Savard/Lee International Symposium on Bath Smelting

Front Cover:
Photos illustrating work by G. Savard and R. Lee
during development of shrouded tuyere.
Top left: Cross section of a typical mushroom formed
on the tip of a concentric injector.
Top right: Illustration of Savard-Lee concentric tuyere as patented in 1966.
Bottom right: One of the first tests of the Savard-Lee tuyere in operation,
Cap-de-la-Madeleine, Quebec, Canada, June 11, 1965.
Bottom left: The K-vessel designed with capabilities for top, for bottom
and for side oxygen injection (1965).

Proceedings of the
Savard/Lee
International Symposium
on Bath Smelting

Proceedings of an International Symposium sponsored jointly by
The Minerals, Metals & Materials Society, The Iron & Steel Society
and The Metallurgical Society of CIM.
The International Symposium was held October 18-22, 1992
at the Radisson Gouverneurs Montreal Hotel, in Montreal, Quebec, Canada.

Edited by
J. K. Brimacombe, P. J. Mackey, G. J. W. Kor,
C. Bickert and M. G. Ranade

A Publication of

TMS
Minerals • Metals • Materials

A Publication of The Minerals, Metals & Materials Society
420 Commonwealth Drive
Warrendale, Pennsylvania 15086
(412) 776-9024

Printed in the United States of America
Library of Congress Catalog Number 92-61283
ISBN Number 0-87339-191-8

If you are interested in purchasing a copy of this book, or if you would like to receive the latest TMS publications catalog, please telephone 1-800-759-4867.

Savard/Lee International Symposium on Bath Smelting

Organizing Committee

General Meeting Chairman:
J. K. Brimacombe
University of British Columbia
Vancouver, Canada

Technical Program Chairmen:

Iron and Steel
G. J. W. Kor
The Timken Company
Canton, Ohio

Madhu G. Ranade
Inland Steel Company
East Chicago, Indiana

Non-Ferrous Metals
Phillip J. Mackey
Noranda Technology Centre
Pointe Claire, Quebec

Light Metals
Christian M. Bickert
Pechiney Corporation
Greenwich, Connecticut

Committee Members:

Alejandro A. Bustos
Canadian Liquid Air Ltd.

John M. Floyd
Ausmelt Pty. Ltd.

Helge O. Forberg
Hatch Associates Consultants, Inc.

Moto Goto
Mitsubishi Materials Corporation

Pierre Homsi
Pechiney LRF

Hiroyuki Katayama
Nippon Steel Corporation

Sydney R. Leavitt
Altimag Consultants Inc.

Theo Lehner
Boliden Minerals AB

Dieter Neuschuetz
Technical University Aachen

Claudio H. Queirolo
Codelco-Chile

Greg G. Richards
University of British Columbia

Nickolas J. Themelis
Columbia University

K. Torssel
ABB Powdermet AB

Companion's Program:

Angele Mackey

Anne Marie Savard

VI

Session Chairmen:

Session 1 - Overview of Bath Smelting

R. J. Fruehan
Carnegie Mellon University
Pittsburgh, PA

Peter Tarasoff
Beaconsfield
Quebec, Canada

Session 2 - Innovative Bath Smelting Processes

Alejandro Bustos
Canadian Liquid Air Ltd.
Montreal, Quebec, Canada

P. J. Koros
LTV Steel Co.
Independence, OH

Session 3 - Bath Smelting Fundamentals

Herbert H. Kellogg
Columbia University
New York, NY

G. J. W. Kor
The Timken Co.
Canton, OH

Session 4 - Mechanisms and Models in Bath Smelting

Phillip J. Mackey
Noranda Technology Centre
Pointe Claire, Quebec, Canada

Howard K. Worner
University of Wollongong
Wollongong, Australia

Session 5 - Injection into Baths

Christian M. Bickert
Pechiney Corporation
Greenwich, CT

Madhu G. Ranade
Inland Steel Co.
East Chicago, IN

Session 6 - Emerging Bath Smelting Processes

Paul E. Queneau, Sr.
Dartmouth College
Hanover, NH

Karl Brotzmann
Atzevus
Sulzbach-Rosenberg, Germany

Preface

In Celebration of Dreamers and Innovators

Progress in any industry is not continuous but is punctuated by key discoveries and innovations that accelerate change and frequently transform the industry. One need only think of the transistor, laser and super-conductor to appreciate the truth of this statement. In the making of steel, the Bessemer converter, the top-blown LD furnace and the Savard/Lee bottom injection tuyere similarly stand out because each changed the course of the industry. The Bessemer converter profoundly accelerated the rate at which steel could be made through the bottom injection of air and the use of the carbon and silicon dissolved in the liquid metal as a fuel to render the process autogenous. The LD furnace exploited developments in tonnage oxygen to accelerate steel refining further by injecting pure oxygen into the bath from the top. The Savard/Lee concentric tuyere permitted the bottom injection of oxygen by shrouding the gas with a hydrocarbon to protect the adjacent refractory. And early developments by Lee and Spire on the bottom injection of inert gas through porous plugs into steel were the harbinger of ladle metallurgy. Thus the names of Bessemer, Dürrer, (instigator of the LD process), Savard and Lee figure prominently amongst the dreamers and innovators who have transformed the steelmaking landscape. Interestingly none of these inventors originated from the steel industry itself.

In the view of the importance of the invention of Savard and Lee, not only to steelmaking processes but increasingly to the nonferrous industry as well, it is appropriate to honour these men with a named International Symposium on Bath Smelting at which their story could be told. There are lessons to be learned from their creativity, courage and persistence in the face of numerous obstacles. The insight of luminaries, especially Horace Freeman and Karl Brotzmann, who facilitated the early invention and commercialization of the Savard/Lee concentric tuyere, is also an inspiration to those who would make change. In many respects, then, this Symposium is a celebration not only of the achievements of Savard and Lee, but of all dreamers and innovators who have sought to transform the metallurgical landscape.

The papers assembled in this Symposium have all been invited from prominent inventors, researchers and practitioners to cover the many fundamental and applied aspects of bath smelting with an emphasis on injection. Papers were invited purposely from the steel, base metals and light metals industries which traditionally have stood as three solitudes. It is the fervent hope of the Organizing Committee that this symposium will be the first of many which aim to break down the walls of these solitudes to discover what we can learn from each other.

J. K. Brimacombe
General Meeting Chairman

Table of Contents

Session 1
Overview of Bath Smelting

Session 2
Innovative Bath Smelting Processes

Session 3
Bath Smelting Fundamentals

Session 4
Mechanisms and Models in Bath Smelting

Session 5
Injection into Baths

Session 6
Emerging Bath Smelting Processes

Session 1

Overview of Bath Smelting

P.J. Mackey
Noranda Technology Centre
240 Hymus Boulevard
Pointe Claire, Quebec Canada H9R 1G5

J.K. Brimacombe
The Centre for Metallurgical Process Engineering,
The University of British Columbia
Vancouver, B.C. Canada V6T 1Z4

Abstract

It was a dream to remove the dead hand of nitrogen from the Bessemer converter and to inject pure oxygen through the bottom of a furnace to make steel, without destroying the refractory lining. But it was to take over a century until Savard and Lee discovered how to turn the dream into a reality simply by shrouding the discharging oxygen stream with hydrocarbon which, at steelmaking temperatures, cracks and provides local cooling. Their invention of the concentric tuyere, so elegant in concept, transformed the steelmaking landscape spawning new processes, much as did Bessemer's discovery that pig iron contained sufficient fuel in its dissolved silicon and carbon to be refined autogenously with air. Now the concepts of the concentric tuyere and gas shrouding are being applied in process developments for non-ferrous metals production like lead and copper. The birth of their invention and its evolution to create new metallurgical processes are traced in this paper.

Proceedings of the
Savard/Lee International Symposium on Bath Smelting
Edited by J. K. Brimacombe, P. J. Mackey,
G. J. W. Kor, C. Bickert and M. G. Ranade
The Minerals, Metals & Materials Society, 1992

Introduction

The huge, pear-shaped OBM converter* tilts into blowing position, driven by motors with a combined power of 450 KW. The operator activates the computer controlled oxygen system to commence injection of pure oxygen through a unique and ingenious submerged tuyere system, invented by Savard and Lee, using a hydrocarbon shroud for injector protection. Quietly and efficiently, the 300-tonne melt of steel is refined in less than 40 mins., producing, through computer-monitored quality control techniques, the precise composition, which would have taken many hours and required numerous individual heats three or four decades ago.

The quietly efficient OBM converter of the nineteen-nineties has an ancestral resemblance to its earlier counterpart, the Bessemer converter (Fig. 1), which dates from the last century and even continues into the present [1]-[2]. The elegant invention of Guy Savard and Robert Lee, whom we recognize and celebrate at this International Symposium, had its beginnings in Montreal over forty years ago. These inventors are, to quote author Richard Preston [3], truly "Hot Metal Men". To fully appreciate this development, it is necessary to look back on the metallurgical industry, especially the steel industry in the nineteen forties, fifties, and sixties (and perhaps even earlier). It was into this industry that Guy Savard and Robert Lee, from the outside, transformed the steelmaking landscape. The birth of their shrouded tuyere invention and its evolution to create new metallurgical processes are sketched in this paper.

Background - The Iron and Steel Industry in the 1940's to 1960's

The iron and steel industry has been long recognised as the key driver of the Industrial Revolution in Britain (Fig. 2) in the period 1720-1850 [2]. The production of steel on a massive scale really began in the second half of the last century in Britain and the U.S. and was possible due to the early contributions notably of William Siemens, Henry Bessemer and Sidney Gilchrist Thomas in Britain, Pierre and Emil Martin in France and William Kelly in the U.S.A. Bessemer's discovery in 1855 that pig iron could be autogenously blown with air led to the Bessemer converter for steelmaking, and spurred the development of pneumatic processes for other metals, particularly for copper and metal converting. Henry Bessemer's engineering background developed through his father's type-foundry business interests in Britain and France. Though not a metallurgist by training, Bessemer's early curiosity in metal-making and fabrication processes led to one of the most significant discoveries in industrial history [4]. Almost one hundred years later, in Montreal, Canada, Guy Savard and Robert Lee, with a board mandate to find new markets for Canadian Liquid Air's industrial gases and looking at the steel industry from this perspective and therefore from the outside, somewhat like Henry Bessemer, have similarly produced an invention which significantly altered the direction of metallurgical processing. It is interesting to comment that the zeal of the "outside" inventor may sometimes bring more creative skills in process development than is often mustered directly from within an industry itself.

In the nineteen forties and early fifties, the technical literature on the Bessemer process was generally concerned with steel quality, converter performance and related mechanical piping and tuyere systems. Thus in a 1941 paper [5] discussing Bessemer converter operation, J.S. Fulton recommended several improvements to the air delivery systems of the typical Bessemer plant of the period. The author was associated with Ingersoll Rand Co., blower

---

* The submerged oxygen injection technology invented and pioneered by Savard and Lee spawned a whole new family of pneumatic processes generally referred to as OBM (Oxygen Bottom Metallurgy Maxhütte) or Q-BOP (Quick-Basic-Oxygen-Process), and other variants such as K-OBM, KMS, KMS-S, KVA and K-ES.

4

Figure 1 - Bessemer converter blowing a heat of steel at Workington Iron and Steel Works, early 1930's. Rails for the Canadian Pacific Railway were rolled at Workington [1].

Figure 2 - Images of metallurgical operations at early iron and steel plants of the Industrial Revolution in Britain - "Coalbrookdale by Night", by P.J. de Loutherbourg from the Science Museum Collection, London.

and compressor manufacturers, and it is of interest to note the improvements introduced to an industry itself (the steel industry) by outsiders, a pattern which was to repeat itself again later in the present history. Typically, such Bessemer converters in use at the time (Fig. 3) were about 4 m (13 ft.) dia. and 7.6 m (25 ft.) high and operated with a typical steel charge of 22 tonnes blown with about 18,000 Nm³/h of air distributed through some two hundred tuyere holes. (Larger converter units were also in operation). The actual tuyere blocks in use at the time generally contained seven holes, the same as in one of Bessemer's original designs (the tuyeres consisted of six holes on a circle of a given diameter in the block, and one in the centre). Fulton's paper recommended a series of pneumatic and instrumentation improvements to enhance blowing efficiency. The use of pure oxygen in the tuyeres would cause severe erosion of the bottom, and although oxygen was included in patents by Bessemer in 1855 [6], pure oxygen injection was not used in commercial vessels until Savard and Lee's breakthrough.

Testing of oxygen enrichment of the tuyere air for the Bessemer converter and enrichment of open hearth furnace burners was carried out in the 1930's and 1940's principally in Europe. It is interesting to note that by 1938, Maximilianshütte, the company which would play a pivotal role in producing the first commercial heat of steel using the Savard-Lee shrouded tuyere, was using oxygen enrichment for its Bessemer steel plant [7]. The early history of oxygen steelmaking is outside the scope of this paper but details are available in several reviews [7-9]. However a brief description of the key events leading up to the 1960's, and a review of the industry in Canada at that time is useful to help place the development of the shrouded tuyere in perspective.

Professor Robert Dürrer of Switzerland experimented with the application of oxygen for steelmaking in the 1930's and 1940's. In 1948 at the Gerlafingen Works in Switzerland, he succeeded in producing steel by top blowing in a 2.7-tonne pilot unit. This was followed by successful testing of the process by the Linz-Donawitz companies in Austria who were the first to commercialize oxygen steelmaking by top blowing in 1952 [8]. The process became known as the LD Process after the first initial of Linz where testing began, and both Donawitz where the first testing also occurred, and Dürrer the originator. Over the ensuing ten years or so, more than 12 million tonnes of steel capacity worldwide were introduced using this process. Improvements in open hearth technology with oxygen enrichment also continued. In the U.S., oxygen consumption in the steel industry increased nearly six fold from about 1 million tonnes in 1956 to 5.6 million tonnes in 1965 [7]. It was during this period of remarkable change and growth in the steel industry that Guy Savard and Robert Lee launched their quest.

In the United States in 1959, up to 4% of steel output was produced by the newly introduced basic oxygen process (total U.S. steel production was slightly over 100 million tonnes, or about one-quarter of world output). Open hearth furnaces, many equipped with oxygen enrichment facilities, accounted for 87% of steel output, while electric furnace output accounted for 9% of the total. At that time, Canada produced about 5 million tonnes, or a little over 1% of world steel output. Stelco, the largest [10] produced nearly 2 million tonnes, followed by Algoma at about half that amount. (A history of Canada's steel industry up to about the 1960's is included in the monograph edited by Wayman [11]). Dofasco (the No. 4 plant in Canada after Dominion Steel and Coal Corporation in Nova Scotia), had commenced experimenting in 1952 with oxygen steelmaking based on the process developed at Linz and Donawitz in Austria. Dofasco produced the first heat from a 40-tonne commercial installation on 13 October 1954 [12]. The new furnace was expected to have a lining life of 300 heats with a charge of about 40 tonnes per heat. A few years later, on

6

18 November 1958, oxygen steelmaking commenced with a larger furnace design at Algoma Steel Corporation Ltd., in Sault Ste. Marie, Ontario, making 516,000 tonnes in 1959 by this process [13]. Stelco, the nation's largest steelmaker, continued using open hearth furnaces and improved their productivity by perfecting oxygen enrichment [14]-[15] which was first tested there in 1947 [7]. This program was enlarged when in 1961 a fifth furnace of 450-tonne capacity was brought on line at the Hilton Works in Hamilton. Canadian Liquid Air participated in this development work at Stelco. By 1965, all Stelco furnaces were using oxygen enrichment and a paper published in that year [15] summarized the company's faith in these developments. Within a few years, the improved cost effectiveness and productivity of the new oxygen steelmaking process would result in Stelco commencing its first installation of this process in 1971 [16]. The interest in oxygen steelmaking in the mid-1950's could perhaps be characterized by the title of the Dofasco paper [12] describing its first LD plant as follows: "Oxygen Steel Produced at Dofasco can Compete with Open Hearth" - a prediction which turned out to be true. In a lead article five years later in the Journal of Metals [17], it was predicted that LD steelmaking capacity would double in two years, a forecast which was confirmed by actual events. It was against this background that Guy Savard and Robert Lee introduced a new tuyere for oxygen steelmaking.

Early Experimentation by CLA on Submerged Gas Injection

In the late 1940's, the Metallurgical Section of the Research and Development Department of Canadian Liquid Air Company Limited (CLA) was headed by Etienne Spire. The company's mandate was to extend the use of industrial gases into Canada's expanding industrial and metallurgical industries. Thus Spire and co-workers started investigations on submerged gas injection. Initially the use of the porous plug injector was identified as a candidate for a variety of metallurgical operations. Robert Lee, who graduated in metallurgical engineering from McGill University in 1947 joined CLA the same year . He began working on these gas applications initially with Etienne Spire. Although Spire would leave the company a few years later in 1951, a mission for finding new techniques for oxygen and gas injection in metallurgy appears to have been established. At about this time, Guy Savard, a graduate of the Royal Military College, Kingston, Ontario, returned to the Montreal office after an overseas posting, and the department was consolidated over the next few years, and re-named the Research Department. Except for a brief period while serving the Quebec sales region, Guy Savard headed the department.

Initial tests of the porous plug injector on 200-kg iron, steel and non-ferrous melts were carried out at the Mines Branch Ottawa (Booth Street foundry) in 1947-1951 by Lee and Spire. At the time, it was difficult to commercially obtain suitable test plugs; however with the aid of the Mines Branch, which was also working on Kilmar refractory ores, Robert Lee was able to secure sufficient quantities of refractory porous plugs for the injection tests. The porous plug injectors were found to work effectively when injecting with nitrogen and argon in steel melts, but when oxygen was used the plugs disintegrated. The intense local heat created by the reaction of oxygen with the melt destroyed the refractory plugs. The CLA investigators certainly experienced first hand in these early tests the self-destructive capability of oxygen in an uncontrolled injector; yet they also began to appreciate the enormous refining potential of submerged injection. Obviously, the injector was the key to success if bottom oxygen injection was to work; and no doubt, ideas to avoid the injector erosion began to develop during this early phase.

The porous plug results with inert gases, however, were promising and pilot tests began using a 7-tonne pilot unit (Fig. 4) built with the co-operation of Canadian Car and Foundry Company of Montreal (later called Canadian Steel Foundries). In these pilot tests, argon gas

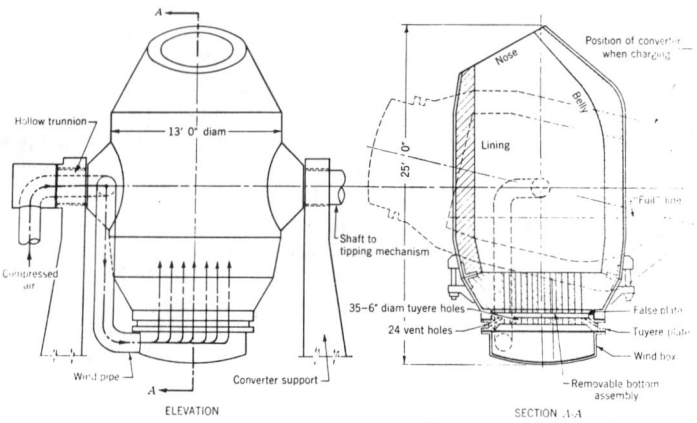

Figure 3 - Illustration of typical 22-tonne Bessemer converter in use in 1940's and 1950's [21]. Normally such converters would have about 30 "tuyere blocks" each containing seven to ten 16-mm holes [5,21].

Figure 4 - Portable test vessel built in 1951 for inert gas flushing tests originally using the porous plug injector [20].

was injected through the porous plug to homogenize ferro-alloy additions to a high-carbon "dead melt down" steel. The technique used was probably one of the earliest modern applications of what is referred to today as Ladle Metallurgy. While the technique was successful, the Montreal foundry did not adopt it at the time. This process was successfully commercialized some years later as the Gazal Process, a proprietary technology of CLA. Meanwhile, CLA also pursued oxygen injection (Table I).

High-Pressure Oxygen Injection Tests by Savard and Lee

Collaborative tests with oxygen also began in 1951 at DOSCO (Dominion Steel and Coal Corporation, now Sydney Steel Corporation) on the desiliconizing of hot metal in the transfer ladle by jetting oxygen onto the melt surface - a variant of the newly-introduced LD concept. This approach unfortunately led to the generation of a large amount of red iron-oxide fumes,, and work commenced by CLA and DOSCO to find a solution. In first-class collaborative work, which became their legend, and which was to eventually lead to their first joint patent [18] on high-pressure oxygen injection, Guy Savard and Robert Lee helped solve DOSCO's dilemma by successfully introducing bottom oxygen blowing which was more effective metallurgically and reduced the amount of fuming. The pilot unit used for these early tests was an 18-tonne ladle as illustrated in Fig. 5; typical test results are presented in Table II [19]. The oxygen was introduced via four injectors, with an average flow of about 95 Nm^3/hr. per injector. Experiments were carried out with high-pressure oxygen flow to confirm the effectiveness of Joule-Thompson cooling from the expanding jet. At this time, testwork on high-pressure oxygen injection was also carried out on different melts (from 0.5 up to 35 tonnes) at several foundries including B and T Foundry (Richmond, Quebec), Eastern Electro Castings (Lachine, Quebec) and Quebec Iron and Titanium Corp. (QIT) at Tracy, Quebec.

Following these tests and the successful trials at DOSCO, similar tests were subsequently undertaken in 1957 at the foundry of Sorel Industries Limited, at Sorel, Quebec, on decarbonizing QIT iron (Table I). The objective was to lower the carbon content initially for electric-arc furnace refining prior to casting for gun manufacture. Successful tests using the same 18-tonne ladle as employed at DOSCO demonstrated the validity, in principle, of high-pressure oxygen injection. These tests also confirmed to CLA the superiority of bottom injection. North American steel executives were invited to view these demonstration tests at Sorel; however, sadly for Savard and Lee and for CLA, there was no commercial interest in using the high-pressure injectors.

The injector which they developed employed heavy-walled copper tubing imbedded in a refractory plug [18]. The amount of "cooling" was in part a function of the integrity of the "mushroom" cap which formed at the injector. It was found that the injector burn-off rate was primarily a function of the copper wall thickness, and oxygen pressure, and also, in part, was related to the type of refractory. Reasonable results were found with oxygen pressures in excess of 2760 KPa with a 1.6 mm orifice having a wall thickness greater than 2.4 mm imbedded in a burned magnesite refractory plug. Higher oxygen pressures were found to improve injector life. Such pressures were, however, well beyond normal accepted operating limits. As noted above, this work led to the award of their first joint patent in 1958 [18].

Oxygen injection tests were also conducted in the early 1960's at Strategic-Udy Metallurgy Ltd., Niagara Falls, Ontario, Table I. These tests were initially carried out with the same transfer ladle test unit used previously, but with an extension added to the bottom for greater depth. This unit held about 7 tonnes of ferro-chrome melt for these tests. The tests evidently confirmed that ultra high-pressure injection (~ 8300 Kpa, 1200 psig) prolonged injector life.

9

Table I Summary of Early Joint Test Work Undertaken by G. Savard and R. Lee of Canadian Liquid Air

Company or Organization	Location	Period	Objective	Test Facility	Result
Mines Branch	Ottawa, Ontario	1947-51	To introduce gas into melt using porous plugs	200-kg melt	Led to successful development of CLA "Gazal" Process
Canandian Car and Foundry	Montreal, Quebec	1951	To test homogenization of alloy additions using porous plug	7-tonne vessel	Technical success - led to first use of Ladle Metallurgy
DOSCO	Sydney, Nova Scotia	1951-52	Desiliconizing hot metal with top oxygen lancing using consumable steel pipe	35-tonne heat (80 tests)	"Too many fumes, too labour intensive, must find an easier way"
B and T Foundry	Richmond, Quebec	1954	Desiliconize cupola iron with a refractory coated (top) lance, high-oxygen pressure (~ 2760 kPa)	1-tonne ladle	Top lancing resulted in erosion of refractory coating
Eastern Electro Castings	Lachine, Quebec	1954	Decarbonizing iron, initially top lancing, later bottom injection ($O_2 \sim$ 2760 kPa)	1 to 5 tonne ladle	Discovered bottom injection more acceptable, but injector eroded at this pressure
Quebec Iron and Titanium Corp.	Tracy, Quebec	1954-55	Testing various designs of submerged injectors at higher pressures	- 0.5 to 5-tonne ladle from previous tests - 15-35-tonne QIT vessels	Higher pressures appeared promising. First "mushroom" recovered at 4825 kPa indicated sign of cooling [25]
National Research Council	Ottawa, Ontario	1955	Modelling of submerged injector using steam injection into water	0.5 m^3 water tank	Revealed shape of gas (steam) entering and disappearing into liquid (water). "Onion" shaped plume gave first clue to injector protection
DOSCO	Sydney, Nova Scotia	1956	Desiliconizing hot metal using submerged high-pressure oxygen (4135 kPa)	18-tonne transfer ladle fitted with 4, and later 6 bottom injectors	Tests successful, not commercialized

cont'd...

10

Table I Summary of Early Joint Test Work Undertaken by G. Savard and R. Lee of Canadian Liquid Air (cont'd)

Company or Organization	Location	Period	Objective	Test Facility	Result
Sorel Industries Ltd.	Sorel, Quebec	1957	Decarbonizing QIT iron to steel via submerged oxygen injection at 8270 kPa	18-tonne transfer ladle fitted with bottom injector	Improvement of high-pressure techniques; quality of refractory at injector important
IRSID	Maizieres-les-Metz, France	1961	Demonstrate high-pressure submerged injection of oxygen	2.5-tonne vessel, two injectors	Confirmed refractory surrounding injector was critical
Strategic-Udy	Niagara Falls, Ontario	1962	Submerged high-pressure oxygen injection in ferro-chrome melts, moisture spray to enhance injector cooling agent.	Savard-Lee vessel (4-11 tonne), 8 injectors, 369 heats	Tests successful, not commercialized
Mines Branch	Ottawa, Ontario	1964	High-pressure oxygen submerged injection	150-kg K-vessel (Figure 7)	Refined iron and ferro-nickel melt, need to move test facility to Freeman Corp.
Freeman Corporation	Cap-de-la-Madeleine, Quebec	1964-68	Test concentric injectors	K-vessel (Figure 7)	- Invention of double-pipe tuyere having oxygen in centre with protective hydrocarbon in annulus space (1964-65) - Tested many variables related to injector life
Freeman Corporation	Cap-de-la-Madeleine, Quebec	1967	Demonstration for Prof. K. Brotzmann	K-vessel	Two demonstration heats (150 kg) successfully carried out
Eisenwerk-Gesellschaft Maximilianshutte, mbH	Sulzbach-Rosenberg, Germany	1967	Scale-up submerged Savard-Lee injectors for use in a Thomas converter	20-tonne heat	Tests successful, OBM process born

*Initial work was conducted by E. Spire and R. Lee

11

Table II Early Trials on Submerged Oxygen Injection -
Desiliconizing at DOSCO STEEL [19]

Weight of metal	13.1 tonnes
Melt depth (approximately)	< 1 m
Duration of injection	14.4 mins.
Rate of oxygen flow*	380 Nm³/h
Specific oxygen usage	0.5 Nm³/min. tonne

	Temperature	Melt Analysis Content, %				
	°C	C	Si	Mn	S	P
METAL						
Before	1269	4.09	0.85	0.74	0.045	1.10
After	1340	3.81	0.20	0.37	-	-

FINISH SLAG	Analysis, %				
CaO	SiO₂	FeO	MnO	MgO	P₂O₅
23.8	52.8	4.13	12.75	5.79	0.15

*Four injectors with approximately 95 Nm³/h flow per injector

One early memorandum of this period, dated 10 January 1962 referred to the tests on the "Savard-Lee Vessel" which was a new dedicated test unit built by Strategic-Udy for the work. The new vessel, shaped somewhat like an elongated Bessemer converter with the high-pressure injectors, was about 1 m I.D. at the bottom, 2 m I.D. at the top and nearly 3 m high.

A conducive atmosphere for testing, an important ingredient for success, existed at the time at Strategic-Udy. Dr. N.J. Themelis, currently Stanley-Thompson Professor of Chemical Metallurgy at Columbia University, was the Research Engineer at the plant, and he evidently brought unique creative skills to the investigations. It would be less than three years later when he would move to Noranda Research Centre to co-invent the Noranda Process. The tests at Strategic-Udy, while generally successful, did not lead to further commercial development because of business reasons related to a drop in the price of ferro-chrome. The company later ceased operations as a result of this decline.

Perhaps as a compromise approach between the known, but not trusted, technique of indirect water-cooling of the tuyere members, such as had been identified in previous patents (see patents cited in ref. (18)), and their newly developed use of high pressure with heavy-wall copper injectors, Savard and Lee also undertook tests at Strategic-Udy in which a water mist was introduced into the high-pressure oxygen distributor prior to injection. The use of oxygen-enriched air mixed with steam had been pioneered earlier on Thomas-converters (basic lining) since 1947 at the Hagen-Haspe Works in Germany operated by Klöckner [22]. This was pursued as a means to minimize nitrogen in the blast to produce low-nitrogen steel, yet have high oxygen enrichment plus a reasonable tuyere life. A plant in the UK later operated in this mode at Port Talbot [21]. Thus in the tests in Canada, the intention of the moisture spray was to try and remove some of the intense heat from the injector zone. In these tests, water spray additions up to 4 or 5% by weight of the gas flow were made. These amounts of water improved the performance of the injector, and the burn-off rate was evidently less. This approach helped provide an understanding of the heat effects; however, the 1958 patent on the heavy-wall copper tubing with high-pressure oxygen remained unchanged.

Data taken from this patent illustrate the injector burn-off rate as a function of the wall-thickness of the copper tubing with an orifice diameter of 1.6 mm (Fig. 6). The patent claims covered up to 25-mm wall thickness when the burn-off rate was considerably improved. Prior art cited in the patent referred to the LD concept for top lancing; there was also reference to the use of indirect water-cooled tuyeres, but for practical reasons, these were not considered for commercial use on account of the ever present danger of hot metal-water (steam) explosions. However, even with the evident technical success of the new thick-walled copper injector, widespread application of CLA's device appeared to be limited - there remained a concern about ultimate injector and furnace life such that apparently world-class companies were just not interested in the development for commercial operation. Nevertheless, Guy Savard and Robert Lee continued their search; perseverance obviously is an important ingredient for success.

Birth of an Idea - The Shrouded Tuyere

The good working relationship which CLA established with the Mines Branch during the porous plug injection work continued with the high-pressure injection tests using inert gases and oxygen for the specific purpose of stirring and degassing molten metals. Co-operation and, in today's vocabulary, entrepreneurship, were some of the hallmarks of Savard and Lee's working style, and this was no doubt an important factor in their ultimate success. At about

13

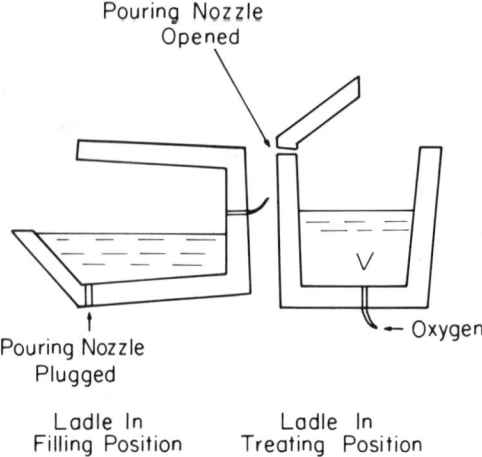

Figure 5 - The 18-tonne ladle lined with standard fireclay brick as employed by Guy Savard and Robert Lee for early pilot tests on bottom blowing with oxygen [19].

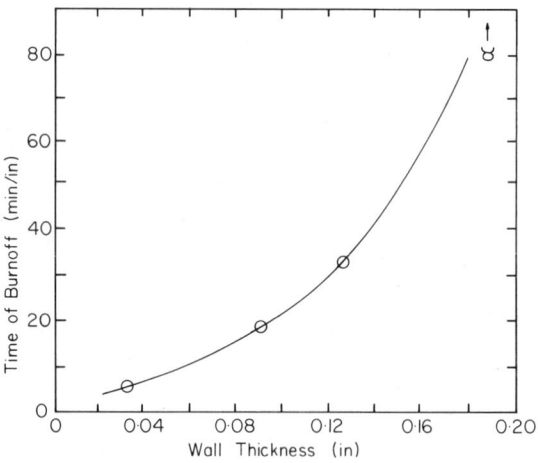

Figure 6 - Burn-off rate of early design of heavy wall copper tubing injector of 1.6-mm orifice diameter imbedded in fireclay refractory. Oxygen pressures in range 2760-4140 kPa were employed [18].

14

this time, R. Bergman of Falconbridge, who later made distinguished process developments in nickel smelting, indicated interest in oxy-refining ferro-nickel in connection with lateritic nickel production. This test work was carried out at the Mines Branch using a new pilot converter named the K-vessel, after the brilliant designer and technician at CLA, the late Robert Kottmeier who evidently was an indispensable partner in all the Savard-Lee tests. This converter, which was approximately 0.6 m O.D., 0.3 m I.D. and about 1.2 m high excluding the "nose" (Fig. 7), was equipped for bottom, side and top blowing. It was in fact, almost a precursor of today's modern steel converter. This vessel was successfully tested with a variety of high-pressure gas injection runs involving 100-150 kg of ferro-nickel and other melts.

As the test program expanded, it became evident that the wooden structure housing the Mines Branch foundry could constitute a fire hazard for the work that was proposed, and an alternative test site for the K-vessel was sought. An interested and willing friend to provide test facilities was found at the small foundry of Freeman Corporation at Cap-de-la Madeleine, about 150 km east of Montreal. Here, in the shadow of the famous Basilica built on the site (founded in the seventeenth century after the first settlers were saved from certain drowning in the St. Lawrence river by a miraculous "bridge of ice"), and also close to the site of one of Canada's first and most famous ironmaking facilities (Les Forges du Saint-Maurice), the "Hot Metal Men" of the twentieth century successfully conceived and experimented with the revolutionary new shrouded tuyere injector.

The plant operated by Freeman Corporation produced iron powder for metal cutting operations. The owner, Horace Freeman, a successful and innovative entrepreneur had himself introduced several new technologies to industry, including the development of a safe Fe-Al alloy powder for special metal cutting applications and, a hydrocyclone to clean pulp slurry, while taking an interest in direct smelting of iron ores by a type of spray smelting technique. Apparently the latter had evolved from testwork conducted on the spray burning of black Kraft mill liquor in a boiler for chemical recovery. Evidently on personal terms with Maurice Duplessee, Horace Freeman had at this time pursued direct smelting of Quebec iron ores by spraying a pre-mixed slurry of iron ore, coal and oxygen into a vertical tower furnace. It would be several years later in 1968 that the CIM would honor Horace Freeman with the Inco Medal Award, for ".... his early work on flash roasting of sulphides and for his pioneering work on iron powder metallurgy". CLA, and specifically Robert Lee, was acquainted with Horace Freeman from the time that CLA supplied the oxygen for his experimental program. Thus with Freeman Corporation providing all test facilities and necessary personnel in a conducive atmosphere, Guy Savard and Robert Lee continued work on the K-vessel.

Initial tests were carried out to study the moisture addition concept. This approach had been found to improve injector life on account of both the dilution of the oxygen, but principally due to the heat of evaporation which caused a cooling effect. It was considered that the previous method of addition that was tested (spray addition to oxygen distributor) actually conditioned the entire jet and not specifically the critical zone at the injector-refractory interface where wear was occurring. If the modifier flow (water or other medium) could be adjusted independently of the bulk oxygen flow the wear might be better controlled. This notion of control at the wear site led to the first idea of a nest of concentric pipes as the injector, such that the flow of the modifying medium could be adjusted independently of the oxygen. It was reasoned that by shrouding the skin of the oxygen jet with a medium that was non-reactive with the melt, but reactive with the oxygen itself, the injector would be protected [25].

15

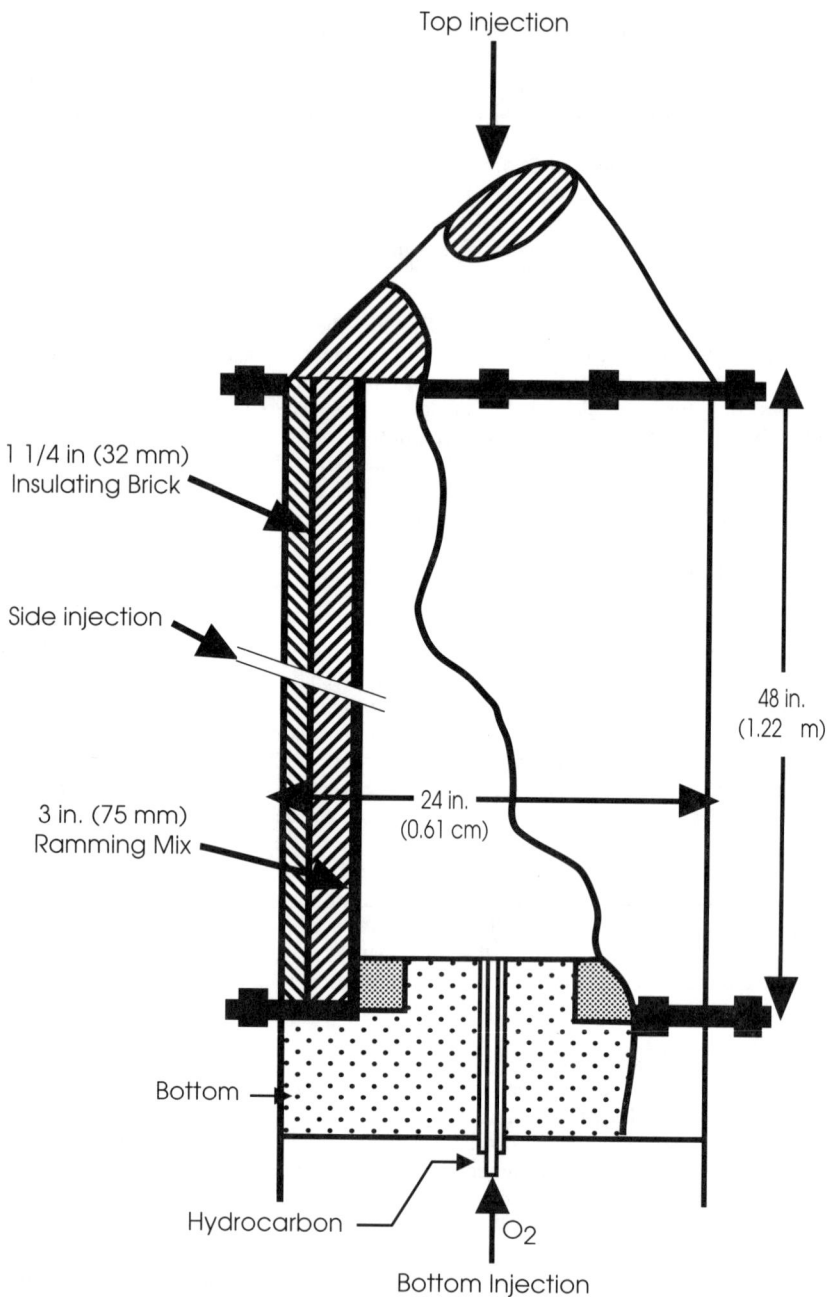

Top injection

1 1/4 in (32 mm)
Insulating Brick

Side injection

48 in.
(1.22 m)

3 in. (75 mm)
Ramming Mix

24 in.
(0.61 cm)

Bottom

Hydrocarbon

O₂

Bottom Injection

Figure 7 - Experimental K-vessel designed in early 1960's by R. Kottmeier of CLA to test injection techniques. The unit was fitted with top, side and bottom injection ports and operated with a 100-200 kg melt added via a transfer ladle [39].

16

The first tests at Freeman Corporation in 1964 were carried out using propane as the shrouding fluid, followed by later tests on many other shrouding media, including water, methane, carbon dioxide, nitrogen and argon. Since top lancing of steel with oxygen was known to generate iron oxide fumes, Savard and Lee also considered that the concentric pipe concept operating with oxygen/propane gases could, in fact, suppress fuming. During 1964, several tests using the K-vessel also explored these ideas; however, the focus of the effort was soon given to the bottom-mounted concentric injector (Fig. 8).

Significant tests carried out on 10-11 June 1965 with 110 kg of QIT iron confirmed the approach. The injector in these first tests was made of 3.2 mm I.D., 8.7 mm O.D. copper tubing for the oxygen and 11.9 mm I.D., 15.9 mm O.D. stainless steel tubing for the natural gas; a design of a similar test injector is given in the paper by Savard and Lee [25] which concludes this symposium. Typical flows for one test (K213) were oxygen, 50 Nm^3/h (pressure about 1220 kPa) and natural gas, 10 Nm^3/h (pressure about 125 kPa), representing a relatively high specific total gas rate of 9 $Nm^3/min.$ tonne. Figure 9 taken from a report on these tests [24] shows this test in operation at Cap-de-la-Madeleine on 10 June 1965 - the world's first OBM/Q-BOP!

The Idea Takes Hold

The tests at Cap-de-la Madeleine were continued successfully until about 1968. During these trials, the various parameters were tested and the relative flows and pressures were refined. The first patent on the submerged concentric injector was granted in 1966 [27]. The parent company of CLA, L'Air Liquide of Paris, evidently did not see special merit in the Canadian work as presented to them at the time, nor did any of the French steelmaking companies who were approached. Most of the plants were evidently ready for change, and at the time had not yet switched to the new LD steelmaking technology (France was one of the last European countries to adopt oxygen steelmaking (Fig. 10)). In 1965, submerged oxygen injection with shrouded tuyeres, thus, was not part of their modernization plans. By this time (1965), oxygen steelmaking using the LD approach began to be utilized widely in Japan, Austria, and was making in-roads in countries like the U.S., Canada and the United Kingdom (Fig. 10), and at 81 million metric tonnes worldwide, represented 18% of the world steel production (Fig. 11). Many plants were evidently reluctant to install a new improved system.

As mentioned above, the first patent [27] covering the shrouded tuyere was granted in 1966, (Fig. 12). The developments were no doubt watched by Eisenwerk-Gesellschaft Maximilianshütte (Maxhütte) in Sulzbach-Rosenberg, Germany. In the 1950's and 1960's, Maxhütte was operating six, 15-tonne Thomas converters treating high-phosphorus hot metal. As noted earlier, this plant had also been one of the pioneers of oxygen enrichment in steel converters. In late 1967, Dr. Karl Brotzmann, Director of Research for Maxhütte, expressed interest to CLA concerning commercial application of the shrouded tuyere technology for Maxhütte's plant. A license agreement which included a cross-licensing arrangement with CLA was finalized [29]-[30] and the first industrial heat in one of Maxhütte's old Thomas converters was poured on 17 December, 1967 [28]-[29], just four months after the demonstration tests on a small scale at Freeman Corporation (Table I). Immediately, there was an increase in bottom life and the operating metallurgical results were excellent. The process was called the OBM process (for Oxygen Bottom Metallurgy Maxhütte) and a joint patent with Maxhütte was subsequently granted [31].

17

Figure 8a - K-vessel (0.6 m O.D., 1.2 m high) fitted with top, side and bottom injectors in operation at Freeman Corporation, Cap-de-la-Madeleine, Quebec, December 1964, [24].
(a) Approximately 100 kg of QIT Metal (4% C), 3 kg graphite and 3 kg of 50% ferrosilicon were melted in indirect arc furnace and transferred to K-vessel by fork lift carrying transfer vessel.

Figure 8b - K-vessel (0.6 m O.D., 1.2 m high) fitted with top, side and bottom injectors in operation at Freeman Corporation, Cap-de-la-Madeleine, Quebec, December 1964, [24].
(b) The top lance shown here employed one of the first designs using concentric pipes for oxy-hydrocarbon injection. The operators are positioning the lance prior to test (17 December, 1964). Robert Lee (partially obscured) in check jacket, H. Freeman on right, others in photo were operators at Freeman Corporation.

18

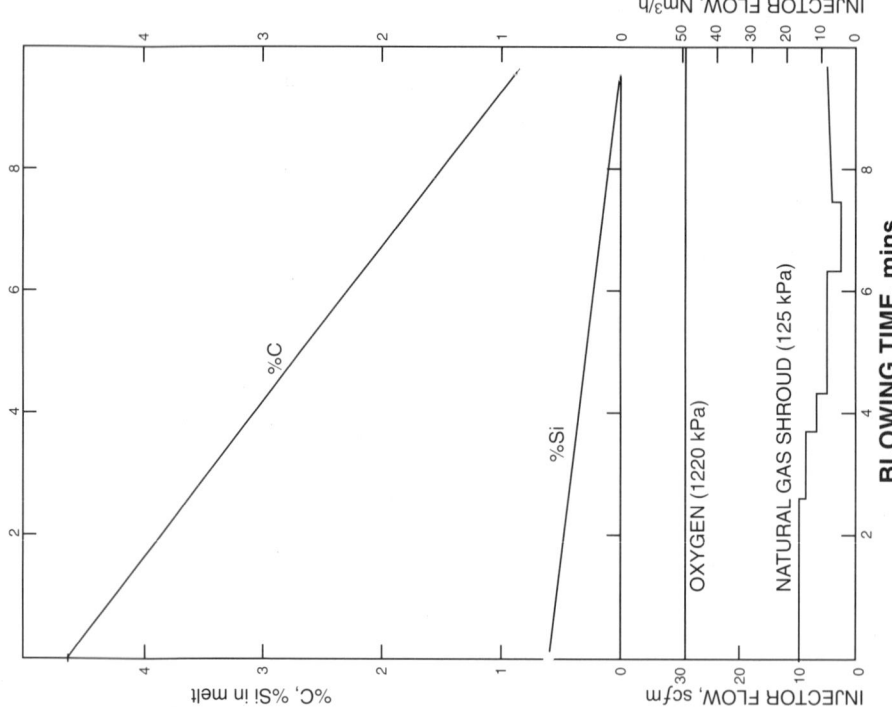

Figure 9 -

One of the first tests of the Savard-Lee shrouded tuyere injector in operation during initial trials at Freeman Corporation, 11 June 1965 on K-vessel. At the time of photo, the test had been in operation for 1 min. 50 sec. Graphs at right illustrate reduction in C and Si of melt during blow. Starting melt approximately 115 kg QIT metal, 3 kg graphite, 3 kg (50%) ferrosilicon. Oxygen and sheld natural gas flows also shown at right correspond to specific gas rate of about 9 Nm³/min. tonne [24].

19

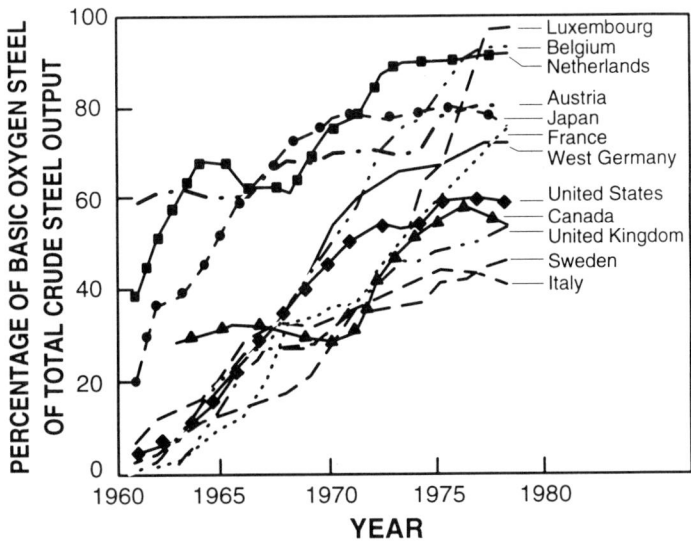

Figure 10 - The diffusion of oxygen steel making for 12 countries, 1961-1978 [26].

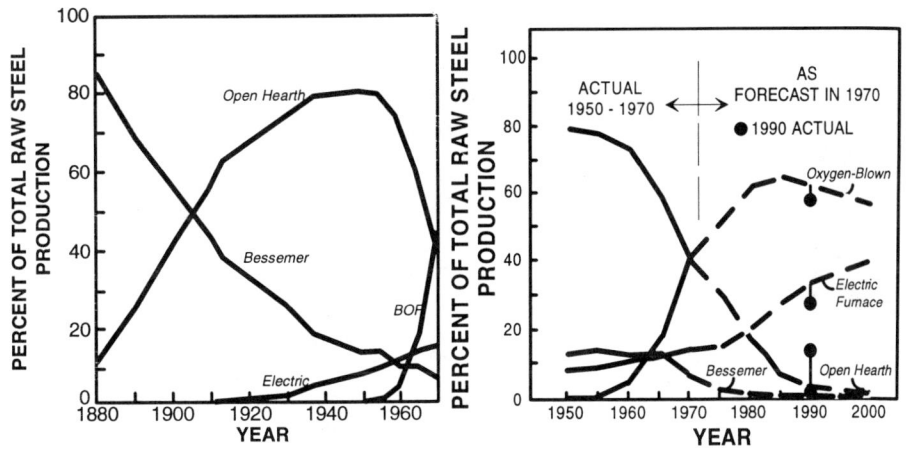

Figure 11a - World steel production by process, 1880-1970 [7].

Figure 11b - World steel production by process (1950-2000), as forecast in 1970 with 1990 actual data added [7].

20

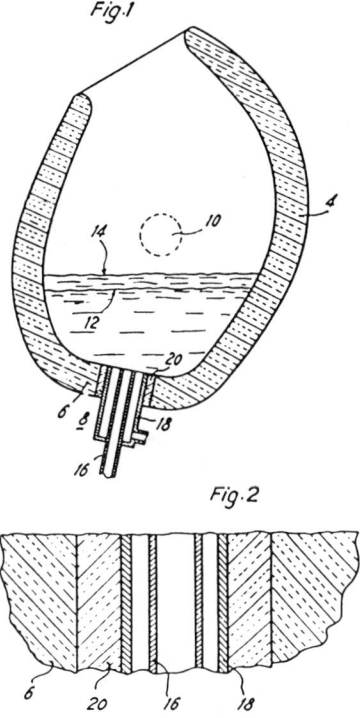

Fig.1

Fig.2

Figure 12 - First shrouded tuyere design as shown in French Patent No. 1,450,718, 18 July 1966 [27]. Numbers refer to following components

4	Refractory lining	14	Slag level
6	Lower section of surface	16	Inner copper tube for
8	Reference location		oxygen (7 bars), 4.8 mm
10	Trunnions		I.D., 9.5 mm O.D.
12	Metal level	18	Outer stainless steel tube for
			Methane, (1.8 bar), 12 mm I.D.
		20	Refractory Plug

The next few years saw adoption of this process with remarkable speed [32]-[33]. Following successful implementation elsewhere in Germany and also in France, U.S. Steel became interested in the process in 1971. In December 1971, U.S. Steel announced that following conclusive test work with Maxhütte, 200-tonne bottom blown oxygen converters were proposed for their Gary, Indiana plant [29]-[32], [37]. The U.S. Steel press release was quite unusual in that there was no reference to Savard and Lee, whose development was clearly at the heart of the U.S. Steel announcement. The largest OBM-type installation at the time commenced operation on 200-tonne units in 1973. These converters were about 10 m diameter compared to the earlier Bessemer units of 4 m diameter. U.S. Steel called this version of the process Q-BOP (the origin of the "Q" is unclear; some have suggested it comes from quiet, quality and quick, while "BOP" was the U.S. Steel acronym for "basic oxygen process"). Over the next four or five years, the OBM/Q-BOP process was widely implemented, and flexibility extended with flux handling and injection systems. Apart from significant metallurgical improvements, capital cost savings resulted from the lower and less expensive OBM/Q-BOP building. By this time, the steel industry was the largest end user of industrial oxygen, requiring some 9 million tonnes in the U.S. in 1973, or about 70% of all oxygen demand.

By 1977, there were some 20 OBM/Q-BOP installations worldwide representing 23 million tonnes of capacity; this reached 27 million tonnes by 1980, and doubled again to about 50 million tonnes by 1992. The chronology of the first twenty installations is shown in Table III [33], while Fig. 13 illustrates a modern OBM/Q-BOP plant. A modified version and close copy called the LWS process, using liquid fuel oil shrouding, was pursued in France (LWS after Creusot-Loire and Wendel-Sidelor, the developers).

At about this time, it was discovered that the refining ability of top-blown converters could be enhanced by retrofitting the new bottom blowing injectors to BOP-units. This introduction of combined top and bottom blowing processes, which incorporated the advantages of both techniques, was possible in its current mode of operation due to the Savard-Lee tuyere. The concentric tuyere idea has been adopted for other processes; for example the AOD process now incorporates argon shrouding for some operations. These developments have been discussed by Ishihara of Nippon Steel in a 1982 paper when the benefits of the different operating modes were evaluated [34]. Figure 14, taken from Ishihara's paper, illustrates the relationship between bottom gas blowing rate in different vessels and the calculated time for complete mixing in the melt [36]. The faster mixing time promotes better slag-metal reaction allowing equilibrium conditions in the bath to be achieved. The family of processes spawned by the original Savard-Lee development is also indicated in Fig. 13. The technique is also spreading to applications in the electric arc furnace improving the performance of this process [20].

By 1992, the steel capacity available worldwide by the submerged oxygen process employing the Savard-Lee tuyere had increased to about 50 million tonnes annually, or approximately 9% of world oxygen steelmaking capacity. (This capacity equalled 541 million annual tonnes in 1992). A further 262 million tonnes of world oxygen steelmaking capacity (or 48%) was provided by combined top and bottom blowing processes in which bottom injection, with relatively inert gases[*], is implemented to improve mixing and

---

[*]When bottom oxygen injection is applied only Savard-Lee shrouded injectors are employed. When inert or non-active gases (e.g. CO_2, CO, N_2, Ar, etc.) are used for stirring, a variety of other injectors are now employed including porous plugs, directional pore plugs, single tubes or concentric pipes patterned after Savard-Lee, but with the center plugged with refractory and inert gas in the annulus, and other configurations. Many of these injectors were pioneered by CLA as described in this paper.

Table III Chronology of OBM/Q-BOP Installed Between 1967 and 1977 [33]

Name Of Company	Location	Start-Up	Number Of Furnaces	Heat Size Tonnes	Annual Production Capacity. Million Tonnes
Germany					
Eisenwerk-Gesellschaft Maximillianshütte, mbH	Sulzbach-Rosenberg	1967 / 1976-77	6 / 3	30-32 / 54	closed in 1976 / 1.35
Rochlingsche Eisen-und Stahlwerk Gmbh	Volkingen	1969	2	34-40	0.5
Metallhüttenwerke Lubeck GmbH	Lubeck	1970	1	45	0.035
France					
Acieries et Treflieries de Neuves-Maison	Chatillon	1969	4	30-32	0.55
Union Siderurgique du Nord et de l'Est de la France (USINOR)	Valencienne / Longwy	1970 / 1970	3 / 2	69-72 / 36-41	0.8 / 0.35
Luxembourg					
Miniere & Metallurgique de Rodange	Rodange	1970	2	30	0.27
Belgium					
Cockerill-Ougree Providence	Marchienes	1971	2	32-34	0.4
Forges de Thy-Marcinelle et Monceau	Monceau / Marcinelle	1971 / 1976	4 / 3	30-32 / 135	0.6 / 2.7
Sweden					
Surshammar Bruks AB	Surahammar	1974	1	32	0.22
Stora Kopparberg	Donnarvet	1974	1	32	0.27
United States					
United States Steel Corporation	South Chicago / Gary / Fairfield	1971 / 1973 / 1974	1 / 3 / 2	27 / 180-200 / 180-200	5.0 / 3.2
Others					
Republic Steel Corp.	Chicago	1976	2	180-200	2.0
Kawasaki Steel Corp.	Japan	1976	2	210	2.8
Italsider	Comigliano	1977	2	220	2.1
			TOTAL INSTALLED CAPACITY (ROUNDED)		23.1

Figure 13 - OBM/Q-BOP furnace installation at Geneva Steel [35].

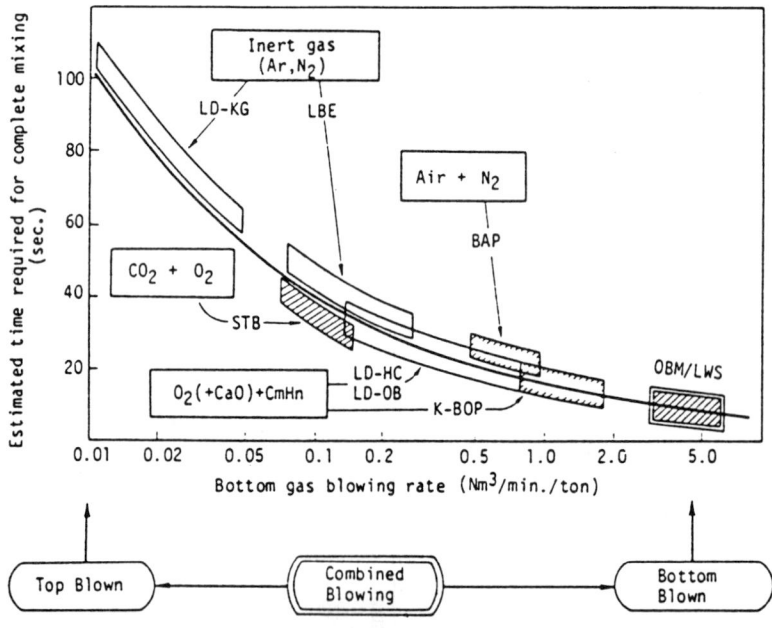

Figure 14 - Top and bottom blowing processes illustrating the influence of
bottom gas blowing rate on mixing time, after Ishihara [34].

converter performance. Out of Canada's 13.2 million annual tonnes of oxygen steelmaking capacity in 1992, 8.5 million tonnes were produced by combined blowing processes, out of which 2.4 million tonnes (or 18%) were made by oxygen processes using the Savard-Lee tuyere. The latter tonnage includes the most recent converter installations in Canada, (Dofasco, Hamilton, Ontario in 1987 and QIT, Tracy, Quebec in 1986), both of which employ the K-OBM process with the Savard-Lee tuyere [37].

While perhaps the forecasts made in 1970 (Fig. 11) on the future trends in steelmaking processes did not exactly match actual events, the immense benefits brought by the Savard-Lee development certainly contributed to the oxygen blowing units remaining supreme amongst steelmaking processes. Perhaps also what was not fully recognised back in 1970 was the significant strides that would be made in direct steelmaking processes using bath smelting techniques. Many of these developments have also been made possible because of the effectiveness of the shrouded tuyere technique invented by Guy Savard and Robert Lee back in 1964.

The new tuyere has found applications in non-ferrous processing as well. However, the utilization of this device has been slower than in the iron and steel industry, in part due to the slower growth of oxygen usage in non-ferrous metallurgy [38]. Also, a high-oxygen tuyere for conventional converting, for example, is not a critical barrier to development (as existed in the steel industry in the 1960's), since todays batch converting operations are essentially autogeneous at moderate levels of oxygen enrichment with conventional single-pipe tuyeres. With continued development in bath smelting processes, the Savard-Lee concept will, however, find new applications, as has occurred for example, in the QSL Process [40]. As non-ferrous converting operations eventually become continuous processes [38], it is expected that the shrouded tuyere design will likely find additional applications. The use of the porous plug injector first pioneered by Lee and Spire in 1947 may also find new non-ferrous applications, as ladle refining techniques employed in the steel industry spread to non-ferrous metallurgy.

On Reflection

History teaches us lessons that we would do well to heed. There are the lessons of fresh thought, new insight and all that is possible. There are the lessons of stagnation, comfort and linear change. Creativity draws upon the spirit of man and fathoms his soul. This symposium is about that human dimension - and demands an appreciation of the intellect and perserverance of great men and women who overcome conventional barriers. This is the essence of the achievement of Savard and Lee.

Acknowledgement

The authors are indebted to Guy Savard and Bob Lee who have been so helpful in the preparation of this paper, by supplying their knowledge and by checking the accuracy of our presentation.

References

1. J.Y. Lancaster and D.R. Wattleworth, The Iron and Steel Industry of West Cumberland, (The British Steel Corporation, Teeside Division, 1977), 198 pp.

2. R.F. Tylecote, A History of Metallurgy, (The Metals Society, London, 1976), 182 pp.

3. Richard Preston, American Steel - Hot Metal Men and the Resurrection of the Rust Belt, (Prentice-Hall Press, New York, 1991), 278 pp.

4. Sir Henry Bessemer, F.R.S., An Autobiography, published by Offices of Engineering, London, 1905, (The Institute of Metals, London, 1989), 381 pp.

5. J.S. Fulton, "Analysis of the Generation and Delivery of the Blast to the Metal in a Bessemer Converter", Technical Publication TP1344, (The American Institute of Mining Engineers, 1941).

6. H. Bessemer, "Manufacture of Iron", British Patent No. 2, 768, 7 December, 1855.

7. BOF Steelmaking, Volume I, (The Iron and Steel Society of AIME, Warrendale, PA., 1974), ibid, Volume V, 1977.

8. E.W. Starratt, "LD ..... In the Beginning", J. Metals, 12(7), (1960), 528-529.

9. P.H. Dauby: "The Multiple Facets of Integrated Steelmaking from 1925 to the Present Day: A Pattern of Continuous Improvement", in Steelmaking in the 20th Century - From Black Magic to Technology, (Iron and Steel Society, Warrendale, PA., 1992), 1-30.

10. W. Kilbourn, "The Elements Combined - A History of the Steel Company of Canada", (Clarke, Irwin and Company Limited, Toronto, 1960), 335 pp.

11. M.L. Wayman, "All that Glitters: Readings in Historical Metallurgy", (The Metallurgical Society of Canadian Institute of Mining and Metallurgy, Montreal, 1989), 197 pp.

12. F.J. McMulkin, "Oxygen Steel Produced at Dofasco Can Compete with Open Hearth", J. Metals, 7, (4), (1955), 530-534.

13. C.C. Benton, "Notes on Algoma's LD Steel Plant", J. Metals, 12(7), (1960), 548-570.

14. G.A. Ferris, "Oxy-Fuel Increases Open Hearth Potential", J. Metals, 13(4), (1961), 298-299.

15. F.W. Irwin, "The Impact of Tonnage Oxygen on Steelmaking at Stelco", CIM Bulletin, 58, (1965), 838-842.

16. G. Newton, "Stelco's First BOF", J. Metals, 24(10), (1972), 15-19.

17. A.B. Wilder, "The Place of Basic Oxygen Steelmaking", J. Metals, 12(7), (1960), 529-530.

18. G. Savard and R. Lee, "Method and Apparatus for Treating Molten Metal with Oxygen", U.S. Patent No. 2, 855, 293, 7 October 1958.

19. G. Savard, R. Lee and M.R. Campbell, "Bottom Blowing With Oxygen", J. Metals, 12(7), (1960), 566-569.

20. R. Lee, "Innovations in Ferrous Pyrometallurgy - A Canadian Perspective", CIM Bulletin, 83, (1991), 125-131.

21. J.L. Bray, Ferrous Process Metallurgy, (John Wiley and Sons, New York, 1954), 414 pp.

22. A. Jackson, Oxygen Steelmaking for Steelmakers, (Newnes-Butterworths, London, 1969), 358 pp.

23. H. Bessemer, "Manufacture of Iron", British Patent No. 2, 768, 7 December 1855.

24. R. Lee, "Test K-205 on Top Lancing of Iron with Propane and Oxygen, 17 December, 1964", Canadian Liquid Air, Montreal; Ibid, "C.L.A. Experimental Converter: Oxy-Fuel Bottom Injector Tests, June 10-11, 1965", (Canadian Liquid Air, Montreal, 1965).

25. G. Savard and R. Lee, "Submerged Oxygen Injection for Pyrometallurgy", Proc. Savard-Lee International Symposium on Bath Smelting, (TMS, Warrendale, Pa, 1992).

26. "Technology and Steel Industry Competitive Mess", U.S. Office of Technology Assessment, Report No. PB80-208200, (U.S. Department of Commerce, Washington, D.C., 1980).

27. G. Savard and R. Lee, "Improvements in Metallurgical Process", French Patent No. 1,450,718, 18 July, 1966.

28. K. Brotzmann, "The Bottom Blown Oxygen Converter - A New Method for Steelmaking", Technik und Forschung, 41 (1968), 718-720, BISI Translation No. 7255.

29. "Q-BOP: from Blow to Go in 90 Days", J. Metals, 24(3), 1972, 31-37.

30. B. Crosariol, "Men of Steel, Wills of Iron", Financial Times of Canada, (Oct. 28 - Nov.4, 1991), 16-18.

31. H. Kauppel, K. Brotzmann, H.G. Fassbieder, G. Savard and R. Lee, "Method for Refining Pig-Iron into Steel", U.S. Patent No. 3,706,549, 19 December, 1972.

32. "Q-BOP: Year II", J. Metals, 25(3), (1973), 33-41.

33. "Bottom-Blown Q-BOP Furnaces See a Major Revival for Tonnage Production", Metals and Materials, The Metals Society, London, (1975), 19-25.

34. S. Ishihara, "Latest Advances in the Oxygen Steelmaking Processes", Iron and Steelmaker, 9(7), (1982), 43-48.

35. American Metal Market, Feb. 19, 1992.

36. K. Nakanishi, T. Fujii and J. Szekely, "Possible Relationship Between Energy Dissipation and Agitation in Steel Processing Operations", Ironmaking and Steelmaking, 3(1975), 193-197.

37. L-D Process Newsletter, (Voest-Alpine Industrieanlagenbau, Ges. m.b.H., Linz, Austria, May 1992).

38. P.J. Mackey, "Oxygen in Non-Ferrous Metallurgical Processes - Past, Present and Future", in The Impact of Oxygen on Productivity of Non-Ferrous Metallurgical Processes, (The Metallurgical Society of Canadian Institute of Mining and Metallurgy, Montreal, 1987), 1-30.

39. R. Lee, Canadian Liquid Air, Montreal, Private Communication, 1992.

40. P. Arthur, A. Siegmund and M. Schmidt, "Operating Experience with QSL Submerged Bath Smelting for Production of Lead Bullion", Savard-Lee International Symposium on Bath Smelting, (TMS, Warrendale, PA., 1992).

FROM OBM TO HISMELT: A RETROSPECTIVE AND A VISION

K. Brotzmann

Almost exactly 25 years ago, a novel process for steel production was developed. It took only half a year to realize its technical application on production scale - an incredibly short time for the development of a new complex metallurgic process. What was the situation at the time?

Maxhütte, a company which produced steel mainly according to the Thomas process, was facing a rather difficult situation: The smelting of their own ores resulted in a pig iron of a high phosphorus and silicon content and the refining of that pig iron according to the Thomas process had the well-known disadvantages - high nitrogen content and a low scrap rate. Moreover, it had another great disadvantage with regard to the Maxhütte's intentions to restructure their installations, - the temperature could not be increased to such an extent that the steel could subsequently be cast in a continuous casting plant. Likewise, it was impossible to process this pig iron as per the LD-AC process which had just been developed in those days. These aspects left little hope for Maxhütte's survival.

In the course of the planning stage, it became evident that the only possibility for a successful general restructuring of Maxhütte rested in achieving a considerably higher output of the Thomas converters with improved steel qualities. A more theoretical chance was seen in the development of a bottom-blowing oxygen converter, which, however, appeared to be an illusion in view of the results known so far. It was the general opinion in those days that, due to the high temperatures and the formation of iron oxide, it was basically impossible to operate a bottom-blowing oxygen converter with a sufficient bottom life.

It was probably this apparently hopeless situation which induced us to tackle once again the problem of developing a bottom-blowing oxygen converter. While looking for possible solutions, we got some hints that in Canada tests were carried out using nozzles where the oxygen was injected under the surface of an iron bath.

Proceedings of the
Savard/Lee International Symposium on Bath Smelting
Edited by J. K. Brimacombe, P. J. Mackey,
G. J. W. Kor, C. Bickert and M. G. Ranade
The Minerals, Metals & Materials Society, 1992

This is why I went to Canada to obtain further information. I saw a small converter in which for about ten minutes oxygen was injected by a bottom tuyere into an iron bath. I also observed that the formation of red fume could be reduced considerably by jacketting the oxygen with propane. Upon completion of the test, the nozzle was pulled out. Its point had a small metal deposit, which was of utmost significance as it implied that there was a chance for the nozzle not to wear. I brought the nozzle to Germany. In the course of six weeks, a converter bottom was built using about 50 nozzles similar to those employed in Canada, and on 17th December 1967, the first melt was blown. There were more men from the fire brigade than metallurgists at the site, as nobody could fancy that oxygen was less dangerous in connection with an iron bath if propane were added. A most astonishing thing happened: After about 20 minutes, the converter was turned down, and, after having carried out relevant modifications, the first melt was cast.

In the following three months, the bottom of the Thomas converter was for certain test periods removed and replaced by a new bottom with oxygen nozzles. During these tests a few heats were produced, and afterwards the Thomas converter was operated furtheron in the normal production. By mid-March of 1968, enough experience was gained allowing a continuous operation of the converter on a production scale. A quarter of a year later, it was already decided to restructure Maxhütte as a whole on the basis of the new process. Two continuous billet casters were installed at Sulzbach-Rosenberg, the open hearth shop and the electric arc furnace at Haidhof plant were shut down, three obsolete sheet rolling mills were scrapped, and two bar mills were installed. Indeed, everything "was staked on one throw", which meant that a reliable operation of the new steel process had to be assured. Since everything sounded somewhat unbelievable, the board had postulated to install an electric arc furnace as well, so that, if necessary, the melts coming out of the OBM converter, as it was called by then, could be subjected to an after-treatment in that furnace. No melt, however, had to undergo such an after-treatment.

The first step, i. e. to employ the injector developed by Savard and Lee on production scale, was only made with a view to solving the problems which had to be coped with at Maxhütte. Therefore, we were surprised to learn that very quickly other Thomas plants disposing of a more suitable pig iron were interested in rearranging their converters accordingly. Thus, many of the Thomas plants existing in Western Europe were adapted to the new process.

To our surprice the OBM process could be operated with lump lime just as the Thomas process, whereas the LD-AC process needed powdered lime in order to refine high phosphorus pig iron.

THE REFINEMENT OF HEMATITE HOT METALL

As is well-known, the LD-process refined hematite hot metal more easier than high phosphorus hot metal. Hence, we were of the opinion, that it should be particularly easy to refine hematite hot metal in a bottom-blowing converter. In a first trial, hot metal was synthetically produced in an electric arc furnace and charged into the converter. To our utmost surprise, however, the blowing behaviour was so bad that the refining of the charge could hardly be finished. The necessity arose, therefore, to develop lime powder injection into the bottom of the converter. It turned out that hematite hot metal could be refined by this method without any problems and that also the blowing behaviour for refining of high-phosphorus hot metal could be improved furtheron. From that point on, oxygen bottom-blowing was principally used together with the injection of lime powder - a vision that had always been dreamt of by Thomas converter metallurgists; but because of the wind box at the bottom of the converter, it could not be realized.

At that time, the CRM at Liege had also investigated on the refining of hematite hot metal by bottom oxygen injection. They found that the blowing behaviour could also be improved considerably by adding powdered lime to the top of the hot metal. A cooperation was agreed on, which, for two decades, has proved fruitful results for both companies.

In 1971, the first new steel plant with a 30 tonne bottom-blowing oxygen converter was commissioned at Surahama-Bruck in Sweden. Not long afterwards, US Steel restructured the first open hearth shop at Fairfield, installing two bottom-blowing converters in the existing halls. This is where large converters with a capacity of 180 tons were used for the first time. At the same time, US Steel restructured an LD steel mill with 3 converters of 180 tons each, which were under construction to adapt the bottom-blowing process. It soon turned out that the bottom-blowing process offered considerable metallurgical advantages due to reaching equilibrium created by the strong stirring motion.

On the basis of experience gained by US Steel, Kawasaki Steel erected a new steel mill at their Chiba plant. Investigations carried out later on by Kawasaki Steel clearly demonstrated that mainly for the production of low carbon grades, the bottom-blowing process had essential advantages compared to the LD process.

Surprisingly however, the approach to the metallurgical equilibrium in case of the LD process could be improved considerably by bottom stirring with small quantities of inert gases. These improvements were so distinct that it became difficult to install the bottom-blowing process in existing LD-plants. On the other hand, it also became evident that in case of the bottom-blowing process, it was not necessary to inject the whole amount of oxygen through the bottom.

Investigations carried out at Maxhütte and later on by Kawasaki Steel at their Mizushima plant have demonstrated that the same results could be obtained with bottom-blowing rates of 30-40%.

Nippon Steel have reduced the bottom-blowing rate further to 5-7% and found out that there was still enough mixing energy. For these low bottom-blowing rates only very few nozzles are needed, so that it was no longer necessary to use an exchangeable bottom.

As is well known, all of these different developments were influencing the converter processes now resulting in almost all oxygen blowing converters use a certain bottom-blowing rate with oxygen or inert gas.

POST-COMBUSTION OF THE CONVERTER OFF GASES

New areas for metallurgical developments could be forseen, when the phenomenon of the so-called post-combustion was incidentally discovered in 1978. Tests were made with a bottom-blowing converter to oxidize the slag by top-blowing of oxygen, thus inducing dephosphorizing in case of catch-carbon heats. Surprisingly, the final temperature of these melts was extremely high. The anticipated metallurgical reactions had however not occurred.

As found by investigations carried out later on, the following discoveries were made during this tests: The high bottom-blowing rate had caused an intensive stirring of slag and bath with high amounts of iron droplets being simultaneously ejected into the gas zone. The soft top-blowing jet was sucking off gases into the converter post-combusted them, and transfered heat due to the contact with the iron droplets, while, at the same time, part of the reaction gases were reduced again. An evaluation showed that the high heat transfer rate could neither be explained by radiation nor by the contact of the top-blowing jet with a flat iron melt. The conditions of the top-blowing jet were of such a nature that, in case of a converter without bottom-blowing, a strong formation of foaming slag had to be expected, which was most probably prevented by the strong stirring action. The oxidation rate of the converter off gases was about 20%.

Already at a post-combustion degree of 20% it was worthwile to use coke or coal as an energy source in the converter. Therefore, a combined blowing converter of Maxhütte was restructured for the injection of coal powder. By injecting of approximately 40 kg coal, the scrap rate was increased by about 200 kg per ton of steel. For several years the Maxhütte converters were operated according to this process which was later on called KMS process.

Experiences gained during these tests served as a basis to install a converter with a capacity of 100 tons in an existing OH plant at Georgsmarienhütte in 1983, with the goal to melt 100% scrap. After an initial phase during which the converter was operated according to the KMS process, the whole production of the plant was operated with a scrap rate of 100% (KS process) for about one and a half years. It turned out, however, that the new process was too complex and not yet sufficiently controllable. When the difference between the costs of pig iron and scrap was reduced to approximately 200,-- DM, it was decided to reoperate the blast furnace and to operate the converter again according to the KMS process.

In general, the problems with the KS process had a negative influence on the application of post-combustion. There was even a general opinion stating that post-combustion was only a phantom, some publications alleging that post-combustion was theoretically impossible.

POST-COMBUSTION AND SMELT REDUCTION

Investigations on the process of post-combustion carried out by Maxhütte using a KMS converter operated on a technically large scale had two essential results:

a)   Post-combustion does exist and is technically controllable
b)   The energy transfer rate is higher than in any other metallurgical process.

These findings were the basis of first tests for smelt reduction on a large scale in 1982, using an existing 60 ton KMS converter. Within one hour, about 20 tons of lignite coke and 25 tons of ore were blown through the bottom of this converter. It turned out that there were no limits with regard to reaction kinetics, and that, also for post-combustion, the thermal balance was calculable and controllable.

Hence it was proved that it is possible to carry out smelt reduction of iron by means of a converter process. To ensure a technical application, however, the thermal efficiency had to be further improved. For this purpose, more knowledge had to be gained on processes taking place during post-combustion. In 1984, a test converter with a liquid steel capacity of 10 tons was installed at Maxhütte for investigations on post-combustion. By means of this converter, detailed investigations were carried out on the influence of the bottom-blowing rate, the coal type, the nature of the post-combustion gas and the fluid dynamics on post-combustion and on thermal efficiency. It turned out that, using oxygen under optimal conditions, a post-combustion degree of 35% with a thermal efficiency of 85% could be achieved. Using hot air of 1.200 °C, a post-combustion degree of about 60% at also about 85% thermal efficiency was attained. A further improvement by about 8 percentage points could be achieved by modifying the structure of the top-blowing jet.

It also turned out that the high bottom blowing rates usually employed are unnecessary and unfavourable for a high post-combustion. The investigations demonstrated as well that the individual process parameters could be kept under complete controll.

CRA and Midrex formed a joint venture for the further development of this so-called HIsmelt process. A demonstration plant with a capacity of about 100.000 tons per year is being established in Perth/Western Australia at present. Preheated fine ore is blown into an iron bath of about 30 tons. The coal consumption will be at 630 kg per ton of pig iron.

NEW DEVELOPMENTS AND VISIONS

In the course of the past 20 years, the metallurgy of the bottom-blowing and the combined bottom-blowing oxygen converter have been investigated in detail. There are not many surprise findings to be expected in the future in this field. In my opinion, new fields of application might still be possible with regard to two special areas which have not found the required attention: The first area covers the combined blowing with a bottom-blowing rate of about 6% oxygen. The conversion of existing converters to this process type would be simple. It is not necessary to provide an exchangeable bottom. The nozzle life is improved by using smaller nozzles and by increasing the hydrocarbon rate to such an extent that the nozzles will last for 2.000 melts. In case of this bottom-blowing rate, the injection of lime powder is not necessary. The metallurgial results, however, correspond rather closely to the known results obtained with the combined blowing converter. So far, the findings of post-combustion have not yet been applied for this mode of operation. There are some indications that a bottom-blowing rate of about 6% would be sufficient to achieve favourable post-combustion conditions. The pretreatment of the pig iron on the one hand and the high tapping temperature for steel which is continuously cast on the other hand might require a reinvestigation of this technique.

Investigations have demonstrated that, using fine-grain powdered lime with a grain size below 0.03 mm, effective dephosphorizing at high carbon contents of the melt can be achieved. This effect has not been investigated in detail. An interesting field of application is the dephosphorisation at catch-carbon heats in combined blowing converters.

In general, it can be expected that new applications for the bottom-blowing oxygen converter in the future will be in such fields where post-combustion will be applied. The application for the smelt reduction is already recognized at present. To achieve an effective post-combustion, the bottom-blowing rate has to be sufficiently high so that oxygen will have to be used as stirring gas for economical reasons.

On the other hand, the gas content of the coal types generally used is that high, that enough gassing will occur during the blowing-in through the bottom. For process technological reasons, bottom tuyeres alternately operated with coal and oxygen will find an optimal field of application in this respect.

The development of smelt reduction processes will be determined by a multitude of parameters, the consequence of which have not yet been sufficiently known so far. Finally, the use of fine-grain ore and coal powder as feed stock will be determined by their applicabilities, as to how these new treatment steps could be optimized to an integrated process. A lot of investigation will still be necessary in this field.

The fact that in two areas of the world, i. e. in Japan and in Western Australia, demonstration plants will be commissioned in due course demonstrates that this interaction of individual steps will now be investigated on a semitechnical scale and that - as both sites will use different process principles - very important new experiences will certainly be gained soon. Nevertheless, it goes without saying that this new complex process will still require a longer time till the great goal to replace the blast furnace with its sintering and coking plants can be realized by installing small production plants.

It appears to be easier to realize those steps that are based on the knowledge gained at the smelt reduction developments and to apply them for the melting of scrap or sponge iron. If a converter for the melting of iron was operated in the same way as the vessel for the smelting reduction, only approximately 100 kg coal would be needed for the melting of one ton of iron. Complicated devices for the preheating of the scrap would not be necessary. It is easy to imagine that sponge iron or shredded scrap are continuously charged in such a converter. Existing converters could simply be restructured and the existing supply and gas cleaning facilities could widely be used. Interesting applications would mainly be found where existing converter shops are intended to be transformed to electric furnace plants. The consumption of primary energy is only 60% as in case of coal being transformed into electrical energy in a power plant and then used for the melting of scrap. In this field, it will perhaps be possible to promote new developments rather quickly, as it is possible to revert to the comprehensive understanding of the elementary processes.

There are also novel applications within sight for the melting of dust or hazardous materials from gas cleaning facilities if such powders are blown into an iron bath and were the experiences of the smelt reduction development would be applied.

The preceding part of my talk gives a general view on the different developments which have been made since the introduction of the bottom-blowing converter, and their chances for the future. It was a report on the technical possibilities and their realization. Such developments often depend on casual events and their sucess is always unpredictable. They are, however, decisely influenced by the question how personages are showing great zeal for them and how much unusual men of talent have sustained these developments. I would, therefore, like to remind you of those who have contributed to the phenomenon that a new element could give birth to those many novel metallurgical applications. The beginning was the excellent performance, of course, for which we are gathering today, and which is the idea of our symposium to recognize this. Without this first step taken by Savard and Lee all the above events would not have been imaginable. Nevertheless, we should also give credit to the performance of those persons who have participated by their efforts and outstanding distinction to the effect that also the following steps could be realized.

In this respect, Professor Helmut Knüppel and Dr. Hans-Georg Faßbinder of Maxhütte should be mentioned in the first place. Professor Knüppel, the former head of the technical management of Maxhütte believed in the vision that a bottom-blowing oxygen converter was imaginable and that it was possible to solve the problems of Maxhütte by means of this device. He had shown great courage by combining the future of the company with the success of the new development. And it was Dr. Hans-Georg Faßbinder's special merit to tackle the technical realization of this development. Very well informed about the individual steps of the new process he developed the required installations. He was able to see through the problems always occuring at the beginning of such a development to the effect that his constribution represented an essential prerequsite for the final success at Maxhütte.

Sven Fornander, who had built the first new steel mill for the refining of hematite hot metal according to this new process, and Ed Speer, who was reponsible at US Steel to use this bottom-blowing process in large converters, have essentially contributed to the further acceptance of this process. Finally, Mr. Kawana of Kawasaki Steel has promoted the acceptance of the new process in Japan, which formed the basis of further efforts in investigating in detail the application possibilities of this process and in introducing it into Japan's steel industry.

As to the second important step which will even be more important for the future, I would like to point out the activities of John Innes of the Australian Raw Material Company CRA. He was the first to recognize the great chances of this new process for the reduction of iron ores and he had exposed himself for this process at his company, thus enabling to take this decisive step of the development at all.

There are of course no visions as to which personages will decide on the progress in the future. But it will only be possible to realize it if persons do really engage themselves in these new opportunities and also the future will be mainly determined by those who will expose themselves to the new tasks.

To complete my description, I would like to finally revert to Guy Savard and Bob Lee once again. Apart from the great technical achievement which we are celebrating today, I would like to point out that it was not only 25 years of a succesful technical cooperation but also 25 years of an unspoiled friendship. This is a special aspect of our relationship which cannot be taken for granted.

BATH SMELTING PROCESSES IN NON-FERROUS PYROMETALLURGY:

AN OVERVIEW

H.H. Kellogg * and C. Díaz **

* Stanley-Thompson Professor Emeritus
Columbia University
New York, New York, U.S.A. 10027

** Section Head, Pyrometallurgy
J. Roy Gordon Research Laboratory
Inco Limited
2060 Flavelle Boulevard
Mississauga, Ontario, Canada L5K 1Z9

Abstract

Operational data for thirteen non-ferrous bath smelting processes are analyzed with respect to oxygen utilization, smelting intensity, and other features related to metallurgical efficiency and environmental control. These processes include both old and new designs, batch as well as continuous reactors, tuyere and lance injection, and reduction as well as oxidation chemistry. The smelting intensity of these very different reactors is compared by a measure of their specific oxygen consumption rate: SBSR, specific bath smelting rate, expressed as Nm^3 O_2 used$/h/m^3$ of bath volume. On this basis, processes are grouped into high (SBSR > 250), intermediate (SBSR 100–200) or low (SBSR < 100) intensity.

Process features that determine reactor intensity are discussed. Comparisons of the intensity of non-ferrous bath smelting with flash smelting and with oxygen steelmaking are presented.

Proceedings of the
Savard/Lee International Symposium on Bath Smelting
Edited by J. K. Brimacombe, P. J. Mackey,
G. J. W. Kor, C. Bickert and M. G. Ranade
The Minerals, Metals & Materials Society, 1992

Introduction

Bath smelting has a long history in metallurgical processing. The successful application of the bottom-blown Bessemer converter for steelmaking in 1860 at Sheffield, England, marks the beginning, but not many years went by before copper matte was converted to blister, first (circa 1885) in converters similar in shape to the Bessemer vessel, and later (circa 1905) in the horizontal, cylindrical Peirce-Smith converter. Fuming of zinc from lead blast furnace slags began in 1927 at the Anaconda plant in East Helena, Montana, employing a stationary, rectangular furnace. All of these processes used submerged tuyeres, either bottom or side blown.

These early processes possess the unique characteristics that we have come to associate with bath smelting — heat generation results from oxidation of metallic sulfides, impurity elements or fuel by *submerged combustion* within a turbulent molten bath; air/oxygen is supplied to the bath by tuyeres or lances; and they are *intensive* processes, characterized by very high throughput compared with the processes they displaced.

Successful processes for direct bath smelting of copper concentrate were not commercialized until the early 1960s. It may come as a surprise, therefore, that extensive experimentation on direct smelting of copper concentrates in a Great Falls converter was conducted as early as 1907, as described by Wheeler and Krejci (1). Concentrates were added to the converter, both by dumping them into the mouth and by injection through the tuyeres. Wheeler and Krejci explain that trials of the process, although encouraging, were discontinued, but they prophetically conclude that "it should be borne in mind that this was an experimental run with only a small amount of previous trial of the manipulation of the process, and was carried out in a small converter of a class which has since been abandoned. With the present basic-lined converters of a much larger size, there is no doubt that much better results can be obtained, particularly with practice".

The potential for intense continuous smelting processes, either bath or flash smelting, tantalized design engineers since the turn of this century, but such processes were only reduced to practice in the period following World War II. The availability of tonnage oxygen at reasonable cost was a key factor in spurring this development. No invention in the history of metallurgy rivals the spectacular success and rapid worldwide adoption of the top-blown converter for basic oxygen steelmaking. Savard and Lee, who we honour at this symposium, perfected the bottom-blown tuyere which could employ pure oxygen, protected against severe overheating and refractory attack by use of a shrouding gas. Use of oxygen in non-ferrous metallurgy began in 1952 with Inco's oxygen flash smelting of copper concentrate (2), and has since spread to a wide variety of new, intensive bath smelting processes that employ tuyeres or lances.

This paper offers an overview of non-ferrous applications of bath smelting, with emphasis on a comparison of the various means to achieve submerged combustion — submerged tuyeres (bottom and side blown), lances above the bath, and submerged lances. We have limited our attention to processes that are in commercial operation or are about to be commercialized, and because of the many different processes that must be considered, only the principal characteristics of each will be discussed. The details of chemistry, which are specific to the metal being recovered, are left to other papers that deal with that application. Here we will emphasize reactor design — whether it lends itself to batch or continuous processing, how it contributes to the ability of the process to achieve intensive oxidation by accepting high oxidant injection rates and/or high levels of oxygen enrichment of the blast while attaining acceptable reactor campaign life and meeting current workroom and external environmental standards.

Whether tuyeres or lances supply gaseous oxidant to a bath smelting process, the physical interaction between gas and molten bath — the nature of the gaseous plume and its dispersion in the bath, the resulting agitation, mixing, splashing or frothing of the bath, and the effect of these on the wear of refractories, tuyeres or lances — plays a vital role in determining the limitations on the operating parameters of the process. Experimental and theoretical studies of the gas/bath interaction over the past twenty years have contributed significantly to our understanding of this complex phenomenon (3,4,5,6). These have already contributed to process improvements, and it seems likely that such studies will continue to play a major role in the future as is evidenced by some of the papers in these proceedings. The nature of this review precludes detailed consideration of this fundamental aspect of bath smelting, but this in no way implies a lack of importance.

Oxygen Utilization in Bath Smelting

The high intensity (high specific-capacity) of bath smelting can best be measured by the *rate at which the process consumes oxygen*. Consumption of oxygen, from air, oxygen-enriched air or "pure" oxygen, relates directly to the rate of smelting. This is true whether the smelting reactions involve the oxidation of metal sulfides (as in copper or lead sulfide smelting) or reduction of metal oxides. In the latter case the oxygen may first react with excess reducing agent (coal, fuel oil, or reducing gas) to produce heat for the process, and CO and H_2, which then act to reduce metal oxides; or the primary reaction may be between hydrocarbon and metal oxide, to produce CO, H_2 and metal, with oxygen reacting with a portion of these products to produce heat for the process.

By using oxygen consumption rate as a measure of smelting rate one gains the advantage that the resulting measure is independent of the smelting task, whether this be smelting of lean or rich feed materials, or which metal (copper, nickel, lead or zinc) is being smelted.

Oxygen Utilization Efficiency

Not all oxygen blown into a bath smelting process is *consumed*. That portion which leaves the furnace as free oxygen does not contribute to smelting reactions, and should not be counted in any measure of smelting intensity. The measure of the total oxygen input that is consumed, expressed as a percentage, is the *oxygen utilization efficiency*, OUE. Process engineers attempt to measure or estimate the OUE, but inaccuracy in metering, loss of oxygen at tuyere valves, and infiltration of ambient air into process gas, among other factors, represent sources of error in measurements and estimates of this quantity. Values of OUE, reported in this paper range from 70 to 100%, with most values being 95% or greater. We estimate that reported values of OUE are, for the most part, uncertain to ± 5%. Clearly, bath smelting processes are capable of very efficient use of oxygen.

The rate at which a process consumes oxygen may be calculated from the metered rate of air, of known percentage of oxygen, blown into the process multiplied by the OUE, expressed as a fraction, to yield the rate of oxygen consumed in units of Nm^3/h. Aside from inaccurate metering, another problem arises from different conventions regarding the standard conditions chosen for reporting metered values. Here we adopt zero degrees centigrade and one atmosphere pressure as standard conditions for calculation of Nm^3, and we have attempted to adjust all of our data to this standard. Selection of one bar as the standard pressure results in a difference of only 1.3% in calculation of Nm^3, but a metering standard of 70°F, rather than 0°C, causes a difference of 7.7% in the value of Nm^3.

As a check on the rate of oxygen consumption calculated in the above manner, we have also calculated this parameter from reported data on feed rates and compositions of process input and output streams. This method of calculation of consumption of oxygen suffers from inaccuracies resulting from incomplete chemical analyses, uncertainty about the state of oxidation/reduction of some feed and product materials, and inaccuracies in reported feed and product rates. However, we were gratified to find that the agreement between oxygen consumption rates calculated in this manner with those calculated from the air/oxygen blast seldom exceeded 5%.

Measures of Specific Oxygen Consumption Rate

The rate of oxygen consumption can be converted to a more meaningful measure of smelting intensity by relating it to some measure of the size of the smelting vessel or the volume of the liquid bath contacted by the oxygen blast. We offer here two different measures of *specific oxygen consumption rate*, as follows:

Specific Bath Smelting Rate, SBSR: Nm^3 O_2 used/h/m^3 of bath volume
Specific Furnace Volume Smelting Rate, SFVSR: Nm^3 O_2 used/h/m^3 of furnace volume

41

The bath rate, SBSR, is the true measure of the intensity of reaction within the bath; this rate can only achieve high values if a vigorous blast is well dispersed throughout the bath, without significant "dead volume". In the examples that follow we record values of this measure that range from 33 to 504 $Nm^3/h/m^3$. Factors that lead to low values of SBSR are: the need to provide a region of quiescent bath for settling of matte from slag; the necessity to avoid intense agitation so as to minimize back-mixing in a process where a gradient of bath composition along the length of the reactor must be maintained; and low reaction rates which require increased residence time in the reactor.

The furnace volume rate, SFVSR, is always lower than SBSR because the furnace volume is necessarily larger than the bath volume. The furnace must provide some volume for the gas phase above the bath surface, for control of splash, disengagement of liquid droplets, and, in some cases, for postcombustion of CO, H_2 and metal vapours that issue from the bath. Furnace geometry and provision for feeding and tapping also influence the furnace volume relative to the bath volume it contains. A high value of SFVSR is desirable because this will result in a smaller and less costly reactor, and smaller heat losses from the smaller reactor shell, for the same smelting capacity.

The ratio Furnace Volume/Bath Volume, F/B, is a measure of the excess furnace volume necessary to provide gas space and address the other factors noted above. This ratio should be kept as small as possible in order to yield a smaller, less costly reactor. As shown in the examples that follow, the ratio varies in the range 2.7 to 8.3.

There follows a summary for each process we have reviewed, including the data made available to us, calculations of SBSR, SFVSR and F/B, along with a brief analysis of the results. Finally, we present a comparison between the processes and our thoughts regarding future developments in bath smelting.

Injection Through Submerged Tuyeres

In his paper "Copper Converting — A Historical Perspective" (7), S. Marcuson reminds us that the pneumatic steelmaking process, developed simultaneously in 1855–56 by Sir Henry Bessemer in the U.K. and William Kelly in the U.S.A., was the precursor of the early and many of the modern non-ferrous bath smelting processes in which the oxidant is injected through submerged tuyeres. These processes are discussed in this section.

Peirce-Smith Copper Converters

Submerged oxidation of iron and sulfur from primary smelting mattes in Peirce-Smith converters has been the standard route to blister copper for almost a century. During this period, increasingly large converters have been built and operated; mechanized tuyere punching has been adopted in most smelters; oxygen enrichment of the blast has become a widespread practice; better refractories have been introduced; and other operating practices have been improved. However, the original, basic converter design of William H. Peirce and Elias A.C. Smith (8) has stayed the same. Moreover, some modern copper bath smelting processes, such as Noranda and El Teniente, use tilting, horizontal, submerged tuyere vessels and can, accordingly, be considered Peirce-Smith progeny.

Current Peirce-Smith converting practice is practically the same as that reported in the 1979 worldwide survey by Johnson et al. (9). This batch process is carried out in horizontal, cylindrical vessels, typically 9.1 m by 4 m diameter, although both larger and smaller sizes are in use. Air is supplied, at about 1 kg/cm^2 gauge pressure, to tuyeres (40 to 50) mounted along one side of the vessel. The tuyeres can be submerged into or raised above the bath by tilting the vessel about its cylindrical axis. Details of this well established process may be found in the above mentioned survey and throughout the metallurgical literature.

The information collected by Johnson et al. covered 47 smelters from 14 countries. The data showed a reasonable correlation between blowing rate and converter internal volume, with a value of 482 Nm^3 of air/h/m^3 (8.5 SCFM/ft^3) representing average late 1970s industrial practice. Interestingly, similar specific blowing rates were achieved in the various versions of the Great Falls converters used in Montana in 1892–1913 (1,7).

These specific blowing rates continue to be standard industry practice. However, following the 1979 survey, progress has accelerated in other areas of converter technology in the continuous pursuit of higher equipment productivity and improved environmental and workroom practices. For instance, the use of mechanical punching, quite extensive nowadays, has greatly contributed to limiting fluctuations in blowing rates during a converting cycle due to tuyere blockage. The adoption of puncher-silencers has resulted in much decreased air losses during punching (10), still as high as 5–10% in smelters which do not use these devices. In addition, mass flow meters with the capability to totalize air (and oxygen) injected into converters during fixed periods of time are now in use in a number of plants. As a result of these developments, more accurate and reliable data on blowing rates and other converter operating parameters are becoming available. The development of computerized converter models, already employed in some smelters as an effective process control technique, has emphasized the need for quality operating data.

Blowing rates in submerged tuyere converters are limited by bath slopping and splashing (3,5,11). "Slopping" refers to the wave action which occurs in a converter as a result of the transfer of momentum from the blast to the molten bath. Vigorous slopping may contribute to splashing. Excessive splashing results in high dusting and rapid formation of accretions on the converter mouth and in the hood. In turn, removal of these accretions has a negative effect on equipment life and maintenance cost. It is customary not to install tuyeres in the area immediately under the converter mouth to minimize these problems.

Richards et al. (5) investigated the mechanism of slopping and splashing using a plexiglass-water model. They concluded that blowing rate, bath volume (mass) and tuyere submergence are critical parameters. They suggested that slopping and splashing could be reduced by maximizing tuyere submergence and converter percent fill, thus allowing an increased blowing rate. However, no industrial testwork appears to have been conducted to explore this possibility.

Based on the foregoing and additional experimental information on the fluodynamics of Peirce-Smith converters (12), which indicates that the transfer of momentum from a tuyere jet to its corresponding segment of molten bath is a localized phenomenon, and using plant data published by Noranda (11,13) for three different sized vessels,* a rough "blast-bath momentum transfer" parameter can be calculated, expressed in normal cubic meters of blast per hour per cubic meter of bath directly influenced by the tuyeres. The value of this parameter varies from about 1,700 for the 4.3 m by 9.75 m Noranda converter to about 1,900 for the Noranda reactor. In the case of the 4.3 m by 9.75 m converter, the parameter was calculated assuming 35% bath fill and using a blowing rate of 47,900 Nm^3/h (29,800 scfm), the "highest average blowing rate for a single month above which material slopping out of the mouth occurs" (11). This blowing rate corresponds to about 550 Nm^3 of blast/h/m^3 of converter volume, a value some 15% higher than the late 1970s industrial average (9). Evidently, a true blast-bath momentum transfer parameter should reflect the effect of factors such as blast injection velocity, tuyere angle and submergence, among others (5,6). More industrial data are required to establish better correlations between slopping and splashing and converter operating parameters.

Forty one out of forty seven smelters provided information on OUE for the 1979 Johnson et al. survey (9). There was considerable scatter in the data, but 70% of the smelters quoted OUEs higher than 80%, with 30% achieving better than 90%. Attempts to correlate OUE with operating parameters such as tuyere submergence and blowing rate failed. This apparent lack of correlation may have been due to inconsistency in the data provided by the various smelters. Considering that the residence time of the oxidant in the bath is extremely short, one second or less (14), deeper tuyere submergence would be expected to increase OUE and vice versa. In fact, it is still industry practice to "blow shallow" to avoid bath overheating when operators are short of cold dope. Operating practices such as this may explain low OUEs. Similarly, a relationship between OUE and blowing rate appears to exist, as was observed in one of the few reported plant efforts to directly measure converter OUE (15). More data on current, improved converter practice are required to clarify these as well as

---

* 4 m by 9.1 m (13'x30') and 4.3 by 9.75 m (14'x32') Peirce-Smith converters and the 5.2 m by 21.3 m (17'x70') Noranda reactor .

other questions, such as the impact of air inleakage, related to OUE. In any case, the available information indicates that, at the blowing rates practised in Peirce-Smith converting, OUEs well above 90% are achievable both in the slagging and copper blows, even at high levels of oxygen enrichment of the blast. *The barrier to increased blowing rates, as discussed above, is physical rather than chemical.*

Possibly the single most important development in copper converting during the 1970s and 1980s has been the increase in primary smelting matte grades resulting from the rapid substitution of flash furnaces or bath smelting processes for reverberatories. The decrease in primary matte latent chemical heat (less iron sulfide available for combustion) has been compensated by oxygen enrichment of the blast in order to maintain converter capability to digest smelter reverts. Oxygen enrichment of the converter blast, typically 25 to 30%, is widespread practice today.

Based on the foregoing, we have used a blowing rate of 500 Nm^3 of air/h/m^3 of internal converter volume and a rather conservative 90% OUE to calculate typical Peirce-Smith, furnace and bath (35% converter filled) specific oxygen consumption rates for blowing with plain air and with 30% oxygen enriched air. These values are:

		Plain Air	30% O$_2$–Air
SFVSR	Nm^3 O$_2$ used/h/m^3 furnace volume	95	135
SBSR	Nm^3 O$_2$ used/h/m^3 bath volume	270	385
F/B		2.85	2.85

These values place Peirce-Smith converting in the group of high intensity bath smelting processes.

As mentioned in the introduction, concentrate smelting in submerged tuyere converters was attempted in the early years of this century (1). The mechanical problems experienced by Anaconda's pioneering metallurgists, in particular during the injection of concentrate through tuyeres, deterred operators from other attempts, with few exceptions (16,17), for many years. However, with the growing availability of tonnage oxygen in the late 1940s, concentrate smelting in converters, using oxygen enriched air, came to be considered a simple, low capital approach to increasing smelter capacity (18). This type of operation was practised in several smelters (19,20,21,22) during the 1960s and 1970s and well into the 1980s. At Hitachi, where concentrate smelting in converters was conducted for many years, the blast oxygen content was normally in the range 35–40% during the slagging blow. Apparently, such high levels of oxygen enrichment did not shorten refractory life "provided the heat balance was properly maintained" (19). High dusting of the concentrate, even when previously pelletized, and the above mentioned increase in primary smelting matte grades combined to discourage further spread of this technology. Instead, those who persisted in exploring concentrate smelting in converters ultimately evolved modern continuous bath smelting techniques such as the Noranda and El Teniente processes.

Noranda Process

The Noranda process was developed in the late 1960s. The first commercial reactor was commissioned at Noranda's Horne smelter on March 1st, 1973. Several descriptions are available of the Noranda reactor and the Horne smelter (23,24,25).

The essential features of the Noranda reactor and process are shown in Figure 1. The reactor consists of a cylindrical, tilting vessel equipped with tuyeres located between the mouth, which serves as exit for the off-gas, and the feeding-end riding ring.

The charge, which consists of a blend of fresh concentrate, reactor slag concentrate, reverts, recycled dust and flux, with a moisture content varying from 6 to 8%, is slingered onto the molten bath. The charge is smelted-converted to a high grade matte, analyzing 70 to 75% copper, and a slag with a composition similar to converter slag. Under normal operating conditions, the bath occupies about 28% of the reactor internal volume. The matte tapping and the slag skimming holes are located, respectively, in the mantle and in the wall opposite to the feeding end. The slag is slow-cooled and milled to recover its copper value. Dust carry-over by the off-gas is about 2% by weight of concentrate.

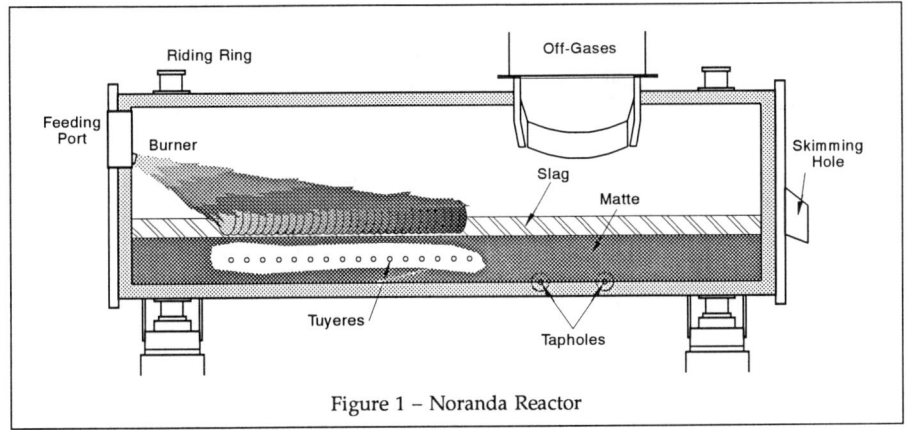

Figure 1 – Noranda Reactor

The Noranda reactor can accept a wide range of feed compositions and sizes. The feeding system and the reactor can handle feed sizes up to about 100 mm with moisture contents up to about 14%. This versatility is one of the important merits of the process.

A small burner, natural gas at the Horne smelter, located in the feeding wall and a small amount of coal added to the feed, about 1% by weight of total solid charge, generate heat in the free-board and on the surface of the bath to partly offset the cooling effect of localized endothermic reactions, such as water evapora-tion and decomposition of sul-fates. A portion of the labile sulfur also burns in the reactor free-board. The charge is readily digested into the turbulent bath where most of the oxidation reac-tions and heat generation take place.

At today's high feeding and blowing rates and high oxygen enrichment of the blast, the reac-tor requires a dedicated process control strategy (26) involving frequent, periodic matte samp-ling and temperature measure-ments through tuyeres, to limit changes in matte composition or bath temperature.

The main Noranda process and reactor characteristics are given in Table I (13,27). The reactor specific smelting rate is 11.7 tonnes of dry solid charge/m³ of

Table I – Noranda Reactor (Horne Smelter) *		
Charge, tonnes/h, (% Cu):		
Concentrate	97	(21.9)
Slag concentrate, reverts, scrap, etc.	43	(28.0)
Total charge including flux	154	(21.7)
Products, tonnes/h, (% Cu):		
Matte	35	(72.4)
Slag	80	(5.7)
Fuel: Coal, tonnes/h		2
Natural gas, Nm³/h		0–560
Reactor: External diameter, m		5.2
Length, m		21.3
Internal volume, m³		316
Number of tuyeres		56
Tuyere diameter, mm		54
Bath volume, m³		88
Blast, Nm³/h, (% O₂):	72,340	(35.8)
Velocity at tuyere tip, m/s		157
Gauge pressure, kg/cm²		1.27
Utilization of Oxygen		
Oxygen utilization efficiency, %		97.5
SBSR, Nm³ O₂ used/h/m³ bath volume		286
SFVSR, Nm³ O₂ used/h/m³ furnace volume		80
F/B		3.6

* Data from the literature (13) and from Noranda engineers (March–April 1992).

furnace/day, one of the highest in the industry. However, the SBSR and SFVSR values for the Noranda reactor are not significantly different from those typical of Peirce-Smith converters operated with plain air. The reactor's higher blowing rate per tuyere and higher oxygen enrichment of the blast are offset by the smaller proportion of the vessel equipped with tuyeres. There are no tuyeres below the mouth or at the skimming end of the vessel, thus providing an approximately 10 m long matte-slag settling zone.

The Horne reactor campaign life is about 300 days. The current, typically high levels of oxygen enrichment of the blast, 35 to 40%, appear to have little effect on refractory wear.

Another Noranda reactor, 4.5 m diameter by 17.5 m long, was recently commissioned by Southern Copper (formerly ER & S) of Australia. Kennecott has apparently decided to replace the Utah smelter Noranda reactors with Outokumpu technology.

El Teniente Converter

The development of the El Teniente converter was spurred by an unsuccessful attempt in the late 1960s to expand the capacity of the El Teniente Caletones Smelter by processing concentrates in Peirce-Smith converters. The first commercial El Teniente converter was commissioned in January, 1977. In the El Teniente process, seed primary smelting matte and fresh concentrate are converted-smelted continuously in a modified Peirce-Smith converter (Figure 2) to produce high grade matte, analyzing about 75% copper, and a converter type slag.

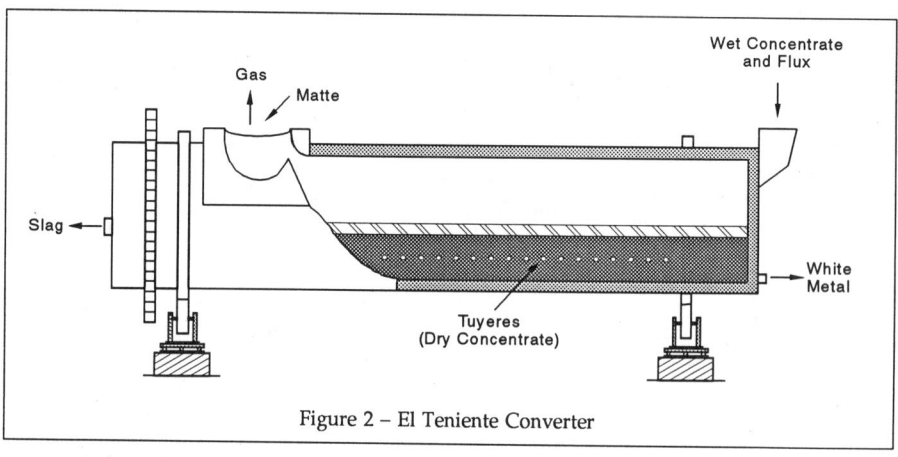

Figure 2 – El Teniente Converter

Descriptions of the El Teniente process and converter are available in the literature (28,29,30). Feeding of the solid charge, mainly a blend of fresh, wet concentrate (containing about 8% moisture) and flux, is done continuously, either through a hole in the converter mantle or by Garr gun. In recent years, the trend at the Caletones Smelter has been to inject an increasing proportion of the concentrate, bone dried, through tuyeres. Primary smelting matte, reverb matte at Caletones, is added intermittently through the mouth with the converter off tuyeres. Tapping of high grade matte and skimming of slag are done through holes located in the opposite end walls of the vessel. Bath occupancy is about 37% of the internal volume of the converter. Dust carry-over by the off-gas is similar to the Noranda reactor, about 2% by weight of dry solid charge.

The relative proportions of primary smelting matte and dry and wet concentrates fed to the modified converter are dictated by heat balance considerations. The main variables are matte and concentrate composition and oxygen enrichment of the blast. No fossil fuel is used in the El Teniente converter. Key process and vessel characteristics are presented in Table II. These data correspond to the largest El Teniente converter currently in operation (there are two such at the Caletones Smelter) in the so called high injection mode (90% of the concentrate injected through tuyeres). Under these conditions, only a small amount of seed reverb matte is required. Autogenous operation, with no seed matte, is planned for the near future (31).

The current SBSR and SFVSR of the El Teniente converters, presented in Table II, are relatively low for submerged tuyere vessels. Peirce-Smith converter and Noranda practice indicate that El Teniente converters should be capable of operating at higher blowing rates and higher levels of blast oxygen enrichment.

Campaign life for an El Teniente converter is about 300 days, similar to that of the Noranda reactor. Stack time throughout one campaign is about 90%. There are El Teniente converters of various sizes operating in three Chilean smelters in addition to Caletones.

Vanyukov Process

The Vanyukov process was developed in the former USSR in the late 1960s and early 1970s and was commercialized at the Norilsk copper and nickel complex in 1977 (32). There are six commercial Vanyukov furnaces operating today: four in Norilsk and two in Balkhash, also smelting copper. The process employs a rectangular stationary furnace with tuyeres located along the side walls. The portion of the side walls in direct contact with the agitated upper region of the bath consists of water (or steam) cooled panels. Schematic transversal and longitudinal cross-sections of the Vanyukov furnace are presented in Figure 3.

Table II – El Teniente Converter (Caletones Smelter) *		
Charge, tonnes/h, (% Cu):		
Bone dry concentrate	72	(30.1)
Wet concentrate	8	(30.1)
Reverb matte	4.2	(47.1)
Flux	9	
Products, tonnes/h, (% Cu):		
Matte	29.9	(75.1)
Slag	45.4	(6.8)
Fuel:		None
Converter:		
External diameter, m		5
Length, m		22
Internal volume, m³		293
Number of tuyeres		47
Tuyere diameter, mm		59
Bath volume, m³		108
Blast, Nm³/h, (% O₂):	60,000	(30.1)
Velocity at tuyere tip, am/s		79
Gauge pressure, kg/cm²		1.27
Utilization of Oxygen		
Oxygen utilization efficiency, %		95
SBSR, Nm³ O₂ used/h/m³ bath volume		159
SFVSR, Nm³ O₂ used/h/m³ furnace volume		59
F/B		2.7

* Data supplied to the authors (March 1992) by El Teniente engineers.

Highly oxygen-enriched air, 70 to 75 volume %, is injected into the slag layer, about 50 cm below the surface of the bath. Wet feed — concentrate, flux and reverts — is dropped into the furnace through roof ports. The solid materials are readily digested into the turbulent slag, where the oxidation of labile sulfur and iron sulfide generates most of the heat required by the process. Combustion of natural gas, also injected through the tuyeres, supplies the remainder of the heat. Matte and slag are discharged continuously from siphons located at opposite ends of the furnace. The off-gas exits through an uptake located in the furnace roof. The furnace atmosphere is reducing and the off-gas typically contains 2 to 3% elemental sulfur and minor amounts of carbon monoxide and hydrogen. Dust carry-over is 0.5 to 1.5% by weight of dry solid charge.

The Vanyukov furnace accepts feeds with up to about 10% moisture and sizes up to about 50 mm. At Norilsk, crushed massive nickel-containing copper sulfide ore appears to be an important component of the charge.

The location of the tuyeres in the Vanyukov furnace results in the formation of a vigorously agitated region in the upper portion of the slag layer with a relatively quiescent zone underneath. Consequently, fast smelting kinetics and good slag-matte separation are achieved simultaneously within the furnace. This separation is further enhanced by the slag matte siphon discharge arrangement.

47

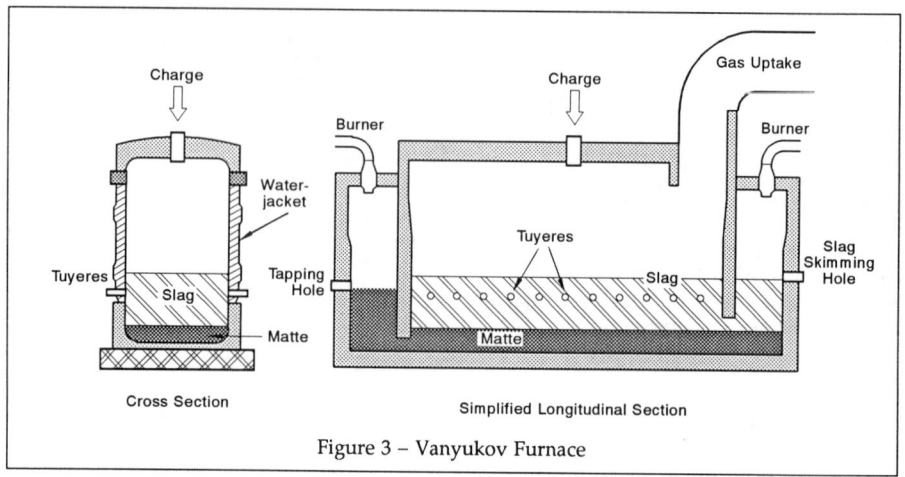

Figure 3 – Vanyukov Furnace

The dimensions and main operating parameters of two different sized Norilsk Vanyukov furnaces are given in Table III. The specific smelting rates of these furnaces, 15.4 and 17.4 tonnes of concentrate /m³/day respectively, are higher than those of the Noranda reactor. The SBSR and SFVSR values for the bigger Vanyukov furnace are also higher. This probably reflects the fact that the active portions of furnace and bath volumes in the Vanyukov process are larger. However, the Norilsk Vanyukov furnaces produce a much lower matte grade, 50% Cu versus 70–75% for the Noranda reactor.

Norilsk experience indicates that such a furnace can operate steadily for two years without complete draining of the bath to allow repairs. Furnace campaign life between major rebuilds is estimated to be longer than 5 years.

Table III – Vanyukov Process (Norilsk) *		
	Furnace 1	Furnace 2
Charge, tonnes/h, (% Cu):		
Concentrate	50 (20.0)	130 (20.0)
Converter slag		25
Total charge including flux	63	195
Products, tonnes/h, (% Cu):		
Matte	18 (50.0)	50 (50.0)
Slag	35 (0.6)	90 (0.6)
Fuel: Natural gas, Nm³/h	500	600
Furnace:		
Hearth area, m²	20	36
Volume, m³	130	230
Number of tuyeres (operating)	16	24
Tuyere diameter, mm	–	–
Bath volume, m³	42	72
Blast, Nm³/h, (% O₂):	16,500 (68)	37,000 (75)
Velocity at tuyere tip, m/s	200	200
Gauge pressure, kg/cm²	0.8–1	0.8–1
Utilization of Oxygen		
Oxygen utilization efficiency, %	100	100
SBSR, Nm³ O₂ used/h/m³ bath volume	244	369
SFVSR, Nm³ O₂ used/h/m³ furnace volume	79	116
F/B	3.1	3.2

* Data supplied to the authors (December 1991) by Intermet Engineering, Moscow.

<u>Baiyin Process</u>

This process (33), developed in China in the 1970s, is practised in a rectangular furnace divided into two sections by a transversal partition. The smelting section is equipped with tuyeres, located along the two side walls, which inject oxygen-enriched air into the matte layer. The wet charge is fed into this section through holes in the roof. Matte and slag flow through a tunnel in the partition into the settling section of the furnace which also receives converter slag. Heat to the settler is supplied by combusting coal. The smelting section and the settler have separate off-gas uptakes. There are two of these furnaces operating at Baiyin, with total hearth areas of 44 and 100 m² respectively. The smelting section occupies about 40% of the hearth in the larger furnace.

Only partial information is available on the operation of the Baiyin process. The feed is low grade concentrate which yields low to medium grade matte, i.e. 35 to 50% Cu. Oxygen enrichment of the blast is about 40%. Our guestimated value of the SBSR for the smelting section of these furnaces is in the range 75 to 90 Nm³ of O_2 used /h/m³ of bath, well below that of the processes previously discussed. About half of the process heat requirement is supplied by combustion of coal.

<u>Conventional Zinc Fuming Furnace</u>

Fuming of zinc from lead blast-furnace slag has been practised for more than sixty years. The furnace is a rectangular steel shell, entirely constructed of water jackets, with horizontal tuyeres along its length. The process is operated batch-wise: 40-50 tons of slag (mostly molten, but sometimes accompanied by ladle skulls and frozen slag) is charged to the furnace, blown with a mixture of powdered coal and air (the coal/air ratio is sufficiently high as to provide a strongly reducing

Table IV – Zinc Fuming Furnace (Submerged Horizontal Tuyeres, Batch Process)		Kellogg *	Richards et al. **	No Coal ***
Slag charge:	Weight, tonnes	45.4	45.0	47.2
	Volume, m³	11.9	11.8	12.4
Fuel:	Powdered coal, tonnes/h	4.63	3.90	0.0
Furnace:	Hearth area, m²	15.6	11.3	15.6
	Volume, m³	98.6	(80.0)	98.6
	Number of tuyeres	(40)	30	(40)
Blast:	Nm³/h	23,970	19,850	26,160
	% O₂	20.6	20.6	20.6
	Velocity at tuyere tip, m/s	–	~100	–
	Gauge pressure, kg/cm²	–	~0.97	–
Utilization of Oxygen				
Oxygen utilization efficiency, %		~98	~92	99
SBSR, Nm³ O₂ used/h/m³ bath volume		405	318	430
SFVSR, Nm³ O₂ used/h/m³ furnace volume		49.0	47.1	54.1
F/B		8.3	6.8	8.0

* El Paso furnace, standard run (36).
** Standard conditions (35).
*** Trail, B.C., 5-minute operation without coal (34). In this test, oxygen reacts with Fe^{2+} and sulfide in the slag to form Fe^{3+} and SO_2.
() Values in parentheses are estimated.

environment); the blow continues for 90-120 minutes, until the slag analyzes about 2% zinc, then the tap-hole is opened and the depleted slag is discharged; zinc vapor, emanating from the slag during the blow, is oxidized to ZnO fume by "tertiary air" introduced above the slag bath, and the oxide fume is recovered from the process gas by a bag house.

Table IV lists some characteristics of zinc fuming furnaces that have been reported in the literature (34,35,36). All cases show very high specific bath-smelting rates. It is particularly instructive to note the high values for the "No Coal" run, during which the air is oxidizing the slag, rather than reducing zinc oxide, as in the normal operation. This oxidizing run operates at very much the same specific capacity as the reducing runs, illustrating the inherent ability of the furnace to use oxygen regardless of the chemistry involved.

Submerged horizontal tuyeres, blown into a relatively shallow slag bath, produce a violent agitation (34), with ejection of large slugs of slag into the gas space above the bath, and cataracting of these back into the bath. The gas/slag interface area created in this manner, must be very large to account for the very high values of SBSR shown in Table IV.

The violent agitation in this furnace, with virtually no dead space at the walls or bottom, would probably result in excessive wear of refractories, thus the water-jacket design of the furnace, with frozen slag on the jackets acting as self-healing refractory. Heat-loss rates from water jackets are six to seven times larger than those from refractory-lined furnaces, but this is a small price to pay for this very intensive furnace design.

Although the values of SBSR are among the highest found in any non-ferrous bath smelting process, the values of SFVSR for slag fuming are much lower, with values of F/B in the range 6.75 to 8.25. This reflects the unusually large gas space, and associated wall area, required to oxidize the CO, H_2 and Zn(g) that issue from the bath, and to dissipate the heat released by this combustion.

The authors believe that the fuming furnace behavior can be studied to advantage by designers of slag-cleaning processes for lead or other non-ferrous slags. New slag-cleaning processes, including zinc fuming, need not follow the batch design of traditional fuming; design of continuous processes seems entirely feasible.

QSL Lead Smelting Process

The QSL process features a single, continuous reactor for smelting lead sulfide concentrates and lead-bearing secondaries to crude lead bullion (37,38). At the time of writing, only one plant, in Stolberg, Germany, was operating, but three others have been built and are expected to come on-stream in 1992.

A schematic diagram of the QSL reactor is shown in Figure 4. The Stolberg reactor consists of a long cylindrical vessel (32 m long by 2.74 m dia. at the oxidation end, and 2.25 m dia. at the reduction end) that can be tilted about its axis so as to raise the Savard-Lee bottom injectors out of the molten bath. The oxidation zone (11.6 m long), at the feed end of the reactor, is separated from the slag-reduction zone (20.3 m long) by a partition. Feed material, consisting of damp pellets of concentrate, recycle dust, flux and coke (if required) are fed to the oxidation zone. Oxygen is introduced through the injectors which are shrouded with moist nitrogen containing a small percentage of natural gas (0 to 25%) for fine control of the bath temperature. Slag from the oxidation zone, containing 25-30% Pb, flows through an underpass in the partition to the reduction zone. There, fine lignite coke, transported in a stream of air, is blown with oxygen through the injectors into the slag to reduce the lead.

Table V summarizes the operating parameters of the plant at Stolberg. The oxygen utilization efficiency is said to approach 100% (shown as 98% in Table V). The SBSR in the oxidation zone, 157 Nm^3 O_2 used/h/m^3 of bath volume, is similar to the intermediate range found for other bath smelting processes. In contrast, the SBSR for the reduction zone has the very low value of 33 Nm^3 O_2 used/h/m^3 of bath volume. It should also be noted that the reduction zone uses both 96% oxygen and air, so that the average oxygen content in the tuyere air is only 58%.

50

Table V – QSL Process (Stolberg) *
(Savard-Lee Bottom Injectors)

	Oxidation Zone		Reduction Zone	
Charge, tonnes/h, (% Pb):				
Raw feed	20.75	(50.0)		
Recycle dust	5.50	(50.0)		
Lignite coke	0.75	(0.0)	0.50	(0.0)
Moisture	3.00	(0.0)		
Products, tonnes/h, (% Pb):				
Pb bullion	9.0	($\geq$ 98)		
Discard slag			3.5	(2.0)
Dust	5.5	(50.0)		
Furnace internal volume, m³:		68		81
Bath volume, m³		18		13
Savard/Lee injectors, number		3		5
Total blast: Oxygen, Nm³/h, (% O₂)	3000	(96)	375	(96)
Velocity, m/s		~250		~250
Gauge pressure, kg/cm²		9–16		7–14
Coke carrier air, Nm³/h				375
Shroud gas	N₂/H₂O/CH₄		N₂/H₂O/CH₄	
Shroud gas volume, Nm³/h		450		375
Utilization of Oxygen				
Oxygen utilization efficiency, %		~98		~98
SBSR, Nm³ O₂ used/h/m³ bath volume		157		33.0
SFVSR, Nm³ O₂ used/h/m³ furnace volume		41.5		5.3
F/B		3.8		6.2

* Data supplied to the authors (February 1992) by Lurgi.

The low intensity in the reduction zone undoubtedly results from the need to avoid intense agitation and back-mixing, so as to permit a gradient of slag composition along the length of the reduction zone. Low SBSR in this zone is responsible for excessive length of the zone, large heat losses from the long shell and high capital cost of this section of the reactor. These disadvantages may be justified by a principal advantage of the QSL process — the single vessel design that greatly improves the environmental control, so important in lead smelting. The QSL process is still in the early stages of commercialization, and it is too soon to predict whether a significant increase in process intensity in the reduction zone will become possible. The many years of experience with the highly intensive zinc-fuming process in which lead, as well as zinc, is reduced from the slag may provide clues to improvement in the intensity and efficiency of reduction in the QSL reactor.

The QSL process has encountered the usual number of start-up problems, most of them associated with the operation of the reduction zone. The original design called for a reactor slope of 0.5% downward toward the oxidation zone, and this, coupled with a shell diameter 0.5 m larger in the oxidation zone, resulted in operation without a lead layer in the reduction zone. However, it was found that a lead layer at least 20 cm deep was necessary to facilitate lead reduction. The reason for this is uncertain. Lurgi engineers state that the lead layer enhances the rate of combustion of the fine coke, as well as of natural gas, an alternate reductant. This important finding might result either from the effect of the lead layer on the injector plume, so as to minimize agitation and mixing in the slag layer, and/or from its acting as a collector for the metallic lead nuclei formed in the slag. In any case, a small ring dam has been installed in the Stolberg reactor to maintain the required lead layer

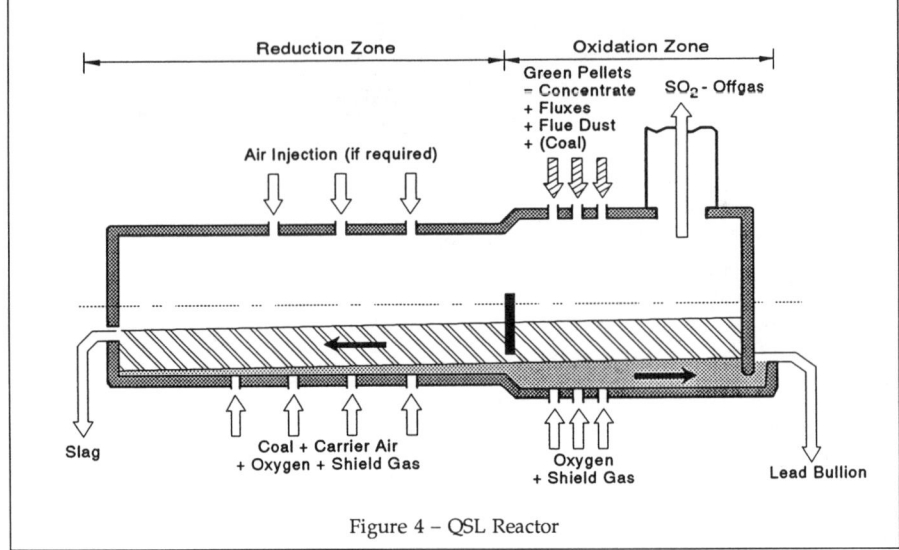

Figure 4 – QSL Reactor

in the reduction zone. Other problems in the reduction zone have arisen from attempts to use natural gas in place of powdered coke or coal as the reductant. Slow and incomplete combustion of natural gas has been observed in efforts to use this fuel in zinc fuming and in poling of anode copper, and the same problem may explain the difficulty encountered in the QSL reduction. The new finding, regarding the enhancement of natural gas combustion by the lead layer, has encouraged Lurgi engineers to plan further trials of natural gas reduction at Stolberg.

The metallurgical community looks forward to the outcome of these first plant trials of QSL, a lead smelting process with the potential to meet stringent environmental and workroom standards along with high energy efficiency.

Injection Through Lances

Lance injection was first used in steelmaking (Kaldo and LD processes) in the 1950s (39). The adaptation of this technology to non-ferrous pyrometallurgy led to the development of the bath smelting processes which are reviewed in this section.

Top Blown Rotary Converter (TBRC)

The TBRC or Kaldo furnace is essentially a cylindrical vessel, equipped with a mouth in one of the end walls and capable of rotating about its longitudinal axis. A lance, inserted through the mouth, is used to inject oxidizing or reducing gases or as a burner, or a separate lance and burner can be provided. During blowing, the vessel is tilted at an angle of about 20° from the horizontal. The gas jet, impinging obliquely on the surface of the bath, and the rotation of the vessel impart a stirring motion to the melt thus enhancing mass and heat transfer with the gas.

The process used at Inco's Copper Cliff operations to prepare nickel granules for high pressure carbonylation (40,41,42) — incidentally, the first (September 1972) non-ferrous commercial application of the TBRC — illustrates the flexibility and also the shortcomings of this technology. First, a 70 to 120 tonne bath is established from a blend of nickel-copper metallics, nickel sulfides and oxides and various reverts. Heat for smelting is supplied by an oxygen-natural gas flame. Oxygen is then blown to lower the sulfur content of the melt to provide a Cu/S ratio of 3.5. The blowing rate, 4,000 Nm³/h, is limited both by bath splashing and by OUE which at this rate is about 70%

(desulfurization efficiency is only 55%). The heat generated by the oxidizing reactions raises the temperature of the bath from about 1450°C, at the start of blowing, to about 1600°C, thus avoiding excessive formation of nickel oxide. Finally, coke is added to the TBRC to reduce nickel oxide mush. Once the melt specifications are met, the material is transferred to a granulation sluice to produce the feed for carbonylation. The vessel is rotated at 10 rpm during smelting and blowing and 15 rpm during reduction.

Using an average internal volume of 38 m³ (31 m³ for a newly relined vessel and 45 m³ at the end of a campaign) and an average bath volume of 15 m³ and the above blowing rate and OUE, the TBRC's SFVSR and SBSR are 70 Nm³ of O_2 used/h/m³ of furnace and 180 Nm³ of O_2 used/h/m³ of bath. These values are substantially lower than those for Peirce-Smith converters. Moreover, the OUE drops to the low 50s in processes generating substantial amounts of slag. This factor and the high mechanical maintenance cost of TBRCs (43) combine to discourage the spread of this technology. The low OUE is probably due to the rather mild blowing and bath stirring conditions of the TBRC. Mass and heat transfer occur mainly at the bath surface. In this respect, the TBRC deviates from typical bath smelting.

TBRCs are currently used in circumstance where operating flexibility over a wide range of temperatures and atmospheres offers a unique advantage. Such processes include the treatment of complex and highly contaminated feeds (44,45) and the recycle of secondary materials, in partic-ular copper scrap (46). Small TBRCs are also used in the production of precious metals (47).

<u>Mitsubishi Continuous Copper Smelting</u>

The Mitsubishi continuous copper smelting process has been in commercial operation for about twenty years, with two plants operating today (48,49). The two principal furnaces employed in the process design — the smelting furnace and the converting furnace — are both continuous bath smelting processes. These furnaces are circular in design, and feature top-blown lances, with lance tips about 50 cm above the bath surface. Dry concentrate, fuel (if required) and flux are introduced through the smelting furnace lances, along with oxygen-enriched air; limestone flux and oxygen-enriched air are fed to the converting furnace lances. Matte and slag from the smelting furnace overflow, together, through a launder to a slag-cleaning furnace (not a bath smelting process); matte from the slag-cleaning furnace flows through another launder to the converting furnace; blister copper and slag separately overflow from the converting furnace. Converter slag is granulated and returned to the smelting furnace. A schematic flowsheet of the Mitsubishi process is shown in Figure 5. A comprehensive discussion of the process metallurgy can be found elsewhere (50).

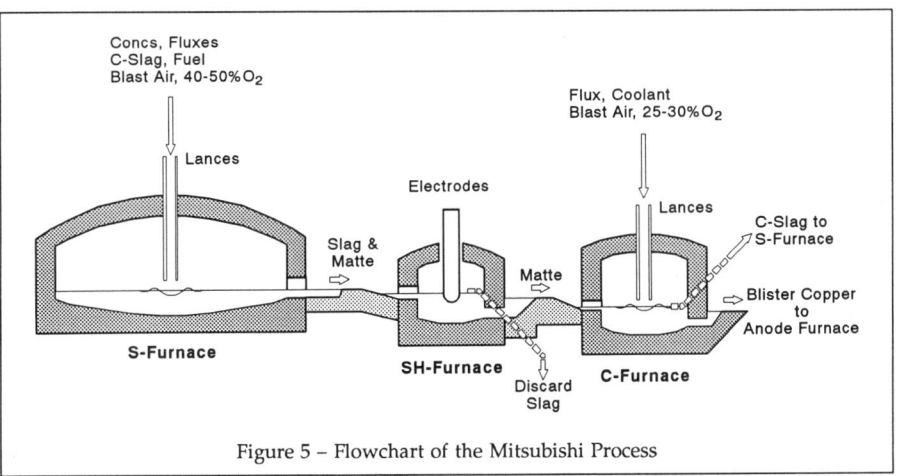

Figure 5 – Flowchart of the Mitsubishi Process

Table VIA – Mitsubishi Continuous Copper Smelting Furnace *
(Top Lances, Above the Bath)

	Naoshima		Kidd Creek	
Charge, tonnes/h, (% Cu):				
Concentrate	78.3	(34.2)	65.0	(25.0)
C-slag, scrap, reverts, etc.	13.1	(24.3)	12.0	(26.7)
Total charge including flux	109.5	(27.4)	97.0	(19.8)
Products, tonnes/h, (% Cu):				
Matte	45.7	(67.0)	27.0	(68.0)
Slag	57.1	(0.6)	58.0	(0.9)
Fuel: Coal, tonnes/h		4.5		0.0
Furnace: Internal volume, m^3		305		220
Number of lances		9		10
Lance height above bath, cm		50		30–50
Bath volume, m^3		97	74	(37) **
Blast, Nm^3/h, (% O_2):	40,800	(45.4)	28,000	(47.0)
Velocity at lance tip, m/s		177		171
Gauge pressure, kg/cm^2		0.7–1.5		2.5
Utilization of Oxygen				
Oxygen utilization efficiency, %		99		98
SBSR, Nm^3 O_2 used/h/m^3 bath volume		189	184	(368) **
SFVSR, Nm^3 O_2 used/h/m^3 furnace volume		60		62
F/B		3.1		3.0

* Data supplied to the authors (November–December 1991) by engineers of the respective companies.

** These values are calculated on the assumption that 50% of the nominal bath volume is occupied by magnetite build-up.

Tables VIA and VIB list the characteristics of the two plants that employ this process. The Naoshima plant treats a charge that is much richer in copper than the Kidd Creek plant. Since the Naoshima feed has less iron and sulfur to oxidize, a small amount of fuel (coal) is added to the feed to supply the necessary heat for the smelting furnace. Both plants report very high oxygen utilization efficiency for the smelting furnace, and the specific smelting rates are in the intermediate range for bath smelting processes. High velocity blast from top lances results in depression of the bath surface and intense agitation directly under the lances. This can cause erosion of the refractory furnace bottom if the lance velocity is too high or the bath depth too shallow, particularly if the blast is loaded with solids, as in the smelting furnace (51). These factors probably limit the permissible blast velocity and the resulting specific smelting rate that can be achieved in a given furnace.

The Kidd Creek plant reports that magnetite build-up may reduce actual bath volume of the smelting furnace by 60% of its nominal value (74 m^3). In Table VIA, we have shown in parenthesis a bath volume of 37 m^3 — 50% reduction in the nominal volume due to magnetite build-up. With such a small bath volume, SBSR is increased to a value characteristic of the more intense bath smelting processes — 368 Nm^3 O_2 used/h/m^3 of bath volume. The new plant at Naoshima has not been in operation long enough to determine the magnetite build-up, but for the old plant the magnetite build-up was much less than at Kidd Creek — reduction of only 10% in the nominal bath volume. The difference in magnetite formation in the two plants must be related to differences in the chemistry of the feed materials; in particular, the Naoshima plant believes that the small amount of coal added to the feed materials helps to prevent magnetite formation.

	Table VIB – Mitsubishi Converting Furnace			
		Naoshima		Kidd Creek
Charge, tonnes/h, (% Cu):				
	Matte	45.7 (67)		27.0 (68)
	C-scrap, slag, reverts, etc.	3.0 (99.3)		3.5 (22.6)
	Limestone	2.8		2.0
Products, tonnes/h, (% Cu):				
	Blister	31.8 (98.6)		17.5 (99.2)
	C-slag	8.7 (13.7)		8.0 (18.0)
Furnace:	Internal volume, m³	182		140
	Number of lances	10		6
	Lance height above bath, cm	50		30–50
Bath volume, m³		46		34
Blast, Nm³/h, (% O₂):		29,970 (30.3)	16,500	(32.0)
	Velocity at lance tip, m/s	127		167
	Gauge pressure, kg/cm²	0.5–0.7		2.5
	Utilization of Oxygen			
Oxygen utilization efficiency, %		83		92
SBSR, Nm³ O₂ used/h/m³ bath volume		164		143
SFVSR, Nm³ O₂ used/h/m³ furnace volume		41		35
F/B		4.0		4.1

Compared to the smelting furnaces, the converting furnaces of both plants show slightly lower specific smelting rates, use of less oxygen enrichment of the blast, and lower oxygen utilization efficiency. We offer the following possible explanations for these differences. Lower values of specific smelting rate in converting may result from the need to employ a less vigorous blast so as to avoid refractory wear by excessive splashing of the highly corrosive lime-ferrite slag. The degree of oxygen enrichment is probably dictated by furnace heat balance considerations. The lower value of oxygen utilization efficiency in converting is more difficult to understand, but it may result from the slow rate of sulfur diffusion within the molten copper (0.6% S) to the reacting interface.

Isasmelt (Sirosmelt) Copper, Lead

A pilot plant for Isasmelt smelting of copper concentrate began operating in 1987, and has smelted over 512,000 tonnes of concentrate between April 1987 and May 1992 (52). It was designed to operate at 15 tonnes of concentrate/h, but has recently

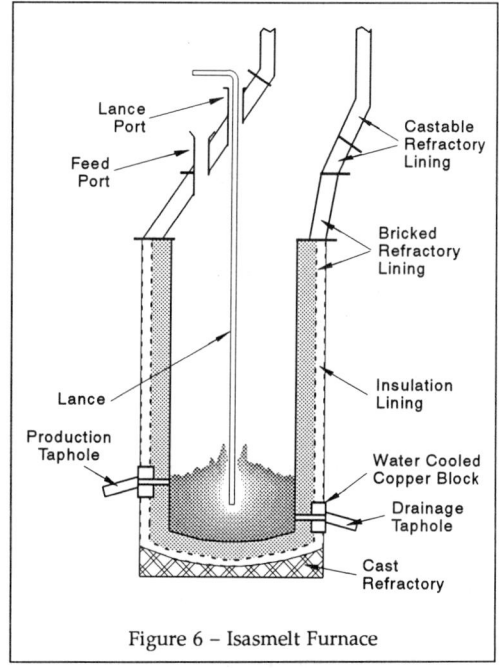

Figure 6 – Isasmelt Furnace

Table VII – Isasmelt Processes (Sirosmelt Lance, Submerged in Bath)

	Copper Smelt		Lead Smelt			
			Oxidation		Reduction	
Charge, tonnes/h, (%Cu or %Pb):						
Concentrate	18.8	(~30)	20	(50)		
Recycle fume			3			
Slag from oxidation					19	(~50) *
Products, tonnes/h, (%Cu or %Pb):						
Matte, metal	10.8	(51)			9.2	(99)
Slag		–	19	(~50) *	8.5	(~4)
Fuel, tonnes/h, (type)	2.24	(coal)	0.9	(coke)	2.0	(coal)
	0.28	(oil)			0.56	(oil)
Furnace:						
Internal volume, m³		45.2		49.1 *		77 *
Lance submergence, cm		10–60		30.0 *		30.0 *
Bath volume, m³		12 *		13 *		15 *
Blast, Nm³/h, (% O₂):	18,930	(25.3)	25,560	(27)	6,000	(21)
Gauge pressure, kg/cm²		0.7–1.5		1.38		1.38
Utilization of Oxygen						
Oxygen utilization efficiency, %		100		95 *		100
SBSR, Nm³ O₂ used/h/m³ bath volume		400 *		504 *		84 *
SFVSR, Nm³ O₂ used/h/m³ furnace volume		110		134 *		16.4 *
F/B		3.8 *		3.8 *		5.1 *

* Values are estimates by the authors.

been re-rated to 20 tonnes/h as a result of the regular use of oxygen enrichment (53). The operational data shown in Table VII are mostly from the literature (53,54), and are the average for a 28-day period in September-October 1989. The smelting furnace is an upright cylinder, 2.4 m.i.d. and 10 m high (54). A schematic of the furnace is shown in Figure 6 (55). Feed consists of concentrate, flux and lump coal in the form of damp pellets (10% moisture) which are fed through the top of the reactor. Oxygen-enriched air is fed from a Sirosmelt lance (56) submerged in the bath. A small amount of oil is fed through the lance for fine control of temperature. Matte and slag are batch tapped together, and flow via a launder to a pre-existing reverberatory furnace. Matte grades have normally ranged from 45 to 51% copper. The pilot plant copper reactor has been run for short periods with up to 40% O_2 enrichment of the blast. Under these conditions, the operation has been autogenous.

The Sirosmelt lance is equipped with internal, helical vanes which impart a swirling motion to the air. This enhances heat transfer between lance and air and promotes the formation of a protective layer of frozen slag on the lance. The need for air cooling of the lance may limit the blast oxygen enrichment. Lance life in the copper reactor is typically about 150 hours. Reactor refractory life is currently in excess of one and a half years. Copper pilot plant availablity, i.e. actual hours of operation based on 28 day periods, ranges from 85 to 95%.

At the time of writing, Mt. Isa Mines was constructing a commercial furnace with a capacity of 730,000 tonnes of concentrate per year; and a second Isasmelt furnace, rated at 600,000 tonnes of concentrate per year, was under construction at Miami, Arizona, for Cyprus Miami Mining Corp. Both plants are expected to be commissioned by mid 1992 (52).

The data for Isasmelt lead smelting (Table VII) are also from the literature (57). It represents the projected performance of the 20-tph plant commissioned in March 1991, but the estimated performance has been backed up by nine years of experience with a 5-tonne/h pilot reactor which has been operated continuously for extended periods. The process consists of two separate continuous reactors, connected by a launder. Both reactors are upright cylinders. The oxidation reactor is 2.5 m.i.d., and the reduction reactor, 3.5 m.i.d., with estimated heights of 10 and 8 m, respectively. Damp pellets of concentrate, flux, and coke breeze are fed through the top of the oxidation reactor, and the feed is completely oxidized to a high-lead slag by the oxygen enriched blast from the submerged Sirosmelt lance.

The slag from the oxidation reactor flows to the reduction reactor, where lead is reduced from the slag by the action of fine lump coal (−12 mm) and the lance air (coal/air ratio is maintained at a high level so as to yield strongly reducing conditions). A small amount of fuel oil is fed through the lance for fine control of temperature. Additional air is fed above the bath to burn the combustible gases coming from the bath reactions. Slag and lead bullion flow continuously to a forehearth for separation.

The values of SBSR and SFVSR for copper smelting, shown in Table VII, are among the highest for the processes reviewed here. They are based on a reported OUE of 100% (52) and an estimated average bath volume of 26.5% of the internal volume of the reactor (F/B of 3.8). In the Mt. Isa pilot plant, the volume of the bath fluctuates because of the batch nature of the operation.

The values of SBSR and SFVSR for the oxidation stage of lead smelting are even higher than for the Isasmelt copper process. As mentioned above, they correspond to the projected performance of the commercial plant and are derived from an estimated 26.5% bath volume occupancy and an OUE of 95%. The latter estimate is based on the fact that slightly more oxygen than stoichiometrically required is used in the smelting reactor to guarantee complete elimination of sulfur as SO_2 and complete oxidation of lead (52,57).

The SBSR for the reduction reactor is based on a reported OUE of 100% and an estimated bath volume of 15 m^3, including the lead heel (corresponding to a bath depth of about 1.5 m). The estimated bath volume results in a slag retention time about twice longer than that required in the pilot plant reduction reactor to achieve a slag PbO content of about 2 wt% (57). In any case, the SBSR for the reduction reactor appears to be much smaller than that for the oxidation reactor. Possible entrainment of coal near the slag surface, less agitation of the bath due to the lower blast rate, and the larger diameter are probably responsible for the lower intensity.

It will be of interest to see whether experience with operation of the reduction reactor leads to changes in the mode of operation from that reported in Table VII. Sixty years of experience with zinc fuming furnaces suggests that lead, like zinc, is rapidly reduced from slag by an intense blast of *powdered coal* and air, with the coal/air ratio maintained at about 0.195 kg coal per Nm^3 of air. In contrast, the Isasmelt reactor operates with fine lump coal (−12 mm) on the coal even if it were powdered at a coal/air ratio of 0.333, high ratio for appropriate gasification of the coal even if it were powdered. Isasmelt appears to rely on the contact of fine lump coal with slag, which provides limited surface area for reaction. The zinc fuming furnace is thought to operate either by reaction of powdered coal with air to yield CO and H_2, which then react with the slag (36), or by penetration of powdered coal into the slag (35), followed by rapid reaction because of the large surface area presented.

Inco Top-Blowing, Bottom-Stirring Process

Inco developed an oxygen top-blowing, nitrogen bottom-stirring process to convert semiblister, i.e. sulfur-saturated, nickel-contaminated (about 5%) copper, to low sulfur (< 100 ppm) and low nickel (< 0.7%) blister. Following laboratory and pilot plant testwork (58), the technology graduated to the smelter floor and will become a key component of the new Copper Cliff smelter flowsheet (59).

At Copper Cliff, a 10.7 m long by 4 m diameter Peirce-Smith converter was adapted for this process. The new vessel (Figure 7) has no tuyeres. Instead, it is equipped with five refractory porous plugs for nitrogen injection. Oxygen is blown through water cooled lances installed in the end walls. The

57

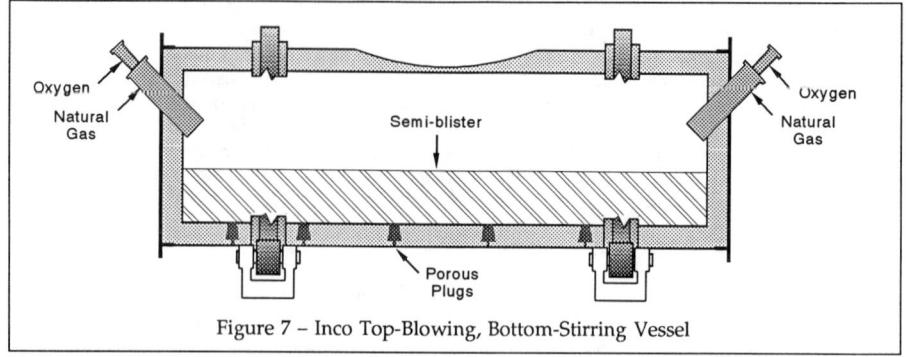

Figure 7 – Inco Top-Blowing, Bottom-Stirring Vessel

lances are capable of acting as burners when required. During commercial testing of the technology, batches of about 135 tonnes of semiblister were converted at oxygen blowing rates of 5330 Nm^3/h while bottom stirring the bath with about 14 Nm^3 of nitrogen/h/plug. Oxygen utilization efficiencies of about 85% were achieved. Sufficient excess heat was generated during blowing to consume scrap at a rate of 20% by weight of semiblister. Using the preceding numbers, the SFVSR and SBSR for the Inco top blowing-bottom stirring vessel are 61 Nm^3 of O_2 used/h/m^3 of furnace and 257 Nm^3 of O_2 used/h/m^3 of bath. These values are slightly lower than those typical of Peirce-Smith converters.

The potential for independent inert gas sparging is an additional benefit of the technology. At Copper Cliff, this capability has permitted sulfur and nickel elimination from copper close to thermodynamic equilibrium and has contributed to improved control of the process end-point.

This lance injection technology is applicable to the pyrometallurgical refining of copper and to the processing of secondary materials such as copper scrap, among many other processes. In addition, it has the same flexibility as the TBRC to operate over wide ranges of temperature and atmosphere.

Discussion

Smelting intensity, as measured by the parameters used in this paper, SFVSR and SBSR, describes one essential quality of a process. However, the authors recognize that there are many other factors which contribute to the success of a commercial smelting process. Among these are ease of operation; adaptability to a variety of feed materials, including external reverts; high energy efficiency and metal recoveries; good elimination of impurities; low maintenance costs; and amenability to current workroom and external environmental standards. Nevertheless, the recent history of process development shows a strong tendency toward more intensive smelting, and this should justify the attention it has received here.

The specific bath smelting rates of the processes considered here fall naturally into three groups that we will call High Intensity, Intermediate Intensity and Low Intensity. Table VIII shows these three groups, along with the SBSR, F/B, OUE and degree of oxygen enrichment employed by each process.

The data summarized in Table VIII indicate that high intensity processes employ either side-blown tuyeres or submerged lances, or lances combined with inert gas stirring of the bath. They include both oxidation smelting and reduction smelting (zinc fuming), and continuous as well as batch processes. Interestingly, the two oldest processes, Peirce-Smith converting and zinc fuming, are in this group.

The case of the Peirce-Smith converter is particularly outstanding. Despite its obvious shortcomings (batch process, high air dilution of the off-gas, source of fugitive emissions, high refractory wear), this converter has been around for almost a century, surviving the profound changes which have occurred in primary smelting, just across the converter aisle. It may still be some time before the obituary of the Peirce-Smith converter is written, despite the challenges from solid matte converting

58

Table VIII – Comparison of Bath Smelting Processes				
	SBSR	F/B	OUE	% O$_2$
High Intensity				
P-S converter	270–385	2.9	90	21–30
Noranda	286	3.6	97.5	36
Vanyukov	244–369	3.2	100	68–75
Zinc fuming	318–405	6.8–8.3	92–98	21
Isasmelt, Cu *	400	3.8	100	25
Isasmelt, Pb oxidation *	504	3.8	95	27
Inco top blown, bottom stirred	257	4.0	85	96
Intermediate Intensity				
El Teniente	159	2.7	95	30
QSL, Pb oxidation	157	3.8	98	96 ***
Mitsubishi, smelting	184 ** –189	3.0	98–99	45–47
Mitsubishi, converting	143–164	4.0	83–92	30–32
Low Intensity				
QSL, Pb reduction	33	6.2	98	58 ***
Isasmelt, Pb reduction *	84	5.1	100	21

*   Values for SBSR and F/B are estimates by the authors.
**   This value becomes 368 (High Intensity) if the magnetite build-up at Kidd Creek is taken into account.
*** Shrouding gas not included.

processes (60,61,62,63). The critical issue in the development of these newer processes is to match or, better, surpass the Peirce-Smith's intensity and flexibility while overcoming its drawbacks.

Some of the newer processes of intermediate and low intensity are likely to become more intensive as experience is acquired. For example, both stages of the Mitsubishi process have undergone considerable increases in intensity since the first commercialization twenty years ago. The two low-intensity lead reduction processes, in particular, are likely to achieve greater intensity as they undergo design modifications and climb the learning curve.

It would appear that high oxygen enrichment is not a prerequisite for high intensity. Besides, the degree of oxygen enrichment depends on process heat balance considerations. Some of the high intensity processes in Table VIII employ less than 30% oxygen in the blast, whereas one of the least intensive process (QSL lead reduction) uses 58% oxygen. Added oxygen does, however, contribute to high intensity, as witness the cases of the Peirce-Smith converter and the Vanyukov process.

The high values of OUE for all processes in Table VIII indicate that neither reaction kinetics nor mass transfer of oxygen to the reaction zone limits the smelting intensity, at least for the blast rates currently employed by these processes. One can expect, however, that such limitation may occur for some of these processes if higher blast rates are employed, with a concurrent drop in the OUE. The tuyere or lance design that most effectively promotes rapid mass transfer of oxygen to the reaction zone, by providing an intimate dispersion of gas throughout the bath, without creating excessive splashing, slopping or refractory erosion, will have the highest probability of achieving increased blast rate without sacrifice of OUE.

Table VIII shows that most of the high intensity processes are either batch processes (Peirce-Smith converting, zinc fuming, Inco top-blowing, bottom-stirring) or continuous processes that operate without slag-matte separation (Isasmelt lead oxidation, Isasmelt copper smelting). Three of these (zinc fuming, Isasmelt lead oxidation, and the Inco process) are, in essence, single phase processes. The need for quiescent conditions to promote slag-matte separation is met in the Peirce-Smith converter by turning the tuyeres off the bath, and in Isasmelt copper smelting by transferring the two liquids to a settling furnace. Of the high intensity group in Table VIII, only the Noranda and Vanyukov processes achieve the separation of two liquid phases in a single continuous reactor.

F/B values in Table VIII are mostly in the range 2.7–4.0. These values are determined by the need to control splashing and disengagement of liquid droplets, and are further influenced by reactor shape — horizontal or vertical cylinder, rectangular or circular hearth-furnace. The zinc fuming furnace requires a much larger F/B because of the need for post-combustion of CO, H_2 and Zn(g), and the dissipation of this combustion heat to the water jackets above the bath level. It seems unlikely that reactor designs can be expected to achieve F/B values below the range 2.5–3.0.

Comparison with Flash Smelting

We propose here that a meaningful comparison of the intensities of bath and flash smelting can be made by equating the SBSR for bath smelting to SSSR (Specific Shaft Smelting Rate expressed as Nm^3 of oxygen used/h/m^3 of flash-shaft volume) for flash smelting. To calculate this parameter, we have used data published by Themelis et al. (64) for six Outokumpu type flash smelters, four of which have used various degrees of oxygen enrichment, thus providing ten different sets of operating conditions. Assuming 100% OUE (which is characteristic of flash smelting) in each case, the range of SSSRs is 41 to 111 Nm^3 O_2 used/h/m^3 of shaft volume, with a strong correlation between SSSR and oxygen enrichment of the air. For normal moist air (20.6% O_2), the average SSSR of four smelters was 55; for oxygen enrichment between 35 and 50%, the average SSSR for four smelters was 89. All of these SSSR values are significantly lower than the SBSR values for the bath smelting processes in the high and intermediate ranges of Table VIII.

The lower intensity of flash smelting, compared to bath smelting, can best be understood by another measure used by Themelis et al. (64), the percentage of the shaft volume occupied at any instant by concentrate particles. They found loadings, for the same ten operating regimes considered above, to range from 0.004 to 0.01 percent: the dispersed solids occupy less than one part in ten thousand of the shaft volume. Evidently, the flash shaft contains an extremely dilute suspension of reacting particles, and even though the average residence time of particles in the shaft is very short (1.5 to 4.7 seconds), the ability to consume oxygen and to smelt sulfides is limited by the very dilute suspension. In contrast, the intensive bath smelting processes operate with far more concentrated suspensions of gas in the liquid bath; the volume percentage of entrained gas in the liquid for the Peirce-Smith converter is reported to be in the range 8–10 percent (14). In addition, most of the heat of reaction is generated within the bath in which it is immediately consumed and which acts as a flywheel to limit fluctuations in temperature and product composition.

Cyclone reactors (Kivcet, Contop) can achieve smelting intensities several times higher than the Outokumpu shaft (65,66). Combustion of the feed in a cyclone occurs in a much more confined space. This results in higher reaction and wall temperatures, faster reaction rates, and a need for a substantially increased rate of heat transfer through the reactor wall (66).

Comparison with LD Steelmaking

Typical LD steelmaking, a bath smelting process employing a top-blown oxygen lance, operates at an SBSR of 900–1100 Nm^3/h/m^3 of bath (steel plus slag) volume based on a value of 3.0–3.5 Nm^3/min/tonne of steel (67). This specific rate is more than twice that of the highest rate we report here for non-ferrous processes. Indeed, the LD process possesses special characteristics that make such a high rate possible. Pure oxygen is blown at a supersonic rate into the bath, causing the unique phenomenon of gas-slag-metal emulsion and thus generating a very large surface area for reaction. An OUE of 100% is achieved during decarburization. An F/B volume ratio of about 5.5 is necessary in order to contain the frothing and splashing action caused by the very intense lance action. Bottom-blown steelmaking, OBM, is reported to operate at even higher SBSR than LD steelmaking.

For any given non-ferrous bath smelting process, there may be special reasons why the specific oxygen rate cannot approach those of the steelmaking processes. Nevertheless, the experience of the highly successful oxygen steelmaking processes does offer a challenge to non-ferrous process designers. There exists a window of opportunity for processes of even higher productivity in non-ferrous bath smelting than those we accept today.

Future Directions

As the newer bath smelting processes gain operating experience, and as more understanding of the fundamentals of gas injection into molten baths is acquired, we can expect to see more processes operating in the SBSR range of 300–500. Intensities beyond this range are certainly possible, as oxygen steelmaking teaches, but they may require some of the same unusual conditions that apply in steelmaking. New designs to achieve high-intensity non-ferrous smelting will bring with them other advantages — improved energy efficiency, higher productivity of capital and labour, and, most necessary, improved environmental control.

Acknowledgements

The authors express their appreciation to the engineers of the various companies who provided data and comments on their respective processes. Their contributions are acknowledged at the appropriate places in the text.

Thanks are also due to Lucille Green of Inco's J. Roy Gordon Research Laboratory for her assistance in editing this paper.

References

1. A.E. Wheeler and M.W. Krejci, "Great Falls Converter Practice," Trans. Amer. Inst. Min. Eng., XLVI (1914), 486–561.

2. "Oxygen Flash Smelting Swings into Commercial Operation," J. Met., 7 (6) (1955), 742–750.

3. N.J. Themelis, P. Tarassoff and J. Szekely, "Gas-Liquid Momentum Transfer in a Copper Converter," Trans. Metall. Soc. AIME, 245 (1969), 2425–2433.

4. J.K. Brimacombe, A.A. Bustos, D. Jorgensen and G.G. Richards, "Toward a Basic Understanding of Injection Phenomena in the Copper Converter," Physical Chemistry of Extractive Metallurgy, ed. V. Kudryk and Y.K. Rao, (Warrendale, PA: The Metallurgical Society, 1985), 327–351.

5. G.G. Richards, K.J. Legeard, A.A. Bustos, J.K. Brimacombe and D. Jorgensen, "Bath Slopping and Splashing in the Copper Converter," The Reinhardt Schuhmann International Symposium, ed. D.R. Gaskell et al., (Warrendale, PA: The Metallurgical Society, 1987), 385–402.

6. Jong-Leng Liow and N.B. Gray, "Slopping Resulting form Gas Injection in a Peirce-Smith Converter," Metall. Trans. B, 21B (1990), 657–664 and 987–996.

7. S.W. Marcuson, "Copper Converting – A Historical Perspective" (Paper submitted for publication in CIM Bull.).

8. W.H. Peirce and E.A.C. Smith, "Method of and Converter Vessel for Bessemerizing Copper Matte," U.S. Patent 942,346, 7 December 1909.

9. R.E. Johnson, N.J. Themelis and G.A. Eltringham, "A Survey of Worldwide Copper Converter Practices," Copper and Nickel Converters, ed. R.E. Johnson, (Warrendale, PA: The Metallurgical Society, 1979), 1–32.

10. J.A. Vogt, P.J. Mackey and G.C. Balfour, "Current Converter Practice at the Horne Smelter," Copper and Nickel Converters, ed. R.E. Johnson, (Warrendale, PA: The Metallurgical Society, 1979), 357–390.

11. G.C. McKerrow, "A 14 Ft. x 32 Ft. Converter at the Noranda Smelter," Pyrometallurgical Processes in Nonferrous Metallurgy, ed. J.N. Anderson and P.E. Queneau, (Warrendale, PA: The Metallurgical Society, 1967), 247–258.

12. E.O. Hoefele and J.K. Brimacombe, "Flow Regimes in Submerged Gas Injection," Metall. Trans. B, 10B (1979), 631–648.

13. P.W. Godbehere, "An Outline of Operations at the Horne Smelter" (Paper presented at the Canadian Smelter Group, Timmins, Ontario, Canada, May 1990).

14. G.G. Richards, private communication.

15. D.W. Rodolff and I.A. Rana, "Converter Practice at Magma Copper Company in 1978," Copper and Nickel Converters, ed. R.E. Johnson, (Warrendale, PA: The Metallurgical Society, 1979), 81–109.

16. F.J. Longworth, "Smelting Copper Concentrates in a Converter," Trans. Amer. Inst. Min. Metall. Eng., 71 (1925) 969–971.

17. G.E. Beavers, "Smelting Copper Concentrates in a Converter," Trans. Amer. Inst. Min. Metall. Eng., 106 (1935) 149-150.

18. T. Tsurumoto, "Copper Smelting in the Converter," J. Met., 13 (11) (1961), 820–824.

19. T. Tsurumoto, "Improvements on the Oxygen Smelting Process at Hitachi Smelter," Pyrometallurgical Processes in Nonferrous Metallurgy, ed. J.N. Anderson and P.E. Queneau, (Warrendale, PA: The Metallurgical Society, 1967), 291–305.

20. R. Saddington, W. Curlook and P. Queneau, "Tonnage Oxygen for Nickel and Copper Smelting at Copper Cliff," J. Met., 18 (4) (1966), 440–452.

21. M.E. Messner and D.A. Kinneberg, "Direct Converter Smelting at Utah Using Oxygen," J. Met., 21 (7) (1969), 23–29.

22. S. Edlund and S. Lundquist, "Copper Converter Practice at the Roennskaer Works," Copper and Nickel Converters, ed. R.E. Johnson, (Warrendale, PA: The Metallurgical Society, 1979), 239–256.

23. L.A. Mills, G.D. Hallett and C.J. Newman, "Design and Operation of the Noranda Continuous Smelting Process", Extractive Metallurgy of Copper, ed. J.C. Yannopoulos and J.C. Agarwal, (Warrendale, PA: The Metallurgical Society, 1976), 458–487.

24. J.B.W. Bailey, G.D. Hallett and L.A. Mills, "The Noranda Smelter – 1965 to 1983," Advances in Sulfide Smelting, Vol. 2: Technology and Practice, ed. H.Y. Sohn, D.B. George and A.D. Zunkel, (Warrendale, PA: The Metallurgical Society, 1983), 691–707.

25. P.J. Mackey, J.B.W. Bailey and G.D. Hallett, "The Noranda Process – An Update," Copper Smelting – An Update, ed. D.B. George and J.C. Taylor, (Warrendale, PA: The Metallurgical Society, 1981), 213–236.

26. Y. Prevost and D. Verhelst, "Metallurgical Conrol of the Noranda Process Reactor," Copper 91 – Cobre 91, Vol. 4: Pyrometallurgy of Copper, ed. C. Díaz et al., (Toronto: Pergamon, 1991), 583–597.

27. D.G. Pannell and P.J. Mackey, "Noranda Process Operations 1988 and Future Trends" (Paper presented at the Copper Committee Meeting of the GDMB, Antwerp, Belgium, 27–29 April 1988).

28. R. Campos and C. Queirolo, "Improvements to the Oxygen Smelting Process," Copper and Nickel Converters, ed. R.E. Johnson, (Warrendale, PA: The Metallurgical Society, 1979), 257–273.

29. G. Vera and R. Campos, "Codelco-Chile Copper Concentrates Smelting Technologies," Extraction Metallurgy '85, (London: Institute of Mining and Metallurgy, 1985), 117–147.

30. C. Queirolo, L. Torres and A. Tapia, "Commercial Operation of Submerged Smelting of Concentrates in Teniente Converters at Caletones Smelter," Today's Technology for the Mining and Metallurgical Industries, (London: Institute of Mining and Metallurgy, 1989), 597–602.

31. R. Campos, J. Achurra and O. Rojas, "Teniente Converter: A Leading Pyrometallurgical Technology," Copper 91 – Cobre 91, Vol. 4: Pyrometallurgy of Copper, ed. C. Díaz et al., (Toronto: Pergamon, 1991), 229–246.

32. V.P. Bystrov, "The Vanyukov Process: A New Pyrometallurgical Technology" (Paper presented at Copper 91 – Cobre 91, Ottawa, Canada, 18–21 August 1991).

33. B. Q. Chen, Q. X. Huang and Y. Z. Gao, "Development of Baiyin Bath Smelting Process with Oxygen Enrichment," Copper 91 – Cobre 91, Vol. 4: Pyrometallurgy of Copper, ed. C. Díaz et al., (Toronto: Pergamon, 1991), 259–268.

34. H.H. Kellogg, "A New Look at Slag Fuming," Eng. Min. J., 158 (3) (1957), 90–92.

35. G.G. Richards and J.K. Brimacombe, "Kinetics of the Zinc Slag-Fuming Process: Part III. Model Predictions and Analysis of Process Kinetics," Metall. Trans. B, 16B (3) (1985), 541–549.

36. H.H. Kellogg, "A Computer Model of the Slag-Fuming Process for Recovery of Zinc Oxide", Trans. Metall. Soc. AIME, 239 (1967), 1439–1449.

37. P.E. Queneau, "The QSL Reactor for Lead and Its Prospects for Ni, Cu and Fe," J. Met., 41 (12) (1989), 30–35.

38. K. Mager and A. Schulte, "Process and Technological Aspects of the First Four Commercial QSL Plants," Primary and Secondary Lead Processing, ed. M.L. Jaeck, (Toronto: Pergamon, 1989), 3–14.

39. Steelmaking in the 20th Century, (Warrendale, PA: Iron and Steel Society, 1992).

40. M.D. Head, V.A. Englesakis, B.C. Pearson and D.H. Wilkinson, "Nickel Refining by the TBRC Smelting and Pressure Carbonyl Route" (Paper presented at the 105th AIME Annual Meeting, Las Vegas, Nevada, 22–26 February 1976).

41. W.J. Thoburn and P.M. Tyroler, "Optimization of TBRC Operation and Control at Inco's Copper Cliff Nickel Refinery", Copper and Nickel Converters, ed. R.E. Johnson, (Warrendale, PA: The Metallurgical Society, 1979), 274–290.

42. G.P. Tyroler and G.E. Cuthbert, "Nickel Smelting Rates in Top-Blown Rotary Converters," Symposium on Quality Control in Non-Ferrous Pyrometallurgical Processes, (Montreal, PQ: The Metallurgical Society of CIM, 1985) 175–194.

43. J.D. Guiry and A.D. Dalvi, "P.T. Inco's Indonesian Nickel Project: An Update," International Seminar on Laterite, (Tokyo: The Mining and Metallurgical Institute of Japan, 1985), Paper B-2-7, 141–158.

44. G. Lundqvist, P.-L. Nystedt and S. Petersson, "Application of the Kaldo Process at the Copper Smelter of the Ronnskar Works, Boliden Metall AB, Skelleftehamn, Sweden," Copper Smelting – An Update, ed. D.B. George and J.C. Taylor, (Warrendale, PA: The Metallurgical Society, 1981), 41–49.

45. L. Hedlund and L. Johanson, "Recent Developments in the Boliden Lead Kaldo Plant," Recycle and Secondary Recovery of Metals, ed. P.R. Taylor et al., (Warrendale, PA: The Metallurgical Society, 1985), 787–796.

46. W.S. Nelmes, "Current Trends in Smelting and Refining of Secondary Copper Materials," Trans. Inst. Min. Metall., Sect. C, 96 (1987), C151–C155.

47. J.G. Cooper, J.W. Matouek and J.G. Whellock, "Bulk Oxygen Use in the Refining of Precious Metals," The Impact of Oxygen on the Productivity of Non-Ferrous Metallurgical Processes, ed. G. Kachaniwsky and C. Newman, (Toronto: Pergamon, 1987), 107–119.

48. T. Shibasaki, K. Ochichi, K. Kanamori and T. Kawai, "Construction Of New Mitsubishi Furnaces for Modernization of Naoshima Smelter and Refinery," Copper 91 – Cobre 91, Vol. 4: Pyrometallurgy of Copper, ed. C. Díaz et al., (Toronto: Pergamon, 1991), 3–14.

49. C.J. Newman and A.G. Storey, "Productivity Improvements in the Kidd Creek Copper Smelter," Copper 87, Vol. 4: Pyrometallurgy of Copper, ed. C. Díaz et al., (Santiago, Chile: Universidad de Chile, 1988), 123–138.

50. T. Shibasaki and M. Hayashi, "Top-Blown Injection Smelting and Converting: The Mitsubishi Process," J. Met., 43 (9) (1991), 20–26.

51. M. Goto and T. Echigoya, "Effect of Injection Smelting Jet Characteristics on Refractory Wear in the Mitsubishi Process," J. Met., 34 (11) (1980), 6–11.

52. C.R. Fountain, private communication.

53. C.R. Fountain, M.D. Coulter and J.S. Edwards, "Minor Element Distribution in the Copper Isasmelt Process," Copper 91 – Cobre 91, Vol. 4: Pyrometallurgy of Copper, ed. C. Díaz et al., (Toronto: Pergamon, 1991), 359–373.

54. R.L. Player, C.R. Fountain, T.V. Nguyen and F.R. Jorgensen, "Top-Entry Submerged Injection and the Isasmelt Technology," Savard-Lee International Symposium on Bath Smelting (this volume), (1992).

55. W.J. Errington, J.H. Fewings, V.P. Keran and W.T. Denholm, "The Isasmelt Lead Smelting Process," Extraction Metallurgy '85, (London: Institution of Mining and Metallurgy, 1985), 199–218.

56. J.M. Floyd and B.W. Lightfoot, "Sirosmelt and the Wide World of Opportunity," Eng. Min. J., 186 (6) (1985), 52–56.

57. S.P. Matthew, G.R. McKean, R.L. Player and K.E. Ramus, "The Continuous Isasmelt Lead Process," Lead-Zinc '90, ed. T.S. Mackey and R.D. Prengaman, (Warrendale, PA: The Metallurgical Society, 1990), 889–901.

58. C. Díaz, S. Marcuson, H. Davies and R. Stratton-Crawley, "Conversion of Nickel and Sulfur-Containing Copper to Blister," Copper 87, Vol. 4: Pyrometallurgy of Copper, ed. C. Díaz et al., (Santiago, Chile: Universidad de Chile, 1988), 293–304.

59. C.A. Landolt, A. Fritz, S.W. Marcuson, R.B. Cowx and J. Miszczak, "Copper Making at Inco's Copper Cliff Smelter," Copper 91 – Cobre 91, Vol. 4: Pyrometallurgy of Copper, ed. C. Díaz et al., (Toronto: Pergamon, 1991), 15–29.

60. C.E. O'Neill, H. Davies, C. Díaz and S. Marcuson, "Top Blowing of Copper Mattes with Simultaneous Overflow of Copper and Slag" (Paper presented at the 23rd Annual Conference of Metallurgists, Québec, Québec, 19–22 August 1984).

61. K.J. Richards, D.B. George and L.K. Bailey, "A New Continuous Copper Converting Process," Advances in Sulfide Smelting, Vol. 2: Technology and Practice, ed. H.Y. Sohn et al., (Warrendale, PA: The Metallurgical Society, 1983), 489–498.

62. J.A. Blanco, T.N. Antonioni, C.A. Landolt and C.M. Mitchell, "Productivity Improvements at Inco's Copper Cliff Smelter," (Paper presented at the 115th AIME Annual Meeting, New Orleans, Louisiana, 2–6 March 1986).

63. M. Reist, H. Persson and D. Poggi, "Process for Converting of Solid High-Grade Copper Matte," U.S. Patent 5,007,959, 16 April 1991.

64. N.J. Themelis, J.K. Makinen and N.D.H. Munroe, "Rate Phenomena in the Outokumpu Flash Smelting Reaction Shaft," Physical Chemistry of Extractive Metallurgy, ed. V. Kudryk and Y.K. Rao, (Warrendale, PA: The Metallurgical Society, 1985), 289–309.

65. G. Melcher, E. Müller and H. Weigel, "The Kivcet Cyclone Smelting Process for Impure Copper Concentrates," J. Met., 28 (7) (1976), 4–8.

66. N.J. Themelis, "Transport Phenomena in High-Intensity Smelting Furnaces," Trans. Inst. Min. Metall., Sect. C, 96 (1987), C173–C226.

67. A. Chatterjee, C. Marique and P. Nilles, "Overview of Present Status of Oxygen Steelmaking and its Expected Future Trends," Ironmaking Steelmaking, 11 (3) (1984) 117–131.

TRENDS IN ALUMINUM SMELTING TECHNOLOGY

Nolan E. Richards
Reynolds Metals Company
Manufacturing Technology Laboratory
3326 East Second Street
Muscle Shoals, Alabama 35661-1258

Introduction

The electrolyte, or bath, is the heart of the Hall-Heroult process for the production of aluminum. The production and quality of aluminum from a plant is influenced by the total quality management practices concerning control of bath and its interfaces.

While there may be what appears to be a wide range in the compositions of electrolytes used world wide in the industry, several basic commonalities prevail. For instance, despite the modifying or fine tuning anyone may deduce from physicochemical properties which can lead to cost effective improvements, the essence of that change was embodied in Hall's patents (1). Secondly, no alumina, our basic raw material, is electrochemically decomposed that has not dissolved in the cryolitic solvent flux. The dynamics of dissolution of aluminum and, as Dr. E. Dewing has indicated in his presentation at this symposium (2), the loss of Faradaic efficiency - current inefficiency - is so dependent upon the balancing, control, and indirect means of sensing the kinetics of reactions and mass transfer between the electrodes, both the theoretical and practical aspects of understanding and utilizing our flux remains both intriguing and challenging. Presenting the basics for this appreciation, some pragmatics and scope for opportunities, is the purpose of this presentation.

The Scope of Bath in Production

World production of primary aluminum for 1990 was estimated to be 17,817,000 MT (3). There are two basic types of cells, differentiated according to whether the consumable, carbonaceous anode is brought in after being baked at a separate facility - prebaked anode pots - or using

Proceedings of the
Savard/Lee International Symposium on Bath Smelting
Edited by J. K. Brimacombe, P. J. Mackey,
G. J. W. Kor, C. Bickert and M. G. Ranade
The Minerals, Metals & Materials Society, 1992

a combination of process and electrical energy to bake a plastic mixture of coke and pitch to form a monolithic anode - Soderberg anode pots. About 23% of the world's capacity is in Soderberg cells, and 77% prebake. These technologies differ in several aspects. For example, prebakes make more efficient use of bath. The ratio (weight of liquid flux)/(unit of current) -- a reflection of the intensity of production per unit volume -- is shown in relation to the current in the potline in Figure 1. There is a trend for the Soderberg cells to utilize about 1.7 times as much bath as prebakes. (There is ±8% standard deviation to the points.) This larger volume is attributable to factors such as methods and frequency of adding alumina and the greater distances for transport into the central zone of the interelectrode distance necessitating a larger inventory of solvent, and to deeper cavities. These have an impact on moderating the intensities of fluid flow from both the bubbles of CO_2 and electromagnetic forces in the aluminum cathode. In prebaked anode cells, the preheated alumina-rich crust, by means of which the solute is fed to the bath, surrounds each of multiple anodes so mass transfer should be more efficient and the volume needed for supply-depletion of alumina, much less.

Some of the scatter in the graph indicates a trend embarked upon by the whole industry: reduction of unit energy and costs. By enlarging anodes, thereby reducing current densities, anodic overpotential, ohmic drops, and enhancing utilization of the volume of the cavity - with improved alumina management - Soderbergs can equal prebakes on this particular criterion. Most recent designs of prebaked cells exploit great improvements in bus work for compensation of magnetic forces resulting in significantly reduced flow vectors in the metal pad, together with point or virtually continuous feeders that add alumina in 2 to 5 kg increments. There is decreased risk of intermixing liquid metal and bath components, and potential for oscillating waves in the surface of the cathode. These improvements have allowed dramatic reduction in the inventory of flux.

A bit of trivia connecting world aluminum production to bath can be offered. The Hall Heroult process is characterized as "extensive." Given the apportioning of technologies above, the integrated area of bath for the world's capacity of aluminum is 920 million m^2 (227,400 acres).

Role of Bath

In the electrolytic extraction of aluminum, bath meets the following important functions:

-- Solvent and electrolyte medium for the solubilization and chemical reactions of Al_2O_3 with Na_3AlF_6 to form cations and oxy-

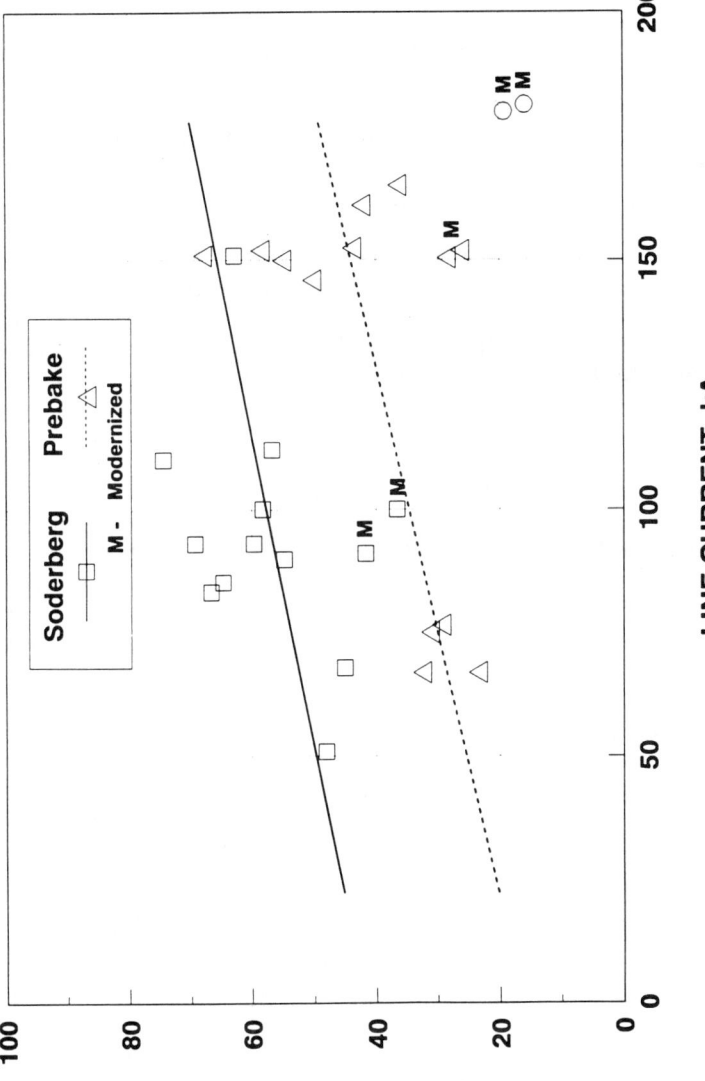

Figure 1 - The comparison of flux ratio with cell capacity of aluminum reduction cells.

fluoride anions. The key properties for understanding this are alumina solubility and phase diagrams.

-- Source of thermal energy as an internal resistive component of the electrical circuit. The IR drop across the bath is typically 1.6 ±.2 volts, or 35-40% of the total cell voltage. Electrical conductivity must be known for predictive changes in heat production over an adjustable interelectrode distance.

-- A layer of frozen bath against sidewalls of carbon or silicon carbide is the primary and durable containment for the flux. Because the velocity vectors in the bath and metal range from 3 to 12 cm/sec, implying strong convective mass accompanied by radiative heat transfer, then the temperature of the operation (bath) cannot be very much above the primary freezing point, e.g., 10 to 20°C. Typically, the frozen cryolite-rich ledges occupy 20 ±5% of the initial cavity. Besides understanding the dynamics and maintenance of the heat balance, we depend upon the fundamentals of phase diagrams, alumina solubility, knowledge of heat transfer coefficients, and thermal conductivities in accounting for and predicting the relationships between energy distribution and status of the apportionment of solid and liquid flux.

-- Bath, depending upon the temperature and proportions of the components in the commonly used quaternary, $NaF-AlF_3-CaF_2-Al_2O_3$, can have a vapor pressure 5-15 torr. The sparging action of the (CO_2 + CO) bubbles helps transport both major vapor species, HF (gases) and $NaAlF_4$ (particulate). Although 5-15 kg F/MT Al may be transported from a cell (4), efficient capture and/or recycle systems reduce these emissions by 90-95%. Models for vapor pressure of the bath are important for evaluating modifications in composition so that continued compliance with environmental regulations can be assured.

-- As Dr. Dewing notes (2), the electrolyte is also a solvent for aluminum and co-products, such as Na. The consequence of this chemical reaction along with transport of CO_2 gas bubbles across the horizontal surface of the anode is reoxidation - and loss - of products (e.g., Al, Na, and Li) that must be reduced again -- our major source of current inefficiency. This inefficiency is one of the more intriguing problems of the Hall Heroult process and consequently has been and continues to be researched in detail. When formulating changes to electrolytes, a quantitative measure of the impact on the amounts of Al and alkali or reactive metals dissolving/distributed in the bath is essential.

-- Components of the electrolyte enter into the electrode reactions, e.g., at the cathode, sodium is co-deposited to the extent that virgin aluminum can contain 100 ±60 ppm Na and, when the bulk concentration of Al_2O_3 decreases below 1.8 to 1.2 wt.%, only then does the flouride ion discharge at a higher anodic potential to form a proportion of CF_4 and C_2F_6 in the anode gases (anode effect). The trend in the industry is to decrease the frequency of these events through control of dissolved alumina at appropriate and more consistent levels.

--   Bath soaks into the carbon lining and thermal insulation. All of the sidewalls, carbon hearth, and thermal insulation have porosity. Bath, and particularly Na-enriched components, percolate into, accumulate in, and exert chemical and physical forces on the lining. This initially rapid absorption of flux, the continual vaporization of HF and $NaAlF_4$, contributions of sodium and calcium by way of 0.5 wt.% Na_2O, and 0.03 wt.% CaO in metallurgical alumina, requires an ongoing intelligence and adjustment on the composition and volume of bath. The trends are towards predictive control of the additions to maintain the specified chemistry, increased automation of chemical analyses, and more stringent specifications of raw and cathode construction materials.

Industrial Baths

The choice and design of electrolytes in recent years has been leveraged by three major independent factors: effect an increase in the current efficiency, decrease the energy requirement, and decrease emissions. Lowering the temperature of operation can help the first and last, but cannot improve electrical conductivity.

Increased concentration of AlF_3 and adding other thermodynamically acceptable fluorides such as LiF or MgF_2 depress the 1011°C freezing point of pure cryolite. Depending upon the small amount of CaF_2, due to the unavoidable component CaO in the Al_2O_3, CaF_2 equilibrates to 3.5 to 7 wt.% in an industrial bath affording 7 to 19°C reduction in liquidus temperature. Alumina, itself, lowers the freezing point about 5°C/wt.% and in a well adjusted cell with a "steady" balance between dissolved-decomposed Al_2O_3, the temperature will track the primary liquidus.

Differing compositions of baths that have evolved since the outset of the Pittsburgh Aluminum Company are exemplified in Table I (5). Those relevant to prevailing operations would permit operation at 945 to 970°C. The significance of density difference is that this should be greater than 0.15 gm/cm^3 to minimize risk of encapsulation or upward excursion of the liquid aluminum into the bath.

The quantity and quality of fundamental research since the mid 1950s contributing to the inventory of scientific information has been both remarkable and inspirational. As a consequence, almost every important aspect of the alumina reduction process can be modeled

TABLE I. BATH COMPOSITIONS ENCOUNTERED IN INDUSTRY

Company	Composition, wt.%				Calculated Freezing Point,°C at 3% Al_2O_3	Value at 20°C Above Freezing Point		
	AlF_3	CaF_2	LiF	MgF_2		Conductivity mho⁻¹ cm⁻¹	Density Difference g/cm³	Al_2O_3 Solubility wt.%
Early USA	28	15.6	--	--	783.5	1.27	0.05	--
"Classical"	3	7	--	--	970.9	2.41	0.16	7.2
Low Ratio, 1940	7	5	--	--	967	2.29	0.19	7.2
Pechiney (F Cell)	13	5	--	--	936.9	2.03	0.19	6.2
Nippon Light Metals	2	5	4.3	--	940.1	2.57	0.16	6.9
Alusuisse	7	4	--	--	970.0	2.31	0.20	7.6
	1-2	4	4	--	945.8	2.61	0.17	7.4
	1-2	4	2.5-3	3	939.7	2.41	0.16	6.8
Revere	7	2	2.5	--	953.7	2.39	0.21	7.7
Intalco	7	6	--	--	963.8	2.27	0.18	6.8
	6	4.8	2	--	952.8	2.36	0.18	6.9

mathematically with enough reliability for helping make decisions industrially.

From the above and related sources, I have extracted some useful equations for calculating or predicting selected properties of bath. These demonstrate today's trend away from the old empiricism to the extent that the design of the cell, embracing electromagnetics, thermal insulation, heat energy balance, voltage distribution, bath specifications and management, and current efficiency may all be computed.

Calculation of Bath Properties

Properties noted above that are perceived as essential to the intelligence-based selection or modification for a flux are conveniently reviewed and collated in several invaluable sources, particularly by Grjotheim and colleagues (6) and in the proceedings of The Metallurgical Society (7,8). The latter has become an international forum for sharing basic and practical information concerning the aluminum industry.

Freezing Point

The equations best accounting for the liquidus temperature deriving from comprehensive experimental data have been published by Peterson and Tabereaux (9) and Rostrum, Solheim & Sterten (10). The latter's inclusive equation accounting for normally encountered compositions in a quintary system is:

$$t \quad = \quad 1011 - 0.072 \cdot (AlF_3)^{2.5} + 0.0051 \cdot (AlF_3)^3$$

$$+ 0.14 \cdot (AlF_3) - 10 \cdot (LiF) + 0.736 \cdot (LiF)^{1.3}$$

$$+ 0.063 \cdot [(LiF) \cdot (AlF_3)]^{1.1} - 3.19 \cdot (CaF_2)$$

$$+ 0.03 \cdot (CaF_2)^2 + 0.27 \cdot [(CaF_2) \cdot (AlF_3)]^{0.7}$$

$$- 12.2 \cdot (Al_2O_3) + 4.75 \cdot (Al_2O_3)^{1.2} \tag{1}$$

where t is °C and the components are in wt.%.

Alumina Saturation

Combinations of the rate of dissolution, rate of depletion, and mass transfer normally do not let the amount of dissolved alumina exceed 70% $(Al_2O_3)_{sat}$. However, for projecting the likely behavior of a different electrolyte, a particular time-profile or feeding mechanism for alumina and thermal management, to avoid unnecessary polarization, anode effects, and sludging (freezing out from hypereutectic solutions), knowledge of the saturation level of alumina is desirable, and can be obtained from (11):

$$[Al_2O_3]_{sat} = exp[A + B \cdot (1000/t - 1)]$$

where:

$$A = 2.464 - 0.007 \cdot (AlF_3) - 1.13 \cdot 10^{-5} \cdot (AlF_3)^3$$

$$- 0.0385 \cdot (Li_3AlF_6)^{0.74} - 0.032 \cdot (CaF_2)$$

$$- 0.040 \cdot (MgF_2) + 0.0046 \cdot [(AlF_3) \cdot (Li_3AlF_6)]^{0.5}$$

and

$$B = -5.01 + 0.11 \cdot (AlF_3) - 4.0 \cdot 10^{-5} \cdot (AlF_3)^3$$

$$- 0.732 \cdot (Li_3AlF_6)^{0.4}$$

$$+ 0.085 \cdot [(AlF_3) \cdot (Li_3AlF_6)]^{0.5} \tag{2}$$

Electrical Conductivity

In the very near future we can expect what should become the most definitive data on electrical conductivity through the research of Wang, Peterson & Tabereaux who, through improved techniques seem to have overcome some of the complexities and sources of error inherent in obtaining direct measurements (12). Meanwhile, the industry continues to use a regression equation (13) based on the better data available up to 1971. Numbers derived from this empirical correlation were within ±4% of the experimental database. However, the trend in composition of electrolytes has been to higher proportions of AlF_3 and LiF and lower temperatures and, as Tabereaux et al. (12) point out, the predictions of the Choudhary equation (13) are 6-8% too low in those cases. Until the improvement is published, workable conductivities (K) can be obtained from:

$$\ln K \, (ohm^{-1}cm^{-1}) = 2.0156 - 0.0207 \, (\%Al_2O_3) - 0.005 \, (\%CaF_2)$$

$$- 0.0166 \, (\%MgF_2) + 0.0177 \, (\%LiF)$$

$$+ (0.00623) \, (\%NaCl)$$

$$+ 0.4349 \, (Bath \, Ratio) - 2068.4/(t+273) \tag{3}$$

where t is °C.

It is not unusual for a molten salt equilibrated with a common cation metal to exhibit some electronic conductivity. Through the studies of

74

Haarberg, et al., evidence for electronic conductance of about 0.3 $ohm^{-1}cm^{-1}$ at saturation has been presented (14). The project is continuing and we will be hearing more about this because, if directly applicable to the reduction process, this process should lead to an automatic 6% decrease in current efficiency. Since there are cells operating at almost 97% CE, it is plausible that the "electronic circuit" attributable to this parallel path is interrupted by the reoxidation of any dissolved metal by the CO_2 immediately adjacent to the surface of the anode.

Vapor Pressure

Aluminum producers have, for decades, recycled fluoride compounds for economic reasons and, today, through efficient fume capture and hooding systems, have improved this phase of the whole reduction process for reasons of both conservation and compliance with air quality standards. The main vaporizing species are HF, from the reaction of any source of moisture (including the small residual in metallurgical alumina) with AlF_3, and $NaAlF_4$, a discrete compound evolved from Na_3AlF_6-AlF_3, at partial pressures dependent upon the NaF/AlF$_3$ ratio and temperature of thermodynamic activities of the fluoride complexes. In both dry alumina and wet scrubbed technologies for fluoride capture, the adsorption-capture efficiency must be carefully balanced and monitored. Changes in flux composition can have a large effect on vapor pressure and awareness of these impacts must be known. Fortunately, reliable data has been produced by researchers (e.g., 15-18) and this has been incorporated into a mathematical model by Haupin (19).

$$P - Exp\ (B - A/T)$$

where:

$$A = 7101.6 + 3069.7(R_b) - 635.77(R_b)^2 + 51.22\ (\%LiF) - 24.638(R_b)$$

$$(\%LiF) + 764.5\ (\%Al_2O_3)/[1 + 1.0817\ (\%Al_2O_3)] + 13.2(\%CaF_2)$$

$$B = 7.0184 + 0.6844(R_b) - 0.08464(R_b)^3 + 0.01085\ (\%LiF)$$

$$- 0.005489(R_b)\ (\%LiF) + 1.1385\ (\%Al_2O_3)\ /$$

$$[1 + 3.2029\ (\%Al_2O_3)] + 0.0068\ (\%CaF_2) \tag{4}$$

R_b is the bath ratio, components are in wt.% and T is °K.

Aluminum Solubility

Current <u>inefficiency</u> is possibly the focal point of the alumina electrolyte reduction process. Dr. Dewing has addressed this topic for the audience. The apparent simplicity of the reaction,

$$3 CO_2 (g) + 2 Al_{(1)} \circledR 3 CO (g) + Al_2O_{3(1)} \tag{5}$$

continues to intrigue the curious and courageous researchers because while the outcome is easily measurable, the chemistry, mechanisms, kinetics, intricacies of what is rate determining, and nature of the structural or ionic species, are not. In time, with the improvement in experimental and analytical procedures building upon vast cumulative experiences on this topic, an even more complete story will be told about current inefficiency. From a purely pragmatic point of view we need to be able to predict what most probable trends in current efficiency will be if and when we want to change the flux. Subsequent to the excellent work of Odegard (20), cited by Dewing, my colleagues undertook a rigorous determination of the individual solubilities of Al, Na, and Li in cryolitic melts because it was suspected that these varied differently over the significant ranges of composition of choices of additives being used in our industry today (21,22). The regression equations generated for each of the dissolved metals, in wt. %, were:

$$\log (Al)_1 = 10.129 - 1.525 \log(CR) - .0452 LiF$$
$$- 14.738 \times 10^3/T + 0.1317(Al_2O_3)^{1/2} \cdot \log(CR)$$

$$\log (Na)_1 = 10.464 + 0.4024 CR - 16.324 \times 10^3/T$$
$$+ 0.1610(LiF)^{1/2} + 0.198(Al_2O_3)^{1/2}$$
$$- 0.1247(LiF)^{1/2} \cdot Al_2O_3/CR$$

$$\log(Li)_1 = 7.683 - 0.6453CR + 6.077\log(CR)$$
$$+ 0.4039LiF - 0.01998Al_2O_3$$
$$- 14.789 \cdot 10^3/T - 0.4190LiF \cdot \log(CR)$$
$$- 0.007269(LiF)^2/\log(CR) \tag{6}$$

The components are in wt.%, CR is the molar cryolite ratio and T is temperature °K.

The most modern magnetically compensated cells are fitted with bar and point feeders. The results of Peterson and Wang, in Tables II and III, Figures 2 and 3, are worthy of reproduction because they strongly reflect and correlate with actual plant results. This type of modelling for

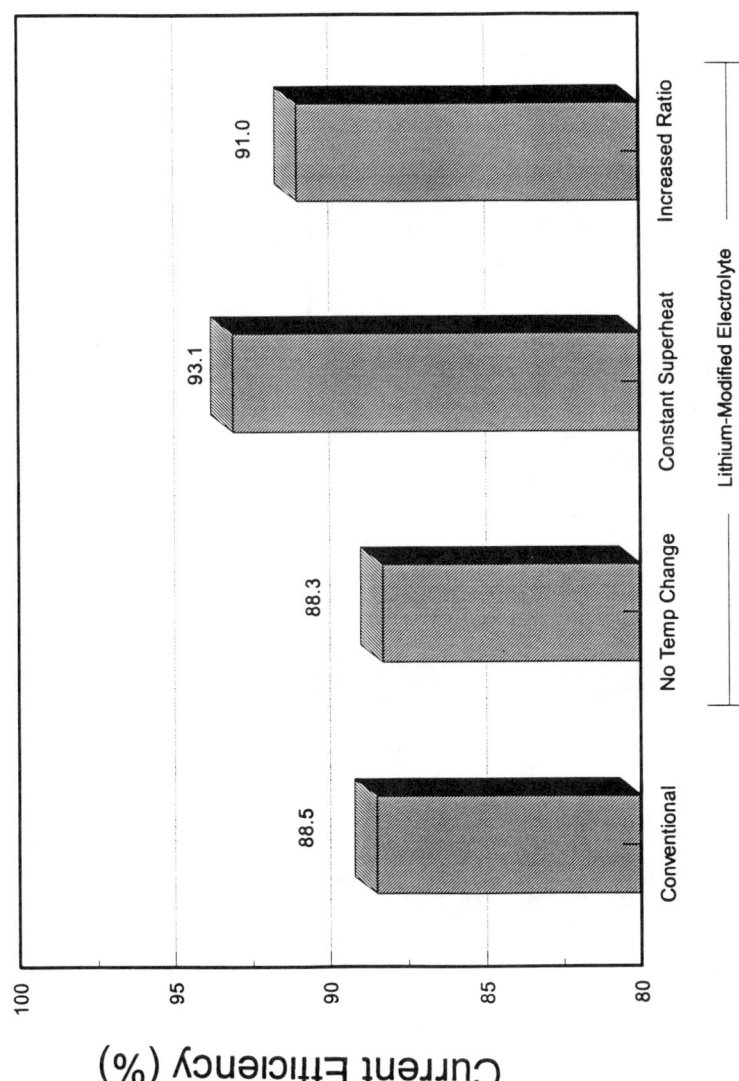

Figure 2 - Current efficiency predictions for a center-break prebake cell with various electrolyte compositions.

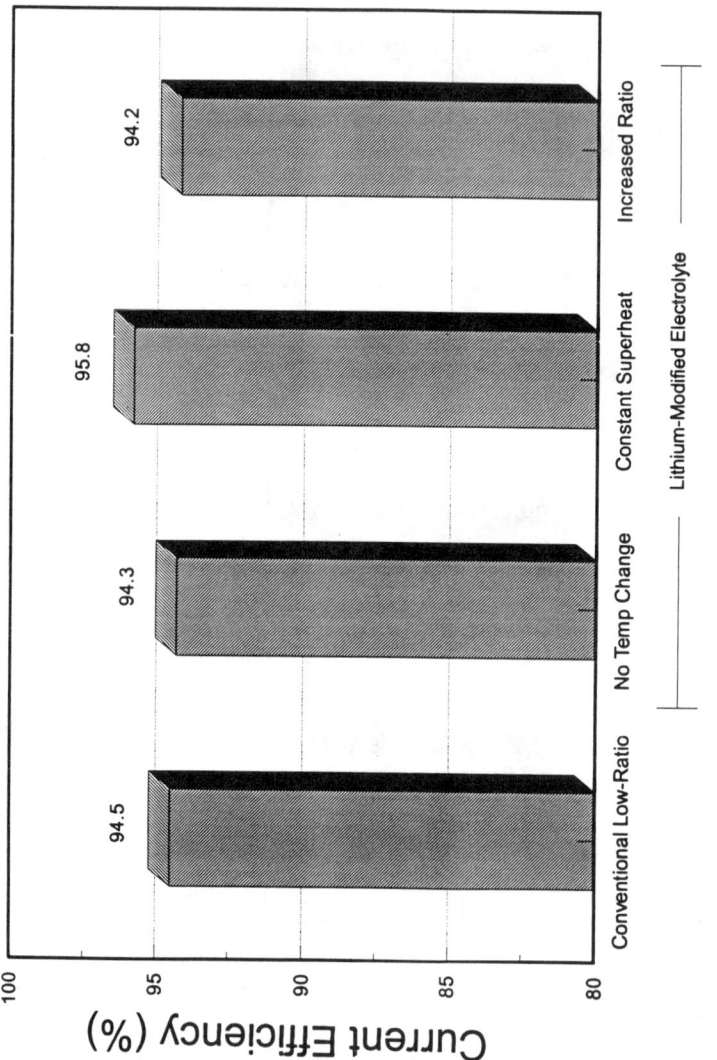

Figure 3 - Current efficiency predictions for a point-feed prebake cell with various electrolyte compositions.

Table II. Electrolytes for a Center-Break Prebake Cell

Molar Cryolite Ratio	% Al_2O_3	% CaF_2	% MgF_2	% LiF	Freezing Point °C	Operating Temp. °C	Metal Solubility %
2.60	3.5	6.0	0.5	0.5	960	975	0.0311
2.60	3.5	6.0	0.5	3.0	939	975	0.0305
2.60	3.5	6.0	0.5	3.0	939	954	0.0184
2.80	3.5	6.0	0.5	3.0	943	958	0.0243

Table III. Electrolytes for a Point-Feed Prebake Cell

Molar Cryolite Ratio	% Al_2O_3	% CaF_2	% MgF_2	% LiF	Melting Point °C	Operating Temp. °C	Metal Solubility %
2.32	2.5	5.0	0.5	0.5	957	967	0.0204
2.32	2.5	4.0	0.5	2.0	945	967	0.0207
2.32	2.5	4.0	0.5	2.0	945	955	0.0155
2.40	2.5	4.0	0.5	2.0	955	966	0.0215

electrolytes clearly shows the impact on current efficiency of choices for both the electrolyte and the regimen of the heat balance.

The 'total aluminum' in the bath is total equivalents of all the light metals. This data quantitatively confirms Haupin's early indications (23) that sodium is the larger proportion of the equivalent dissolved aluminum.

Wang and Peterson showed how this solubility data could be coupled with mathematical models for current efficiency to anticipate the consequences of changes in fluxes, cell design and operational parameters (22). From a model proposed by Lillebuen, et al. (24), using the above equations, they calculated, a priori, current efficiencies for two generations of prebaked anode cells equipped for additional alumina between the rows of anodes, one equipped with a center breaker and the other, a point fed cell.

Summary

The Hall Heroult process for extraction of aluminum is an extensive process producing 62 kg $Al/m^2/day$. The electrolyte, based on the original $NaF-AlF_3-Al_2O_3$ system, is and will continue to be a focal aspect. Within this bath or flux, separating the horizontal electrodes by 3.5 to 5.5 cm, desirable and undesirable chemical and electrochemical reactions occur. An excellent database exists for the key properties such that, for any given type of cell -- older, retrofitted, or modern -- the properties, chemistry, and management of the molten bath can be prescribed for maximizing production and minimizing energy and costs. Although plants are increasing the use of automation for control of the interelectrode distance, dissolved alumina, and current distribution, there is still room for further improvement. In particular, mathematical models for the whole process leading to more sensitive algorithms will improve, and the need for cost effective means for on-line sensing of important temperatures, in situ chemical compositions, and current distribution will become available.

Acknowledgement

I thank my colleagues for invaluable discussions leading to this review, and Dr. Alton Tabereaux, especially. I am also grateful to Reynolds Metals Company and interactions with people of the aluminum industry around the world who have made learning about the Hall Heroult process, still incompletely understood, highly enjoyable.

References

1.   C. M. Hall, U. S. Patent 400766, Line 15 (1889); U. S. Patent 400664, Lines 55-60 (1889); U. S. Patent 400665, Lines 5-15 (1889).

2.   E. W. Dewing, "The Thermochemistry of Aluminum Smelting," Savard-Lee Symposium (1992).

3.   "Aluminum, Bauxite and Alumina," U.S. Bureau of Mines Report, (1991), 6.

4.   Warren Haupin, "Chemical and Physical Properties of the Electrolyte," (Paper presented at the International Course on Process Metallurgy, Trondheim, Norway, 1991).

5.   N. E. Richards, "Evolution of Electrolytes for Hall-Heroult Cells," (Paper presented at the Hall Heroult Centennial, 1986) ed. Warren Peterson, The Metallurgical Society, 114.

6. K. Grjotheim et al., <u>Aluminum Electrolysis</u>, 2nd edition, (Aluminium-Verlag, Dusseldorf, 1982).

7. "Extractive Metallurgy of Aluminum," <u>Aluminum</u>, vol. 2, ed. G. Gerard (John Wiley, New York, 1963).

8. <u>Light Metals</u> series, (Warrendale, PA: The Metallurgical Society, 1971-1992).

9. R. D. Peterson and A. T. Tabereaux, "Liquidus Curves for the Cryolite AlF_3-CaF_2-Al_2O_3 System," <u>Light Metals</u>, (1983), 383.

10. A. Rostrum, A. Solheim and A. Sterten, "Phase Diagram Data in the System Na_3AlF_6- Li_3AlF_6-AlF_3-Al_2O_3," Part I, <u>Light Metals</u>, (1990), 311.

11. E. Skybakmoen, A. Solheim, and A. Sterten, "Phase Diagram Data in the System Na_3AlF_6-Li_3AlF_6-AlF_3-Al_2O_3," Part II, "Alumina Solubility," <u>Light Metals</u> (1990) 317.

12. Xiangwen Wang, R. D. Peterson and A. T. Tabereaux, "Electrical Conductivity of Cryolitic Melts," <u>Light Metals</u> (1992) 481.

13. G. Choudhary, "Electrical Conductivities in the Aluminum Extraction Cell," <u>Journal of Electrochemical Society</u>, 120 (1973) 381.

14. G. M. Haarberg, et al., "Measurement of Electronic Conduction in Cryolite-Alumina Melts and Estimation of Its Effect on Current Efficiency, <u>Light Metals</u>, (1991) 283.

15. U. Kuxmann and U. Tillessen, "Dampfdruckmessungen in den Systemen NaF-AlF_3 und LiF-NaF-AlF_3," <u>Erzmetall</u>, Band XX (1967) 147.

16. H. Kvande, "Thermodynamics of the System NaF-AlF_3-Al_2O_3-Al, Studied by Vapour Pressure Measurements," (Dr. Techn. thesis, Institute of Inorganic Chemistry, University of Trondheim, Norway, 1979).

17. J. Guzman, K. Grjotheim and T. Ostvold, "The Influence of Different Fluoride Additions on the Vapour Pressure of Molten Cryolite," <u>Light Metals</u> (1986) 425.

18. K. Grjotheim, H. Kvande and K. Motzfeldt, "Vapor Liquid Equilibria in the System NaF-AlF_3-Al_2O_3," <u>Light Metals</u>, vol 1, ed. A. Sterten and J. Thonstad, (1975) 125.

19.   W. E. Haupin, "Chemical and Physical Properties of the Electrolyte," (Paper presented at the 10th International Course on Process Metallurgy of Aluminum, Trondheim, May 1991).

20.   R. Odegard, "On the Solubility of Aluminum in Cryolitic Melts," Metall Trans. B, (19B) (1988), 449.

21.   Xiangwen Wang, R. D. Peterson and N. E. Richards, "Dissolved Metals in Cryolitic Melts," Light Metals, (1991) 323.

22.   R. D. Peterson and Xiangwen Wang, "The Influence of Dissolved Metals in Cryolitic Melts on Hall Cell Current Efficiency," Light Metals (1991) 331.

23.   W. E. Haupin, "Metal Mist and Aluminum Losses in the Hall Process," Journal of Electrochemical Society, 107, (1960), 232.

24.   B. Lillebuen et al., "Current Efficiency and Back Reaction in Aluminum Electrolysis," Electrochimica Acta, vol 25 (1980), 131.

CONTINUOUS FERROUS AND NON-FERROUS BATH SMELTING

Howard K. Worner

Consultant
c/- The Illawarra Technology Corporation Limited
University of Wollongong
PO Box 2112
Wollongong NSW 2500 Australia

Abstract

The discovery in 1961, while with BHP, that fine lump ore could be continuously smelted in a slowly flowing stream of blast furnace hot metal which was sequentially lanced with oxygen, launched the author into bath smelting in both ferrous and non-ferrous metallurgy. Under the name WORCRA, zoned horizontal furnaces with both tuyeres and lances were evaluated. Problems and advantages of each will be discussed, as will the subtle interplay of kinetics and thermodynamics in systems involving turbulent and quiescent zones and slag moving generally countercurrent to matte or metal. Metal was produced in the same furnace as smelting and slag cleaning were achieved continuously. With a change in top management, CRA discontinued the developments in the early seventies, but aspects of WORCRA technology continue in other processes. Currently, the author and his colleagues are using WORCRA principles in the smelting of composites of steelworks dusts and a variety of carbonaceous wastes to produce a foundry type iron, phosphorus-containing slag and zinc oxide in the off-gases.

Proceedings of the
Savard/Lee International Symposium on Bath Smelting
Edited by J. K. Brimacombe, P. J. Mackey,
G. J. W. Kor, C. Bickert and M. G. Ranade
The Minerals, Metals & Materials Society, 1992

Introduction

When our guests of honour, Guy Savard and Robert Lee, conceived of their shrouded tuyeres they were no doubt thinking of applications in oxygen steelmaking. They could hardly have dreamed that 28 years later we should be celebrating their important contributions to pyrometallurgy in an international symposium on bath smelting and in both ferrous and non-ferrous metal production. Several of us participating in this meeting have experienced that kind of serendipity, starting out with one invention and, either ourselves, or others, discovering new and unexpected applications or adaptations.

I am honoured that in my eightieth year I should be included amongst the speakers, not only to represent the older members in paying tribute to Guy and Robert but to be able to share with a younger generation of metallurgists work on continuous integrated smelting-converting-slag cleaning that I began thirty years ago.

In some of the applications of that concept my colleagues and I thought to use tuyere injection but we did not have the benefit of the ideas that in two years time would be germinating in the minds of Savard and Lee. As I shall explain later, our early experience with conventional tuyeres (of course, small for a pilot plant) were most disappointing so I decided to revert to lances. I have often pondered the question, how might my concepts and practices have developed if my inventions had followed rather than preceded those of Savard and Lee? We did not get to know about shrouded tuyeres in Australia until later in the sixties when we learned what Karl Brotzmann was achieving with them at Maxhütte in Germany. By that time our technology was fairly firmly tied to fixed furnaces and lances. I was then with CRA Ltd which, if you like, had "hedged its bets" by entering into a collaborative agreement with Noranda, who were developing technology involving tilting furnaces with conventional tuyeres.

I hope that I may live long enough to see shrouded tuyeres used in the scale-up of another of my more recent inventions in which iron bath melting/smelting is used for the processing of contaminated scrap, steel plant dusts and a whole spectrum of carbonaceous wastes.

Before I recount briefly the story of how I got into bath smelting, I want to thank our General Chairman, Keith Brimacombe, and Chairman of the non-ferrous program, Phillip Mackey. It was Keith's kind thought that I should participate and Phillip who, via faxes, has brought me up-to-date with the world scene in sulphide smelting. The papers he sent me included three excellent general reviews (1, 2, 3).

Beginnings for the Author

My preparation for bath smelting took place in 1960 when I was Director of Research for the big Australian steelmaker, The BHP Co Ltd. While gardening one Saturday afternoon, I suddenly conceived of how continuous iron making might be integrated with continuous steelmaking using, for the latter, sequential oxygen lancing in an elongated hearth. The concept is reproduced in Figure 1 exactly as it was drawn in late 1960 and used in the May 1961 "Annual Address" to the Australian Institute of Metals Annual General Meeting. (4).

I envisaged a start-up with molten pig iron from a cupola (not shown); the hot metal would flow into the elongated hearth from the left and would trigger off the reduction-smelting of the burden descending in the shaft. Hot carbon monoxide from the sequential lancing would speed-up the reduction of the closely sized ore lumps or pellets and also assist in the final smelting in the slowly flowing bath. The countercurrent upward flow of hot gases would be aided by a slight negative pressure in the gas handling system.

As you will imagine, my submarine-like furnace generated vigorous debate. More sober reflection after the meeting caused me to trim my objectives and start by testing the concept of sequential refining of blast furnace hot metal in a small (2.0 x 0.3 metre internal dimensions) launder furnace. Oxygen would be jetted into the flowing molten pig iron by four vertical lances.

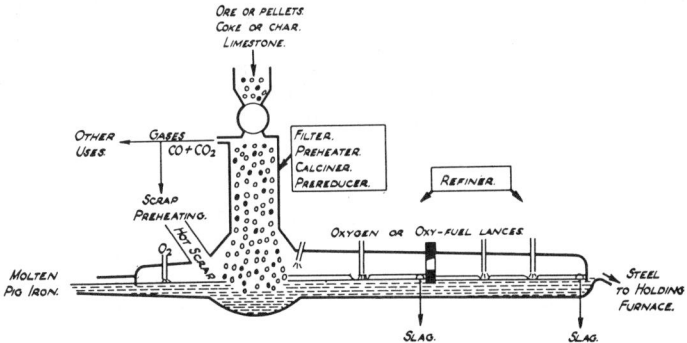

Figure 1 - A 1960 concept of a possible integrated continuous iron and steelmaking operation

Most of my colleagues were highly sceptical, believing that the heat losses from such a small furnace would be so great that the bath would progressively freeze. The opposite proved to be the case. Feeding the little furnace with normal hot metal at about 1500°C (at the beginning of the run) and with a flow rate as close to 5 ton per hour as our ladle tilting mechanism would permit, we found that the exiting metal got hotter and hotter. Within twenty minutes we were tapping a rimming type steel and the temperature was approaching 1750°C. We frantically searched for every nut and bolt that we could find to add as coolant but, of course, the slag at that temperature soon eroded our final slag dam.

For the next run I decided to include an ore plus limestone bin. The endothermic reduction and calcination reactions would be more effective than "scrap" additions. The arrangement is shown in Figure 2, the only photographs retained in BHP archives.

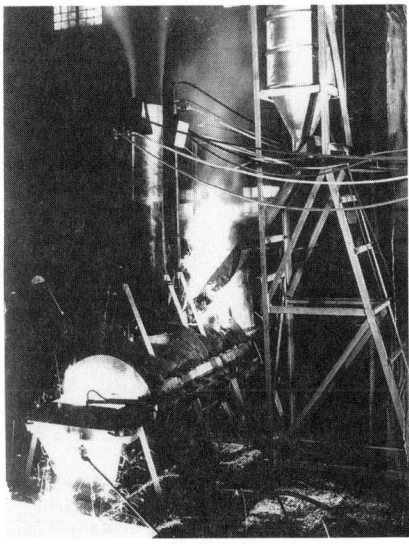

Figure 2 - Small continuous steelmaking furnace operated in late 1961 and early 1962. On the left, launder being prepared for run; on the right in the early stages of a trial at night time. Desiliconized semi-steel is flowing from the furnace after six minutes blowing. Ore bin is clearly visible well above the furnace.

Rimming steel production was achieved in less than twenty minutes and continued until we had run out of hot metal. (Colour slides will show the evidence.)

Even with 15% ore addition and 5% limestone our steel tapping temperature was almost 1550°C. Clearly, in a large launder furnace refining, say, 200 tons per hour, there would be the capacity to incorporate a considerably higher proportion of ore.

As I watched through a feed-end sight port, I could discern that the whole bath seemed to be "boiling" and there was a particularly vigorous action slightly downstream from the point where the ore and limestone were falling into the slowly flowing bath.

I had many times witnessed an ore boil in an open hearth. Here clearly we were generating the same reactions continuously. I recognised that silicon and carbon were not only acting as "internal fuels" but there was an on-going reaction between the oxygen in the ore and the carbon in the slowly flowing metal. It was, indeed, a "continuous bath smelting operation".

It dawned on me that if we could add more carbon with the ore, we could still further enhance the bath smelting reactions. In later runs we tried adding some Newcastle coke but at that time it was too light and was carried downstream on the bubbling slag. It did not work for us where we needed the carbon up-take, between lance 1 and lance 2. Of course, in early 1962 we were many years ahead of Karl Brotzmann's brilliant use of bottom tuyeres for coal injection.

<u>Non-Ferrous Possibilities</u>

My mind was tremendously stimulated; I began to think of applications to lead, nickel and copper concentrates and, possibly, "bulk" concentrates, where sulphur and iron would be the internal fuels.

Taking chalcopyrite-rich concentrates as an example, I thought about smelting the finely particulate sulphides by injecting them into a furnace with supplementary fuel and some oxygen. The matte generated would flow along an elongated hearth and be sequentially jetted with oxygen, or air enriched with oxygen, and I envisaged the composition changing as indicated by the heavy line in the ternary diagram of Figure 3.

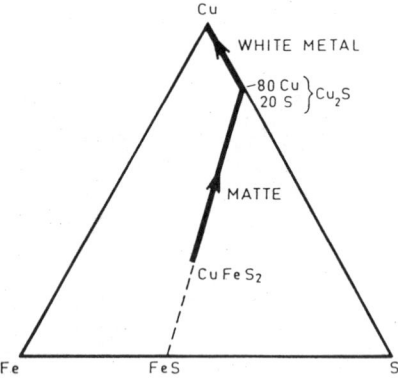

Figure 3 - Ternary diagram showing expected change in composition (heavy line) of chalcopyrite melt progressively oxidised while flowing along an elongated hearth furnace (concept of early 1962).

As 1962 progressed, I became persuaded that bath smelting might be easier to develop with non-ferrous feed than with iron ore and coal. My problem was that I was working for an iron and steelmaking company and my General Manager, Mr (later Sir Ian) McLennan, could see no reason why his Director of Research should be concerned about copper, nickel or lead smelting. He was not even enthused about ferrous bath smelting, so I took the bold decision to resign my position in BHP as from the end of 1962.

Departure from BHP and New Beginnings

I left BHP with very mixed feelings for my seven years with the "big Australian" had been both challenging and exciting and I had an excellent staff. Now, however, I would be free to explore a wider spectrum of extractive metallurgy.

Simultaneously with thinking about continuous non-ferrous, as well as ferrous, bath smelting, I began to cogitate about the advantages of having the slag flow in the opposite direction to matte or metal. But there was no point in sharing those thoughts with a company that, at that time, could see no future in bath smelting. As 1962 progressed, I perfected my ideas during weekends while throwing all the effort I could into completing the projects we had going at the BHP Central Research Laboratories, Shortland, near Newcastle, New South Wales.

Having lodged my patent applications, I went overseas early in 1963 to "sell" my continuous processing concepts to whomsoever would listen. Several companies in Europe and the UK were interested but in the end I responded to the invitation of Mr (later Sir Maurice) Mawby, Chairman of Conzinc Riotinto of Australia Ltd. (later to shorten its name to CRA Ltd.) to return to Australia to take up the position of Director of New Process Development. Mr Mawby proposed that I continue to live in Newcastle and do the pilot scale work at Cockle Creek (also near Newcastle) where Sulphide Corporation had recently commissioned the Imperial Smelting Furnace (ISF) for simultaneous production of zinc and lead. My continuous processes would be named WORCRA.

It is relevant to record that one of my young colleagues who had been with me at Shortland and had resigned shortly after me to join CRA, was John Innes. After working with me on a variety of continuous processes, he was later to become associated with Karl Brotzmann in the development of HIsmelt of which we shall hear more at this meeting. Another young colleague was Barry Andrews, now Managing Director of Southern Copper Pty Ltd (formerly ER & S) where the Noranda reactor is performing well. He had a thorough training in continuous copper pyrometallurgy with me.

The WORCRA Processes - Continuous Countercurrent Flow Technology

The first WORCRA furnace built at Cockle Creek was a little smaller than the first continuous steelmaker at BHP steelworks but it proved to be large enough to demonstrate the concepts that I had already covered with patent applications.

The furnace was "zonalised"; smelting took place in a more or less central zone; oxidation of the "internal fuels" was achieved in the directly-connected converting (or refining) zone where the majority of the flux was added. By the use of an appropriately cooled slag dam or end wall, the slag was forced to flow generally countercurrent to the matte or metal and be "cleaned" by the incoming sulphide concentrates (or coal in the application to oxidic ores of such as iron, manganese and chromium) in the vigorously mixed smelting zone before flowing on through the relatively quiescent slag settling zone to continuous tapping. The metal was tapped continuously from a deeper taphole at the opposite end.

Selected references reporting progress with the evaluation of the continuous countercurrent flow technology are listed at the end (5-16).

Schematic drawings of vertical cross sections through WORCRA furnaces for (a) iron and steel and (b) copper metal productions are shown in Figure 4.

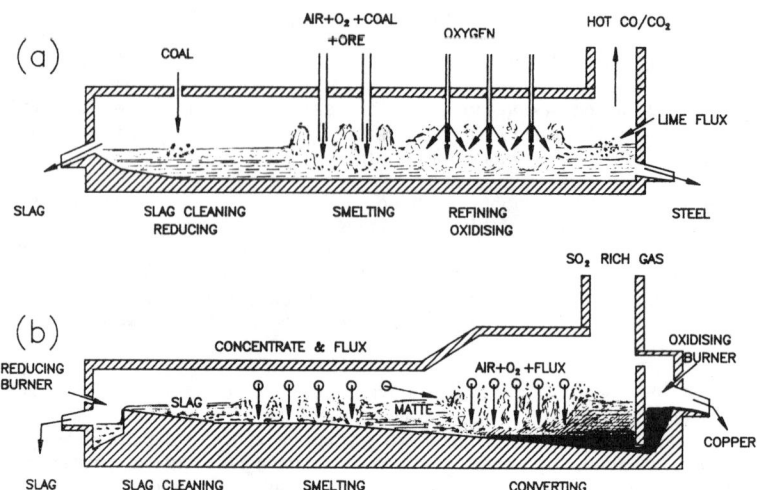

Figure 4 - Schematic vertical sections of the fixed forms of WORCRA furnaces
(a) for iron and steel production;
(b) for metallic copper production.

I emphasise that the drawings are schematic, but they convey the essential nature of a three-zone furnace producing metal at one end and a "throw-away" slag at the other. I am aware that there are still pyrometallurgists who feel that we must have used some trickery for "you cannot make copper in one and the same furnace as concentrates are being smelted to matte". The WORCRA counterflow technology no more "flies in the face of thermodynamics" than a reflux distillation column does. We generate and maintain a steady state dynamic reaction system and do not aim at achieving equilibria throughout the furnace.

We progressed from a pilot furnace at Cockle Creek, treating only a few hundred kilograms per day, to iron and steelmaking furnaces treating eight to ten tonnes per hour and, in the case of copper and nickel concentrates, treating three to five tonnes per hour. During 1963, we completed some promising trials with lead concentrates but, having the relatively new Imperial Smelting Furnace at Cockle Creek, there was little incentive at that time in pursuing bath smelting for lead production. Later, in the larger pilot plant at Port Kembla we did ten weeks work with Kambalda (WMC) nickel concentrates and proved that the technology works just as well in making a nickel-rich matte as it does for copper production.

Tonnage Oxygen

In our iron and steelmaking work, we always used oxygen but there was at that time a feeling in CRA that oxygen was too expensive in non-ferrous smelting, so we used only air in our pilot plants right up to the latter stages of Campaign 3 in the WORCRA copper smelter at Port Kembla. Only then was my personal conviction confirmed, that oxygen enrichment of air would have greatly increased throughput, given us a richer SO_2 gas and saved hydrocarbon fuels.

Tuyeres and Lances

I remind you that the evolution of the WORCRA furnaces took place a few years before Savard and Lee made their important discovery of the shrouded tuyere. As I said in the Introduction,

WORCRA technology, at least in some aspects, might have been different had I known four or five years earlier what Guy and Robert were going to invent.

Some people imagine that we ignored tuyere injection. That is not so. Indeed, one of the early non-ferrous pilot plants, built in late 1963, was a tilting barrel fitted with two "blocks" of tuyeres, one for injection of concentrates and air in the smelting zone and the other for air injection in the converting zone.

Unfortunately, I cannot show you a real photograph of that early furnace. The only record of it in CRA Archives is a photocopy and by including it as Figure 5, I know I am disobeying the instruction of the TMS Publishing Services Department. I crave indulgence in respect of some poor later illustrations for the same reason.

Figure 5 - A tilting cylindrical furnace built in October-November 1963. By the time the (original) photograph was taken, two tuyere blocks had been removed and angled lances were being installed.

That particular barrel-shaped pilot plant suffered from a number of disabilities:

(i)   the tuyeres were small and, as we were using conventional converter air pressure, we experienced frequent blockages (oxygen enrichment would have helped);

(ii)   vigorous punching was necessary and that damaged the tuyere blocks;

(iii)   the furnace was neither long enough nor did it have sufficient slope to minimize back-mixing; consequently, the slag tapped (at the higher end) often contained over 1.5 percent copper. We were aiming for an average of 0.5 per cent copper in slag.

By the end of February 1964, we had decided to use lances rather than tuyeres and it was natural to opt for fixed furnaces. That decision also seemed appropriate in the light of a discovery, some twelve months later, that Noranda had begun to develop a continuous copper smelting-converting furnace. They were using a tilting furnace with tuyeres. Our respective senior executives were long-time friends, so the companies decided on the collaboration that was mentioned in the Introduction. Noranda would develop a tilting furnace with tuyeres and CRA would continue with its fixed furnaces and lances.

It turned out, that for reasons I shall explain later, the WORCRA developments were stopped in the early 1970s, and in recent years, CRA, having acquired the Port Kembla ER&S smelter, opted to install the Noranda reactor (rich matte mode) with an electric furnace slag cleaner. I have already mentioned that the continuous smelter is performing well and has exceeded the capacity of the two existing converters to turn the rich matte into blister copper.

That lancing also has an established role in copper pyrometallurgy is attested by the success of the Mitsubishi process and also Ausmelt and Isasmelt, both of which utilise Sirosmelt type lances.

It is interesting to note the wide variety of engineering designs that are used in bath smelting operations. Some features are listed in Table I and it will be seen that there is almost an even split between tuyeres and lances.

Table I Design Features of Some Developed Non-Ferrous Bath Smelters

Process Name	Country of Origin	Furnace Type	Gas Injection System	Solids Feeding System
NORANDA	Canada	Tilting barrel	Tuyeres	Green, slung onto bath
MITSUBISHI	Japan	Fixed hearth	Vertical lances	Dried, via lances
VANUKOV SMELT IN MELT	Russia	Fixed hearth	Side tuyeres in water cooled copper jackets	Green, from top onto emulsion
BOLIDEN (for lead)	Sweden	Rotating	Angled lance	Angled lance
Q S L	USA/Germany	Tilting barrel	Shrouded tuyeres	Green pellets from top
EL TENIENTE	Chile	Tilting barrel	tuyeres	Dried, via tuyeres; also green added to bath
AUSMELT	Australia	Fixed, tall cylindrical	Sirosmelt "swirling" lance	Green, lightly agglomerated, from top
ISASMELT	Australia	Fixed, tall cylindrical	Sirosmelt "swirling" lance	Green pellets from top
Emerging ferrous bath smelting				
HISMELT	Germany/ Australia	Tilting cylindrical	Shrouded tuyeres	Top and via tuyeres

Ferrous and Other Oxidic Smelting - Refining Applications

While discussing appropriate applications of tuyeres and lances, I should make the point that shrouded tuyeres have advantages if the reductant is comminuted bituminous coal: (I have a different view, as I shall briefly discuss later, if the reductant is very low rank coal or peat or equivalent carbonaceous material which can be used in a "composite" with the oxide).

Top jetting of bituminous coal particles into a hot bath of slag and metal is prone to lead to high dust carry-over for a number of reasons, amongst which are:

90

(a) the lower specific gravity of coal means that particles do not enter the bath with the same momentum as ore or concentrate/particles;

(b) that lower SG means that the coal particles are more readily entrained in the off gases;

(c) the almost explosive pyrolysis of the coal particles contributes to their "blasting" out of the bath.

These problems are greatly reduced if coal enters the bath from the bottom; a foaming slag on the top assists in "catching" ash particles and unabsorbed carbon.

For the above reasons, we did not progress very far with the technology implied in Figure 4a. I go along with what Karl Brotzmann, John Innes and their colleagues are doing with the HIsmelt operation. I am also in general agreement with what Paul E Queneau has written (17,18) about ferrous applications of Savard/Lee bottom tuyeres. I must, however, comment, as I did in a letter to Egil Aukrist in October 1989 (19) in relation to the AISI "Direct Steelmaking" project; long experience has taught me that it is extremely difficult to control carbon levels if steel is being made in a "zone" of the same furnace that much higher carbon hot metal is being generated. I prefer to "stretch out" the steelmaking part, as we did in the WORCRA continuous steelmaking pilot plants at Cockle Creek (8, 10, 11) and later at MEFOS, Lulea, Sweden and also at Oregon Steel Mills plant, Portland, Oregon (10, 12).

Chemistry plays an important role in the design of reactor systems. For copper, the "end of the road" for converting is blister copper and it does not matter what the shape is or whether the furnace is tilting or fixed. For steel, on the other hand, there are many possible "stopping places" in the refining "road". Furthermore, very violent boils can be generated if the furnace shape allows uncontrolled mixing of very hot FeO-rich slag (coming from a low carbon steelmaking zone) with hot high carbon iron.

Some WORCRA Ferrous Developments in the 1960s and Early 1970s

Because of the problem with dust carry-over, already mentioned, and difficulties in controlling bath carbons, we did not pursue on any significant scale the lance injection of fine coal and ore. We did, however, confirm in pilot plants up to eight tonnes per hour, with counter-current slag flow, what we had observed back in 1961-62 at BHP with co-current slag flow. Small lump ore could be "bath smelted" in the slowly flowing molten stream. At Lulea, with the hot metal generated in the electric arc melter/smelter, and containing about 4 per cent carbon and 1.2 per cent silicon, some 20 per cent of our final eight tonnes per hour of steel came from bath smelted slightly preheated ore.

In each of the continuous iron and steelmaking plants operating with countercurrent slags, the vigorous but controlled "boiling" lead to foaming slags; these acted as marvellous "scrubbers" for both fume and dust (as will be shown in colour slides). I shall refer here only briefly to the pneumatic system used at Cockle Creek; electric melting/smelting was employed at Lulea and Portland. In each of the pilot plants the steelmaking section was "stretched out" as a launder where refining took place by a combination of sequential lancing and countercurrent slag flow.

The principles and achievement of the Cockle Creek L-shaped furnace operating at 5 tonnes per hour, are illustrated in Figures 6, 7 and 8. Regularly 95 per cent or more of the phosphorus was removed and between 75 and 85 per cent of the sulphur. Even more remarkable refining was achieved in the Lulea 8 tonne per hour pilot plant. Extra clean steel was made from very dirty granulated pig iron (20). Over 99 per cent of phosphorus was removed and 95 per cent of the sulphur. I am pleased to learn (21) that in USA David Robertson at Rolla and Julian Szekely at MIT are doing modelling work on the continuous counter current (CC) refining of steel. This work assumes that Savard/Lee tuyeres will be used for oxygen and lime injection.

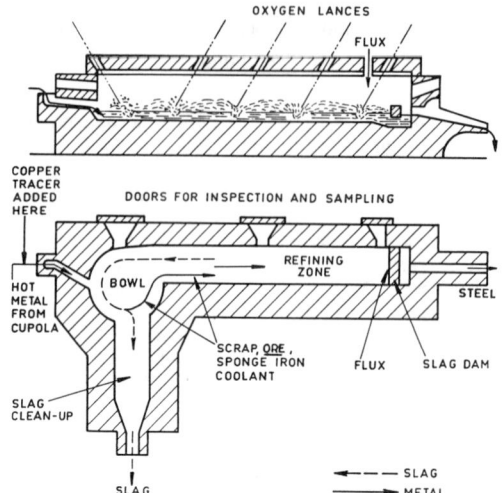

Figure 6 - Sectional plan and elevation of L-shaped pilot WORCRA Steelmaking furnace at Cockle Creek, near Newcastle NSW

Figure 7 - L-shaped furnace operating at 5 t.p.h. Steel is flowing on right and slag on left, both into granulating launders.

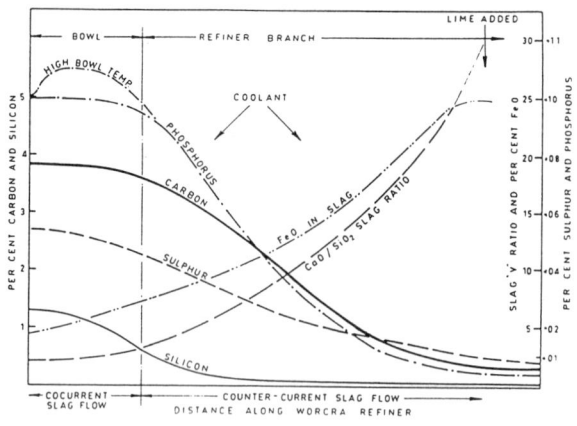

Figure 8 - Typical composition gradients in the launder and "bowl" of both slag and metal when operating at 5 t.p.h. steel production.

92

While thus far a single furnace operation for continuous smelting-converting-slag cleaning/settling has not become commercial, the potential is still there for such technology to:

(i) eliminate equipment for hot liquids transfer;

(ii) lower overall energy requirements, in part by utilising efficiently the heat in the slag flowing back from the converter zone to the smelting zone;

(iii) achieve high SO_2 recovery in a "single" relatively rich continuous stream.

I realise now that there are a variety of furnace designs in which those goals could be achieved. However, it will be fruitful to review briefly what we did at Cockle Creek and Port Kembla in the 1960s and 1970-71, frankly admitting our mistakes as well as sharing information on our successes.

When in early 1964 we decided to move from the tilting furnace to the fixed hearth furnace, we made a decision which was later to plague us. Because of limited site availability, we elected to use a U-shaped furnace, as shown schematically in Figure 9.

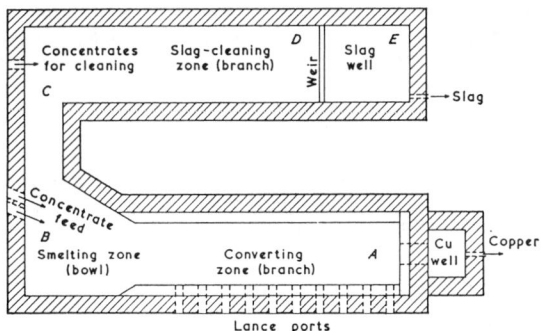

Figure 9 - Plan view of U-shaped WORCRA furnace used at Port Kembla

It had some good features as well as some bad ones. Good design aspects were:

(a) the proximity of the slag and copper tapholes made it possible for one furnaceman to keep his eye on each continuously flowing stream; the "throw-away" slag was granulated while the copper was cast into large iron moulds carried on a moveable platform on rails;

(b) the sloping inverted arch hearth led down to an underpass into a "copper well" where the sulphur level in the metal could be oxidized to any desired level.

There was a significant temperature as well as a composition gradient from top to bottom in the converter zone. At the deepest end where the copper flowed through the underpass, the metal temperature was between 1100 and 1120°C. At that temperature, it contained about 0.9 per cent sulphur. The oxidising burner in the copper well could bring the sulphur level down to 0.4 to 0.5 per cent without difficulty. We stopped at that level because the operators at the anode furnace preferred flat cakes such as those shown in Figure 10. They would "sit" without trouble on fork lift trucks because they did not have the "blistering" characteristic of conventional converter copper.

Figure 10 - Flat topped 3500lb (1600kg) "cakes" of copper stacked near the copper taphole of the Port Kembla furnace. The continuously flowing streams of both copper and slag are just visible in the, admittedly, poor photograph reproduced from a 1970 paper (9).

Unsatisfactory Features of the Port Kembla pilot plant were:

(i) the re-entrant angles in the brickwork in the smelting zone (bowl). (Those corners were continuously "washed" with a mixture of hot fayalite-type slag and freshly melted matte. The chrome-magnesite bricks tended to become dislodged after wear and would float out into the slag-matte mixture. Repair was required after five or six weeks);

(ii) the unnecessarily large slag cleaning-settling branch. It needed much oil fuel to keep it hot and progressively we came to realise that a well reduced fayalite type slag (with 5 to 7.5 per cent CaO) was quite fluid in the temperature range 1220° to 1250°C. Matte droplets would settle rapidly in it and there was no need for most of that part of the "U".

Lance Life Improvment. In Campaigns 1 and 2 at Port Kembla, we were not satisfied with the lives of lances in the converting zone. They were refractory coated steel with about 45mm internal diameter (ID). Frequent punching was required with the normal converter air pressures and, when the ends were damaged, matte attack of the steel would occur. Half way through Campaign 3, we "took the plunge" and began to use larger, rugged, heavy gauge steel pipe; we also adopted a simple but clever tip cooling system and our lance lives immediately increased by a factor of five or more. Days became weeks and, furthermore, we could achieve the same copper production (about 1 tonne per hour without added oxygen) with one third the number of lances.

Figure 11 shows two large lances (refractory coated steel pipe with ID of 85mm) which were still in the converting zone of the Port Kembla furnace after drain-out at the end of Campaign 3. Those two lances had been achieving the same degree of oxidation as five or six of the 45mm ID lances.

It will be noted that the two large lances were set quite low in the converting zone. They were, indeed, in the white metal (Cu_2S) layer. Experience, in the latter stages of Campaign 3, had taught us that when that was done, there was a small but significant lowering of the magnetite levels in the slag.

As others have found, the reactions between oxygen and sulphur in molten sulphides is extremely fast and so, before the bubbles travelled twenty or thirty centimeters, most of the oxygen in the air had been converted to SO_2 and , therefore, less likely to lead to high magnetite levels in the converter zone slag.

94

Figure 11 - Two large, rugged refractory coated lances remaining in converter zone in Port Kembla furnace after drain-out at the end of Campaign 3. They were in excellent condition and probably would have lasted many more weeks. Some of the previously used smaller pipes, 45mm ID, are shown, uncoated, in the bottom right-hand corner.

The schematic diagram shown in Figure 12 represents the data collected when we were operating with smaller lances having their tips somewhat higher in the bath. We did not have time to collect a new set of data when we switched to the larger and deeper-set converting lances.

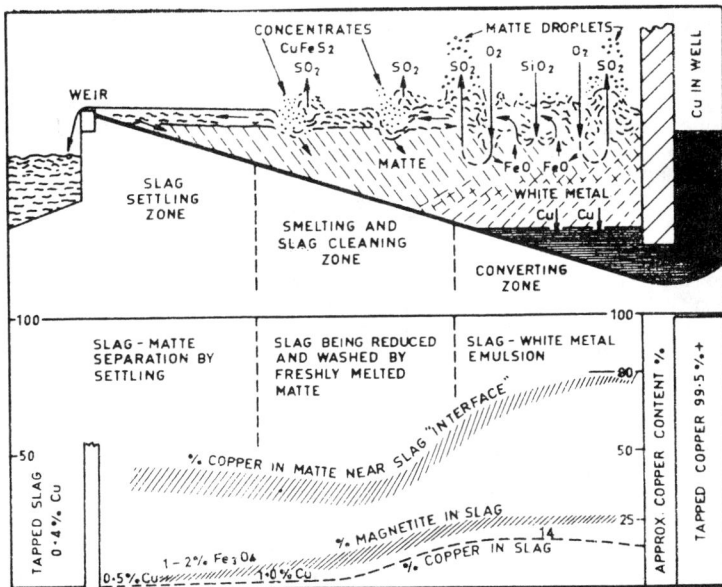

Figure 12 - Schematic diagram illustrating the changes in the bath of the WORCRA furnace producing copper. The data were collected before changing to larger and deeper set lances in the converting zone. That lowered the magnetite levels.

By the end of Campaign 3, by which time we were using fewer, larger and deeper-set converter lances and were also achieving better "mixing" of matte and slag in the smelting zone, we had no difficulty in achieving low copper levels in slag (averaging 0.5 per cent and frequently down to 0.3 per cent) flowing into the slag settling branch. That confirmed that with well reduced slag we did not need the slag settling branch of the "U" and could have saved a large proportion of our fuel bill. Over the last month of operation, we achieved a 99.12 per cent recovery of copper from the Cobar NSW concentrates, which were being used at that time.

Benefits of Oxygen Enrichment

The advantages which oxygen enrichment could confer only became evident when we brought in tankers of oxygen for the last few weeks of operation in Campaign 3. These benefits have already been mentioned but they bear repeating:

(i)   higher throughput;

(ii)   reduced fuel requirements;

(iii)   higher SO_2 tenor in the (single) furnace off-gas.

There was a fourth significant advantage; we could dispense with the drier and sling the slightly moist concentrates onto the vigorously stirred and circulating smelting zone liquids. Noranda had been teaching us that from the beginning of our collaboration.

Although our experience with oxygen enrichment was limited, we did enough to indicate that a higher level of oxygen would be appropriate in the smelting zone than in the converting zone. Many of the other well-established bath smelting technologies have demonstrated that and, particularly, Mitsubishi with their three-furnace operation (22). Of course, Mitsubishi use an electric furnace to clean their unique calcium-copper ferrite type converter slag.

At one stage, in our Port Kembla work, we used lime sand flux in our converter zone with beneficial results; thereafter, we always maintained between 5 and 7.5 per cent CaO in the slags as finally tapped.

Reverting to oxygen enrichment, perhaps amongst bath smelters (as distinct from flash smelters) Vanukov and his Russian colleagues have used the highest level of oxygen; it has been reported (23) that oxygen levels up to 90 per cent have been used in some Smelt-in-Melt operations, presumably, while still achieving less than 0.6 per cent copper in their throw-away slags.

Termination of WORCRA Projects

Many people have asked the question, why did the WORCRA developments stop when results were becoming so promising? The reasons are somewhat complex but, basically, were due to a change in CRA's top management. Sir Maurice Mawby, as Chairman, had been a consistent supporter. He stepped down as Chief Executive just when major improvements were being achieved. His successor, the former Chief Accountant, argued that it did not matter how successful our results were, it was wrong for CRA to continue spending money on radically new technology when we had, at that time, neither a copper smelter nor a steelworks.

The company should have entered into joint ventures with other interested companies, of which there were several. However, another senior director was such an enthusiastic supporter that he insisted that CRA should retain 100 per cent ownership of the technology. My colleagues and I were caught between the two protagonists and there was little we could do but bring the projects to a halt and disperse WORCRA staff into other parts of the Company.

The Future

A few of us continued some cold and hot modelling experimental work at Cockle Creek. As a result of that work, supplementing the experience gained in the Port Kembla furnace, we designed a straight line commercial furnace for copper (or rich nickel matte) production. it is shown schematically in Figure 13.

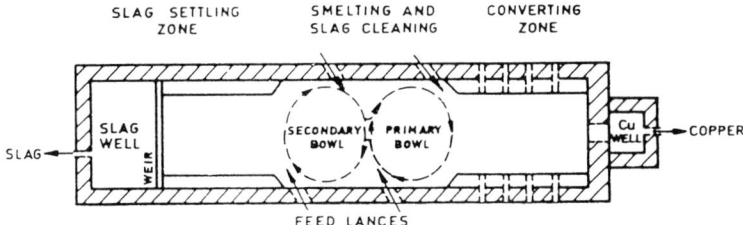

Figure 13 - Plan of a proposed commercial WORCRA smelter-converter-slag settler. The hearth slopes continuously from the slag weir down to the underpass into the copper well.

The modelling work indicated that an enlarged smelting zone with two circulating patterns of fresh matte and slag would:

(i) damp out any slopping or "foam" formed during vigorous lancing in the converting zone;

(ii) ensure thorough "washing" and chemical reduction of the slag flowing back from the converting zone.

The slag settling zone may not need to be as large as is shown in Figure 13 but, because the bath is shallow at that end, the actual volume of slag at any moment in time would not be large and so residence times would not be as long as a cursory glance may imply.

I am not wedded to that particular design. Indeed, with the passing of years and the recent study of other two- or three- furnace technologies, I can see advantages in incorporating several of their features into a future one furnace smelter-converter-slag cleaner.

The relatively recent papers (17, 18) of the doyen of American pyrometallurgy, Paul E Queneau, help to strengthen my conviction that the goal is worth pursuing in both ferrous and non-ferrous metallurgy.

Whatever the engineering design, any future single furnace for smelting-converting (or refining)-slag cleaning should be energy efficient and environmentally "friendly".

Iron Bath Melting/Smelting for Treatment of Wastes

Having moved from Melbourne to Wollongong in early 1986, the opportunity presented to reactivate my long-term interest in ferrous bath smelting and, particularly, using low rank coals or peat as the binder-fuel-reductant. Two of my Melbourne colleagues (one University and one CRA) had discovered that those low rank materials, even sawdust, turned into a sticky plasticine-like mass when comminuted in a shearing type mill. I found that such "pastes" became admirable binders and fuels when incorporated with ore fines to make "composites"; if burnt lime is also added, they dry readily to form hard but micro-porous and self-fluxing feed to a smelt reduction furnace.

Using foundry facilities kindly made available to me by the Wollongong Technical & Further Education (TAFE) College, plus the services of a capable staff member (Len Reilly), we adapted one of the furnaces for bath smelting. First we used hematite ore fines, kindly sent to us by CRA Ltd and then we switched to steelplant dusts supplied by BHP. The composites made from each smelted very well and in the latter case, in addition to the foundry type iron and a blast furnace type slag, we were able to collect zinc oxide from the furnace off-gases.

At the beginning of 1987, the University of Wollongong had asked me to found a Microwave Applications Research Centre (MARC). One of its early projects was the collaboration with an industrial partner in developing a continuous microwave sterilizer of sewage sludge. This work was successful and a staff member suggested that we try the sterilized sludge in place of brown coal paste in our smelting composites. This was also successful but it triggered off in my mind the use of steelworks dusts plus lime plus a trace of flocculent for rapid settling and substantially sterilizing the solids (and most of the pathogens) in raw, heavy metal contaminated sewage. This was so promising that we concentrated on that program instead of the microwave sterilizing.

The settled mixed sludges lent themselves to ready dewatering and, when more dusts and burnt lime were added, the mixture could easily be agglomerated by granulating or extrusion in brick making type equipment. Drying of these composites requires some odour control in certain weather but the dried composites possess the same desirable characteristics (strength and micro-porosity) for feeding to the bath smelting furnace and are quite safe to handle.

As phosphate detergents seem to be ubiquitous in sewage these days, we have had to incorporate systems for dephosphorizing the hot metal. Once again, I have called on countercurrent slag flows plus top and bottom (Savard/Lee) injection of oxygen with lime (plus a little coal, if required, to keep bath carbon levels high).

My patents of the late eighties cover not only the sewage settling benefits but both pneumatic and electric furnace bath smelting. Figure 14 shows schematically how we propose to modify an electric arc furnace to achieve both efficient bath smelting/melting and dephosphorization and desulphurization. If the phosphorus levels in slag are high enough, granulation will be used and the slag will make a useful soil conditioner and fertiliser.

We have discovered that several "waste" carbonaceous materials, including pulped paper, lake weeds and algae, can be used along with or in place of sewage solids. Equally, finely shredded plastics, particularly polyethylene, can beneficially be incorporated in the composites.

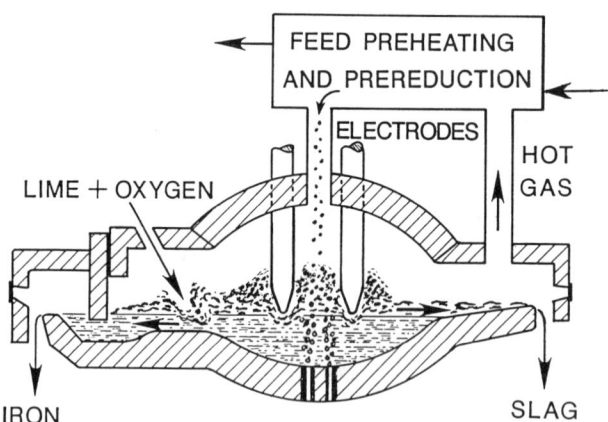

Figure 14 - Schematic vertical section through proposed electric smelter/melter for wastes and contaminated scrap.

An electric furnace will operate more efficiently, in terms of power consumption per tonne of iron, if appropriately sized scrap is charged along with the dried composites. The scrap we propose to add is that disliked by ordinary EAF shops, namely, oily swarf, shredded steel pieces to which plastics, cloth or paper are attached and shredded or crushed drums which contain residual organic liquids. The operation is therefore envisaged as a combination of melting and smelting as the heading implies.

Since more detail of our R & D program has recently been presented (24) at a Melbourne Symposium to honour Mervyn Willis, a long-time leader in Australian chemical metallurgy, I shall here confine myself to drawing attention to the valuable contribution which the Savard/Lee bottom tuyeres can make. They are an efficient method of ensuring that the lime plus a little oxygen gets mixed with all of the bath metal; they also will enable fast addition of coal if bath carbons need "topping up". The other potential is the use of the shrouding annulus to inject hazardous, even so-called "intractable" liquid organic wastes. We had thought about this, but had not gone as far as Molten Metals Technologies in USA in application of this novel approach to waste processing and recycling. MMT have done most of their work in modified OBM type vessels.

Guy Savard and Robert Lee have had yet another application added to their clever original objective of oxygen steelmaking.

Our "wastes" bath smelting work has been supported by the Sydney Water (& Sewage) Board, the New South Wales electricity authority , with help also from BHP and a smaller iron and steel foundry to all of whom The Illawarra Technology Corporation is indebted.

<u>Conclusion</u>

Work on bath smelting has occupied just under one third of my long and rewarding career. It continues to provide us all with stimulating challenges, for it makes possible the achievement of high rate chemical reactions between solids and liquids, liquids and liquids, liquids and gases and even solids and gases.

The work I have described demonstrates also the generation within baths of steady state temperature- and composition-gradients which make possible widely different types of reactions in one and the same reactor. Oxidation can be taking place in one zone while reduction is being achieved in another liquid phase in a zone only a few metres away. That such is possible is hard for some people to believe but it has, for me, made the achievement the more satisfying.

In a sense, my inventions of the early 1960s may have been ahead of their "commercial" time but the science cannot be disputed for we made several thousand tonnes of copper in the same furnace that matte was being generated and slag was being tapped continuously with 0.5 per cent or less of contained copper. Likewise, many hundreds of tonnes of low carbon steel were continuously flowing from furnaces in which, in another zone, high carbon hot metal was being generated or was being fed. Simultaneously, by the use of countercurrent (slag) flow, low iron slags were being tapped at an appropriately located taphole.

With all the "tools" and materials that are available to pyrometallurgists today, such as mathematical modelling and novel, fast response sensing devices, it should be even easier to achieve and sustain the conditions I have described.

I shall feel rewarded if this paper stimulates others to take up the challenge of developing single reactors in which high energy efficiency, high metal recovery and low emissions of all kinds are achieved.

Thank you Professor Brimacombe for the pleasure and privilege of participating.

REFERENCES

1. P.J. Mackey and P. Tarassoff, "New and Emerging Technologies in Sulphide Smelting", Advances in Sulphide Smelting, Vol 2, Technology and Practice, Eds. H.Y. Sohn, D.B. George and A.D. Zunkel, (The Metallurgical Society of AIME, Warrendale, PA, USA 1983), 399-426.

2. P.J. Mackey, "Sulphide Smelting – The Science and Technology from the Perspective of SO_2 Control", Workshop on SO_2 Control, CIM Conference of Metallurgists, Toronto, Canada, 1987, pp 59 (includes update of reference 1).

3. P.J. Mackey, "Trends in Copper Processing", Metals Week Copper Conference, Jan 7-8, 1991, Orlando, Florida, USA, pp 21 plus overheads.

4. H.K. Worner, "Revolutions in Iron and Steel Making", The Australasian Engineer, June 1961, 60-71. Also Jl. of Aust. Inst. Metals, 6, No 3, (1961) 167-185.

5. H.K. Worner, "Continuous Iron and Steel Making", Bulletin of The AusIMM, No 252, (Nov 1962), 6-7. (Abstract only.)

6. H.K. Worner, "Continuous Copper Converting", Letter to Editor, Journal of Metals, (AIME, New York, NY, 16, 1964) p 614.

7. H.K. Worner, "The WORCRA Processes for Continuous Smelting and Refining", Proceedings of 1967 Symposium on Advances in Extractive Metallurgy, (The Institution of Min. and Met., London, 1968), 245-263.

8. H.K. Worner et al., "WORCRA (Continuous) Steelmaking", Journal of Metals, (AIME, New York, NY, 21, 1969) 50-56.

9. H.K. Worner, J.O. Reynolds and B.S. Andrews, "WORCRA Copper Smelting", Copper Metallurgy, Ed. R.P. Ehrlich, (The Metallurgical Society, AIME, New York, NY, USA, 1970), 198-219.

10. F.H. Baker and H.K. Worner, "WORCRA Iron and Steelmaking", Alternative Routes to Steel, (The Iron & Steel Institute, London, 1971), 99-106.

11. T.W. Jenkins, N.B. Gray and H.K. Worner, "Application of the Levenspiel Dispersion Model to Metal Flow in WORCRA Continuous Steelmaking", TMS Metallurgical Transactions, 2, (1971) 1258-1259.

12. T.A. Engh et al., "A One-Dimensional Model of the WORCRA Steel Refining Launder", Jernkontorets Annaler, 155, No 9, (1971), 553-564.

13. H.K. Worner et al., "Developments in WORCRA Smelting-Converting", Proc. of 1971 Symposium on Advances in Extractive Metallurgy and Refining, (The Institution of Min. & Met. London, 1972), 19-38.

14. H.K. Worner, "WORCRA Countercurrent Pyrometallurgy", Proc. First Heat and Mass Transfer Conference, Melbourne Australia, May 1973, 33-40.

15. H.K. Worner and B.S. Andrews, "Integrated Smelting-Converting-Slag Cleaning in a Single Furnace", Proc. AIME TMS Symposium, Dallas, Texas, Feb 1974, (Paper A74-19) pp.11.

16. H.K. Worner, "Some Pioneering Pyrometallurgy", The AusIMM Bulletin and Proceedings, 293, No 8, (1988) 36-43.

17. P.E. Queneau, "The Q S L Reactor for Lead and Its Prospect for Ni, Cu and Fe", JOM, (TMS, AIME) 41, No 12, (1989), 30-35.

18. P.E. Queneau, "Direct Steelmaking - Quo Modo?", Iron & Steelmaker (AIME) 17, No 12, (1990), p.76.

19. H.K. Worner, Letter of 21 August, 1989, to Dr. Egil Aukrist, Leader AISI Direct Steelmaking Project.

20. H.K. Worner, "Extra Clean Steel Production via the WORCRA Continuous Steelmaking Process", Iron & Steelmaker (AIME) 15, No 1 (1988), 23-24.

21. D.G.C. Robertson private communication with the author, 27 January, 1992.

22. T. Shibasaki and M. Hayashi, "Top-Blown Injection Smelting and Converting: The Mitsubishi Process" JOM (TMS, AIME) 43, No 9 (1991) 20-26.

23. V.P. Bystrov, Unpublished paper on "Vanukov Smelting - A New Technology for Processing Sulphide and Oxide Minerals", communicated to author by J. Cucvara of Kaiser Engineers, February, 1992.

24. H.K. Worner, paper entitled "Iron Bath Melting/Smelting in the Treatment of Wastes", presented at Mervyn Willis Symposium, Melbourne, Australia, 6-8 July, 1992.

SIROSMELT - THE EMERGING ROLE OF NEW BATH SMELTING

TECHNOLOGY IN NON-FERROUS METALS PRODUCTION

J M Floyd

Ausmelt Pty Ltd
(A.C.N. 005 884 355)
2/13 Kitchen Road
DANDENONG VIC 3175
AUSTRALIA

Abstract

Ausmelt's development of top submerged lancing reactor systems over the last ten years has covered the spectrum of non-ferrous metal production from sulphide and oxidic resources. From laboratory studies, pilot plant trials and plant design and commissioning applications of this technology have been established for smelting ores, concentrates, residues, slags, fumes, drosses and many other sources of base and precious metals.

This technology represents more than just another bath smelting process; it also introduces opportunities for carrying out separations between elements in feed sources which have not been achievable in other systems. Top submerged lancing has been commercialised for many different purposes and there are many opportunities which have not yet been tapped. Some of these opportunities are outlined and Ausmelt's furnace system is compared with alternative approaches which have been or could be considered.

Proceedings of the
Savard/Lee International Symposium on Bath Smelting
Edited by J. K. Brimacombe, P. J. Mackey,
G. J. W. Kor, C. Bickert and M. G. Ranade
The Minerals, Metals & Materials Society, 1992

103

Brief Description of the Technology

Details of the Sirosmelt reactor system are shown in Figure 1. The refractory-lined vessel would normally be stationary, but for small-scale operations where very short cycle-times for batch treatments are involved, such as in copper converter slag reduction, a tilting furnace on trunnions would be employed. The flue offtake acts as an after burner in many applications and for this reason it is often refractory-lined, but designs incorporating cooled steel offtakes are appropriate for some applications.

The vessel may be refractory-lined and insulated for applications where heat losses must be limited and where refractory life is acceptable. Where very aggressive slags or high temperatures are employed, water cooling is required behind refractories to ensure long campaigns between vessel relines.

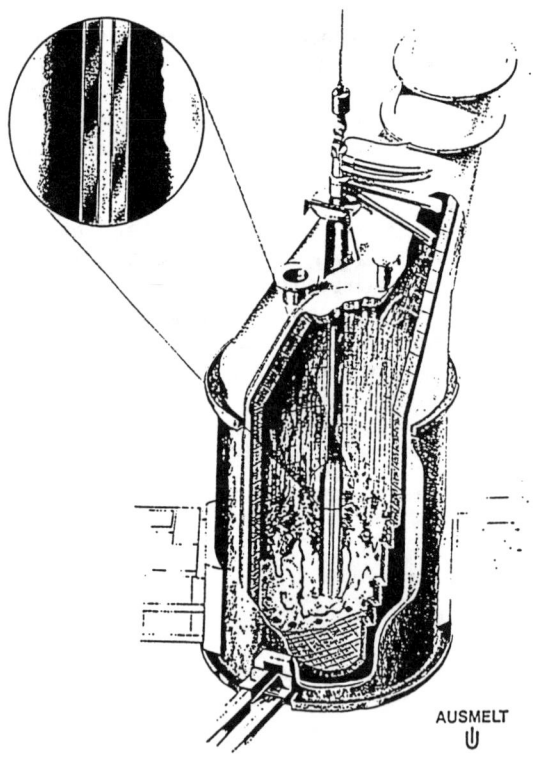

Figure 1 - Sirosmelt Reactor System

The heart of the reactor is the Sirosmelt lance. This is non-consumable and constructed of steel pipe. Air blown into the furnace cools the outer steel pipe. By the manner of its operation, the lance is coated by solid slag before immersion into the bath. The solid slag layer protects the steel from attack by the liquid slag bath.

Injected gas from the lance creates a cascade of slag above the surface as it rises through the slag layer. The lance effectively pumps slag from the lance tip through the top surface. For a central lance operating in a cylindrical vessel, there is a relatively quiescent region beneath the lance. Droplets of slag are ejected into the gas space above the cascading bath, and , for this reason, a high freeboard is provided to contain splashed slag within the vessel. Guides provide support and control of the lance position during raising and lowering by hoist.

Oxygen-enrichment of process air to high oxygen levels can be used with specially designed lances. Fuel and fine materials are conveyed down the lance through inner tubes. Internal facilities are provided to ensure turbulence of air at the outer wall of the lance in order to achieve high heat transfer rates. Coarse materials are dropped into the furnace through a feeding chute and are rapidly incorporated into the bath on the cascading slag surface.

The system may be operated with continuous feeding and a constant bath volume that is maintained by continuous tapping. Alternatively, the feed can be interrupted to remove products during a tapping or pouring operation, leaving a heel of slag in the vessel for a following feed cycle. Figure 2 shows a flowsheet for a two stage continuous operation for processing zinc plant residues.

Top Submerged Lancing Technology - Its Uniqueness

The Sirosmelt development has been carried out over the last twenty-two years, and it has now become more accepted as a viable smelting unit. However at this stage we should not lose sight of the fact that there was a great deal of skepticism about the capability to maintain a lance in a smelting environment where it is submerged in a bath of liquid slag. Early advice and comments from some of the technological leaders of the industry was that submerged lances had been tried and failed, and this development was just a laboratory curiosity.

In the first five years of the development work, from 1970 up to 1975 there would have been only a few people in industry who accepted the viability of the technology for industrial applications. By 1980 the number of believers would have doubled, but was still quite small.

By the mid-1980s, possibly a quarter of the industry recognised the viability of what Ausmelt was doing. Thanks to major successes in Industrial operations, by 1990 the believers probably greatly out numbered the non-believers.

The technology has gone through the experimental, developmental and prototype stages to being accepted as industrially viable. The commercial viability for a wide range of applications and situations is also becoming accepted.

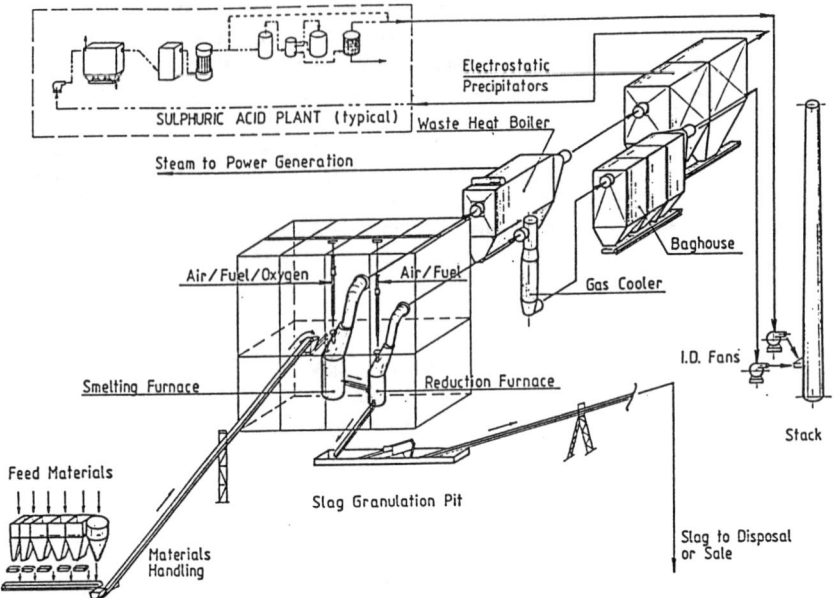

Figure 2 - Schematic Flowsheet for Two Stage Residue Smelter

Reasonably large industrial trials were carried out after less than five years and a commercial prototype was in operation within seven years for tin slag reduction at Associated Tin Smelters in Sydney. However the industrial acceptance has been slow compared to expectations. With the benefit of hindsight, it is suggested that this is because of the following:

1. The technology is radically different to any other technology in existence. It is amongst those processes which are revolutionary, rather than evolutionary, and this introduces an additional level of caution in its acceptance.

2. The basic furnace system has applications over a wide range of materials and for the complete range of operating scales. This means that the development is targeted at different benefits and compared with a variety of alternative reactors. This tends to diffuse the path of development. Table I lists the applications and the alternatives in use.

3. There are superficial similarities between Sirosmelt top submerged lancing and other developments which might have resulted in an expectation of a similar outcome whereas in practice the differences give a much different outcome. This is discussed below.

4. The advantages offered by Sirosmelt over alternative technologies were not clearly perceived by potential users.

Table I Applications of Ausmelt Technology and Alternative Reactors

Application	References	Alternative Reactor
Tin Smelting	1,2,3	Reverberatory, Rotary, Electric Furnace
Tin Concentrate Fuming	1,2,5,6	Box Fumer
Tin Slag Fuming	2	Box Fumer
Tin Slag Reduction	1,2,3	Reverberatory, Rotary, Electric Furnace
Copper and Nickel Sulphide Smelting	1,4,7	Large Range of Smelting Furnaces eg. Outokumpu, INCO Flash, Noranda, Mitsubishi, El Teniente
Copper and Nickel Oxide Smelting	8	Kiln for Reduction and/or Drying Blast Furnace, Electric Furnaces
Copper and Nickel Matte Converting	7	Pierce Smith Converter, Hoboken, Mitsubishi, Flash Converters
Secondary Copper Smelting		Reverberatory, Rotary, Shaft Furnace, TBRC
Primary Lead Smelting	9,10,11	Sinter Plant/Blast Furnace, QSL, Kivcet, TBRC
Secondary Lead Smelting	11	Rotary, Reverberatory, Shaft Furnace
Zinc Leach Residue Smelting	9,12,13, 14,16,24	Flash, Electric, Plasma Furnaces
Zinc Concentrate Smelting	15	Flash, Electric, Plasma Furnaces
EAF Dust Smelting	9,13,14,24	Flash, Electric, Plasma Furnaces, Waelz Kiln
Solid Slag Processing	17,18	Rotary or Blast Furnaces
Liquid Slag Processing	1,2,3, 4,19	Zinc and Tin Fuming Furnace, Electric Furnace, TBRC
Precious Metal Smelting	20	Rotary, Reverberatory, TBRC, BBOC Furnace
Precious Metal Cupellation	21	Crucibles, TBRC, BBOC Furnaces
Fume Smelting	22	Rotary, Reverberatory Furnaces, TBRC
Complex Sulphide Smelting	23,24	TBRC
Complex Oxide Smelting	23,24	TBRC

5.   The original development was not an "in-house" development targeted at a specific resource by a company, but by a government research organisation which had to "sell" its suitability to the industry.

6.   The development was not "adopted" by a major mining organisation until quite recently.

7.   Some of the early applications had engineering faults introduced during installation because of inadequate understanding of the quite specific needs of the technology compared to conventional systems (ie. the technology is not as simple as it seems and close involvement of experts in the technology is needed at all stages of the implementation to avoid possible problems).

Key Features of the Sirosmelt System

Table II lists the important features of the Ausmelt top submerged lancing technology. Each of the points listed has to be considered in relation to a specific material and operation sequence. A full explanation in comparison with the alternatives is beyond the scope of the present paper, but there have been many technical articles written on specific applications which explain these factors[1,23,24,25].

There are other innovations carried out by Ausmelt which have improved on these key features. For example Ausmelt's patented shrouded lance system[28] provides for very close and direct control of afterburning in the above bath zone (one of the multiple reaction zones included in Item 8 of Table II). The shrouded lance provides an improved lance system for many applications.

Important commercial opportunities are presented by the process benefits of low operating cost, low capital cost, process flexibility and environmental benefits. These open up opportunities for small scale as well as large-scale smelting operations. One example of the capital cost advantages from the process over conventional technology is the replacement of an electric furnace/Pierce Smith Converter system for sulphide smelting by an Ausmelt smelter and converter, with a capital requirement of less than half the conventional system.

Origins and Development of the Technology in the 1970s

Development of top submerged lancing was started in CSIRO in the early 1970s for tin smelting applications. The development has been described in a paper on the first ten years of Sirosmelt by Floyd and Conochie[1]. Table III lists the companies involved in industrial developments during the 1970s.

During the early investigation of the idea and its applications, it was called "Submerged Combustion Smelting Technology". The name "Sirosmelt" was coined by the inventor and leader of the development team in the second half of the 1970's in order to distinguish it as a new technology.

Table II Key Features of the Sirosmelt Furnace System

1.	Enclosed refractory-lined furnace facility with top submerged lance technology using slag coated lance cooled by injected air.
2.	Efficient heat and mass transfer from combustion of injected fuel and air/oxygen within a slag bath while feeding materials into the bath.
3.	Low energy cost and small furnace system result from fuel efficiency and rapid smelting rates.
4.	Feed preparation is minimal and feed preparation and handling is inexpensive. (wet or coarse feed suitable).
5.	Any fuel is suitable (Coals, natural Gas, LPG, Liquid Oil, Heavy Oil, Carbon in feed, Iron Sulphide in feed).
6.	Very little carry over of feed material into flue lines results in undiluted volatilised products in fume.
7.	Efficient operation on small scale as well as large scale.
8.	Multiple reaction zones in the furnace allows for efficient operations.
9.	Oxidising neutral or reducing conditions can be provided on call by operator.
10.	Metal or matte phases can be used as collecting agents by control of conditions or by reagent addition.
11.	Environmentally acceptable discard slags are achievable.
12.	Emissions from the furnace are avoided due to small size and tight sealing of ports.

Table III List of Australian Companies or Organisations Involved in Developments of Sirosmelt During the 1970s

COMPANY (ORGANISATION)	START DATE
1. CSIRO - Division of Chemical Engineering (later Mineral Engineering), Melbourne. (Invention, Development Team, Lab Work, Pilot Plant Work, Development of Scientific and Engineering Knowledge and Know-how).	1970
2. ASSOCIATED TIN SMELTERS, Sydney, New South Wales. (a) One tonne Pilot Plant for Slag Reduction.	1974
(b) 5-6 tonne Production Prototype Slag Reduction Unit.	1977
3. BROKEN HILL ASSOCIATED SMELTERS, Port Pirie, South Australia. (Operation of CSIRO's Transportable 50kg Pilot Plant for Antimonial Slag Reduction).	1974
4. ER&S, Port Kembla, New South Wales. (One tonne Capacity Pilot Plant). (Mount Isa Mines also sent materials for testwork to ER&S).	1975
5. COPPER REFINERIES, Townsville, Queensland. (50 kg Transportable Rig and one tonne Capacity Pilot Plant for Anode Furnace Slag Reduction).	1976
6. ABERFOYLE, Melbourne, Victoria. (a) Development of Process for Tin Fuming at CSIRO	1977
(b) 5 tph Pilot Plant for Tin Fuming at Kalgoorlie Nickel Smelter, Kalgoorlie, Western Australia.	1979
7. MOUNT ISA MINES, Mount Isa, Queensland. (a) One tonne Pilot Plant for Converter Slag Reduction.	1978
(b) Lead Smelting Laboratory Testwork at CSIRO.	1979

Acceptance and Application of the Technology in the 1980s

After over ten years of work on the technology, CSIRO made the decision in 1981 to stop further research and development and to allow industry to take up the technology, or to drop it, as it saw fit. The companies who were undertaking major developments were Associated Tin Smelters and Aberfoyle Limited (Table III). Mount Isa Mines had built a 120 kg capacity pilot plant for the lead smelting process called "Isasmelt", which it was developing using the "Sirosmelt" smelting system.

There was much assistance which was still needed to transfer the technology to industry and to have it fully developed and commercialised, and the inventor and team leader left CSIRO to form Ausmelt in 1981 to provide a vehicle for the technology transfer, and to allow further development of the technology and its applications. Ausmelt worked with the above organisations to complete the technology transfer started from CSIRO. Many other companies were encouraged to consider and evaluate the technology, as a result of the promotional and marketing work of Ausmelt, and operations outside Australia were developed from an early date.

Ausmelt carried out the role of technology marketer, promoter and commercialiser as an independent, self funding company which provided access to the lance patent owned by CSIRO through an agreement reached with the organisation in 1982. This agreement was extended in 1985 when the venture capital company, Australian Pacific Technology, became a shareholder of Ausmelt. In 1989 CSIRO provided an opportunity for Mount Isa Mines to enter the market place in supplying the technology through an agreement and Ausmelt's agreement was extended to include for the first time the right to sub-license directly to end users. Figure 3 details the major developments and change of status of the various organisations with respect to the technology which occurred during the 1970s and 1980s.

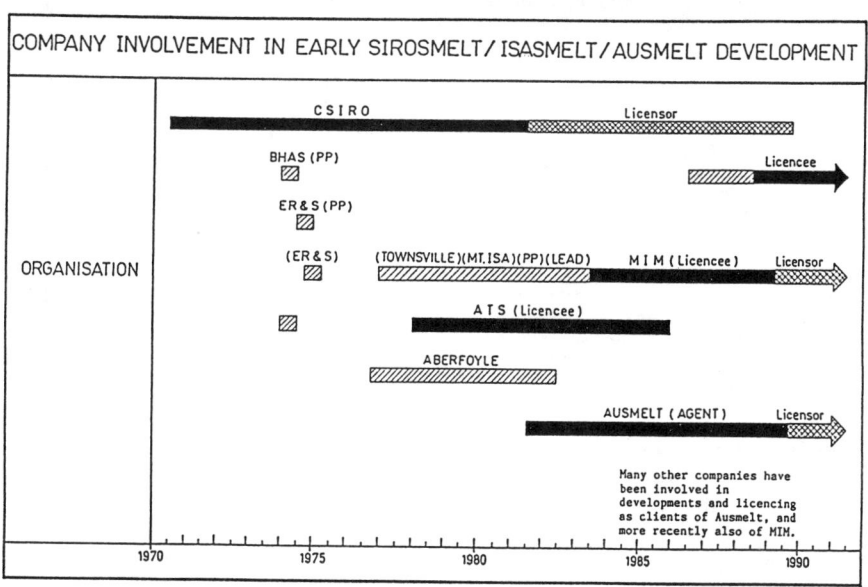

Figure 3 - Major Developments and Change of Status of the Main Organisations Involved in the Technology Since the 1970s

Since 1989 the technology has been available from either MIM and Ausmelt. The transferors and licensors for plants using the technology are shown in Table IV. Mount Isa Mines now has three smelting plants using the technology for three different applications. The original lead smelter has been replaced by a larger unit and the original copper smelter is about to be replaced by a larger unit. Mount Isa Mines uses and sells the technology under the trade name "Isasmelt".

Table IV Current Status of Licences of Top Submerged Lancing Systems

	LICENSEE	TRANSFEROR	LICENSOR
1.	Associated Tin Smelters (Tin Slag Reduction)	CSIRO	CSIRO (Closed in 1986 due to state of tin industry)
2.	Mount Isa Mines (i) Lead Smelting (ii) Copper Smelting (iii) Secondary Lead Smelting (Britannia Refined Metals)	CSIRO/Ausmelt MIM* MIM	CSIRO CSIRO/MIM MIM
4.	Broken Hill Associated Smelters Silver Retort Bullion Cupellation	Ausmelt	CSIRO
5.	Sulphide Corporation Zinc Slag Fumer	Ausmelt	CSIRO
6.	HMIB (Arnhem) Tin Smelter	Ausmelt	CSIRO
7.	Agip Radio Hill Nickel Smelter	MIM	MIM
8.	Rio Tinto Zimbabwe Nickel Residue Smelter	Ausmelt	Ausmelt
9.	Cyprus Metals Copper Smelter	MIM	MIM
10.	Korea Zinc (i) Zinc Slag Fumer (ii) Zinc Residue Smelter	Ausmelt	Ausmelt

* Follows earlier large pilot scale development by Ausmelt[7].

In its development and commercialisation of the technology since 1981 Ausmelt has carried out many improvements, extensions and modifications of the technology. Where major inventions have given opportunities for commercial developments these have been patented but many improvements have remained in the areas of know-how. Patents have been taken out by Ausmelt for both equipment and process which have been developed. Presumably Mount Isa Mines also have know-how developments as well as the patents which they have taken out and it is these difference which distinguish Ausmelt and Isasmelt technology from each other and from the original Sirosmelt system developed in CSIRO during the 1970s.

Ausmelt Technological Base

Ausmelt's business is dedicated to the development, application, commercialisation and use of top submerged lancing technology. The core of the Company is a very strong group of process engineers and metallurgists with a great depth of experience in pyrometallurgical process development, design and operation. An expert and experienced Engineering Division provides the engineering design and installation facilities needed to establish operational smelters. The company has a marketing department and a group aiming at establishing equity ventures using the technology.

As well as servicing clients needs and developing commercial operations, this team is involved in a very wide range of research and development activities involving improvement and extension of the plant and equipment being used for the smelting operations, as well as providing new processing opportunities. The research work is generally carried out at outside laboratories such as the G K Williams Centre of the University of Melbourne, whilst development work is carried out 'in-house' by the Ausmelt team.

Table V lists some of the equipment developments which have been carried out by Ausmelt.

Table V Plant and Equipment Developments by Ausmelt

1.	Special lance for powder and dust injection using an injector (together with Aberfoyle Limited) (Patented).
2.	Lance for high levels of oxygen enrichment of combustion and process air.
3.	Lance system using shroud air for improved performance as well as process advantages. (This lance system has been patented by Ausmelt and is a major extension of the capabilities of the original Sirosmelt system).
4.	Improvements and modifications in refractory and furnace shell design.
5.	Liquid transfer facilities between furnaces.
6.	Materials transfer systems for feed into the furnace.
7.	Lance handling facilities.
8.	Furnace design and construction methods.
9.	Furnace tapping facilities and arrangements.
10.	Control systems and sensors.
11.	A new lance and furnace system for iron smelting applications. (Patented).

Table VI lists some of the more notable achievements in new processing opportunities which have been created by Ausmelt.

Table VI Major Process Developments By Ausmelt

1.	Smelting copper concentrates and ores
2.	Batch and continuous converting of copper and nickel matte.
3.	Stibnite smelting and fuming for gold and antimony separation and recovery (Patented).
4.	Gold and silver recovery from a range of primary and secondary materials.
5.	Zinc - lead - silver concentrate smelting. (Patented).
6.	Smelting zinc plant wastes and similar materials. (Patented).
7.	Zinc slag smelting and fuming. (Patented).
8.	Ferro-nickel production from laterites. (Patented).
9.	Iron-production from low grade as well as high grade resources. (Patented).
10.	Desulphurising nickel matte.
11.	Processing complex lead-zinc-copper-precious metals concentrates and other materials.
12.	Processing "dirty" or complex copper concentrates and other materials.
13.	Smelting of various fumes and dusts.

The company is proud to have been recognised in 1991 by the Award for Excellence in Waste Technology given to Ausmelt by the Australian Federal Department of Industry, Technology and Commerce and the magazine "Waste Management and Environment".

The technology being installed in Korea with the Korea Zinc Company for processing zinc plant waste materials such as jarosite, goethite and primary leach residue is one application of Ausmelt's developments in that area. Table VII lists these and other wastes being worked on by Ausmelt.

Ausmelt emphasises the need for a "system" approach to establishing a new smelting furnace system. A synergy between concept, design, installation, training, start up and commissioning, and operations is essential to being certain of success in a new installation. Project failure can result from inadequate attention to any one of these areas and great attention to detail is needed to maximise the efficiency of the resultant project.

Table VII Typical Analyses of Metallurgical Wastes

Material	Primary Leach Residue	Jarosite	Goethite	ISF Slag	Pb Blast Furnace Slag	Red Mud	Steel Plant Dusts
Zn %	16.9	5	10	6	19.8	-	20
Pb %	9.9	5.2	2	0.5	4.1	-	3.3
Fe %	28.9	25	40	35.4	12.4	38.5-42	31.6
S %	12.5	13	4	0.3	0.8	-	0.6
SiO_2 %	6.1	3	2.5	17.8	27.2	-	3.7
Al_2O_3 %	-	-	-	-	-	12-15	1.2
TiO_2 %	-	-	-	-	-	4-5	-
CaO %	-	-	-	-	-	4-5	3.6
Na_2O %	-	-	-	-	-	2	1.0
Cu %	1.1	0.2	1.5	0.1	-	-	0.2
Cd ppm	8100	200	500	-	-	-	400
As ppm	1800	600	2000	1000	900	-	-
Ag ppm	1800	80	80	5	24	-	-
Sb ppm	1700	100	10	10	1200	-	-

A caution should be given at this point about a piece-meal approach to a project. Many people in engineering companies or engineering departments, or in research or operations departments will claim to have sufficient expertise and experience in smelting plant development, design, construction, start up, commissioning and operations to carry out part or all of the steps needed without involvement of experts in the technology. This will be supported by arguments relating to "saving money". However, one significant mistake in any of the areas will risk having an unprofitable operation.

Another way to look at this is to consider the normal development path for achieving a profitable operation. Figure 4 illustrates this path and it can be seen that starting at year zero is an expensive and time-consuming task which is not in the interest of a commercial project. Ausmelt offers its clients the opportunity of starting projects at or near point A, which is a much more attractive proposition for an investor.

Comparison with Alternative Bath Smelting Systems

The Ausmelt system is illustrated schematically in Figure 5 which shows the different zones which can be utilised for controlling reaction conditions in the furnace. It can be regarded as a "black box" smelting furnace, but its capabilities cannot be described simply by equilibrium calculations for a "black box".

115

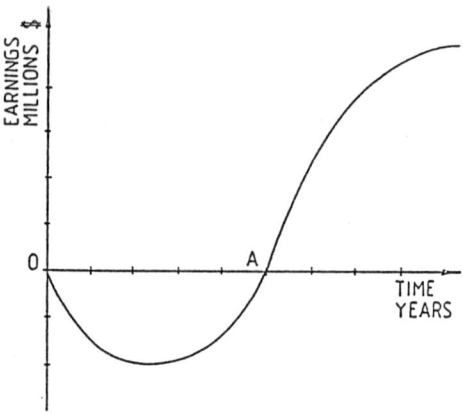

Figure 4　　- Development Path for a Project
　　　　　　(Earnings from Project Development)

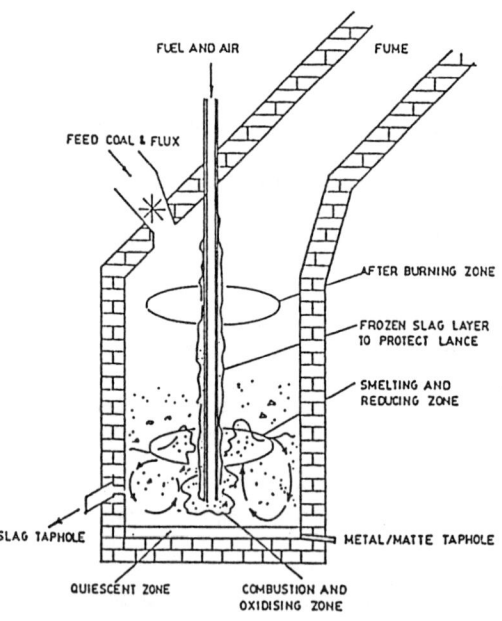

Figure 5 - Different Zones in the Ausmelt Furnace System

There are, in effect, four main reaction regions:

♦ at the lance tip - combustion and reactions involving injected material and the bath;

♦ on the top surface - reaction between reductant coal or feed materials and the bath;

♦ at the furnace bottom - normally a quiescent region where settled metal or matte accumulates and may or may not react further with the bath. However, the lance can be lowered to stir this region to enhance reactions;

♦ above the bath - where gas reactions such as afterburning of volatiles, CO, etc., can be carried out under controlled conditions.

A fifth region can be identified in the plume of gas where reaction between the combustion gases and the slag bath occurs.

The two main reaction sites are at the lance tip and on the slag surface. The conditions at these locations can be oxidising, neutral or reducing and it is control of these conditions which provides the flexibility and controllability in relation to metal separation in the system.

The main advantages of the technology are its intense and controllable stirring and its ability to accept a wide range of feeds ranging from run of mine crushed ore to mud and fine concentrates injected through the lance. These features, together with the simple design of the furnace, allow plants to be built for much lower capital costs than competing technologies. The same features also ensure lower operating costs.

Ausmelt technology is particularly robust with respect to process excursions since the top entry lance can be raised or lowered to control the rate of reactions and to make operational changes.

The technology occupies a special position because it is equipment oriented, rather than process oriented. The bath smelting technology is very well suited to a range of different feed materials because the basic principles of the intense mixing and heat transfer offer improvements to most smelting processes.

In contrast alternative technologies such as flash smelting were developed for specific types of feed and have had limited application for different processes[25,26].

Table VIII gives a brief description of the bath conditions for various types of bath reactors. The geometry involved in the various pneumatic injection reactors is illustrated in Figure 6 for tuyeres and Figure 7 for lances. The top submerged lancing system is the only one which gives full control of the degree of stirring and mixing in the reactor and this provides it with a large degree of flexibility in processing options.

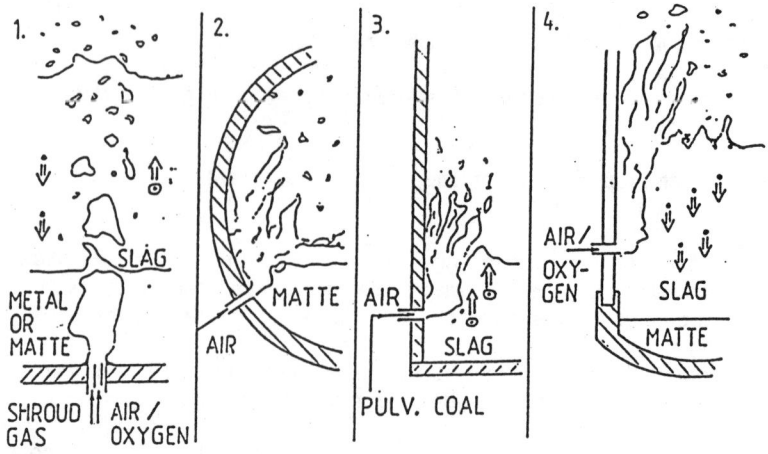

1. BOTTOM BLOWN SHROUDED TUYERE
2. SIDE BLOWN - PLAIN TUYERE CONVERTER
3. SIDE BLOWN - PLAIN TUYERE FUMER
4. SIDE BLOWN - PLAIN TUYERE - SMELT - IN - MELT

Figure 6 - Pneumatic Injection Reactors (Tuyeres)

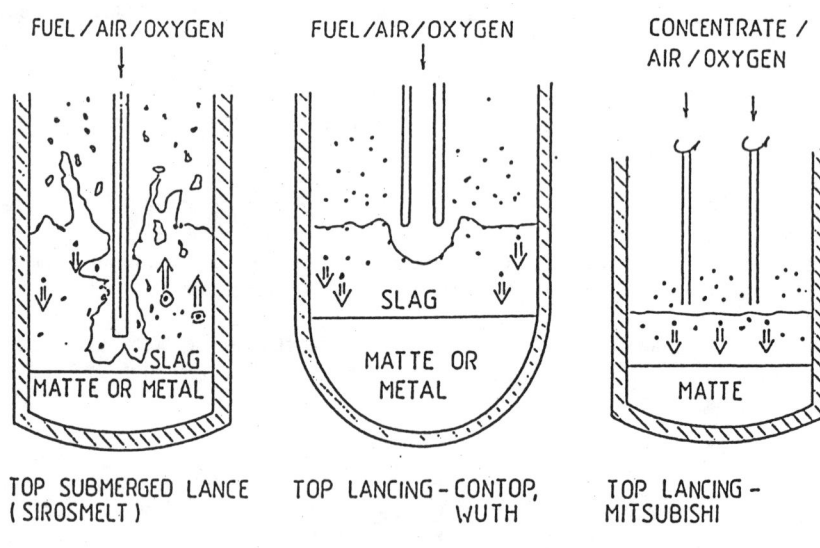

Figure 7 - Pneumatic Injection Reactors (Lances)

Table VIII Bath Conditions in Various Reactors

Reverberatory Furnace:	Little stirring of bath (except for open hearth steelmaking where the carbon boil gives reactions/stirring) oxidation from top surface, slow reactions. Melting and reaction during smelting only.
Electric Furnace:	Weak convection stirring of bath. Surface strongly reducing during reduction reactions. Electrodes also reducing. Slow reactions with the bath.
Rotary Furnace: Waelz kiln Short rotary furnace	Rotary action gives some stirring of the bath which allows reactions between phases, such as metal and slag. For a liquid slag, poor mixing of coal at top surface with slag gives quite slow reactions.
Tuyere Injection: P.S. Converter OBM or Q - BOP Zinc or Tin Fumer, Noranda Reactor El Teniente Reactor QSL Vanyukov	Oxidising, neutral or reducing conditions are possible and can be changed with time. Fast reactions and good mixing above the tuyeres. Little mixing and slow reaction below the tuyeres (for side tuyeres) and at significant distances from the area affected by the gas movement and splashing/slopping of the bath surface.
Top Jetting: LD/BOF Contop/Wuth Mitsubishi	Oxidising, neutral or reducing conditions are possible. Interaction of injected material and bath depends on jet position with respect to the bath, which is not always easy to determine, especially with consumable lances. Reaction between two bath phases is not rapid, since there is little stirring. Oxidation during slag formation is a problem with iron silicate slag because of magnetite formation and viscous slags.
Top Blown Rotary Converter:	Oxidising, neutral or reducing conditions are possible. Mixing of the bath is controllable depending on the rotation speed. Either slow reaction between different bath phases or fast reactions can be allowed for by no rotation or fast rotation. Oxidation during slag formation is a problem with iron silicate slag because of magnetite formation and viscous slags.
Ausmelt:	Oxidising, neutral or reducing conditions are possible and can be different in different regions of the reactor. Lance position adjustment allows for mixing or non-mixing of bath phases. Afterburning is controllable for degree and position.

The processing conditions in the Ausmelt system are completely variable and controllable. The oxygen potential, temperature, slag composition and presence or absence of matte or metal collecting phase are the important factors controlling the distribution of various elements during a smelting operation. Control of these variables and use of multiple operations at different times in one batch reactor, or in different continuous or consecutive reactors provides the opportunity of chemical optimisation of operations using the Ausmelt reactor[22,23,24,27].

Maintenance of Refractories

The top submerged lancing system has advantages over other bath smelting systems because the point of injection is remote from the walls and because the whole bath region is at essentially the same temperature. Furthermore, stationary furnaces are used and external cooling of the walls can be used to extend refractory life.

Tuyere injection results in refractory wear problems due to the severe thermal, mechanical and erosion conditions adjacent to the tuyere.

Top injection can give a hot region at the top surface of the slag and flow of this to the refractory walls can give severe conditions at the top surface.

Rotatable furnaces such as TBRC, rotary furnaces, Noranda reactors and Pierce-Smith converters cannot use external cooling, so that refractory life is limited by the properties of the materials involved.

Maintenance of Injector

Ausmelt's lance systems are easily replaced and repaired without substantial downtime for the reactor. Tuyere systems are repaired by rolling out the reactor or lowering the level of the furnace contents to allow replacement or repair. The replacement is not rapidly or easily achieved because the Tuyere must be sealed in the brickwork or, in some cases, it is simply a hole in the brickwork.

The Future of the Technology

Currently, the top submerged lance technology has commercial plants built, or being built, for the following applications:

1. Tin smelting.
2. Retort bullion smelting and cupellation.
3. Zinc slag fuming.
4. Smelting nickel leach residue.
5. Smelting zinc plant waste and steel plant dust.
6. Copper concentrate smelting.
7. Lead concentrate smelting.
8. Secondary lead smelting.
9. Nickel concentrate smelting.

Each one of these applications involves a single unit, except for zinc slag fuming and copper smelting which have two units each. The operations involved cover the spectrum of non-ferrous smelting applications. The scale of operation varies from a furnace capable of smelting close to a million tonnes per annum of feed materials, to the smallest, smelting a thousand tonnes per annum. The range of temperatures involved in the above list is 1050°C to 1400°C. Pilot plant operations have been carried out at temperatures as low as 800°C and as high as 1700°C, and the system can be designed to operate at these extremes.

Operations can include continuous steady state smelting or multiple processing in a single furnace by smelting followed by further processing the material retained in the furnace. Two-stage processing in adjacent continuously operating furnaces is also used in the units listed above and a third unit can be used for further metal recovery operations in some circumstances.

Thus, the technology provides a cost-effective and efficient universal reactor which finds applications, not only in the complete range of smelting and processing operations at present performed for non-ferrous and precious metal materials, but it also has the ability to open up new opportunities for achieving metal separations and processing steps which are not possible or readily achievable using other technology.

Acknowledgment

The author acknowledge the major contribution of his colleagues in Ausmelt, in academia and in industry to the technical and commercial developments which are discussed in this paper.

References

[1] J M Floyd and D S Conochie, "Sirosmelt - The First Ten Years", The AusIMM Melbourne Branch Symposium on Extraction Metallurgy, November 1984, 1-8.

[2] G P Swayn, K R Robilliard and P J King, "Recent Developments in Tin Smelting with Reference to Ausmelt's Sirosmelt Technology", (Paper being Presented at the 1993 Annual Tin Conference, Phuket, Thailand.

[3] J M Floyd, K W Jones, W T Denholm, R N Taylor, R A McClelland and J O'Shea, "Large Scale Development of Submerged Combustion Lancing Sirosmelt Tin Processes at Associated Tin Smelters", The AusIMM Melbourne Branch Symposium on Extraction Metallurgy, November 1984, 25-33.

[4] J M Floyd and B W Lightfoot, "Direct Smelting of Complex Copper Ore Using Sirosmelt Technology", 23rd CIM Annual Conference of Metallurgists, Quebec, August 19-22, 1984, Vol II, 25pp.

[5] K A Foo and B W Lightfoot, "Operation of a 4 tph Matte Fuming Pilot Plant", Pyrometallurgy '87 Institute of Mining and Metallurgy, 1987, 389-418.

[6] R L Biehl, D S Conochie, K A Foo and B W Lightfoot, "Design, Construction and Commissioning of a 4 tph Matte Fuming Pilot Plant", AusIMM Melbourne Branch Symposium on Extractive Metallurgy, November 1984, 9-15.

[7] L E Anderson, J M Floyd, B W Lightfoot and R Mullar, "Smelting of Olympic Dame Copper Concentrates Using Sirosmelt Technology", Complex Sulphides: Processing of Ores, Concentrates and By-Products, TMS-AIME, 1985, 69-76.

[8]   J M Floyd, B W Lightfoot, K R Robilliard and G P Swayn, "Smelting of Nickel Laterite and Other Iron Containing Nickel Oxide Materials", International Patent Application, WO91/05879, 2 May 1991.

[9]   K R Robilliard, G A Guorgi, S K Wu, P J King and J M Floyd, "Sirosmelt for Solving Environmental Problems of Lead-Zinc Production", The International Lead and Zinc Study Group, Rome, June 1991, 15 pp.

[10]   P J King, G A Guorgi and J M Floyd, "The Role of Sirosmelt in the Lead and Zinc Industry", Metals Week Conference, Orlando, Florida, 11-12 March 1991, 19 pp.

[11]   J M Floyd, B W Lightfoot, K R Robilliard and G P Swayn, "Lead Smelting Using the Sirosmelt Furnace System", Lead and Zinc in the 1990's, Lead and Zinc Study Group Conference, Sao Paulo, 5-7 February 1991, 167-195.

[12]   J M Floyd, P J King, G A Guorgi, "Sirosmelt Technology for Environmental Compliance in Zinc Plants", The International Lead and Zinc Study Group, Geneva, 1990, 24 pp.

[13]   S K Wu, K R Robilliard, P J King and J M Floyd, "Sirosmelt - Modern Smelting for Environmental Compliance", Minerals, Metals and the Environment, IMM Conference, Manchester, 4-6 February 1992, 18pp.

[14]   K R Robilliard, P J King and J M Floyd, "Sirosmelt Technology for Solving the Lead and Zinc Industry Waste Problem", TMS Symposium on Processing of Residues and Effluents, Processing and Environmental Considerations, 1992, 331-348.

[15]   J M Floyd and B W Lightfoot, "Top Submerged Reactor and Direct Smelting of Zinc Sulphide Materials Therein", Australian Patent 592398, 19 November 1986.

[16]   B R Baldock, J M Floyd, B W Lightfoot and K R Robilliard, "Smelting of Metallurgical Waste Materials Containing Iron Compounds and Toxic Elements", International Patent, WO91/02824, 7 March 1991.

[17]   J M Floyd, B W Lightfoot, J P Bultitude-Paull, "Recovery of Metal Values from Metal Bearing Materials", Australian Patent Application 61094/90.

[18]   J M Floyd, J M Bultitude-Paull and B W Lightfoot, "Recovery of Zinc and Other Valuable Metals from Slag Dumps by Use of Sirosmelt", Pyrometallurgy 1987, Institute of Mining and Metallurgy, London, 1987, 725-741.

[19]   M Waladan, G R Firkin, J M Bultitude-Paull, B W Lightfoot and J M Floyd, "Top Submerged Lancing for Recovery of Zinc from ISF Slag", Pb-Zn-90, TMS-AIME Conference Volume, Annaheim, California, 18-21 February 1990, ed, TS Mackey 607-628.

[20]   B R Baldock and G P Swayn, "Recovery of Gold from Complex Materials Using Sirosmelt Technology", Contemporary Gold AusIMM Conference, Bendigo, Victoria, Australia, 17-29 September 1988, 10 pp.

[21]   J M Bultitude-Paull, J M Floyd and G P Swayn, "Development of a Submerged Injection Process for the Production of Doré Silver at BHAS", Non Ferrous Smelting Symposium, AusIMM, 1989, 255-261.

[22]   B R Baldock, K R Robilliard, J M Floyd and B W Lightfoot, "Top Submerged Lancing (Sirosmelt) for Achieving Difficult Separations in Smelting Complex Materials", International Conference on Advances in Chemical Metallurgy, ICCM -91, Bombay, India, 1991, 12 pp.

[23]   J M Floyd and B W Lightfoot, "Sirosmelt and the Wide World of Opportunity", Versatile System has Down Stream Possibilities at Small Mines, Engineering and Mining Journal, June 1985, 52-56.

[24]   K R Robilliard, B W Lightfoot, B R Baldock and J M Floyd, "Sirosmelt Top Submerged Lance Reactor for Multi Processing in Base Metal Extraction", International Symposium on Injection in Process Metallurgy, TMS-AIME, 1991, 181-198.

[25]   J M Floyd, "Bath Smelting of Base Metals", Technological Advances in Extractive Metallurgy Conference, Bulea, Sweden, September 1988, Ed. MEFOS and Boliden, 25 pp.

[26]   J M Floyd, "Base Metal Smelting and Refining", The Howard Worner Symposium - Frontiers in Pyrometallurgy, AusIMM Newcastle and District Branch Symposium, August 1988, 35-54.

[27]   B W Lightfoot and J M Floyd, "Chemical Optimisation of Smelting Processes", AusIMM North Queensland Branch, Smelting and Refining Operators Symposium, Townsville, Queensland, Australia, 27-31 May 1985, 69-76.

[28]   J M Floyd, "Top Submerged Injection with a Shrouded Lance, International Patent Application, WO91/05214, 18 April 1991.

Session 2

Innovative Bath Smelting Processes

OPERATING EXPERIENCE WITH QSL SUBMERGED BATH SMELTING

FOR PRODUCTION OF LEAD BULLION

Philip Arthur, Andreas Siegmund and Manfred Schmidt

LURGI AG
Lurgi Allee 5
W-6000 Frankfurt am Main
Germany

Abstract

A QSL demonstration plant was operated within the Berzelius Lead/Zinc smelter in Duisburg, Germany, from 1981 to 1986. During its operation a total of 100,000 tonnes of both high and low grade lead bearing feed materials (varying from 30-70% Pb in fresh feed) were treated. Since the completion of the demonstration plant phase four commercial plants have been designed and built. One of these plants, operated by Berzelius Stolberg in Germany, commenced operation at the end of August 1990. This paper concentrates on the experience gained in the Stolberg plant since startup.

Experience on a commercial scale has shown that the process concept is viable but that certain modifications were necessary to the original design which was based on the limited results obtained from the demonstration plant. Major progress has been made in improvement of submerged injector design and in the understanding of factors affecting the reduction of PbO slag.

Proceedings of the
Savard/Lee International Symposium on Bath Smelting
Edited by J. K. Brimacombe, P. J. Mackey,
G. J. W. Kor, C. Bickert and M. G. Ranade
The Minerals, Metals & Materials Society, 1992

127

Introduction

The QSL process, named after the inventors Professors Queneau and Schuhmann and the process developer Lurgi AG, is a continuous direct lead smelting process. The entire process takes place in a single reactor as shown in figure 1. Four commercial scale units have been designed and built to date. The statements in this paper are based on experience made at the Berzelius plant located in Stolberg, Germany. This plant has been in operation since the end of August 1990.

The reactor, which is divided into an oxidation zone and three reduction zones, is a slightly sloped (0.5%) refractory-lined cylinder, which can be rotated through 90 degrees when operation is interrupted. Concentrates, residues, fluxes, recirculated flue dust and additional solid fuel are agglomerated and charged through feed ports located in the roof of the oxidation zone without any further treatment. The agglomerates fall into a melt consisting of primary slag and lead bullion. Industrial grade (min 96%) oxygen is blown into the melt through submerged gas-cooled injectors based on the Savard-Lee patents. Currently three oxidation injectors are employed in the Berzelius plant. The roasting reaction takes place at 1100-1150°C producing metallic lead, primary slag with a lead oxide content of 25-30%, and a sulphur dioxide-rich offgas containing the flue dust.

The lead bullion is discharged via a syphon, whereas the primary slag passes into the first reduction zone which is separated from the oxidation zone by a partition wall. Pulverized coke is injected into the reduction zones through submerged gas-cooled injectors, together with carrier air and oxygen. Currently five reduction injectors are employed in the Berzelius reactor, although up to eight injectors could be operated simultaneously in up to twelve different locations on the reactor shell. The lead is gradually reduced out of the slag as it flows to the opposite end of the reactor. The low-lead final slag is tapped at the end of the third reduction zone, whereas the lead settles to the bottom and flows back towards the oxidation zone to combine with the primary lead bullion.

Figure 1 - Overview of QSL Reactor Berzelius Stolberg

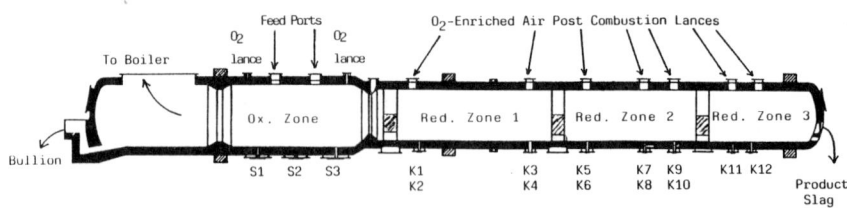

The offgas, which contains a high concentration of sulfur dioxide due to the use of tonnage oxygen, leaves the reactor through a vertical uptake. It passes a waste heat boiler for heat recovery and an electrostatic precipitator for dedusting before flowing to the sulfuric acid plant. The precipitated flue dust is recirculated to the feed mixing system.

Process Philosophy and its Effect on Reactor Geometry

The initial concept of the QSL process was to operate the reduction zone with as little metallic lead as possible, while retaining a lead sump above the injectors in the oxidation zone. The intention was to inject the reduction media directly into the slag bath, which would act as a molten slag gasifier. Although a lead layer covered the floor of the reduction zone in the demonstration plant, due to the fact that the reactor was not sloped and had a constant diameter over its entire length, this was not believed to have an influence on the rate of slag reduction. Therefore all four commercial plants were designed with a 0.5% slope and with a step in the reactor shell, the oxidation zone having a larger diameter than the reduction zone, while the reduction zone submerged injectors were angled 12 degrees from the vertical. This was to ensure that any lead reduced from the slag would drain from the reduction zone as quickly as possible and would not be reentrained into the slag by the submerged gas jets.

Experience has shown that this concept does not work as it was intended, especially when natural gas or lignite coke is used as the reductant. The cold reductant must be heated, together with the oxygen, carrier air and a limited amount of shroud gas, as quickly as possible on entry to the molten bath. It is believed that the heat transfer rate from the slag to the fuel is insufficient to facilitate gasification of the coke or reforming of the natural gas. As the slag is only about 100°C above its freezing point it is possible that a frozen slag layer coats some of the coke particles reducing the heat and mass transfer rates futher still. The problem has been further intensified by the fact that only relatively coarse coke has been used to date in the Berzelius plant, due to problems handling fine coke dust in the solids injection system.

It has also been shown that injection directly into a slag bath can lead to massive accretion formation on the reactor floor, which hinders lead flow out of the reduction zone, and complicates the plant operation generally.

It has been found both on the commercial Berzelius reactor and in laboratory scale (50kg) tests in Aachen University that the rate of gasification of coal or coke, or reforming of natural gas can be accelerated greatly if a molten lead layer covers the injector. It is now believed that a lead layer of approximately 250-300mm depth is necessary over the entire reduction zone if the desired reduction

129

rates are to be achieved. In order to incorporate this concept into the existing reactor design, without it leading to excessive lead bath depths in the oxidation zone, it was necessary to install a 300mm high refractory ring dam (see figure 1) adjacent to the step in the shell diameter. This measure has proven successful not only in improving the reduction efficiency of the reactor, but also in simplifying the reactor operation as problems with massive accretion formation in the reduction zone no longer occur, the contact area of slag with the refractory lining being minimised due to the presence of the lead layer in between.

The reactor is divided into the oxidation zone and the three reduction zones by three refractory partition walls. These reduce back-mixing effects caused by the high bath turbulance. The partition walls have an underflow for the exchange of slag and metallic lead between each zone, and a gas opening for passage of the process offgases (figure 2).

Figure 2 - Partition Wall Design

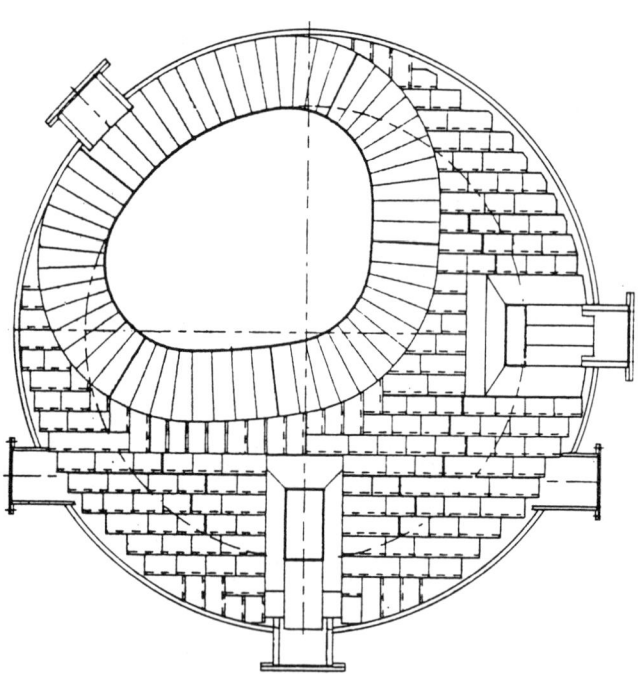

In the initial design the feed ports were located directly above the oxidation injectors. This led to major problems with feed port blockages caused by splashing from the injectors. Consequently, the oxidation injectors are now offset from the feed ports, in order to minimise the amount of splashing. They are placed such that their gas jets surround the fresh feed as it falls into the bath, thereby ensuring that there is sufficient heat and turbulence to melt and oxidise the feed immediately.

Oxygen or oxygen-enriched air is injected through lances in the roof of the reactor into the reactor gas atmosphere over the whole reactor length in order to optimise the heat balance of the process through post combustion of remaining combustibles.

Refractory

Various refractory materials were tested in the demonstration plant and as a result a special chrome-magnesite based brick (see table I) was adopted for the commercial plants. It is a direct-bonded chrome-magnesia brick, made from chrome ore and fused magnesia, with low silica content, good slag resistance, and very good thermal shock resistance.

Table I Refractory Brick Analysis

RADEX DB505B		
Chemical Analysis		
MgO	49	%
Cr_2O_3	26	%
Fe_2O_3	15	%
Al_2O_3	8	%
CaO	1	%
SiO_2	0.8	%
Physical Data		
Bulk Density	3.32	g/cm3
Open Porosity	<18	%
Cold Crushing Strength	>30	N/mm2
Refractoriness under Load	>1700	ta(C)
	>1700	tb(C)
Specific Heat	0.85	kJ/kgK
Thermal Expansion	11	10-6K-1
Thermal Conductivity (600C)	2.0	W/mK
(1200C)	2.0	W/mK

131

Lifetimes of more than one year have been achieved with this material in the general reactor shell lining.

The wear rate of the bricks surrounding the injectors is much higher due to the aggressive and turbulent nature of the bath in their vicinity. Provision is made for straightforward replacement of the bricks in the areas adjacent to the injector positions, in the "rolled-out" position without having to empty the reactor. The brick arrangement is shown in figure 3. The injectors are cemented into a conical brick while the cone brick is surrounded by four "collar" bricks. Depending on the wear observed, either the cone brick alone, or the cone and collar bricks can be replaced with an injector. A brick frame surrounds the collar bricks in order to avoid collapse of the surrounding shell lining while replacing the injector bricks. The frame bricks form a flat arch in the injector vicinity.

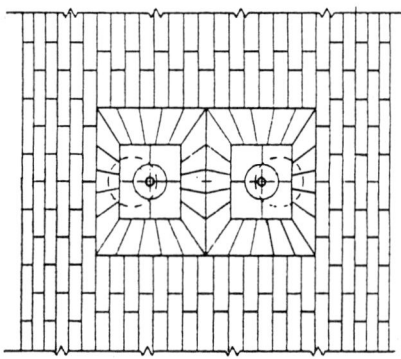

Figure 3 - Brickwork in Injector Vicinity

The lifetime of the partition walls is somewhat shorter than that of the shell lining because they are heated on both sides and there is little chance for heat transfer away to the shell. An original design with tongue-and-groove construction resulted in excessive wear at the brick joints. As a result a ground-and-glued construction has been adopted which, it is hoped, will extend the lifetime significantly.

132

Control of Oxidation and Reduction Zones

It is of utmost importance to control the flowrates of all media introduced to the process very closely as the following process parameters are determined by them:

* Heat balance and temperature profiles throughout the reactor

* Lead deportment between primary bullion and primary slag

* Degree of reduction

Oxidation Zone

Current practice is to aim to keep the PbO content of the oxidation zone slag at approximately 25-30% and the bath temperature approximately 1150°C. The more stable the composition of the oxidation zone slag, the easier it is to maintain a stable low PbO level in the final tapped slag. In normal operation the oxygen input to the bath through the oxidation injectors is interlocked with the total feed rate charged to the reactor. Should the feed composition remain constant, the oxygen rate, and hence the lead deportment in the oxidation zone is controlled automatically. However, some fluctuation in feed composition must be tolerated and this can be compensated for by adjustments to the specific oxygen rate. Regular oxidation zone slag analyses are used as the main control parameter for this purpose.

The bath temperature is controlled mainly via the variation of the amount of solid fuel added to the raw feed mixture, with short term alterations possible by addition of gaseous fuel through the submerged injectors. The relative change in bath temperature is monitored online via thermocouples installed in the floor of the reactor. The absolute bath temperature is measured regularly with a disposable thermocouple. Should the fuel input be altered, the specific oxygen rate must be adjusted accordingly.

In the Berzelius Stolberg plant the flue dust from the reduction zone passes into the oxidation zone where it combines with the dust from the oxidation zone bath and then passes into the waste heat boiler. The recycled flue dust is fed into the feed mixture automatically. Total dust rates are in the order of 20% of total mixed feed.

Reduction Zone

The main factors to be controlled in the reduction zone are the bath temperature and the reduction capacity. During its passage through the reduction zones the slag's temperature must be raised to about 1250°C and its lead oxide content lowered to 2%. In order to achieve this,

the amount of reductant and the oxidant:reductant ratio are adjusted
for each of the submerged reduction injectors. Slag samples are taken
regularly from each of the reduction zones as are disposable thermocouple temperature measurements. The relative change in bath temperature is monitored via thermocouples installed in the floor of the
reactor.

Reduction Mechanism

One advantage of the QSL process is its ability to produce a large
amount of metallic lead directly from the lead-bearing feed materials
in the oxidation zone. A certain amount of lead is unavoidably oxidised through to lead oxide, however, and passes in the slag into the
reduction zone. The task of the reduction zone is to clean the slag of
this lead and return the reduced lead to the lead syphon. This is
necessary to make the process economic and to ensure that an environmentally acceptable discard slag is produced. It is currently achieved
in a three stage process. The three reduction zones are separated from
the oxidation zone and from each other by refractory partition walls.
The partition walls restrict the amount of back-mixing of slag between
the individual zones and thus aid the attainment of a steep concentration gradient along the reactor's length. Figure 4 shows an example of
lead flows in slag and bullion through the individual zones, along
with the lead concentration in the feed, dust and the various slags
for the Berzelius reactor.

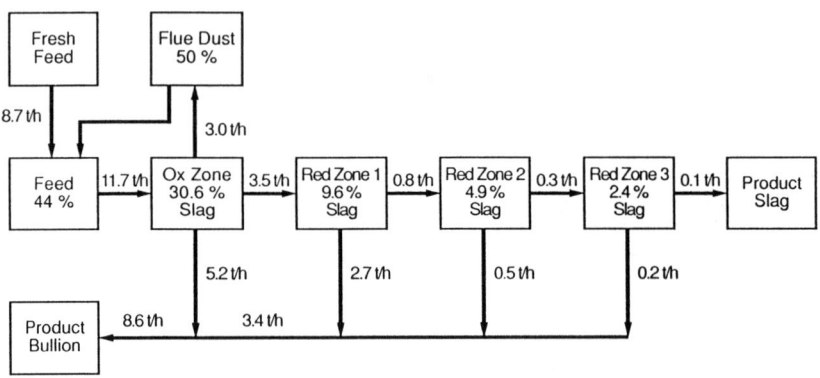

Figure 4 - Example of Lead Balance QSL Berzelius
Stolberg

Five reduction injectors are installed in the reduction zone of the Berzelius Stolberg reactor. Two are located in the first reduction zone, two in the second reduction zone and one in the final zone. Most experience to date has been obtained with lignite coke as the reductant, but test campaigns with natural gas have proven it to be of equal quality. Reducing agent, with carrier air if necessary, oxygen and shroud gas are injected into the bath from below. As mentioned previously, experience in the Berzelius reactor has highlighted the importance of the presence of a metallic lead layer at the tip of the submerged injector to achieve good slag reduction. The reduction reactions are believed to function along the lines illustrated schematically in figure 5.

Step 1: The oxygen, coke and air are heated quickly on introduction to the lead continuous layer, due to the high heat transfer properties of the molten lead, and the coke is gasified by the oxygen to form carbon monoxide. The oxidant:reductant ratio is adjusted automatically to obtain the desired gas mixture at the injector tip.

Step 2: The carbon monoxide rises into the slag continuous layer and reacts with the oxides in the slag to reduce iron, and form metallic phases, eg. lead and zinc, and carbon dioxide. The lead forms as small droplets and descends through the slag layer, finally reaching the lead continuous zone below. Any zinc fumed passes from the slag into the gas continuous zone.

Step 3: The zinc, the carbon dioxide and any unreacted carbon monoxide rise into the gas space above the slag layer. The zinc is totally, and the carbon monoxide partially, post combusted by the oxygen-enriched air injected through lances in the roof.

In table II the amounts of reductant used for each of the aforementioned reactions are shown for the Berzelius reactor.

Table II Example of Carbon Usage in the Reduction Zone

PbO --> Pb	30%
ZnO --> Zn	10%
Fe3+ --> Fe2+	15%
Total Reduction	55%
Heating	45%

135

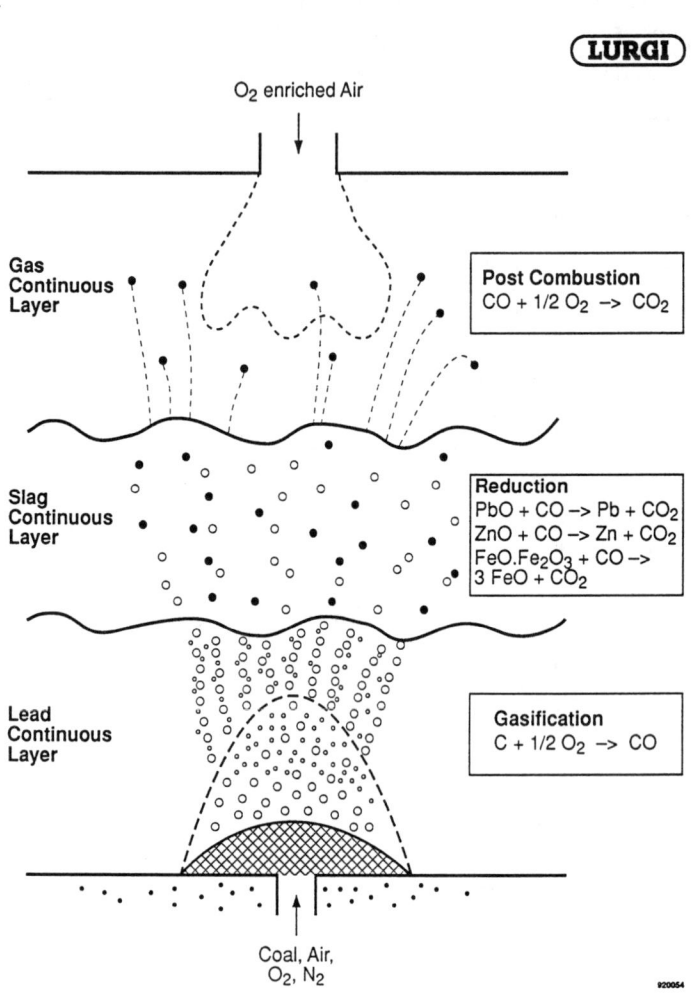

Figure 5 – Slag Reduction Mechanism

Overall reduction efficiency can be impaired when an oxidising post combustion jet impinges into the slag layer. The post combustion jet should optimally be so adjusted that it can oxidise the gaseous species exiting the bath and transfer as much as possible of the resultant heat back to the bath, without allowing the reduced species in the slag to be reoxidised. However, if excessive air and oxygen rates are employed, reoxidation of reduced species can occur and the overall reduction capacity of the reduction zone can be decreased dramatically. This has occurred on occasions with individual lances and also at times throughout the entire reduction zone.

Figure 6 shows the lead content of the oxidation zone slag and the final tapped slag over a ten day period. Initially the total post combustion gas rate was approximately 2.5 times too high due to an error in the flowrate measurement. Subsequently the error was corrected and the flowrate reduced accordingly. This led to a significant improvement in the reduction efficiency of the reduction zone.

Figure 7 shows a similar albeit milder effect brought about by a post combustion lance installed in the reactor end wall immediately above the slag tap. Initially the lead content of the final slag varied between 2 and 10%. Once the lance was replaced with a burner operating under slightly reducing conditions (end of day 5) the lead content stabilised.

Submerged Injectors

Two basic types of submerged injectors are employed in the QSL process. The oxidation, or so-called S-injectors (Sauerstoff=Oxygen), are installed in the oxidation zone. Here oxygen is injected to oxidise the feed materials and shroud gas is used to protect the injector tips from excessive burn-back. The reduction, or so-called K-injectors (Kohle=Coal), are employed in the reduction zones. In this case reducing agent is injected in addition to the oxygen and the shroud gas.

S-Injectors

The development of the S-injector has produced a variety of cross-sections. The best proven design to date can be seen in figure 8. The actual dimensions of the channels vary from plant to plant depending on the local requirements. The data for the current injectors in the Berzelius plant and that of Korea Zinc can be seen in table III.

Oxygen is injected through the central bore and two of the surrounding rings of channels. It has been found that the impulse of the jet is reduced if it is split up into a number of channels with resultant decrease in the amount of splashing produced above the bath surface. This is especially important in the case of the S-injectors due to the vicinity of the feed ports. The small bore in the centre of the injector allows regular measurement of the injector length when the

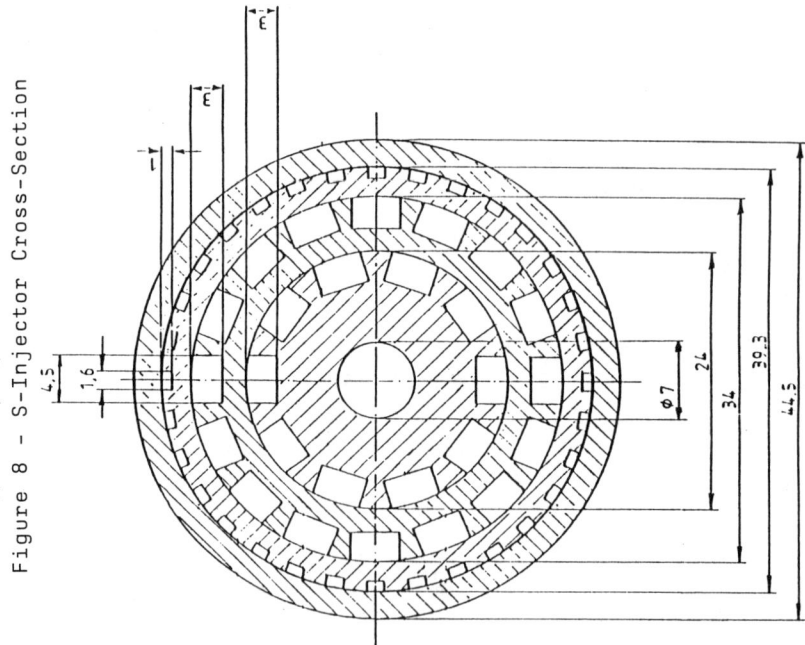

Figure 8 – S-Injector Cross-Section

Table III – S-Injector Flowrate and Pressure Data

	X-Sectional Area mm^2	Nominal Flowrate Nm3/h or L/h	Nominal Pressure barg
Berzelius Stolberg			
Oxygen			
Total	242	1350	12
Central Bore	38.5		
Inner Ring	60		
Outer Ring	144		
Shroud			
Total	51	180–260	10–15
Water		20–30	
Hydrocarbon Gas		0–50	
Korea Zinc Corporation			
Oxygen			
Total	405	2000	12
Central Bore	38.5		
Inner Ring	145		
Outer Ring	232		
Shroud			
Total	51	180–260	10–15
Water		20–30	
Hydrocarbon Gas		0–50	

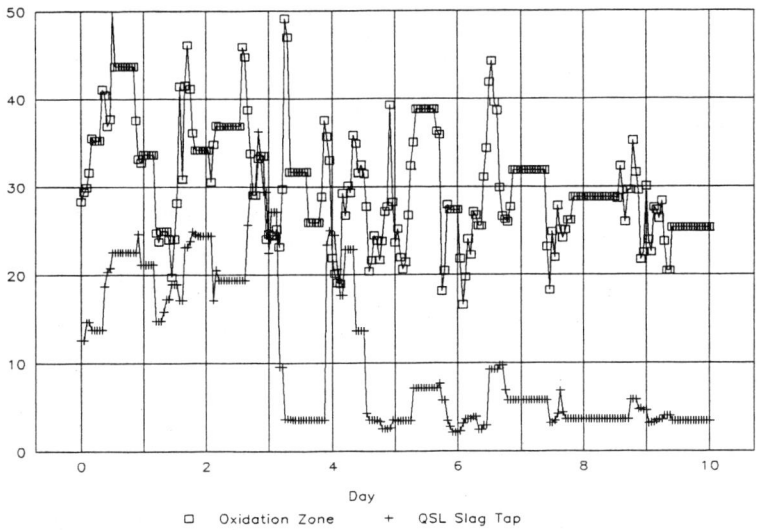

Figure 6 - Effect of Post Combustion Jet on Slag
Composition.

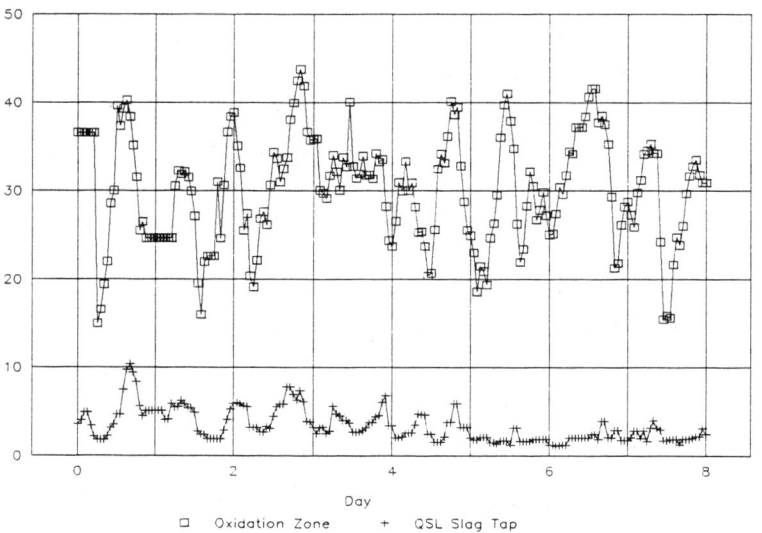

Figure 7 - Effect of Post Combustion Jet on Slag
Composition.

139

reactor is in the standby position using a hooked wire. This is essential to keep a track on injector wear rates and in order to allow preventative maintenance to be carried out.

The oxygen is surrounded by a ring of channels through which shroud gas is injected. It has been found that the amount and type of shroud gas is of utmost importance for ensuring long injector lifetimes, especially considering the high oxygen flowrates through the centre of the injector. Currently a mixture of nitrogen, atomised water and hydrocarbon gas is used to obtain the cooling intensity required.

A schematic diagram of the control system for an S-injector is seen in figure 9. The oxygen rate required for the individual S-injectors is adjusted automatically using a specific oxygen rate based on the total feed rate. A nitrogen back up system ensures that a minimum pressure is maintained at all times on the oxygen channels even when the oxygen is shut off. This reduces the likelihood of run-back of liquid lead and/or slag into the injector. The shroud gas total pressure is held constant as are the flowrates of water and hydrocarbon gas. The nitrogen flowrate is adjusted automatically to make up the difference.

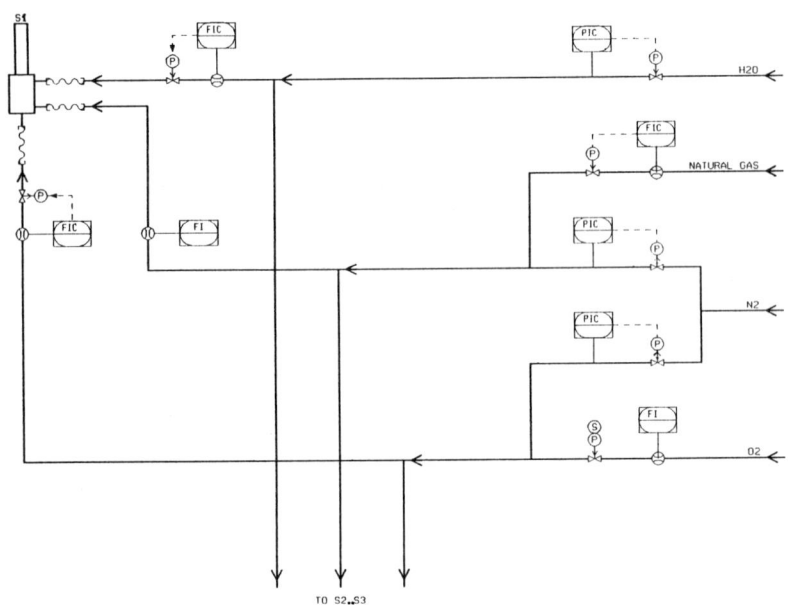

Figure 9 - S-Injector Control System

Experience has shown that the best lifetimes are obtained when a ferritic stainless steel is used for the injector pipes. Because of the high solubility of nickel in molten lead austenitic grade stainless steels are not suitable. The grade of steel currently in use is the German standard 1.4762, similar to the SAE 51446. The composition of this steel is shown in table IV.

Table IV Composition of Steel Used for Injector
Pipes

Element	%
C	<= 0.12
Cr	23.0 - 26.0
Si	0.7 - 1.4
Al	1.2 - 1.7
Mn	1.0 Max
P	0.040 Max
S	0.030 Max
Ni	---

The injector is cemented into a refractory block prior to installation in the reactor. This ensures that the gap between the injector and the surrounding refractory is sealed as well as possible at all times. This is especially important in the case of superheated liquid lead due to the low viscosity. If the gap is not sealed lead can flow between the injector and brick, thus accelerating the injector and refractory wear substantially. The seal also improves the heat transfer between the cooled injector and the surrounding brick, which aids the formation of a protective "mushroom" around the injector tip.

S-Injector wear rates of about 0.3mm per operating hour are currently achieved. This results in effective injector lifetimes (given a useable length of 200mm) of up to 4 operating weeks. The intense cooling brought about by the shroud gas mixture is utilised to form a protective mushroom in the vicinity of the injector tip and the surrounding refractory. A photograph of a typical S-Injector mushroom can be seen in figure 10. The mushrooms are typically of cow pat shape and vary in diameter from 300 to 500mm. It is believed that extended injector life can only be obtained when a stable mushroom of this type is formed. Major efforts are being concentrated on improving the reliability and lifetimes of the injectors further as injector replacement still represents a major operating cost.

141

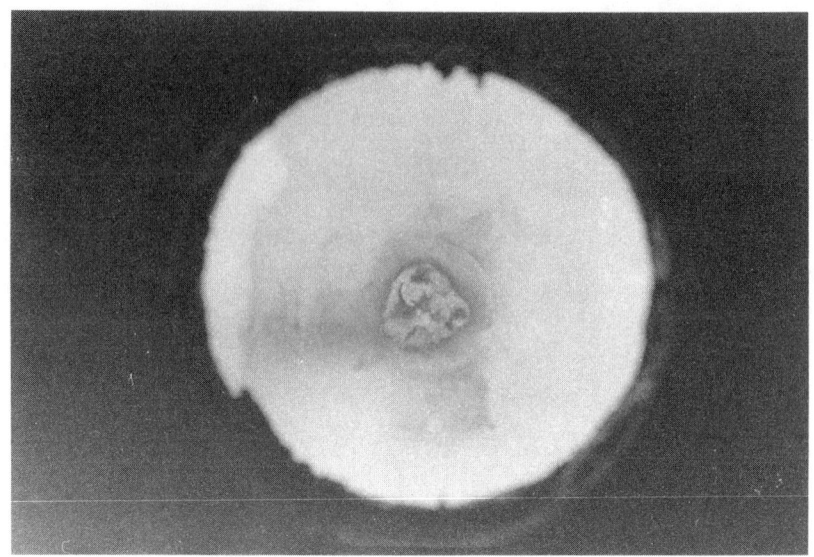

Figure 10 - Typical "Mushroom" on Injector Tip

Although it would be preferable to use more hydrocarbon gas in the shroud gas to intensify the injector cooling, it has been found that once a mushroom forms, a certain amount of shroud gas is forced back through the gap between the injector and the refractory brick towards the reactor shell, even when the gap is minimised using a cement seal. The resultant reducing atmosphere between the refractory lining and the steel shell can lead to refractory attack. The "vagabond" gas can also reemerge in the gas space above the bath surface, where it combusts, leading to unnecessary overheating of the offgas. Therefore hydrocarbon gas in the shroud mixture is kept to a minimum, normally being used just for short-term bath temperature corrections.

K-Injectors

The best proven K-injector design to date can be seen in figure 11. The actual dimensions of the channels also vary from plant to plant depending on the local requirements. The data for the current injectors in the Berzelius plant and that of Korea Zinc can be seen in table V.

Solid fuel is injected through the central bore of the injector along with carrier air. Due to the abrasive nature of the solid fuel, it is necessary to install a ceramic pipe in the innermost steel pipe. The ceramic pipe and surrounding steel pipe can be removed from the injector when the reactor is in the standby position. This allows measurement of the injector length and replacement of the entire coal pipe should it block.

Table V – K-Injector Flowrate and Pressure Data

	X-Sectional Area mm^2	Nominal Flowrate Nm3/h, L/h or kg/h	Nominal Pressure barg
Berzelius Stolberg			
Solid Fuel Pipe	50		4.5
Lignite Char		100	
Carrier Air		60	
Oxygen	28	75	8
Shroud	14		6–8
Total		40–60	
Water		0–10	
Hydrocarbon Gas		0–50	
Korea Zinc Corporation			
Solid Fuel Pipe	78		4.5
Lignite Char		200	
Carrier Air		90	
Oxygen	48	180	14
Shroud	39		6–8
Total		70–90	
Water		0–10	
Hydrocarbon Gas		0–50	

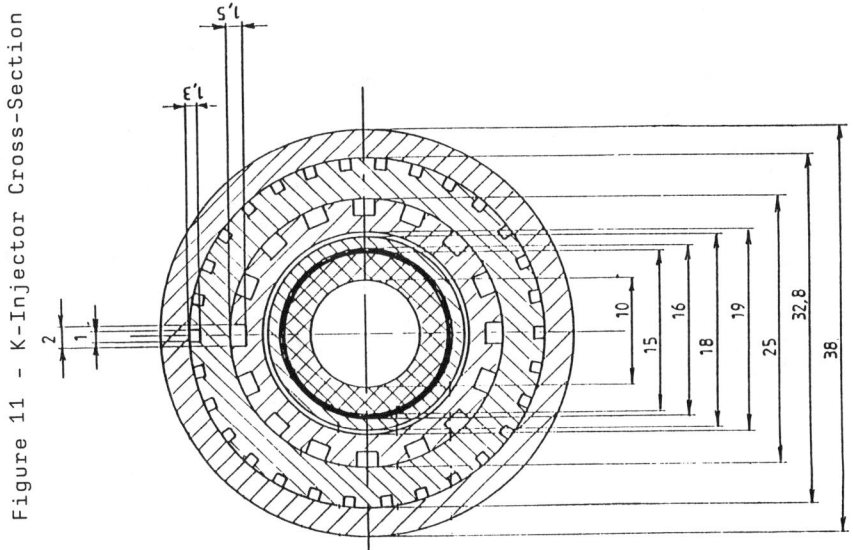

Figure 11 – K-Injector Cross-Section

143

Oxygen is injected through a ring of channels arranged concentrically around the central pipe.

The oxygen channels are surrounded by a further ring of channels through which shroud gas is injected. A mixture of nitrogen, atomised water and hydrocarbon gas is again used to obtain the cooling intensity required to protect the injector tip.

A schematic diagram of the control system for a K-injector is seen in figure 12. The solid fuel flowrate for each K-injector is set by the operator and the carrier air pressure maintained constant to ensure that run back cannot occur. The oxygen rate required to achieve the desired oxidant:reductant ratio is calculated automatically for each injector by the control system, based on the solid and gaseous fuel rates and the rates of oxygen and carrier air. The oxygen flow is adjusted correspondingly. Individual nitrogen backup lines ensure that the oxygen channels do not run back should the oxygen pressure fall below a minimum value. The total shroud gas pressure is held constant as is the total flowrate of hydrocarbon to all K-injectors. The water flowrate to the individual injectors is held constant and the nitrogen flowrate adjusted automatically to make up the difference. The distribution of the nitrogen and hydrocarbon to the individual injectors is determined by the resistance at the individual injector tips.

As with the S-injectors, experience has shown that the best lifetimes are obtained with ferritic stainless steel injector pipes. The same grade of steel is used as with the S-injectors.

The K-Injectors are also cemented into a refractory block prior to installation in the reactor.

K-Injector wear rates of about 0.2-0.3mm per operating hour are currently achieved. This results in effective injector lifetimes (given a useable length of 200mm) of up to 6 operating weeks. Mushrooms are also observed on the K-injector tips. Controlled mushroom formation is made more difficult however, due to the injection of solid fuel through the mushroom. If the mushroom is too large the solid particles can blind the pores and lead to blockage of the injector. If the fuel is coarse, as in the case of the Berzelius reactor, or contaminated with foreign matter, the situation is worsened. This means that the cooling intensity has to be balanced so that the mushroom is large enough to afford reasonable protection to the injector tip, but not so great as to provoke injector blockages. As with the S-injectors, major effort is being expended on extending the lifetimes of the K-injectors and making them more reliable.

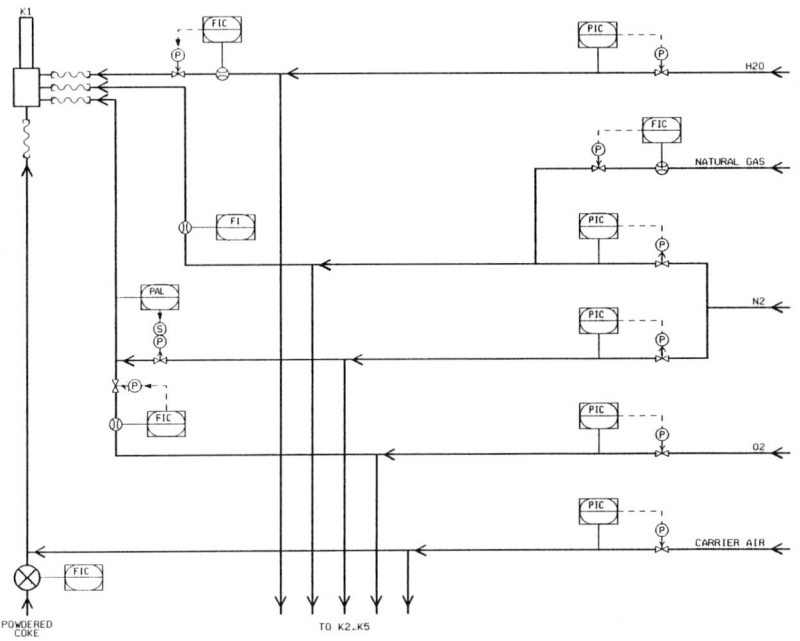

Figure 12 - K-Injector Control System

Summary

Fifteen months operation of the Berzelius Stolberg QSL plant have shown the viability of the overall process concept and the reactor design. It is possible to produce lead bullion and low lead slags from a wide range of feed materials (concentrates and secondaries) as well as sulphuric acid from the cooled, dedusted SO_2-rich offgases. The lifetime of the shell refractory is acceptable but there is potential for further improvements.

Experience has shown that a lead layer along the entire length of the reduction zone aids the slag reduction process significantly. The reactor design has been modified to facilitate this major change in operating philosophy.

Injector wear rates are still one of the major problems affecting the reliability of the process. Major efforts are continuing to be concentrated on extending injector lifetimes in order to reduce operating costs.

145

DEVELOPMENT OF LARGE SCALE

MITSUBISHI FURNACE AT NAOSHIMA

T. Shibasaki, K. Kanamori, and M. Hayashi

Mitsubishi Materials Corporation
Naoshima Smelter & Refinery
4049-1 Noashima-cho,
Kagawa 761-31, JAPAN

Abstract

Since the first commercial plant of Mitsubishi Process started its operation in 1974 at Naoshima, various development works were carried out to improve the productivity and process efficiency. The hearth productivity was increased almost twice by increasing the degree of oxygen enrichment. The high grade matte operation expanded the flexibility of the process to the variation of concentrate grade and enabled to treat larger proportion of higher grade concentrates. Impurity distribution was analyzed for and a process was established to control the impurity level in anodes. It was proven that the process can smelt anode scraps, purchased copper scraps and other miscellaneous materials. The furnace design has been developed to attain at least three years campaign. Finally it was enabled to design a larger Mitsubishi furnace unit with over 200,000 mtpy anode production capacity. The new plant commenced its operation in May 1991. This paper discusses the various development works and the design and earlier operation of the new plant.

Proceedings of the
Savard/Lee International Symposium on Bath Smelting
Edited by J. K. Brimacombe, P. J. Mackey,
G. J. W. Kor, C. Bickert and M. G. Ranade
The Minerals, Metals & Materials Society, 1992

Introduction

Mitsubishi Process is a continuous copper smelting and converting process that utilizes the injection and bath smelting technology. The process is being practiced in a full industrial scale for almost two decades, since the first commercial plant commissioned at Naoshima in 1974. The second plant was built in 1981 at Timmins, Ontario for Texasgulf Canada Ltd., now Kidd Creek Mines Division of Falconbridge Ltd. In May 1991, Mitsubishi built the third and larger plant at Naoshima in purpose to modernize and to make more efficient the overall smelter operation.[1]

Process Description

Mitsubishi Process utilizes three furnaces which are mutually linked with launders as shown in Figure 1.

The first one is S-furnace for smelting. Dried concentrates are injected through vertical lances into the molten bath in admixture with pulverized fluxes, reverts and coal, with oxygen enriched air. The molten bath is mostly matte with a thin layer of slag over the top of it. It is considered that the solid particles are captured in the molten bath while their temperature is not high enough for them to be involved of any metallurgical reaction, and that all the metallurgical reactions proceed within the bath. Resulting slag and matte are taken out by overflow and are sent to the second SH-furnace for separation. Matte grade is controlled at a fixed level in the range from 65 to 72 %. Typical slag analysis is 0.6 % Cu, 34 % SiO_2 and 6 % CaO. Slag is discarded. Matte is siphoned out and is further sent to the third C-furnace.

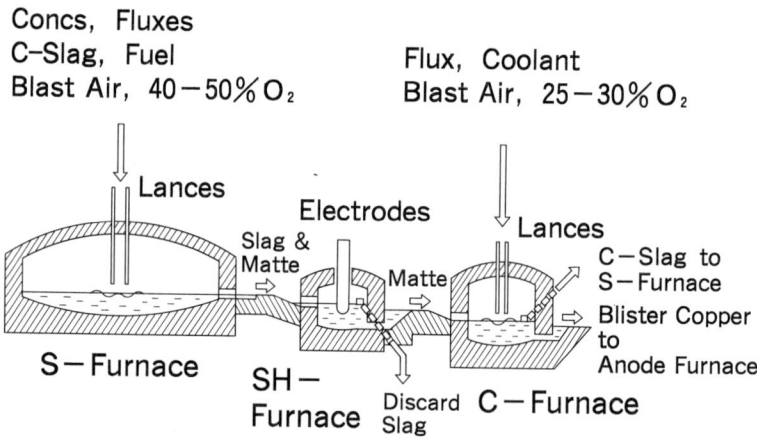

Figure 1 - Schematic Flowsheet of the Mitsubishi Process.

The C-furnace is for converting in which fed matte is converted to blister copper. Oxygen enriched air and calcareous flux are blown into the molten bath through vertical lances in a similar manner to S-furnace. Special lime slag has been developed to prevent thickening of the slag caused by magnetite saturation. Typical C-slag analysis is 15 % Cu in Cu_2O, 15-18 % CaO, and the balance is mostly magnetite. There exists in the furnace only two phases of molten copper and slag. White metal phase dose not normally coexist, mainly due to slightly overoxidized conditions.

148

Advantage of the Process

As the metallurgical features of the process were discussed in a previous paper[2], only its advantages are summarized below :

* As the raw material copper concentrates are finely pulverized and injected directly into the molten bath, the smelting reaction proceeds very rapidly and is considered to be finished almost instantaneously just at the vicinity of the point of injection.

* Major part of heat evolution during smelting is the heat of metallurgical reactions, eg., oxidation of iron and sulfur, and thus the transfer of heat from combustion flame is not the rate controlling factor for the smelting reaction.

* Therefore, furnace throughput could be increased by a substantial degree just by increasing the degree of oxygen enrichment. Only the autogenous point would be the limitation.

* The furnaces and related facilities can be designed compact.

* The man power requirement is smaller due to continuousness of the operation and spontaneousness of the molten metal flow between the furnaces.

* It can be operated with a minimum spillage of SO_2, which could be captured rather easily when compared with the spill gas associated with the PS converter operations.

Development Works

Two smelter lines, a conventional reverberatory furnace line and an older Mitsubishi Process line, were in operation at Naoshima during 80's. The operating cost of the Mitsubishi Process was much lower than the conventional process, and therefore it was obvious that the operation efficiency could be remarkably improved by integrating the entire operation into a single larger Mitsubishi Process line. The technical feasibility of such a project was studied from various point of view and was confirmed by a series of plant tests.

High Grade Matte Smelting

High Grade Concentrates There are two different types of concentrates available at Naoshima as shown in Table 1. One is chalcopyrite type, grading 25 to 33 % Cu, and the other is high grade concentrates, grading 35 to 47 % Cu. The proportion of the latter has gradually been increased.

The smelter was originally designed to treat mainly the chalcopyrite type concentrates and it relies on the heat of oxidation of iron and sulfur in the concentrates to save fuel requirements. So the low calorific value high grade concentrates causes an unbalance of heat requirements between S and C-furnace when they are treated alone or in larger proportions, i.e., the fuel requirement at S-furnace and the lance burden at the C-furnace will be increased.

Therefore the furnace throughput would remarkably be reduced due to increase of fuel rate and resulting increase of offgas volume. The limitation in the converting furnace could also impose the limitation on the smelter throughput. The common measure to prevent such inconvenience

149

is to blend the concentrates and to minimize the fluctuation in grade of furnace feed. Such strategy, however, requires large concentrate storage and is not economical.

Table I. Typical Analyses of Copper Concentrates

		Chalcopyrite type	High grade type
Analysis,			
Cu	wt %	28	42
Fe	wt %	27	15
S	wt %	32	28
SiO_2	wt %	5	7
Heat of reactions, Mcal/mt			
at 68 % matte		557	300
at 76 % matte		635	401

High grade matte smelting was proposed and tested at the actual plant. The calorific value of the concentrate is increased substantially by increasing matte grade to about 75 % as shown in Table 1. The anticipated difficulties of such process scheme were the increase of slag loss and slag thickening due to high magnetite.

Table II. Magnetite Balance

	Reverberatory (Calcine)			S-Furnace Chalcopyrite Matte 68 %			S-Furnace High Grade Conc Matte 76 %		
	Mass kg	Fe_3O_4 %	Fe_3O_4 kg	Mass kg	Fe_3O_4 %	Fe_3O_4 kg	Mass kg	Fe_3O_4 %	Fe_3O_4 kg
Input									
Concentrates	100		–	100	1.0	1.0	100	1.0	1.0
Calcine	90	15.0	13.5	–	–	–	–	–	–
Reverts	3	15.0	0.5	3	15.0	0.5	3	15.0	0.5
C-Slag			–	9	60.0	4.5	4	60.0	2.4
Total(a)			14.0			6.9			3.9
Output									
Matte	62	4.0	2.5	47	1.5	0.7	56	0.5	0.3
Slag	38	8.0	3.0	67	10.0	6.7	49	15.0	7.4
Total(b)			5.5			7.4			7.7
(a)-(b)			+8.5			-0.5			-3.8

Magnetite Balance Excess magnetite, either fed back to the smelting furnace as contained in the converting slag or formed while the concentrates are being oxidized at temperatures below the melting point, is normally reduced to attain equilibrium by the lower grade matte during the course of smelting process, but could cause slag thickening when the matte grade is higher.

At S-furnace however, the concentrates are not oxidized at solid state, and thus the excess magnetite is not formed during smelting. The magnetite contained in the return C-slag is rather smaller due to smaller amount of C-slag. Then the magnetite will be formed to attain the equilibrium as shown in the Table 2. The situation does not change even at the higher grade matte smelting, at 75 % Cu for example, and it is expected that the smelting operation is free from so called magnetite problem.

Slag Loss Figure 2 shows the plot of slag loss at the higher matte grade range. The slag loss increased up to 1.5 % at 75 % matte grade range. At the intermediate matte grade range of 70 - 73 %, the slag loss is in the order of 0.7 - 0.9 %. There was no change in the fluidity of the slag.

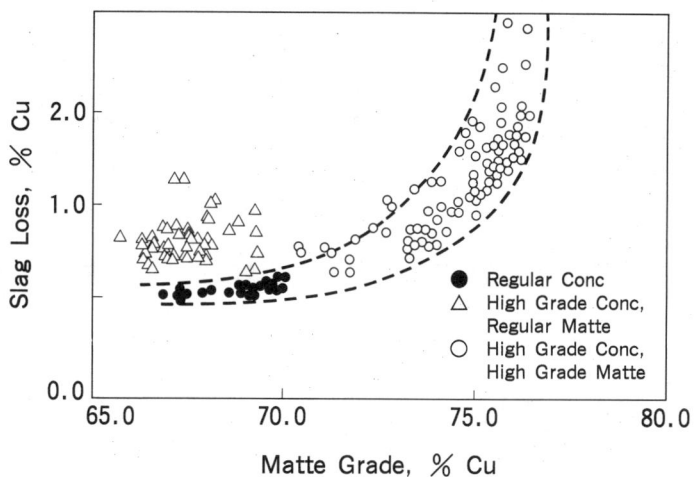

Figure 2 - Slag Loss at High Matte Grade Range.

High grade matte smelting Two alternatives were proposed for the smelting of the higher grade concentrates :

I - High grade matte smelting at 75-76 % Cu. Due to high copper loss in the slag, it is not economical provided that the slag is not milled for flotation.

II - Intermediate matte grade smelting at 70 - 72 % Cu. Considering relatively lower copper loss and smaller slag fall, it is still economical to discard the slag simply.

The flexibility of the Mitsubishi Process to the variation of the concentrate grade have been developed very wide with such provisions, and as a result the higher grade concentrate could be smelted at any proportion in the furnace feed.

Secondary Materials

Classification The secondary materials are classified into the following five groups as shown below, in which groups 1-3 are purchased materials, and groups 4-5, inplant reverts.

1. Fine materials such as sludges, cement copper, and screen under from

151

miscellaneous copper bearing secondary materials.

2. Low grade scraps such as scraps of electronics parts, copper printed circuit board and others.

3. Regular copper scraps

4. Non-metallic reverts such as boiler clinker, used copper soaked bricks, and solidified chunks of slag and matte.

5. Metallic reverts such as anode scraps, reject anodes, impure cathodes from liberator cells, and reject molds.

While Naoshima had two smelter lines, only a part of the type 1 materials were treated at the Mitsubishi Process line and all the others were fed to the PS converters in the conventional line. Aiming at the realization of the integration into a single smelter line system with a larger Mitsubishi Process unit, it was studied to treat all the materials in the Mitsubishi Process. Anticipated amount of materials of respective type and test conditions are shown in Table 3.

Table III. Anticipated Amount of Secondary Material Treatment

Type	Material	Anticipated amount		Equivalence at Old line	Test at Old line	
		mtpm	mtpd	mtpd	mtpd	Furnace
2	Low Grades, Sredded	700	24	12	24	S-furnace
4	Crushed Rverts, -50mm	–	–	–	1t-batch	S-furnace
3	Pressed Scrapcubes,	2,400	80	40	40	S-furnace
	(500kg)				10	C-furnace
5	Anode Scraps, Shredded	3,000	100	50	1t-batch	C-furnace

Low Grade Scraps Low grade scraps of type 2 are normally shredded to -50 mm. Anticipated amount is 700 mtpm, or 24 mtpd. Considering that the furnace capacity of the older Mitsubishi Process line was half of the expected new line, the equivalent feed rate was 12 tpd. A double gate feeding hopper was installed above the drop hole at the old S-furnace and the raw grade shredded scraps were fed at 24 mtpd rate. So far as they were fed in smaller batches, there could not be observed any overflow of unsmelted material. Any noticeable change (increase) of the fuel requirement was not observed.

Crushable inplant reverts, type 4 materials, were normally pulverized to -3 mm when they were treated in the Mitsubishi Process line. Considering the favorable test results for low grade scraps, it was tested to smelt the reverts crushed to -50 mm, similar size to them, by feeding through the drop chute. It was found that the batch as large as 1 metric ton could be smelted without causing overflow of unsmelted chunks.

Regular Scraps Regular scraps pressed into cubes of 400 to 500 kg and were fed to PS converters in the past. There are two alternatives to treat them at the new Mitsubishi Process line. One is the C-furnace and the other is the S-furnace. It is more natural to smelt metallic copper scraps at the C-furnace rather than at the S-furnace. However some of the purchased scraps occasionally contain high level of lead in them, and the lead level in the blister copper could go high when they are treated without any control. So the tests were mainly conducted at S-furnace to smelt them at 40 mtpd rate for one week. Cubes dropped onto the molten bath stay afloat for a while, around 20-30 min, then disappeared. Major operation data are summarized for the test period in comparison with normal operation. Estimated unit consumption of coal was 144 kg/mt-scraps, and the actual figure was only 5 kg/mt. Smaller fuel requirement at the test was presumably due to associated combustibles or high calorific constituents such as zinc and aluminum in the scraps. The effect on metallurgical control could be compensated during regular feed back controls.

Table IV. Fuel Increase by Charging Scrap into S-furnace

		Normal Operation	Scrap Test
Concentrate	t/h	39.2	39.5
Scraps	t/h	0	1.8
Lance Blowing	Nm^3/h	23,100	23,100
	% O_2	39.4	39.7
Coal	kg/h	1,158	1,178
Total Oxygen	Nm^3/t.conc	232	232
Matte Grade	% Cu	67.2	67.6

Anode Scraps Anode scraps should preferably be treated at the C-furnace. As it was unable to install an adequate drop hole for an anode scrap in the original shape at the older C-furnace due to the limitation in the access to the roof area, the test was conducted with the shredded anode scraps. There was not any metallurgical limitation to treat them but to compensate the heat requirement to melt them.

Secondary materials treatment flowsheet finalized though a series of plant tests is shown in Figure 3. Very large scraps and used molds are fed to the anode furnace.

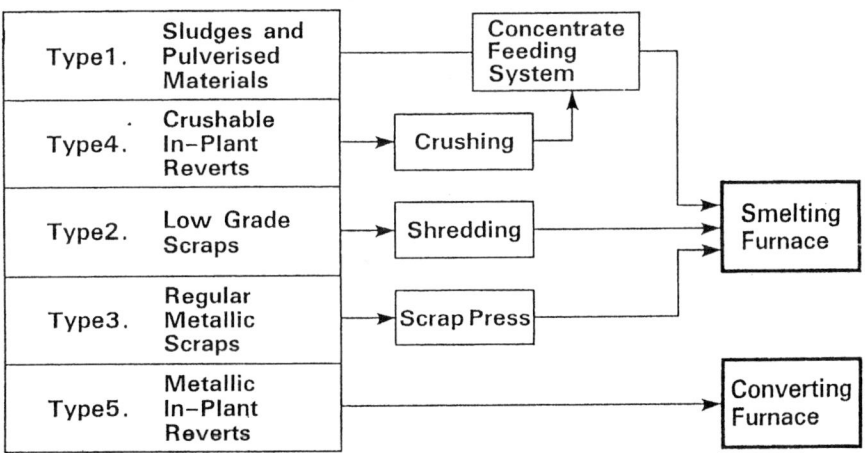

Figure 3 - Schematic Flowsheet of Secondary Materials Treatment.

153

Impurity Control

Distribution of impurities Distribution of impurities to anodes in the Mitsubishi Process is shown in Table 5 in comparison with the conventional process. In the conventional process, the precipitator dust of the PS converters is bled off for separate treatment, moreover the converter slag is slow cooled and milled. The recovery of such impurities as lead and nickel are rather poorer than in the direct recycle of converter slag to the smelting stage. In the Mitsubishi Process the precipitator dust of the S-furnace is bled off and all the C-slag is directly recycled to S-furnace. Elimination of lead in the Mitsubishi Process is slightly poorer than the conventional, and the recovery of nickel much better. Elimination of antimony is much better in the Mitsubishi Process, due to larger slag matte distribution coefficient at higher matte grade.

Table V. Impurity Distribution to Anodes

Process	Distribution to anodes, %				
	Pb	As	Sb	Bi	Ni
Reverb/PS converter(1)	8-13	9-11	28-34	14-26	45-50
Mitsubishi Process (2)	15-19	4- 6	15-20	15-25	65-75

(1) Converter EP dust is bled off for separate treatment.
    Converter slag is slow cooled and milled.
(2) S-furnace EP dust is bled off for separate treatment.

Lead Control Lead level in the anodes is controlled by bleeding off and separate treatment of the EP dust. As the cost of such operation is relatively higher, it is favorable as costwise to minimize the amount of dust bleed off. On this reason, a model was developed to evaluate the lead % in the anodes from the lead % in matte as shown in Figure 4. Assuming the highest limit of lead in anode as 0.25 %, the lead in matte should be controlled at a level lower than 0.35 %. Thus the lead in matte is monitored hourly, and the timing of dust bleed off is decided.

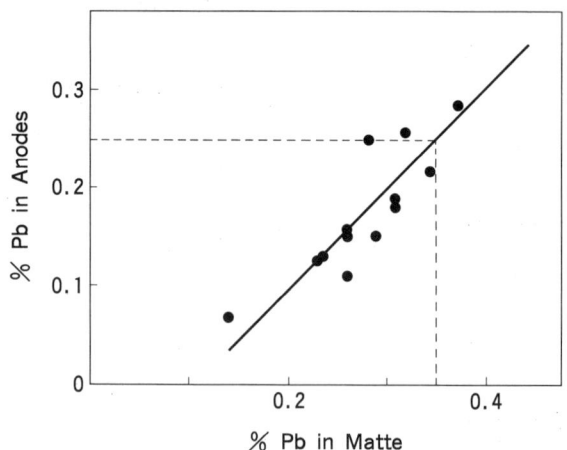

% Pb in Matte

Figure 4 - Relation between Lead in Matte and Lead in Anode

154

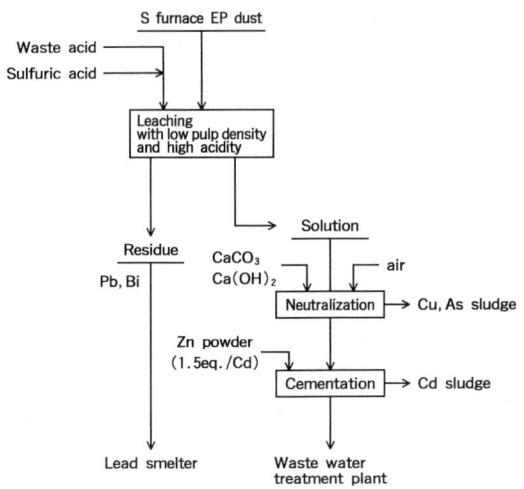

Figure 5 - The Proposed Flowsheet for Dust Treatment

Dust Treatment A simplified hydrometallurgical process was developed for treatment of the dust as shown in Figure 5 [3,4]. At first dust is leached at relatively lower pulp density and higher acidity. The lead residue containing some bismuth and antimony is sent to lead circuit. The solution is neutralized at pH 4.5 to remove copper and arsenic, then cadmium is removed by zinc dust cementation. Copper residue is recycled back to the copper circuit, and the cadmium residue is sent to the zinc plant. The final solution still containing zinc and some cadmium is sent to the existing waste water treatment for further cleaning.

Typical anode analyses are shown in Table 6. Original intention was to control lead at 0.25 %. As it was confirmed later that the refinery could meet with higher lead levels in anodes without deterioration of the cathode quality, dust bleed off ratio was reduced so as to allow 0.3-0.4 % lead in anodes.

Table VI. Typical Anode Analysis

| | Old Scheme | | New Scheme (Mitsubishi Furnace) | |
	Reverbline	Mitsubishi Furnace	Intended	Present
Cu %	99.4	99.3	99.3	99.2
Pb %	0.15	0.25	0.25	0.3-0.4
Ni %	0.15	0.10	0.20	0.2-0.3
As ppm	700	1000	1000	1000
Sb ppm	350	300	300	300
Bi ppm	100	100	100	100

155

Design of a Large Scale Mitsubishi Furnace

Unit Capacity

It was intended to achieve a smelting rate of almost 2,000 mtpd concentrates in moderate sized furnaces, 10mϕ for S-furnace and 8mϕ for C-furnace for example, being similar size to those as Kidd Creek smelter. At Kidd smelter, however, the smelting operation of S-furnace is autogeneous at 1,500 mtpd of smelting rate at 48 % oxygen enrichment for lance blast.

The limitation on the increased throughput due to attainment of autogeneous point has been overcome by the following counter measures;

S-furnace :
* The carolific value of the concentrates available at Naoshima is relatively lower due to larger proportion of higher grade concentrates.
* Purchased copper scraps are fed to S-furnace and that helps to compensate the excess heat, if any.
* Oxygen enrichment was limited at a degree lower than 50 % and lance blow rate is increased, if required.

C-furnace :
* Increased matte feed requires larger total oxygen supply. The lance blast burden is limited within a tolerable range by increasing the degree of oxygen enrichment.
* Excess heat generated due to higher oxygen enrichment is compensated by feeding anode scraps as coolants.

Major furnace design parameters are summarized in Table 7 in comparison with those for the Kidd Smelter and old plant at Naoshima. Concentrate feed rate was increased to twice that of the old plant and anode production to three times.

Table VII. Major Furnace Design Parameters

		Naoshima (Old)	Kidd Creek	Naoshima (New)
Furnaces				
S-Furnace,	mϕ	8.25	10.3	10.1
C-Furnace,	mϕ	6.50	8.2	5.05
SH-Furnace,	KVA	1,800	3,000	3,600
S-Furnace Feed				
Concentrate,	tpd	900	1,500	1,900
Conc. grade,	Cu %	30	25	31
Scraps,	tpd	–	–	90
C-Furnace Feed				
Anode scraps,	tpd	–	–	100
Anode production,	tpd	260	380	750

Lances		S	C	S	C	S	C
Number,		8	5	10	6	9	10
Diameter,	in ϕ	3	3.5	3	3	4	3.5,4
Lance blast,	Nm^3/Hr	22400	12000	29000	16000	40000	24000
Oxygen,	%	42	28	48	33	45	32
Hearth efficiency,	Nm^3/Hr/m^2	420	360	350	300	500	480

Operation Data

The metallurgical data are summarized in Table 8 for March, 1992. Concentrate feed rate was slightly lower than the design, but the anode production was higher mainly due to higher concentrate grade.

Table VIII. Metallurgical Data in March, 1992

	Mass mtpm	Cu %	Fe %	S %	SiO_2 %	CaO %
S-Furnace Input :						
Concentrates	52820	32.8	24.0	29.1	8.6	0.5
Pressed Scraps	1380	85.5	5.0	-	-	-
Shredded Scraps	750	40.1	13.3	1.5	12.3	1.3
Silica Flux	8990	-	4.0	-	86.0	0.2
Limestone Flux	1160	-	-	-	0.9	53.9
Return C-slag	4970	13.9	43.8	-	0.1	17.1
Return Dust	2990	36.9	4.5	10.7	1.0	0.6
Inplant Reverts	1030	9.6	1.5	-	0.5	12.0
Total Solid Charge	74090	27.9	21.0	21.2	15.5	2.6
S-Furnace Output :						
Matte	29180	68.3	7.6	21.1	-	-
Slag	37860	0.7	33.9	0.5	32.7	5.1
Total	67040	30.1	22.5	9.5	18.5	2.9
C-Furnace Input :						
Matte	29180	68.3	7.6	21.1	-	-
Anode Scraps	2740	99.2	-	-	-	-
Limestone	1590	-	-	-	0.9	53.9
Coolants (C-Slag)	4240	13.9	43.8	-	0.1	17.1
Total Charge	37750	61.6	10.8	16.3	-	4.2
C-Furnace Output :						
Blister Copper	21640	98.7	-	0.6	-	-
C-slag	9210	13.9	43.8	-	0.1	17.1
Total	30850	73.4	13.1	0.4	-	5.1
Anode Production	22460	99.2	-	-	-	-

Summary

The Mitsubishi process has various advantages over the conventional process. Since the start up of the first commercial plant at Naoshima in 1974, various development works were carried out as described above, such as increasing hearth productivity, high grade matte operation, treatment of secondary materials, longer furnace campaign life and impurity control.

Such developments were incorporated into the design for the new big continuous smelter which started its operation in May 1991 after shutdown of both plants, old continuous and reverb line.

By overcoming the difficulties during the earlier start up operation, now it was proven that the process, as a stand alone custom smelter, can treat any kind of copper concentrate, secondary materials and inplant reverts like spent anode and boiler chunks, while keeping high productivity and economic superiority.

Reference

1) T. Shibasaki et al., <u>Pyrometallurgy of Copper. Copper '91</u> (Pergamon Press, 1991), 3-14.

2) T. Shibasaki and M. Hayashi, "Top-Blown Injection Smelting and Converting : The Mitsubishi Process", <u>Journal of Metals</u>, Sept. 1991, 20-26.

3) T. Shibasaki and Y. Tosa and N. Hasegawa, "Impurity Control strategy at Integrated Operation of Naoshima Smelter into a single Larger Mitsubishi Furnace Line", <u>Met. Rev. of MMIJ</u>, Vol.8, No.1. (1991), 119-129.

4) T. Shibasaki and N. Hasegawa, "Combined Hydrometallurgical Treatment of Copper Smelter Dust and Lead Smelter Copper Dross," Paper to be presented at the Ernest Peters International Symposium on Hydrometallurgy Theory and Practice, to be held at Vancouver, B.C., June 14-17, 1992.

The COREX[R] Technology

From Theory to Commercial Reality

Michael Lemperle

Deutsche Voest-Alpine Industrieanlagenbau GmbH

From the original innovative idea to a new technology on an industrial scale the COREX process for the production of hot metal passed through many different development phases.

In the beginning was the technical concept, consequently analysed on a theoretical basis and tested practically in the laboratory and on a pilot scale. Finally 1989, there was the commercial scale COREX plant with a hot metal production of more than 300,000 t per year. Using coal instead of coke, a COREX plant additionally produces high-quality export gas of approx. 8000 kJ/m^3 (STP) which has a tremendous potential for power generation, especially in gas turbine based combined cycle schemes. To the integrated steel plant, COREX offers different variants of application depending on infrastructure, gas and energy demand, and electric power credits, respectively. High flexibility with respect to gas quantity and quality combined with environmental compatibility show substantial benefits for numerous applications.

Proceedings of the
Savard/Lee International Symposium on Bath Smelting
Edited by J. K. Brimacombe, P. J. Mackey,
G. J. W. Kor, C. Bickert and M. G. Ranade
The Minerals, Metals & Materials Society, 1992

INTRODUCTION

The shortage of coking coals and high cost of coke and natural gas has given rise to worldwide efforts in recent years to smelt ore on the basis of raw coals.
Attempts to gasify coal by conventional methods and to use this gas for reduction processes previously fueled by natural gas have failed for economic reasons. Direct reduction although having gained a remarkable market potential suffers on high prices for natural gas or low demand for sponge iron, if cheap gas is available.

All these processes could not offer a true alternative for the blast furnace until 1970 when processes for the production of liquid iron without using coke were patented.

Today hot metal production with a technic different from the conventional blast furnace-coke oven route has become a cost and energy saving reality. On 10th November 1989 the first large-scale smelting reduction plant of the world was put on stream in the metallurgical works ISCOR, Pretoria, S.A., and since then has been in continuous operation.

The COREX® technology was jointly developed by DEUTSCHE VOEST-ALPINE INDUSTRIEANLAGENBAU GMBH, Düsseldorf/Germany and VOEST-ALPINE INDUSTRIEANLAGENBAU GES.M.B.H., Linz/Austria.

The advantages compared to the blast furnace are obvious:

-   The coal based COREX® process does not require coking- and sinterplants which cause high investment and pollution problems.

-   High operational flexibility with respect to production capacity, raw material change and stopping times allow for a lot of applications in integrated steelworks and satisfy best the needs of the international iron and steel market. Thus the COREX® process fills two gaps within today's industrially available steelmaking processes:

-   Mini mills can now produce pig iron in small capacities at low cost, which is a precondition for these works to successfully enter into the flat product market. For the large integrated steelproducer it allows to have a small capacity, cost efficient unit as a swing producer of hot metal at up to some 800,000 tons per year to complement the large blast furnaces operating at up to 5 million tons per year. This allows the large modern blast furnaces to operate at constant optimum conditions without being forced to adjusts production rates to the ups and downs of the market.

FIRST STEPS AND PILOT PLANT

The possibility that a smelting-reduction process could be based on the use of coal alone was first investigated by KORF STAHL in 1977. The basic idea consisted in the separation of iron ore reduction and melting steps into two reactors:
- Generation of reducing gas and liberation of energy from coal for melting occurs in the melter gasifier;
- Reduction of iron ore occurs in a shaft furnace.

In 1978, tests were conducted using a melter-gasifier of 760 to 1000 mm diameter, with promising results. This work led to the first application for a COREX® patent in October 1978.
In a development contract, which was signed in 1979, KORF STAHL and VOEST ALPINE INDUSTRIEANLAGENBAU agreed to work jointly on the development of a smelting-reduction process.
The basic and detailed engineering aspects of a pilot plant were established, and a large plant with a melting capacity of approximately 60,000 t/a was planned. By May 1981, a melter-gasifier of 3,5 to 5 m diameter, and a reduction shaft of 2 m diameter, had been built as a joint venture. After cold tests, the first campaign was started in July 1981.
The pilot plant was a semi-industrial plant, and incorporated features that were used later in the design of the production plant at ISCOR's Pretoria Works. It was situated at Kehl-am-Rhein, Germany, near a steelworks that could use the liquid iron produced during campaigns.

Figure 1 shows the basic flowsheet of the COREX® plant.

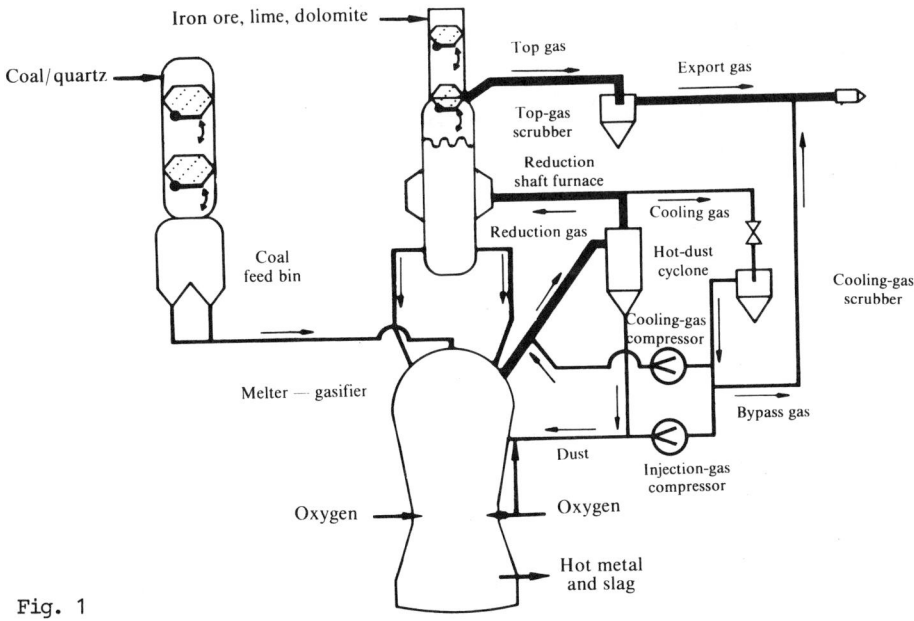

Fig. 1

The main components of the pilot plant were:

- a melter-gasifier
- a reduction shaft
- a dust-recycling system
- gas scrubbers for the cooling gas and the top gas
- a system for water treatment and clarification
- a coal-drying plant
- material-handling systems for coal, ore, and additives
- a supply system for oxygen and nitrogen
- a granulation plant for slurry
- a runner system for the liquid iron and the slag.

The dimensions of the process tower were approximately 10 by 15 m, and its total height was more than 50 m. The area of the site, which included the systems for storage of the raw materials, treatment and clarification of the water, and drying of the coal, was about 15,000 m^2.

INVESTIGATIONS AND PATENTS

The Fluidized Bed

The first COREX® patent, applied for in October 1978, was for a fluidized bed (Fig. 2). The main claims of the patent for this fluidized bed are as follows:

- The fluidized bed uses fine coke.
- The DRI is molten in front of the tuyere
- The grain size of the coke is 2 to 3 mm (1 to 4 mm).
- The maximum grain size of the coal charged is 12 mm.
- The maximum height of the fluidzed bed is 3,5 m.

It was found that liquid iron could be produced from non-coking coal and iron ore but, although the overall results were encouraging, they were not overwhelmingly convincing.

162

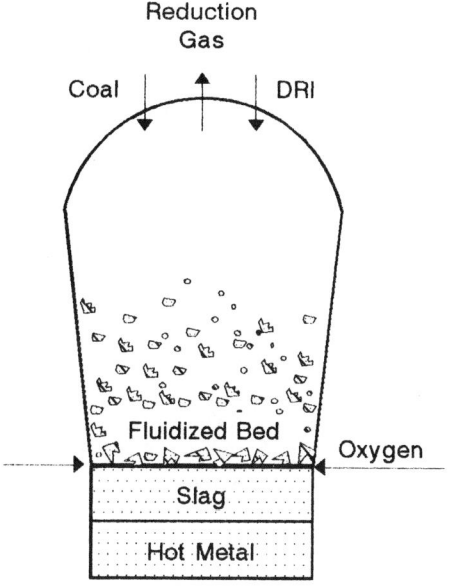

Fig. 2 - The COREX® fluidized bed in the melter-gasifier.
A patent was applied for in October 1978.

The results of the first campaigns showed that the liquid iron had a
carbon content of 2,5 to 3,5 per cent, and a sulphur content between 0,2
and 0,4 per cent. No silicon was present. The tapping temperature was
1300 to 1350°C, and the highest temperature under the dome was above
1200°C. The dust losses into the clarifier were high.
Examination of the above-mentioned results led, between 1982 and 1984, to
further development work and applications for patents.

The Fixed Bed

The application for the first COREX® fixed-bed patent describes a fixed
bed above the level of the tuyeres (see Fig. 2).
By control of the average grain size of the coal, a fixed bed can be
formed that increases the residence time of the DRI, and also of the
additives in the high-temperature zone above the tuyeres. Metallurgical
reactions such as carburization and the reduction of silica are therefore
possible and, furthermore, the contact time between the molten iron and
the hot char or coke is increased, which leads to higher tapping temperatures.

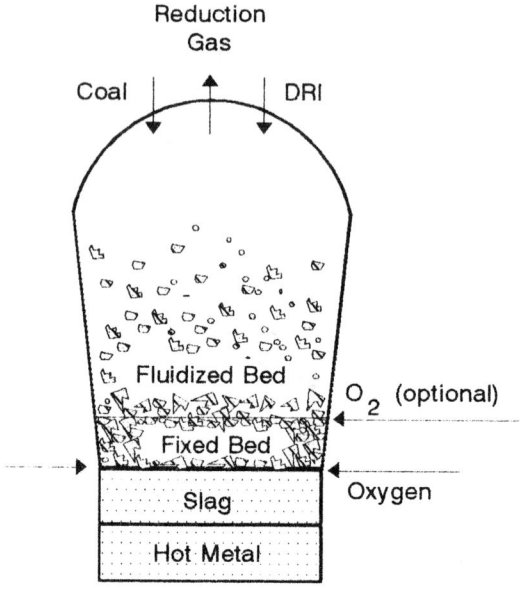

Fig. 3 - The COREX® fixed bed in the melter-gasifier (Patent No. 1)

The following reactions were noted

$$SiO_2 + 2C \longrightarrow Si \quad + 2CO + \Delta H, \tag{1}$$
$$MnO \ + C \longrightarrow Mn \ + CO \ + \Delta H, \tag{2}$$
$$FeO \ + C \longrightarrow Fe \ + CO \ + \Delta H, \tag{3}$$

$$3Fe \ + C \longrightarrow Fe_3C \quad + \Delta H. \tag{4}$$

The application for the COREX® fixed-bed patent (No. 2) describes a second fixed bed below the tuyeres (Fig. 4). The upper fixed bed is above the level of the tuyeres. The lower fixed bed, which is a part of the deadman, is very important, since several reactions influence the purity and quality of the liquid iron. Since not all the FeO is reduced above the tuyeres, the slag and the liquid iron must be allowed additional residence time so that the sulphur reaction can approach equilibrium. The DRI contains the sulphur as FeS, and the desulphurization can be represented as follows:

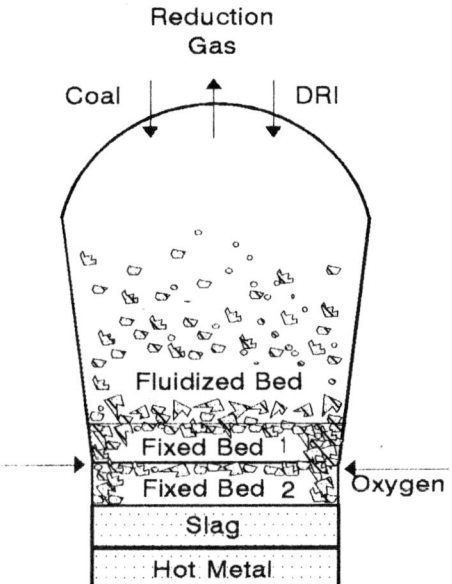

Reduction Gas

Coal | ↑ | DRI

Fluidized Bed

Fixed Bed 1

Fixed Bed 2 — Oxygen

Slag

Hot Metal

Fig 4 - The COREX® fixed bed in the melter-gasifier (Patent No. 2)

$$FeS + CaO \rightarrow CaS + FeO. \tag{5}$$

A sulphur distribution $(S)/[S] = 20$ can be obtained where:

(S) is the sulphur in the slag, and
$[S]$ is the dissolved sulphur in the liquid iron.

The FeO also participates in the silicon equilibrium reaction:

$$2FeO + SiO_2 \rightarrow 2FeO + Si. \tag{6}$$

The thermodynamic equations are as follows:

$$\frac{(S)}{[S]} = K_s \cdot \frac{a_{(CaO)}}{[O]} , \tag{7}$$

where $a_{(CaO)}$ is the CaO activity in the slag, and [0] is the oxygen potential in the metal bath.

$$[0] = K_{Si}^{1/2} \cdot \frac{a_{(SiO_2)}}{[Si]}^{1/2} , \tag{8}$$

where $a_{(SiO2)}$ is the SiO_2 activity in the slag, and [S] is the dissolved silicon in the metal.

$$K_{Si,S} = \frac{K_S \cdot a_{(CaO)}}{K_{Si}^{1/2} a_{(SiO_2)}^{1/2}} , \tag{9}$$

$$\frac{(S)}{[S]} = K_{Si,S} [Si]^{1/2} \tag{10}$$

It follows that, only when the silicon content of the liquid iron is sufficiently high, and the basicity of the slag is high (i.e. the concentration of CaO-MgO in the slag is high), will the sulphur content of the liquid iron decrease. The behaviour of manganese in the liquid iron is similar to that of silicon. A certain amount of fixed bed must be provided in the COREX® gasifier if liquid iron of the quality required for the making of steel is to be produced.

The work done by that stage resulted in the production of COREX® liquid iron that was comparable in quality to liquid iron produced in a blast furnace.

The Dust-recycling system

The next important step in the development of the COREX® process was the incorporation of a dust-recycling system.

During devolatilization and gasification of the coal, fines are created that are entrained in the gas leaving the gasifier. On the one hand, this dust represents a considerable loss of carbon. On the other, it has a dramatic influence on the operation of the reduction shaft. Extensive development work was carried out in an effort to find a reliable system that can handle fine dust at temperatures higher than 800°C.

Particular attention was paid to the determination of the most suitable location for the burners in the melter-gasifier. The principal arrangement of the dust-recycling system, as patented, is shown in Fig. 5.

This system not only prevents the build-up of dust, but can have a very beneficial influence on the temperature under the dome and the composition of the reduction gas. A temperature of at least 1000°C is required in the dome to prevent the condensation of hydrocarbons in the reduction gas. The additional energy from the dust burners allows the temperature in the dome of the melter-gasifier to be controlled within a certain range. Hence, the dust-recycling system is also an important factor in pollution-free operation of the COREX® process.

Tests on the above-mentioned improvements were carried out, and the improvementes were patented in 1983 and 1984. At the same time, tests were being done using South African raw materials.

166

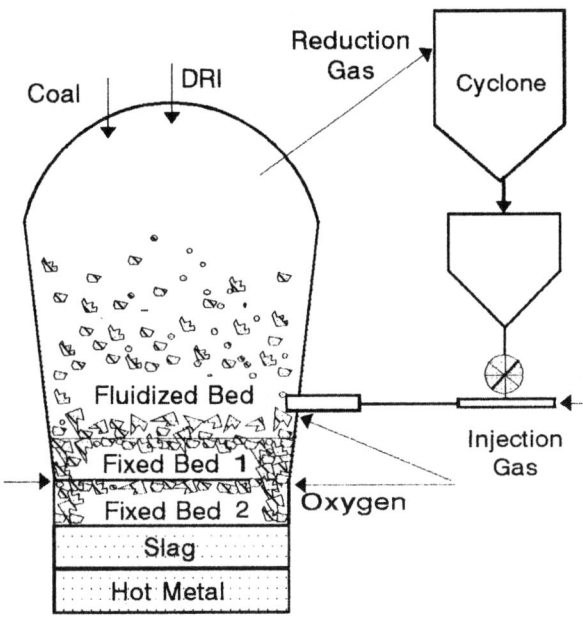

Fig. 5 - The COREX® dust-recycling system

Discharge Screws for DRI

A further patent of great importance in the efficient operation of the
COREX® process is that for the discharge screws for DRI.
This exclusive design was developed in 1980/1981, and the discharge
screws in the pilot plant did not fail during the lifetime of the plant.
This was proved conclusively when the plant was dismantled in 1988, and
no signs of wear were observed.

Reduction shaft

A very important matter was the configuration of the reduction shaft. The
COREX® shaft differs from conventional direct-reduction shafts in several
respects. In the COREX® shaft,

-   gas can contain dust
-   gas contains 60 to 70 per cent carbon monoxide
-   100 per cent lumpy ore is used
-   DRI is discharged by means of screw conveyors
-   fluxes and the ore are both in the burden.

All these aspects had to be born in mind when the layout of the plant was
being planned. An accompanying test was done on a 1:5,5 scale model at
Linz, Austria.

The following parameters were studied:
- the isobars in the shaft
- the influence of the burden on the gas flow
- the influence of the dust load on the gas flow
- the influence of the descending rate of the burden
- methods for "checking" of the shaft.

The results are presented in Fig. 6.

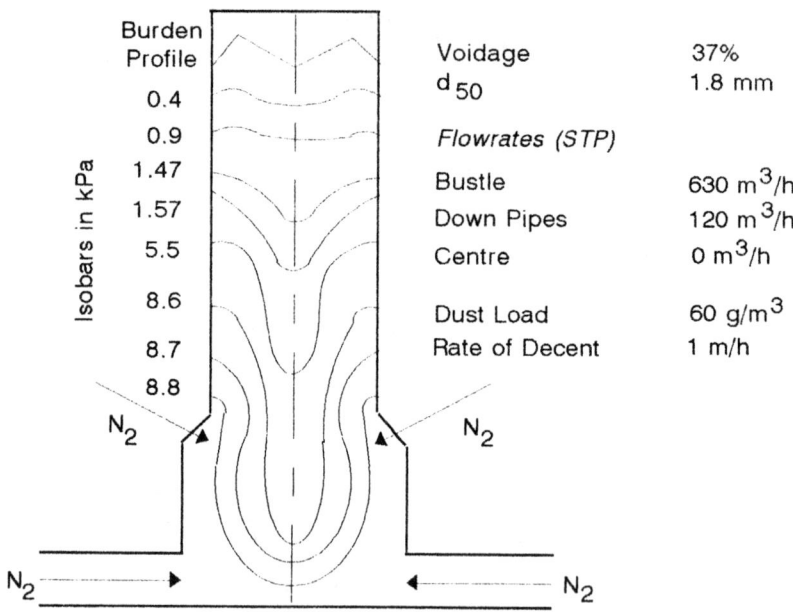

Fig. 6 - Results of tests in the COREX® reduction shaft

EXPERIENCE GAINED ON THE PILOT PLANT

As was mentioned earlier, the pilot plant at Kehl was a semi-industrial plant, which, because it included all the auxiliary facilities, was comparable to a commercial plant.
The modifications introduced as a result of the experience gained on the pilot plant could therefore be adopted with confidence in the scale-up of a production plant. The main improvements in the process were as follows:

- A broad range of coals and ores can be used.
- The dust-recyling system on the pilot plant was optimized, and the optimal location of the dust burner was determined. (The location of the dust burner has a strong influence on the composition of the reduction gas and the temperature in the dome of the melter-gasifier.)
- Reliable gates for the hot gas were developed.
- The gas scrubbers for the cooling gas and the top gas were optimized.
- The start-up procedure which was developed exclusively for the COREX® process, was optimized.
- The tuyeres were optimized.

SCALE-UP OF THE COREX® PROCESS TO 300,000 T/A

It was clear that the risks involved in the scale-up of the COREX® process from the pilot-plant scale by a factor of 5 would be minimal. Accordingly, a COREX® production plant with a nominal capacity of 300.000 t/a was installed at ISCOR's Pretoria Works, and completed in November 1987.

The melter-gasifier was designed in such a way that the gas velocities are comparable to those in the pilot plant. The dust losses therefore remain the same.

The tuyeres are of the same shape as those in the pilot plant, but the diameter of the nozzle was enlarged to accommodate the higher flow of oxygen. The number of tuyeres was increased from 12 to 20.

The hearth was designed after very extensive studies. Because higher specific amounts of slag were to be expected, the design of the hearth differs slightly from that of a blast furnace. Also, the movement of the deadman during tapping had to be considered in the layout.

The principle of dust recycling remained unchanged, but the dust bins and the burners were enlarged.

The cyclones were improved because the cyclones in the pilot plant were too small, and their cones too short, for efficient operation.

The design and number of DRI screws remained unchanged because of the excellent results obtained on the pilot plant, but their discharge capacity was increased.

Scale-up of the reduction shaft presented the biggest problem. Extensive experience had been gained on the pilot plant using pellets, but experience with lumpy ore and sinter was lacking. The diameter of the shaft needed to be increased by a factor of more than 6, whereas the height of the shaft was to be increased by only about 50 per cent. This led to considerable differences in the gas distribution. However, the shaft could be optimized by the use of information gained in tests on the 1:5,5 scale model.

The water system had been tested on the pilot plant by flotation and chemical treatment, and the results of those tests showed that the system was suitable for an industrial plant.

SCALE-UP OF THE COREX® PROCESS TO 600,000 T/A

The successful operation of the production plant with a rated capacity of 300,000 t/a led to the next step in the development of the COREX® process. This step, namely the design of a plant of 600,000 t/a, has already been taken. The main dimensions of the pilot plant at Kehl, the production plant at ISCOR's Pretoria Works and the C2000 of today are compared in Fig. 7.

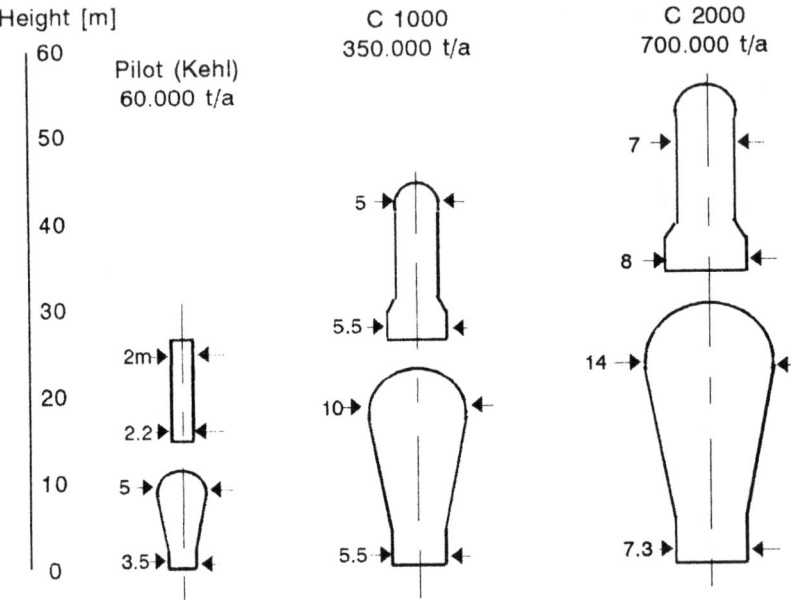

Fig. 7 - Comparison of the main dimensions of COREX® plants at Kehl, C1000 (ISCOR) and C2000 respectively

OPERATIONAL RESULTS

The nominal melting capacity of the COREX® plant at ISCOR is 40 tons of hot metal per hour.

This performance figure was already achieved in the first three months and also exceeded by up to 15 % with pure lump ore charging.

Based on the melting capacity achieved, the nominal performance of 300,000 tons has been reached and even sometimes exceeded.

Campaigns with 100 % CVRD blast furnace pellets, yielded melting figures of over 53 tons of hot metal per hour, exceeding the nominal capacity, by more than 25 %.

These typical production data achieved with 100 % Thabazimbi lump are highlighted in Figure 8 and 9 which indicate that the production rate and the consumption figures are better than anticipated.

The analysis of the hot metal as it is shown in Figures 10 - 15 produced in the COREX® process is identical with that produced in a blast furnace provided that the raw materials do not have an exceptional composition.

The hot metal currently produced according to the COREX® process in Pretoria has an average C content of 4.3 %, a Si content of 0.3 % and a S content of 0.055 %.The N_2 content is substantially lower than that of hot metal from the blast furnace, whereas the tapping temperature is slightly higher.

The originally higher Si and S contents are to be attributed to the higher acid contents of the slag deliberately chosen during the initial phase of operation to obtain a fluid slag, and also represent the successful improvement along with the learning curve of the operators.

At ISCOR the hot metal produced is charged into 125 t electric arc furnaces with scrap at a ratio of 1:1. Steels are produced for high-quality long and forged products.

The COREX® export gas features a mean calorific value in the range of 7000 kJ/Nm^3 and a high degree of cleanliness. It is exclusively used for the heating of metallurgical furnaces of the Pretoria works. As a result, the maintenance expenditure for the burners are reduced, the surface quality of rolled and forged products improved, and the heating times are shortened as compared to the coke oven - blast furnace gas mixture previously used.

The availability of the plant - referring the process and machine-related downtimes was about 95 %. It is to be noted that during the total continuous operation from Nov. 10th, 1989 until October 1991 only 2 out of the 20 tuyeres nozzles had to be exchanged because of damage.

As in a blast furnace, thermocouples or permanent temperature monitoring are installed in different levels of the melter gasifier lining. A computer analysis has shown that the brickwork reached a steady state of slagging within a few weeks of operation. The isotherms calculated from the temperature records have practically remained the same, although the nominal melting capacity was often exceeded during this period.

CONCLUSIONS AND OUTLOOK

In a lecture held at the SRNC '90 Conference at Pohang (Republic of Korea) in Oct. 1990 Mr. R.S. Fruehan, Professor at the Carnegie Mellon University, Pittsburgh, PA (U.S.A.), stated 8 principles which a new direct smelting processes shall meet.

These are the principles that a new direct smelting process shall fulfill:

171

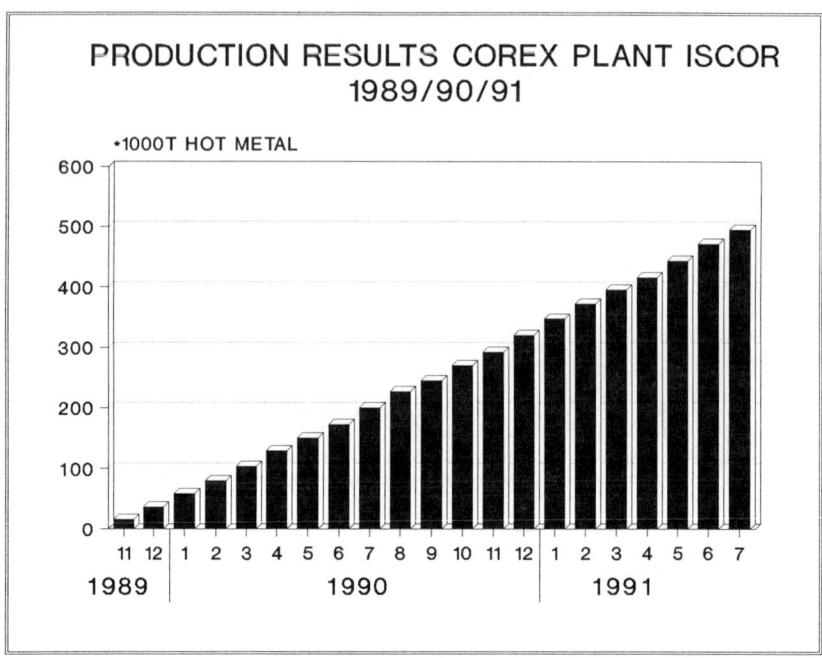

Fig. 8

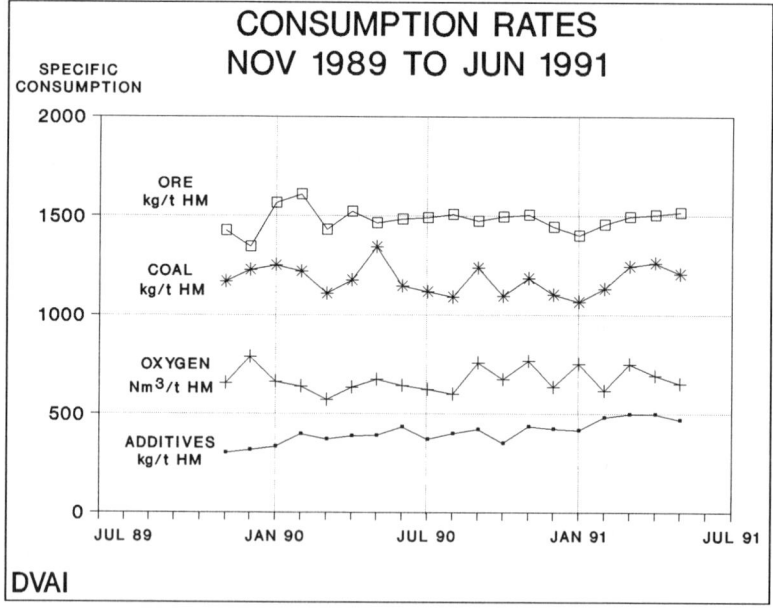

Fig. 9

172

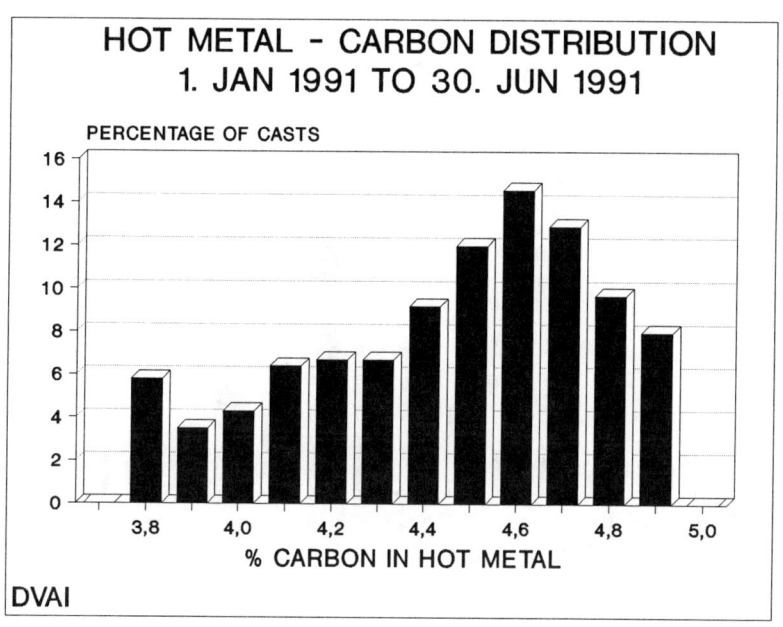

Fig. 10

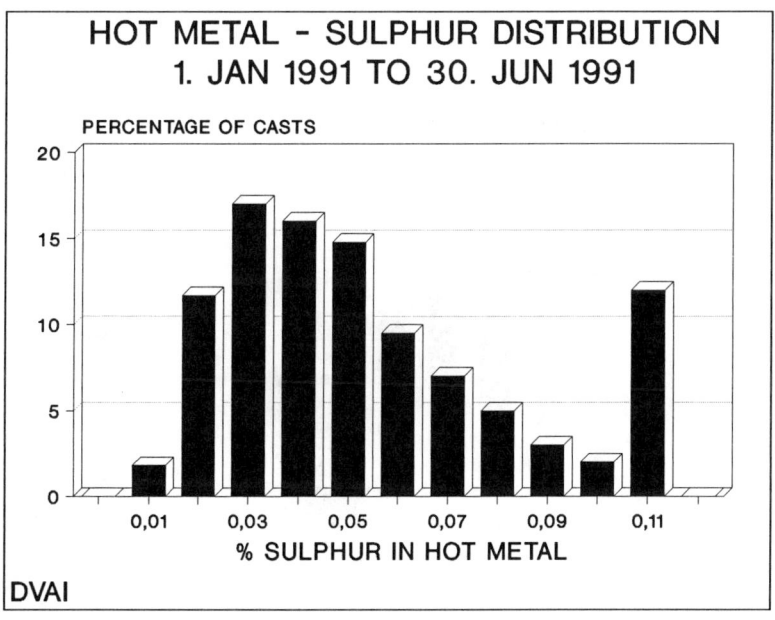

Fig. 11

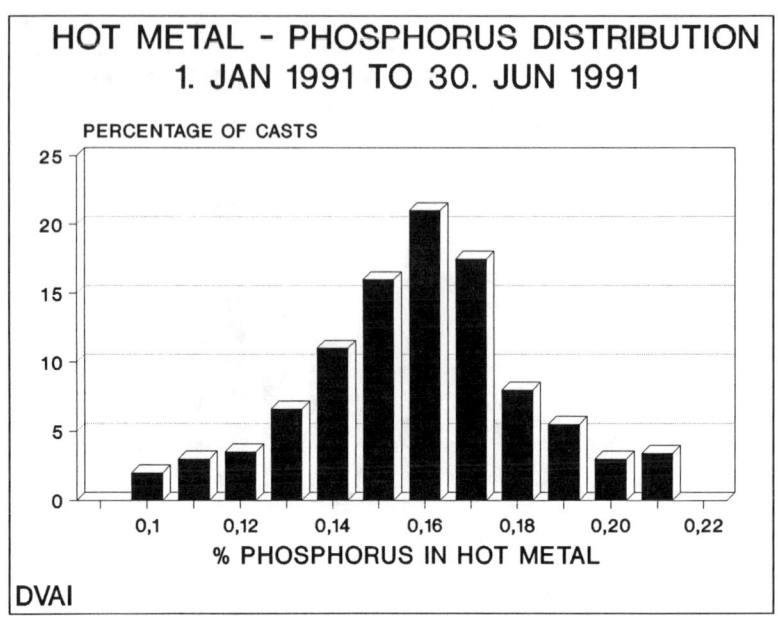

Fig. 12

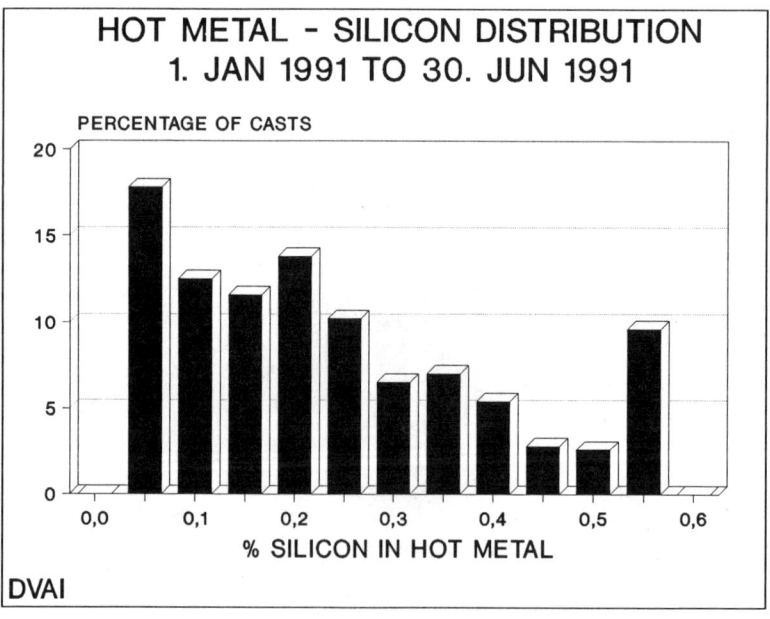

Fig. 13

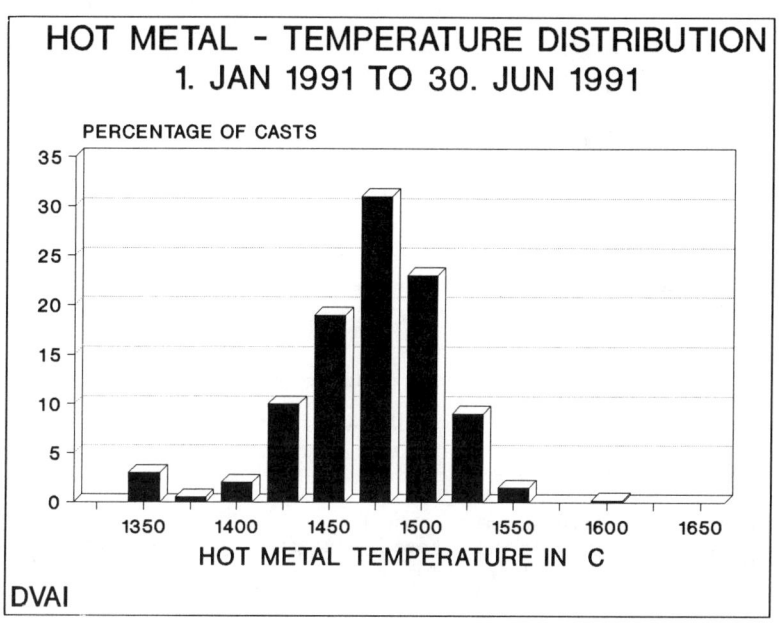

Fig. 14

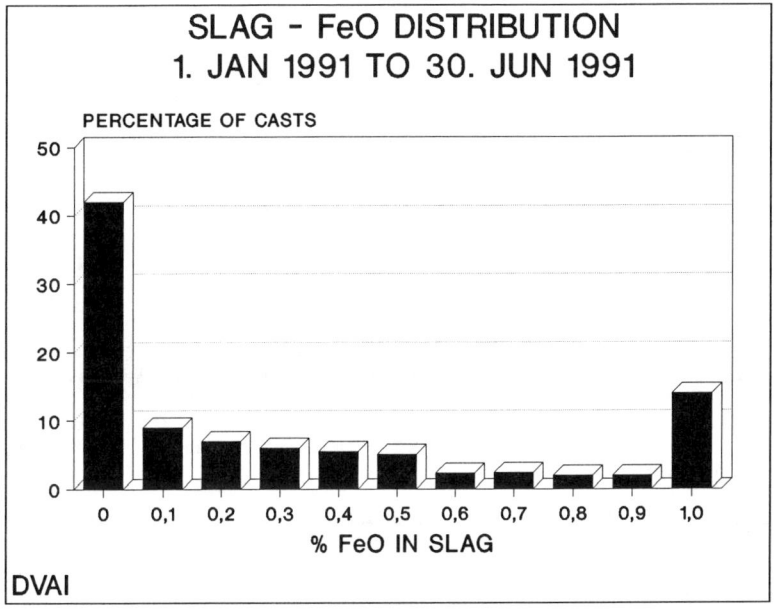

Fig. 15

I. No coke plant necessary.
   - COREX® does not require a coke plant.-

II. High specific melting capacity, i.e. better than a blast furnace
   - COREX® has already exceeded it by approx. 50 %.-

III. No agglomeration, i.e. Sinter or Pellets, necessary
   - Although COREX® can not yet be operated with fine ore COREX®
   allows the 100 % lump ore operation, thus not being absolutely
   dependent on agglomeration plants.-

IV. Low Capital expenses
   - The investment cost for a COREX® plant with an oxygen plant are
   substantially lower than for a conventional coke oven/blast furnace set up.-

V. No coking coal quality necessary
   - COREX® permits operation with a wide range of non-coking coals.-

VI. Low operational cost
   - The operational cost for COREX® is substantially lower than for
   the coke plant/blast furnace route.-

VII. Economical operation of small capacity
   - COREX® allows economical operation already with small scale
   plants like 300,000 t/y. This offers a great potential for mini
   mills wit high quality requirementes for their final product.-

VIII. Flexible operation
   - COREX® allows a higher flexibility than a blast furnace. -

We have completed Prof. Fruehan's principles by two further principles, which we deem necessary not only for reasons of analogy.

IX. High degree of environmental compataibility
   - COREX® avoids environmental pollution at the source of origin.-

X. Hot metal quality corresponding to that of the blast furnace.
   - COREX® produces the same quality as from a blast furnace.-

Nine out of these ten principles as you can see are already fulfilled by
the COREX® technology. We are working on principle III to make the use of
fine ore in COREX® possible.

FUTURE DEVELOPMENTS

Further plans include optimization of the energy consumption, on the assumption that larger units will have lower specific heat losses, which
will result in reduced energy consumption.
A further step will be the design of a plant of the next module size,
namely about 1,000,000 t/a. The COREX® plant will then be the size of
standard blast furnaces.

Apart from the capacity increase, our short and medium-term goals are the
minimization of carbon consumption, the production of different hot metal
grades and ferro-alloys, and the optimization of integrated power systems, e.g. a combined cycle power station.

SUBMERGED SMELTING OF CONCENTRATES IN TENIENTE CONVERTERS
AT CALETONES SMELTER - AN UPDATE

Claudio H. Queirolo

CODELCO-CHILE, División El Teniente
Millán 1040, Rancagua, Chile

ABSTRACT

In order to optimize the Teniente Converter process heat balance, in the year 1988 there started the feeding of bone-dry concentrate through special design tuyeres submerged in the reactor molten bath, replacing almost all wet concentrate (7-8% moisture) added directly over the bath. This new method of feeding concentrate has permitted the Teniente Converter to increase its smelting capacity up to 2000 tons of concentrate per day as well as an important increase of the concentrate/matte ratio until making matte addition unnecessary to balance the process heat requirement. The Teniente Converter operating with bone-dry concentrate is being consolidated as a highly productive technology with low cost investment.

Proceedings of the
Savard/Lee International Symposium on Bath Smelting
Edited by J. K. Brimacombe, P. J. Mackey,
G. J. W. Kor, C. Bickert and M. G. Ranade
The Minerals, Metals & Materials Society, 1992

Introduction

The Teniente Converter is a technology developed at the Caletones Smelter of CODELCO-CHILE, División El Teniente to process copper concentrates and is classified as a bath smelting processes for copper concentrates, with intensive use of oxygen. It was initially developed to extend the capacity of an existing smelter or to improve its efficiency. Since industrial scale commissioning at Caletones Smelter in 1977, very successful results have been attained[1,2], so its application has been extended to other three smelters. Besides, there exists projects to use this technology in two smelters more. [3]

The strengths of this technology are based on its high level of autogenous smelting of concentrates, a large converting capacity, a high and steady sulphur dioxide off-gas concentration and a low amount of suspended dust carried away through the off-gas. The smelters where this technology has been applied have significantly reduced their energy consumption and operational costs and they have improved environmental control, all of which has been attained with a low capital cost investment.

Notwithstanding the success of this process, CODELCO-CHILE División El Teniente has continued making its best efforts in research and development to attain all the potential benefits of this technology. As a result of this work, a new technological innovation called "Submerged Smelting of Concentrates" has been generated. This technology applied to a Teniente Converter permits an increase of its smelting capacity with an optimization of its energetic balance.

Teniente Converter - Reactor and Process Characteristics

The original pyrometallurgical process carried out in the Teniente Converter, shown in Figure 1, mainly consists in the simultaneous converting of copper matte, by blowing air or oxygen enriched air, and the smelting of concentrate using heat generated by the matte converting reactions. This process produces high grade matte or "white metal" (74-76% Cu) and slag (6-8% Cu).

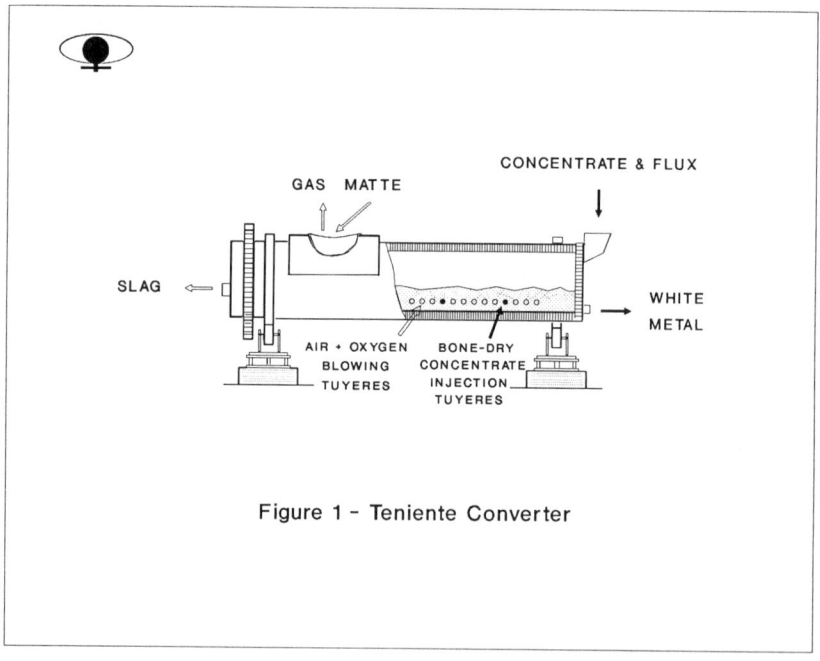

Figure 1 - Teniente Converter

The Teniente Converter is basically a cylindrical reactor, refractory lined, with its off-gas mouth located near one of the vessel end plates . It is provided with a concentrate and flux feeding garr-gun located at the most distant end plate from the off-gas mouth, and slag and "white metal" tap holes located at different levels, one at each end plate. The "white metal" tap hole is located at the reactor garr-gun end plate.

A typical operational practice considers matte addition through the converter off-gas mouth at regular intervals, a continuous feeding of "green" charge composed of wet concentrate (7-8% moisture) and required flux (3-4% moisture) through the garr-gun and the continuous converting of the matte charged, when necessary, as well as matte generated by concentrate smelting.

"Green" charge fed over the molten bath surface is firstly heated and then digested in the bath under agitated conditions produced by the blown air through the tuyeres and afterwards smelted. "White metal", as well as slag, are tapped at regular intervals without interrupting blowing operation and charge feeding.

"White metal" (74-76%) is processed in conversion units (Peirce-Smith, Hoboken or other alternatives) to produce blister copper. The slag generated in the Teniente Converter, which normally contains 6-8% Cu, 25% SiO_2 and 18% Fe_3O_4, is separately treated to recover copper content.

Concentrate Moisture Effect in a Teniente Converter Operation

As described before, the process carried out in a Teniente Converter, conceptually implies the use of heat generated by matte oxidation reactions to smelt wet concentrate (7-8% moisture). Under this conventional operating condition, an important part of heat generated by matte converting reactions is lost, when used only to evaporate and heat the water contained in the concentrate from ambient temperature to 1.200°C., which is the temperature of the off-gases at the reactor mouth. This consideration was the initial reason to study the specific effect resulting by feeding a drier concentrate on the heat and mass balances of the Teniente Converter.

Figure 2 shows in a schematic form, theoretical heat and mass balances of a 5 m ϕ X 22 mL Teniente Converter for a determined operational condition in relation to a specific mineralogical and chemical concentrate composition, to the matte chemical composition and to the blowing air flow and its oxygen enrichment. Said balances are presented as a function of the fed concentrate moisture in the range of 0% to 8%.

From the above mentioned figure, it can be clearly seen the moisture effect on the concentrate smelting capacity of the Teniente Converter. For wet concentrate condition (8% moisture), mass and energy balances indicate that concentrate smelting capacity is 1.217 tpd, being 969 tpd of matte required to thermally balance the smelting and converting process. As concentrate moisture decreases, the reactor smelting capacity is increased up to 1.791 tpd for bone-dry concentrate and on the other hand, matte requirement decreases until almost disappearing. Thus, the
relation concentrate/matte increases significatively while moisture of fed concentrate decreases as less energy is consumed to heat the water contained in concentrate fed to the reactor.

These mass and energy balances, as a function of the moisture of wet concentrate fed to the Teniente Converter, can be also carried out for different mineralogical and chemical compositions of concentrate, blowing air flow rate, and in special, air blown oxygen enrichments, reaching similar trends. Moreover, referred balances indicate that a Teniente Converter can theoretically attain steady operational conditions without matte requirement, to thermally balance the smelting-converting process, by mostly feeding bone-dry concentrate with high oxygen enrichment in blown air.

179

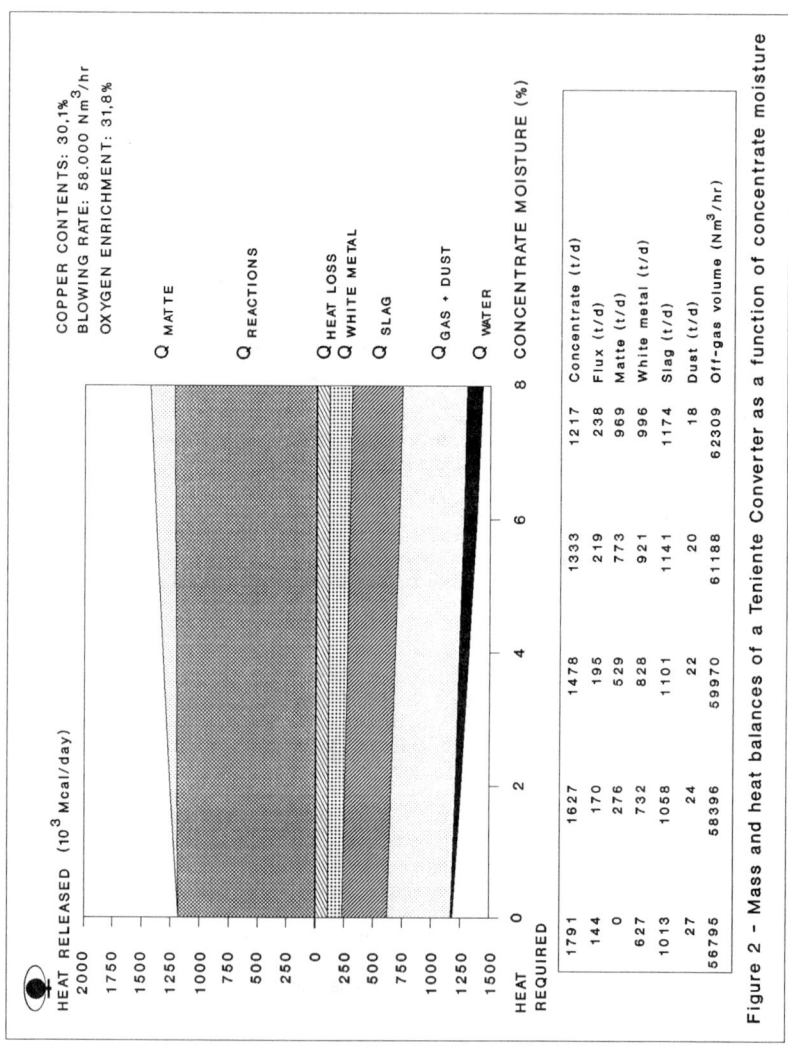

Figure 2 - Mass and heat balances of a Teniente Converter as a function of concentrate moisture

This theoretical analysis derived in the idea of investigating the technical feasibility of feeding bone-dry concentrate to the Teniente Converters in 1981. After having finished laboratory scale tests (1984), pilot tests were initiated in which a 4 m ϕ x 17 mL Teniente Converter was partially fed with bone-dry concentrate[4].

From the beginning of this technological research, it was evident the need to look for a system to feed bone-dry concentrate to the reactor without an important increase of the amount of dust carried over by the reactor off-gas, which could be expected if using the same garr-gun system. It should be taken into account that a low dust carry over is one of the main advantages of the Teniente Converter technology. On the other hand, due to the fact that the drying of concentrate was a technology already available in the market, it was only required to choose an adequate type of dryer. Therefore, pilot scale tests were basically centred in testing different ways to feed concentrate to the reactor, standing out among these, the use of lances and tuyeres with different designs and locations. After a long theoretical and practical work, based on considerations related with dust carried over in the reactor off-gas, a decision was taken in order to use a pneumatic conveying system with specially designed tuyeres intercalated among the air blowing tuyeres.

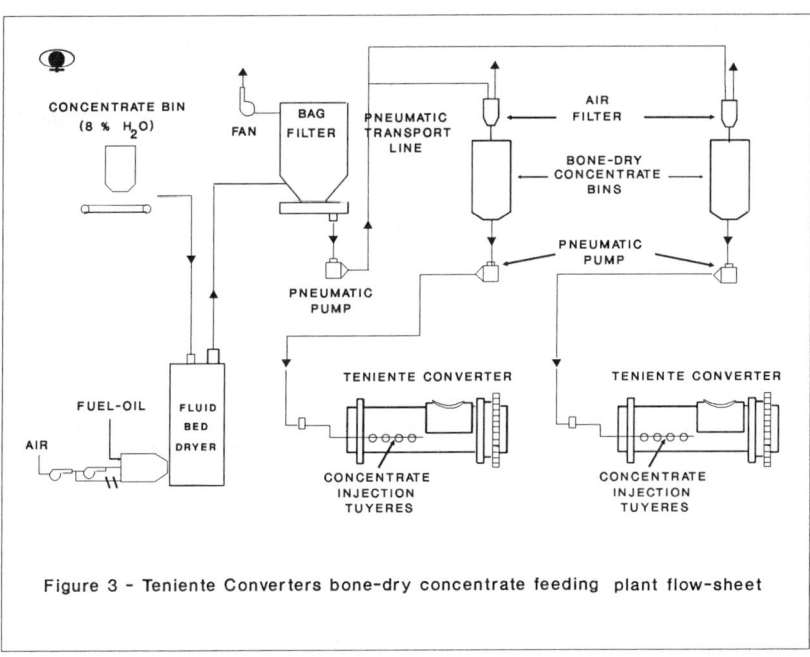

Figure 3 - Teniente Converters bone-dry concentrate feeding plant flow-sheet

Industrial Plant

Once having successfully finished the pilot scale tests and after taking a decision to apply this technological development in a commercial level at the Caletones Smelter, a plant started its operation on July 1988. Figure 3 shows a schematic process diagram of this plant. The concentrate is bone-dried in an 80% tph capacity fluid bed dryer, from 8% moisture down to 0,2% moisture. Afterwards, bone-dry concentrate is pneumatically conveyed from the dryer to feed-bins located close to the Teniente Converters. The concentrate feeding system to each reactor basically comprises dosage equipments, pneumatic pumps, pipe lines, a splitter and the specially designed injection tuyeres.

The first stage of the application of the Submerged Smelting of Concentrates [5] on one of the Teniente Converter , considered to feed only half of the concentrate with 0,2% of moisture through the specially designed tuyeres and the rest in the conventional way, that is, through the garr-gun located at one of the reactor end plates. This operational scheme, which permitted to attain a global moisture equivalent to a 4% approximately, was based on practical considerations making advisable a progressive application of this technological innovation in an important production unit as it was the first 5 m ϕ x 22 mL Teniente Converter commissioned at the Caletones Smelter in 1987.

On January 1990, that is, 18 months after the industrial commissioning of this technological innovation and having overcome some of the initial difficulties, the second 5 m ϕ x 22 mL Teniente Converter was also fed with bone-dry concentrate under the same operational scheme. In this way, the total capacity of the fluid bed dryer was filled and both reactors began to operate similarly. For the Caletones Smelter, this operating scheme represented the second stage of the commercial application of the

Submerged Smelting of Concentrates, permitting thus, an important improvement in its concentrate smelting capacity, with a significant decrease of the specific energy consumption[6,7].

Since June 1991, a third stage is being performed in the development of this technology. A more intensive application of the Submerged Smelting of Concentrate is being put in practice in order to attain the operation of one of the Teniente Converters without matte addition. To this effect, it has been necessary to increase both the oxygen tonnage and proportion of the bone-dry concentrate until thermically balance the process, without making use of heat supplied by the oxidation of added matte. This has been attained by increasing the feeding of bone-dry concentrate over 1.200 tpd which represents at least 75% of the total smelted concentrate in one reactor. It has been necessary to put all the bone-dry concentrate produced by the fluid bed dryer to only one Teniente Converter, so the other reactor returned to the conventional operation, that is, with wet concentrate and matte addition.

Results

The results attained when operating a Teniente Converter under the three mentioned modes can be seen on Table I. First, fed only with wet concentrate, then fed with a mixture of 50% wet concentrate and 50% bone-dry concentrate, and finally the present stage under experiment, in which the Teniente Converter is being fed with a high proportion of bone-dry concentrate through the injection tuyeres and the rest as wet concentrate.

These results have indicated that based on a mixed operation, an increase of near 16% of the autogenous smelting capacity of the Teniente Converter was achieved and matte requirements diminished in 45%, both figures with respect to the conventional operation mode. This means, in practical terms, that the mass ratio smelted concentrate to matte, was doubled. Afterwards, when increasing the proportion of bone-dry concentrate to 87%, the smelting capacity of the reactor increased in 46%, avoiding in practical terms the matte requirements

Figure 4 shows results obtained by the Teniente Converters of the Caletones Smelter operating with intensive application of the Submerged Smelting of Concentrate during the period june 1991-march 1992, in terms of concentrate smelting capacity and matte required. It can be seen that a concentrate smelting rate level between 1600 and 1950 tpd on a monthly basis has been achieved, being matte addition almost negligible. Nevertheless, during this period there are some days in which capacities over 2.000 tpd and with no matte addition have been reached.

The high specific smelting capacity obtained in the Teniente Converter through this technological innovation, has been favoured by the achievement of high reactions velocities, which are attained due to two basic reasons. First, when fed bone-dry concentrate is conveyed in dilute phase by an oxygen enriched air stream, a high specific surface of reaction is achieved, and second by the great energy of mixture generated in the liquid bath by the air jet injected through the reactor tuyeres[8].

Table I - Teniente Converter Operational Results

Input/Output		Operation Mode		
		Wet Concentrate (8%)	Wet/Dry Concentrate (4%)	Dry Concentrate (0,3%)
Feeding				
Bone-Dry Concentrate	(t/d)	--	750	1.572
Wet Concentrate	(t/d)	1.237	690	231
Total Concentrate	(t/d)	1.237	1.440	1.803
Flux	(t/d)	238	192	154
Matte	(t/d)	964	528	57
Concentrate/Matte	(t/t)	1,28	2,73	31,63
Blowing Conditions				
Blowing Rate	(Nm³/hr)	58.000	58.000	58.000
Oxygen Enrichment	(%)	32	32	32
Blowing Time	(%)	90	87,5	92
Products				
White Metal	(t/d)	999	815	668
Slag	(t/d)	1.178	1.079	1.053
Dust	(t/d)	25	22	18
Gases				
Flow rate (mouth)	(Nm³/hr)	61.243	58.488	55.661
SO_2 (mouth)	(% d.b.)	21,3	22,3	23,4

Concentrate Chemical Composition (%)

Cu	Fe	S	SiO_2	Al_2O_3	CaO	MgO	Others
30,1	25,3	31,1	5,8	2,7	1,5	1,5	2,0

Besides the excellent results achieved in terms of smelting and converting capacities, it is important to point out the continuous generation of a gas stream with a high and steady SO_2 concentration, due to the intensive use of oxygen and to the high blowing time of the reactor. Figures obtained are in the range of 21.3% to 23.4% of SO_2 at the off-gas mouth of the reactor, facilitating the sulphur dioxide fixation in an acid plant.

Another attractive result of this technological innovation has been the low amount of dust carried over in the reactor off-gases. When operating the Teniente Converter with a high proportion of bone-dry concentrate, measurements performed have indicated that carried out dust is less than 1% of the total fed concentrate. This figure represents by itself an outstanding advantage of this technological development in comparison to other pyrometallurgical processes that treat bone-dry concentrate.

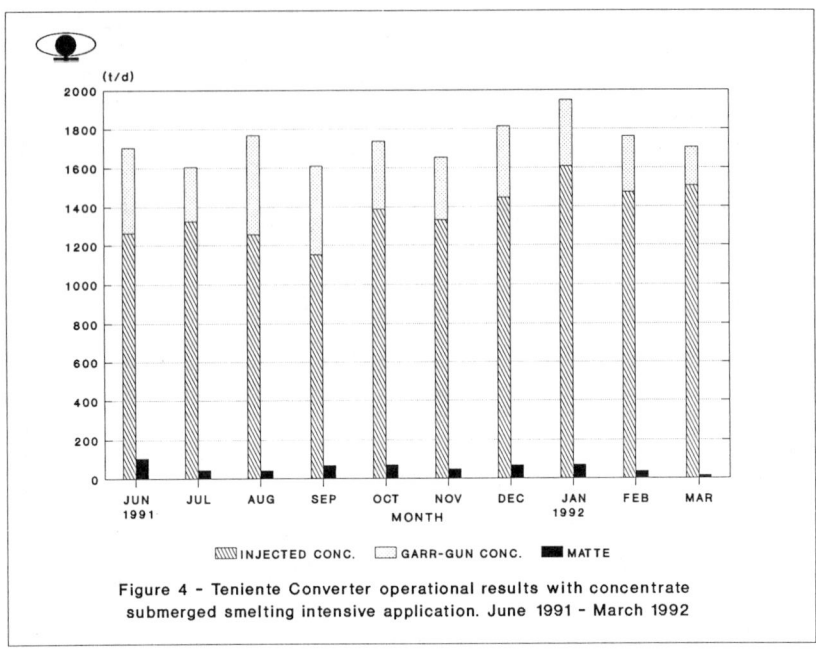

Figure 4 - Teniente Converter operational results with concentrate submerged smelting intensive application. June 1991 - March 1992

Nevertheless, this technological development has not been free from difficulties in its scaling up from pilot to commercial level and to overcome them, an important amount of resources in technological and operational research have been spent. Efforts have mainly been directed to ensure a uniform distribution of the concentrate among the injection tuyeres, to the use of abrasive highly resistant materials, to define adequate control operational strategies and to determine the best maintenance programs for the dry concentrate conveying and handling system, among other aspects. The experience achieved during these three years of industrial operation have meant an important contribution to solve faced problems.

Future Trends

After the stage where a large concentrate smelting capacity without matte requirement has been attained, the future development of this technology aims to ensure the operativeness of a Teniente Converter as a primary smelting unit, that is, to enable its operation in a smelter without any previous smelting units, such as reverberatory furnaces, blast furnaces or others. To attain this goal at the Caletones Smelter, it will be necessary for the Teniente Converter to process a high concentrate tonnage as well as part of secondary materials that are generated throughout the smelter, especially those with low copper content that are not processed in the conventional converters. With this purpose, División El Teniente has started studies and plant tests to determine levels of blowing rates, oxygen enrichments and concentrate moisture contents that should make this goal possible.

Figure 5 shows calculated smelting throughputs of a 5 m ϕ x 22 mL Teniente Converter for a defined blowing air rate, as a function of oxygen enrichment of the blowing air and the moisture of fed concentrate. It can be seen that by increasing the oxygen enrichment and decreasing moisture content of concentrate, the smelting capacity increases and matte requirements diminishes until reaching a point in which matte addition is not necessary to thermically balance the process. Furthermore, for less moisture content in the concentrate and higher oxygen enrichments in the blowing air, there is an excess of heat generation that should make it possible to smelt secondary materials. At present, the operational range indicated at the right side of Figure 5 is being a matter of deeper studies.

184

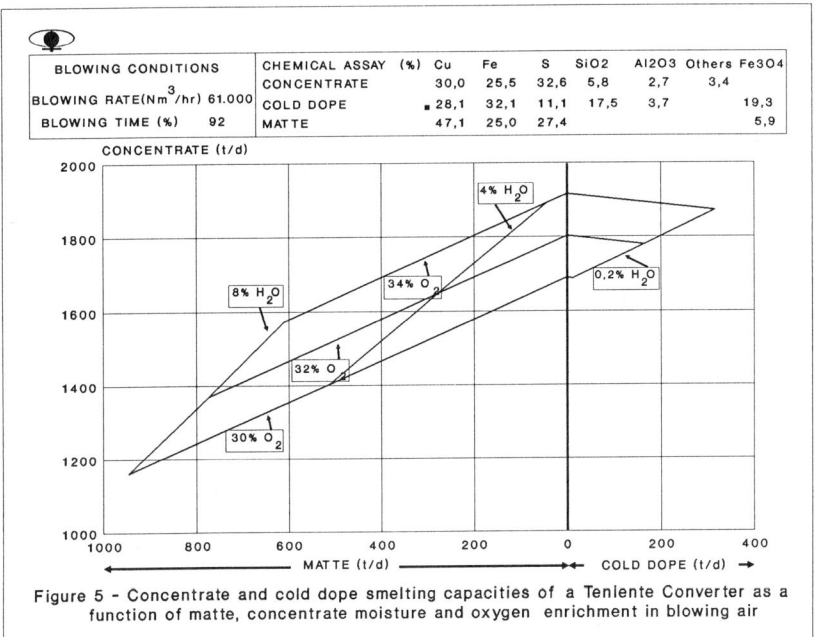

BLOWING CONDITIONS	CHEMICAL ASSAY (%)	Cu	Fe	S	SiO2	Al2O3	Others	Fe3O4
	CONCENTRATE	30,0	25,5	32,6	5,8	2,7	3,4	
BLOWING RATE(Nm³/hr) 61.000	COLD DOPE	28,1	32,1	11,1	17,5	3,7		19,3
BLOWING TIME (%) 92	MATTE	47,1	25,0	27,4				5,9

Figure 5 - Concentrate and cold dope smelting capacities of a Teniente Converter as a function of matte, concentrate moisture and oxygen enrichment in blowing air

Figure 6 shows theoretical smelting capacity of concentrate and secondary materials of a Teniente Converter operating without matte addition under the same operational parameters than those in Figure 5, for different levels of oxygen enrichment in the blowing air and of concentrate moisture. This graphic shows that the Teniente

Converter has the capacity to smelt an important amount of secondary materials, that can reach up to an equivalent of 15% of the fed concentrate, without affecting the autogeny of the process. A sequence of practical tests is going to be carried out soon at the Caletones Smelter to confirm these theoretical results.

In this way, the submerged smelting technology is rapidly advancing to the specific aim proposed at the beginning of its development and thus, the Teniente Converter developed firstly to retrofit conventional smelters with an existing smelting unit for matte generation, is evolving towards a primary smelting-converting technology, adequated to partially or totally replace primary smelting units of existing smelters or to be considered in new smelter projects.

Due to the results attained and to the projection of this technological innovation, División El Teniente of CODELCO-CHILE has started a project to install a second fluid bed dryer of 100 tph capacity, in such a way that in the short term both Teniente Converters could be simultaneously fed with bone-dry concentrate.

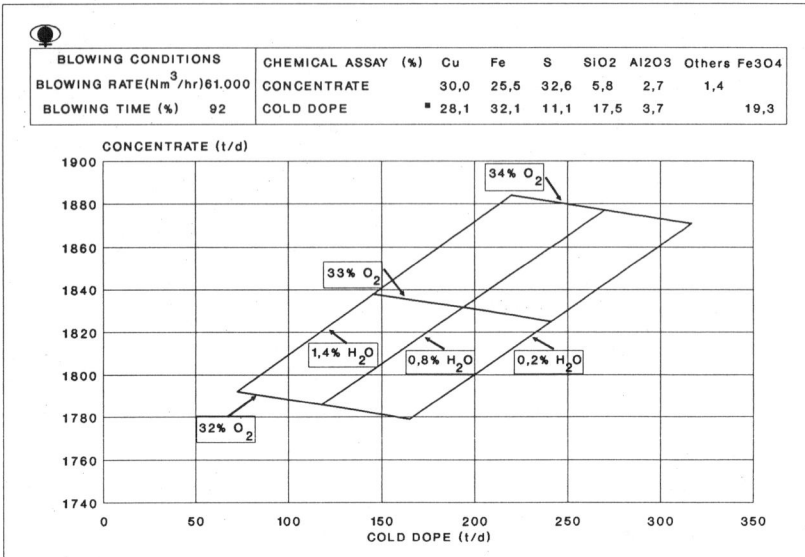

Figure 6 - Concentrate and cold dope smelting capacities of a Teniente Converter operating without matte, as a function of concentrate moisture and oxygen enrichment in blowing air

Conclusions

Mass and energy balances of the smelting-converting process carried out in a Teniente Converter, indicate that when reducing the moisture content of the concentrate charged to the reactor an increase of its autogenous concentrate smelting capacity is attained, and consequently decreasing matte requirements to thermically balance the process.

The feeding of bone-dry concentrate to the Teniente Converter through submerged tuyeres has shown its technical feasibility at commercial scale in the new 5 m ϕ x 22 mL reactors installed at Caletones Smelter. During the first industrial application stages, that considered 50% of bone-dry concentrate fed through tuyeres and the rest by means of the conventional way, the Teniente Converters attained an increase of a 16% of their smelting capacity and a decrease of 45% of its matte requirement.

A more intensive application stage of this technological innovation, which is presently in a very advanced consolidation stage on commercial scale, is practically resulting in an increase of 46% of the Teniente Converter smelting capacity, achieving up to 2.000 tons/day, with an almost negligible matte requirement.

The Teniente Converter running with the Submerged Smelting of Concentrate is evolving towards a highly productive technology fitted to be considered in smelters that need to completely discontinue the operation of low efficiency primary smelting furnaces or in a project for a new smelter, involving low capital investment and operating costs.

References

(1)   Campos R. and Queirolo C., "Improvements to the Oxygen Smelting Process at Caletones Smelter", 108th AIME Annual Meeting, New Orleans, USA, 1979.

(2)   Vera G., Campos R., "Codelco-Chile Copper Concentrates Smelting Technologies", Extraction Metallurgy, London, September, 1985.

(3)   Campos R., Strachan M., Cross A., "Teniente Converter Technology at Caletones and Nkana Smelters". TMS Annual Meeting, New Orleans, USA, February 1991.

(4)   Vera G., "Recent Developments in the Teniente Modified Converter Operation and in Converter Slag Cleaning at the Caletones Smelter". Pyrometallurgy 87, The Institution of Mining and Metallurgy, Pag. 1031-1045. London 1987.

(5)   Queirolo C., Torres L., Tapia A., "Commercial Operation of Submerged Smelting of Concentrates in Teniente Converters at Caletones Smelter", MMIJ/IMM Joint Symposium, Kyoto, Japan, 1989.

(6)   Queirolo C., "Present Status and Future Out Look of the Submerged Smelting of Concentrates in Teniente Converters", 41 Annual Meeting Chilean Institute of Mining Engineers, Santiago, September 1990.

(7)   Campos R., Achurra J., Rojas O., "Teniente Converter: A Leading Pyrometallurgical Technology", International Conference Copper'91, Ottawa, Canadá, August 91.

(8)   Mackey P.J., "Trends in Copper Processing", Metals Week Copper Conference, January, 1991.

APPROACHING TWENTY YEARS OF OPERATION OF THE NORANDA

PROCESS REACTOR AT THE HORNE DIVISION OF NORANDA MINERALS INC.

Yves Prévost
Process Metallurgist
Noranda Minerals Inc. - Horne Division
101, avenue Portelance
Rouyn Noranda, Quebec
J9X 5B6

Abstract

The Noranda Continuous Smelting Process at the Horne Smelter was commissioned in March 1973. Since that time, the capacity of the vessel has been increased by a factor of about four and it is now the Horne's principal smelting unit. This evolution has been achieved by a steady stream of technical developments that has allowed the vessel to smelt several types of metal bearing materials while maintaining an adequate control of metallurgical parameters.

This paper presents an overview of the main Noranda reactor improvements that have occurred and the effects they have had on the capacity and control of the Noranda Process.

Proceedings of the
Savard/Lee International Symposium on Bath Smelting
Edited by J. K. Brimacombe, P. J. Mackey,
G. J. W. Kor, C. Bickert and M. G. Ranade
The Minerals, Metals & Materials Society, 1992

Introduction

The Horne smelter was commissioned in 1927 to smelt 910 mtpd of ores and concentrates, mainly produced by the newly discovered Horne Mine. The original smelter operation used eight multiple hearth roasters, two reverberatory furnaces and two converters. By 1964, a decision was taken to develop a new technology that would allow the production of copper from concentrates in a single vessel, while producing a low copper content slag and a concentrated SO_2 gas stream. Developed in laboratory tests at the Noranda Research Center, the technology was piloted at the Horne Division during 1967-1972[3]. The full scale Noranda Continuous Smelting Process, also known as the Noranda Reactor, was commissioned in March 1973. Since then, continuous improvements have allowed the instantaneous feed rate to increase from the design capacity of 726 mt NRPM (metric tonnes of new revenue producing material) per day to the present input in excess of 2700 mt NRPM per day.

These improvements also permitted an increase of the campaign life, a reduction in the fuel ratio and better metallurgical control. At the eve of the reactors' 20th anniversary, this paper will review the major changes that have occurred at the reactor and their effects on production.

General Description of the Noranda Continuous Smelting Process

The Noranda Process has been described previously[1-2]. The reactor is a 5,2m diameter by 21,3 m long refractory-lined cylindrical vessel. Copper bearing feed, flux and coal, used as a fuel, are fed into the vessel using a high speed slinger while oxygen enriched air is blown into the liquid matte using a series of fifty four submerged tuyeres.

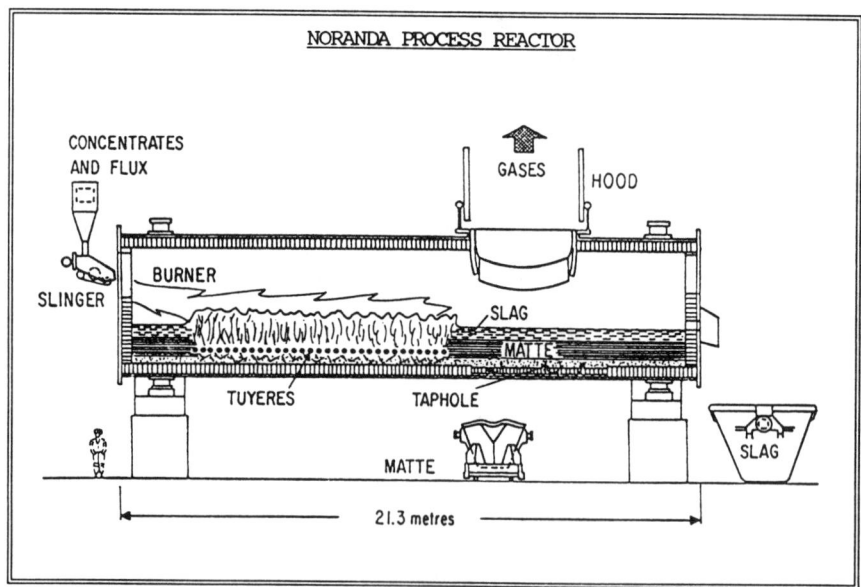

The relatively high blowing rate(59,000 to 76,000 Nm³/hr) produces an excellent mixing of the liquid matte, that is almost homogeneous along the length of the reactor. The effect of the high blowing rate combined with the even spreading of the concentrate by the slinger allows the feed to be rapidly absorbed and oxidised within the molten bath. Most of the heat required to melt the feed is generated by the oxidation of sulphur and iron. The rate of coal addition is influenced by the composition and rate of feeding and the oxygen enrichment of the blowing air. Operating without coal or natural gas for hours has become a routine operation. However, coal is commonly used to improve the capacity of the reactor to melt low energy materials.

A settling zone of 8,9 m in length between the last tuyere and the slag end of the reactor allows the separation of the slag and the matte. The slag is tapped from a taphole located in the end plate opposite the feed end plate. It is tapped into 33 mt capacity ladles and sent to the slag cooling area where it is water-cooled for 29 hours. It is then broken and sent to the concentrator where this 5.5 % copper content slag is processed to produce a 38 % Cu slag concentrate and a 0.35 % Cu tailing. The slag concentrate is recycled to the reactor and the tailing is discarded to an impoundment area.

The matte grade target may be varied from 55 % to 75 % copper when operating on the matte mode (production of copper sulphides). The present matte grade target is 70 % Cu. Matte is tapped from a taphole located on the barrel of the vessel. It is tapped into 18 mt capacity ladles and sent to the converter aisle where the smelting is completed. Refined copper is cast as anodes which are sent to Noranda's CCR refinery in Montreal.

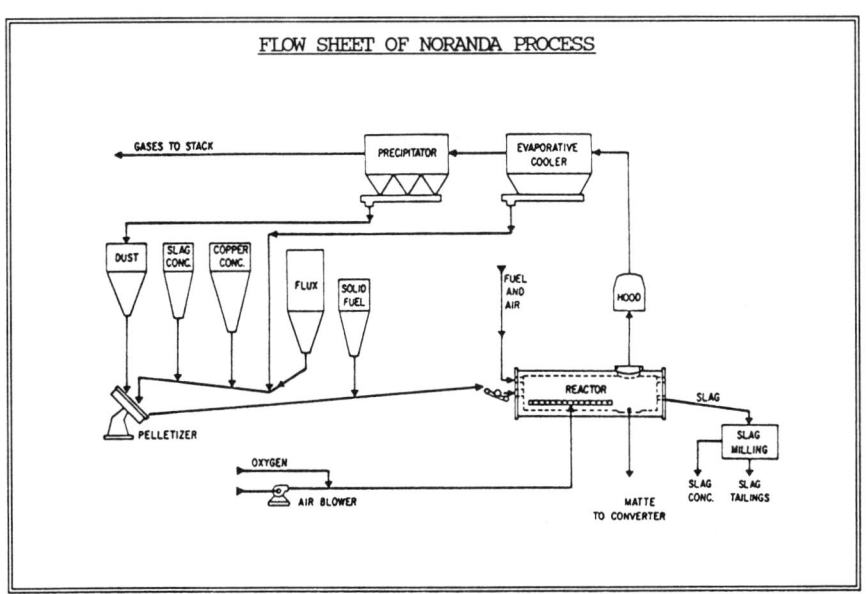

The outlet gases are collected by a hood, water-cooled in a spray chamber and cleaned in electrostatic precipitators. Most of the dust is captured by the electrostatic precipitators and recycled to the reactor. A small fraction of this dust is sent to Brunswick Mining and Smeltings' lead smelter in Belledune where the dust is processed. The SO_2 from the gases is treated in an acid plant to produce sulphuric acid.

The Development of the Process[3]

The origin of the Noranda Process lies in an initial objective at the Horne in the 1960's to decrease the copper losses in the reverberatory furnace slag [3]. Soon afterwards, other reasons emerged to spur the development of a new method to produce copper:

-       To reduce the matte and slag transfer between the reverberatory furnaces and the converters.

-       To provide for a method to eventually increase the strength of the SO_2 in the off-gases.

-       To improve the heat efficiency of the process by using the energy generated by oxidation to melt the feed.

-       To replace the existing batch process by a more efficient continuous process.

At an early stage of the development, it was established that the reactor would essentially resemble an elongated Pierce-Smith converter. By using a technology that was already known as a base, the developers were able to concentrate on the metallurgy of the process in the knowledge that the mechanical features of the vessel would be sound.

The following years were mainly focused on testing the new process in both a small modified reverberatory experimental furnace with a capacity of 2 tpd and a pilot plant fabricated from an existing converter, having a capacity of 100 tpd. Supporting cold modelling studies were conducted concurrently.

In March 1971, the decision was taken to build the prototype plant and two years later, in March 1973, the prototype plant commenced operation.

The first reactor campaign lasted 59 days, with a blowing time of 69.4 %, while producing blister copper. Copper production continued for a total of 10 campaigns during two years before switching to the high matte grade mode. One of the key metallurgical improvements that occurred during that period was the use of solid fuel added with the feed to control the heat balance of the vessel. Prior to the use of coal, the temperature control was performed only by the feed end and slag end burners; such a technique had limitations and did not easily allow the smelting of significant amounts of low-energy content materials and at times the flame damaged the bricks. Both coal and petroleum coke were used as solid fuel. It was eventually found that petroleum coke, because of a lower ash content, allowed for the production of a better quality slag.

The year 1975 was an important year for the Noranda Reactor. In order to improve the minor element elimination, it was decided to produce a high grade matte at 75 % Cu rather than blister copper [1]. The control of the matte grade was then based on matte sampling instead of the level of copper in the bath.

During that same year, a small second hand oxygen plant was installed, having a capacity of 82 mtO$_2$/day. While increasing the reactor capacity by virtue of a greater oxygen input, the use of oxygen enriched air of up to 25 % O$_2$ also allowed a decrease in the fuel ratio.

It was also in 1975 that ladles were first used for slag cooling replacing the pit cooling technique previously adopted. These ladles enabled a more controlled cooling resulting in improved metal recovery in the slag flotation process reduced the risk of slag explosions. A more definite matte/slag segregation was also achieved, allowing a better recovery of matte and an easier handling of the slag.

During the next few years, the two burners were shut down and a small gas lance was introduced to keep the feed port open. This reduced the brick wear above the bath level.

By 1982, the reactor was smelting 1300 mt of NRPM daily per 24 hours of production, for a total charge of 1731 mt\day. A new 463 mt/day oxygen plant had been commissionned, as well as three new feed bins giving a total of six. These changes occurred under the program Horne Metallurgical Improvement Project (HMIP), designed to improve the efficiency of the Horne smelter during a period of concern for the future of fossil fuel prices.

The newly available oxygen allowed a significant increase of the reactor capacity and a decrease of the fuel ratio as predicted. It was also an opportunity to build a control strategy based on the oxygen demand of each material and their proportion in relation to the total reactor feed, leading to a better control of the matte grade. This control, combined with the ability to mix different materials in the six available bins, is the basis of the reactor's flexibility to handle a wide variety of materials.

A year later, a two wavelength pyrometer (see drawing on next page), developed by Noranda was installed at the reactor for temperature control[4]. Using a fiber optic to transmit the light emission from the liquid matte to a detector, the patented tuyere pyrometer allowed the continuous monitoring of the melt temperature via the tuyeres, where the oxidizing reactions makes the temperature critical for the refractory lining. The operators became so confident of the new instrument that the hood radiamatic pyrometer, previously used for the temperature control was eventually removed while a second tuyere pyrometer was installed as a back-up.

Over the next couple of years, Horne personnel improved established techniques for minor element control and the smelters' ability to handle increasingly complex feeds. Combining the efficiency of the last remaining reverberatory furnace to eliminate minor elements such as bismuth, coupled with the efficiency of the reactor in regards to other elements such as nickel, control recipes and guidelines were created and applied. It became possible to treat an increasing quantity of minor elements and by directing feed sources based on composition to the respective smelting vessels, a high quality copper could be produced. Most of these techniques were later proven to be of considerable value when eventually the sole remaining reverberatory furnace was shut down in 1989.

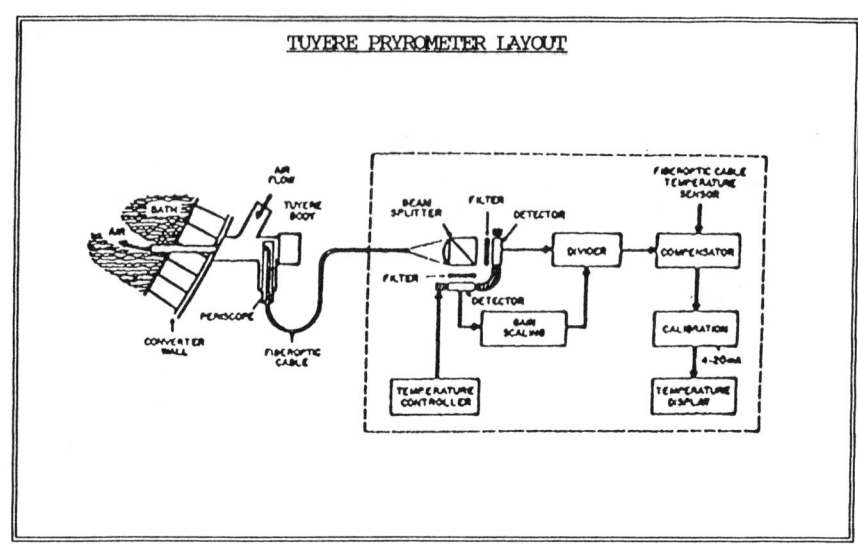

TUYERE PRYROMETER LAYOUT

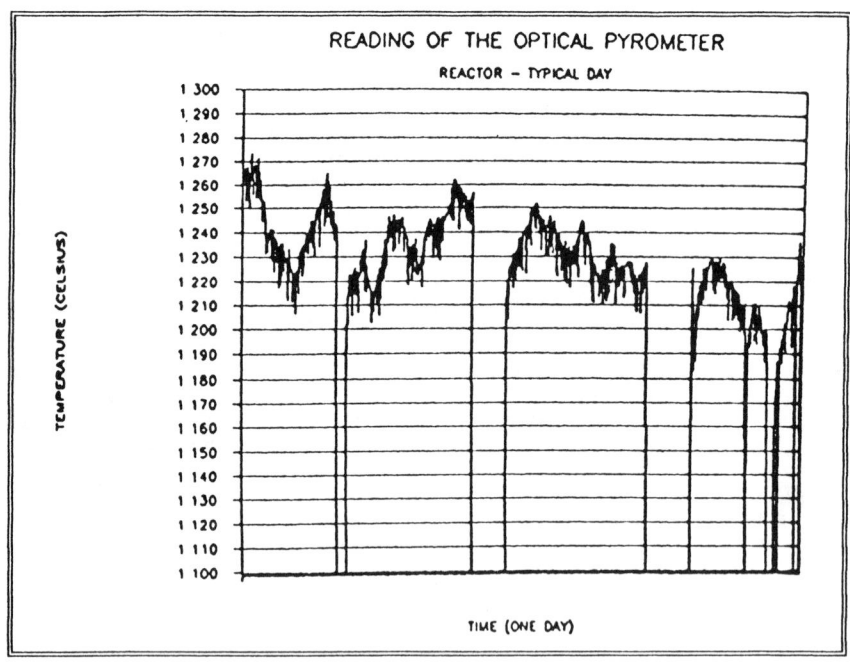

READING OF THE OPTICAL PYROMETER

REACTOR – TYPICAL DAY

By 1988, the reactor had already reached two and a half times the original design capacity and further increases were projected. Unfortunately, the control strategies remained the same as in 1982 and were not adapted to these high production rates. Concurrently, the campaign length of the vessel began to decrease. From a maximum 400 day campaign in 1984, the campaign life dropped to 150 days by 1988. Most of the shut downs occurred due to brick wear around the tuyere line.

Controlling the reactor at the end of the 80's

A task force was formed with the objective to improve both the control of the reactor and increase campaign life. Physical and metallurgical areas of control were studied. The tuyere pressure and blowing rate were the main physical problems considered. Both of these were solved in 1988 when the first automatic Gaspé puncher replaced the previous manual Gaspé puncher. Controlled by a computer, the puncher will start or stop punching every time the blowing rate or the pressure is outside the established limits, or will punch with a predetermined sequence. The computer controls the movement of the machine, allowing it to position itself in front of the tuyeres and unblock them. The gain in stability of the blowing rate became immediately visible.

The introduction of punching bars with a flat instead of round head occurred at about the same time. These flat head bars allow the blowing air to flow over the sides of the bar head when the bar is retracted from the tuyere, while with the round head bars, the non-aerodynamic head creates a vacuum that tends to suck the liquid matte into the tuyeres. The flat bars were therefore less efficient for punching (the head surface is smaller) but a lot better for the tuyere life.

Three different metallurgical controls were identified as key to good reactor control; the matte grade, temperature, and slag fluidity. A fourth one, minor elements control, was not deemed critical since it had no effect on the tuyere line wear.

These three controls are obviously linked together. To understand the links involved, a certain knowledge of the reactor control is necessary.

The Horne reactor is always expected to operate at maximum production. Therefore, a maximum blowing rate and the use of all the oxygen available is usually required. It is therefore non productive to vary the oxygen flowrate to control the matte grade if other more productive approaches are available.

Since one has to maintain the balance between the total oxygen demand of the feed and the oxygen blowing rate in order to control the matte grade, and if the oxygen is to be maintained at a maximum, then any change of the matte grade will have to be followed by a change of the feed rate in order to alter the total oxygen demand. The matte grade can also be controlled by varying the proportions of the different materials in the feed, allowing a change of the total oxygen demand with a constant feed rate. The decision to use one or the other will usually have to be made by the operator, using his knowledge and experience, but supplemented with daily advice from the reactor metallurgist.

There are several ways to vary the major operating parameters - matte grade temperature and the slag quality. They are listed in the table below, along with the effects on the other metallurgical parameters.

PARAMETER	CHANGE	EFFECT
Change of matte grade	Modify feed rate	-A higher feed rate will cool the vessel -A higher feed rate means more silica flux. The iron/silica ratio of the slag is not maintained. -A cold slag with a incorrect Fe/SiO2 ratio is more viscous.
	Modify proportions of feed	-Even if the feed rate is the same, some materials are more exothermic than others. The temperature will vary. -The flux feed rate will have to be readjusted to maintain the Fe/SiO2 ratio.
	Decrease air/O2 blowing	-The production is not optimum. -Less reactor heat is generated. The temperature will vary.
Change of temperature	Increase coal feed rate	-The oxygen used by the coal is not available for the feed. The feed rate must be modified. -Less FeO will be produced. The flux rate must be modified.
	Decrease air enrichment	-This is done by decreasing plant O2. Once again, the feed rate and the flux rate must be modified. The plant O2 not used is wasted. The operation is not optimum.
	Increase coolant feed rate	-The best solution when the coolant is available and the converters can smelt the extra production. Coolant will not affect the matte grade nor the slag Fe/SiO2 ratio.
Change of slag quality	Change temperature	-A higher temperature makes for a more fluid slag. On the other hand, it is harder on the refractory and less energy efficient. This change is controlled by one of the ways described above.
	Change flux feed rate	-A correct iron/silica ratio for a fluid slag is between 1,6 and 2,0. The flux requires energy to smelt. An increase of the flux feed rate will decrease the temperature.

A decision was taken to focus only on one of the parameters: the matte copper grade. It was expected that a more stable matte grade would reduce the need to change the feed rate and, therefore, would also smooth out the temperature and slag quality variations. In addition, it would improve minor element control and converter operation by producing a matte with more uniform characteristics.

The approach used was to create an oxygen balance around the reactor. Using a computer, the balance would be recalculated each time a factor would be changed (oxygen availability, coal feed rate) and would suggest to the operator the best feed rate to maintain the matte grade[5] constant. Because the computer would be linked with the instrumentation panel, it would use the actual feed rate and total oxygen blown to perform the calculations. It would also allow to make corrections for the deviations between the target and the actual rates.

A decision was also taken to allow the operator to decide wheather he would or not accept the computer suggestion. The operator could therefore override the computer suggestions, if knowing additional operating information unavailable to the computer, such as a mechanical failure in part of the feed system or an inadvertant mix up with the feed materials.

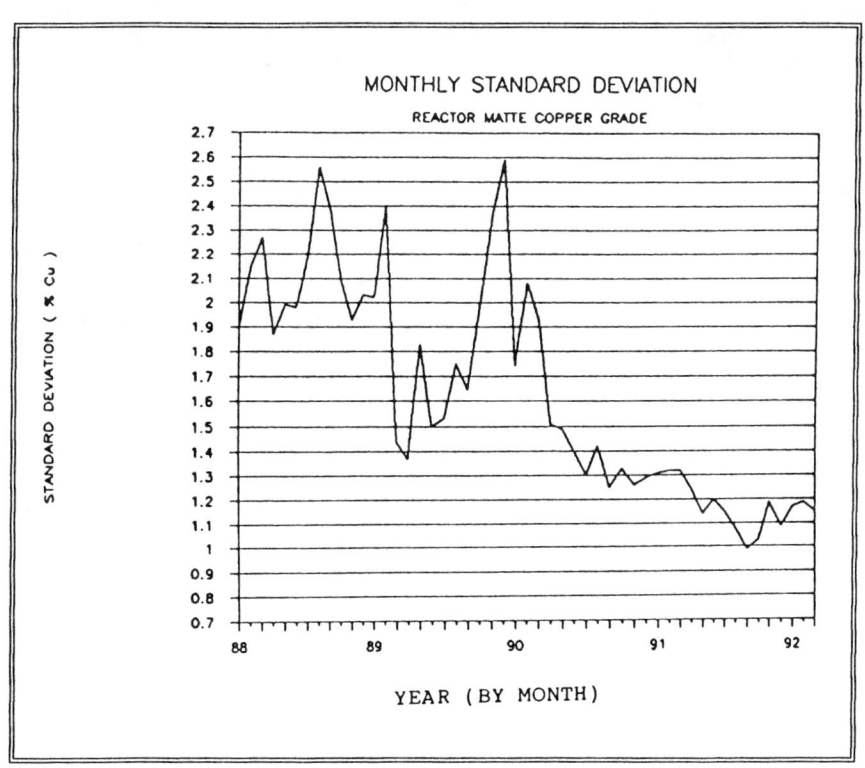

In January 1990, after a one day course was given to all operators and foremen on how to use the computer, the system was installed at the reactor and was followed for three weeks with shift technical supervision. The hardware adopted was simply a lap top computer with a monochrome monitor linked to the existing microprocessor PM 550. The software was based on a algorithm written in C language.

Even though operators were experiencing an on-line computer for the first time, this use of an advisory computer was immediately accepted. During the first few weeks, the acceptance rate of the computer suggestions exceeded 70 %. Eventually, the operators became more confident using the computer and started making fewer input mistakes. Concurrently, metallurgical staff involved with reactor control used the data memorized by the computer to modify the program in order to limit the risks of computer misuse.

One year after, the installation of the system, the acceptance rate of the computer suggestions was 90 % and has remained in the 90-95 % range. From an original standard deviation of the matte grade of 2,11 in 1988 and 1,87 in 1989 (when the sampling rate was increased from every hour to every half hour) the standard deviation decreased to 1,50 in 1990 and to 1,20 in 1991 and for 1992 to date. Concurrently, the campaign life of the reactor increased to 330 days and shutdowns are no longer due to tuyere wear. At the present time, achieving a 400 day campaign life is no longer the primary priority. Rather, reactor shutdowns are now planned ahead to minimise costs and metal inventories. Indeed, the tonnage smelted during the last campaign was 40 % higher than ever achieved during a previous reactor campaign.

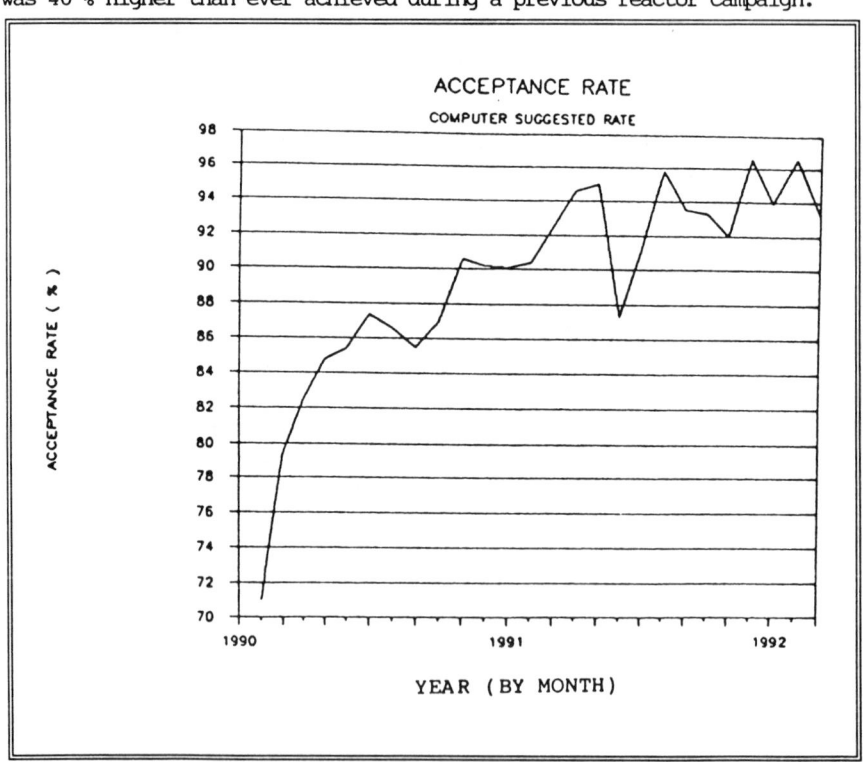

The 90's: Reactor Only

In September 1989, the temporary closure of the last reverberatory furnace at the Horne made the reactor the sole vessel at the plant for concentrate smelting. This closure was made necessary due to the low energy efficiency of the reverberatory furnace, the increasing smelting capacity of the reactor, the inability to produce a high strength SO_2 gas stream from the furnace and inadequate gas cleaning capacity.

The reverberatory closure allowed the reactor to use the 90 tons of oxygen previously consumed in the reverberatory furnace. Once again, the effect on the reactor was to increase both the production rate and the fuel efficiency of the vessel.

At the end of 1989, the Horne acid plant was commissionned treating reactor gas, allowing a reduction of the SO_2 emissions by more than 50 % compared to that in 1980. Noranda also announced the reduction would be of 70 % by the year 1995. This reduction implied that the reactor would remain the main smelting vessel. However, the reliance on a single smelting vessel was judged to be imprudent. For this reason a system of concentrate injection in converters was developed and commissionned in December 1990.

This process involves drying concentrate and injecting it into a converter via special tuyeres with oxygen enriched air. Such a process is heat efficient, easy to operate, and provides alternative smelting capacity when the reactor is unable to smelt all the available concentrate and during its annual shutdown for repairs.

Conclusion

Through twenty years of operation, the development of the Noranda Reactor has led to a unique smelting vessel that is highly productive, energy-efficient and easy to operate and control. The process provides flexibility to handle a wide range of feed materials within a framework of tightly controlled emission standards.

Reactor operating personnel in conjunction with metallurgical staff of the Horne and the Noranda Technology Centre actively continue the challenge of process development that started almost thirty years ago.

Acknowledgement

The author wishes to express his thanks for Noranda Minerals' permission to publish this paper and to recognise the contribution of P.W. Godbehere and Dr. P.J. Mackey in editing the text, as well as those who have contributed to the success of the reactor process over the years.

SUMMARY OF REACTOR OPERATING PARAMETERS AND DEVELOPMENT MILESTONES

OPERATION	COPPER MODE	MATTE MODES		
		BEFORE HMIP	AFTER HMIP	REACTOR ONLY
Campaign Number	1 - 10	11 - 25	26 - 36	37 - 38
Year	1973 - 1974	1975 - 1982	1982 - 1989	1989 - 1991
Average length of campaign (available days)	46,8	153,7	195,7	292,4
Instantaneous feed rate (mt NRPM)	715	1 146	1 953	2,515
Blowing rate (Nm3/h)	60,000	67 212	69 190	74 363
Blowing time (%) including I.C.S.	83,2	86,2	72,9	76,7
Blowing time without I.C.S.	83,2	90,1	86,6	80,5
Fuel ratio (GJ/mt NRPM)	4,52	2,34	1,76	0,78
O2 enrichment (%)	21	23,6	32,2	36,3
Main cause of the shut down	Tuyere brick wear	East wall brick, mouth and tuyeres	Tuyere brick wear	Planned shut down

NRPM: New revenue producing material
I.C.S.: Intermittant Control System; system involving mandatory production cutbacks to control ambient SO2 levels in the community.

OPERATION	COPPER MODE	BEFORE HMIP	AFTER HMIP	REACTOR ONLY
Campaign Number	1 - 10	11 - 25	26 - 36	37 - 38
Year	1973 - 1974	1975 - 1982	1982 - 1989	1989 - 1991
Key Improvements	1973: Reactor start-up	1975: Reactor on matte mode	1982: Horne Metallurgical Improvement Program (HMIP) Large O2 plant (463mt O2 per day)	1989: Installation of the on-line advisory system
	1974: Use of coal for heat balance	1975: Slag cooled in ladles	1983: Installation of an optical pyrometer	1989: Start-up of the Horne acid plant
		1975: Small O2 plant (82mt O2/d)	1988: Matte sampling rate increases to every half an hour	1989: Closure of the last reverberatory furnace
		1976: Feed end burner stopped	1988: Automatic Gaspé Puncher	1990: Start-up of Concentrate Injection Process
		1976: Intermittent Control System (I.C.S.) for SO2		
		1976: Closure of Horne Mine		
		1977: Slag end burner stopped		

201

References

1. L.A. Mills, G.D. Hallett, and C.J. Newman, <u>Design and Operation of the Noranda Continuous Smelting Process</u> (Extractive Metallurgy of Copper, edited by J.C. Yannopoulos and J.C. Agarwal, The Metallurgical Society of AIME, New York, 1976) pp. 458-487.

2. G.D. Hallett, "The Noranda Process", (Paper presented at the Copper Smelting and Refining Technologies Seminar, organized by the B.C. Ministry of Energy, Mines and Petroleum Ressources, Vancouver, British Columbia, 5-6 november 1980).

3. P. Tarassoff, "Process R & D - The Noranda Process", (The 1984 Extractive Metallurgy Lecture, The Metallurgical Society of AIME).

4. J. Lucas, "A Fiber-Optic Pyrometer for Tuyere Temperature Measurement, (Paper presented at the ISA 1987 conference).

5. Y. Prévost, D. Verhelst, "Metallurgical Control of the Noranda Process Reactor", (Paper presented at the Copper-Cobre 1991 Conference in Ottawa, Ontario, August 1991).

6. D.G. Pannell, P.J. Mackey, "Noranda Process Operations 1988 and Future Trends", (Paper presented at the Copper Committee Meeting of the GDMB, Antwerp, Belgium, April 1988).

7. P.W. Godbehere, B.W. Brooks, "Milling of Copper Slags from the Noranda Coninuous Smelting Process", (Paper presented at the 106th AIME annual meeting, Atlanta, Georgia, March 1977).

USE OF OXYGEN IN REVERBERATORY FURNACE AND TENIENTE CONVERTER

AT LAS VENTANAS SMELTER.

Jaime Olguín, Sadi Medina, Hernán Cuadro, Ricardo Bassa[*]

ENAMI-CHILE, Las Ventanas Smelter
[*]Mining and Metallurgical Research Center

Abstract

In September 1990, Las Ventanas Smelter began the intensive use of oxygen, in order to increase its concentrates smelting nominal capacity from 325,000 to 480,000 t/year. This project also included a new 88,000 Nm^3/h Sulfuric Acid Plant, to improve environmental conditions. The new Oxygen Plant has a nominal capacity of 315 t/d. The Reverberatory Furnace operation with oxy-fuel burners and original frontal burners, has permitted an important decrease in fuel oil consumption. The industial oxygen consumption varies between 180 and 200 t/d. On the other hand, the use of oxygen at the Teniente Converter has increased its concentrate smelting nominal capacity from 45,000 to 200,000 t/year. This increase has been attained with oxygen enrichment level between 28 and 32% in blowing air, which means industrial oxygen consumption between 100 and 120 t/d.

Proceedings of the
Savard/Lee International Symposium on Bath Smelting
Edited by J. K. Brimacombe, P. J. Mackey,
G. J. W. Kor, C. Bickert and M. G. Ranade
The Minerals, Metals & Materials Society, 1992

Introduction

In 1964, Las Ventanas Smelter of the Empresa Nacional de Minería, began to operate using traditional smelting and converting technology for copper concentrates in Reverberatory Furnace and Peirce Smith Converters.

Holding these technologies, introducing new production units and groups of support, plus improving their operational production, Las Ventanas Smelter attained 258,000 t concentrates smelting in 1983. With the setting up of a Teniente Converter with air in 1984, and the raising of the suspended roof of the reverberatory furnace in 1985, the Smelter capacity further increased to 325,000 ton/year. Starting September 1990, the Teniente technology was totally adopted in the Smelter. The use of oxygen in Reverberatory Furnace and Teniente Converter, increased concentrates smelting nominal capacity to 480,000 t/year.

The project considered a total inversion of US$ 85,000,000 for:

- A 315 t/d, 95% purity and 350 kPa g pressure, Oxygen Plant.

- A reception equal to 88,000 Nm³/h of dry gases, at the Acid Plant.

- And a series of other neccesary projects to confront new production levels. These are called complementary projects and include among others: 60 t cranes, 6 m³ ladles, oxygen distribution, converting building reinforcement, plus adequate and modern concentrates reception and blends preparations.

The objective of this work is to give a general idea about Las Ventanas Smelter while operating with Teniente technology, permitting the knowledge of plant facilities, capacities, operational results and long and short term proyections.

General description of main equipment.

Figure 1 shows a general pyrometallurgical flowsheet process (smelting - converting) at Las Ventanas Smelter.

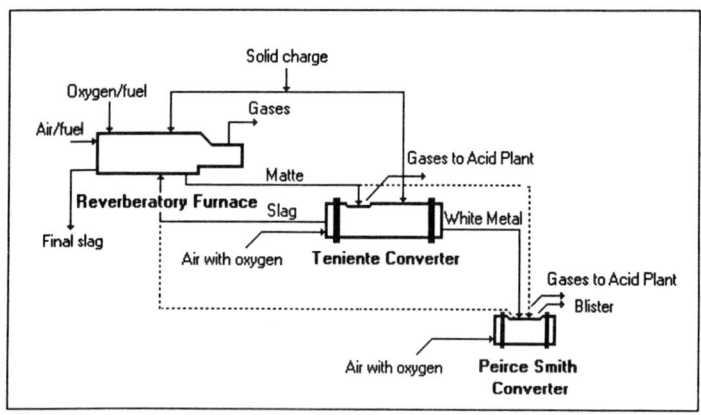

Figure 1.- General pyrometallurgical flowsheet process at Las Ventanas Smelter.

Table I shows the principal characteristics of the Reverberatory Furnace, Teniente and Peirce Smith Converters.

204

Table I.- _Smelting and converting characteristics of the reactors at Las Ventanas Smelter._

Reactor	Units	Principal characteristics
Reverberatory Furnace	1	Width 8 m, length 35 m, height (crisol) 3.8 m Suspended roof thickness, 305 mm (12") 2 frontal burners for fuel or coal, with air 14 oxygen/fuel burners 3 matte and 2 slag tap holes at each side 1 slag return chanel at the burners wall Blend charge through 26 tubes per side 3 boilers producing steam at 470°C and 3,550kPa g
Teniente Converter	1	Lenght 15 m, diameter 4 m 50.8 mm (2") nominal diameter tuyeres x 48, distributed in 6 hot repair desmountable compartments Injector charge at head Automatical punching machine 1 white metal tap hole 76.2 mm (3") diameter 1 slag tap hole 177.8 mm (7") diameter Refractory brick thickness, 457.2 mm. (18") Mouth area, 4.2 m²
Peirce Smith Converter	3	Lenght 8 m, diameter 3 m 50.8 mm (2") nominal diameter tuyeres x 28, distributed in, 4 hot repair desmountable compartments Refractory brick thickness, 381 mm (15") Mouth area, 4.0 m²

Teniente and Peirce Smith Converters can send process gases to the Acid Plant through the hot gases cleaning unit which basically consists of cooling towers (one per Converter), a high-speed duct, an electrostatic precipitator and several blowers, or to the principal stack through the balloon flue.

Results and operation with Teniente technology at Las Ventanas Smelter

In September 1990, Las Ventanas Smelter adopted completely the Teniente technology in its smelting processes. Nonetheless it is important to mention here that each Smelter has its own characteristics, limitations and resources which makes them differ among them and each one must adopt its own operating mode in the whole and the processes in particular.

One of the principal characteristics in Las Ventanas Smelter is that it is a multurer unit processing copper, silver and gold concentrates and copper precipitates coming from 120 different clients approximately. So the products receptioned have different and complex chemical and physical characteristics which permanently require control and operational planning.

This control and operational planning is done with mathematical models designed and validated at the Smelter and its principal objective is to predict the behaviour of the mining products before treatment.

Since working with oxygen, the use of these models has allowed to achieve and maintain operational equilibrium at the Smelter in the short term: Smelting concentrates have been maximized, matte production has been adequate in regards to quality and quantity, revert materials production has been reasonable and fuel, oxygen, air and flux consumption has been rationalized.

Results provided by mathematical models are used to daily program liquid movements and estimate blister copper production. (See figure 2).

Hours								
	1	2	3	4	5	6	7	8

Liquid	Number of ladles							
Matte	3		3		3		3		3
Converting slag	4	3		4		3		3	
White metal			3	3		2		1	
Final slag		7		6		7	6		5

Liquid	Ladle volume (m3)	Net weigth (t)	
Matte	6	15	Blister
Converting slag	6	12	production
White metal	6	18	370 t/d
Final slag	6	10	

Figure 2.- *Liquid movements program (8 hours) and estimate blister copper production at Las Ventanas Smelter.*

Smelting capacity at Las Ventanas Smelter as well as type of concentrates to feed reactors are basically defined by three conditions:

a) *Reverberatory Furnace working with Teniente and 3 Peirce Smith Converters.*

b) *Reverberatory Furnace working with Teniente and 2 Peirce Smith Converters.*

c) *Reverberatory Furnace working with 3 Peirce Smith Converters.*

In general, smelting capacity at Las Ventanas Smelter depend on the converting capacity. Then, operational condition :

a) *Allows maximum smelting capacity in Reverberatory Furnace because it has maximun converting capacity thus permitting conventional cycles with remaining matte that is not possible to charge into Teniente Converter.*

b) *Makes that all matte coming from Reverberatory Furnace has to be blown into Teniente Converter. The blends are planned in such a way so as to maximize smelting concentrates.*

c) *Is the lowest smelting concentrate capacity while the converting capacity is the limiting. So blends are planned to obtain matte law close to 52% Cu.*

All this must be conjoined with a series of other restrictions such as: Emissions, air and oxygen availibility, conditions and availabilities of support equipment, concentrate impurities, yearly production program and mining products supply.

Operational results for each reactor, operational conditions and future projections within Las Ventanas Smelters are provided below.

Reverberatory furnace.

The Reverberatory Furnace at Las Ventanas Smelter operates introducing energy in two ways. That is, about 33 % by means of fuel/air frontal burners and the remaining 67 % by oxygen/fuel vertical burners.

It is possible to locate these oxygen/fuel burners in 27 different positions on the suspended roof of the furnace, thus allowing to operate only with a maximum of 14 units simultaneously. In this condition, they do operate with frontal burners.

206

Figure 3 shows the most frequent oxygen/fuel burners disposition while operating with frontal and vertical burners.

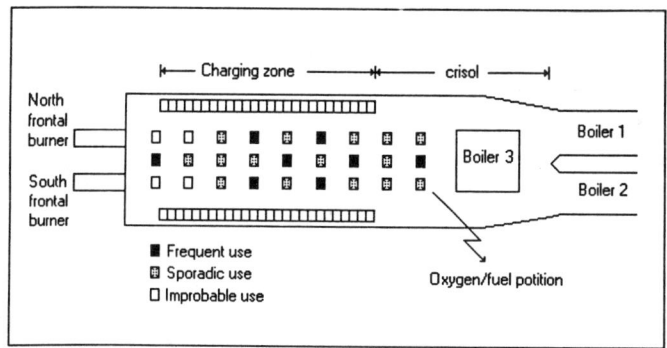

Figure 3.- *General view of different oxy-fuel positions on the suspended roof of the Reverberatory Furnace at Las Ventanas Smelter.*

In normal operating conditions (frontal and 6 a 8 oxygen/fuel burners), the energy supplied to the furnace varies between 32,000 and 36,000 Mcal/h. This means a reduction close to 38% of combustion gases volume with respect to conventional operating with air.

Steam production has dicreased, because part of the sensible heat gases evacuated from Las Ventanas Reverberatory Furnace, is recovered in the boilers, affecting directly the electric energy production.

The use of oxygen provides the Reverberatory Furnace with higher operational flexibility because it can return to its steady state condition faster than conventional air operation when unwanted disturbances are present, besides, it allows to smelt concentrates with bigger caloric requirements and it keeps the thermal intensity close to its steady state condition in case of emergency (repairs). Lastly, when the boilers fail, there comes a strong gas restriction evacuation, then it can operate with more oxygen, in order not to lose smelting concentrates capacity.

Mixed operation with frontal and vertical burners have the following difficulties:

— *It creates zones of high thermal density inside the furnace and the effect on the refractory is highly harmful.*

— *The dust carried away is high due to interactions between vertical flames and the charge bank.*

Oxygen/fuel burners used in Las Ventanas Smelter Reverberatory Furnace obey to a special design for mixed operations. This burners were designed so that heat transference to the charge bank would be efficient, and this is attained when the vertical flame traspasses the frontal flame, permitting thus to easily reach the base of the charge bank. The design conditions for these burners are the following:

- *Oxygen flow rate, 10 Nm^3/min*
- *Oxygen pressure, 350 kPa g*
- *Fuel flow rate, 4.7 1/min.*
- *Fuel pressure, 3,550 kPa g*

Table II shows the typical mineralogic compound of the blend feed to Reverberatory Furnace and Table III shows the operational results by using oxygen and its comparison with the conventional air operation.

207

Tabla II.- *Mineralogic composition of the blends feed to Reverberatory Furnace and Teniente Converter at Las Ventanas Smelter.*

Mineralogic Compound	Reverberatory Furnace, (%)	Teniente Converter, (%)
Calcopyrite	30 - 35	20 - 30
Calcosine	6 - 10	10 - 14
Coveline	1 - 2	3 - 6
Pyrite	10 - 12	20 - 26
Bornite	6 - 8	2 - 4
Enargite	0,1 - 0,2	1 - 2
Sílica	8 - 12	12 - 18
Alúmina	1 - 2	1 - 2
Calcite	1 - 2	1 - 2
Magnetite	0,5 - 1	0,1 - 0,6
Magnesite	0,1 - 0,2	0,1 - 0,2
Hematite	3 - 5	2 2 - 3
Pyrrotite	2 - 3	0,2 - 0,8
Copper	1 - 2	0,01 - 0,02
Galena	0,1 - 0,3	0,1 - 0,3
Esfalerite	0,5 - 1	0,5 - 1
Fayalite	1 - 3	0 ,1 - 1,5
Cuprite	5 - 10	0,2 - 2

Tabla III.- *Operational results of the Reverberatory Furnace at Las Ventanas Smelter, before and after oxygen.*

Operational Index		Before Oxygen	After Oxygen
- Dry charge,	t/d	942	1,012
Reverts materials,	%	16	21
Flux,	%	4	5
Humidity,	%	8-10	8-10
- Matte,	t/d	581	622
Cu,	%	47	48
- Final slag,	t/d	453	814
Cu,	%	1.0	1.1
- Converter slag,	t/d	364	700
Cu,	%	8.3	9.4
Fe_3O_4,	%	22.3	23.5
SiO_2,	%	23.1	24.6
- Energy (fuel),	Mcal/h	54,000-56,000	32,000-36,000
- Thermal efficiency, %		35	41
- Oxygen,	t/d	0	180-200
- Fuel index,	kg fuel/t	135	79
- Refractory index,	kg/t	1.74	2.34
- Dust in waste gas,	t/d	8-12	12-16

The principal differences with the conventional air operation are:

- *Increase in thermal efficiency, which has caused to diminish energy and specific fuel consumption.*

- *Increase in final slag and its copper content, principally due to an important increase in returning slag.*

- *Increase in returned magnetite which has obliged to keep a permanent control on furnace floor and treatment with eucaliptus sticks and iron scrap. Excellent results have been obtained using permanent oxygen/fuel in the crisol which has permitting to maintain an adequate temperature in liquids avoiding an accelerated banking with magnetite. All slag coming from the Teniente Converter is returned to the Reverberatory Furnace.*

- *Increase in specific refractory consumption due to high thermal density zones and an increase of the temperature inside the furnace.*

In 1992 we hope to reach a smelting concentrates capacity equal to 274,400 t, and a total smelting capacity of 369,400 t, including fluxes and revert materials.

The long and short term projections of the Reverberatory Furnace at Las Ventanas Smelter, that point out to improve smelting capacity and also to increase their thermal efficiency, are :

- *Replacement of the present combustion system for air/petroleum/oxygen vertical burners. This will allow a better distribution for the thermal energy inside the furnace and minor dust carried away in the gases.*

- *Methods development for refractory repairs that would minimize thermal energy variations and maximize brick waste recovery to maintain the useful volume of the crisol.*

- *The possibility to install a slag treatment furnace to diminish slag return at Reverberatory Furnace. This will increase metal recovery, treatment capacity of concentrates and will diminish revert materials production.*

Teniente Converter

The Teniente Converter is the most important reactor in Las Ventanas Smelter because although not being selfsufficient (it requires matte to hold FeS equilibrium), it determines the rate of the smelter operations.

Once the operational conditions are defined: Types of concentrates constituting the charge, amount and chemical composition of the matte, and oxygen air enrichment, then the blowing flow by the tuyeres determines the kinetic of the process and indirectly the rate of the smelter operations.

This air and oxygen blowing flow is variable in time and requires permanent and strict process control to keep it within reasonable variations and at an adequate operational level.

The most important parameters which must be considered and controlled are: Total liquid level, metal phase level, liquids composition and process temperature. These parameters are more sensible when the reactor is small and the blowing pressure is low.

In Las Ventanas Smelter in which the Teniente Converter is relatively small (4 m diameter and 15 m in length) and the available pressure at the tuyeres is only 100 kPa g, it is fundamental to keep these parameters at an adequate range for the operation. The use of oxygen accentuates the sensibility of these parameters, each time that the kinetic of the process increase.

This present year, Las Ventanas Smelter Teniente Converter has reached with 28% oxygen enrichment a very good production level, operating under permanent operational control of these parameters and materializing a series of actions tending to maximize the blowing flow and the availability of the reactor. Table IV shows results and compare them with air operation.

Tabla IV.- *Operational results of the Teniente Converter at Las Ventanas Smelter, before and after oxygen.*

Operational Index		Before Oxygen	After Oxygen
- Dry charge,	t/d	246	685
Reverts materials,	%	9	2
Flux,	%	27	18
Humidity,	%	8-10	8-10
- Matte,	t/d	477	539
Cu,	%	46.5	48
- Others liquids, (1)	t/d	18	42
- Slag,	t/d	308	632
Cu,	%	9.2	9.4
Fe_3O_4	%	22.6	21.7
SiO_2	%	23.0	25.6
- White metal,	t/d	278	470
Cu,	%	75.9	75.4
- Flow,	Nm^3/min	500-550	500-600
O_2,	%	21	28
- Oxygen,	t/d	0	100-120
- Load efficiency, (2)	%	60	80
- Blown efficiency, (2)	%	70	92
- SO_2 after dilution, %		4-6	6.5-7.5
- Dust in waste gas, t/d		6-10	14-18

These values (t/d) are daily averages which do not take into account time lost when Converter is out of due to refractory repairs.

(1)   Anode furnace slag and Peirce Smith Converter slags (white metal blown)

(2)   It does not consider time lost by Teniente Converter due to refractory repairs. .

The principal modifications with respect to the air operation, are the following:

—   Modification of the charge injector hopper in order to keep the concentrates feeding when the Converter turns to charge matte.

—   Modification of the hood hatch in order to easy up emptying matte and avoid cutting the blowing flow during this operation.

—   Modification of the tuyeres piping in order to diminish pressure losses and avoid air escape.

—   Punching machine automation.

—   Improvements in liquids extraccion systems.

—   Air blowing pressure increase, from 99 to 106 kPa g (at the blower).

Las Ventanas Smelter Teniente Converter campaign is indicated below:

- *Total repair (18 days.)*
- *70 days operation.*
- *1st refractory hot repair in tuyere line (2 days.)*
- *70 days operation.*
- *2nd refractory hot repair in tuyere line (2 days.)*
- *70 days operation.*
- *3rd refractory hot repair in tuyere line (2 days.)*
- *70 days operation.*

Las Ventanas Smelter Teniente Converter blowing time has been 92 %, not considering these repairs.

In 1992 we hope to reach a concentrates smelting capacity equal to 168,000 t, meaning total smelting capacity of 207,000 t, including flux and revert materials.

The future projections of this reactor in Las Ventanas Smelter, are basically the following:

- *Increase of blow pressure*
- *Automatic operation*
- *Set-up bigger size Converters*
- *Inject dry concentrate by tuyeres*

All these projections lead to obtain a stable, secure and efficient operation that will translate into higher productivity.

Peirce Smith Converter

Las Ventanas Smelter has 3 Peirce Smith Converters which are 3 m diameter and 8 m lenght.

The operational sequence with matte and white metal cycles, is the following:

Cycles with matte   :   Feeding with 2 matte laddles
Overfeeding with 1 matte laddle

Cycles with white metal :   a) Feeding with 2 white metal laddles
Overfeeding with 1 white metal laddle
b) Feeding with 3 white metal laddles

A Converter can do 3 to 5 cycles per day depending on the ruling operational conditions.

Oxygen use in Peirce Smith Converters has been sporadic and used only for short periods at an enrichment level of 24% when the white metal has high copper concentration (>77%) or when it becomes necessary to increase of the revert materials smelting.

Las Ventanas Smelter Peirce Smith Converters campaign is the following:

- *Total repair (13 days).*
- *200 cycles operations*
- *1st refractory hot repair in tuyeres line (2 days).*
- *150 cycles operations.*
- *Partial repair (12 days).*
- *2nd refractory hot repair in tuyeres line (2 days).*
- *150 cycles operations*
- *3rd refractory hot repair in tuyeres line (2 days).*
- *100 cycles operations*

All these cycle produces 25,000 to 30,000 tons of copper.

The yearly availability of these Converters is:

- *a)   Operation with 3 Converters during 214 days*
- *b)   Operation with 2 Converters during 151 days*

Today operating with 3 white metal laddles produces too much splashing and solid material accumulation at the Converter mouth and it requires much Converter and crane time. We are studying, as a future projection, the possibility to enlarge them, separate mouth from the liquid level or replace them for others with bigger diameter.

Table V shows present operational results obtained since the oxygen was set up.

Tabla V.- *Operational results of the Peirce Smith Converter since Oxygen Plant was set up at Las Ventanas Smelter.*

Operational Index		Results
- Availability,	units/month	2.58
- Matte,	t/d	83
Cu,	%	48.6
- White metal,	t/d	470
Cu,	%	75.4
- Blister,	t/d	371
Cu,	%	98
- Smelting solid charge,	t/d	101
- Matte cycles,	cycles/d	1
- White metal cycles,	cycles/d	3
- Flow,	Nm^3/min	250-350
O_2,	%	21-24
- Oxygen,	t/d	0-1

Operation with Acid Plant

Las Ventanas Smelter diminished 20% of its environmental sulphur outcome rather than operating with air, regardless increasing the smelting concentrates by using oxygen. Now the acid plant captures 50% sulphur income.

Using oxygen at the Smelter has allowed a decrease in production costs due to the Acid Plant and at the same time produces a high SO_2 concentration in the converting processes gases. SO_2 concentrations after dilution are adequate to feed a conventional Acid Plant without installing high inversion systems at the hoods in order to decrease air filtrations.

Las Ventanas Acid Plant feeds itself principally with 75,000 Nm^3/h dry and diluted gases provided by the Teniente Converter. The total capacity of the Plant (88.000 Nm3/h) is completed with gases coming from Peirce Smith Converters. The draft at the hood Converter is very important to obtain reasonable dilutions (120%) and are automatically regulated between -30 to -50 Pa g.

The gases temperature is automatically regulated at 350°C using spray cooling. This is the adequate temperature so that the gases at the electrostatic precipitation entry would be at 320 to 340°C.

Figure 4 shows a general flowsheet of hot gases cleaning system.

The dust dragged by process gases, specially in the Teniente Converter has not been an impossible problem to solve for its best operation with the Acid Plant. This has been possible by keeping the gases cooling system in optimum operational conditions, so that the dust attached to the ducts remain reasonable and requires cleaning only each 12 days during 4 to 8 hours.

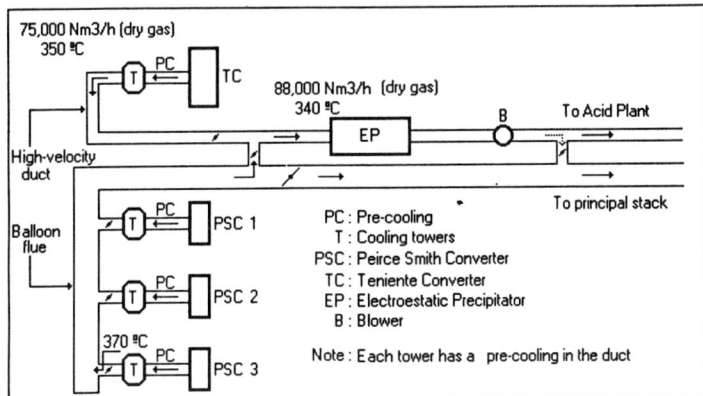

Figure 4.- General flowsheet of the hot gases cleaning to Acid Plant at Las Ventanas Smelter.

Coordinating converting operations is vital to maximize the SO_2 capture, the Teniente Converter operating in series with 2 Peirce Smith Converters programmed in order not to work simultaneously, is a good example.

CONCLUSIONS

The Teniente process adopted by Las Ventanas Smelter has proved to be an efficient technology, at a low inversion rate, easy to operate and able to produce acid in a conventional Sulphuric Acid Plant without inconvenience.

Slag coming from the Reverberatory Furnace is discarded and does not require a flotation plant for metals recovery.

ACKNOWLEDGEMENTS

The authors wish to gratefully acknowledge to the Management of the Empresa Nacional de Minería for allowing this publication at the Savard Lee International Symposium on Bath Smelting.

TOP-ENTRY SUBMERGED INJECTION AND

THE ISASMELT TECHNOLOGY

R.L. Player[1], C.R. Fountain[1], T.V. Nguyen[2],
and F.R. Jorgensen[2]

[1]Process Development Department
Mount Isa Mines Limited
Mount Isa, Qld 4825, Australia.

[2]CSIRO Division of Mineral and Process Engineering
P.O. Box 312, Clayton, Vic 3168, Australia.

The Isasmelt Process is a new lead smelting process, developed jointly by Mount Isa Mines Ltd (MIM) and an Australian Government research body, the Commonwealth Scientific and Industrial Organization (CSIRO). The Isasmelt process is based on the top-entry, submerged combustion Sirosmelt lance developed by CSIRO in 1973.

The technology developed for lead smelting has also been applied to smelting lead dross, copper and nickel-copper concentrates, and various waste materials, such as flue dusts, refining residues and battery paste.

MIM has operated Lead Isasmelt furnaces in its Mount Isa lead smelter since 1983. A Lead Isasmelt continuous smelting and reduction plant was commissioned at Mount Isa in 1991 to produce 60,000 t/y of crude lead bullion.

A 15 t/h Copper Isasmelt pilot plant was commissioned in the Mount Isa copper smelter in April 1987. By March 1992, it had treated over 490,000 tonnes of copper concentrate. A furnace to treat up to 115 t/h of concentrate is currently under construction.

Isasmelt furnaces have a very rapid mixing rate due to the injection of large volumes of gas deep into the bath. The furnace can be treated as a continuous stirred tank reactor.

The flushing action of the gas ensures high elimination rates for such minor elements as arsenic, lead, zinc and bismuth.

The paper considers the development of the Isasmelt technology from the bench scale through to plants producing over 180,000 t/y of contained copper. The role of injection modelling is also considered.

Proceedings of the
Savard/Lee International Symposium on Bath Smelting
Edited by J. K. Brimacombe, P. J. Mackey,
G. J. W. Kor, C. Bickert and M. G. Ranade
The Minerals, Metals & Materials Society, 1992

INTRODUCTION

We in the minerals industry are faced with a continuing real decline in the prices we receive for our products. In 1900, the price of a pound of aluminum was equivalent to about 160 minutes' labor for the average American wage earner. Today, a pound costs around 3.1 minutes' work. Similarly, a pound of copper in 1900 was worth about 84 minutes' labor, while today that pound could be had for a mere 5.8 minutes' toil.

The decline in the real prices for commodities, goods and services that has occurred over the past century is known as the "rising standard of living." It is the result of increasing productivity. Companies are able to stay in business in the face of falling prices only by decreasing their operating costs and increasing their output per employee. It took more people back in 1900 to produce a pound of aluminum or copper than it does today.

While falling prices benefit the population as a whole, it does have a downside for producers and manufacturers. Companies must continually seek further increases in productivity or they will cease to be competitive. It is the old story of running hard just to stay in the same place. Any company that steps off the treadmill for too long becomes a footnote in history.

It was in such an effort to remain competitive that Mount Isa Mines Limited (MIM) began a survey of available and potential lead smelting technologies in the 1970s. MIM operated a lead smelter based on an updraft sinter plant and a single blast furnace. The company wished to achieve metallurgical, cost and hygiene benefits.

None of the technologies investigated fully satisfied MIM's requirements, so a project was jointly undertaken with the CSIRO[*] to develop a new smelting technology: the top-entry, submerged combustion Lead Isasmelt process. The result of the work was construction and commissioning of a $A65 million Isasmelt plant, designed to produce 60,000 metric tons per year of crude lead bullion.

The technology developed for the Lead Isasmelt process has also been applied to other feed materials, including lead dross, lead battery paste, copper concentrate, and nickel concentrate. MIM has now been marketing Isasmelt technology for three years.

A small Isasmelt furnace has been commissioned at Britannia Refined Metals (London, England) to treat an average 4.5 tph of battery paste[1]. Copper Isasmelt plants capable of producing matte with 150,000 tpy and 180,000 tpy of contained copper are currently under construction at the Cyprus Miami smelter (Miami, Arizona) and MIM's own copper smelter (Mount Isa, Australia). An Isasmelt furnace was commissioned at AGIP Australia's Radio Hill Mine (Karratha, Western Australia)[2], but the mining and smelting complex has since been placed on care and maintenance due to low nickel prices.

Isasmelt technology has been the subject of much interest over the past few years. It has the potential to make substantial reductions to smelter operating costs while improving smelter hygiene. This article will outline some of the steps taken in the development of the process and present some modelling results.

---

*   Commonwealth Scientific and Industrial Research Organization, an Australian government body.

216

The Anatomy of the Technology

Figure 1 is an illustration of the Isasmelt concept. An Isasmelt furnace is an upright, cylindrical reactor. The steel shell is protected by a lining of refractory bricks.

A single lance passes through a hole in the roof of the reactor. The lance tip is submerged in the molten bath at the bottom of the furnace. Process air, fuel (such as oil, coal, or natural gas), and possibly concentrate, are injected into the bath through the lance. The usual mode of operation is to add the concentrate as pellets through a second hole in the roof.

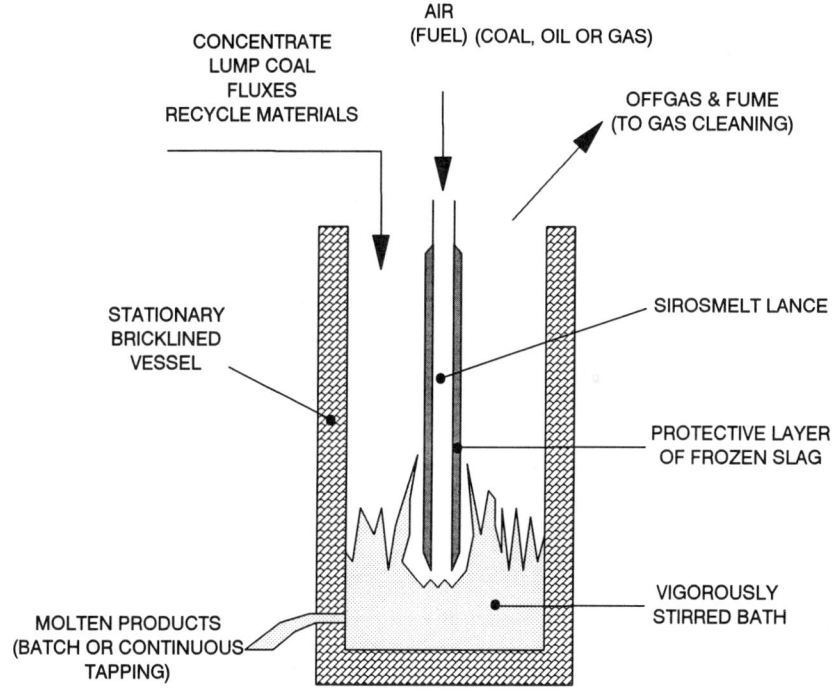

Figure 1 - The Isasmelt concept

The lance is protected by a layer of slag that builds around it. This layer is formed and preserved by the cooling action of the process air. The whole system is made practical by the presence of internal swirlers, increase the heat transfer rate.

The submerged injection of large volumes of air into the bath makes it extremely turbulent. This ensures rapid incorporation of the feed materials into the bath, and a high rate of reaction, allowing a low residence time.

The products of the reaction are removed through one or more tap holes in the furnace.

The Lead Isasmelt process comprises two furnaces (see Figure 2). The first smelts lead concentrate to a lead-oxide slag. The slag is transferred via a launder to the second furnace where the lead is reduced to metal. The metal and slag are tapped from the reduction furnace through separate holes.

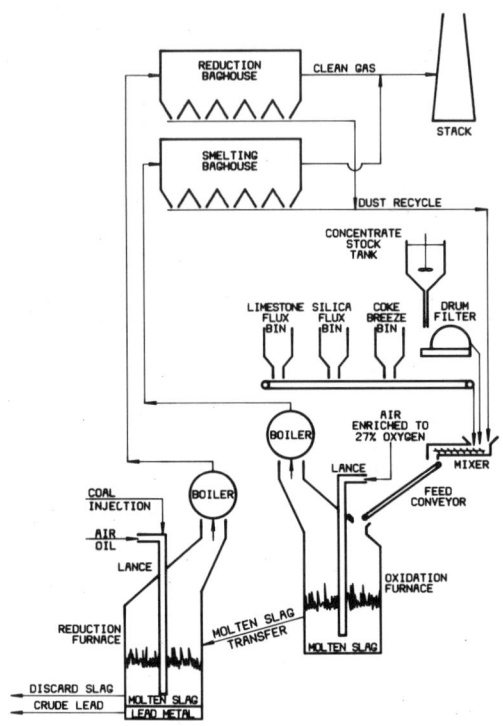

Figure 2 - Flowsheet for MIM's 60,000 tpy (of lead bullion) Isasmelt plant.

The process is continuous, with an overflow taphole on the smelting furnace and a weir on the reduction furnace, although difficulties can be encountered with this system at low feed rates. Under these circumstances, the smelting furnace could be fed continuously but tapped intermittently. Reduction in the second vessel would then be a batch process.

The Copper Isasmelt process currently operates with a single furnace. Copper concentrate is smelted in this reactor to form a matte. The matte and slag are tapped through a single hole and are transferred to a settling furnace for separation. A discard slag containing about 0.5% copper is thus produced[3].

Again, the process is designed to be continuous but practical difficulties at the scales so far operated have led to a continuously fed, batch tapped furnace.

THE DEVELOPMENT OF ISASMELT TECHNOLOGY

THE LEAD ISASMELT PROCESS

The philosophy of the development of the Isasmelt technology has been to make the large pilot plants self-funding by using them to increase the company's smelting capacity. The initial justification for the development was the need to improve the productivity of MIM's lead smelter. The company had contributed to industry-funding of early CSIRO work with the Sirosmelt lance[4], and from this gained a clear insight into the potential of the submerged combustion lance.

Crucible scale tests

Crucible scale tests and thermodynamic modelling work were undertaken in Melbourne by CSIRO and MIM personnel. This work, carried out during 1978 and 1979, provided fundamental data on the bath smelting of lead concentrate.

The early modelling work at CSIRO revealed that both lead metal and a lead oxide slag could be produced in a smelting furnace when fed with high grade lead concentrate (say, 65-75% lead). However, this strategy would cause excessive fuming with MIM's lower grade concentrates (45-50% lead)[5]. The crucible work confirmed this, and showed the best results for MIM would be obtained in the two-stage process that was eventually developed.

250 kg pilot plant tests

A 250 kg/h scale pilot plant was constructed in the Mount Isa lead smelter in 1980. This plant was used during the next two years to develop first the smelting stage and then the reduction stage. The pilot plant was located in the lead smelter to take advantage of existing gas handling equipment.

Up until this point, the costs of the research were relatively small and were primarily the wages of the people involved. The next step was a further scale-up of the process. The capital cost of the next scale was close to $A2 million and additional labor had to be hired to operate the plant. The justification for the expenditure was the use of the pilot plant to increase the capacity of the lead smelter.

The 5 tph Lead Isasmelt pilot plant

The first 5 tph Lead Isasmelt plant was constructed in August 1983, using a furnace purchased from Aberfoyle Limited. Aberfoyle had built the furnace at the Kalgoorlie Nickel Smelter (Kalgoorlie, Western Australia) to recover tin by matte fuming[6].

The 5 tph pilot plant was first used to demonstrate the smelting step and to gain operating experience. Lead sulfide concentrate was smelted to produce a lead oxide slag. The product slag was tapped, granulated and fed through the lead smelter sinter plant.

Lead sinter plants are characterized by large recirculating streams. Experience has shown the plants to operate best when the sulfur content of the sinter feed is in the range of 6.5 to 7.5%. A high proportion of the product sinter is crushed and mixed with concentrate to dilute the sulfur. The level of sulfur in MIM's concentrate is of the order of 22 to 25%, requiring a 3:1 to 4:1 sinter return ratio.

219

The lead oxide slag from the pilot plant was used to replace some of the return sinter, reducing the size of the recycle stream. This practice increased the productivity of the lead smelter by about 15%. The pilot plant was indeed self-funding.

A second furnace was commissioned in 1985. It included modifications based on operating experience. The two vessels were then used to demonstrate both batch and continuous smelting and reduction of concentrate. Levels of lead as low as 2% in the discard slag were achieved during continuous operation[7].

A process for treating lead dross was also developed and put into production.

Improvements, both to the pilot plant and in operating know-how, increased its capacity from 5 tph to 10 tph of concentrate. By the time it was shut down in January 1991, the pilot plant had treated over 175,000 t of lead concentrate and over 40,000 t of lead dross.

The 60,000 tpa Lead Isasmelt plant

Based on the success of the 5 tph pilot plant MIM built an Isasmelt plant to produce 60,000 tpa of crude lead. Commissioning of the two furnaces began early in 1991, and the plant is currently being optimized. The plant was designed to produce a discard slag containing 4% lead[7].

The lead oxide is reduced by coal in the reduction furnace. This coal is injected down the lance.

The plant has achieved lead levels in the discard slag of 1.2% while operating continuously. The lance wear in the smelting furnace is low, with the same lance being used over a period now exceeding 12 weeks.

THE COPPER ISASMELT PROCESS

The initial phase

The development of the Copper Isasmelt process began with trials in the 250 kg/h pilot plant. The next step was the construction and commissioning in April 1987 of a 15 tph pilot plant in MIM's copper smelter. The layout of the plant is shown in Figure 3.

The pilot plant consists of a single furnace with an internal diameter of 2.4 meters and a height of 10 meters. The offgas is cooled in a water spray gas-cooler and passes through an electrostatic precipitator before being vented up the Copper Smelter's stack. The matte and slag are tapped through a single tap hole and transferred to the smaller of MIM's two reverberatory furnaces for settling. Coal and oil are used to fuel the plant.

One of the major advantages seen with the Isasmelt process is the minimal feed preparation required. The concentrate is filtered by a drum filter that produces a filter cake with a moisture content of 13 to 15%. The moisture in the feed is reduced to 9 to 10% by blending the filter cake with dry concentrate and fluxes. No further drying is required.

The philosophy of self-funding furnaces used in the development of the Lead Isasmelt pilot plant was continued. The Copper Isasmelt furnace was installed to increase MIM's copper smelting capacity. The pilot plant cost an initial $A6 million ($US4.5 million at the time) to build. This cost was recovered after 14 months of operation.

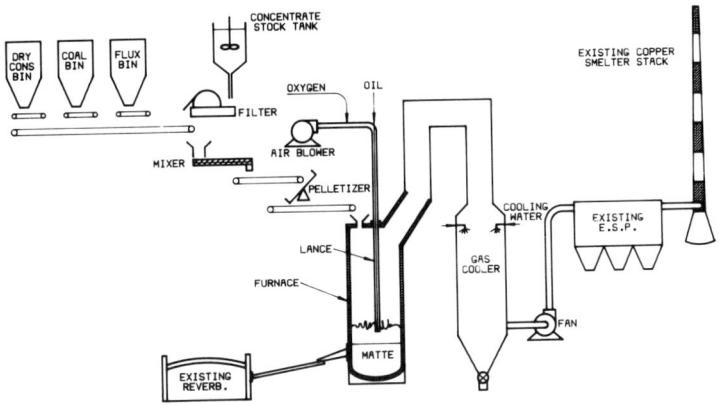

Figure 3 - The Copper Isasmelt pilot plant

The scale-up of the concentrate treatment rate was, at sixty times, relatively large. MIM was able to contemplate this because of the experience gained with the Lead Isasmelt pilot plant. There are considerable differences involved in operation of Sirosmelt lances on the 200 to 250 mm scale when compared with the 25 to 50 mm scale used in the 250 kg/h pilot plant. Lance lives were initially low, sometimes as low as 20 minutes, but with experience and some redesign, lances now typically operate for a week. The record Copper Isasmelt lance life stands at 506 operating hours. Repairs usually consist of cutting off the worn tip and welding on a new section.

Despite the normal mechanical problems encountered when commissioning a new plant, the Copper Isasmelt furnace performed largely as expected and achieved its design throughput of 15 tph within the first two months of operation.

Copper Isasmelt feeds

The Copper Isasmelt furnace was used to treat a wide variety of feed types. These included MIM's own chalcopyrite concentrate, a concentrate recovered from converter slag, and various purchased concentrates.

Table I gives analyses for some of the concentrates successfully treated in the Isasmelt furnace. Note that some concentrates lack sufficient sulfur to allow the formation of a matte. These were blended with others to give an acceptable feed composition.

The flexibility of the Copper Isasmelt process is demonstrated in the treatment of feeds with high magnetite contents. At times, converter slag concentrate has comprised half the concentrate mix being fed to the furnace. The turbulent bath ensures there is no magnetite precipitation on the walls of the furnace. This is a marked contrast with the difficulties experienced when feeding similar proportions of magnetite into a reverberatory furnace.

221

Table I Concentrates treated by the Copper Isasmelt furnace

Component (wt.%)	MIM conc	Converter slag conc	Purchased concentrates		
			A	B	C
Copper	24.7	40.5	21.4	24.7	15.5
Iron	28.9	26.1	10.2	33.5	29.3
Sulfur	31.4	14.7	9.9	33.7	9.0
Silica	8.86	7.42	22.2	1.95	20.8
Magnesia	0.88	-	2.57	0.81	1.51
Alumina	0.63	0.61	2.64	0.83	4.9
Lime	0.32	0.2	-	-	-
Arsenic	0.19	0.23	0.82	0.015	< 0.01
Cobalt	0.12	0.16	0.004	0.022	0.007
Lead	0.10	0.50	1.67	0.06	< 0.01
Zinc	0.095	0.32	3.92	< 0.01	< 0.01
Bismuth	0.007	-	0.09	0.37	< 0.01
Magnetite	-	15.0	-	-	-

The use of oxygen

A 60 tpd oxygen plant was installed to increase the capacity of the pilot plant. The purchase was preceded by a series of trials to establish the abilities of the lances to cope with oxygen enrichment of the process air. There was the worry that additional oxygen could significantly shorten the lance life.

The oxygen trials showed there was no measurable shortening of lance life with oxygen reaching levels of up to 40% in the process air.

The use of oxygen enrichment substantially increased the concentrate treatment rate of the plant. During one trial, a feed rate of 48 tph of concentrate was achieved for a period of five hours. A 50 tph feed rate was reached but the oxygen ran out, so this could not be sustained.

The oxygen enrichment substantially reduced the fuel requirement of the furnace and, given enough oxygen, it was possible to operate the furnace autogenously. Table II gives some figures for fuel consumptions with and without oxygen.

The first two columns of data in Table II describe average operating conditions over 28 days, the first without oxygen enrichment, and the second with 26% oxygen in the process air. The third column of data represents the operating conditions during a five hour trial with 35% oxygen in the process air.

Note that the ratio of converter slag concentrate to chalcopyrite concentrate in the feed mix varied between the cases, as did the total feed rate. The converter slag concentrate provides little energy to the process and so its use increased the gross energy figure. Counteracting this, to some extent, the matte grade was lower during the high oxygen trial.

Table II The Effect of Oxygen Enrichment

| | 28 Day averages | | Trial |
	Feb/Mar 1989	Apr/May 1989	May 1989
Concentrate (t/h)	15	18	48
Ratio slag conc/chalcopyrite conc	0.43	0.51	0
Matte grade	48	50	43
% O_2 in smelting air	21	26	35
Oxygen use (t/t Cu)	0	0.32	0.5
Coal (t/t Cu)	0.68	0.48	0.09
Oil (kg/t Cu)	79	31	4
Air $(Nm^3/t Cu)^*$	5140	3570	1641
Gross energy $(GJ/t Cu)^+$	25.6	17	4.1

*1 Nm^3 = 1 m^3 at 101.3 kPa and 0°C.

$^+$Gross energy includes coal, oil, and the electricity used in the blower and oxygen plant.

The performance of the Copper Isasmelt plant

Figure 4 shows the number of tons of concentrate treated through the Copper Isasmelt plant during each four week accounting period since the furnace was commissioned. Between April 1987 and February 1992, the Copper Isasmelt furnace treated 490,000 t of concentrate.

Figure 4 is characterized by a steady increase in the four weekly throughput. This increase was initially due to an increase in the availability of the plant, and later to the introduction of oxygen. Drops in the throughput were largely due to either a furnace reline or a lack of oxygen (when smelter economics demanded that it be used elsewhere).

Figure 5 is a graph of the availability of the plant. The plant is said to be available when it is smelting concentrate. Figure 5 shows a trend of increasing availability during the first two years of operation, followed by a plateauing, with the availability usually in the range of 85 to 95%. The major drops in availability correspond to shut-downs to reline the furnace when the refractory was worn. The record availability over four weeks was 95.9%.

Refractory life was a major concern during the first two years of the Copper Isasmelt plant's operation. The first lining failed after 14 weeks, the second after 17 weeks, and the third (which used some very cheap, second hand bricks) after 12 weeks. Since then, major changes have been made to the bricking of the furnace, and a 90 week lining life has been achieved.

223

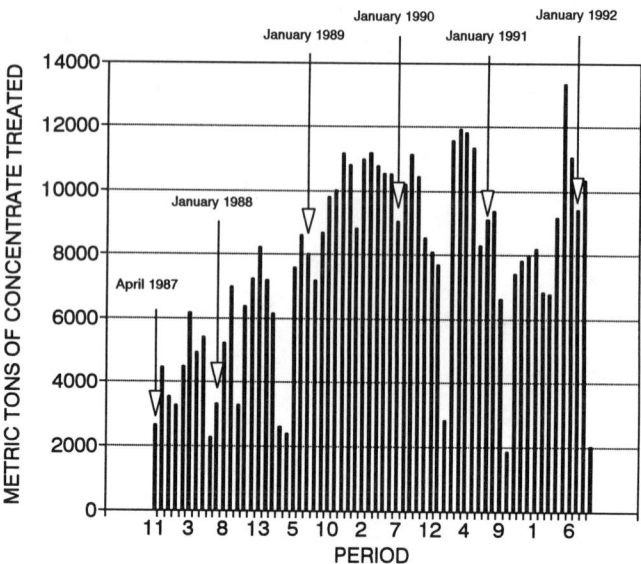

Figure 4 - The number of metric tons of concentrate treated during each four week accounting period since the Copper Isasmelt plant was commissioned.

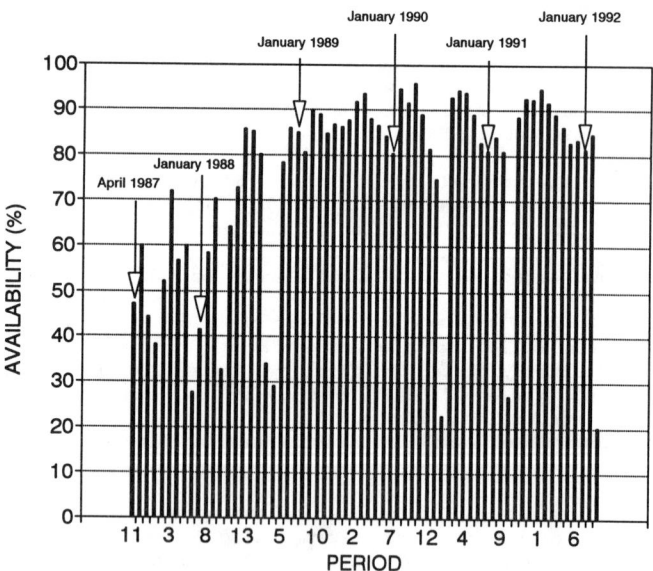

Figure 5 - The availability of the Copper Isasmelt plant each period since the plant was commissioned.

The pilot plant has been used as a test bed for further changes to the refractory lining in preparation for the construction of other, larger plants. Unfortunately, recent experimental modifications to the lining seem to have been backward steps, with each of the past two campaigns being 8 to 9 months in length.

The regular use of oxygen has led to the rerating of the 15 tph plant to 20 tph. The construction of the plant has increased the copper smelter's capacity by 20%, converting MIM from a seller of surplus copper concentrate to a purchaser.

Minor element distributions

A feature of the Copper Isasmelt plant is a relatively high elimination of such minor elements as arsenic, bismuth, lead, and zinc from the matte[8].

The Copper Isasmelt furnace consistently removes over 90% of the arsenic from the matte (the arsenic content of the feed has ranged from <0.2 to 0.8%). 80 to 90% of the bismuth, 70 to 80% of the zinc, 60 to 80% of the antimony, and 50 to 75% of the lead are removed from the matte. Arsenic, lead and bismuth report mainly to the offgas, while the zinc is retained mainly in the slag.

About 99% of the gold, over 95% of the silver and 67% of the cobalt are retained in the matte.

Table III presents the averages of measured distributions of some minor elements.

Table III Minor element distributions in the Copper Isasmelt furnace

Element	Feed (%)	Distribution (%)		
		Matte	Slag	Balance
Arsenic	0.39	6.2	5.1	88.7
Cobalt	0.11	72.4	28.9	-1.3
Lead	0.17	38.2	9.8	52.0
Zinc	0.29	28.6	55.4	16.0
Bismuth	0.057	13.9	0.1	86.0
Antimony	0.39	29.3	39.0	31.7
Silver (g/t)	120	101.4	1.4	-2.8
Gold (g/t)	44	98.8	0.5	0.7

A 180,000 tpy Copper Isasmelt

The Sirosmelt injection technology - first developed with 25 and 37 mm lances in Melbourne, and now operated using 250 mm lances - is about to take another giant leap. Based on the success of the Copper Isasmelt pilot plant, MIM is constructing a furnace to treat up to 115 tph of concentrate. The contained copper will be about 180,000 tpy. Lances with diameters up to 500 mm will be used.

The new Copper Isasmelt plant will eventually replace MIM's existing roaster and two reverberatory furnaces. It is due to begin commissioning in July 1992.

At a total cost of $A100 million, the 180,000 tpy plant will have a number of innovative features over and above the Isasmelt furnace: the concentrate is to be filtered using Andritz pressure filters; the offgas is to be cooled with an Ahlstrom Flux-Flow recirculating fluidized bed boiler; and the matte and slag will be separated in a rotary holding furnace.

MODELLING THE ISASMELT FURNACES

The development and scale-up of the Isasmelt furnaces could not have proceeded with confidence without extensive modelling work, needed to develop a fundamental understanding of the processes. Some of this work has been carried out by MIM personnel, but much of it was undertaken by CSIRO staff in Melbourne.

There are essentially two approaches which must be taken when modelling a new smelting process. The first is to build an understanding of the metallurgy of the process. This involves thermodynamic modelling of the reactions that take place inside the furnace. The second is to model the fluid flow and heat transfer involved. How does the gas interact with the bath? And, in the case of the Sirosmelt lance, how do you design the swirler to develop the maximum possible cooling.

We will very briefly outline some of the modelling work undertaken to date. Unfortunately, much of the knowledge gained is commercial, so this paper is unable to give more than a flavor of what has been done.

THERMODYNAMIC MODELLING

An Isasmelt furnace is essentially an equilibrium reactor. The reactions are very fast, and the bath is well mixed. It is a good example of the classic chemical engineering "continuous stirred tank reactor".

Two approaches have been taken to the thermodynamic modelling of the Isasmelt processes. The first is the crude heat and mass balance approach based on empirical measurements. This has worked very well. A model of the Copper Isasmelt furnace has been developed, and can quickly and accurately predict the fuel requirements for a given copper concentrate - provided it does not deviate too far from a standard Mount Isa chalcopyrite copper concentrate.

The second approach is the free energy minimization path using programs such as CSIRO's *Thermochemistry* system or *Smelt_2*. These programs take longer to run, but they are better able to handle new types of concentrate, and can predict the distributions of minor elements.

All models, both empirical and free energy minimization, have been verified against plant operations. They have been found to produce good, consistent agreement with plant results.

FLUID FLOW AND HEAT TRANSFER MODELLING

The modelling of fluid flow and heat transfer has involved both physical models and mathematical models using packages such as *PHOENICS*[9]. The predictions of the models were then verified against plant data.

Areas investigated have included the behavior of the process air in the Sirosmelt lances. Measurements were first made on small scale lances, and later on full-scale lances operating in Mount Isa. The results from this have been incorporated into the design of lances and their swirlers.

The interaction of the process air with the bath is another important area of research. Figure 6 shows the output from a *PHOENICS* model run showing the direction of the gases injected into an Isasmelt bath.

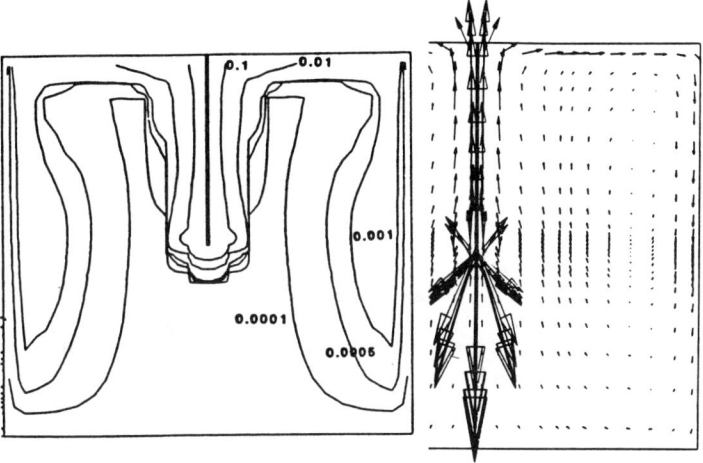

Figure 6 - Predictions of the turbulent kinetic energy distributions and the computed velocity fields in a water model.

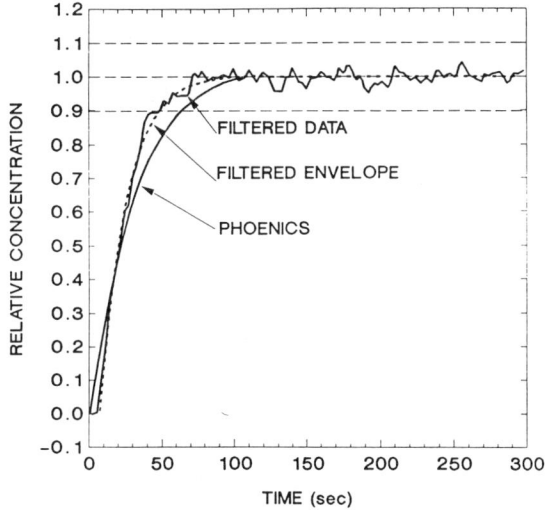

Figure 7 - Calculated and predicted tracer concentration curves for the injection of air at 5.5 Nm^3/h into water.

227

Figure 7 is an example of the verification process. It shows the calculated changes in the dimensionless tracer concentration with time, as predicted by a *PHOENICS* model run. In the case shown, the air flow rate was 5.5 Nm³/hr. Verification of the predicted mixing behavior was by injecting air into water at 5.5 Nm³/hr. A tracer was used, and its concentration was measured using an electrode located at mid-height at the point diametrically opposite to the tracer injection point. There is a time delay of several seconds in the measured curve which represents the flowing time between the tracer input position and the detector.

Unfortunately, the scatter of experimental mixing time measurements makes comparisons with numerical results difficult. In the present work, a statistically refined method has been used to process the results from tracer injection tests in order to extract mixing time estimates from the raw data. Figure 7 shows the experimental data after is has been filtered by this statistical method.

CONCLUSIONS

- The past 15 years have seen the development of a new base-metal smelting technology: the Isasmelt process. This technology has been successfully applied to smelting lead, lead dross, lead battery paste, copper and nickel-copper concentrate.

- The Isasmelt process relies on the Sirosmelt submerged injection lance, which creates a highly turbulent bath and promotes rapid smelting.

- Confidence gained through extensive mathematical and physical modelling of the Isasmelt furnaces and processes has allowed scale-up steps as large as sixty times.

- The development of the Isasmelt technology has been a collaboration between Mount Isa Mines Limited and the CSIRO.

REFERENCES

(1)   R.B.M. Brew, C.R. Fountain and J. Pritchard, "Isasmelt for secondary lead smelting," Lead 90: 10th International Lead Conference, Nice, France, 29-31 May 1990, (Lead Development Association, London, 1991), 170-181.

(2)   P. Bartsch, C.R. Fountain and B. Anselmi, "The Radio Hill project," Pyrosem WA, ed. E.J. Grimsey and N.D. Stockton, (Murdoch University Press, Perth, Western Australia, 1990), 232-250.

(3)   M.D. Coulter and C.R. Fountain, "The Isasmelt process for copper smelting," Non-ferrous Smelting Symposium, Port Pirie, South Australia, 17-21 September 1989, (Aus.I.M.M., Melbourne, 1989), 237-240.

(4)   J.M. Floyd, G.J. Leahy, R.L. Player, and D.J. Wright, "Submerged combustion technology applied to copper slag treatment," Australasian Institute of Mining and Metallurgy, North Queensland Conference, (Aus.I.M.M., Melbourne, 1978), 323-327.

(5)   W.J. Errington, J.H. Fewings, V.P. Keran, and W.T. Denholm, "The
       Isasmelt lead smelting process," Transactions of the Institution of
       Mining and Metallurgy, Section C, 96 (1987), 1-6.

(6)   R.L. Biehl, D.S. Conochie, K.A. Foo, and B.W. Lightfoot. "Design,
       construction and commissioning of a 4 tonne/hour matte fuming pilot
       plant," Extractive Metallurgy Symposium, Melbourne, 12-14 November
       1984, (Aus.I.M.M., Melbourne, 1984), 9-15.

(7)   S.P. Matthew, G.R. McKean, R.L. Player, and K.E. Ramus, "The
       Continuous Isasmelt Lead Process," Lead-Zinc '90, ed. T.S. Mackey and
       R.D. Prengaman (The Minerals, Metals & Materials Society, Warrendale,
       PA, 1990), 889-901.

(8)   C.R. Fountain, M.D. Coulter, and J.S. Edwards, "Minor element
       distribution in the Copper Isasmelt process," Copper '91, Volume IV,
       ed. C. Diaz, C. Landolt, A. Luraschi, C.J. Newman, (Pergammon Press,
       New York, 1991), 359-373.

(9)   W.J. Rankin, F.R.A. Jorgensen, T.V. Nguyen, P.T.L. Koh, and R.N.
       Taylor, "Process engineering of Sirosmelt reactors: lance and bath
       mixing characteristics," Extraction Metallurgy '89, (The Institution
       of Mining and Metallurgy, London, 1989), 577-600.

229

Session 3

Bath Smelting Fundamentals

Reaction Rates and Rate Limiting Factors

in Iron Bath Smelting

R. J. Fruehan

Department of Materials Science and Engineering
Carnegie Mellon University
Pittsburgh, PA

Abstract

In several versions of the iron bath smelting processes ore, coal and oxygen are added, or injected, into an iron and slag bath. The reduction reactions occur by primarily two mechanisms; the reaction of Fe-C drops and of coal char with FeO dissolved in the slag. The fundamentals of these reactions are reviewed and a simple reduction reaction model for the process is presented. Another limiting phenomenon in the process is slag foaming. The fundamentals of slag foaming and predictions of slag foaming in the actual process are presented. A reduction model using laboratory data is capable of predicting the total production rate reasonably well. The reduction is about equally divided between that by the Fe-C drops and that by the char in the slag. The limiting production rate considering reduction and slag foaming for a hypothetical reactor is also discussed. Work to date indicates that the processes are able to produce iron using as much, or less, coal than required for coke for a blast furnace and the smelting intensity tonnes/m^{3}day could be three times.

Proceedings of the
Savard/Lee International Symposium on Bath Smelting
Edited by J. K. Brimacombe, P. J. Mackey,
G. J. W. Kor, C. Bickert and M. G. Ranade
The Minerals, Metals & Materials Society, 1992

Introduction

There are world-wide efforts to develop a new ironmaking process which would eliminate the coke ovens, blast furnaces and, in some cases, the sinter or pellet plants. The blast furnace process has been significantly improved in the past decades and further advances, particularly in the area of reducing coke requirements by coal injection, are expected. However, due to environmental concerns, and capital restraints, the new processes are being developed. The incentive for these processes has been discussed elsewhere in detail.[1] Briefly, they include:

Elimination of coke plants

Elimination of the need for coking coal

Higher smelting intensity

Possible elimination of pellet or sinter plants

Lower capital costs

Lower operating costs

Economical on a small scale

Flexibility in production

Of the cokeless processes, the COREX process is the furthest advanced and a 300,000 tonne per year unit is operating at ISCOR in South Africa.[2] The process, shown in Figure (1), uses the off gas from the melter-gasifier to prereduce pellets, or lump ore, to 90% metalization. Whereas, this process eliminates the coke plant, and has been reasonably successful, the developing bath smelting processes have a greater potential in that they do not produce excess gas, have lower coal consumption, a higher smelting intensity and may not require pellets, or lump ore, as compared to COREX.

The processes which are classified as bath smelting include:

- **Hismelt**

- **DIOS**

- **AISI-Direct Steelmaking**

In these processes coal and partially reduced iron oxide are introduced into an iron and slag bath where melting and final reduction takes place. The reaction gases, CO and H_2, are then postcombusted with oxygen or air. The postcombusted off gas is used for prereduction, generally about 25-30% (Fe_2O_3 to FeO). Hismelt[3] (Figure 2) differs from DIOS and AISI in that air is used for postcombustion and reactions and heat transfer take place via the metal phase. In the DIOS and AISI processes, coal is top fed or injected into the slag. There is a deep foamed slag where reduction and heat transfer from post combustion take place. The energy and material balances for these processes have been developed, for example by Fruehan et al[4] for the AISI process.

234

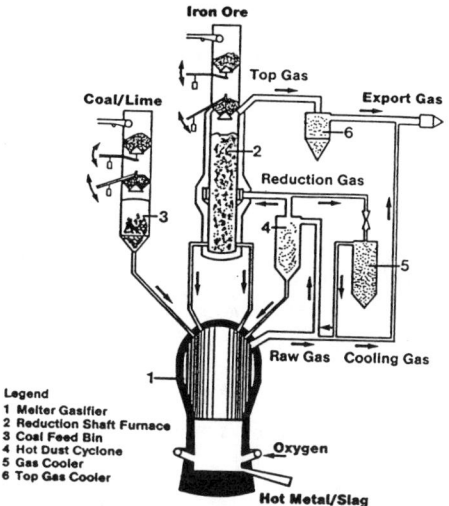

Figure 1. Schematic diagram of the COREX process

With postcombustion of 50%, and using FeO, the coal consumption is 600-700 kg per tonne of metal depending on the coal type. The energy in the off gas will be similar to that in blast furnace off gas.

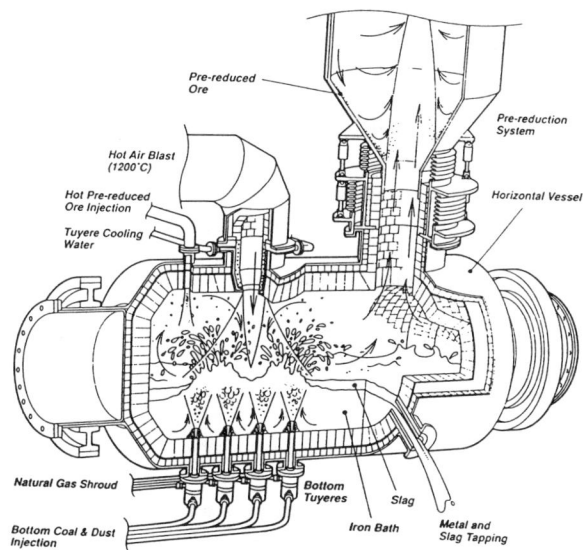

Figure 2. Schematic diagram of the Hismelt process.

Possible rate limiting phenomena in bath smelting include:

Solution of ore and coal

Reduction of FeO

Slag foaming or gas removal

Heat transfer

The rates of solution of FeO in slag[5] and coal char in the metal[6] bath are fast and not rate limiting. For example, it was estimated that it takes only a few seconds (1-3) for an FeO or Fe_2O_3 pellet to dissolve in the slag.[5] Heat transfer from postcombustion of greater than 90% has been reported at moderate production rates.[7] However, heat transfer in larger vessels and higher production rates may be less and possibly insufficient.

This paper will discuss the fundamentals of the reduction reactions and of slag foaming. A simple reduction model is presented. From this model and the results of foaming studies, the maximum production for the process is predicted.

Reduction

A schematic of the bath smelting process, as practiced in DIOS and AISI, is shown in Figure (3). The reaction takes place via two reactions:

$$(FeO) + \underline{C} \ (\text{in Fe}) = CO + Fe \tag{1}$$

$$(FeO) + C \ (char) = CO + Fe \tag{2}$$

Min and Fruehan[8] investigated the rate of reaction (1) for metal droplets (1-5 grams) using the experimental equipment shown schematically in Figure (4). The rate was determined by the rate of gas evolution; typical results are shown in Figure (5). In addition, an x-ray video of the metal droplet reacting with the slag was taken. The x-ray indicated a gas halo developed around the drop and the drop remained suspended in the slag for the fast reaction period.

The presence of the gas halo indicated reaction (1) took place via of two reactions:

$$CO + (FeO) = CO_2 + Fe \tag{3}$$

$$CO_2 + \underline{C} = 2CO \tag{4}$$

The overall rate of the reaction decreased significantly with increasing sulfur content in the metal indicating that reaction (4) was significantly affecting the rate. The rate of reaction (4) has been measured by several investigators and sulfur, which is surface active on liquid iron, retards the rate. The overall rate of reaction (1) was calculated using the independent data for the rate of reaction (4)[9-10] and the equilibrium CO_2 content for reaction (3). As indicated in Figure (6) the agreement is excellent indicating that the reaction sequence is reasonable and reaction (4) is the

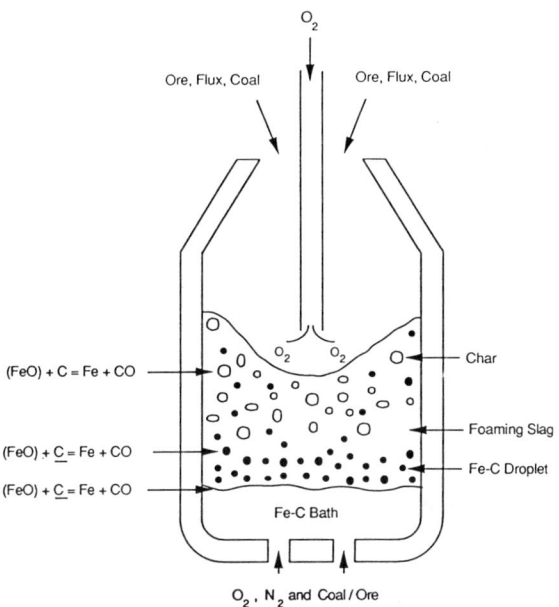

Figure 3. Schematic diagram of bath smelting indicating reduction reactions

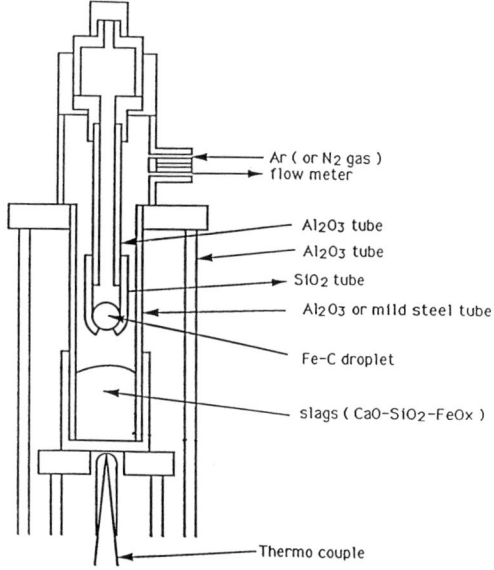

Figure 4. Experimental arrangments to study the reaction of iron drops in slag.

major rate limiting step at high sulfur contents (>0.02%).

There are still some important unanswered questions regarding reaction (1). In particular, the rate appears to decrease significantly at low carbon contents. If the rate is strictly limited by

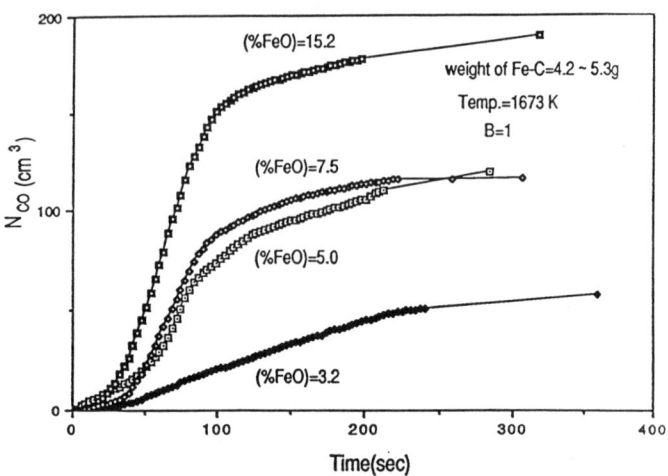

Figure 5. Typical results for reduction of FeO in slag by Fe-C drops.

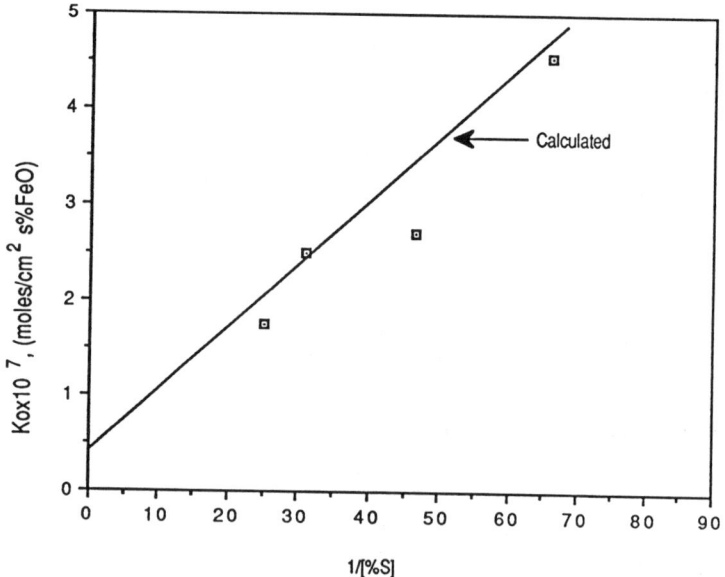

Figure 6. Rate of reduction of FeO in slag by Fe-C drops as a function of sulfur content.

reaction (4) the rate should be independent of carbon content.

Reaction (2) has been recently and extensively studied by Sarma, Cramb and Fruehan.[11] The rate was measured for rotating rods, disks and spheres and the reaction was observed by x-ray video. In this case, as well, a partial gas film existed around the char. Typical results for disks are shown in Figure (7). The gas coming off the sample significantly increased the rate of mass transfer and, therefore, the rate reaction increases with the FeO content more than linearly. The reaction, in a way, is self catalyzed with the evolution of CO gas providing additional stirring. Assuming only mass transfer is controlling the rate the calculated mass transfer coefficient using existing correlations is compared to the observed value as a function of the volume of gas generated in Figure (8). It was concluded that the rate is primarily controlled by mass transfer but reaction (3) may also be influencing the rate.

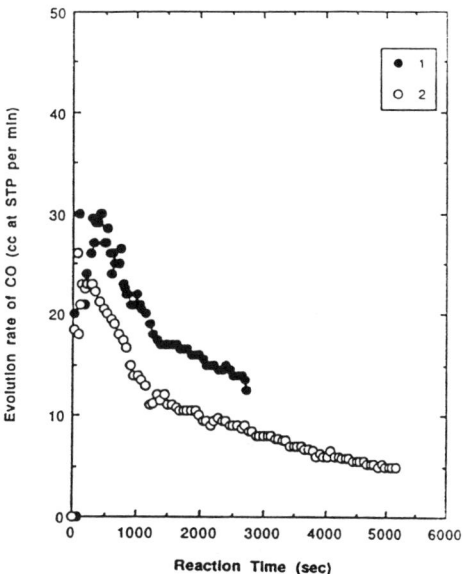

Figure 7. Typical results for the rate of reduction of FeO in slag by graphite(1) and coke disks(2), 6.4 cm².

Reduction Model for Smelting

Sarma, Cramb and Fruehan[11] developed a very simple model for bath smelting. The model uses the fundamental rate constants determined from laboratory studies and pilot plant data on amounts and the sizes of metal droplets and coal char in the slag. The total rate of reduction (R) of FeO in the slag is the sum of the reactions for the metal bath surface (R_{SM}), the char (R_{SC}) and metal drops (R_{SD}). Each of the reduction reactions is approximately proportional to the FeO content and the total rate is given by (6).

$$R = R_{SM} + R_{SC} + R_{SD} \tag{5}$$

$$R = (k_{SM} A_B + k_{SC} A_C + k_{SM} A_D) (\%FeO) \tag{6}$$

where k_{SM} and k_{SC} are the experimental rate constants for the slag metal and slag char reactions and A_B, A_C and A_D are the bath area, char area and metal droplet areas respectively.

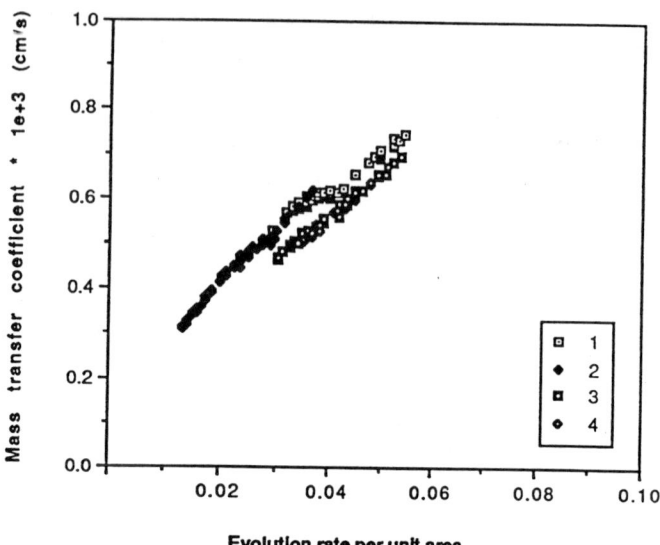

Evolution rate per unit area

Figure 8. Mass transfer coefficient as a function of gas evolution rate.
1. Coke sphere (1.5 cm) 2. coke sphere (1.42 cm)
3. Graphite sphere (1.33 cm) 4. Graphite sphere (1.29 cm)

Nippon Steel conducted 170 ton bath smelting trials and reported on the size and amount of metal droplets and char in the slag along with the production rate.[7] Using equation (6) the production rate was computed to be 42.37 tons/hour compared to 41 tons/hour measured (Table I). Even though the area of the metal bath was taken as three times the planar surface this mechanism contributes less than 1% to the total rate. The major part of the reduction is about equally divided between the char-slag and metal drop-slag reactions.

In general, the weight and, therefore, the surface area of the char and metal drops in the slag is proportional to the slag weight. Also, the slag-bath surface reaction is small and can be neglected. Therefore, to a first approximation the rate can be given by:

$$R = (k'_{S-m} + k'_{S-C}) \, W_S \, (\%FeO) \tag{7}$$

where k'_{S-M} and k'_{S-C} represents the rate constant and the surface area of drops and char per unit weight of slag and W_S is the weight of the slag. Both NSC and AISI indicate the rate is proportional to slag weight and FeO content supporting equation (7).

Slag Foaming

It was recognized early that slag foaming could limit the rate of production in smelting. Ito and Fruehan[12] examined foaming and defined the foaming index (Σ) which is given by:

Table I

Comparison between the calculated production rates from individual reactions with respect to the total production rate
(Actual production 41 tons/hour)

Reaction	Production Rate (ton/hr)	% of total production
Slag-metal bath	0.21	0.5
Slag-char	25.21	59.5
Slag-metal droplet	16.95	40.0
Total production	42.37	100.0

$$\Sigma = \frac{H_f}{V_g} \qquad (8)$$

or,

$$\Sigma = \frac{V_f}{Q} \qquad (9)$$

where H_f and V_f are the increased height and volume of slag due to foaming respectively, V_g is the superficial gas velocity and Q is the volumetric flow rate of gas through the slag. Typical results for the determination of the foam index are shown in Figure (9).

Jiang and Fruehan[13] conducted experiments on the foaming of bath smelting type slags, the foam index is shown in Figure (10). It was found that the foaming index was a function of the properties of the slag. A correlation was developed through dimensional analysis using the data for a large number of slag. The relationship is shown in Figure (11) where $\Pi_1 = \frac{\Sigma g \mu}{\sigma}$ and $\Pi_2 = \frac{\rho \sigma^3}{\mu^4 g}$

The resulting correlation is given by:

$$\Sigma = 115 \frac{\mu}{(\rho g \sigma)^{1/2}} \qquad (10)$$

where,

μ = viscosity
ρ = density
σ = surface tension
g = gravitation constant

In these experiments the foam bubbles were generated by gas injection through an orifice and were about 5-10 mm diameter.

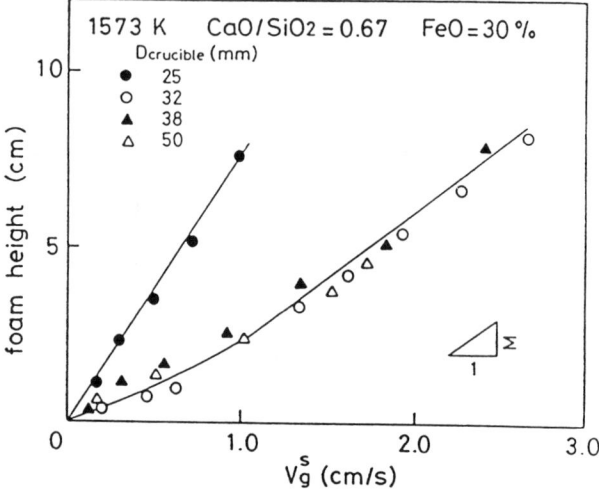

Figure 9. Determination of the foam index.

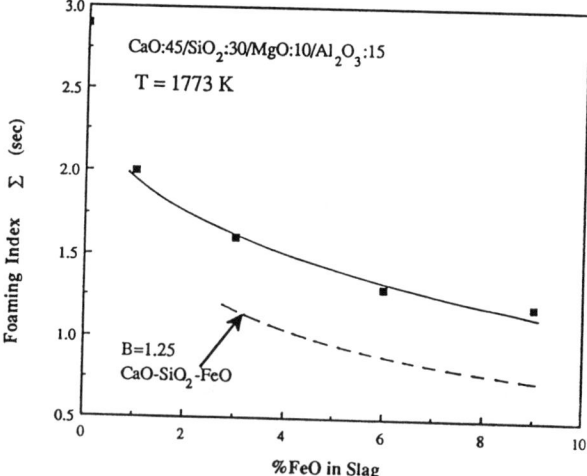

Figure 10. The foam index for bath smelting slags at 1773 K.

In a recent study, Zhang and Fruehan[14] found that the foam index was a function of the bubble size and was inversely proportional to the bubble diameter as shown in Figure (12). It was

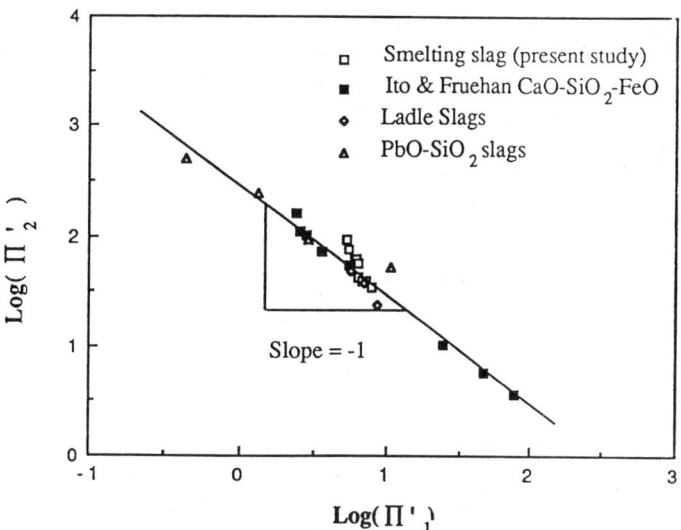

Figure 11. Dimensional analysis relationship of the foaming index to the slag properties.

also found that the foam bubbles generated from reaction (1) were small and depended on the sulfur content of the metal. The small bubble foam was very stable, with a foam index more than 10 times that for 5-10 mm diameter bubbles.

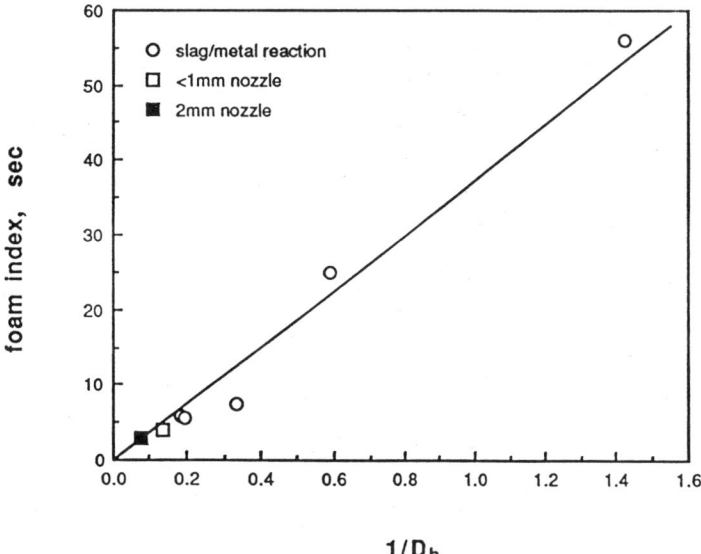

Figure 12. Effect of bubble size on the foam index.

In smelting, the sulfur content is about 0.03-0.15% and the bubble size from reactions (1-2) was found to be 5-10 mm. In addition, gas bubbles result from other sources such as reaction (1)

243

and char oxidation. Therefore, in a real smelting process, the foam will be made up of bubbles of different sizes. It was found in laboratory experiments that the large bubbles pass through the stable small bubble foam. In a first approximation, the foam index is an average of the foam indexes for the different size bubbles. This is the subject of current research.

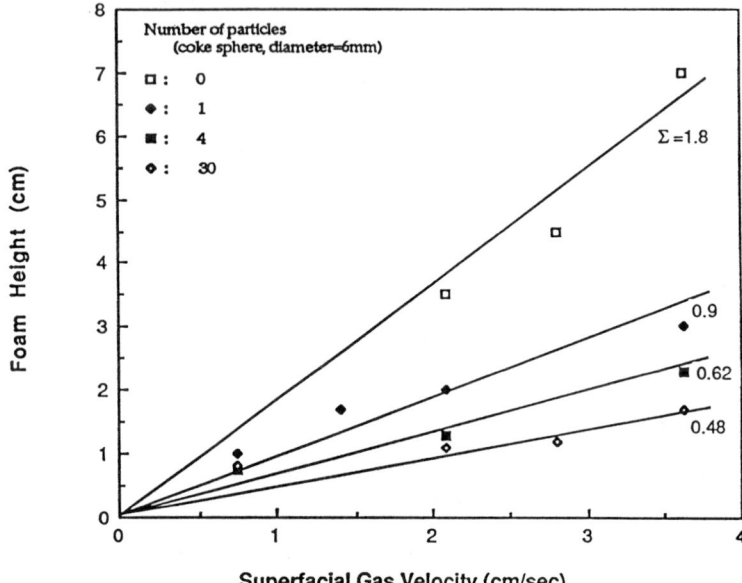

Superficial Gas Velocity (cm/sec)
Figure 13. Effect of coke particles on slag foaming (Crucible 50 mm).

Several investigators had reported that coke and coal char could reduce foaming. Jiang and Fruehan have investigated the fundamentals of this phenomena. The effect of coke is shown in Figure (13). It was found that other non-wetting substances, such as FeO and Al_2O_3, did not affect foaming as shown in Figure (14).

The laboratory measurements of the foam index can be used to predict foaming in the actual smelting operation. AISI measured the foam height and the total gas flow rate. Using equation (9) and the measured foam index for sufficient amounts of char in the slag, the predicted foam height was calculated and is compared to the measured values in Figure (15). The AISI slags were oversaturated with MgO. In computing the bulk viscosity and the foam index, it was necessary to estimate the amount of the solid phase in the slag. The calculation of the foam height was made using the bulk viscosities and resulting foam indexes calculated assuming 10% and 20% of second phase particles in the slag. The agreement is excellent considering the extrapolation from the laboratory experiments to the smelting operation. Also shown in Figure (15) is the height required in OSM as a function of flow rate. Slag foaming at a given gas generation rate is less in bath smelting than in OSM (e.g., BOF) because the char reduces foaming. In the AISI pilot plant when the amount of char is below the critical amount, about 25%, the slag foamed to a greater level and, in some cases, out of the vessel.

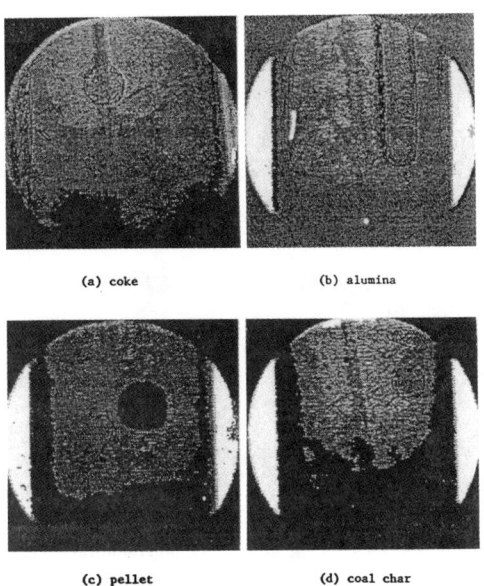

(a) coke   (b) alumina

(c) pellet   (d) coal char

Figure 14. Effect of coke and other particles on slag foaming.

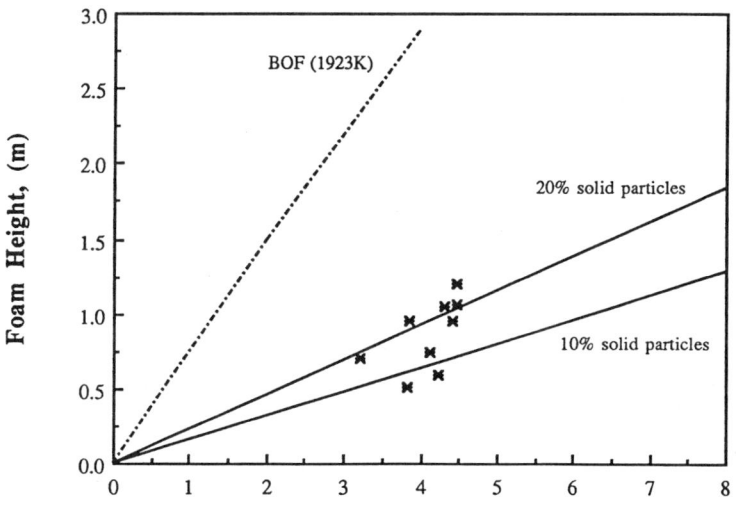

Gas   Velocity at 1873 K, (m/sec)

Figure 15. Calculated and observed foam heights in bath smelting.

245

Theoretical Production Rate

As discussed earlier, reduction, foaming or heat transfer may limit the production rate in bath smelting. NSC was able to smelt over 40 tonnes per hour without any heat transfer limitation. However, for high specific production rates tonnes/m³ day) it was not certain if heat transfer may be a problem. For the present discussion it is assumed that heat transfer will not be limiting and only reduction and foaming are considered.

Zhang and Fruehan[14] have developed a model to predict the smelting rate for foaming or reduction control. Briefly, the available reactor volume (V_A) for foaming is related to the total available volume (V_T), the volume of slag (V_S) and the volume of metal (V_M).

$$V_A = V_T - V_S - V_M \qquad (11)$$

In this calculation, the volume of char is neglected. It is unclear in the NSC correlation, between the reduction rate and slag weight, if the slag weight also includes the char. If the char volume needs to be considered this will reduce the volume available for foam by about 10-15%.

The volume of foam is given by equation (9) using the volume of gas computed from the energy and material balance.[4] For 40% post combustion, 95% heat transfer and smelting FeO pellets at one and three atmospheres the foam volume is given as a function of production rate in Figure (16). The foaming is for a typical slag containing over the critical amount of char for control. The foam volume decreases with pressure simply because the volume or superficial gas velocity decreases with pressure.

According to equation (7), and pilot plant results, the rate of reduction or production is related to the amount of slag, i.e. faster rates of production require more slag. Therefore, to sustain a higher production rate since V_S increases and V_A must decrease. NSC has published data on the effect of slag weight on reduction. Kor[15] found a similar relationship for the AISI pilot results.

For a hypothetical generic smelter with a total volume of 100 m³ of which 65 m³ can be used effectively. The volume of slag (V_S) required for a given production rate was calculated using the NSC and AISI data (3% FeO). A typical metal weight of 12 tonnes was assumed. Using these values, the available volume was computed and plotted in Figure (16). The intersection of the lines gives the theoretical maximum production rate. Using the more conservative NSC reduction data the production rates are 7.2 and 10.3 (t/m³d); if AISI data are used even higher production rates are predicted. It is interesting to note the maximum production rate is not proportional to the pressure as might be expected. This is because higher production rates require more slag reducing the volume available for foam.

These production rates are considered to be conservative since only 65% of the vessel volume and the slower reduction rates were used. The production rates are for the smelter only; obviously the prereducer volume should be included. However, it is reasonable to conclude smelting intensities of 7-10 t/m³d, or more, about 2-3 times that of the blast furnace are possible.

246

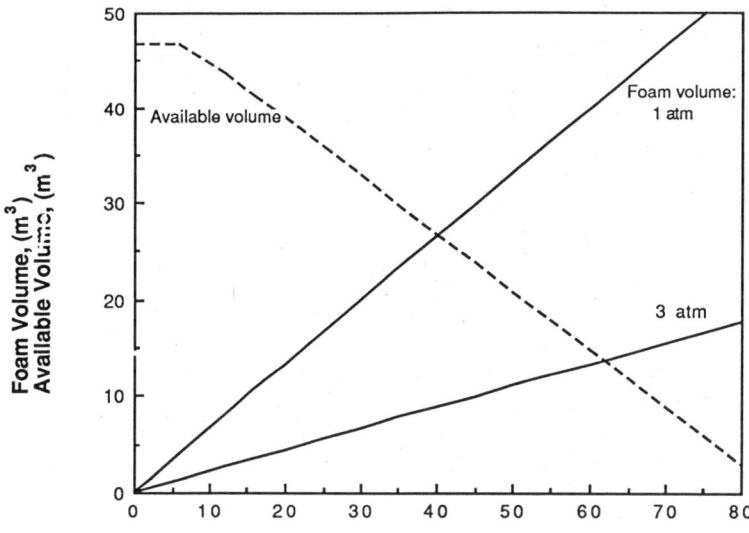

Production Rate, (tonne/hr)

Figure 16. Foam volume and available volume for generic 100 m³ vessel using 65 m³
for smelting FeO. (Coal consumption 700 kg per tonne)

Conclusions

Due to economical and environmental incentives, new ironmaking processes based on bath smelting are undergoing worldwide development. The possible production limiting processes include postcombustion, heat transfer, solution of pellets and coal, reduction and slag foaming. The solution reactions are fast and, most likely, not production limiting. High degrees of postcombustion and heat transfer efficiency have been achieved on a large scale at moderate production rates. Reduction and foaming have been studied in the laboratory and in pilot plants. The reduction model is able to predict the total production rate which is about equally divided by reduction by Fe-C drops and char in the slag. Based on reduction, and foaming limiting production, it is concluded that production intensity in the smelter could be up to 10 tonnes/m³ day.

Acknowledgements

The author acknowledges the AISI Direct Steelmaking Project, funded 77% by the Department of Energy, for primary support of this work. Support by the Center for Iron and Steelmaking Research (CMU) is also acknowledged. The technical input from various members of the AISI Direct Steelmaking research team is gratefully acknowledged.

References

1. R. J. Fruehan, Proceedings of the Elliott Symposium, (Boston MA, 1989, ISS-AIME), 1.

2. L. W. Keppling, New Smelting Reduction and Near Net Shape Casting Technologies, (Pohang Korea, 1990), 373-398.

3. J. V. Keough, G. J. Hardie, D. K. Philip and P. D. Burke, Proceedings of 50th Ironmaking Conference, (ISS-AIME), 635-641.

4. R. J. Fruehan, K. Ito and B. Ozturk, Trans ISS, (I&SM, 1988), 83.

5. B. Ozturk and R. J. Fruehan, Trans ISIJ, (1992), 538-544.

6. L. Zhang and F. Oeters, Steel Research, 62, 3, (1991), 95-106.

7. T. Ibaraki, M. Kanemoto, S. Ogato, M. Matsuo, H. Hirata and H. Katayama, SNRC 90, (Pohang Korea, 1990), 351-361.

8. D.-J. Min and R. J. Fruehan, Metall Trans B, 23B, (1992), 29-37.

9. F. J. Mannion and R. J. Fruehan, Metall Trans B, 20B, (1989), 853.

10. D. R. Sain and G. R. Belton, Metall Trans B, 7B, (1976), 95.

11. B. Sarma, A. W. Cramb and R. J. Fruehan, to be submitted to Metall Trans B.

12. K. Ito and R. J. Fruehan, Metall Trans B, 20B, (1989), 509-521.

13. R. Jiang and R. J. Fruehan, Metall Trans B, 22B, (1991), 481-489.

14. Y. Zhang and R. J. Fruehan, to be published in Metall Trans B.

FUNDAMENTALS OF IN-BATH
SMELTING WITH POST-COMBUSTION

F. Oeters

Berlin Technical University
Berlin, Germany

Introduction

In-bath smelting with post-combustion is known as a concept for the energy-saving production of hot metal and steel. Possible process routes have frequently been described [1 - 7]. The main advantages to be expected from this new technology are

- direct use of fine ore,

- use of coal instead of coke,

- a decrease of the primary fuel consumption up to autothermic production.

The knowledge of the new processes has considerably increased in the last few years and various investigations into the process fundamentals have also been carried out. In this paper, fundamental aspects of in-bath smelting with post-combustion are presented.

Heat and mass balances with post-combustion

The processes of in-bath smelting are in most cases divided into two stages, namely pre-reduction of iron ore with gas in the solid state and subsequent smelting reduction with coal in the liquid state. In the smelting stage, gas is produced from the reaction of coal with iron ore and oxygen. In the pre-reduction stage, this gas is consumed. For a minimum fuel consumption of the entire process, the gas production and the gas consumption should be equal. Otherwise either excess gas is produced or the lacking gas must be produced separately. This leads to a certain operating point of the process at which the two stages are

Proceedings of the
Savard/Lee International Symposium on Bath Smelting
Edited by J. K. Brimacombe, P. J. Mackey,
G. J. W. Kor, C. Bickert and M. G. Ranade
The Minerals, Metals & Materials Society, 1992

connected. Fig. 1[8] shows the gas production in the smelting reduction stage as a function of the degree of pre-reduction for the case that pure pre-reduced iron oxide is charged, that pure carbon is the reducing agent and that no heat losses occur. The gas production rises as the pre-reduction sinks. The figure also shows the gas consumption for the counter-current pre-reduction of hematite as a function of the degree of pre-reduction. It was assumed here that at the stage where the reduction of wustite to iron starts, the wustite-iron equilibrium with the gas phase is established. This equilibrium is defined as % CO_2/(% CO + % CO_2) = 0.31 at 900 °C. The gas consumption rises with the degree of pre-reduction. At the point of 74 % pre-reduction the two lines in Fig. 1 cut, which means that gas production and gas consumption are equal. This point is, therefore, the operating point of the entire process. The point may shift under various influences. If the reductant is, instead of

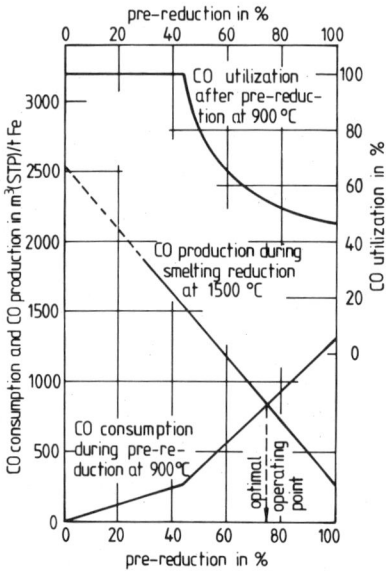

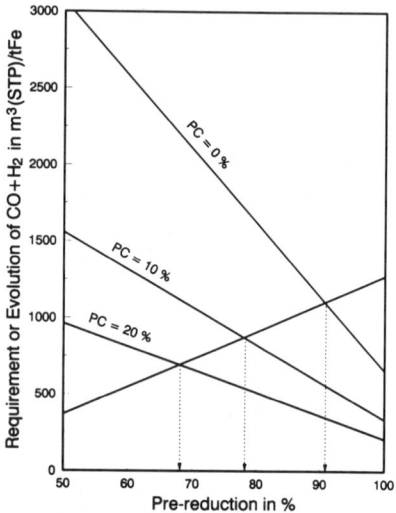

Fig. 1: Determination of optimal operating point in a two-stage reduktion process with counter-current pre-reduction [8]

Fig. 2: Influence of post-combustion degree on the operating point of a two-stage reduction process with counter-current pre-reduction

pure carbon, a real coal which contains ash and hydrogen from the volatiles, the gas production increases and the operating point shifts to higher pre-reduction degrees. If the gas composition remains below equilibrium at the point where iron formation starts, gas consumption increases and the operating point shifts to lower pre-reduction degrees.

The shift to higher pre-reduction at the use of coals mostly prevails. It is considerable at coals with medium or high volatile contents. The operating point may then theoretically be at more than 100 % pre-reduction. This means that excess gas is produced in the smelting reduction stage which, even at 100 % pre-reduction, cannot fully be utilized for pre-reduction. As a further circumstance it has to be attended to that the generation of hot gaseous hydrogen from the volatiles of coal in the smelting reduction stage is heat consuming. If the consumed heat is subtracted from the heat liberated by oxidation of the carbon content of the coal to carbon monoxide, it may happen with high volatile coals that the remaining heat is not sufficient to melt the charged iron even if it is 100 % pre-reduced. Coals that have such properties cannot be used, unless additional heat is supplied by post-combustion [8, 9]. The combustion energy of hydrogen to water vapor can then be utilized and hydrogen-rich coals are usable.

Fig. 2 [8] shows the influence of post-combustion on the position of the operating point for a coal with 6.2 % hydrogen. Whereas without post-combustion the point lies at 90.5 % pre-reduction, it shifts to lower pre-reduction degrees with increasing percentage of post-combustion, as post combustion lowers the gas production in the smelting reduction stage. It was assumed in Fig. 2 that the $CO_2 + H_2O$ is washed out of the gas before introducing the gas into the pre-reduction stage. Such washing must be done at process lines with high degrees of pre-reduction and correspondingly low post-combustion percentages, in order to attain the equilibrium gas composition of $\% CO_2/(\%CO + \%CO_2) = 0.31$ within the pre-reduction stage at the position where reduction of iron starts.

If the degree of post-combustion in Fig. 2 is further increased, the operating point finally shifts to 35 % pre-reduction or less. This is the region where metallic iron is no more generated in the pre-reduction stage. Hence, the thermodynamic limit of $\%CO_2/(\%CO + \%CO_2) = 0.31$ vanishes. The limit now lies at $\%CO_2/(\%CO + \%CO_2) = 0.78$ which corresponds to the equilibrium of wustite and magnetite with the gas phase. The degree of post-combustion must be about 60 % here, in order to supply the energy necessary for the remaining reduction in the smelting stage. However, even a gas with 60 % post-combustion is able to reduce magnetite to wustite in accordance with the above-mentioned equilibrium. The post-combusted gas can be directly used for pre-reduction without an intermediate CO_2-washing process.

A further application of post-combustion is the so-called fully autothermic iron ore reduction process. If the CO_2 of the post-combusted gas is washed out before introducing the gas into the pre-reduction unit, the washing facility may also be used to separate the CO_2 from the top gas of the pre-reduction stage. The remaining gas may then be re-introduced into the bottom of the pre-reduction stage together with the gas from the smelting stage. In this way, the off-gas of the entire process only consists of the $CO_2 + H_2O$ washed out. The process is autothermic. If the same coal is used as in Fig. 2, the coal consumption for such process when carried out at the above-mentioned operating conditions, is 351 kg coal/t Fe [6].

It should be mentioned that a process with about 60 % post-combustion and 35 % pre-reduction also has an almost 100 % gas utilization and is thus nearly autothermic. In this case, no CO_2 separation would be needed which indicates the attractiveness of such procedure.

Process kinetics

The applicability of post-combustion depends on the successful procedure of heat and mass transfer from gas to melt. In the melt itself, the reduction reaction proceeds, and this part of the system is, therefore, governed by the kinetics of reduction. In order to describe heat and mass transfer and reduction kinetics quantitatively, an appropriate process model must be set up.

For modelling the process of in-bath smelting it is suitable to characterize it by the following properties:

1. The process is determined by a number of input flow rates, namely those of ore, coal, and oxygen, and by a number of output flow rates, namely those of iron, slag, and off-gas, which in most cases are constant in time.

2. From the constancy in time of the input and output flow rates it follows that the process is in a steady-state.

3. In accordance with the steady-state condition, the reaction system in the smelting reactor has a number of state variables such as temperature, concentrations and others which are also constant in time.

4. In the reaction system there are at least four different phases present, namely gas, slag, metal, and solid carbon.

5. The phases are intermingled with one another in a number of emulsions and suspensions.

Starting from these properties, the process kinetics can be divided into a number of steps [10 - 13]. The first step is to formulate the equations of macro-kinetics within the in-bath smelting system. Toward this end the system is divided into locally separated reaction sites, namely the gas site, the slag site and the metal site. The concept of reaction sites was early applied to the BOF process[14]. Between the sites there are macro-kinetic heat and mass fluxes. These fluxes are on the one hand interrelated to one another by heat and mass balances which include the input and output flow rates. They are on the other hand determined by kinetic flow equations, in which the flux is expressed by the product of a macro-kinetic rate constant and a driving force. The driving force is the difference of two values of a state variable, for instance, temperature or concentration. The state variables are initially unknown. They establish themselves within the system by the condition that the fluxes given by the heat and mass balances and those given by the macro-kinetic flow equations must be equal. On this basis the state variables and the macro-kinetic fluxes can be determined as functions of the input and output variables and of the macro-kinetic rate constants.

The macro-kinetic rate constants are obtained from a consideration of micro-kinetics. This is the second step of process kinetics, and it is that step which requires most of the labour in the entire procedure of modelling. Micro-kinetics describe the processes taking place within the single reaction sites which, as mentioned above, are structured as emulsions or suspensions of several single phases. According to this structure, the single micro-kinetic processes include the chemical and thermal reactions at the interfaces between the dispersed phases, the physical processes of drop formation and of the establishment of the life times of drops and bubbles and of their size distribution functions as well as flow and mixing processes. Our knowledge of these reactions is good at present. An overview of the physical processes is given in [15]. In some areas, further work has to be done, for instance, about the size distributions of bubbles and drops and about the kinetics of iron ore reduction under in-bath smelting conditions. For an actual modelling of in-bath smelting processes the lack in these areas may be overcome by including observations from pilot-plant and production plant trials into the calculations.

The third step is the numerical calculation. Methods used here are described in [16].

Macro-kinetics of bottom-blowing
in-bath smelting with post-combustion

The process technology of the bottom-blowing reactor with coal injection and post-combustion was developed, using the Savard-Lee tuyere, at the so-called OBM-S and K(M)S steelmaking process in order to attain high scrap rates [17, 18]. It was then transferred to coal gasification and to the development of the HIsmelt process for smelting reduction [19 - 21]. At this technology, pulverized coal is injected with inert gas through the reactor bottom into the melt. Ore is fed from the top. The bottom injection of coal and the strong carbon monoxide evolution from the reaction of iron oxide with the dissolved carbon result in an intense mixing of slag and metal accompanied by the ejection of metal and slag droplets into the gas space above the melt.

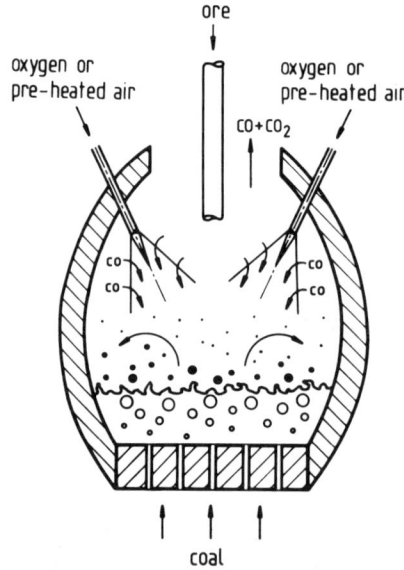

Fig. 3: Iron-bath reactor with post-combustion and heat transfer via suspended droplets. Schematically [28]

Under these circumstances, the entire system may be considered as consisting of two reaction sites: the metal site including the emulsified slag and the gas site with ejected droplets, whereby the gas site is assumed to be completely mixed. This assumption is the optimum possible condition (cf., however, [22 - 24]). A schematic view of the system is shown in Fig. 3 for a converter shape. It is assumed that the oxygen blown in from the top through several nozzles oxidizes the carbon monoxide to carbon dioxide and that the thus produced CO_2-CO mixture is homogeneously distributed over the gas site where the droplets are present. In practice, also hydrogen is combusted to water vapor. The influence of this reaction is discussed later. The heat liberated by the post-combustion reaction is transferred from gas to droplets by convective and radiative heat transfer. The droplets are heated up and, by falling back, give their surplus heat content to the melt. Simultaneously, the carbon dioxide oxidizes the carbon contained in the metal droplets and eventually the iron itself by convective mass transfer. The convective heat and the convective mass transfer from the bulk gas phase to the droplets surface are coupled with one another according to their physical similarity. Therefore, heat transfer is unavoidably accompanied by reoxidation when the heat is directly transferred via iron droplets.

The macro-kinetic heat and mass flow diagram for the process described is presented in Figs. 4 a and b. Fig. 4a gives the diagram for the gas site, Fig. 4b gives that for the metal site, for which complete mixing with the slag is assumed. Fig. 4a at the upper rectangle shows the post-combustion reaction with the corresponding enthalpy change and gas formation rates, the state variables relevant to the gas phase and the input and output flows. The reactions are described by the overall rate equations of heat and mass transfer kinetics on the one hand and by heat and mass flow balances on the other. This is shown in the medium and lower rectangles. The main state variables relevant to the kinetics are also indicated. A large part of them serve to describe the details of micro-kinetics. These details are different at each process type and must be carefully evaluated. As a result of the macro-kinetic flow diagram the steady-state gas temperature and gas composition, the rate of reoxidation $\dot{V}_{CO,reox}$, and the total heat flow from gas to melt $\dot{Q}_{gas-melt}$ yield. The gas leaves the system as off-gas with the steady-state gas temperature and composition. The heat flow $\dot{Q}_{gas-melt}$ is transferred to the melt site. For simplification, the iron oxidation is not considered in Fig. 4a.

The heat and mass flow diagram for the metal site is shown in Fig. 4b. It is similar to that for the gas site. The heat flow $\dot{Q}_{gas-melt}$ coming from the gas serves to meet the demand of heat per time for reduction. The carbon monoxide evolution rate $\dot{V}_{CO,melt}$ results from the reduction reaction and from the carbon oxidation by bottom injected oxygen. Both

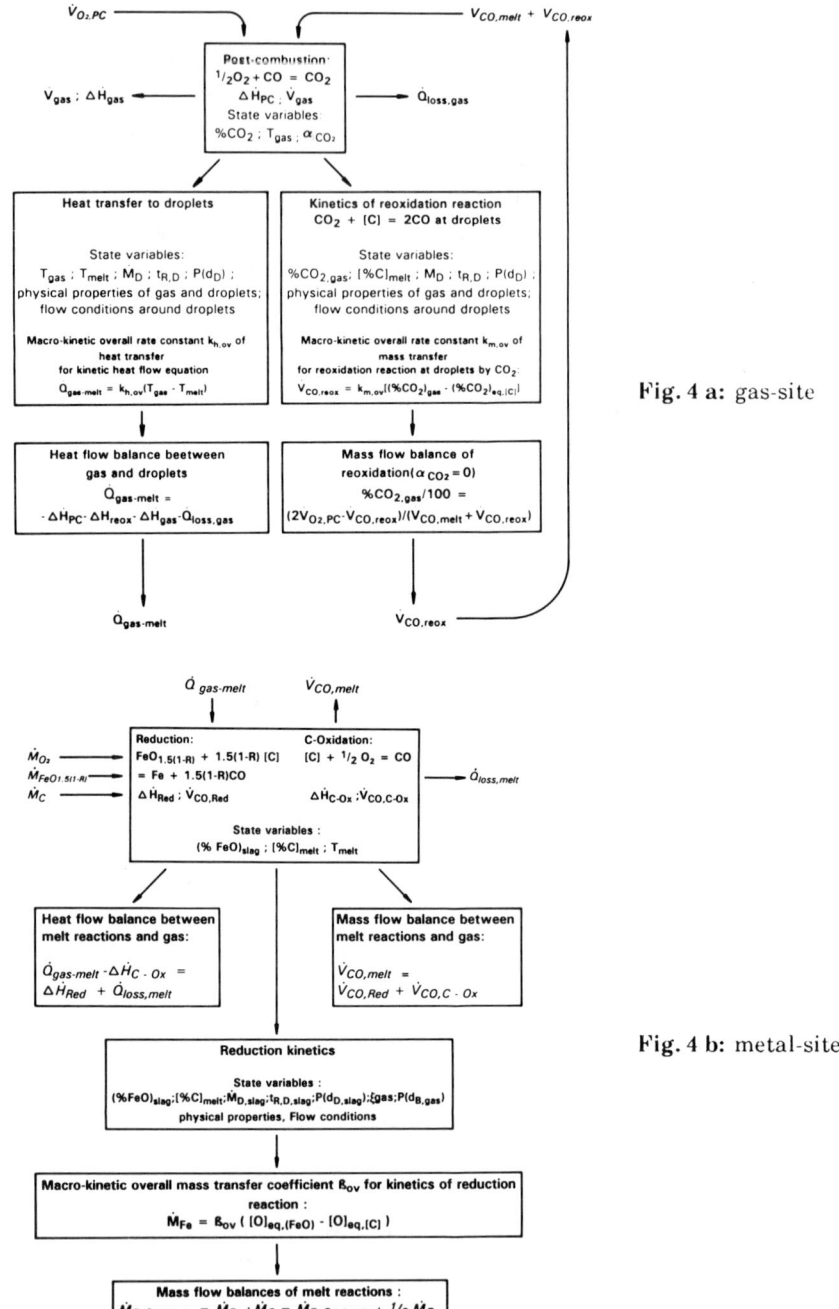

Fig. 4 a: gas-site

Fig. 4 b: metal-site

Caption on next page

Fig. 4: Macro-kinetic heat and mass flow diagram for an iron-bath reactor with post-combustion and heat and mass transfer via suspended droplets

$\dot{V}_{O_2,PC}$	volume flow rate of post-combustion oxygen
$\dot{V}_{CO, melt}$	carbon monoxide evolution rate from melt
$\dot{V}_{CO, reox}$	carbon monoxide evolution rate by reoxidation
$\dot{V}_{gas}$	Off-gas flow rate
$\Delta\dot{H}_{gas}$	of-gas enthalpy flow rate
$\Delta\dot{H}_{PC}$	enthalpy production rate by post-combustion
$\dot{Q}_{loss,gas}$	flow rate of gas heat losses
a_{CO_2}	degree of dissociation of CO_2
T	temperature
$\dot{M}_D$	drop formation rate
$t_{R,D}$	drop residence time in gas space
$P(d_D)$	drop diameter distribution function
$\dot{Q}_{gas-melt}$	heat flow rate from gas to melt
$\dot{M}_{O_2}$	oxygen blowing rate into melt
$\dot{M}_{FeO\ 1.5(1-R)}$	feeding rate of pre-reduced ore
$\dot{M}_C$	coal feeding rate
$\dot{M}_{Fe}$	iron production rate
$\dot{M}_{D,slag}$	slag-drop formation rate
$\dot{V}_{CO,Red}$	carbon monoxide evolution rate by reduction
$\dot{V}_{CO,C-Ox}$	carbon monoxide evolution rate by carbon oxidation
$\Delta\dot{H}_{Red}$	enthalpy production rate by reduction
$\Delta\dot{H}_{C-Ox}$	enthalpy production rate by carbon oxidation
$\dot{Q}_{loss,melt}$	flow rate of melt heat losses
$t_{R,D,slag}$	slag-drop residence time in metal phase
$P(d_{D,slag})$	slag-drop diameter distribution function
ξ_{gas}	gas bubble content in metal phase
$P(d_{B,gas})$	bubble diameter distribution function

reactions are shown in the upper rectangle together with the corresponding enthalpy change and gas formation rates, the state variables relevant to macro-kinetics and the input and output flow rates. The medium rectangles show heat and mass balances between melt reactions and gas. Below, the reduction kinetics are shown. Besides the state variables relevant to macro-kinetics also slag droplet evolution rate, residence time and size distribution as well as gas

257

content and bubble size distribution function are indicated. They are relevant to the micro-kinetics of reduction[25 - 27]. As the main rate equation the reduction reaction is shown. The rate, expressed by the quantity of metallic iron $\dot{M}_{Fe}$ produced per time, is determined by the product of an overall macro-kinetic rate constant and the driving force. The rate constant follows from the details of micro-kinetics. The driving force is the difference of oxygen potentials in equilibrium with the slag iron oxide content and with the melt carbon content at the given carbon monoxide partial pressure respectively. As the oxygen potential in equilibrium with the melt carbon content and the carbon monoxide pressure -- except at carbon contents below 0.5 % -- is small compared to the oxygen potential in equilibrium with the slag iron oxide content, it is the iron oxide of the slag which determines the magnitude of the driving force and, hence, which adjusts itself in such way that the rate of reduction given by the macro-kinetic flow equation and that following from the iron oxide mass balance become equal. The carbon content of the melt is adjusted by the coal feeding rate. As a result, the state variables, the carbon monoxide evolution and iron production rates, and the feeding rates of ore and coal yield. The flow diagrams shown in Fig. 4a and b differ from normal heat and mass balances in such way that the state variables are predicted. This is made possible by the rate equations.

A model of post-combustion for bottom-blowing in-bath smelting

Structure of model

Based on the flow diagram of Fig. 4a, a model of post-combustion in iron-bath reactors was developed [28]. As mentioned above, heat transfer and reoxidation take place at the droplets ejected from the melt into the gas space. The drops have a certain drop formation rate $\dot{M}_D$, whose order of magnitude is obtained from converter experience. The droplets follow ballistic curves which are calculated and from which the droplets residence times in the gas site are obtained. The size distribution function of the droplets is taken from known literature data [29].

Convective heat transfer from the bulk gas phase to the droplets takes place by heat conduction through the gas side temperature boundary layer surrounding each droplet and by heat conduction within the droplet. It can be shown that under the conditions prevailing, heat conduction through the gas side boundary layer is solely rate determining. Conduction within the droplet is so fast that the droplet temperature is homogeneous. The heat transfer rates for this situation were calculated by known methods [30], taking the temperature-dependent properties of gas and droplets and the relative flow conditions between gas and droplets which follow from the ballistic curves and the velocity of the gas into account. Radiative heat transfer is possible by the effect of the sub-microscopic particles of the red iron oxide dust. Dust generates at direct contact of oxygen with liquid iron [5]. The mechanism of radiative heat transfer is shown in Fig. 5. Each droplet present in the gas space is surrounded by a cloud of radiating dust particles. The average volume of the cloud is the reciprocal of the number of droplets per volume. An exchange of radiation takes place between the single droplet and the particle-containing cloud around it. As the particles are at gas temperature, the net heat transfer is directed from the cloud to the droplet. The emission coefficient of the cloud can be estimated from experiences with soot containing flames. It yields that the attenuation depth of the radiation is only a few centimeters, which is in the order of the cloud diameter. This means that radiative transfer between cloud and drop is considerable and that radiation from the wall can be neglected.

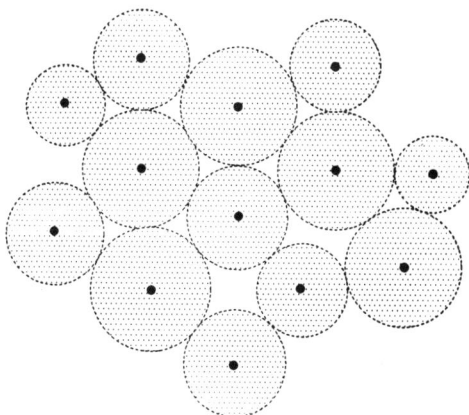

Fig. 5: Droplets suspended in gas space and surrounded by red fume for radiative heat exchange. Schematically

The calculations of convective and radiative heat transfer were carried out separately for each of the 32 classes, into which the size distribution function of the droplets was divided, and then summed up. For the single droplets classes the calculations were performed stepwise over the residence time of the droplets in order to adjust the temperature-dependent physical properties to the instantaneous drop temperature attained at the respective step.

The rate of the reoxidation reaction with CO_2 is determined by diffusion of CO_2 from the bulk gas phase to the drop surface through the gas-side diffusion boundary layer around the drop and by carbon diffusion within the drop. At the surface, the Boudouard reaction CO_2 + [C] = 2 CO is in equilibrium. The counter-diffusion of CO_2 and CO in the gas-side diffusion boundary layer is described by a known mass transfer equation [30]. The gas-side mass transfer coefficient of this transport is calculated in analogy to the corresponding heat transfer coefficient as described before. The carbon diffusion in the drop is calculated by a simplified algorithm for diffusion in spheres. The oxidation of carbon can be followed by an oxidation of iron, if the burn-off time of the carbon is shorter than the residence time of the droplet. Iron oxidation starts when the carbon content of the drop is in the order of 0.01 to 0.02 %.

The calculations were carried out for the cases

- that no reoxidation occurs, i.e. that heat is transferred via slag droplets,

- that reoxidation occurs, i.e. that heat is transferred via metal droplets,

- that instead of pure oxygen pre-heated air is used for post-combustion.

Heat losses are not considered.

Results for post-combustion with oxygen

The following figures show results selected from [28] for post-combustion with oxygen. The values in these figures are in most cases plotted as a function of the so-called theoretical post-combustion degree. That is the ratio of twice the number of moles of oxygen blown in for post-combustion per time to the number of moles of carbon monoxide evolved from the melt per time. This value may also be expressed as $2(P_{O_2}/P_{CO})_P$, where $(P_{O_2}/P_{CO})_P$ means the quotient of the partial pressures of oxygen and carbon monoxide in the gas prior to combustion.

Figures 6 and 7 show the gas temperature and the heat transfer efficiency for the case that no reoxidation occurs, i.e. that heat is transferred via slag droplets. The drop formation rate $\dot{M}_D$ is used as a parameter. It was varied within about one order of magnitude. The values indicated are valid for a 170 t reactor. The size distribution of the droplets obeys an RRS-function with a characteristic diameter d' = 3 mm and a homogeneity exponent n = 1.3 [28, 29]. Fig. 6 shows the steady-state gas temperatures. Depending on the drop formation rate they attain values of up to 2700 °C. The curve for $\dot{M}_D = 114$ kg/s nearly represents the situation of adiabatic combustion. It can be shown that it practically coincides already with the curve for $\dot{M}_D = 0$. No remarkable heat is transferred to the melt in this case. The strong dependence of the gas temperatures on the drop formation rate indicates how much the drop formation is the determining factor for heat transfer between gas and metal and thus for heat transfer efficiency. The efficiency is the percentage of the real heat transfer to the ideal heat transfer which exists, if the gas temperature attains the melt temperature and if the post-combustion reaction $1/2\ O_2 + CO = CO_2$ completely proceeds from left to right. The efficiency, shown in Fig. 7, is highest at the highest drop formation rate. Values of 90 % which are technically satisfying are attained at $\dot{M}_D = 912$ kg/s which means that at the given droplets residence times in the order of 1 second [28] about 0.5 % of the melt weight is steadily suspended within the gas site.

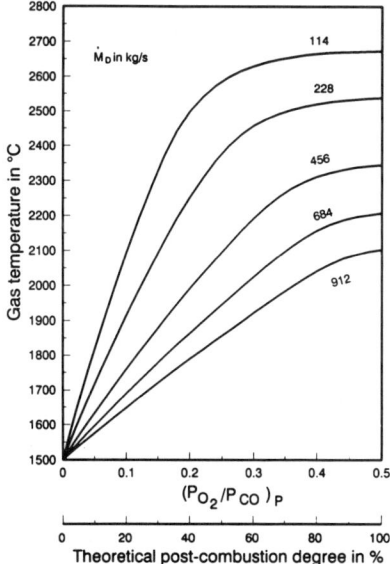

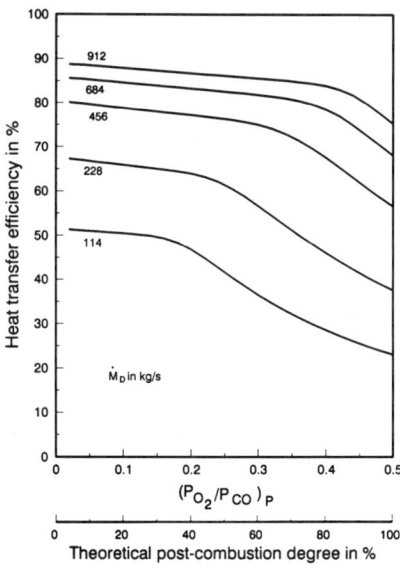

Fig. 6: Gas temperature as a function of the theoretical post-combustion degree for various amounts of slag-drop formation rates

Fig. 7: Efficiency of heat transfer as a function of the theoretical post-combustion degree for various amounts of slag-drop formation rates

261

An additional phenomenon can be observed in Figs. 6 and 7 at high post-combustion degrees. In Fig. 6 the curves change into a more or less horizontal direction, in Fig. 7 the values decrease. This behaviour is caused by the thermal dissociation of carbon dioxide. The dissociation attains a maximum value of 50 % at $(P_{O_2}/P_{CO})_P = 0.5$ and adiabatic combustion [28]. It is negligible in the region where in Fig. 6 the curves are straight and upwardly directed and where in Fig. 7 the curves are almost horizontal. As a consequence of dissociation, less heat is generated and the efficiency decreases. A view into Fig. 6 shows that the dissociation gets remarkable above 1900 °C. This temperature will presumably be the upper tolerable limit for the thermal load of the refractory lining and, hence, should not be exceeded. It means that CO_2 dissociation plays a negligible role under practical conditions.

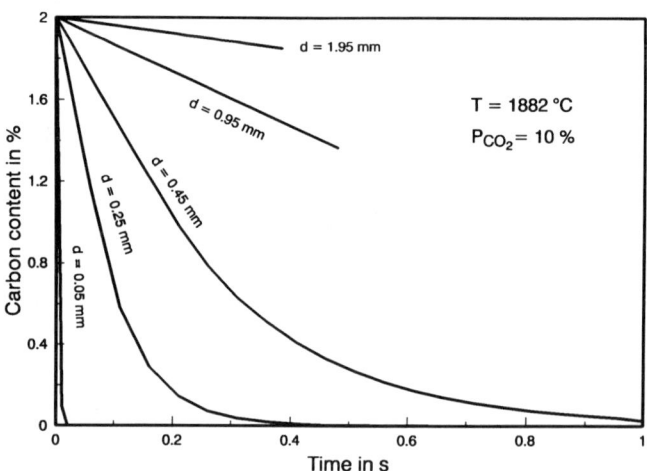

Fig. 8: Carbon concentration in iron droplets as a function of time for various drop diameters

Figs. 8 to 13 show results of model calculations for the case that heat is transferred via metal droplets, i.e. that reoxidation occurs. The results are presented for the same drop size distribution function and the same volumetric drop formation rates as before and, hence, are again valid for a 170 t reactor. The calculations were carried out here until the point where the degree of CO_2 dissociation exceeded 1 %. The curves in the figures cease at this point. In Fig. 8 the decrease of carbon content in the iron droplets during reoxidation is plotted vs. time for various drop diameters, for a gas-phase CO_2 content of 10 % and for an initial carbon content in the droplets of 2 %. Each curve is drawn until the respective

262

residence time of the droplet in the gas space. This time lies between 0.4 s for droplets of 1.95 mm and more than 1.0 s for droplets below 0.45 mm diameter. The decarburization rate increases with decreasing drop diameter. Thus, a drop with 1.95 mm diameter is only weakly decarburized when it falls back into the melt, whereas drops with less than 0.45 mm diameter loose their carbon almost completely. Such drops may subsequently undergo an iron oxidation. The iron oxide produced at the drop surface returns into the melt together with the droplet itself and there reacts with carbon to form carbon monoxide. Thus, iron oxidation finally is also a carbon oxidation.

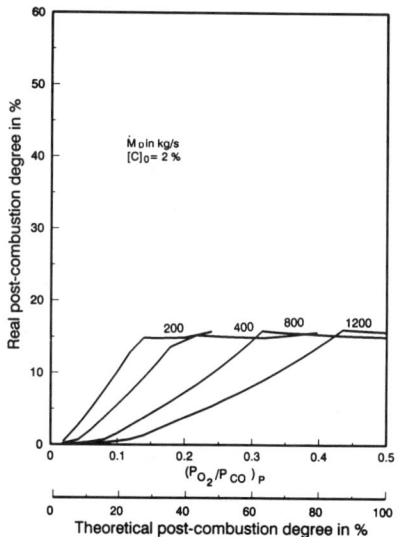

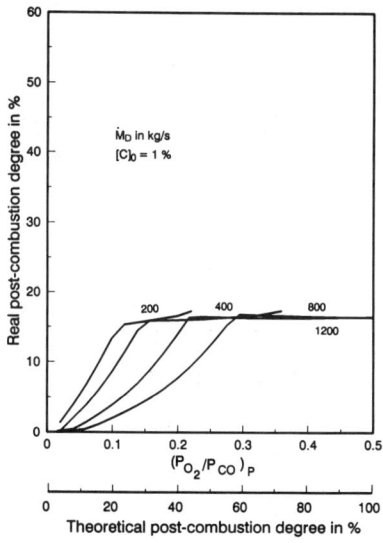

Fig. 9: Bulk gas-phase CO_2 content as a function of the theoretical post-combustion degree for various metal-drop formation rates at 2 % C in the melt

Fig. 10: Bulk gas-phase CO_2 content as a function of the theoretical post combustion degree for various metal-drop formation rates at 1 % C in the melt

The reoxidation reaction at the droplets limits the CO_2 content that can be established in the gas phase at steady-state, as parts of CO_2, initially generated by the post-combustion reaction, are consumed for reoxidation. The result can be seen in Fig. 9. The CO_2 content is shown as a function of the theoretical post-combustion degree for 2 % carbon in the melt and with the drop formation rate as the parameter. The CO_2 content which expresses the real post-combustion degree remains far behind the theoretical post-combustion degree. This effect increases with rising drop formation rate, since more carbon is oxidized and, hence, a higher $(P_{O_2}/P_{CO})_P$ value is needed in order to attain the same CO_2 content in the gas phase. The sharp bends in the curves indicate the begin of iron oxidation. Here, the bulk CO_2 content in the gas phase and thus the CO_2/CO ratio exceeds the value in equilibrium with wustite and iron. The iron oxidation reaction initially proceeds at the smallest droplets. As these droplets have a large surface-to-volume ratio, the reaction is fast and the gas CO_2 content is near equilibrium. The further increase of the CO_2 content is, therefore, rather slow in this region. This leads to the consequence that it is mainly the iron oxidation equilibrium which limits the gas-phase CO_2 content. Fig. 10 shows corresponding results for 1 % carbon in the melt. According to the lower initial carbon content in the droplets the point where iron oxidation starts is reached at somewhat lower theoretical post-combustion degrees than in Fig. 9. The gas CO_2 contents at the start of iron

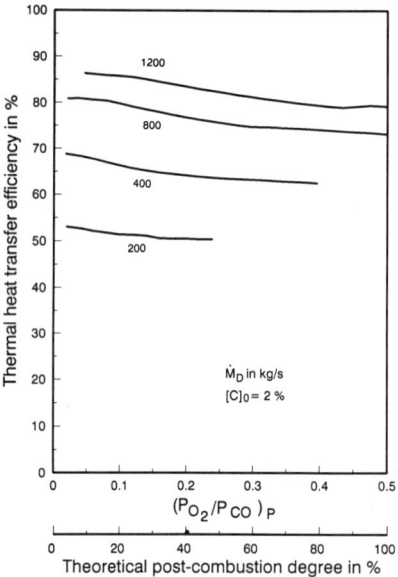

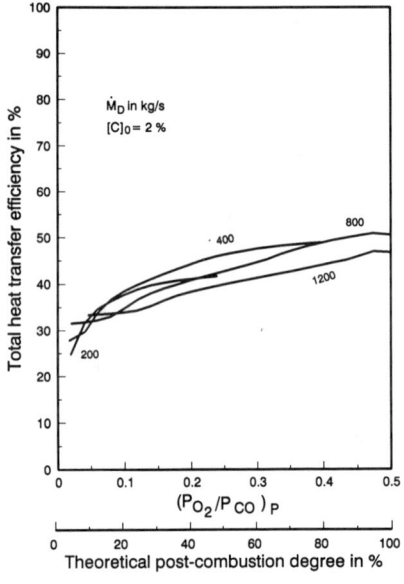

Fig. 11: Thermal heat transfer efficiency as a function of the theoretical post-combustion degree for various metal-drop formation rates at 2 % C in the melt

Fig. 12: Total heat transfer efficiency as a function of the theoretical post-combustion degree for various metal-the drop formation rates at 2 % C in the melt

oxidation are the same as before. Fig. 11 shows the thermal heat transfer efficiency. This efficiency is defined as such that it becomes 100 %, if the gas temperature is equal to the melt temperature. A comparison with Fig. 7 shows that it has only slightly lower values than those attained, if no reoxidation occurs. However, the actual heat generated and transferred to the melt per time by post-combustion at a

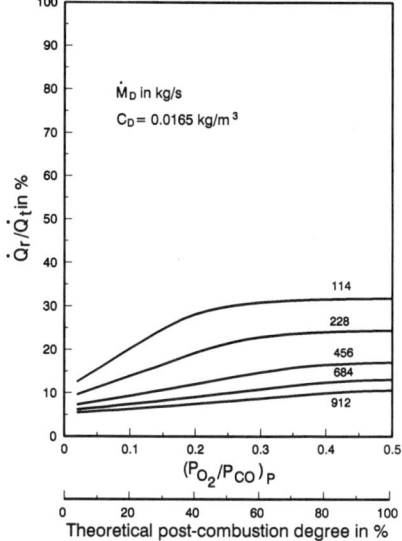

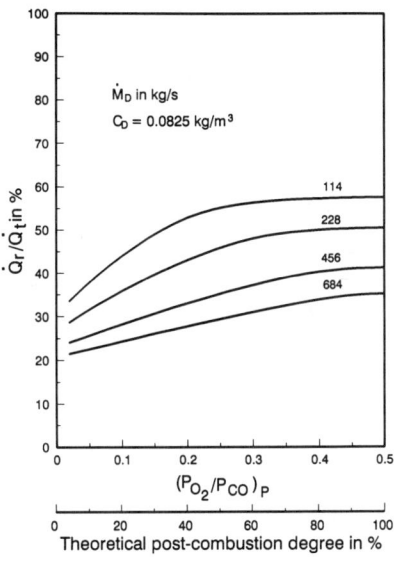

Fig. 13: Percentage of radiative heat transfer from gas to droplets as a function of the theoretical post-combustion degree for various slag-drop formation rates. Fume concentration 0.0165 kg/m³

Fig. 14: Percentage of radiative heat transfer from gas to droplets as a function of the theoretical post-combustion degree for various slag-drop formation rates. Fume concentration 0.0825 kg/m³

given $(P_{O_2}/P_{CO})_P$ value is much smaller with reoxidation than without it, since the post-combustion oxygen blown in is partially consumed for reoxidation instead of post-combustion. Therefore, the total heat transfer efficiency which gets 100 %, if both the real post-combustion degree is equal to the theoretical one and the gas temperature is equal to the melt temperature, is comparatively low (Fig. 12). Contrary to Figs. 7 and 11, the efficiency is lowest at the highest drop formation rate. The reason is that reoxidation is most intense in this case.

As mentioned before, heat is partially transferred from gas to droplets by radiation. Fig. 13 shows the percentage of radiative heat transfer at the total heat transfer, if the red dust concentration in the gas is assumed to be equal to that usual at BOF steelmaking [31]. The percentage lies between 10 and 30 %. Compared to this, at a five times larger dust concentration, the percentage may be between 30 and 55 % (Fig. 14). Radiative heat transfer is useful, as it is, contrary to convective heat transfer, not coupled to the reoxidation reaction.

In total, the results show the narrow range, in which post-combustion is possible, when carbon-containing iron droplets come in direct contact with the oxidizing post-combustion atmosphere.

There are principally two ways to overcome the problem of reoxidation. One is to reduce the oxidizing power of the gas phase by using pre-heated air instead of oxygen for post-combustion, the other one is to transfer heat via the slag instead of directly conveying it to the metal. Both ways may be combined. The first way will be described subsequently. The second way was already described. It will be further discussed in the section on top blowing in-bath smelting.

Results for post-combustion with pre-heated air

Pre-heated air instead of oxygen for post-combustion is used at the HIsmelt process[20, 21]. The dilution of oxidizing gas with nitrogen diminishes the CO_2 concentration in the gas and thus the driving concentration difference of CO_2 between the bulk gas phase and the drop surface. As a consequence, the rate of reoxidation is reduced compared to the use of pure oxygen, whereas the heat transfer rate remains unchanged.

The post-combustion, heat transfer, and reoxidation processes were calculated for the case that post-combustion is carried out with pre-heated air [32]. The temperature of the air was assumed to be 1100 °C. The calculations were carried out in principally the same way as with oxygen. The physical properties of nitrogen were taken into consideration. A fundamental difference between the use of air and of oxygen consists in the diffusion process within the boundary layer at the drop surface. If nitrogen is present, diffusion takes place in a ternary system. However, as nitrogen, carbon monoxide and carbon dioxide have nearly the same diffusion coefficients in the respective binary systems, the diffusion of carbon dioxide in the ternary system may be approximately treated as a binary diffusion process with only a small error. For a numerical calculation the nitrogen concentration at the gas-droplet interface $x^*_{N_2}$ must be known. This concentration lies between zero and the nitrogen concentration in the bulk gas phase x_{N_2}. Calculations were carried out for the limiting cases $x^*_{N_2} = 0$ and $x^*_{N_2} = x_{N_2}$. In the first case, the equilibrium carbon dioxide concentration $x^*_{CO_2}$ of iron oxidation at the interface is higher and, hence, the driving concentration difference $x_{CO_2,bulk} - x^*_{CO_2}$ is smaller than in the second case. As a consequence, iron oxidation is slower at $x^*_{N_2} = 0$ than at $x^*_{N_2} = x_{N_2}$. At carbon oxidation there is no difference between both cases, as $x^*_{CO_2}$ is nearly zero.

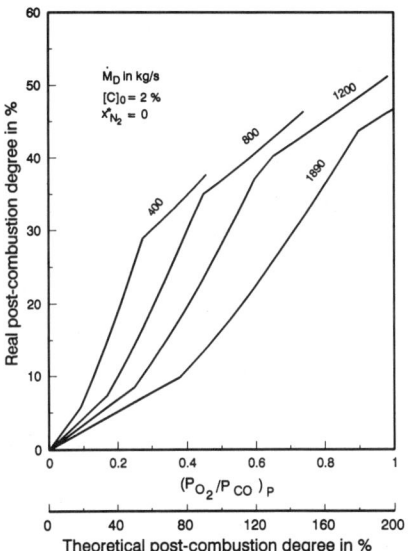

Fig. 15: Real post-combustion degree as a function of the theoretical post-combustion degree for various metal-drop formation rates at 2 % C in the melt and $x^*_{N_2} = 0$

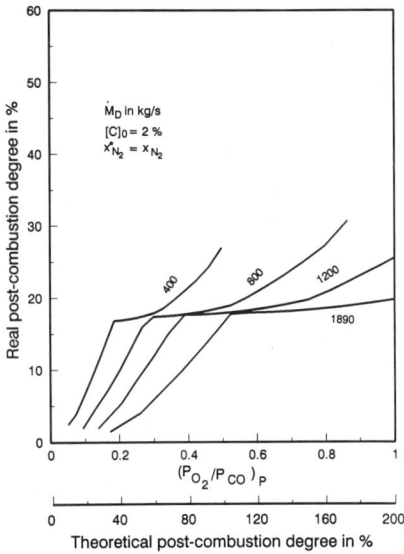

Fig. 16: Real post-combustion degree as a function of the theoretical post-combustion degree for various metal-drop formation rates at 2 % C in the melt and $x^*_{N_2} = x_{N_2}$

The calculations were, contrary to the case of pure oxygen, carried out from $(P_{O_2}/P_{CO})_P = 0$ until $(P_{O_2}/P_{CO})_P = 1$, i.e. until a theoretical post-combustion degree of 200 %. This agrees with the mode, in which the HIsmelt process is actually performed. Oxygen is exclusively blown from the top. Only that oxygen which arises from the ore originally evolves as carbon monoxide from the melt. Under these circumstances, the ratio "moles of oxygen blown in per time"/"moles of carbon monoxide evolving from the melt per time" will have values of up to one for heat balance reasons. The results of calculation are presented in

Figs. 15 to 22. As before, the calculations were finished when a CO_2 dissociation of 1 % was reached. Figs. 15 und 16 show the attained post-combustion degrees for 2 % carbon in the melt and for $x^*_{N_2} = 0$ and $x^*_{N_2} = x_{N_2}$ respectively. The drop formation rate is taken as a parameter. Its values are the same as before. The post-combustion degree is expressed here by $\%CO_2/(\%CO + \%CO_2)$. It appears that the post-combustion degrees are considerably higher than those shown in Figs. 9 and 10 for post-combustion with oxygen. This effect is greater with $x^*_{N_2} = 0$ than with $x^*_{N_2} = x_{N_2}$. The curves in Fig. 16 are similar to those in Fig. 9 with

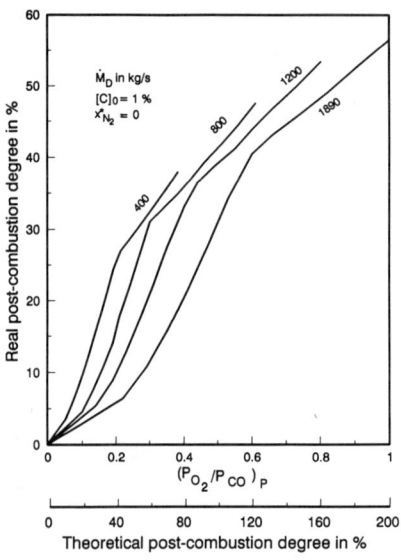

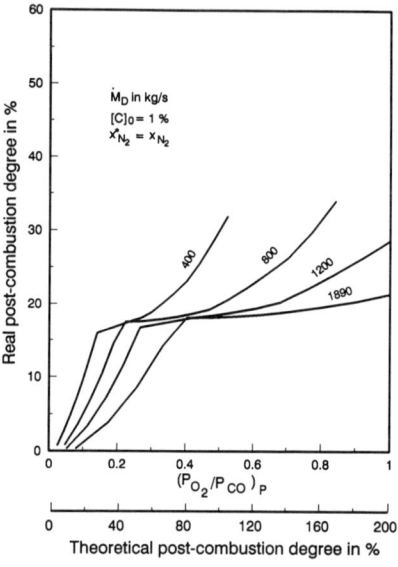

Fig. 17: Real post-combustion degree as a function of the theoretical post-combustion degree for various metal-drop formation rates at 1 % C in the melt and $x^*_{N_2} = 0$

Fig. 18: Real post-combustion degree as a function of the theoretical post-combustion degree for various metal-drop formation rates at 1 % C in the melt and $x^*_{N_2} = x_{N_2}$

the difference that the flat section after the start of iron oxidation at 17 % post-combustion is followed by a relatively steep increase which is not observed in Fig. 9. The reason is that iron oxidation at the droplets gradually extends from small to large droplets. Due to the smaller surface/volume ratio of large droplets, iron oxidation gets slower and less CO_2 is consumed per time. As a consequence, the steady-state bulk gas CO_2 content increases. The same steep increase of the curve would at first sight also be expected at pure oxygen in Fig. 9. There, however, the point of 1% CO_2 dissociation were the curve ceases is reached earlier than at air, as the gas temperature grows steeper. An extrapolation of the curves in Fig. 9 or Fig. 16 to higher $(P_{O_2}/P_{CO})_P$ values would result in a quick flattening of the curves, since CO_2 dissociation makes post-combustion increasingly incomplete. This fact shows that at operating with pure oxygen a further increase of the theoretical post-combustion degree soon becomes useless. A steep increase of the real post-combustion degree as in Fig. 16 is possible only with air because the range where the CO_2 dissociation remains negligible is wider. This fact is an additional advantage of the use of pre-heated air. At $x^*_{N_2} = 0$ (Fig. 15) the attainable post-combustion degrees are higher than at $x^*_{N_2} = x_{N_2}$, since iron oxidation is retarded. Figs. 17 and 18 show comparable results for 1 % carbon in the melt. It is obvious that the curves are steeper at lower carbon contents because less carbon is oxidized. The thermal heat efficiencies are found to be comparable with those shown for post-combustion with oxygen.

Under plant conditions one may expect that the heat transfer between gas and metal site will proceed partially via slag and partially via metal droplets. This mixed heat transfer diminishes reoxidation and leads to a better efficiency and to lower gas temperatures as compared to the calculated results.

The influence of hydrogen on post-combustion

The presence of hydrogen in the gas evolving from the melt influences the reactions in the post-combustion region in three different ways:

1. The molar heat of combustion of hydrogen is smaller than that of carbon monoxide. In presence of hydrogen, therefore, less heat is liberated per volume of gas than in absence.

2. Above 805 °C, the H_2O/H_2 ratio is larger than the CO_2/CO ratio at a given temperature and oxygen partial pressure of the gas phase. Moreover, the H_2O/H_2 ratio in equilibrium with iron oxide and iron increases with rising temperature, whereas the corresponding CO_2/CO ratio decreases. In other words: water vapor acts thermodynamically less oxidizing than carbon dioxide and its oxidizing power on iron decreases with rising temperature while that of carbon dioxide increases. For post-combustion it follows that with larger contents of hydrogen in the gas higher post-combustion degrees, expressed by $(\% CO_2 + H_2O)/(\% CO_2 + CO + H_2O + H_2)$ are to be expected.

269

3. Since hydrogen has a much higher diffusion coefficient than all other technical gases, mass transfer between the bulk gas phase and the surface of iron droplets will in total be accelerated and, hence, reoxidation in the case of mass transfer via metal doplets intensified. This counteracts the thermodynamically lower oxidizing power of hydrogen. How much this effect works quantitatively must be calculated. As at post-combustion temperatures the H_2O/H_2 ratio is about 5 times larger than the CO_2/CO ratio at the same oxygen potential, the remaining hydrogen contents in the gas phase are relatively small. For this reason and due to the counter-action of the thermodynamic and kinetic influences at oxidation, the effect of hydrogen in the gas evolving from the melt on the post-combustion degrees attainable may possibly be small in total. It remains that hydrogen has a lower combustion heat than carbon monoxide.

Top-blowing in-bath smelting with post-combustion.

Besides bottom-blowing in-bath smelting also top-blowing smelting is the object of intense research and development work, mainly in Japan and the United States. Fig. 19 gives a schematic view how the process of reduction and post-combustion may proceed. Lumpy ore is fed from the top and is molten in a hot and foaming iron oxide containing slag, below which the metal melt is located. In many cases it is usual to inject fine-grained coal directly into the slag as the reducing agent. Post-combustion takes place above or within the top region of the slag, whereby heat is transferred to ejected slag droplets or to the fine-dispersed slag in the gas-slag foam respectively. As the heat is transferred to the slag, reoxidation of carbon at this process is excluded. Oxidation of ferrous to ferric iron in the slag must, however, be considered, but its amount is much smaller than the direct oxidation of metallic iron by CO_2. Thus, post-combustion and heat transfer to slag is easier than to metal droplets.

The reduction reaction takes place in the region where the fine-grained coal is injected into the slag. Droplets of metallic iron are produced here and carbon monoxide is evolved in consequence of the reduction reaction. The rising carbon monoxide bubbles generate a circulating flow within the slag. Fig. 19 shows the type of this flow. It should have a pattern that the regions of reduction and of post-combustion are locally separated from one another with respect to the produced iron droplets which means that the flow of iron droplets is directed to the metal bath and, hence, that the droplets do not contact the post-combustion region where they would be oxidized. The slag carries out the complete recirculation, in order to transfer heat from the post-combustion to the reduction region, while the droplets remain deeper under the influence of gravity. This situation is indicated by the small open arrow. The realisation of the described flow pattern requires a careful control of the circulation conditions. A large vertical and a large horizontal distance between post-combustion and reduction region by use of horizontal cylindrical vessels will presumably facilitate the separation of iron droplets. Model experiments may help to find out optimum geometrical conditions.

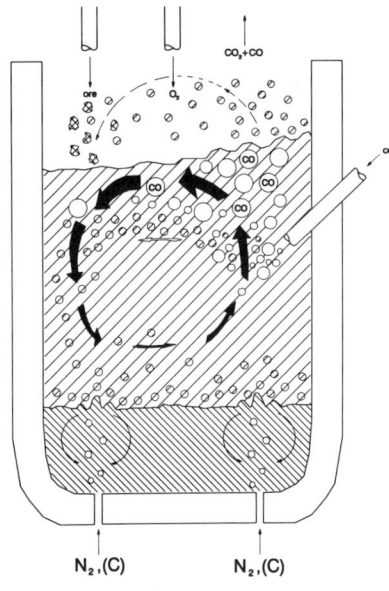

Fig. 19: Schematic representation of the processes taking place at top blowing in-bath smelting with post-combustion

The oxygen potential of the slag in the post-combustion region lies only a little below that of the post-combusted gas itself, since due to the high degree of dispersion, the overall mass transfer resistance is small. Therefore the slag oxygen potential, expressed by the Fe^{3+}/Fe^{2+} ratio, is usually higher in the post-combustion region than in the reduction region where it is near the equilibrium with metallic iron.

The slag heated up in the post-combustion region melts and dissolves the charged ore and transfers heat by the circulating flow to the reduction region where the heat is consumed. Together with the heat, also the oxygen taken up by the slag from the dissolved ore and from the gas is transferred to the reduction region where it is separated by a reduction process. Oxygen-pick up from the gas means that also in this case heat transfer is to some extent accompanied by reoxidation. The heat capacity per volume of the slag and, hence, its ability to transfer heat by recirculation is roughly the same as that of liquid iron, whereas under comparable conditions the oxygen transfer by the slag is less than the corresponding transfer by oxidized iron droplets. Reoxidation is, therefore, minor than with iron droplets. Another object of possible reoxidation by gas or slag is the charged pre-reduced iron ore. If it contains metallic iron, it reoxidizes in the post-combustion region. That can be avoided by directly charging pre-reduced ore into the reduction region. Iron ore which is pre-reduced only until the wustite-magnetite equilibrium and, hence, contains no metallic iron, may be charged into the post-combustion region.

The above presentation describes the case where the post-combustion region and the reduction region are separated from one another with respect to iron droplets and to coal particles. Under these circumstances, the slag is inhomogeneous, a state which has to be maintained during the entire process and which requires that the reactor is geometrically optimized with respect to the separation of metal droplets. Another limiting case would be that the slag is completely mixed and, hence, is macroscopically homogeneous. The driving temperature and concentration differences existing in the slag are then restricted to micro-kinetic dimensions, i.e. to the respective boundary layer thicknesses and drop sizes. Heat transfer would be no problem in this case, but reoxidation would be considerable. The case is, therefore, unfavourable.

For a quantitative description, the system should be divided into the post-combustion site, the reduction site, and the metal site. Post-combustion and reduction site are connected by the recirculatory flow of the slag, reduction site and metal site are connected by the flow of metal droplets. Within the sites, micro-kinetic processes take place, the details of which have to be cleared up and determined in the way described. Besides knowledge in process physics, mixing, and slag foaming, the most important knowledge required for this aim is that of the kinetics of reduction of liquid iron oxide by carbon.

Possible steps in the micro-kinetics of reduction of liquid iron oxide by carbon

The kinetics of reduction of iron oxide and other slag components, such as manganese oxide and silica from liquid slag by carbon was frequently investigated. A systematic evaluation of published results with respect to the dependence of the reduction rate from various parameters was given by Tokuda and Kobayashi [33]. They showed that the overall rate constants of reduction may change over several orders of magnitude. This largely depends on the kinetic conditions which are first and foremost the actual reaction interface, secondly, the stirring conditions. Because the reaction product carbon monoxide acts as the stirring gas, both conditions mutually influence each other.

In a reactor, the stirring effect as well as the steady-state reaction interface - not only of bubbles but also of slag and metal droplets and of carbon particles - are macro-kinetic parameters, while the reaction via the single interface itself is micro-kinetic. For a quantitative description of the entire reaction it should be differentiated between the two types of parameters. Under laboratory conditions which mainly serve the investigation of micro-kinetics, the macro-kinetic parameters should be kept constant as far as possible. Based on earlier publications [25, 33 - 37], a discussion of the possible micro-kinetic reaction steps is given in the following.

The reduction of liquid iron oxide by carbon is written as

$$(FeO) + <C> = [Fe] + CO. \tag{1}$$

At this reaction four phases, namely liquid iron oxide, solid carbon, liquid iron and gaseous carbon monoxide take part. Because mass exchange can occur only at an interface between two phases, the entire reaction must be subdivided into several two-phase partial reactions. In total, oxygen is separated from its bonding to iron in the liquid oxide and entered into the bonding to carbon in the gas phase. Between these two bondings, oxygen must be transported over the respective two-phase reaction interfaces and through the single phases itself. Through the phases, oxygen can be transported in the metal as dissolved oxygen, in the slag via the counter-current flow of ferrous and ferric oxide or alternatively as oxygen ions plus defect-electrons, in the gas via the counter-current flow of carbon monoxide and carbon dioxide. Besides oxygen transport, the transport of carbon must also be considered as carbon dissolves in liquid iron and, hence, is transported in this phase.

All four phases cited are subsequently considered to be present in the reduction region of an in-bath smelting system. Usually the region takes the form of a multi-phase emulsion. Two types of emulsions are possible, one with liquid metal, the other with slag as the continuous phase. According to this classification and the three mentioned modes of oxygen transport, a number of various ways, in which the reduction reaction may proceed, are possible. They are shown in Figs. 20 to 25. For the sake of clarity, the figures are simplified. Not considered are any size distribution, the degree of emulsification, and the size ratios of the dispersed phases. Nor is taken notice of the various chemical and physical factors that may influence the shape of the dispersed entities. At oxygen transport in the slag, only the counter-current flow of ferrous and ferric iron is shown. The alternative transport of oxygen ions plus defect electrons is described by the reactions $(FeO) = [Fe] + O^{2-} + 2e^{\cdot}$ at the slag-metal and $O^{2-} + 2e^{\cdot} + CO = CO_2$ at the slag-gas interface where $e^{\cdot}$ signifies a defect electron. Both reactions together yield the

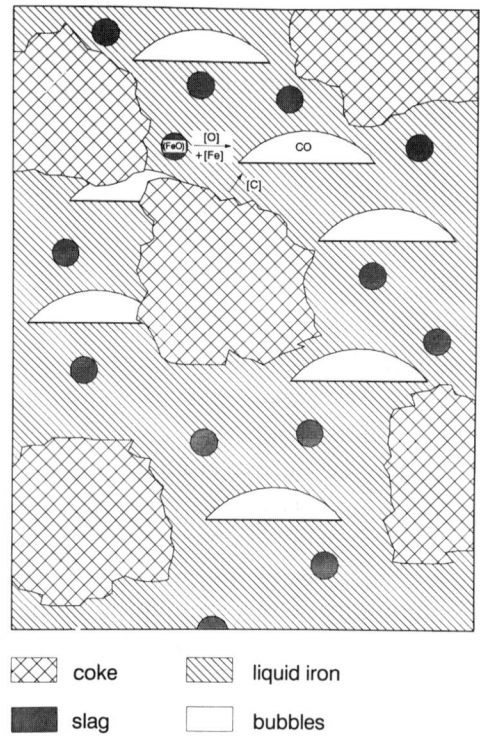

Fig. 20: Distribution of gas bubbles, oxide droplets and coke particles in liquid metal as the continuous phase. Oxygen transfer via the metal phase. Schematically

▨ coke   ▨ liquid iron

■ slag   ☐ bubbles

same result as the counter-current flow of ferrous and ferric iron. All reaction steps in Figs. 20 to 25 are related to one mole oxygen.

Fig. 20 shows the metal phase as the continuous phase. Dispersed in it are coke or char particles, iron oxide droplets and carbon monoxide bubbles. Carbon dissolves from the coke particles into the metal phase.

Iron oxide dissolves from the droplets as [O] + [Fe] also into the metal phase. Dissolved oxygen and carbon move to the bubble-metal interface to form carbon monoxide. If iron oxide droplets are in direct contact with the bubbles (Fig. 21), the reaction may also proceed as shown in Fig. 22

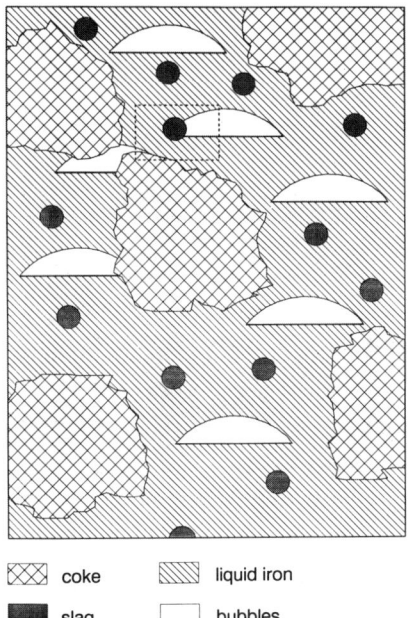

Fig. 21: Distribution of gas bubbles, slag droplets and coke particles in liquid metal as the continuous phase. Bordered section with direkt contact of gas bubbles and oxide droplets. Schematically

XXX coke   \\\\ liquid iron

■■ slag   ☐ bubbles

which presents the bordered section of Fig. 21 in magnification. Within the slag droplet, ferrous and ferric iron move counter-currently. At the slag-metal interface, metallic iron is formed as indicated. Oxygen is withdrawn from the slag droplet at the slag-gas interface by carbon monoxide as the reducing agent. The formed carbon dioxide reacts with dissolved carbon at the gas-metal interface to yield carbon monoxide. The latter reaction may also proceed with solid carbon, if a bubble contacts a coke particle. In the total sequence of the single steps there is one mole oxygen transferred from FeO to CO. Contrary to the way shown in Fig. 20, the transfer here proceeds via two phases: slag and gas phase. Moreover, the reaction can only proceed, if the slag drop is in direct contact with the bubble as the second dispersed phase. Contact between two dispersed phases is less probable than contact between one dispersed and the continuous phase. The first reaction shown in Fig. 20 is, therefore, favoured.

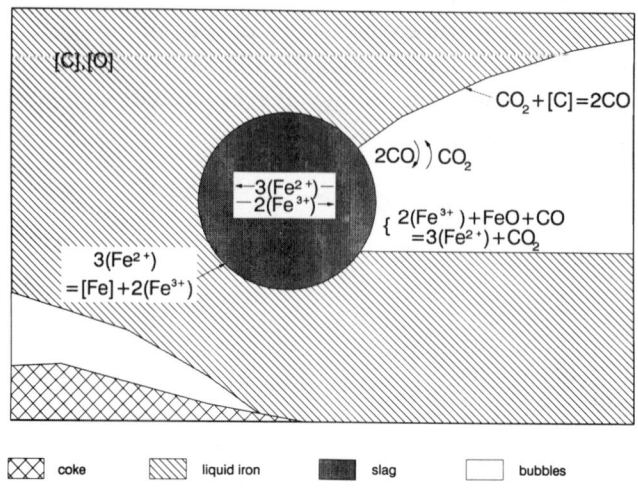

| coke | liquid iron | slag | bubbles |

Fig. 22: Magnified representation of the bordered section of Fig. 21.
Oxygen transfer via the slag and the gas phase. Schematically

Fig. 23 shows the situation, when slag is the continuous phase. Metal droplets, gas bubbles, and coke particles are dispersed. It is assumed here that the coke particles are preferably present within the gas

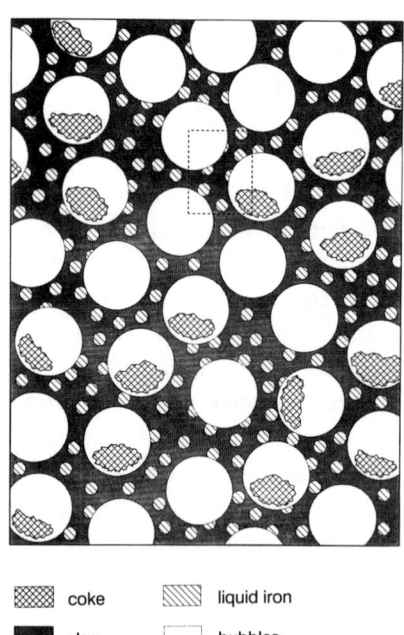

Fig. 23: Distribution of gas bubbles, metal droplets and coke particles in liquid iron oxide containing slag as the continuous phase. Schematically

| coke | liquid iron |
| slag | bubbles |

276

bubbles, as carbon does not wet the slag and as, according to the reaction $CO_2 + <C> = 2CO$ at the gas-coke interface, a gas layer forms between coke particle and slag-gas interface. As a consequence of the preferred presence of coke-particles within the gas bubbles, a direct contact between these particles and the metal droplets produced by the reduction reaction becomes improbable. Hence, a dissolution of carbon into the droplets also becomes improbable. Under these circumstances, the Boudouard reaction $CO_2 + C = 2CO$ will mainly proceed directly at the coke particles within the bubbles and less at the metal droplets. A bordered section, in which all phases partaking are included, is shown magnified in Figs. 24 and 25. Fig. 24 presents the case that oxygen is transferred from FeO to

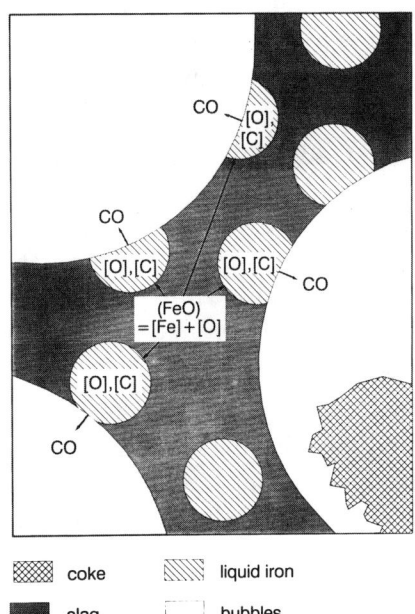

coke   liquid iron

slag   bubbles

Fig. 24: Magnified representation of the bordered section of Fig. 23. Oxygen transfer via the metal phase. Schematically

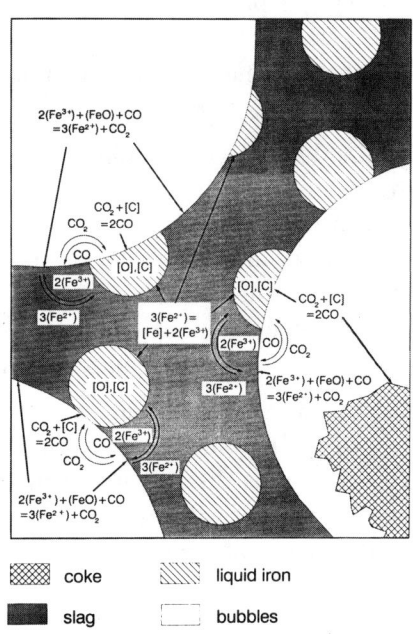

coke   liquid iron

slag   bubbles

Fig. 25: Magnified representation of the bordered section of Fig. 23. Oxygen transfer via the slag and the gas phase. Schematically

CO via the metal phase. Like in Fig. 22, the process is only possible, if the droplet is in simultaneous contact with the gas and the metal phase. Further it is necessary that the droplets have the opportunity to dissolve carbon. As this happens only to a limited extent, this reaction path is unfavourable. Fig. 25 shows the same arrangement for the case that oxygen is transferred via the slag and the gas phase. Here again we have two phases, through which the transport takes place. In Fig. 25, the Boudouard reaction may alternately happen at the gas-coke and at the gas-metal interface. For the reason described, the first way is more favoured than the latter.

The two modes of oxygen transport - one via the metal phase, the other via the slag and the gas phase - are parallel to one another. Within each of them, the transport resistances on the way from FeO to CO are connected in series. Which of the two modes is preferred and which of the various resistances in the sequence of steps is rate determining depends on the type of emulsion, on the actual mass flow densities at the respective reaction interface, on the distances between the single phases, on the local mixing conditions, and on the steady-state sizes of the single interfaces. All these parameters are determined by the process conditions. It is only possible to draw some general conclusions:

1. Under otherwise equal conditions, the steady-state sizes of the single reaction interfaces are larger for interfaces between dispersed and continuous phase than for interfaces between two dispersed phases, since it is less probable, as mentioned, to realize the latter than the former. Numerical modelling of iron oxide reduction requires, therefore, besides size distribution functions and degrees of dispersion, probability considerations on collisions between dispersed phases.

2. The mass flow densities in the two possible modes of oxygen transfer can be estimated. According to Fick's first law, the mass flow density j is considered as the product of the diffusion coefficient D and the driving concentration difference ΔC divided by the thickness of the concentration boundary layer δ_N:

$$j = D\, \frac{\Delta C}{\delta_N}. \qquad (2)$$

For estimation, one characteristic value of the diffusion coefficient is taken in the following for each of the phases considered. The concentration differences are chosen in that way as they may be expected to be typical under in-bath smelting conditions. One value or range is used for each phase. The thickness δ_N of the concentration boundary layer may be approximately expressed for mass transfer at liquid-liquid and at liquid-gas interfaces as

278

$$\delta_N = C \left(D t_R \right)^{1/2} \tag{3}$$

where C is a constant, D the diffusion coefficient and t_R a time that characterizes the details of mass transfer and which depends on the stirring conditions and on the size of the dispersed entity. t_R is fairly independent from the physico-chemical properties of the phases considered and, hence, is assumed to be roughly the same for the three phases metal, slag, and gas. Inserting the lower into the upper equation it yields

$$j \, C \left(t_R \right)^{1/2} = D^{1/2} \Delta C. \tag{4}$$

In this way, the product $D^{1/2} \Delta C$ expresses the influence of the transport properties of, and the typical concentration differences in, the three phases metal, slag, and gas on the reaction rate. The result is as follows:

liquid iron

$D \simeq 10^{-5}$ cm^2/s; $\Delta[O] \simeq 0.1\% \simeq 10^{-5}$ moles/cm^3.

$D^{1/2} \Delta[O] = 3 \cdot 10^{-8}$ moles/cm s$^{1/2}$.

liquid slag

$D \simeq 10^{-5}$ cm^2/s; $\Delta[Fe] = 1$ to $50\% \simeq 10^{-4}$ to $5 \cdot 10^{-3}$ moles/cm^3.

$D^{1/2} \Delta[Fe] = 3 \cdot 10^{-7}$ to $1.5 \cdot 10^{-5}$ moles/cm^2 s$^{1/2}$.

Here, $\Delta(Fe)$ globally expresses the driving concentration difference within the slag.

gas

$D \simeq 1$ cm^2/s; $\Delta CO_2 \simeq 1$ to $10\% \simeq 10^{-6}$ to 10^{-5} moles/cm^3.

$D^{1/2} \Delta CO_2 = 10^{-6}$ to 10^{-5} moles/cm^2 s$^{1/2}$.

It appears that under otherwise equal conditions the mass flow density increases in the sequence metal-slag-gas. The reason for the small value in the metal phase is the low solubility of oxygen in liquid iron. If slag is the continuous phase, it can be concluded that the reduction reaction mainly proceeds via oxygen transport in the slag and in the gas phase. If metal is the continuous phase, oxygen transport in the metal phase, as shown in Fig. 18, may play a role, since in the steady-state the slag-metal interface is greater than the slag-gas interface. In each case, mass transport in the gas phase is fast. Hence, the CO_2 contents at the gas-metal or coke-metal interface will be nearly the same as those at the gas-slag interface. The CO_2 contents at the gas-metal or coke-metal interface are determined by the Boudouard equilibrium. At melt carbon contents of 1% and more, usual at in-bath smelting and at direct contact between gas and coke, the gas CO_2 contents are extremely low. Due to the fast gas transport, this is then also the case at the gas-slag interface. As a consequence, supersaturation of the slag with respect to metallic iron occurs at this interface. Nucleation of iron droplets is enhanced.

3.   For the numerical calculation of the micro-kinetic mass flow densities, the mass transfer coefficients are required. They are obtained by use of known expressions for circumstreamed droplets and bubbles. In order to account for turbulence, the thus calculated values may be multiplied by turbulence factors which for droplets and bubbles are presumably in the order of 2 to 4. Alternately, equations which directly account for turbulence may also be used.

Dissolution of coal

The behaviour of coal at in-bath smelting may be classified into three processes:

-     degassing,

-     gasification,

-     dissolution.

D e g a s s i n g, also called pyrolysis, is a reaction that occurs, if coal particles are brought into a hot atmosphere. The reaction was primarily investigated in connection with the development of new coal gasification processes [38 -40]. A survey of literature is given in [26]. Based on these investigations, a reaction model of degassing at blowing of coal particles into iron melts was developed [26, 27]. If a spherical particle is surrounded by a hot melt, it comes into direct contact with this melt for a few milliseconds. During this time, the surface region of the particle heavily evolves gas and - as a consequence - a gaseous layer between melt and particle develops. The further degassing process is determined by heat transfer through the gas layer and heat conduction within the particle on the one hand and by the temperature dependent rate of the chemical degassing reaction on the other. During degassing, a thin spherical shell, in which the reaction proceeds, moves gradually from the surface to the core of the particle. Fig. 26 shows for Ruhrkohle with 24.8 % volatiles the development of the dimensionless degassed volume as a function of time for various particle diameters. It appears that for usual particle diameters the degassing process needs less than one second.

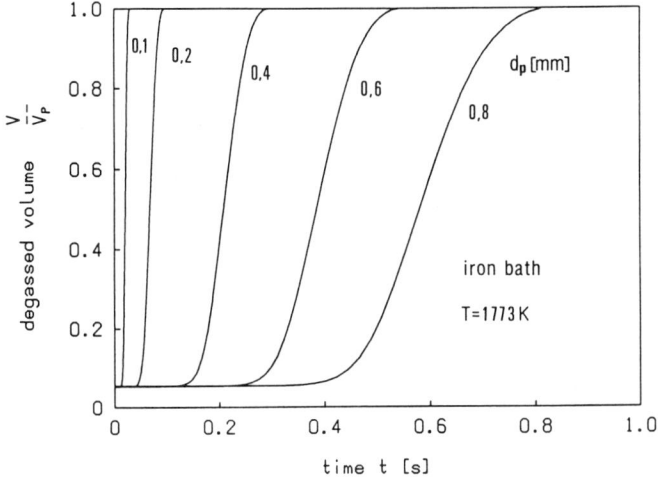

Fig. 26: Course of degassing of coal particles with various diameters after blowing the particles into an iron melt [26]

G a s i f i c a t i o n takes place at coal particles which are blown into iron oxide containing slags after the degassing reaction has finished. The particles are surrounded by a gas layer, whereby, as described, the reaction $CO + (FeO) + 2(Fe^{3+}) = CO_2 + 3(Fe^{2+})$ occurs at the gas-slag interface and the reaction $CO_2 + C = 2 CO$ at the gas-coke interface.

D i s s o l u t i o n of coke particles takes place after the degassing reaction has finished, if coal particles are blown into iron melts . Many investigations were carried out on the dissolution of graphite in liquid iron. The state of knowledge is described in [41]. According to this state, the dissolution of graphite in pure Fe-C melts is transport controlled. Extrapolations of experimental results obtained by means of rotating cylinders to infinitely high rotating speeds revealed the rate constant of the phase boundary reaction of dissolution. This constant is about ten times larger than the transport constants found at usual rotating speeds of a few hundred rpm [41, 42]. Various studies were also made on the influence of sulphur on the dissolution rate of graphite in liquid iron.The sulphur contents of the melt varied up to 1.7 %. Here also, the dissolution is transport controlled. A weak influence of sulphur on the rate is found, as sulphur changes the solubility and the diffusion coefficient of carbon in liquid iron [41]. The thus described influence of sulphur is valid at dissolution times in the order of seconds usual at in-bath smelting processes. If the dissolution needs more than 100 seconds, a retarding effect of sulphur on the phase boundary reaction of carbon dissolution gradually gets remarkable. It works with sulphur in the coal as well as in the melt. Under this influence, the dissolution of coke in liquid iron is more and more retarded with growing time [27]. At the short dissolution times of in-bath smelting, this effect will presumably be insignificant. Nevertheless, high sulphur contents should be avoided.

The dissolution rates of pulverized technical carbonaceous materials in liquid iron are influenced by the behaviour of ash components. It was found in laboratory experiments [27] that the dissolution rate of coke particles is lowered to a certain degree in comparison to ash-free graphite particles, if the ash is liquid, whereas the rate is lowered much more, if the ash is solid. Additions of liquifying slag components to coke particles in amounts sufficient to make the ash liquid at the experimental temperature created a dissolution behaviour like that with a liquid ash.

It is known that carbon is not wetted by liquid slags. If such slag forms during the dissolution of a coke particle, the slag, under the influence of surface forces, tends to contract and approach a spherical shape. This keeps the carbon-melt interface free and the particle continues to dissolve. If a solid oxide forms during dissolution, it maintains its position, since it is not capable to flow. Hence, a dense oxide layer between carbon surface and melt will form and the particle will cease to dissolve. This explanation is confirmed by the observation of Kaask [43] that solid ash layers on coke particles protect the particles against further combustion in an oxidizing atmosphere at 1250 °C.

Productivity

It is known from the blast furnace and from converter processes that productivity is mainly determined by the maximum possible gas through-put. The iron-bath reactor may be compared in this respect with a steel converter. At a blowing time of 12 to 15 min the productivity of the steel converter is usually exhausted [44]. For example, with 40 kg carbon/ton Fe, 12 min blowing time, and a specific converter volume of 1 m^3/t Fe, the maximum decarburization rate is 200 kg C/m³h. This corresponds to a specific gas through-put of 16.7 kmoles gas/m³h. On the other hand, experiments on coal gasification in a 6 t pilot converter type iron-bath reactor in Sweden [45] were carried out with 220 to 300 kg coal/t Fe h. The chemical composition of the used coal (dry) follows from the data published as 64.7 % C; 4.7 % H; 8.6 % O (assumed); 0.7 % S; 12.9 % ash. From this composition and with the assumption that the specific converter volume is - as above - 1 m³/t Fe, the volume-specific gas through-put follows as 17.8 to 24.1 kmole gas/m3h. The lower limit of this range is in good agreement with the above value derived from the steel converter productivity. It seems to be reasonable to take this value as the basis for the gas-through-put that limits the productivitiy of an iron bath reactor, provided the process is run under atmospheric pressure.

The productivity P of a process is generally defined as

$$P = \frac{maximum\ volume\text{-}specific\ gas\ through\text{-}put}{amount\ of\ gas/kg\ fuel \times fuel\ consumption} \tag{5}$$

The amount of gas/kg fuel is specific for a certain fuel. For the coal of the above example it is 0.081 kmoles gas/kg coal. A coal consumption of 600 kg/t of h.m. = 570 kg coal/tFe [46] is considered as typical for a two-stage pre- and smelting reduction process that is run under atmospheric pressure without post-combustion. With this value, taking into account the above-mentioned maximum volume-specific gas through-put of 16,7 kmoles gas/m³h and with 0.081 kmoles gas/kg coal, the productivity results as P = 8.7 t Fe/m³d. This value is in good agreement with published data from pilot plants [47 - 49] related to the above coal consumption of 600 kg/t of h.m. It shows that the limit of productivity seems to be determined by the specific gas though-put in a comparable way as at the oxygen converter process.

Productivities are influenced by various factors. One of them is pressure. It enables a higher volume specific gas through-put and, thus, an increased productivity. In principle, productivity is proportional to pressure. Another factor is decrease in fuel consumption. It reduces the gas production per t Fe and, thus, also increases productivity. For example, at autothermic procedure with post-combustion, CO_2 washing, and gas recirculation a coal consumption of in-bath smelting of 351 kg/t Fe is possible as mentioned above. Under otherwise equal conditions, this would increase the productivity from 8.7 to 15t Fe/m³d.

In this way, post-combustion increases productivity by lowering the coal consumption. The way is important, if high-volatile coals are used, as with these, the gas through-put can considerably be lowered by post-combustion.

Another influencing factor is slag foaming. Strong foaming decreases productivity and foaming increases with the iron oxide content of the slag [50,51].

The total productivity P of a two-stage pre- and smelting reduction process is composed of the productivities of smelting reduction P_{SR} and that of pre-reduction P_{PR} as follows:

$$\frac{1}{P} = \frac{1}{P_{SR}} + \frac{1}{P_{PR}}. \tag{6}$$

Data of P_{PR} are known for counter-current shaft furnaces. The following table presents such data [6]:

Table 1: Productivities of various iron ore direct reduction processes carried out in shaft furnaces with counter-current flow of gas and ore

Process type	Tops pressure bar	Productivity t/m^3d
Midrex [53]	1	8.4
Sumitomo [48]	2	9.6
Nippon-Steel [54] (pilot plant)	5	13.2

In connection with the values for smelting reduction of roughly 8 t/m^3d at 1 bar and 16 t/m^3d at 2 bars or autothermic procedure respectively the data given in table 2 follow from equation(6)[6]. The values of total productivity presented are higher than the productivity of a blast furnace with increased top pressure which amounts 2.4 t/m^3d [52]. This is mainly because at pre-reduction the space required by the coke in the blast furnace vanishes.

In summary, the higher productivities reveal a development potential which is one of the possible advantages of in-bath smelting. It has to be kept in mind, however, that all values cited belong to pilot-plant dimensions, while it is to be expected that with increasing bath height in the reactor which occurs at scaling-up to greater dimensions productivity sinks. Increasing bath height means an increasing running length of the gas and, hence, a greater resistance that

Table 2: Total productivities of two-stage pre- and smelting reduction processes

P_{SR} t/m³d	P_{PR} t/m³d	P t/m³d
8	8.4	4.1
	9.6	4.4
	13.2	5.0
16	8.4	5.4
	9.6	5.8
	13.2	7.0

withstands the gas through-put. The greater resistance may in principle be overcome by an increased pressure drop, but nevertheless, the maximum possible gas through-put $\dot{V}_{max}$ may decrease. The influence of bath height h on $\dot{V}_{max}$ may simply be expressed by $\dot{V}_{max} \propto d^2/h^n$, where d is the diameter of the cylindrically assumed vessel and n an unknown exponent. In order to link gas through-put with productivity, it is required to relate the volume flow $\dot{V}_{max}$ to the reactor volume, whereby the latter is expressed as $V \propto d^2 h$. If it is considered that at scaling-up geometrical similarity is usually maintained, the ratio d/h remains constant. Then it follows.

$$\dot{V}_{max}/V \propto h^{-(1+n)} \propto V^{-\frac{1+n}{3}} . \qquad (7)$$

According to this equation, the maximum volume-specific gas through-put decreases with growing reactor volume, if n > -1. The minimum value of n is zero. If this is the case, there is no growing resistance to gas through-put with increasing bath height. The productivity only sinks because the surface-to-volume ratio becomes smaller at scaling-up. If n

285

> 0, the increasing resistance against gas through-put acts as an additional influence.

For numerical calculation it may be assumed that the volumetric scale-up factor from the pilot to the production plant is 5. According to equation (7), the maximum volume-specific gas through-put of the production plant would then at n = 0 be 0.6 that of the pilot plant, and at n = 1 be 0.35 that of the pilot plant. In the same relation the productivity would decrease. At 2 bar top pressure, the total productivity P of plants for two-stage smelting reduction in table 2 then decreases from 5.8 to 3.5 or 2.03 t/m^{3}d respectively in comparison to 2.4 t/m^{3}d of a blast furnace with top pressure [52].

If in-bath smelting is carried out in one stage with high degrees of post-combustion instead of two stages, a lower productivity P_{SR} than above is to be expected, as the entire reduction now proceeds in one stage with a correspondingly higher gas through-put in this stage. The productivity of a one-stage pilot plant can be estimated from published data of the HIsmelt process [20], if it is assumed that also there the specific reactor volume amounts to 1 m^3/t Fe. The ore feeding rate in the 10 t reactor was 45 kg/min [20]. With 60 % Fe in the ore the productivity then results to 3.9 t/m^{3}d. This value is in the same order of magnitude as the total productivity P of a two-stage process working at 1 bar pressure.

Acknowledgement

The present paper is based on research work, that was supported by the Commission of the European Community of coal and steel. This is greatefully acknowledged.

References

1. Eketorp, S.: Smelting Reduction. Proc. Ironmaking Conf.,Met. Soc. AIME 27(1968), p. 36-39.

2. Steinmetz, E.; Steffen, R.; Thielmann, R.: Stand und Entwicklungsmöglichkeiten der Verfahren zur Direktreduktion und Schmelzreduktion von Eisenerzen. Stahl und Eisen 106 (1986), p. 421-429.

3. Brotzmann, K.: New Concepts and Methods for Iron and Steel production. Proc. Steelmaking Conf. Iron Steel Soc. AIME 70 (1987), p. 3-12.

4. Smith, R.B.; Corbett, M.J.: Coal Based Ironmaking. Proc. Process Technol. Conf. Iron Steel Soc. AIME 8 (1988), p. 49-75.

5. Tokuda, M.: The Development of New Ironmaking Processes in Japan. In: Eichmeyer, H. (ed.): 4. Kohle-Stahl-Kolloquium Berlin, Febr. 21-22, 1989, p. 313-336. Institut für Bergbauwissenschaften der Technischen Universität Berlin.

6. Oeters, F.; Steffen, R.: Entwicklungslinien der Schmelzreduktion und des Einschmelzens. In: Eichmeyer, H. (ed.): 4. Kohle-Stahl-Kolloquium Berlin, Febr. 21-22, 1989, p. 262-310. Institut für Bergbauwissenschaften der Technischen Universität Berlin. Stahl u. Eisen 109 (1989), p. 728-742.

7. Fruehan, R.J.: Iron bath smelting - current status and understanding. Proc. Intnl. Conf. New Smelting Reduction and Near net Shape Casting Technologies for Steel, Rist, Pohang/Korea, Oct. 14-19, 1990, p. 39-56. The Korean Inst. Metals, The Inst. Metals UK.

8. Oeters, F.; Saatci, A.: Mass and Heat Balances during the Reduction of Iron Ores. Düsseldorf, Stahleisen 1987, Proc. Process Technol. Conf., Iron Steel Soc. AIME 6 (1986), p. 1021-1030.

9. Vensel, D.; Henein, H.; Dauby, P.H.: A Thermodynamic Analysis of Decarburization and Post-Combustion in the BOF. Iron Steel Soc. Trans. 9 (1988), p. 5-12.

10. Schenck, H.: Der Einfluß der Verfahrensweise auf den betriebstechnischen Wirkungsgrad der Reaktionen zwischen zwei Phasen, insbesondere Schlacke und Metall. Stahl u. Eisen. 84 (1964), p. 311-326.

11. Ohguchi, S.; Robertson, D.G.C.: Kinetic model for refining by submerged powder injection. Ironmaking Steelmaking 11 (1984), p. 262-282.

12. Oeters, F.: Kinetic treatment of chemical reactions in emulsion metallurgy. Steel res. 56 (1985), p. 69-74.

13. Robertson, D.G.C.; Wei, S.; Raman, A.: Process Dynamics of Metal-Slag and Metal-Slag-Gas Reactions. Proc. Elliott-Symposium on Chemical Process Metallurgy, Cambridge, Ma., June 10-13, 1990, p. 413-445. The Iron and Steel Society.

14. Kootz, Th.; Behrens, K.; Mass, H.; Baumgarten, P.: Zur Metallurgie des Sauerstoffaufblasverfahrens. Stahl u. Eisen 85 (1965), p. 857-865.

15. Brimacombe, J. K.; Nakanishi, K.; Anagbo, P. E.; Richards G. G.: Process Dynamics: Gas-Liquid. Proc. Elliott-Symposium on Chemical Process Metallurgy, Cambridge, Ma., June 10-13, 1990, p. 343-412. The Iron and Steel Society.

16. Szekely, J.; Evans, J. W.; Brimacombe, J. K.: The Mathematical and Physical Modelling of Primary Metals Processing Operations. New York et al. 1988, Wiley .

17. Bogdandy, L. von; Brotzmann, K.; Faßbinder, H. G.; Fritz, E.; Höfer, F.: Verbesserung des Bodenblasens durch die kombinierte Blastechnik und Erhöhung des Schrottsatzes. Stahl u. Eisen 102 (1982), p. 341-346.

18. Bogdandy, L. von.; Geck, H. G.; Großmann, J.R.; Schäfer, R.; Selenz, H. J.; Turner, R.: Oxygen Converter Steelmaking with High Scrap Rates. Proc. Process Technol. Conf., Iron and Steel Soc. AIME 6 (1986), p. 1083-1094.

19. Bogdandy, L. von.; Innes, J.A.: Neuartige Verbundsysteme für Stahl, Chemie und Stromerzeugung auf Kohlebasis. In: 3. Kohle-Stahl-Kolloquium Berlin 1984, p. 313-334. Verlag Glückauf, Essen.

20. Innes, J.A.; Moodie, J. P.; Webb, I. D.; Brotzmann, K.: Direct Bath Smelting of Iron Ores in a Liquid Iron Bath - The HIsmelt Process Proc. Process Technol. Conf., Iron Steel Soc. AIME 7(1988), p. 225-231.

21. Keogh, J. V.; Hardie, G. J.; Philp, D. K.; Burke, P. D.: HIsmelt Process Advances to 100.000t/y Plant. Proc. Iron Steel Soc. Conf. Washington 1991.

22. Gou, H. V.; Lu, W.K.: Mathematical modelling of the combustion chamber in a new metallurgical furnace. In: Szekely, J. et.al. (eds.): Mathematical modelling in materials processing operation, p. 551-564. A publication of The Metals Society Inc. London 1987.

23. Moodie, J.P.; Davies, H.P.; Cross, M.: Numerical Modelling for the Analysis of Direct Smelting Process. Proc. Process Technol. Conf. Iron Steel Soc. AIME 7 (1988), p. 55-64.

24. Hirai, M.; Tsujino, R.; Mukai, T.; Harada, T.; Omori, M.: Mechanism of Post Combustion in the Converter. Trans. Iron Steel Inst. Japan 27 (1987), p. 805-813.

25. Szekely, J.; Todd, M.R.: A Note on the Reaction Mechanism of Carbon Oxidation in Oxygen Steelmaking Processes. Trans. Metallurg. Soc. AIME 239(1967), p. 1664-1666.

26. Orsten, S.: Untersuchung zum Verhalten von Kohleteilchen beim Einblasen in eine flüssige Eisenschmelze - Kohlevergasung im Eisenbad. Dr.-Ing. Dissertation, Berlin Technical University, Berlin 1987.

27. Orsten, S.; Oeters, F.: Behaviour of Coal Particles Blown into Liquid Iron. Proc. W. O. Philbrook Memorial Sympos., Toronto, Ontario/Canada, April 17-20, 1988, p. 27-38. The Iron and Steel Society.

28. Zhang, L.; Oeters, F.: A model of post-combustion in iron-bath reactors. part 1: theoretical basis. part 2: results for combustion with oxygen. Steel res. 62 (1991), p. 95-106 and 107-118.

29. Koria, S. C.; Lange, K. W.: A new Approach to Investigate the Drop Size Distribution in Basic Oxygen Steelmaking. Metallurg. Trans. 15B (1984), p. 109-116.

30. Bird, R.B.; Stewart, W. E.; Lightfoot, E.N.: Transport phenomena. New York et. al. 1960, Wiley.

31. Tsujino, R.; Hirai, M.; Ohno, T.; Ishiwata, N.; Inoshita, T.: Mechanism of Dust Generation in a Converter with Minimum Slag. ISIJ Intnl. 29 (1989), p. 291-299.

32. Zhang, L.; Oeters, F.: unpublished results.

33. Tokuda, M.; Kobayashi, S.: Process Fundamentals of New Ironmaking Processes. Proc. Process Technol. Conf. Iron Steel Soc. AIME 7 (1988), p. 3-11.

34. Robertson, D. G. C.; Deo, B.; Ohguchi, S.: Multi-component mixed-transport-control theory for kinetics of coupled slag/metal and slag/metal/gas reactions: application to desulphurization of molten iron. Ironmaking Steelmaking 11 (1984), p. 41-55.

35. Engell, H.-J.; Vygen, P.: Ionen- und Elektronenleitung in CaO-FeO-Fe_2O_3-SiO_2-Schmelzen. Ber. Bunsenges. Phys. Chem. 72 (1968), p. 5-12.

36. Thielmann, R.; Steffen, R.: Herstellung von Roheisen und Stahlvorschmelzen aus Feinerzen durch Schmelzreduktion. Stahl u. Eisen. 101 (1981), p. 841-851.

37. Koch, K.; Härter, U.; Bruckhaus, U.: Smelting reduction of liquid iron from iron oxide melts with various reducing agents. Proc. Intnl. Conf. New Smelting Reduction and Near Net Shape Casting Technologies for Steel, Rist, Pohang/Korea, Oct. 14-19, 1990, p. 107-119. The Korean Inst. Metals and The Inst. Metals UK.

38. Heek, K.H. van; Jüntgen, H.; Peters, W.G:. Ber. Bunsenges. Phys. Chem. 71 (1967), p. 113-121.

39. Heek, K.H.van: Glückauf, 105 (1969), p. 155-160.

40. Jüntgen, H.; Heek, K. H.van: Brennstoff-Chemie 50 (1969), p. 172-178.

41. Orsten, S.; Oeters, F.: Dissolution of carbon in liquid iron. Proc. Process Technol. Conf. Iron Steel Soc. AIME 6 (1986), p. 143-155.

42. Shigeno, Y.; Tokuda, M.; Ohtani, M.: Influence of Sulphur and Phosphorus on the dissolution rate of graphite into Fe-C alloy. J. Japan Inst. Metals 46 (1982), p. 713-720.

43. Kaask, E.: Slag-Coal Interface Phenomena. Trans. ASME, J. Eng. Power, Jan. 1966, p. 40-44.

44. Kreyger, P. J.: Zusammenhang zwischen Konvertervolumen und Sauerstoffaufblasgeschwindigkeit. Stahl u. Eisen 96 (1976), p. 957-960.

45. Axelsson, C. L.; Kaufmann, D.; Torssell, K.: The CIG Process for Smelting Reduction and Coal Gasification. Proc. Process Technol. Conf. Iron Steel Soc. AIME 6 (1986), p. 1041-1048.

46. Anonymous: The Corex Process. Ed. by VOEST-Alpine Industrieanlagenbau, Linz/Austria.

47. Barin, I.: MIP; Molten Iron Puregas Process. A monograph on coal gasification in a liquid iron bath. KHD Humboldt Wedag Forschungs- und Entwicklungsberichte Nr. 1/1986.

48. Hatano, M.; Miyazaki, T.; Yamaoko, H.; Kamei, Y.: New Ironmaking Process by Use of Pulverized Coal and Oxygen. Proc. Process Technol. Conf. Iron Steel Soc. AIME 6 (1986), p. 1049-1055.

49. Itaya, H.; Katayama, H.; Hamada, T.; Ushijima, T.; Sato, M.; Nakanishi, K.: Development of the XR-Process. Proc. Process Technol. Conf. Iron Steel Soc. AIME 7 (1988), p. 209-216.

50. Ito, K.; Fruehan, R. J.: Slag foaming in smelting reduction processes. Steel res. 60 (1989), p. 151-156.

51. Ogawa, Y.; Tokumitsu, N.: Observation of Slag Foaming by X-Ray Fluoroscopy. Proc. 6th Intnl. Iron Steel Congr. Nagoya, Japan, Oct. 21-26, 1990, Vol. 1, p. 147-152. The Iron and Steel Inst. of Japan.

52. Bonnekamp, H.; Engel, K.; Pfrötschner, G.; Thomas, K.; Winzer, G.: Erfahrungen beim Betrieb eines Hochofens mit hoher Leistung. Stahl u. Eisen 103 (1983), p. 67-73.

53. Förster, E.: personal communication.

54. Muroki, J.; Otsuki, N.; Hachisuka, K.; Nishida, N.; Hara, Y.: Nippon Steel Direct Reduction Process. Nippon Steel Tech. Rep. No. 12, Dec. 1978, p. 114-127.

Recent Advances in Top and Bottom Blowing Converters Based on a Mathematical Model

Yasuo Kishimoto, Toshikazu Sakuraya and Tetsuya Fujii

Iron & Steel Research Laboratories, Kawasaki Steel Corporation
Chiba, 260 , Japan

Abstract

Kawasaki Steel has developed two types of top and bottom blowing converters, namely the K-BOP and LD-KGC. K-BOP blows part of the oxygen gas together with lime through bottom tuyeres. As for the LD-KGC, an inert gas is injected through bottom tuyeres and the oxygen gas is blown through the top lance. Stirring force was intensified by increasing the flow rate of the bottom blowing inert gas in the LD-KGC and by mixing the inert gas with bottom blowing oxygen at the final stage of refining in the K-BOP.

Experiments were carried out in a 5t test converter using five kinds of gases for bottom blowing in order quantitatively to evaluate the role of the bottom blowing gas for the improvement of metallurgical reaction in a converter. These studies revealed that preferential decarburization is more affected by stirring force rather than partial pressure of CO in the gas phase.

In order to explain metallurgical reactions in a converter with bottom blowing gas, a new reaction model was developed theoretically. Taking into consideration oxygen transfer from metal to slag, the model can explain the phenomena that concentrations of carbon and oxygen reduce to those in equilibrium with less than 1.0 atm of CO. The analysis by the reaction model also revealed that the effect of dilution of Pco on behaviour of [O] in the steel and (T.Fe) in the slag is considered small. On the basis of the above results, carbon monoxide to the LD-KGC and carbon dioxide to the K-BOP have been used in practice.

A new parameter for estimation of decarburization characteristics of high Cr steel in combined blowing converters is also theoretically introduced. The parameter can reasonably explain the difference in the decarburization characteristics of high Cr steel in several kinds of combined blowing converters.

Proceedings of the
Savard/Lee International Symposium on Bath Smelting
Edited by J. K. Brimacombe, P. J. Mackey,
G. J. W. Kor, C. Bickert and M. G. Ranade
The Minerals, Metals & Materials Society, 1992

1. Introduction

The introduction of the bottom blowing converter (Q-BOP) at Kawasaki Steel in 1977 and the subsequent improvement of bottom-blowing technology demonstrated that the bottom blowing converter can be operated economically on an industrial scale and that the characteristics of metallurgical reactions in the bottom blowing converter are far superior to those of the top blowing converter (LD converter) [1].

As a result of these efforts, intensive development work was carried out to improve metallurgical reactions in the top blowing converter by introducing bottom blowing technology [2]. Kawasaki Steel developed two types of top and bottom blowing converter, the K-BOP and the LD-KGC [3),4),5)]. The K-BOP is characterized by strong stirring of the molten steel by the high volume of oxygen bottom blowing together with pulverized lime. The LD-KGC, on the other hand, is characterized by its wide range controllability of the flow rate of the bottom blowing gas and moderate investment cost [5].

In this report, recent advances in LD-KGC, K-BOP and Q-BOP technology at Kawasaki Steel based on a mathematical model will be discussed. Then contributions of the stirring force and the dilution of the partial pressure of CO, Pco, of the bottom blowing gas to the metallurgical reaction will be discussed based on experiments conducted with a 5 ton test converter and the mathematical model.

2. Recent Advances in Combined Blowing Converters at Kawasaki Steel

2.1 Development of Top and Bottom Blowing Converter with Wide Range Gas Flow Rate for Stirring

To produce a wide range of steel grades from low carbon to high carbon steel in the same converter, in addition to increasing the durability of tuyeres, wide range control of the flow rate of the blowing gas is necessary from the viewpoint of control of the iron oxide in the slag. To accomplish this, special tuyeres have been used [6]. This type of tuyere was, however, complicated and very expensive. Moreover, the range of the flow rate with this tuyere was not sufficient.

With the establishment of hot metal pretreatment facilities, the need for dephosphorization in the converter has been eliminated. Hence, for low carbon steel, use of a large amount of bottom blowing gas in the final stage of refining is desirable as it permits preferential decarburization. To meet these requirements, Kawasaki Steel developed a new type of top and bottom blowing converter with wide control range of flow rate for bottom blowing gas (LD-KGC).

According to compressive fluid dynamics [7] , the gas flow rate in a tube can be controlled by changing the gas pressure at the inlet of the tube because of the proportionality between the gas pressure and the mass flow rate. To verify this, experiments with water models and hot models were carried out. First, the water model experiments showed that the flow rate of the gas is proportional to the gas pressure of the inlet of the tube from 5 to 100 atm. Figure 1 shows the relation between the flow rate of the bottom blowing gas , Q_{N2} , and the pressure in the inlet of the pipe installed at the bottom of the 5 ton test converter, P . These experiments showed that the flow rate of the bottom blowing gas is proportional to the pressure of the pipe even in hot metal experiments. In this case neither leakage of liquid iron into the tuyere nor the drilling of the gas jet through the bath was observed even when the pressure of the pipe is 100 atm.

Based on the above experiments, a compressor was installed in the bottom blowing gas line of the 150 ton converter at No.2 Steelmaking Shop at Chiba Works in 1984 to realize high pressure gas blowing at rate up to 50 atm. After configuration of the wide controllability of bottom blowing gas flow rates, a similar equipment was installed in the 180 ton converters at Mizushima Works, where the bottom blowing gas was pressurized to 43 atm. Figure 2 shows an outline of the facility. The nitrogen and argon used as the bottom blowing gases are pressurized to 50 atm and stored in receiver tanks.

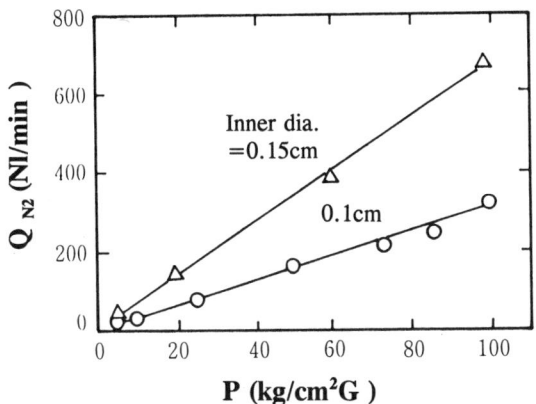

Fig. 1   Relationship between base pressure and gas flow rate in 5 ton experiment

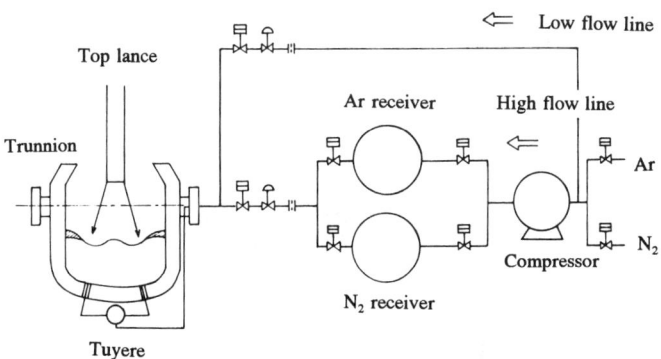

Fig. 2   Conceptional view of LD-KGC

The range of flow rate of bottom blowing gas at Chiba Works is 0.01 to 0.14 Nm3/min per ton of steel. On the basis of the results at Chiba Works, the range of flow rate at Mizushima Works was expanded to 0.005 to 0.2 Nm^3/min per ton of steel.

Adoption of high-pressure gas blowing makes it possible to secure the metallurgical benefits of bottom blowing by expanding the range of flow rate while avoiding problems such as tuyere clogging and tuyere erosion.

In routine operation, the bottom blowing gas flow rate is kept low through the first and middle stages of refining to prevent slopping. The gas flow rate is then increased to the maximum value to prevent excess oxidation of iron and to promote preferential decarburization. By adopting this process, the cost of process gases for bottom blowing can be reduced.

Figure 3 shows the effect of bottom blowing gas flow rate during the final stage of refining on the iron oxide in the slag , (%T.Fe). It was reported previously that the optimum flow rate of bottom blowing gas for improvement of the characteristics of the metallurgical reaction is approximately 0.1 Nm³/min per ton of steel [2] . Nevertheless, it became clear that the characteristics of the metallurgical reaction can be improved by increasing the flow rate of bottom blowing gas above 0.1 Nm3/min per ton of steel. Similar improvements in Mn yield and oxygen content at blow end have been obtained by increasing the gas flow rate .

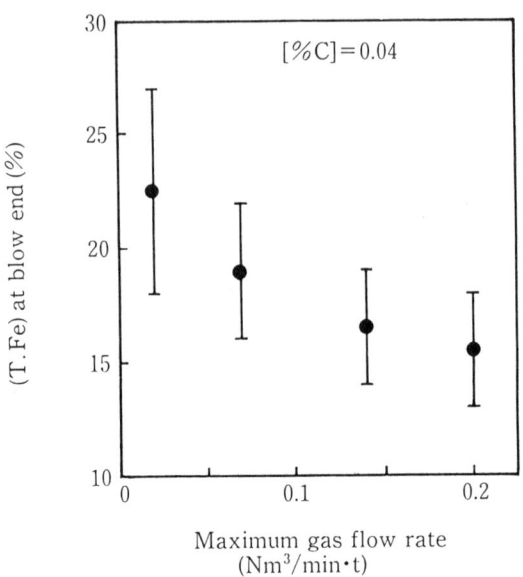

Fig. 3 Relationship between maximum bottom gas flow rate in LD-KGC and (%T.Fe) in slag

The durability of bottom blowing tuyeres has been improved markedly by improving the refractory materials in the bottom area of the vessel, the configuration of the tuyere and slag coating practice. Today, the life of the tuyeres in the 180 ton LD-KGC at Mizushima Works is over 7000 charges without a tuyere change, even with high flow rates of bottom blowing of up to 0.2 Nm3/min per ton of steel.

2.2 Recent Advances in K-BOP and Q-BOP

2.2.1 Development of Mixed Gas blowing Method in K-BOP

A schematic diagram of the three types of the combined blowing converters at Kawasaki Steel, the LD-KGC,K-BOP, Q-BOP are shown in Fig.4. Table 1 shows the characteristics of the converters. One-third of the oxygen necessary for decarburization is blown through the bottom tuyeres in the K-BOP process, hence, a high flow rate of bottom blowing gases can be realized without any cost penalty for stirring gas. The stirring intensity of the oxygen, however, depends on the rate of CO generation in the decarburization reaction, and gradually decreases with decreasing carbon content over the course of refining.

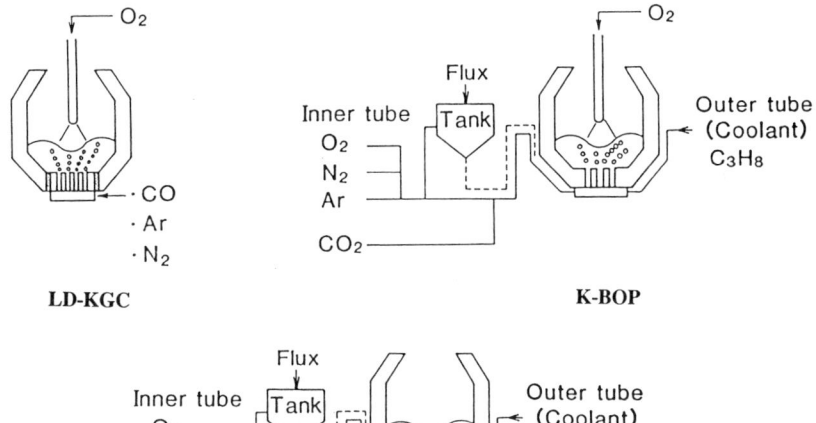

Fig. 4 Schematic diagram of combined blowing converters
at Kawasaki Steel Corporation

Table I Characteristics of Combined Blowing Converters

Item		LD-KGC	K-BOP	Q-BOP
Top blowing	Gas species	Oxygen	Oxygen	-
	Flow rate (Nm³/min/ton)	2.5 ~ 3.5	1.0 ~ 3.0	0
Bottom blowing	Gas species	Ar, N₂, CO	O₂, N₂ , Ar, CO₂	O₂ , Ar
	Flow rate (Nm³/min/ton)	0.005 ~ 0.2	0.25 ~ 0.8	2.5 ~ 4.0

Mixing of inert gas with bottom blown oxygen in the final stage of refining has been successfully practiced with the 250 ton K-BOP to prevent the above-mentioned decrease in stirring intensity [8),9)]. This method, termed IOD (Inert and Oxygen Gas Decarburization), shows the same metallurgical characteristics in the OBM/Q-BOP process.

Fig. 5 a) shows the bottom gas blowing pattern at end point of blowing in K-BOP. Below the critical carbon, [%C] < 0.10~ 0.15, inert gas is injected mixed with oxygen through the bottom and the ratio of the inert gas to oxygen is controlled depending on the carbon content of the melt automatically. Preferential decarburization has been achieved by IOD, resulting in lower [%O] and (%T.Fe) at blow end than with the conventional K-BOP. The IOD process has been used in routine operations and has contributed to the improvement of iron and manganese yield in low carbon steel production.

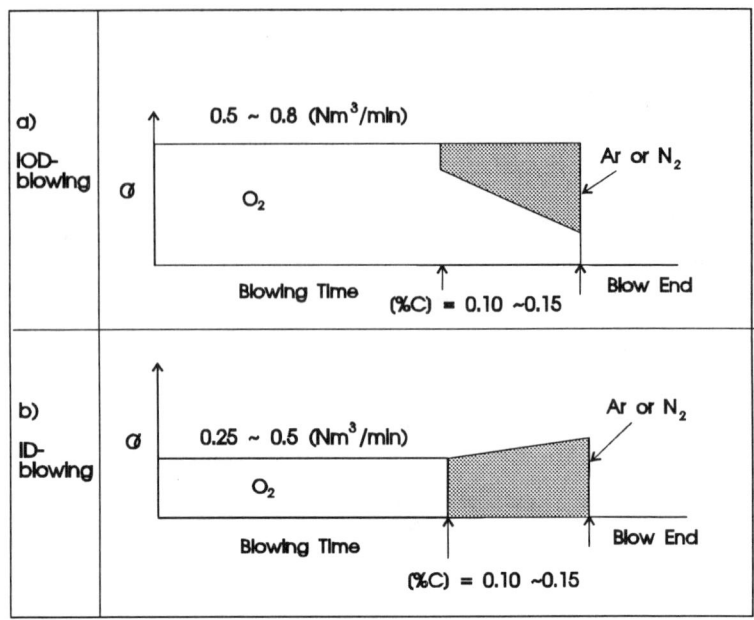

Fig. 5 Bottom blowing gas patterns at end point of blowing
in IOD process and ID process in K-BOP

With the establishment of hot metal pretreatment facilities, the need for dephosphorization in the K-BOP has been removed. Moreover, the decrease in the gas flow rate is desirable from the viewpoint improving tuyere life in the K-BOP. Hence, reducing the bottom blowing oxygen flow rate has been attempted recently [9] from 0.8 Nm^3/min per ton of steel to 0.25 Nm^3/min per ton of steel. On this occasion, inert gas blowing without oxygen through the bottom at the final stage of refining has been practiced to keep the same metallurgical characteristics as with the former K-BOP. Fig.5-b) shows the bottom gas blowing pattern in this process. The inert gas blowing method, termed ID (Inert gas Decarburization) , has shown that the stirring effect of inert gas on the metallurgical characteristics is greater than that of oxygen in the carbon range of less than 0.1 %.

The ID process is being used in the K-BOP and the combination of ID process and the reduction of gas flow rate of bottom blowing gas contributes to the prolongation of tuyere life and the reduction of process gas costs.

2.2.2 Optimization of oxygen flow rate in the final stage in Q-BOP

The establishment of hot metal pretreatment facilities has also enabled us to use dephosphorized hot metal in Q-BOP. Moreover, the production of cold rolled strip sheets with carbon concentration of less than 30 ppm has been increased recently. The decrease of oxidation of iron in the final stage of reining in the converter has been desired from the viewpoint of the improvement of the surface quality of cold rolled sheets [9]. Hence, several improvements have been made for establishing optimum and cost-efficient operations in Q-BOP [10], corresponding with the change in the above-mentioned situation.

Fig.6 shows the influence of oxygen flow rate on oxygen content in steel at blow end, [%O]. The reduction of oxygen flow rate after the sub-lance improves the decarburization efficiency and hence decreases [%O] and (%T.Fe) in the slag. This result shows that the optimization of the oxygen gas flow rate at the final stage of refining is important in the Q-BOP, which can be explained by ISCO values [3] , as shown in Fig.7. The combination of this operation and the KTB method, in which oxygen gas is blown onto the molten steel through the lance in the vacuum chamber in the RH process (Fig.8), realized the effective refining process for ultra low carbon steel [11] .

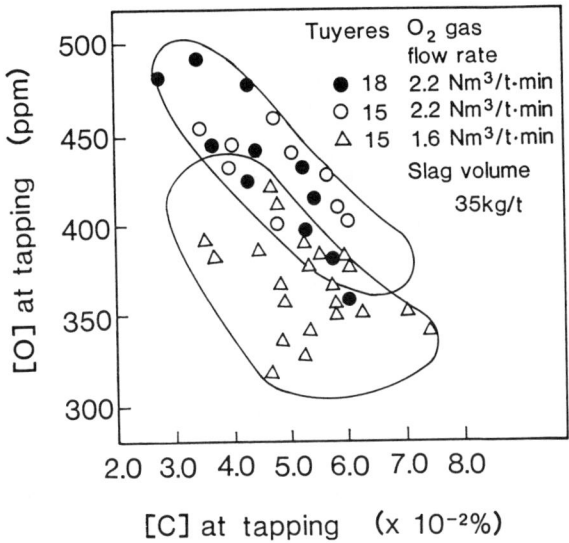

Fig. 6 Influence of oxygen gas flow rate on [%O] at blow end in 230 t Q-BOP [10]

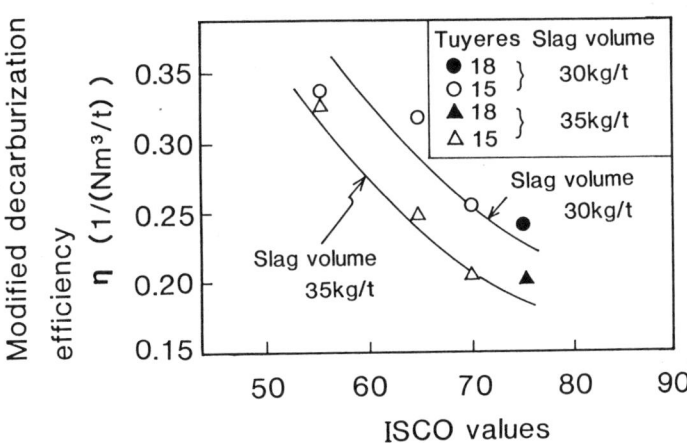

Fig. 7 Relation between modified decarburization efficiency in Q-BOP
and ISCO values[10]

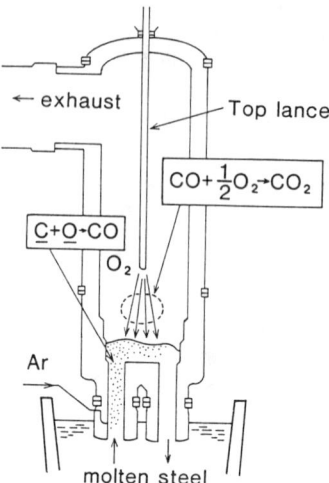

Fig. 8 Schematic illustration of KTB in RH degasser [11]

3. Mathematical Model for Describing Decarburization and Deoxidation in The Converter

3.1 Effect of Bottom Blowing Gas Species on Metallurgical Reaction Characteristics in a Converter [5),12)]

It has been considered that the following two effects of the bottom blowing gas in the combined blowing converter decreases [%O] and (T.Fe): the enhancement of bath stirring and hence decrease in difference between operational values and thermodynamic equilibrium values, and the dilution of the partial pressure of CO gas , Pco, resulting in a change in thermodynamic equilibrium values.

From the latter point of view, inert gases such as Ar or N_2 may be superior to CO or CO_2 gas. According to the operational results of the top and bottom blowing converters, however, improvement of the bottom blowing gas does not depend on the gas species used, but on its flow rate.

Several studies [13),14)] suggested the relation between [%O] in the low carbon range and the distribution ratio of oxygen between the metal and the slag. However, there has been no attempt to explain the phenomena that decarburization and deoxidation in the converter does not depend on the Pco. There is no study performing a quantitative evaluation of the contributions of stirring force and dilution of Pco by the bottom blowing gas to the characteristics of the metallurgical reaction in a converter either.

Experiments were carried out in a 5 ton converter using five kinds of gases as bottom blowing gases in order to clarify these points. On the basis of the experimental results, a new mathematical model is proposed in the latter chapter.

3.1.1 Experimental Method

In experiments , the two types of 5 ton converters, the K-BOP and LD-KGC were used as shown in Fig.4. In the K-BOP, oxygen gas is blown through the inner tube of the bottom tuyeres with pulverized lime in the first and middle stage of refining, and mixed gas in the final stage. Propane gas is blown through the outer annulus of the tuyeres as a protective coolant. Through the top lance, only oxygen is supplied to the iron melt in the bath. In LD-KGC, oxygen gas is supplied through the top lance. The gases used for bottom blowing in the LD-KGC were conventionally either N_2 or Ar. However, CO gas is used for bottom blowing in this research. Table 2 shows the main experimental conditions in the 5 ton converter.

Sampling of metal and slag was carried out during and after blowing. At blow end, measurements of oxygen potential in the metal were carried out with an oxygen sensor in which the solid electrolyte is ZrO_2 stabilized 7%MgO. In some experiments, measurements of oxygen potential in the slag were also carried out.

Table II   Experimental conditions in 5ton converter

Converter type	LD-KGC	K-BOP
Oxygen flow rate of top blowing (Nm3/min)	15	12
Gas flow rate of bottom blowing (Nm3/min)	2	6
Species of bottom blowing gas	CO, N_2	O_2 + Ar O_2 + CO_2
Protective gas for tuyere	-	Propane
The ratio of inert gas to oxygen in bottom blowing gas	-	1/1

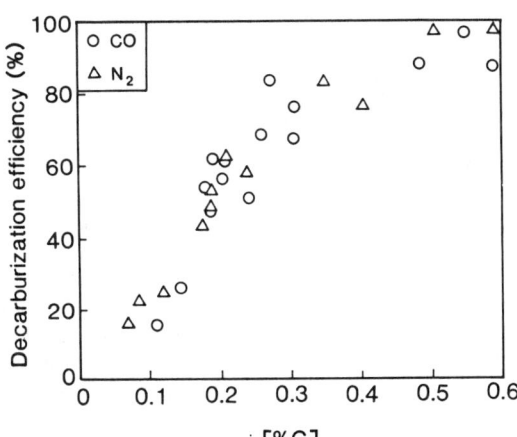

Fig. 9   Relationship between [%C] and utilization efficiency of oxygen gas
for decarburization in 5ton LD-KGC with CO or N_2 bottom blowing

3.1.2 Experimental results
(1) Experiments with CO bottom blowing in 5 ton and 180 ton converter
The relationship between the oxygen utilization efficiency for decarburization

301

and [%C] is shown in Fig.9. The efficiency of CO is equivalent to that of N_2. As shown in Fig.10, the relationship between [%C] and [%O] at blow end for CO blowing is the same as that for N_2 blowing. In addition to the above results, no difference was observed between the metallurgical reaction for CO blowing and that for N_2 or Ar blowing . The relation between [%C] and [%O] for CO blowing in the 180 ton LD-KGC is similar to that seen in the 5 ton converter.

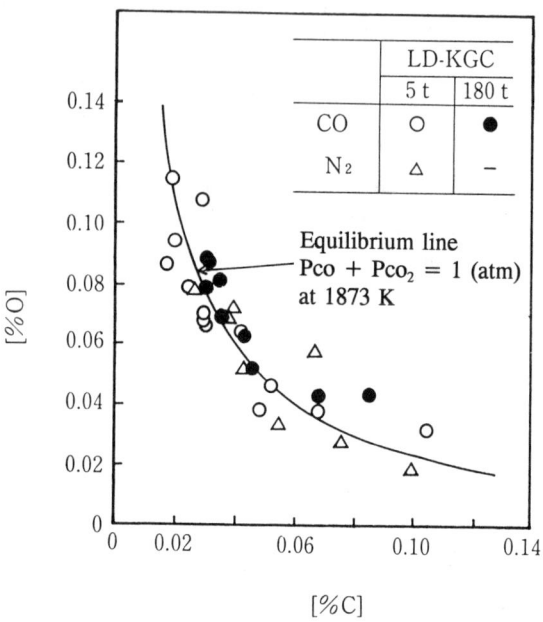

Fig. 10 Relationship between [%C] and [%O] in LD-KGC

(2) IOD experiments in 5 ton converter

Improvements in the efficiency of oxygen for decarburization and a decrease in [%O] and (T.Fe) are observed in IOD practice. The relation between [%C] and [%O] at blow end is shown in Fig.11. The [%C] vs [%O] relationship in the IOD mode is equivalent to the value calculated thermodynamically on the assumption that Pco + Pco$_2$ = 0.5 to 0.6 atm. The [%C] vs [%O] relationship without IOD practice is equivalent to the value on the assumption that Pco + Pco$_2$ = 0.8 atm. It is interesting that there is no effective difference between the [%C] and [%O] relationship in IOD with CO_2 mixing and that with Ar mixing, although a large difference exists between the dilution of Pco by CO_2 and Ar. No difference is observed between the other characteristics of the metallurgical reaction in the IOD mode with CO_2 and that with Ar or N_2 either.

Table 3 shows the comparisons of bottom blowing gas flow rate, Pco calculated from the CO evolution rate and [%O] observed in ordinary blowing (without IOD practice), IOD practice with CO_2 and that with Ar. The calculated value of [%O] which is equilibrium with Pco $_{calc.}$ is also shown in this table. Although the calculated value of [%O] which is in equilibrium with Pco in the gas phase in IOD blowing with CO_2 shows the maximum value, the observed value of [%O] in IOD mode with CO_2 shows the lowest value among the three blowing modes. These results show that the dilution effect of Pco by the bottom blowing gas on the behaviour of [%O] and (T.Fe) is small. In Fig.11, The [%C] vs [%O] relationship in Q-BOP is also shown. The relationship is equivalent to the value calculated thermodynamically on the assumption that Pco + P co$_2$ = 0.5 to 0.7 atm.

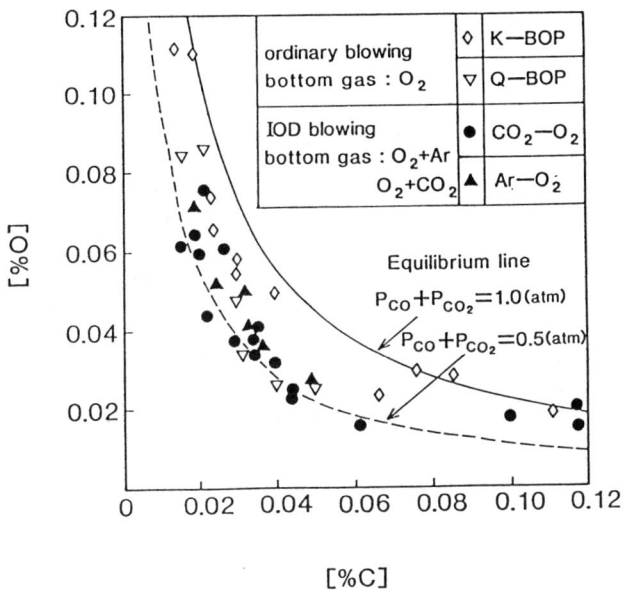

Fig. 11   Relationship between [%C] and [%O] in 5t K-BOP and Q-BOP

Table III   Comparison of bottom blowing gas flow rate, Pco and [%O] in three blowing modes (at [%C] = 0.04 in K-BOP)

	Ordinary blowing	IOD blowing with Ar	IOD blowing with CO$_2$
Flow rate of bottom blowing gas *	1.8 Nm3/min	5.1 Nm3/min	6.15 Nm3/min
Pco $_{calc}$*	0.65 atm.	0.35 atm.	0.86 atm.
[%O] $_{obs.}$ at [%C]=0.04	500 ppm	300 ppm	300 ppm
[%O] equilibrium with [%C] =0.04	383 ppm	209 ppm	503 ppm

* Conversion of O$_2$ to CO is considered by taking account of decarburization efficiency. CO$_2$ is assumed to decompose instantly to CO and O$_2$.

3.2 Mathematical Model Describing Deoxidation and Decarburization in Converter [12]

The above-mentioned experimental results make it clear that the decrease in [%O] and (T.Fe) at blow end in the combined blowing converter is not affected by the Pco of the bottom blowing gas ,but is affected by its stirring effect. To clarify the phenomenon, a new mathematical model for describing decarburization and deoxidation in the converter is proposed below.

The major part of the oxygen supplied to the iron melt bath is consumed for decarburization ; the remainder accumulates as dissolved oxygen in the steel and as iron oxide in the slag. Expanding the theory of Hsieh and Asai [15] for estimating the circulating flow rate of molten steel in the converter, a new reaction model has been developed. The reaction model was theoretically developed on the following assumption:

1) There are two parts in the molten steel in the converter; namely a reaction zone which corresponds to the hot spot formed by the oxygen gas jet, and bulk steel (see Fig.12).
2) The molten steel which is transported to the reaction zone by the circulation flow reacts with the oxygen jet in the reaction zone and, hence, the concentration of component j in the steel , $C_{b,j}$, equilibrates with the concentration of the component j in the reaction zone , C_j *, very rapidly. Then the metal is transported again to the bulk steel by circulation flow.
3) The molten steel transported from the reaction zone is perfectly mixed with the bulk steel.
4) The volume fraction of the reaction zone to the bulk metal is negligible, hence the accumulation of component j in the reaction zone can be neglected.
5) Mixing state of the metal bath is perfect.
6) As excess oxygen exists in the reaction zone, activity of FeO, a_{FeO}, in the reaction zone is taken as a unity (i.e.1).
7) The major part of the oxygen supplied to the iron melt bath is consumed for decarburization ; the remainder accumulates as dissolved oxygen in the steel and as iron oxide in the slag. Owing to the difference between the oxygen potential in the metal and that in the slag, a transfer of oxygen between the metal and the slag occurs.

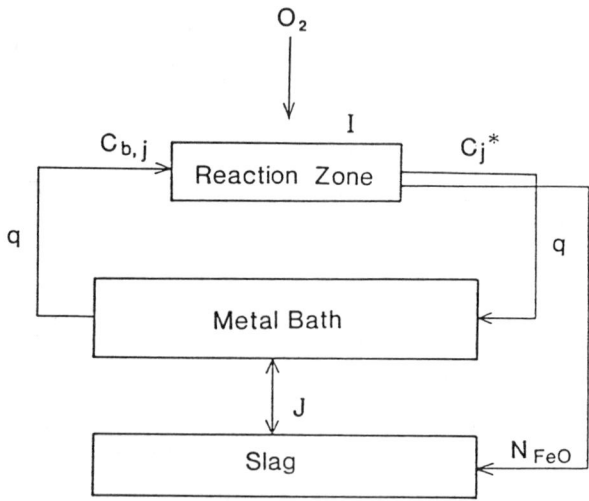

Fig. 12   Illustration of mathematical model describing decarburization and deoxidation in a converter

304

Hsieh and Asai [15] used conditions 1) to 6) and calculated the components of the metal bath in the first and middle stage of refining in the converter. Their calculated values agreed with the observed values. The condition 7) is newly introduced based on the experimental results.

Using these conditions, the mass balances of carbon and oxygen in the bath and that of iron oxide in the slag can be written as :

$$W[dC_{O,b}/dt] = q\ [C_O* - C_{O,b}] + J[C_O** - C_{O,b}] \qquad (1)$$

$$W[dC_{C,b}/dt] = q\ [C_C* - C_{C,b}] \qquad (2)$$

$$(Ws/100)\ [dC_{FeO,b}/dt] = N_{FeO} - (C_{FeO,b}/100)\ dW_S/dt$$
$$+ (71.9/1600)\ J[C_{O,b} - C_O**] \qquad (3)$$

Where, W is mass of the molten steel (kg), q is the circulating flow rate of the molten steel (kg/min), J is a factor which determines the mass flow rate between the metal phase and slag phase (kg/min), N_{FeO} is the production rate of iron oxide (i.e. FeO) in the reaction zone by the supplied oxygen (kg/min), $C_{j,b}$ is the concentration of component j in the metal or slag (%) (the subscripts j denotes C,O or FeO), C_j* is the concentration of component j in the reaction zone (%) and C_O** is the dissolved oxygen concentration in equilibrium with a $_{FeO}$ in the slag, which is calculated by eq. (4) with the thermodynamic value recommended by the Japan Society for the Promotion of Science [16].

$$C_O** = a_{FeO}\ 10^{(-6150/T+2.604)} \qquad (4)$$

The values of the concentration of carbon and oxygen which is in equilibrium with a_{FeO} in the reaction zone (=1) are calculated by eq.(5) and eq. (6).

$$C_{O,e} = 10^{(-6150/T+2.604)} \qquad (5)$$

$$C_{C,e} = (P_{CO} / C_{O,e})\ 10^{(-1160/T-2.003)} \qquad (6)$$

Using the carbon and oxygen concentrations which is in equilibrium with a_{FeO} in the reaction zone, calculated as described above, mass balances of carbon, oxygen , and iron oxide in the reaction zone are given in eq.(7),(8) and (9). Equation (9) is the overall oxygen balance in the reaction zone.

$$q(C_{C,b}-C_C{}^*)=I(C_C{}^*-C_{C,e}) \qquad \text{(7)}$$

$$q(C_{O,b}-C_O{}^*)=I(C_O{}^*-C_{O,e}) \qquad \text{(8)}$$

$$N_{FeO}=(71.9/11.2)\,Q_{O_2}+(71.9/1600)\,I(C_O{}^*-C_{O,e})$$
$$-(71.9/1200)\,I(C_C{}^*-C_{C,e}) \qquad \text{(9)}$$

Where, I is a factor representing the intensity of mixing in the reaction zone (kg/min) and Q_{o2} is the feed rate of oxygen (Nm³/min).

Calculations were carried out under the following conditions:

1) In the case $N_{FeO}<0$, the oxygen supplied to the metal bath was predominantly used for decarburization. The remainder is used for the accumulation of the oxygen in the steel.

2) Temperature was given as the function of time based on the experimental results in the 5 ton converter.

3) The change rate of the mass of the slag, dW_s /dt is decided by taking account of the increase of FeO.

4) All iron oxide is FeO and activity of FeO, a $_{FeO}$, is given as the function of FeO, CaO, SiO$_2$ and temperature, based on the regular solution model by Shim and Banya [16].

Under these conditions, carbon in the metal, C $_{c,b}$, oxygen in the metal , C $_{o,b}$, and iron oxide in the slag , C $_{FeO,b}$ were calculated numerically at a given time step. Other conditions used for the calculations are shown in Table 4.

Table IV   Conditions used in calculation

Initial value	[%C]=0.5,[%O]=0.005, (FeO)=0, T=1823 K, Ws=70 kg/t
Increase rate of temperature	30 K/min
Flow rate of oxygen gas	3.0 Nm³/min/ton
Flow rate of bottom blowing gas	LD-KGC: 0.2 Nm³/min/ton K-BOP : 1.2 Nm³/min/ton Q-BOP : 3.0 Nm³/min/ton
Pco + P co$_2$	1 atm.
Slag basicity	2.5

The circulation flow rate of molten metal, q , is given by eq.(10) based on the study by Sano and Mori [18].

$$q=\frac{60}{\tau}*3\,W \qquad \text{(10)}$$

Where τ is the measured value of mixing time of the bath, which is determined according to calculations by Nakanishi et al [3] .

In these calculations, two parameters , namely I and J, were determined as follows: The first assumption is that J is proportional to I. The values of I for three types of converters were determined to obtain reasonable agreement between the oxygen utilization efficiency for decarburization given by the mathematical model and observed values.

3.3 Results Calculated Using Mathematical Model

Table 5 shows the values of the parameters used in the calculation. Figure 13 shows the behaviour of [%C] and [%O] calculated during blowing in LD-KGC, K-BOP and Q-BOP. As the intensity of stirring by the bottom blowing gas increases, [%O] calculated by the reaction model decreases. The product of [%C] and [%O] calculated by the reaction model is less than the equilibrium value calculated thermodynamically on the assumption that $P_{CO} + P_{CO_2} = 1$ atm.

Table V   Values of parameter used in calculation

	q (kg/min)	I (kg/min)	J (kg/min)
LD-KGC	2.4×10^4	8.0×10^3	5.0×10^4
K-BOP	4.5×10^4	1.0×10^4	6.25×10^4
Q-BOP	6.0×10^4	1.2×10^4	7.5×10^4

These results are explained as follows: With increasing intensity of stirring by the bottom blowing gas, the decarburization rate in the reaction zone increases, while the rate of iron oxide in the reaction zone decreases. Consequently, the oxygen potential in the slag becomes smaller than that of metal, as shown in Fig.14, and oxygen transfer from the metal to the slag occurs. The amount of oxygen transfer increases as the difference between the oxygen potential in the metal and slag becomes greater.

The difference between the two becomes larger with decreasing carbon concentration because of the rapid increases in the oxygen potential in the metal. It also becomes larger as the stirring force of the bottom blowing gas increases. Therefore, the amount of oxygen transfer increases with increased stirring force of the bottom blowing gas or with decreasing carbon concentration in the molten steel.

The fact that decarburization and deoxidation proceed to a point below the values calculated thermodynamically (on the assumption that $P_{CO} + P_{CO_2} = 1$ atm.) can be explained by oxygen transfer from metal to slag. The effect of the partial pressure of CO on the decarburization rate is small because decarburization proceeds in the hot spot, where a_{FeO} is unity (i.e.1).

The behaviour of (T.Fe) in the slag during blowing in the three types of converters was also calculated, as shown in Fig. 15. The (T.Fe) calculated by the mathematical model decreases with increasing stirring intensity, and agrees qualitatively with the observed values in the 5 ton converter and in commercial converters [3] . The mathematical model gives a reasonable explanation of the behaviour of both [%O] in the steel and (T.Fe) in the slag during blowing.

The behaviour of [%O] and (T.Fe) in the IOD mode was also studied, taking into consideration the effect of P_{CO}. Stirring intensity in the IOD mode in the low carbon range improved compared with oxygen bottom blowing in the K-BOP (ordinary blowing).

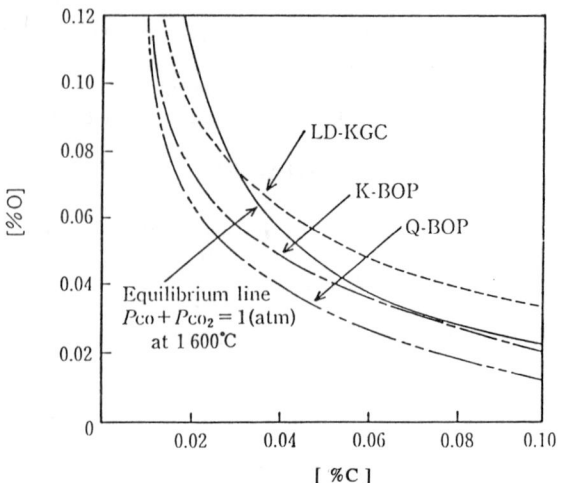

Fig. 13 Behaviour of [%O] calc. during blowing in LD-KGC, K-BOP and Q-BOP

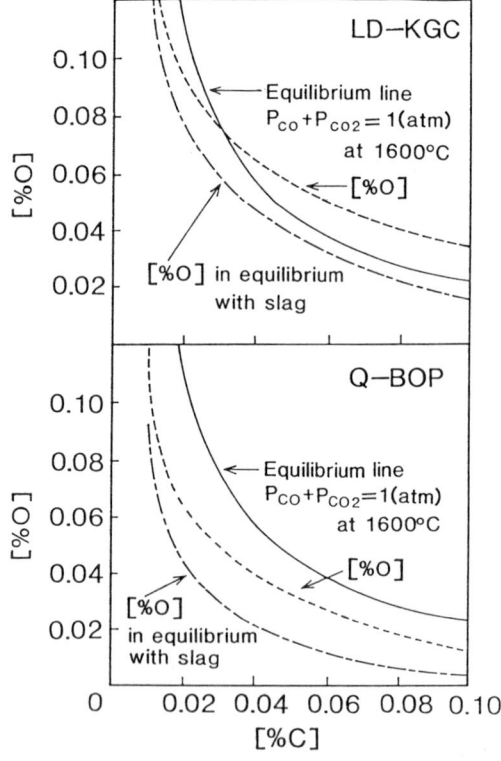

Fig. 14 Behaviour of [%O] calc. in metal and [%O]calc. in equilibrium
with slag during blowing

308

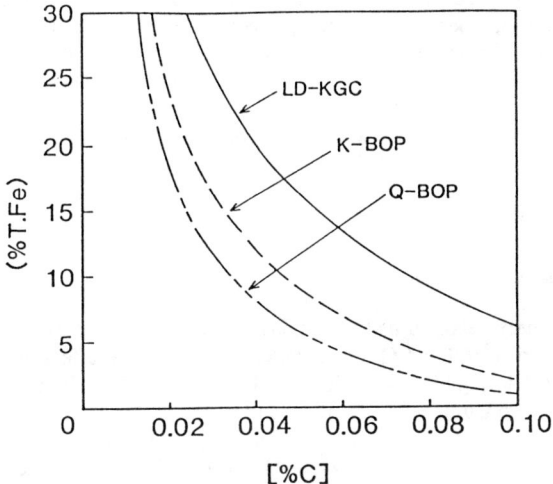

Fig. 15 Behaviour of (%T.Fe) $_{calc.}$ during blowing in LD-KGC, K-BOP and Q-BOP

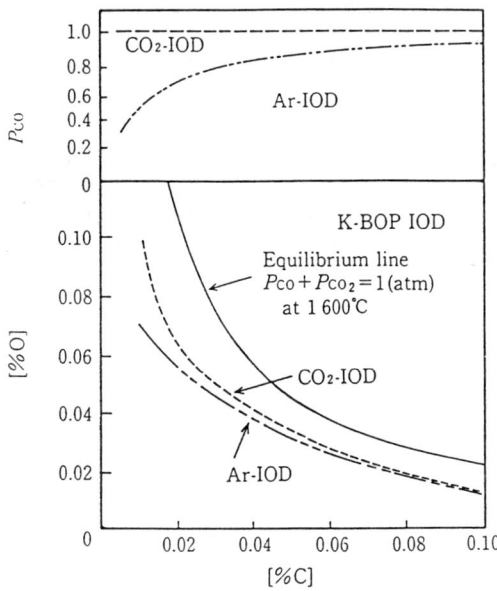

Fig. 16 Effect of Pco on behaviour of [%O] $_{calc.}$ during blowing in K-BOP
     with IOD practice

Hence, assuming that the stirring in the IOD mode is similar to that in the Q-BOP process, q, I and J in the Q-BOP process may be used as a basis for calculations regarding the IOD mode. Figure 16 shows the behaviour of [%O] calculated by the mathematical model during operation in the IOD mode with Ar (Pco + P co$_2$ < 1 atm.) and with CO$_2$ (Pco + P co$_2$ =1 atm.). Figure 16 also shows the change in Pco during blowing. There seemed to be little difference between the two, except that in the ultra-low carbon region [%C] <0.02. Excellent agreement was obtained between these computed results and results observed in the 5 ton K-BOP, as shown in Fig.11. The decrease of oxygen and (T.Fe) in the IOD mode is considered to be the result of the increased stirring intensity obtained by mixing inert gas into the bottom blowing gas.

3.4 Effect of Operational Conditions

It has been reported that operational conditions such as oxygen flow rate and slag volume influenced [%O] and (T.Fe) during blowing [14]. The effect of slag volume on the behaviour of [%O] during blowing in 230 t Q-BOP was calculated by the reaction model. In these calculations, two parameters , namely I and J, were determined to obtain reasonable agreement between the oxygen utilization efficiency for decarburization given by the mathematical model and observed values. Operational conditions used in the calculations are shown in Table 6.

Figure 17 shows the behaviour of [%O] calculated by the mathematical model during operation of the 230 ton Q-BOP. On condition that slag volume is 25 (kg/ton), the values of [%O] calculated by the model become higher than those in the case of slag volume of 50 (kg/ton), and agree well with the observed values in the 230 ton Q-BOP. The (T.Fe) calculated by the mathematical model also agrees qualitatively with the observed values in the 230 ton converter . These results showed that the operational conditions during blowing affect greatly the behaviour of [%O] and (T.Fe). We must therefore take the operational conditions into consideration for evaluation of the metallurgical reaction in a converter.

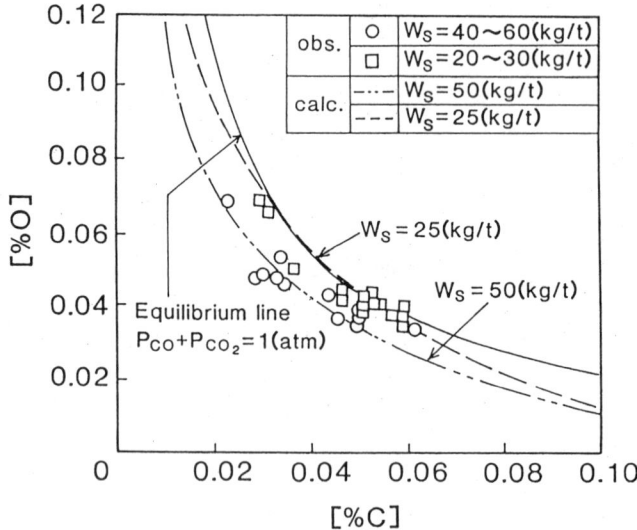

Fig. 17 Behaviour of [%O]$_{calc.}$ during blowing and [%O]$_{obs.}$ in 230 t Q-BOP

Table VI Operational conditions of 230 t Q-BOP and conditions used in calculation

Heat size	230 ton
Oxygen gas flow rate	[%C] > 0.15 : 2.5 Nm3/min /ton [%C] < 0.15 : 2.4 Nm3/min/ton
Slag volume	case (1) : 50 (kg/ton) case (2) : 25 (kg/ton)
q	2.8×10^6 (kg/min)
I	4.5×10^5 (kg/min)
J	3.0×10^6 (kg/min)
Pco + Pco$_2$	1.0 atm.

3.5 Measurements of Oxygen Potential in Metal and Slag in Induction Furnace and 5 ton Converter

To verify the assumption of (7) in the reaction model, the following experiments with an induction furnace containing a 15 kg melt were carried out.

After melting 15 kg of low carbon steel in the induction furnace, CaO -Al$_2$O$_3$ flux, as shown in Table 7, was added to the surface of the molten steel in the furnace. Measurements of oxygen potential of the metal and the slag were carried out after the addition of the flux. A MgO crucible was used for the experiments and Ar gas was introduced to the upper side of the crucible to prevent the entrance of air.

Figure 18 shows the change of [%C] and [%O] after the addition of the flux. After the addition of the flux, the oxygen content decreases. Figure 19 shows changes of the oxygen potential of the metal and the slag during the experiment, demonstrating that the oxygen potential in the slag increases as the oxygen potential in the metal decreases. These results showed that the addition of flux which has low oxygen potential decreases the oxygen potential in the metal .

Table VII Chemical Compositions of Flux and Slag in 230 t Q-BOP

	CaO	Al$_2$O$_3$	SiO$_2$	CaF$_2$	MgO	FeO	Fe$_2$O$_3$	MnO
Flux	50	22	14	9	5	-	-	-
Slag in Q-BOP *	57	0.62	8.8	2.1	3.1	7.6	9.6	8.8

* Q-BOP slag is obtained at blow end at [%C] =0.03.

Second, slag in the 230 ton Q-BOP was added to the metal. The slag was pulverized before addition. The chemical composition of the slag was also shown in Table 7. The change of [%C] and [%O] after the addition of the slag is shown in Fig.18. Deoxidation after the addition of the slag was observed. In this experiment, the oxygen potential in the slag could not measured as the molten layer of the slag is thin. Hence, the oxygen potential in the slag was calculated by the regular solution model [17]. The changes of oxygen potential of the metal and the slag during the experiment were also shown in Fig.19. During the experiment, observed values of the oxygen potential of the metal and of the slag approached each other. These results showed that the addition of slag which has low oxygen potential decreases the oxygen potential of the metal by the transfer of oxygen from the metal to the slag. Hence, the assumption (7) used in the mathematical model is considered to be reasonable.

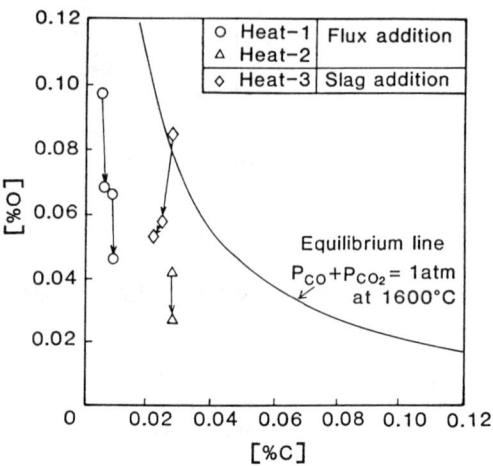

Fig. 18 Changes of [%C] and [%O] in 15kg induction melting furnace

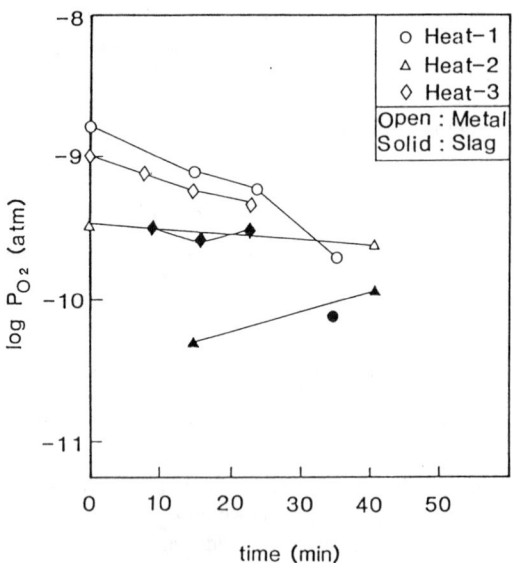

Fig. 19 Changes of oxygen potential in metal and slag
in 15kg induction melting furnace

312

Measurements of oxygen potential of the metal and the slag in the 5 ton converter were also carried out. Fig. 20 shows the relation between [%C] and oxygen potential in the metal and the slag during blowing in the 5 ton LD-KGC. Measurements of the oxygen potential in the LD converter and the 230 ton Q-BOP by Nagata et al. [19] are also shown in Fig.20. The oxygen potential in the slag in the 5 ton LD-KGC is equivalent to that in the metal and increases 10^{-10} atm. to 10^{-8} atm. with decreasing carbon concentration, while the oxygen potential of the slag in the LD converter is $10 \sim 100$ times as large as the oxygen potential of the metal. These results agree well with the calculated results by the mathematical model. However, it is difficult to compare the oxygen potential of the slag with that in the metal qualitatively at low carbon range, considering the accuracy of the oxygen probe. Moreover, several values of the oxygen potential in the slag were sometimes observed in the same experiments (i.e. in the same slag). These results suggested the inhomogeneity of the slag. These problems are to be solved in the future.

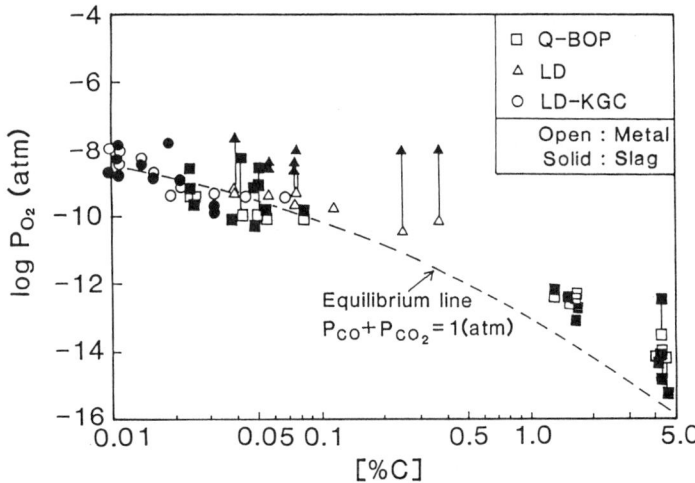

Fig. 20 Relation between [%C] in metal and oxygen potential in metal and slag in LD, LD-KGC, Q-BOP

3.6 Application of the mathematical model to Commercial Plant

The experiment in the 5 ton converter and the mathematical model revealed that the effect of the dilution of Pco by the bottom blowing gas is small. Based on these results, the following two improvements were carried out.

3.6.1 Application of CO bottom blowing to Commercial LD-KGC
The gases used for bottom blowing in the LD-KGC have conventionally been either N2 or Ar. However, nitrogen cannot be used for low nitrogen steel due to nitrogen pick up. The use of Ar is effective for lowering the nitrogen content of the steel, but Ar is expensive.

Research and development to produce CO gas at reasonable cost were therefore intensively pursued. These R&D efforts yield an economical process termed COPISA (CO pressure induced selective adsorption), which makes it possible to purify converter off-gas to a high purity CO gas of ca.99%.

Based on the experiment in the 5 ton converter and the calculation by the mathematical model, commercial use of CO gas as a bottom blowing gas has been adopted at the 180 ton LD-KGC at No. 1 Steelmaking Shop at Mizushima Works since 1985. The flow chart of the system is shown in Fig.21. The CO gas purified by COPISA facility at Mizushima is supplied to a chemical plant for synthetic chemical use and the LD-KGC shop.

Table 8 shows the specifications of the converter off-gas and the CO gas purified by COPISA. The average nitrogen content of the purified CO gas is about 1%, but no difference was observed between the nitrogen content at blow end for CO blowing and that for Ar blowing.

Use of CO gas contributes to improvements in iron and manganese yields by preventing excess oxidation while reducing the operational cost for bottom blowing gas.

Table VIII   Specifications of Gas Composition (%)

	CO	CO_2	N_2	H_2	O_2
BOF gas	>71	<14	<13	<1.2	0.2 ~0.3
Product	>98	<0.4	1.6	-	<1 ppm

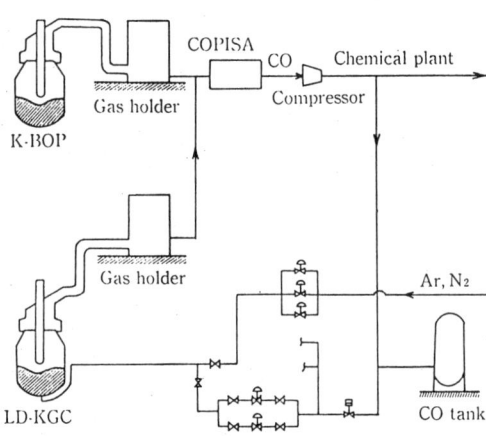

Fig. 21 Schematic drawing of process gas supply system in Mizushima Works

3.6.2 Commercial Use of CO₂ for Operation of IOD in 250 ton K-BOP

Based on the experimental results and calculation, CO_2 has also been used as the mixing gas in the IOD mode with the K-BOP in commercial production. There is no difference between the metallurgical reaction in IOD practice with CO_2 mixing and that with Ar mixing. CO_2 is now being used for ID practice, and contributes to the improvement of iron and manganese yield in low carbon range with the reduction of operational costs for bottom blowing gas.

4. Effect of Stirring Force and Pco on Decarburization Reaction of High Cr Steel[20]

The study in section 3 makes it clear that the effect of Pco on the metallurgical characteristics is small in the refining of carbon steel. Furthermore, the effect of stirring force and Pco on the decarburization reaction of high Cr steel was investigated .

Table 9 shows experimental conditions in the 5 ton LD-KGC and Q-BOP. Below [%C] =0.6, mixed gas is blown through the top lance in the LD-KGC and the bottom tuyeres in the Q-BOP, as shown in Fig.22.

Table IX   Experimental Conditions for Decarburization of High Cr Steel

Converter type		LD-KGC		Q-BOP
Oxygen flow rate (Nm³/min/t)	-	Mixed gas top blowing	O_2 top blowing	Mixed gas bottom blowing
	[%C] >0.6	2.4 ~3.0	2.4~3.0	2.4~3.0
	0.3<[%C]<0.6	1.3	1.3	1.3
	[%C]<0.3	0.67	0.67~1.3	0.67
Ar flow rate (Nm³/min/t)	[%C]>0.6	0	-	0
	0.3<[%C]<0.6	0.67	-	0.67
	[%C]<0.3	1.3	-	1.3
Species of bottom blowing gas		Ar		O_2+ Ar (O_2 +CO_2)
Protective gas for tuyeres		-		propane

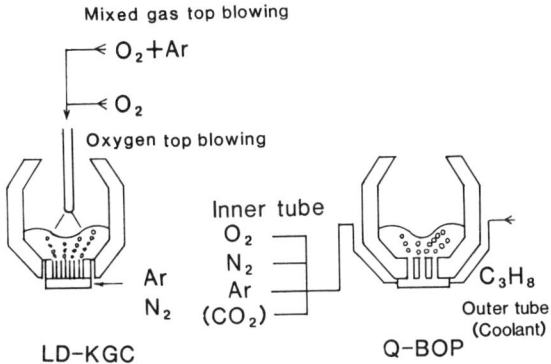

Fig.22 Schematic diagram of 5t converters used for decarburization experiment of high Cr steel

4.1 Effect of Stirring Force on Decarburization Reaction

Fig.23 shows the relationship between the oxygen utilization efficiency for decarburization of 16%Cr steel and [%C] in the LD-KGC and Q-BOP. Decarburization efficiency becomes bigger in the LD-KGC with O_2 top blowing, the LD-KGC with mixed gas blowing and the Q-BOP with mixed gas blowing respectively. As the inert gas mixing ratio in the LD-KGC with mixed gas blowing is same as that in the Q-BOP with mixed gas blowing, the difference in the decarburization efficiency between the two results from the difference of the stirring energy. The relation between gas flow rate,Q, and the rate constant of decarburization,K, is expressed by $K \propto Q^{0.40}$,which is similar to the equation obtained in the decarburization relation of the ultra-low carbon steel in the 5 ton converter, $K \propto Q^{0.33}$ [21].

4 .2 Effect of the partial pressure of CO on Decarburization Reaction

It is well-known that the dilution effect on decarburization in high Cr steel is great [22]. Fig.24 shows the changes in [%Cr] and the oxygen utilization efficiency for decarburization during blowing in the 5 ton Q-BOP with CO_2 /O_2 and that with Ar /O_2 blowing. In the carbon range of less than 0.25%, the decarburization in CO_2/O_2 blowing is stopped and the deoxidation of Cr in steel increases greatly, while the decarburization with Ar/O_2 blowing proceeds to the carbon range of less than 0.05%. A carbon value of 0.25% is consistent with the calculated value [23] , [%C] = 0.27, which is in equilibrium with [%Cr]=16 and Pco = 1 atm. As the stirring energy of CO_2/O_2 mixed gas blowing is equivalent to that in Ar/O_2 mixed gas blowing, the difference in the decarburization reaction in the two methods stems from the difference of the Pco in the two methods.

As mentioned in chapter 3, the effect of Pco on the metallurgical reaction is small in the refining of carbon steel. The difference between the effect of Pco in carbon steel and that in high Cr steel is explained as follows: The carbon concentration in the reaction zone (i.e. hot spot formed by the oxygen gas jet), C_c^* in carbon steel is 0.01% at 1923K on condition that Pco =1 atm, while C_c^* in 16%Cr steel is 0.43% at 1923 K , Pco= 1 atm. C_c^* in carbon steel is so small that the effect of Pco on the driving force of decarburization , I ($C_{C,b}$ -C_c^*), is insignificant. On the other hand, C_c^* in high Cr steel is so large that the effect of Pco on the driving force of decarburization , I ($C_{C,b}$ -C_c^*), is not negligible.

4.3 New Parameter for Evaluation of Decarburization of High Cr Steel

Nakanishi et al. [3] proposed ISCO for the evaluation of the enhancement of decarburization by stirring the steel melt by bottom blowing gas. It is reported that a clear correlation was observed between the ISCO value and iron oxide content in the slag in several combined blowing converters. Kai et al. [24] also proposed BOC for the evaluation of decarburization characteristics in combined blowing converters. However, these parameters were not theoretically introduced, but empirically proposed. Moreover, it is difficult to use these parameters for the evaluation of the decarburization of high Cr steel. In this report, a new parameter for the evaluation of the decarburization of high Cr steel is theoretically introduced.

According to the reaction model described in chapter 3, the ratio of the production rate of Cr oxide to the consumption of oxygen for decarburization is given as follows.

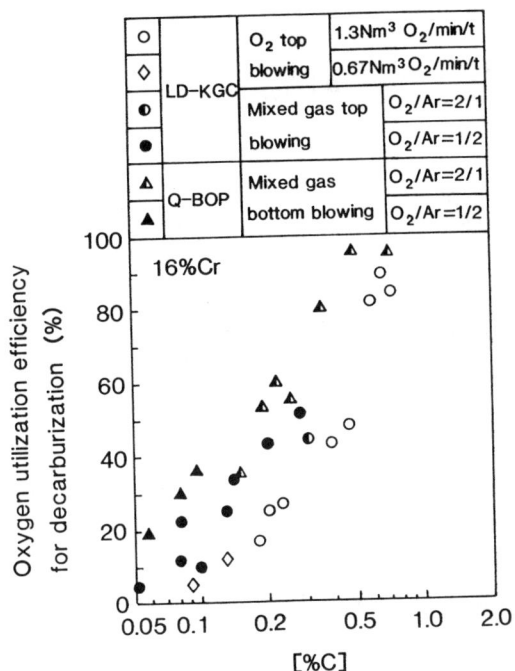

Fig.23 Relationship between oxygen utilization efficiency for decarburization and [%C] in LD-KGC and Q-BOP

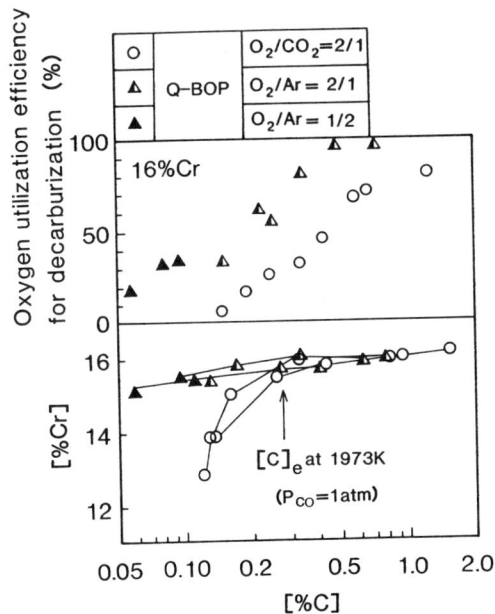

Fig. 24 Changes in [%Cr] and oxygen utilization efficiency for decarburization during blowing in 5t Q-BOP

$$N_{Cr_xO_y} / q(C_{C,b} - C_c*) = \alpha \frac{Q_{O_2}}{q(C_{C,b} - C_c*)} + \beta \frac{(C_{O,b} - C_O*)}{(C_{C,b} - C_c*)} + \gamma \quad (11)$$

Where N_{CrxOy} is the oxidation rate of Cr (kg/min), Q_{O2} is oxygen gas flow rate (Nm³/min), α, β, γ are constants ($\alpha = 4.52$, $\beta = 3.17$, $\gamma = -4.22$).

As the oxygen concentration in the low carbon range of high Cr steel, $C_{O,b}$ is equal to the oxygen concentration which is in equilibrium with $a_{CrxOy} = 1$, C_O*, the right hand second term can be ignored. Moreover, γ is constant and is much smaller than the right hand first term. Hence, the right hand first term is regarded as the main factor which affects the ratio of the oxidation rate of Cr to the consumption of oxygen for decarburization.

As shown in Fig.25, $(1 - C_c* / C_{C,b})$ is proportional to the reciprocal of the partial pressure of CO in the reaction zone, $1/ Pco*$. Therefore, $(C_{C,b} - C_c*)$ is rewritten as follows.

$$(C_{C,b} - C_c*) = C_{C,b}(1 - C_c* / C_{C,b}) \propto C_{C,b} / PCO* \quad (12)$$

Pco* is given by the following equation [25].

$$PCO* = 2Q_{O_2} / (2Q_{O_2} + Q_d) \quad (13)$$

Where Q_d is inert gas flow rate (Nm³/min).

Using eq.(10),(12) and (13), the right hand first term of eq.11) is rewritten as follows.

$$\alpha \frac{Q_{O_2}}{q(C_{C,b} - C_c*)} = \alpha_1 \frac{Q_{O_2} P_{co}*}{(W/\tau) C_{C,b}}$$

$$= \alpha_2 (2Q_{O_2} / (2Q_{O_2} + Q_d)) \frac{Q_{O_2}}{(W/\tau) C_{C,b}} \quad (14)$$

Where α_2, α_3 are constants. The right-hand term of eq.(14) can be used as the parameter for the decarburization of high Cr steel. That is to say, the decrease of the value means that the oxidation of Cr for decarburization is small, hence, preferential decarburization characteristics can be obtained.

New parameters for the evaluation of decarburization of high Cr steel, the Cr Oxidation Index (CROI), is defined as follows.

$$CROI = (2Q_{O_2} / (2Q_{O_2} + Q_d)) \frac{Q_{O_2}}{(W/\tau) [\%C]} \quad (15)$$

Fig.26 shows the relation between BOC values and the ratio of Cr oxide to decarburization, $\Delta Cr / \Delta C$, observed in the low carbon region in several combined blowing converters. Fig.27 shows the relation between CROI values and $\Delta Cr / \Delta C$ in the low carbon region in several combined blowing converters. There seems to be a correlation between BOC values and $\Delta Cr / \Delta C$, however, it is difficult to explain the effect of mixed

318

gas blowing on the decarburization reaction by using BOC values. CROI values enables us to evaluate the decarburization characteristics of high Cr steel in several kinds of combined blowing converters.

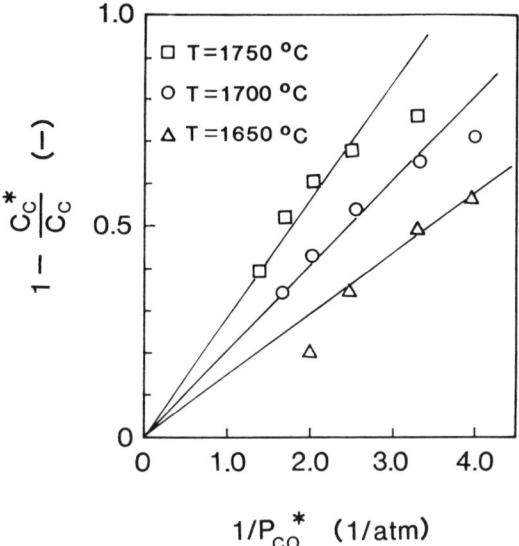

Fig.25 Relation between (1- C_c^* / C_c) and 1/ Pco* ([%Cr] = 16, $C_{c,b}$= 0.3%)

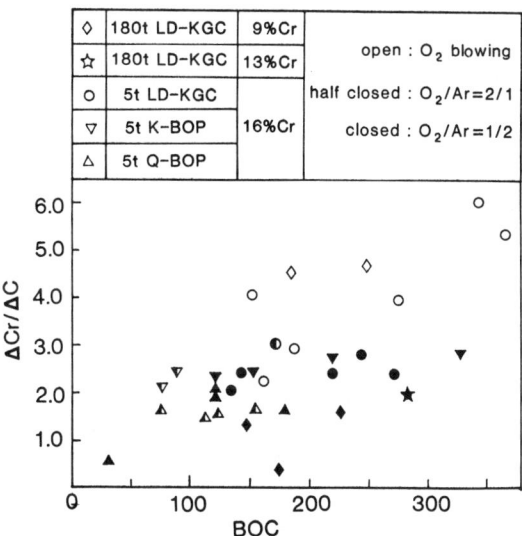

Fig.26 Relationship between ΔCr / ΔC and BOC values in low carbon region in several kinds of combined converters
([%C] =0.05~ 0.3, T =1923~ 1973K)

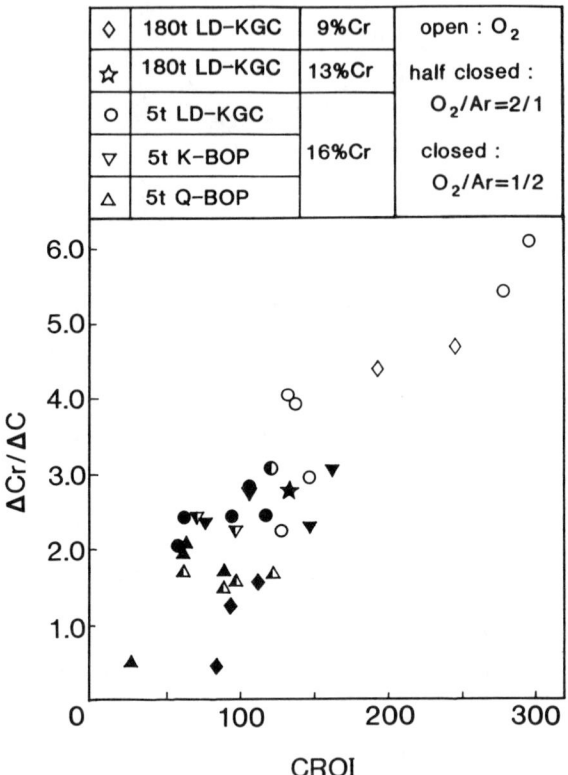

◇	180t LD–KGC	9%Cr	open : O$_2$
☆	180t LD–KGC	13%Cr	half closed :
○	5t LD–KGC		O$_2$/Ar=2/1
▽	5t K–BOP	16%Cr	closed :
△	5t Q–BOP		O$_2$/Ar=1/2

Fig.27 Relationship between ΔCr / C and CROI values in low carbon region
in several kinds of combined converters
([%C] =0.05∼ 0.3, T =1923∼ 1973K)

CROI values calculated in LD-KGC with O$_2$ gas blowing , LD-KGC with mixed gas blowing and Q-BOP with mixed gas blowing are 420, 107, 57 respectively, according to the conditions set forth in Table 9 ([%C] =0.1). This means that the oxygen utilization efficiency for decarburization in the LD-KGC with mixed gas blowing is half as large as that in the Q-BOP with mixed gas blowing, which corresponds with Fig.23, and that the method of mixed gas top blowing in the LD-KGC has a significant effect on the improvement of decarburization characteristics compared with the LD-KGC with O$_2$ blowing. This method permits decarburization of high Cr steel in the conventional combined blowing converter by the minimum alteration.

5. Future Research and Development Needs

The mathematical model made clear the role of stirring force and dilution of Pco of the bottom blowing gas on the metallurgical reaction in a converter. The improvement of the bottom-blowing technology based on the research contributes to the improvement of iron and manganese yield in low carbon range with the reduction of the operational cost of bottom blowing.

However, there appears to be a need for more extensive and accurate information on the effect of the stirring effect on the slag-metal reaction in a converter as well as the oxygen potential in the slag during blowing. Kim and Fruehan[26] observed that the position of the gas input influenced the mass transfer rate between the metal and the slag. Although the mixing state in the bath greatly depends on the gas flow rate or on the specific power input, the details are much more complicated. Moreover, the macroscopic flow pattern is influenced by many factors , such as the gas input position, the ratio of height to the diameter of the vessel[27], etc. Hence, a further study of the mixing state in the bath would be helpful forward understanding the metallurgical reactions in a converter.

Nowadays, the demand for processing clean steel at high productivity levels has been increasing. At the same time, lower levels of iron oxide in the slag at blow end will be demanded to improve the surface quality of cold rolled steel sheets. These seemingly require contradictory objectives in the converter: low carbon in the steel and low iron oxide in the slag. It will take innovative technical development and a new theory to produce steels with such paradoxical objectives in one facility.

6. Conclusions

Recent advances in top and bottom blowing converter technology at Kawasaki Steel based on a mathematical model have been described and may be summarized as follows:

1) Experiments were carried out with a 5 ton converter in order to evaluate the effects of stirring force and the dilution effect of Pco by the bottom blowing gas on the metallurgical reaction in the converter. Experimental results made it clear that the effect of the dilution of Pco by the bottom blowing gas is small.

2) A new reaction model was proposed , and explains the metallurgical reaction in the converter when bottom blowing gas is used. The model suggests that the phenomena that oxygen concentration reduces to a point below the equilibrium value on the assumption that $Pco + Pco_2 = 1$ atm. can be explained by the transfer of oxygen from the metal to the slag.

3) Based on these results, carbon monoxide has been used as a bottom blowing gas to reduce the gas cost for stirring the steel bath in the LD-KGC, with the same metallurgical effects as argon or nitrogen. Mixing of inert gas or carbon dioxide with oxygen in the final stage of refining in the K-BOP has been commercialized and effectively prevents excess oxidation of iron in low-carbon steel production.

4) A new parameter was theoretically introduced to explain the decarburization characteristics of high Cr steel in several kinds of combined blowing converters. The parameter can reasonably explain the difference in the decarburization characteristics in several kinds of combined blowing converters.

Nomenclature

a_{FeO}	Activity of FeO in slag (-)
$C_{j,b}$	Concentration of component j in metal or slag (%)
$C_{C,e}$	Carbon concentration which is in equilibrium with a_{FeO} in reaction zone (%)

$C_{o,e}$	Oxygen concentration which is in equilibrium with a_{FeO} in reaction zone (%)
C_j^*	Concentration of component j in reaction zone (%)
C_U^{**}	Dissolved oxygen concentration in equilibrium with a $_{FeO}$ in slag (%)
I	Factor representing the intensity of mixing in reaction zone (kg/min)
J	Factor which determines the mass transfer rate between metal phase and slag phase (kg/min)
N_{FeO}	Production rate of iron oxide (i.e. FeO) in reaction zone by supplied oxygen gas (kg/min)
N_{CrxOy}	Production rate of chrome oxide (i.e. Cr_xO_y) in reaction zone by supplied oxygen gas (kg/min)
Pco	Partial pressure of CO in gas phase (atm.)
Pco*	Partial pressure of CO in reaction zone (atm.)
Qo_2	Feed rate of oxygen gas (Nm^3/min)
Qd	Feed rate of inert gas (Nm^3/min)
q	Circulating flow rate of molten steel (kg/min)
t	time (sec)
W	Mass of molten steel in converter (kg)
W_s	Mass of slag in converter (kg)
α	Constant in eq. (11)
α_1	Constant in eq. (14)
α_2	Constant in eq. (14)
β	Constant in eq. (11)
γ	Constant in eq. (11)
τ	Measured value of mixing time of bath steel (sec)
suffixes	
j	C, O or FeO

References

1) K.Nakanishi, T.Nozaki, R.Uchimura, T.Ohta, M.Saigusa, J.Nagai and F.Sudo, "Recent Progress of OBM/Q-BOP Steelmaking at Kawasaki Steel Corp.", _Kawasaki Steel Tech Report_, 1 (1980),1-13

2) M.Hanmyou, _The 100 and 101st Nishiyama Memorial Technical Lecture_ (The Iron and Steel Institute of Japan,1984), 201

3) K.Nakanishi, K.Saito, T.Nozaki, Y.Kato, K.Suzuki, and T.Emi, "Physical and Metallurgical Characteristics of Combined Blowing Processes", _Steelmaking Proceeding_ (ISS-AIME, Pittsburgh) 65(1982),101-108

4) R.Tachibana, N.Takashiba, M.Kuwayama, A.Yamane,M.Maeda and H.Osanai, " Combined Blowing System (LD-KGC) with Wide Range of Flow Rate ", _Kawasaki Steel Giho_,17(4) (1985),p357- 364

5) Y.Kishimoto, Y.Kato, T.Sakuraya, T.Fujii, S.Yamada,and S.Omiya, "Recent Progress in Top and Bottom Blowing Converters at Kawasaki Steel Corp.", _Kawasaki Steel Tech Report_, 22 (1990),12-21

6) G.Denier, J.C.Grojean, M.Lemaire, F.Schleimer, R.Henrion and F.Goedert, "Indusrial Development of Bottom Blowing Gas Injection in Top Blown Converters ", _Steelmaking Proceeding_ (ISS-AIME, Washington) 63(1980), 131-138

7) J.Iwamoto, _Attsyuku-sei Ryutai Rikigaku (Compressive Fluid Dynamics)_ (Japan: Kyoritsu Shuppan)

8) H.Osanai, N.Misaki, H.Take, A.Yamane and T.Imai," Improvement of Decarburization Process at Top and Bottom Blowing Converter ", _Tetsu-to-Hagane_,73(1987), s1017

9) N.Kitagawa, H.Osanai, M.Suito, S.Omiya, Y.Kato and Y.Takahashi, "Development of Refining Process for Production of High Purity Ultra-low carbon Steel ",*Tetsu-to-Hagane*, 76(1990),1932-1939

10) H.Nishikawa, H.Kondo, Y.Kishimoto, N.Tamura,R.Asaho and M.Onishi, "Improvement of Refining Technology in Bottom Blowing Converter ",*Tetsu-to-Hagane*, 76(1990),1940-1947

11) K.Yamaguchi, Y.Kishimoto, T.Sakuraya, T.Fujii, M.Aratani, H.Niahikawa," Effect of Refining Condition for Ultra Low Carbon Steel on Decarburization Reaction in RH Degasser ", *ISIJ International*, 32(1992),126-135

12) Y.Kishimoto, Y.Kato,T.Sakuraya and T.Fujii, "Effect of Stirring Force by Bottom Blowing Gas and Pressure of CO in It on Characteristics of Metallurgical Reaction in a Converter",*Tetsu-to-Hagane*, 75(1989),1300-1307

13) T.Usui, K.Yamada, Y.Kawai, S.Inoue, H.Ishikawa, Y.Nimura, " Experiment of Phosphorus and Oxygen Distribution between $CaO-SiO_2-MgO-Fe_tO$ Slag and Liquid Steel and Estimation of Phosphorus Content at End Point of Top and Bottom Blowing Converter", *Tetsu-to-Hagane*,77(1991),1641-1648

14) K.Okohira, S.Tanaka and M.Hirai, *Yutai-Seiren-no Butsuri Kagaku-to -Purosesu Kogaku (Physical Chemistry and Process Engineering for Refining of Melts)*, (The Iron and Steel Institute of Japan), 245-250

15) Y. Hiseh, Y.Watanabe, S.Asai and I.Muchi, "Effect of Recirculating Flow Rate of Molten Steel in Refining Processes ", *Tetsu-to-Hagane*,69(1983),596- 603

16) The Japan Society for the Promotion of Science, The 19th Committee on Steelmaking, *Steelmaking Data Source Book* (2nd ed.,NY: Gordon and Breach Science Publishers,1988), 95,111

17) J.Shim and S.Banya," Distribution of Oxygen between Liquid Iron and Fe_tO -SiO_2- CaO-MgO Slags Saturated with MgO ", *Tetsu-to-Hagane*,67(1981),1745- 1754

18) M.Sano and K.Mori, " Circulating Flow and Mixing Time in a Molten Metal Bath with Inert Gas Injection " , *Tetsu-to-Hagane*, 68(1982),2451- 2460

19) K.Nagata, K.Nakanishi, F.Sudo and K.Goto, " Measurements of Oxygen Potential and Temperature in Liquid Slag, Metal and Gas phase of Q-BOP Converter by Oxygen Concentration Cell " , *Tetsu-to-Hagane*, 68(1982), 277-283

20) Y.Kishimoto, Y.Kato,T.Sakuraya, T.Fujii, H.Osanai, S.Omiya and H.Take, "Refining Technology of High Cr Steel by Mixed Gas Top Blowing in Combined Blowing Converter ",*Tetsu-to-Hagane*, 76(1990),1924-1931

21) S.Takeuchi, Y.Kato, H.Okuda, H.Take and S.Yamada, "Development of Mixed Gas Blowing in Top and Bottom Blown Converter ", *Tetsu-to-Hagane*,70 (1984), A184-186

22) J.M.Saccomano, R.J.Choulet and J.D.Ellis "Making Stainless Steel in the Argon - Oxygen Reactor at Joslyn", *J.Met.*, 21 (2) (1969),59-64

23) D.C.Hilty, H.P.Rassbach and W.Crafts, " Observation of Stainless Steel Melting Practice", *J.Iron Steel Inst.*, 180(1955), 116- 128

24) T.Kai, K.Okohira, M.Hirai, S.Murakami, N.Sato, "Influence of Bath Agitation Intensity on Metallurgical Characteristics in Top and Bottom Blown Converter ",*Tetsu-to-Hagane*, 68(1982), 1946-1954

25) S.Asai and J.Szekely, "Decarburization of Stainless Steel : Part1, A Mathematical Model for Laboratory Scale Results", *Metall. Trans.* 5(1974), 651-657

26) S.Kim and R.J.Fruehan, "Physical Modelling of Liquid/Liquid Mass Transfer in Gas Stirred Ladles" , *Metall. Trans. B*, 18B(1987), 381-390

27) F.Oeters, W.Plushkell, E.Steinmetz and H.Wilhelmi, "Fluid Flow and Mixing in Secondary Metallurgy", *Steel Res.*, 59(1988),192-201

Heat Transfer Between
Molten Cryolite and Solid Phase

A. Warczok and T. Utigard
Department of Metallurgy and Materials Science,
University of Toronto, 184 College Street,
Toronto, Ontario, Canada, M5S-1A4

and

P. Desclaux
Arvida Laboratories and Experimental Engineering Centre,
Alcan International Ltd., P.O. Box 1250(Arvida Sector)
Jonquiere, Quebec, Canada, G7S 4K8

ABSTRACT

The thermal insulation of aluminium reduction cells are designed to promote the formation of a protective layer of frozen cryolite along the carbon based sidewall. This solid freeze preventing intensive corrosion of the carbon by the liquid electrolyte, works also as a temperature buffer. This study was initiated because literature heat transfer coefficient values scatter widely.

The 'cold-finger' technique developed in these laboratory experiments was capable of carrying out heat transfer measurements both during transient and steady state conditions. Nitrogen gas stirring was used to induce fluid flow in the 1.8 kg melts. On a graphite surface, the heat transfer coefficient was found to vary with the bath stirring according to:

$$h(W/m^2K) = 919 + 25.6*Stirring\ Energy(W/ton)^{1/2}$$

This investigation will be extended to more closely simulate the conditions in industrial cells.

Proceedings of the
Savard/Lee International Symposium on Bath Smelting
Edited by J. K. Brimacombe, P. J. Mackey,
G. J. W. Kor, C. Bickert and M. G. Ranade
The Minerals, Metals & Materials Society, 1992

1.0 INTRODUCTION

In the Hall-Heroult process for the electrolytic production of aluminium, a side ledge forms along the side wall of the cell(Fig. 1). This ledge protects the cell lining and is of outmost importance for long service life of the cell. During anode effects the cell heats up and the ledge partially melts/dissolves while for under-cooling of the cell during anode change or alumina feeding, the ledge grows. In this manner the ledge works as an important thermal buffer. Because of differences in i) the fluid flow behaviour and ii) thermal properties of the liquid cryolite and the aluminum, the control of the ledge is difficult.

A review of the literature(1-13) concerned with the i) measurement or the ii) 'calculation/estimation' of the heat transfer coefficient between a cryolite based bath and a solid freeze shows that has been reviewed critically. Reported values scatter widely from 150 to 1050 W/m²*K(12). However, based on a critical analysis of the reported data it is concluded that the heat transfer coefficient during 'steady state' and no forced convection is approx. 608 ± 300 W/m²*K and that it increases from 1,200 to 2,000 W/m²*K with increasing heat flux during non-steady heat transfer upon immersion of a cold specimen in the molten bath.

1.1 Literature Values: None of the experimentally obtained values can be taken as the 'right' value and several of the reported values are of no significance. Further, in most of the publications there are either i) inconsistencies, or ii) poorly described calculation procedure, or iii) lack of experimental and property data.

a) Steady State Measurements: The investigation by Taylor and Welch(7,10) appears to be the only one which has produced fairly reliable data. An internally argon-cooled graphite cylinder(38 mm Outer Diameter) was immersed in a cryolite melt until steady state heat transfer conditions were achieved. The cylinder was pulled out of the melt and allowed to cool down followed by the measurement of the ledge thickness which was typically only 1.0 to 1.5 mm. The heat transfer coefficient was calculated based on the following equation:

$$h = k_L*(T_L - T_C)/(x_L*(T_b-T_L)) \qquad \text{(Eq. 1)}$$

where k_L is the ledge thermal conductivity, T_L is the liquidus temperature, T_C is the temperature at the carbon/ledge interface, x_L is the ledge thickness, and T_b is the bath temperature. Because they did not measure the total heat flux, they depended on the measurement of the ledge thickness as well as assuming a value of the thermal conductivity of the ledge to calculate the heat transfer coefficient. Using a ledge thermal conductivity of 0.759 W/m*K, they calculated the heat transfer coefficient to be approx. 608 W/m²*K. They also reported that the heat transfer coefficient increased with 34 to 45% with a fluid flow of 0.09 to 0.12 m/s.

Mark Taylor(7) carried out an extensive amount of measurements of the heat transfer coefficient in industrial cells during the first 38 days of starting up a new cell, as well as after a period of instabilities of a three year old cell. Based on these measurements which were carried out just above the bath/metal interface, the heat transfer coefficient was calculated to be in the range from 550 to 1820 W/m²*K with an average value of 1,200 W/m²*K(7). It was noted that because of the dynamics of ledge growth and re-melting, the positioning of the thermocouple tip was difficult. Additional uncertainties are caused by fluctuations in the bath- and ledge/bath interface temperature readings, respectively. Based on his data(7), the heat transfer coefficient is plotted versus the sidewall heat flux Fig. 2,

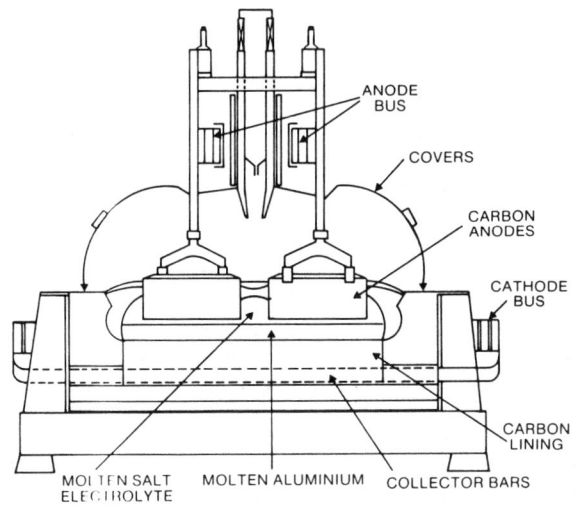

Figure 1. Schematics of a Hall-Heroult cell.

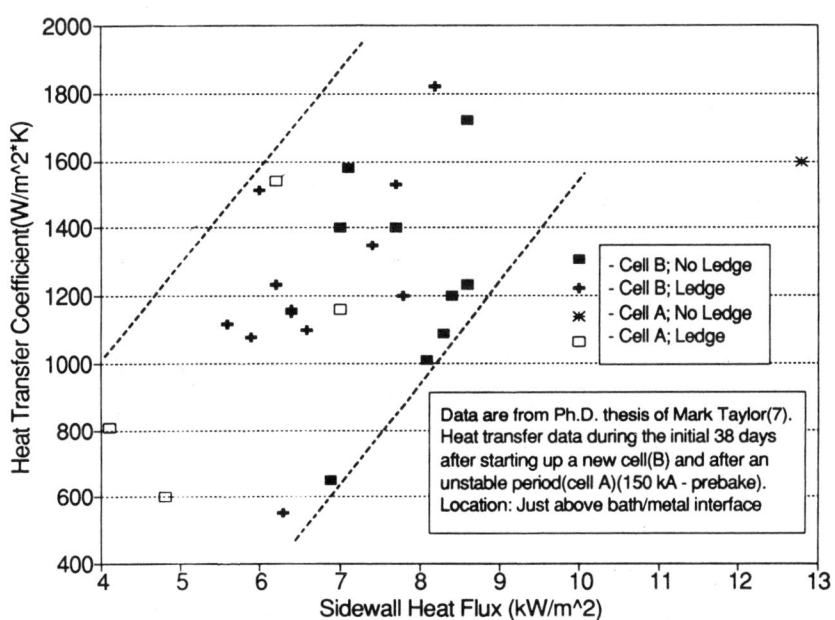

Figure 2. Heat transfer data from industrial measurements(7).

with and without the presence of a frozen ledge. The following observations are made:

1. It appears that the heat transfer coefficient increases with increasing heat flux.
2. There is no significant difference between the heat transfer coefficients measured with and without a ledge.
3. There is significant variance in the data.

b) Transient Measurements: The experimental data by Gan and Thonstad(12) and by Taylor, Welch and McKibbin(10) appear to be sound and representative for rapid freeze formation and remelting upon immersion of a cold specimen in the liquid bath. Figure 3 shows that as a general trend, the heat transfer coefficient increases with the overall heat flux. The values of the heat flux and the heat transfer coefficient given in Fig. 3 have been calculated based on published data(10,12). It is seen that during the rapid heat-up of a cold specimen, the heat flux and the heat transfer coefficient are significantly higher than those during steady state heat flow.

In an industrial cell the heat flux is typically in the range from approx. 1.5 to 5 kW/m^2(2,3,8,9). It is therefore to be expected that the heat transfer data determined in laboratory investigations do not represent those in industrial pots during steady operation and at lower heat fluxes.

c) Heat Transfer Calculations: When a liquid is flowing past a plate of different temperature, a temperature gradient develops. The heat flux from the plate is given by the following equation:

$$q/A = -kdT/dy_{wall} = h(T_{wall} - T_{liquid}) \qquad (Eq.\ 2)$$

where k is the thermal conductivity of the liquid and h is the heat transfer coefficient. A series of expressions based on dimensionless numbers exist for correlating the heat transfer coefficient to the fluid conditions. In the case of natural convection, the following relationship is frequently used(14):

$$Nu_{ave} = 0.59 * Ra_L^{1/4} \qquad (Eq.\ 3)$$

where $Nu_{ave} = h*k/L$ and $Ra_L = Gr*Pr = g*\beta*L^2*(T_b-T_w)/\upsilon*\alpha$. For the cryolite system the following values were used: β(volume coefficient of expansion) = $3.5*10^{-4}$, υ(kinematic viscosity) = $1.4*10^{-6}$ m^2/s, α(thermal diffusivity) = 10^{-7} m^2/s, k(thermal conductivity) = 0.45 W/m*K. The average heat transfer coefficient of a 10 cm vertical surface is shown in Fig. 4 for three different superheats(T_b-T_w). For forced convection, the following equation has been found to describe the heat transfer coefficient fairly well(14):

$$Nu_{ave} = 0.664 * Re_L^{1/2} * Pr^{1/3} \qquad (Eq.\ 4)$$

where $Pr = \upsilon/\alpha$ and $Re = v*L/\upsilon$. The calculated heat transfer coefficient is shown in Fig. 4 versus the bath velocity. It is interesting to note that for velocities less than 2 cm/s, heat transfer by natural convection dominates, while for velocities above 5 cm/s forces convection is the most important. Therefore, it is not obvious how stirring affects the heat transfer coefficient.

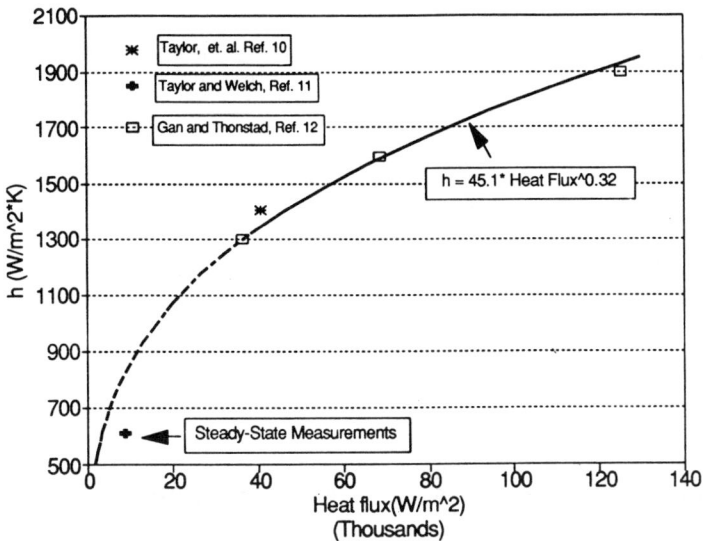

Figure 3. Heat transfer coefficient as a function of the average heat flux. The data for heat fluxes above 20 kW/m² originate from experiments where cold specimens were immersed in liquid cryolite(10,12).

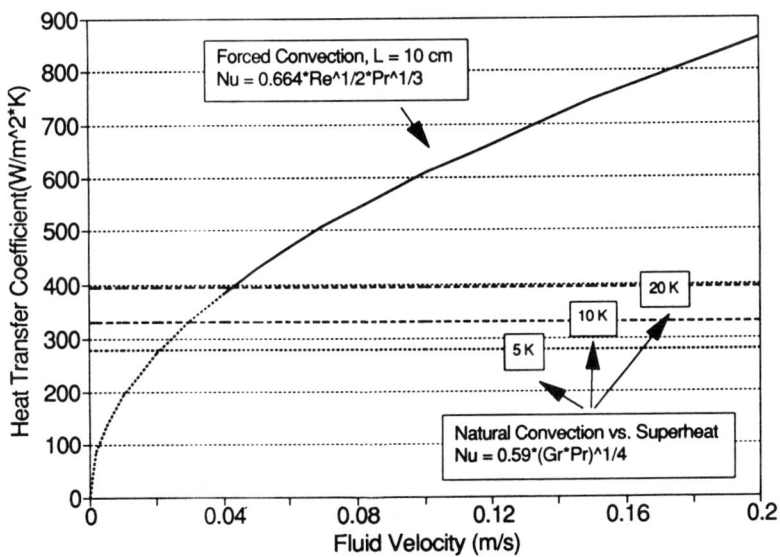

Figure 4. Estimated heat transfer coefficients in cryolite melts for natural- and forced convection.

Based on a similar analysis other researchers(1,5) estimated the heat transfer coefficient to be in the range from 150 to 600 W/m^2*K(1,5). By means of water modelling of gas induced turbulence, a similar range of heat transfer coefficients are reported(5). It must be noted that the fluid flow induced by density gradients caused by composition gradients and also by surface and interfacial tension gradients was not taken into account. This may explain why these model calculations give heat transfer coefficients significantly below experimentally determined values. In addition, the common heat and mass transfer correlations are developed for well established flow conditions over large surface areas (boiler tubes, vertical and horizontal plates). Because of the undefined nature of the fluid flow in the aluminum reduction cell[flow is caused by i) magnetic forces, ii) anode gas evolution, iii) temperature gradients, iv) interfacial tension gradients, and v) alumina feeding], the average velocity does not necessarily give a good picture of the turbulence and the instantaneous flow in a certain volume element.

2.0 EXPERIMENTAL

While reviewing the literature it became apparent that in order to generate reproducible heat transfer data, the experimental technique had to be carefully designed in terms of the following considerations:

2.1.1. Heat flux

In an industrial cell the heat flux is approx. 1.5 to 5 kW/m^2 during steady operation(2,3,8,9). However, most laboratory experiments have been carried out under non-steady conditions and at heat fluxes up to 100 kW/m^2(7,10,12). It was therefore decided to design a 'cold-finger' technique for the continuous measurement of the heat transfer coefficient at heat fluxes more representative to pot conditions.

2.1.2. Na_3AlF_6 - AlF_3 phase diagram

Because of the steepness of the liquidus line it is required to carefully measure the temperature profile in the melt and the ledge to determine the temperature at the interface. Taylor et al.(7,10) found that during rapid heating of a graphite cylinder, the interface temperature was 18 °C below the liquidus temperature.

2.1.3. Ledge formation and melting

Because the heat transfer coefficient may be different during ledge melting than during ledge growth, the technique has to be capable of simulating both situations in a controlled fashion. The values during 'controlled' transient conditions can then be compared to the steady-state values.

2.1.4. Bath flow

The heat transfer coefficient is strongly dependent on the fluid flow within the cell. However, it is very difficult to determine the fluid velocity in an industrial cell as well as in a laboratory crucible. Further, the turbulence of the fluid also affects the heat transfer coefficient. To address this, the experiments were carried under 'controlled' gas stirring conditions by bubbling argon through the melt at a given depth. The energy dissipation in the cryolite melt due to gas injection is given by(13,15):

Power(W) = 0.0123*Q*T(K)*ln(1 + 0.002*L)   (Eq. 5)

where Q is the gas flow rate in standard liter/min and L is the lance immersion in cm. For small immersion depths, this simplifies to(13):

Power(W) = 2.4*10^{-5}*Q*T(K)*L   (Eq. 6)

The energy dissipation due to the anode gas release in a Hall-Heroult cell is approx. 60 to 100 W/ton of electrolyte. For steelmaking processes with gas injection, the corresponding energy varies generally from 500 to more than 10,000 W/ton(15).

2.2 Experimental Set-up

The experimental set-up used in these measurements is shown in Figure 5. The data were continuously recorded using a/D converter DASCON-1 card and a multiplexer/amplifier EXP-16 card. The frequency of data sampling was 3 Hz. Nitrogen gas was used to protect the graphite crucibles.

The cold-finger is made from a 130 mm high graphite crucible with an inner- and outer diameter of 19.8 and 25.6 mm, respectively(Fig. 5). A thermocouple is located at the interface between the crucible wall and the inner thermal insulation as well as in the graphite finger wall. A copper-tube cooling system is located in the centre of the cold-finger and is thermally insulated from the graphite finger by means of an insulating brick. Water flows down into the 'cold-finger' inside one copper tube(6 mm OD) and out through another copper tube. The heat flux is calculated based on the water flow rate and the difference in the water temperature between the inlet and the outlet. Two RTD(Resistance Temperature Detector) sensors are placed inside the copper tubes at the height of the cryolite bath to measure the water temperatures(± 0.2 °C) at the inlet and outlet of the 'finger'. Calibration experiments with thermal insulation around the finger were carried out in air at 1000 °C in order to evaluate the 'back-ground' heat flux. This heat flux(≈8 W) was then subtracted from the heat flux measured(80 - 200 W) upon immersion in cryolite baths.

The graphite crucible(100 mm diameter and 140 mm height) containing approx. 1.8 kg of bath, is located in an inconel container and heated to 960 - 1010 °C in a vertical pot furnace. The stainless steel support for the 'cold-finger' is designed to lock into the top of the graphite crucible. Thermal insulation between the top of the cold-finger and the stainless steel support minimizes the heat loss from the melt. A series of type-K thermocouples are position at determined distances from the graphite cold-finger at a depth of approx. 50 mm below the bath level. These thermocouples are protected using a 1 mm OD inconel sheet placed inside alumina tubes. A stainless steel rod is used for qualitative measurements of the ledge thickness.

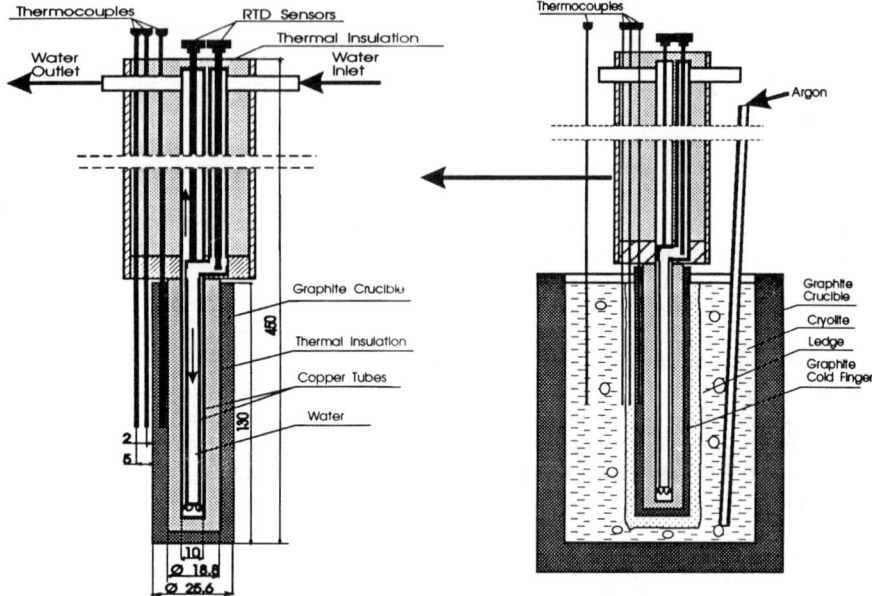

Figure 5. Description of the experimental technique. A data acquisition system was used to record all experimental data. The 10 cm diameter crucible and the cold finger support were placed in a top loading laboratory furnace.

2.3 Procedure

After keeping the melt at temperature for approx. 20 min, the stainless steel support and the cold-finger were lowered into the furnace until the graphite cold-finger reached the set immersion depth(11 to 12 cm). The water flow and the thermocouple readings are continuously determined and stored using a data acquisition system. Typically, the experiments are carried out at one fixed bath temperature until steady state is achieved($\approx$ 10 min) as determined by the temperature readings. The finger is then removed from the bath and allowed to cool down. The ledge thickness profile ie determined and the ledge removed before immersion of the finger in the bath to start a new experiment. In other experiments, gas stirring was introduced and the resulting changes in the temperatures and the heat flux were monitored.

2.4   Variables

Using a liquid bath made up of syntectic cryolite from Bayer containing 1.5 wt% Al_2O_3, 46 wt% F and 0.02 wt% S, the following parameters were varied:

-       gas stirring
-       bath temperature
-       heat flux

2.5 Calculation Procedure

At steady state, the heat flux is given by the following equation:

$$q = \frac{2*\pi*H*(T_{Bath} - T_{Water})}{\dfrac{1}{h_w*r_{Cu}} + \dfrac{\ln(r_{ins}/r_{Cu})}{k_{ins}} + \dfrac{\ln(r_c/r_{ins})}{k_C} + \dfrac{\ln(r_l/r_C)}{k_L} + \dfrac{1}{h_B*r_L}}$$ (Eq. 7)

where q: heat output(W), T_B: bath temperature(°C), T_W: cooling water temperature(°C), H: immersion depth(m), h_w: water-Cu heat transfer coefficient(W/m²K), r_{Cu}: Cu tube radius(m), k_{ins}: insulation thermal conductivity(W/mK), r_{ins}: radius of insulating brick(r), r_C: graphite cylinder radius(m), k_C: thermal conductivity of the graphite(W/mK), r_L: ledge radius(m) k_L: thermal conductivity of solid ledge(W/mK), h_B: bath/ledge heat transfer coefficient(W/m²K). The first three terms in the denominator represent the 'cold-finger' thermal resistance which is independent of the ledge thickness and the fluid flow in the crucible. The heat transfer coefficient can be calculated based on Eq. 7 when knowing each of the resistance terms in the 'cold-finger'. The sensitivity of the heat transfer coefficient to the ledge thickness is given by

$$\Delta h \approx h^2/k_L * \Delta x_L \approx 500*\Delta x_L (mm),$$ (Eq. 8)

where Δx_L is the uncertainty in the ledge thickness, k_L is the thermal conductivity of the ledge(≈ 1 W/m*K), h is in the range of 700 W/m²*K.

To get an accuracy of ± 15%, the ledge thickness would have to be measured to ±0.2 mm and the heat flux and the thermal conductivity of the ledge have to be known very accurately. Also, any thermal resistance such as air gaps will lead to uncertainties in the calculated heat transfer coefficient.

In this initial investigation, only data from tests where no freeze was formed will be presented and analyzed. The reasons for this are:

1)   it is not required to assume a temperature for the bath/ledge interface
2)   it is not required to measure the ledge thickness
3)   the industrial data by Taylor(7) indicate that there is no significant difference between the heat transfer coefficients measured with and without a ledge

Under ledge-free conditions, it was required to measure i) the total heat flux(water temperature and flow-rate), and ii) the carbon wall and bath temperatures. Based on these measurements, the heat transfer coefficient between the bath and the graphite finger was calculated as follows:

$$h = q/r_C*[2\pi*H*(T_B-T_1) - q/k_C*\ln(r_C/r_1)]$$ (Eq. 9)

where q: heat output(W), r_C: graphite cylinder radius(m), H: immersion depth(m), T_B: bath temperature(°C), T_1: graphite temperature(°C) at the location of the thermocouple, k_C: thermal conductivity of the graphite(W/mK), r_1: location of the thermocouple in the graphite wall(m).

3.0 RESULTS AND DISCUSSION

The experimentally determined temperatures upon immersion of the 'cold-finger' in a liquid bath with no forced stirring are shown in Figs. 6a-c. It must be noted that in this case there is no frozen ledge at steady state after approx. 4 minutes. As seen in Fig. 6a), the inlet and outlet temperature of the cooling water slowly increased and then stabilized after approx. 4 minutes. The increase in the inlet temperature is cause by i) the counter-flow down through the 'cold-finger' support and ii) heat-up of the 'cold-finger' support. Even though there is thermal insulation within the support cylinder it gets heated up by the furnace. The heat flux extracted from the bath is equal to the water flow-rate times the difference in inlet and outlet temperature times the heat capacity of the water.

Figure 6b shows the rapid temperature response upon immersion of the cold finger in the liquid cryolite. The thermocouple furthest away from the graphite wall reaches the bath temperature almost immediately. After approx. 2 minutes, the thermocouples in the vicinity of the finger wall also reaches the bath temperature. The wall temperature reaches steady state after approx. 4 minutes and remains below the bath temperature due to the internal cooling. Upon removal of the finger, the temperatures drop rapidly.

Figure 6c show a more detailed view of the changes in the temperature with time. From about 1 to 2.5 minutes after immersion, the overall bath temperature decreases due to the heat load required to heat the cold-finger to steady state. This is followed by a slow increase of all the temperature readings. The very small difference between the three thermocouples located in the bath indicate a uniform bath temperature. Based on the difference between the wall and bath temperatures, the heat transfer coefficient is calculated using Eq. 9. The calculated value is given in Table 1 together with values obtained from other tests.

Figures 7a and 7b demonstrate the effect of bath stirring on the bath and wall temperatures. The following observations are made:

1.   The temperature of the wall and of the liquid in the vicinity of the wall increases upon stirring, while the bulk bath temperature decreases. The opposite occurs when the gas stirring is interrupted.

2.   Figure 7b shows that the difference between the bath and wall temperatures reaches 'steady-state' after 5 minutes of changes to the stirring conditions.

The calculated heat transfer coefficients during the different stages of this experiment are given in Table 1 together with other data. As seen in Table 1, there is no noticeable change of the overall heat flux upon gas stirring. This is because in our experiments, the thermal resistance due to the convective heat flow is very small compared to the overall thermal resistance. Based on the gas flowrate, the specific stirring energy was calculated using Eq. 6. Figure 8 shows that the heat transfer coefficient is strongly dependent on the bath stirring. Based on a regression analysis of our data and by assuming the heat transfer coefficient depends on the stirring power to the half, the following equation was found:

$$h(W/m^2K) = 919 + 25.6*Stirring\ Energy^{1/2} \qquad (Eq.\ 10)$$

where the stirring energy is given in W/ton of bath.

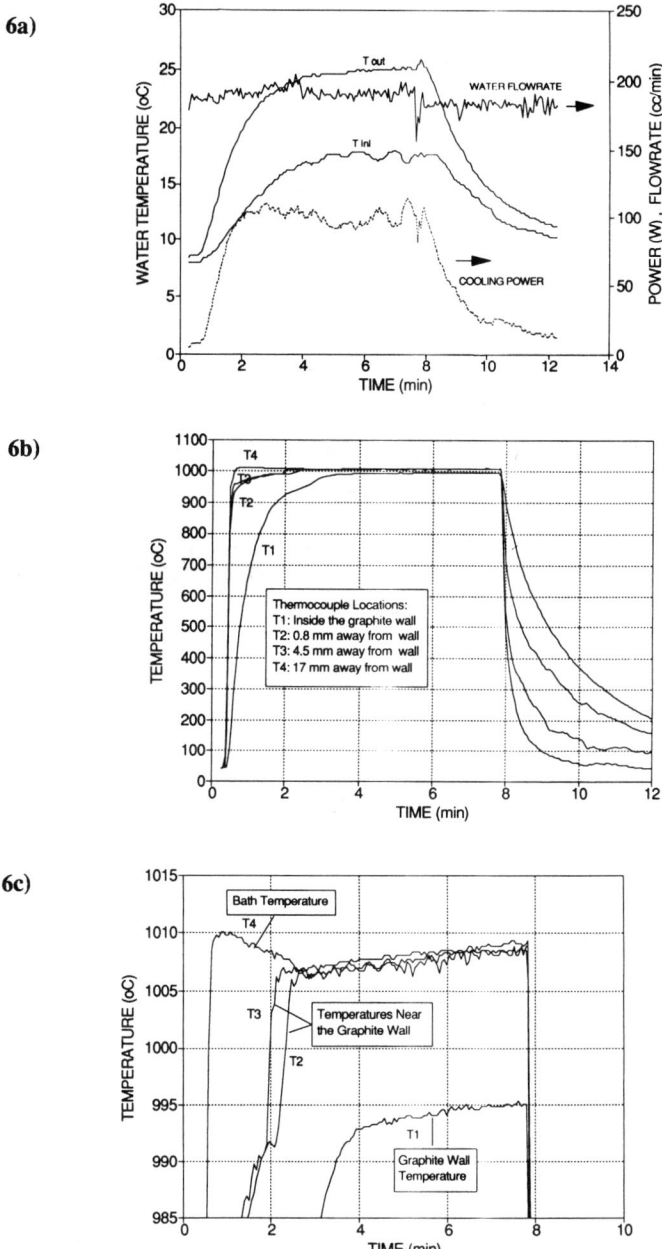

Figure 6. Temperature and heat flux data upon immersion of the 'cold-finger' in liquid cryolite at 1008 °C. In this experiment no ledge was present at steady state.

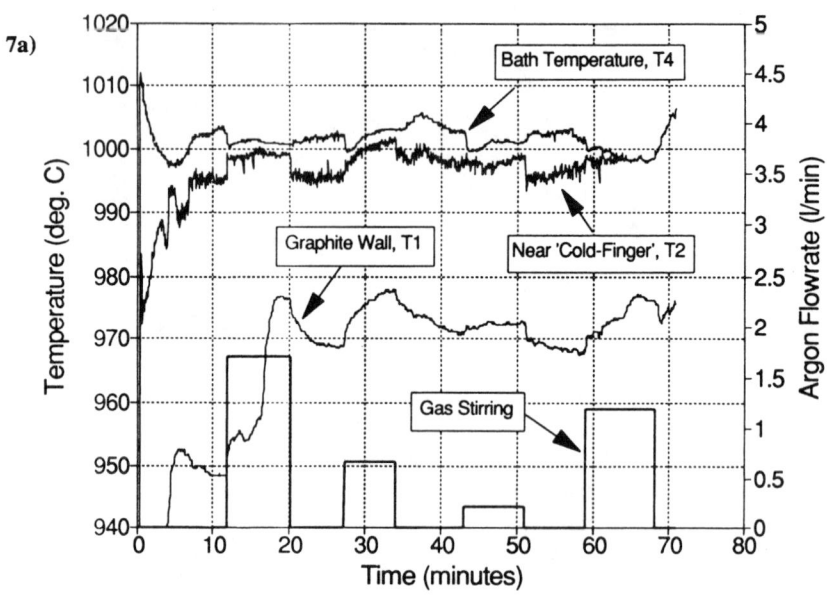

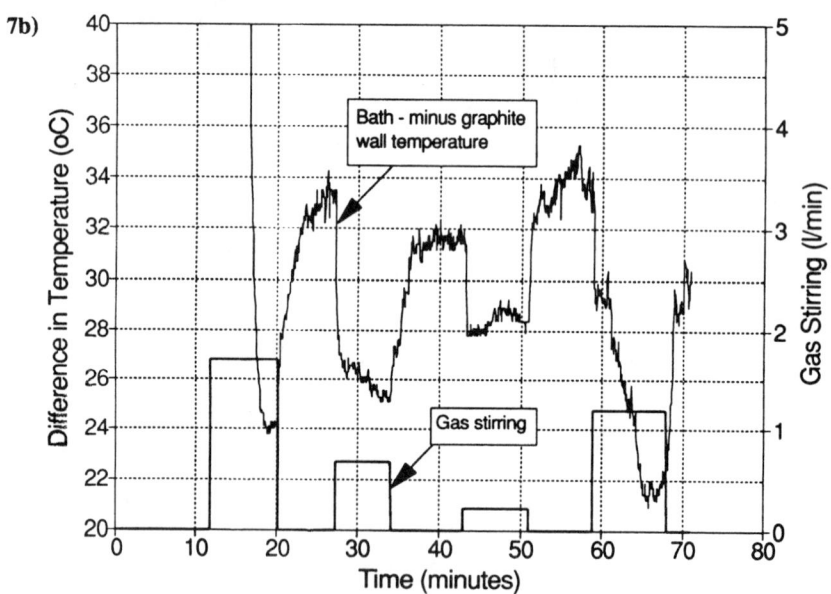

Figure 7. Bath and graphite wall temperatures upon immersion of the 'cold-finger' in liquid cryolite. At various times during this experiment, argon gas was used to agitate the melt.

Table 1. Experimental heat transfer data obtained during four separate experiments
where there was no frozen ledge on the graphite 'cold-finger'. The data given represent
average values during steady state at the particular conditions. In experiment No. 13, a
brick with a low thermal conductivity was used in the construction of the cold-finger.

Test No	Cooling Power W	Temperatures			Heat Flux kW/m2	Argon Stirring l/min	Stirring Energy W/ton	Heat Trans. Coeff.
		T1 oC	T2 oC	T4 oC				
20-1	276.8	976.6	999.2	1000.8	32.3	1.7	361	1336
20-2	274.2	968.9	995.7	1002.0	30.3	0.0	0	916
20-3	276.7	977.0	1000.7	1002.8	31.9	0.7	144	1238
20-4	278.6	971.9	997.9	1003.5	31.0	0.0	0	980
20-5	275.8	972.6	997.8	1001.2	31.2	0.2	45	1087
20-6	280.9	968.5	996.2	1002.7	31.0	0.0	0	905
20-7	270.1	975.6	998.1	998.4	31.8	1.2	255	1394
19-1	280.4	971.2	994.3	998.5	32.0	0.4	74	1173
19-2	275.9	970.6	995.0	998.6	31.3	0.4	74	1117
18-1	265.5	969.6	992.3	998.3	29.8	0.2	45	1037
18-2	275.9	974.6	998.0	1001.3	31.6	0.4	74	1185
13-1	96.8	994.4	1007.7	1008.4	12.1	0.0	0	864

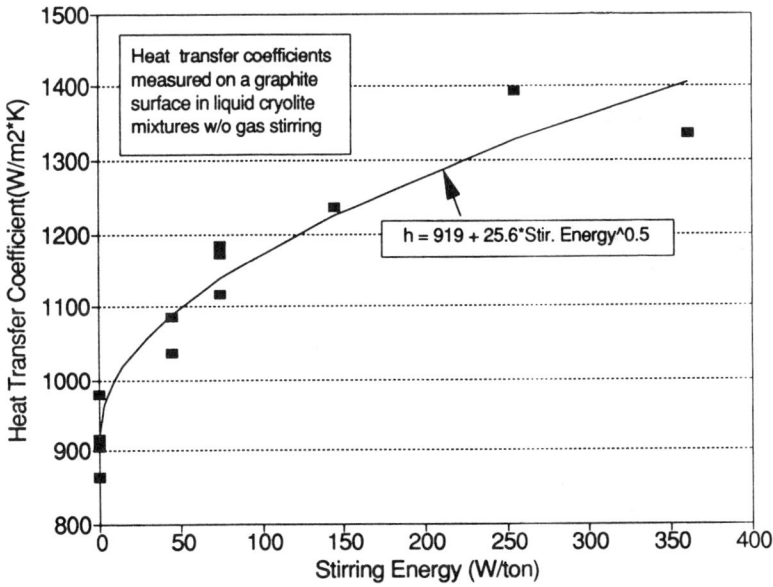

Figure 8. The measured heat transfer coefficients are plotted as a function of the specific
bath stirring energy. The solid line was found by regression analysis of the data. All the
data in this graph were obtained when no ledge was formed on the 'cold-finger' wall.

Using a freeze thermal conductivity of 0.759 W/m*K, Taylor(7) calculated the heat transfer coefficient to be 608 W/m²K. By using the accepted value of 1.1 W/m*K(1,7) the heat transfer coefficient becomes 881 W/m²K which is very similar to our value. It was also reported(7,11) that the heat transfer coefficient increased with 34 to 45% for a fluid flow of 0.09 to 0.12 m/s. Because of the different methods used to induce fluid motion it is not possible to compare the results directly. However, this increase in the heat transfer coefficient are similar in both studies. The range of values obtained in this investigation compares well with the data of Taylor(7) which are shown in Fig. 2. It therefore appears that there is no significant difference between the heat transfer coefficients measured with and without a ledge. However, it must be mentioned that the ledge morphology and roughness may vary significantly and this would affect the heat transfer coefficient.

From an industrial point of view it is not straight forward to apply the above given information. In order to use a heat transfer coefficient to calculate the heat flux through the sidewall, it is required to estimate or measure the bath and interface temperatures. As noted by Haupin(16), there appears to be quite an uncertainty with regards to the measurement of the bath temperature and to the calculated liquidus based on bath analysis. He mentions that in particular for very acid baths, the measured bath temperature is often found to be less than the calculated liquidus. If it is assumed that the ledge/bath temperature equals the liquidus temperature, this would lead to a net heat flux from the ledge into the bath. Obviously, this is not possible of extended periods of time.

It is interesting to note that by using Eq. 6, the stirring energy due to the anode gas generation is typically 60 to 100 W/ton. However, it must be noted that in an industrial pot the gas stirring is most intense around the anodes and that it varies throughout the cell depending on the gas channelling. In addition, the bath movements close to the aluminum level is affected by the metal movements. In general, it is at this level where most of the sidewall erosion/corrosion takes place.

4.0 CONCLUSIONS

An experimental technique capable of carrying out heat transfer measurements both during transient and steady state conditions has been developed and tested. On a graphite surface, the heat transfer coefficient was found to depend on the bath stirring according to:

$$h(W/m^2K) = 919 + 25.6*Stirring\ Energy(W/ton)^{1/2}$$

The results of this investigation compare well with the laboratory investigations of Taylor and Welch(11) and the industrial measurement carried out by Taylor(7).

To extend this investigation to more closely simulate the conditions in an industrial cell, experiments will be carried out at lower heat fluxes and with frozen cryolite covering the graphite 'cold-finger'.

ACKNOWLEDGEMENT

The financial support provided by Alcan International Limited and NSERC is greatly appreciated. The authors are thankful to Alcan International Limited for the permission to make this work public.

REFERENCES

1. W. Haupin, "Calculating thickness of containing walls from frozen melt", Light Metals, 1971, pp. 184-94.
2. J.G. Peacey and G.W. Medlin, "Cell sidewall studies at Noranda Aluminum", Light Metals, 1979, pp. 475-92.
3. H. Tsukahara, N. Ono and K. Fujita, "Establishment of effective operation of prebaked anode pots", Light Metals, 1982, pp. 471-82.
4. J. Thonstad and S. Rolseth, "Equilibrium between bath and side ledge in aluminum cells. - Basic Principles", Light Metals, 1983, pp. 415-24.
5. A. Solheim and J. Thonstad, "Heat transfer coefficients between bath and side ledge", Light Metals, 1983, pp. 425-35.
6. M.P. Taylor, B.J. Welch and J.T. Keniry, "Influence of changing process conditions on the heat transfer during early life of an operating cell", Light Metals, 1983, pp. 437-47.
7. M.P. Taylor, The influence of process dynamics on the heat balance and cell operation in the electrowinning of aluminum", Ph.D Thesis, School of Engineering, The University of Auckland, New Zealand, November 1984.
8. T. Ohta and T. Matsushima, "Thermal analysis of Soederberg pots", Light Metals, 1984, pp. 689-99.
9. W. Schmidt-Hatting, J.M. Blanc, J.C. Bessard and R.v. Kaenel, "Heat losses of different pots", Light Metals, 1985, pp. 609-24.
10. M.P. Taylor, B.J. Welch and R. McKibbin, "Effect of convective heat transfer and phase change on the stability of aluminum smelting cells", AIChE Journal, V32, 1986, pp. 1459-65.
11. M.P. Taylor and B.J. Welch, "Melt/Freeze heat transfer measurements in cryolite based electrolytes", Metall. Trans. B, V 18B, 1987, pp. 391-98.
12. Y.R. Gan and J. Thonstad, "Heat transfer between molten and solid cryolite bath", Light Metals, 1990, pp. 421-27.
13. T.A. Utigard, A. Warczok and P. Desclaux, "Heat and mass transfer between liquid bath and solid cryolite ledge", Extraction, Refining and Fabrication of light Metals, Proceedings Volume 24, CIM Annual Meeting, Ottawa, August 1991, pp. 163-76.
14. D.R. Gaskell, "An introduction to transport phenomena in materials engineering", Macmillan Publishing Company, 1992.
15. N.J. Themelis and P. Goyal, Canadian Metall. Quart., V22, 1983, pp. 313-20.
16. W. Haupin, "The liquidus enigma", Light Metals, 1992, pp. 477-80.

The Thermochemistry of Aluminum Smelting

Ernest W. Dewing

648 Pimlico Place
Kingston, Ontario, Canada   K7M 5T8
(613) 389-9270

Abstract

The current production of aluminum is by the electrolytic reduction of Al_2O_3 dissolved in a molten salt. The thermodynamics and electrochemical kinetics of the process are reviewed. The nature of the molten salt, and the influence on the process of its ability to dissolve also Al_4C_3 and metallic Al, are discussed.

A brief look is taken at the problems and possibilities of direct (non-electrolytic) carbothermic reduction of alumina.

Proceedings of the
Savard/Lee International Symposium on Bath Smelting
Edited by J. K. Brimacombe, P. J. Mackey,
G. J. W. Kor, C. Bickert and M. G. Ranade
The Minerals, Metals & Materials Society, 1992

Introduction

All aluminum produced in the world today is made by the Hall-Heroult process. Invented over a century ago, the process consists of electrolyzing alumina (Al_2O_3) dissolved in molten cryolite (Na_3AlF_6) with a consumable carbon anode at a temperature of about 945 - 975 °C. The overall reaction approximates to

$$2Al_2O_3 + 3C = 4Al + 3CO_2 \qquad (1)$$

Gross cell voltage is 4 - 5 V, and a typical modern cell takes about 200 kA. Figure 1 shows a very schematic diagram of a cell.

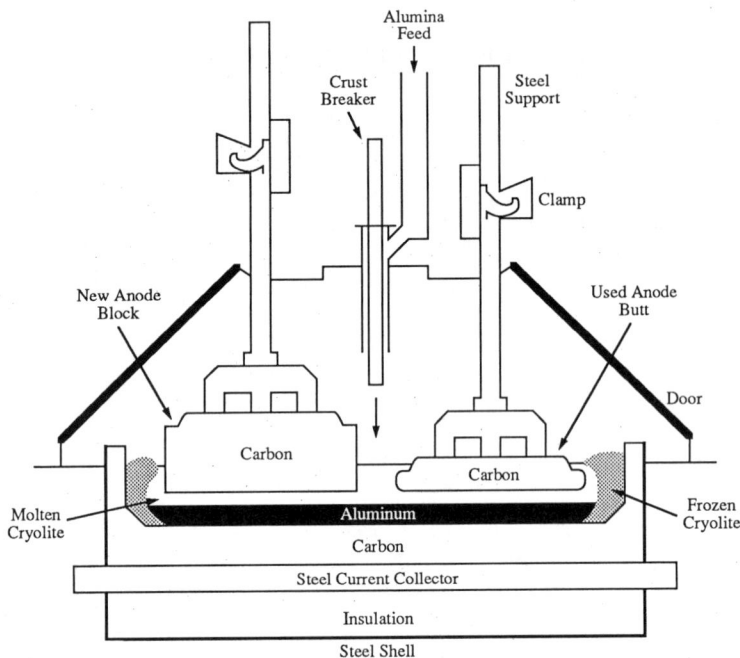

Figure 1 Aluminum Cell (schematic).

Energetics

Although the Hall-Heroult process involves the carbon reduction of alumina, it is an electrolytic process, which means that the passage of the current introduces Gibbs energy and not just ohmic heat. The situation is shown in Figure 2.

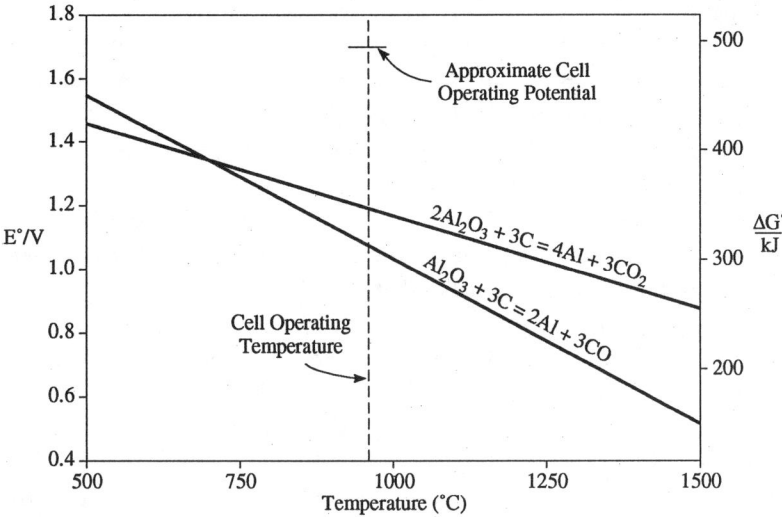

Figure 2 Gibbs energy for reduction of alumina with carbon.
($\Delta G°$ is in kJ per mole of aluminum.)

The conversion factor from $\Delta G°$ (in Joules) to $E°$ (cell emf in Volts) is $1/nF$, where F is Faraday's constant and n is the number of electrons which must be passed to carry out the reaction (= 12 for reaction (1)). At temperatures above 700 °C, formation of CO is energetically more favourable than production of CO_2, but, for kinetic reasons which will be discussed below, the principal product of a reduction cell is CO_2, and reaction (1) is relevant. Gibbs energy may be written as

$$\Delta G°/ J = 2\ 195\ 190 - 661.285\ T \qquad (2)$$

and

$$E°/ V = 1.896 - 5.711 \times 10^{-4}\ T \qquad (3)$$

The first term in both equations (2) and (3) corresponds to ΔH°, and the second numerical term to ΔS°. At a typical operating temperature of 960 °C (1233 K), equation (3) gives a reversible potential of 1.192 V.

The heat balance of an operating cell is of prime importance and, since the current is basically fixed and the thermal characteristics of the cell itself were determined when it was constructed, the situation is controlled by change of cell voltage. This is effected by moving the anode with respect to the cathode, thereby changing the ohmic resistance of the electrolyte. A typical composition has a resistivity of about 0.4 Ω cm, and a typical current density is 0.75 A cm^{-2}, so there is a drop of around 0.3 V cm^{-1}. Anode-cathode distances are usually 4 - 5 cm, which gives 1.2 - 1.5 V as the ohmic drop. To this must be added the reversible emf of 1.2 V, a kinetic overpotential of about 0.5 V (see below) and an additional ohmic resistance of the gas-bubble layer under the anode of 0.15 V, for a total of about 3.2 V dissipated within the inter-electrode gap. The remaining volt or so of gross voltage is lost within anode and cathode connections and in external busbars.

To calculate the <u>heat</u> generated in the inter-electrode gap one must subtract from the total voltage not that corresponding to ΔG° (1.2 V), but that corresponding to ΔH° (1.9 V). The latter figure needs a small modification to allow for loss of Faradaic efficiency (see below) and for the effect of the cold alumina and carbon coming into the cell, but, for the purposes of this paper, it is close enough. An alternative, and equivalent, way of looking at it is to say that energy corresponding to ΔG° (1.2 V) goes to making aluminum, and there is a net cooling effect equal to TΔS° (0.7 V) due to the increase of entropy.

One sometimes sees discussion of the efficiency of electrical energy use in aluminum production, and the way the calculation is done tends to depend on the political stance of the person doing it. Thus a cell running at, say 4.5 V can be held to be 27% efficient or 42% efficient, depending on whether ΔG° or ΔH° is taken as the basis. In fact, one must as a minimum supply ΔH° or the cell, even if perfectly insulated with no heat losses, will freeze. It seems to be the proper measure of 100% efficiency.

Anodic Overpotential

It has already been indicated that the anode is not working close to equilibrium; there is an overpotential of about 0.5 V, and the primary product is CO_2 and not CO. The situation appears to arise because, after an oxygen atom has been discharged onto a carbon atom on the surface, the resulting carbonyl group is fairly stable, and there is an activation barrier of some 330 kJ to breaking the carbon-carbon bonds and liberating CO. That process is slow compared to the rate at which electrolysis is taking place,

344

with the result that a second oxygen is discharged at the same carbon. There is an activation barrier to that process too which needs the 0.5 V overpotential to be surmounted, but with two oxygens now attached to the one carbon the end product is CO_2. A more detailed discussion of this and many other matters can be found in a review shown as reference (1); it gives numerous references to the original literature.

From the point of view of the aluminum industry it is very fortunate that the anode product is CO_2 and not CO; if not, twice as much carbon would be required. It is also interesting to look at the energetic efficiency with which the carbon is burned. The reaction

$$C + O_2 = CO_2 \qquad (4)$$

has a Gibbs energy corresponding to 1.03 V (i.e., that is the emf of a reversible fuel cell burning carbon to CO_2.) In an aluminum cell half of it is lost in the anode overpotential, but that still leaves 50% energy efficiency in the utilization of the carbon. That is much better than can be achieved in a thermal power station.

It is a complete coincidence that the heat generated at the anode surface by the overpotential (0.5 V) is broadly equivalent to the cooling effect of the TΔS tem (0.7 V) at the same location. It certainly simplifies mathematical modelling! The situation has been analyzed in detail by Ødegård et al. (2).

Loss of Faradaic Efficiency

Knowing how many coulombs have been passed through a cell, one can calculate from Faraday's constant how much aluminum should be produced. The real production is invariably less, and the ratio of the two is known as the current efficiency or the Faradaic efficiency. In modern cell operations it is 90 - 95%.

The problem arises because Al metal has a small solubility in molten cryolite — the most recent measurements are those of Ødegård et. al (3) shown in Figure 3. The dissolved metal is carried across the inter-electrode gap and is oxidized by the CO_2 bubbles under the anode. There is a direct relation between the amount of metal lost and the amount of CO appearing in the anode gas, and this is often utilized to determine the efficiency of a cell at a given moment from the analysis of gas samples. What is not clear is the chemical nature of the dissolved metal and the detailed mechanism of the mass transfer (1, 4 - 6).

The impact of loss of Faradaic efficiency on power consumption is obvious — when metal is lost to re-oxidation the energy which went to making it is not recovered.

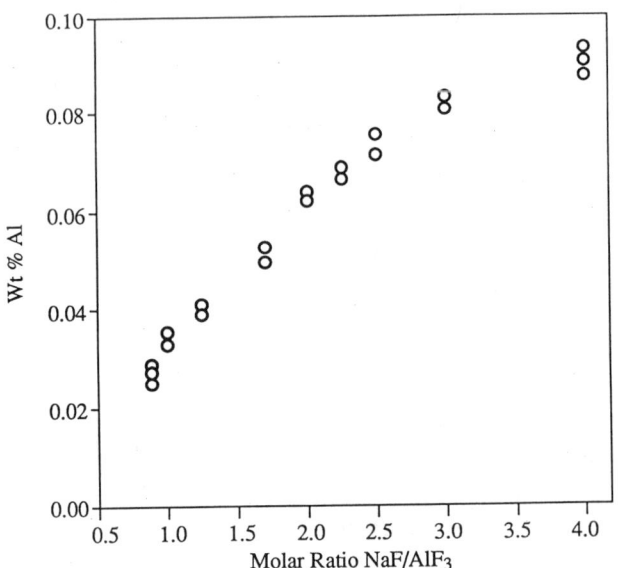

Figure 3 Solubility of Al in NaF - AlF$_3$ melts. (From Ødegård et al. (3).)

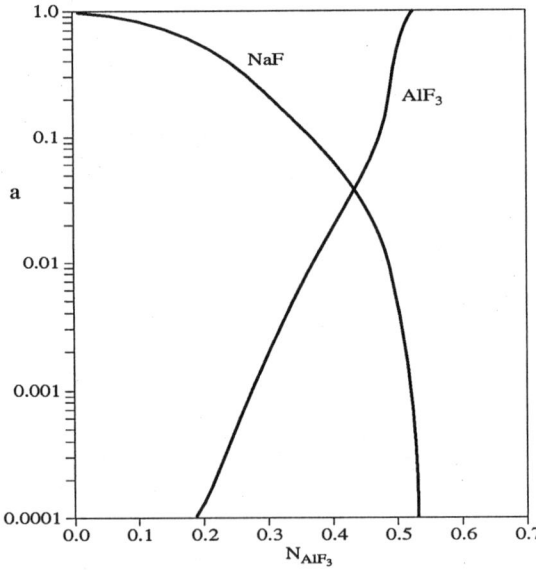

Figure 4 Activities in the system NaF - AlF$_3$ at 1020 °C (7).

346

Cryolite Melts

Energies of mixing in NaF - AlF_3 melts show strong negative deviations from ideality (Figure 4) which can only be accounted for in terms of formation of complex species. The nature of the species has been debated interminably over the last 40 years or so, but it is true to say that the changing views have had no impact whatsoever on the industrial process. The most recent position is that both the activities (7) and Raman spectra (8) are consistent with the existence of major amounts of AlF_5 - species as well as the long-recognized AlF_4 and AlF_6 groups. Alumina dissolves as oxyfluoride species containing only one oxygen atom, and that oxygen seems to be bridged between two aluminums (9). How many fluorines are also attached to the aluminums is not clear, but three on each seems plausible, giving an $(Al_2OF_6)^{2-}$ group.

Solution of Al_4C_3 in Cryolite

Aluminum carbide is soluble in cryolite. This simple, if perhaps somewhat surprising, fact has a major impact on cell operation, since it means that any carbon exposed to cryolite at a cathodic potential will be eroded by formation and dissolution of Al_4C_3. Rates of loss are of the order of a millimetre per day, which begins to be serious if continued for a month or so. Consequently, it is vital to maintain a covering of frozen cryolite on the carbon side walls, and this ties the cell operating temperature very closely to the liquidus diagram for the electrolyte chosen. It also dictates the thermal insulation of the side walls, and makes heat balance the key to successful operation.

The most recent measurements of the solubility (10) are shown in Figure 5. It is remarkable that they can be accounted for by postulating that the dissolved species is $Na_3Al_3CF_8$.

Direct Carbothermic Reduction

Extrapolation of the line for the CO equilibrium in Figure 1 shows that $\Delta G°$ becomes zero at about 2020 °C, and therefore it should be possible, if no complications supervene, to reduce alumina to metal by heating it with carbon to temperatures above that. In fact, it is possible and it has been done, but there are massive difficulties and none of the companies which have undertaken serious work in the field appear to be continuing with it at the present time. This paper can give only the briefest sketch of the situation, and it is a field in which much is known to have been done but little has been published.

347

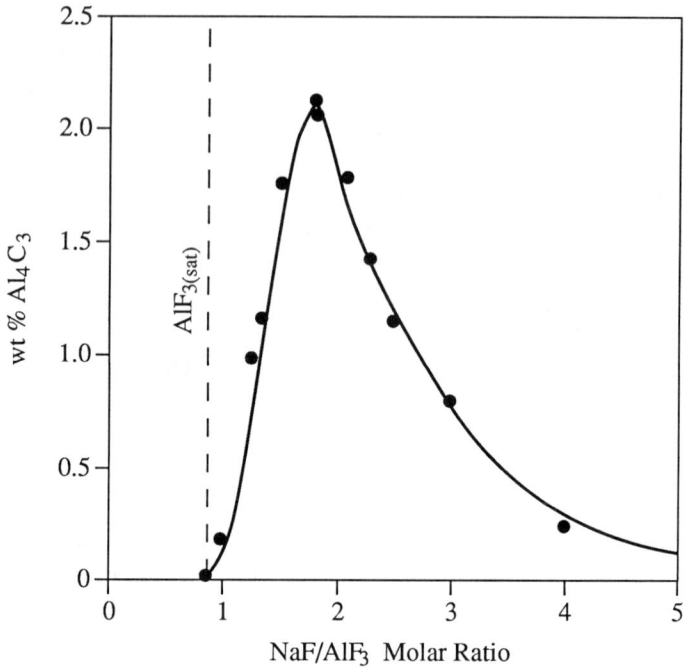

Figure 5 Solubility of Al_4C_3 in NaF - AlF_3 melts. (From Ødegård (11).)

The first complication is the existence of Al_4C_3; the reaction consequently goes in two stages:

$$2Al_2O_3 + 9C = 4Al_4C_3 + 6CO \tag{5}$$

$$4Al_4C_3 + Al_2O_3 = 6Al + 3CO \tag{6}$$

At the temperatures involved, Al_2O_3 and Al_4C_3 form liquid mixtures, so that the product of reaction (5), which is comparatively easy to carry out, is a liquid oxycarbide slag. Reaction (6) involves heating the slag above its formation temperature and boiling out CO; it must be carried out in the absence of free carbon. This is a major constraint.

The second major problem is the formation of the reduced gaseous species Al(g) and $Al_2O(g)$. They represent not only a loss of product but also a loss of the energy which went into making them. To allow them to back-react with CO as the gas cools to re-

348

form Al_2O_3 and carbon is not acceptable; they must be contacted with carbon so that the reversion is only to Al_4C_3.

The third problem is that the Al made contains carbon in solution; it is the equivalent of cast iron. Carbon in aluminum made by the conventional process is not a major problem (although it is there at a level of tens of ppm), but it is limited by the low temperature and the stability of Al_4C_3. In fact, the activity coefficient of C in Al at a given temperature is about the same as that of C in Fe at the same temperature, and at temperatures over 2000 °C it is quite soluble. As the metal cools, the carbon crystallizes as Al_4C_3, which must be separated and returned to the furnace.

Although production of un-alloyed aluminum by direct reduction is very difficult, production of alloys — notably Al - Si — in a submerged arc furnace is well established. The reason is simple. At the reduced activity of Al, Al_4C_3 is no longer stable, so that metal can be produced in the presence of free carbon.

Since Al-Si can be produced in an arc furnace, there have been attempts to make it in a blast furnace with a view to saving electrical energy. Such attempts seem doomed to failure. If one looks at the overall reaction

$$C + O_2 + Al_2O_3 \rightarrow CO(g) + Al(g) + Al_2O(g) \tag{7}$$

with the gas composition in equilibrium with the alloy which is to be made, then that reaction is <u>endothermic</u> (due to the highly endothermic nature of Al(g) and $Al_2O(g)$). This means that, at least in the final stages of the process, <u>one cannot generate heat by burning carbon.</u>

Although work on carbothermic reduction appears dead at the moment, the long-term future is not clear. In a world of ever changing energy costs, capital costs, environmental constraints, and technical possibilities, its day may yet come.

Acknowledgement

The author wishes to thank Alcan International Limited for help with the preparation of this paper.

References

1. E.W. Dewing, "The Chemistry of the Alumina Reduction Cell", Can. Met. Quarterly, 30(1991), 153-161.

2. R. Ødegård et al., "A Thermodynamic and Experimental Study of the Electrochemically Induced Cooling of the Anode in Hall-Heroult Cells", Metall. Trans. B, 22B(1991), 831-837.

3. R. Ødegård, Å. Sterten, and J. Thonstad, "On the Solubility of Aluminum in Cryolitic Melts", Metall. Trans. B, 19B(1988), 449-457.

4. E.W. Dewing, "Loss of Current Efficiency in Aluminum Electrolysis Cells", Metall. Trans. B, 22B(1991), 177-182.

5. E.W. Dewing, "A Degenerate Electron Gas Model for Solutions of Aluminum in Cryolite Melts", Metall. Trans. B, 22B(1991), 669-672.

6. G.M. Hårberg et al., "Electronic Conduction in Molten Cryolite Saturated with Alumina", Proc. 7th Int. Symp. Molten Salts, Pennington, NJ, Electrochem. Soc. (1990), 185-192.

7. E.W. Dewing, "Models of Halo-Aluminate Melts", Proc. 5th Int. Symp. Molten Salts, Pennington, NJ, Electrochem Soc. (1986), 262-274.

8. E.W. Dewing, "Thermodynamics of the System NaF-AlF$_3$: Part VI, Revision", Metall. Trans. B, 21B(1990), 285-294.

9. B. Gilbert and T. Materne, "Reinvestigation of Molten Fluoroaluminate Raman Spectra : The Question of the Existence of Fluoroaluminate (AlF$_5^{2-}$) Ions", Appl. Spectrosc. 44(1990), 299-305.

10. B. Gilbert, G. Mamantov, and G.M. Begun, "Raman Spectra of Aluminum Oxide Solutions in Molten Cryolite and Other Aluminum Fluoride Containing Melts", Inorg. Nucl. Chem. Lett., 12(1976), 415-424.

11. R. Ødegård, "On the Solubility of Aluminum Carbide in Cryolitic Melts", Metall. Trans. B., 19B(1988), 441-447.

Session 4

Mechanisms and Models in Bath Smelting

SOLID-LIQUID AND GAS-LIQUID INTERACTIONS
IN A BATH-SMELTING REACTOR

N.J. Themelis
Henry Krumb School of Mines
Columbia University
New York, NY 10027, U.S.A.

P.J. Mackey
Noranda Technology Centre
Pointe Claire, Quebec, Canada H9R 1G5

ABSTRACT

In bath smelting processes, such as the Noranda Process, concentrates and other materials are fed onto the surface of the bath and are first enveloped in liquid slag. The rate of melting of solid particles is controlled by their size and thermal diffusivity, by the Prandtl number of the liquid slag and the mixing intensity in the bath; the latter is a function of the rate of injection of oxidizing gas through the tuyeres which, also, affects the "plume " velocity and residence time of the oxidizing gas rising through the melt. A method is suggested for estimating the injection-forced heat transfer coefficient between melt and charge particles. The paper also discusses empirical relationships between gas injection rate and reactor volume, which seem to apply both to bath smelting reactors and to conventional Peirce-Smith converters. The analysis made of the solid-liquid and gas-liquid rate phenomena in bath smelting indicates that bath smelting reactors have not as yet reached their full potential in terms of rate of production per unit volume of reactor.

Proceedings of the
Savard/Lee International Symposium on Bath Smelting
Edited by J. K. Brimacombe, P. J. Mackey,
G. J. W. Kor, C. Bickert and M. G. Ranade
The Minerals, Metals & Materials Society, 1992

INTRODUCTION

There are two basic types of smelting processes (1): Flash smelting, where the predominant reactions occur while the concentrate particles are dispersed into an oxidizing gas stream; and bath smelting, where the concentrate particles are enveloped and reacted in a turbulent bath of matte, slag, and gas. The Noranda Process for copper production is a prominent bath smelting process and is presently used at three smelters around the world (2-4). A fourth plant is presently under construction in Canada (5), and a fifth plant is planned in China (23). Another major bath-smelting process is the El Teniente converter (6), which is used at several smelters in Chile, and planned for the modernization of the Nkana smelter in Zambia (7). The smelt-in-melt process developed in Russia (8, 29), and the QSL process for lead production (40-41) are other examples of bath smelting. Approximately 30% of the world's copper is produced by bath smelting processes, 60% by flash smelting and 10% hydrometallurgically (8-9). It is interesting to note that all current international efforts to develop a direct steelmaking process (10-12) are based on the use of bath smelting principles. It is therefore evident that bath smelting is becoming a major processing route for metal production.

In the Noranda bath smelting process (Figure 1), concentrates, flux and other particulate materials are introduced in the horizontal, cylindrical reactor by means of a slinger belt feeder. For concentrate tonnages of 2,500 tonnes per day, the reactor vessel is typically 5 m in diameter by 21 m long (dimensions are inside steel shell). Air enriched to 30-45% oxygen is injected into the melt through a number of tuyeres distributed along about half the length of the reactor.

REACTOR DIMENSIONS AND BLOWING RATE

In some respects, the Noranda Reactor, and also the El Teniente converter, resemble the conventional Peirce-Smith converter; therefore, some of the information available on the dynamics of Peirce-Smith converters (13) can be of value in the analysis of bath smelting vessels. The "standard" Peirce-Smith converter, as used in most smelters in the world, is a 4 m by 9 m long (13 ft. by 30 ft.) cylindrical vessel fitted with 48 tuyeres, injecting 37,000 Nm^3/h (22,000 scfm) under good operating practice. Smaller units are still in use, and larger converters have been in operation for some time (13-15). The largest Peirce-Smith converter (4.57 m by 10.67 m long) in the world, at the San Manuel smelter of Magma Copper, is equipped with 60 tuyeres and has an average blowing rate of about 47,600 Nm^3/h. The range of converter sizes used in copper and nickel smelting is illustrated in Figure 2; some of the bath smelting reactors are also shown on the right-hand of this plot. The length/diameter (L/D) ratio for converters has ranged from about 2 to 3.5. For bath smelting reactors, which smelt and convert continuously, the highest L/D ratio is a little over 4.

Using the 1958 survey data of Lathe and Hodnett (14), Themelis, Tarassoff and Szekely (15) showed that converter blowing rate could be correlated with converter space volume. This trend was confirmed in the 1979 world-wide survey by Johnson, Themelis and Eltringham (13); their results are presented in Figure 3, together with more recent data (21) for large size converters and for bath smelting reactors.

354

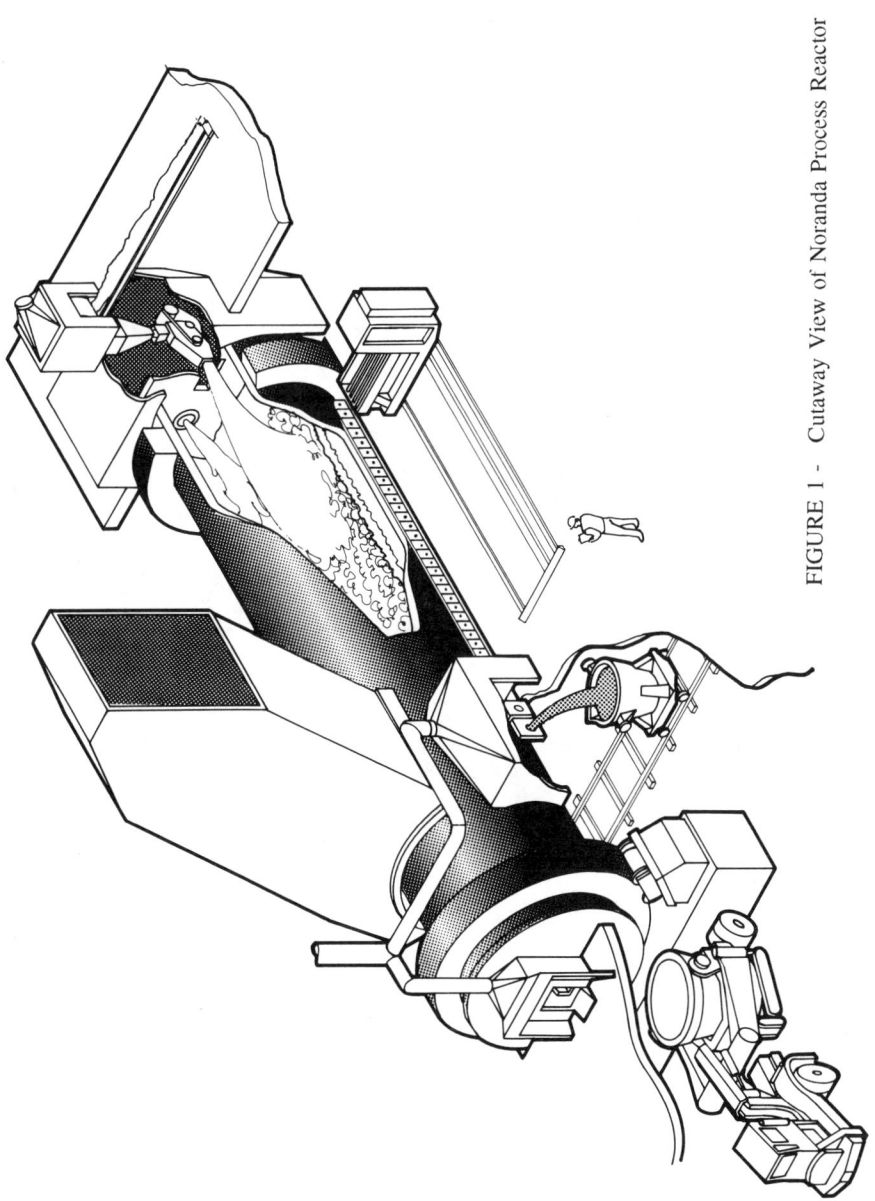

FIGURE 1 - Cutaway View of Noranda Process Reactor

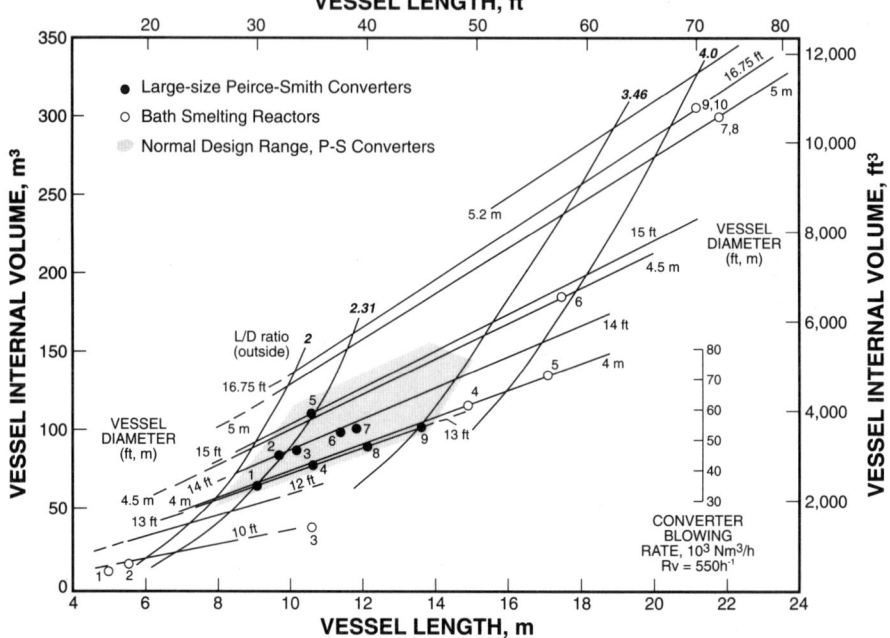

FIGURE 2

Shell dimensions and vessel internal volume for Peirce-Smith converters and bath smelting reactors. Shell dimensions are given in metric and British metric units, refractory lining assumed to be 0.38 m (15 in.). The numbered points refer to the following vessels:

Large-size Peirce-Smith converters (17, 21):
1. Standard, 13 ft. x 30 ft., 37,000 Nm3/h
2. Horne, 14 ft. x 32 ft.
3. Norddeutsche Affinerie, 4.2m x 10.3 m
4. Mt. Isa, 13 ft. x 35 ft.
5. San Manuel, 15 ft. x 35 ft.
6. Saganoseki, 4.2 m x 11.51 m
7. Toyo, 4.2 m x 11.9 m
8. Projected 13 ft. x 40 ft. unit
9. Inco stretch, 13 ft. x 45 ft.

Bath smelting reactors (4, 6, 8, 19, 20):
(ET = El Teniente, N = Noranda)
1. Saganoseki converter smelting
2. Hitachi converter smelting
3. Horne pilot plant (N - copper mode)
4. Las Ventanas (ET)
5. Potrerillos (ET)
6. Southern Copper (N)
7. Chuquicamata (ET)
8. Caletones (ET)
9. Horne (N)
10. Kennecott (N)

This correlation can be expressed in terms of the volumetric ratio, R_v

$$R_v = v_{gas}/V_C$$

where,

v_{gas} = flowrate of air injected, Nm^3/h

V_C = reactor (converter) internal volume, m^3

The "best fit" straight line through the performance data of 56 smelters, as analyzed by Johnson *et al.* (13) yielded a value of $R_v = 510$ h^{-1}. This line is shown on Figure 3 (Line A), along with performance data from Peirce-Smith converters built since the earlier survey. Projecting this line to even larger reactors volumes assumes that,

$$v_{gas} = 510 \ L\left(\frac{\pi \ d^2}{4} \right) \tag{1}$$

where, L and d are the inside length and diameter of the reactor.

However, McKerrow (16), in 1965, analyzed operating data from the 12 ft. and 13 ft. diameter converters at the Horne smelter of Noranda and concluded that converter blowing rate for larger converters should be scaled up according to the following equation,

$$v_{gas} \propto L d^{1.6} \tag{2}$$

In effect, the McKerrow correlation indicates that only part of the converter circular cross-sectional area is utilized in the blowing action, possibly due to the skewed positioning of the tuyeres along one side of the converter; therefore, production capacity does not increase proportionally to the square of the diameter. Table I shows the dimensions of selected Peirce-Smith converters (and bath smelting reactors), and compares the projected blowing rate for converters based on performance for the "standard" converter, according to the direct volume correlation (Equation(1), but with $R_v = 550$ h^{-1}, Line B) and the McKerrow correlation, Equation (2).

The present authors examined the theoretical basis for correlating the gas injection rate to the inside dimensions of the reactor. It is obvious that as the length of the vessel (converter) is increased and more tuyeres are added, v_{gas} should be proportional to L, as assumed by both Equations (1) and (2). When it comes to scaling up the diameter, the choice is not too clear. If the reactor capacity is limited by the gas injection rate per unit volume of the melt, then v_{gas} should be proportional to the square of the diameter, as predicated by Equation (1).

However, as will be discussed in later sections of this paper, it is more likely that the converter capacity will be limited by the superficial velocity of the gas rising through the molten bath, which affects the gas-liquid reactions within the melt, and the effective exhausting of these gases with minimum dust carryover. Or, in the case of bath smelting reactors, the smelting capacity may ultimately be limited by the ability of the bath to absorb the charge per unit surface area of the bath, i.e. by the rate of the charge-liquid reactions near the surface of the bath.

Table I - Analysis of Blowing Rates of Peirce-Smith Converters and Bath Smelting Reactors

PARAMETER	PEIRCE-SMITH CONVERTERS								BATH SMELTING REACTORS				
Plant (13,21)	Home 1964	"Standard" PS	Mt. Isa	Inco	Home	Nord-deutsche Affinerie	San Manuel	Home Pilot	Home	Southern Copper	El Teniente	Potrerillos	Las Venteas
Dimensions inside shell													
Diameter, m	3.66	3.96	3.96	3.96	4.27	4.20	4.57	3.05	5.11	4.50	5.00	4.00	4.00
Length, m	9.14	9.14	10.67	13.72	9.75	10.30	10.67	10.67	21.34	17.50	22.00	17.00	15.00
Dimensions inside brick													
Diameter, m	2.90	3.20	3.20	3.20	3.51	3.44	3.81	2.29	4.35	3.74	4.24	3.24	3.24
Length, m	8.38	8.38	9.91	12.96	8.99	9.54	9.91	9.91	20.58	16.74	21.24	16.24	14.24
Actual internal volume, m³	55.3	67.5	79.8	104.4	86.9	88.9	113.1	40.7	305.1	183.9	300.0	133.9	117.4
BATH SMELTING REACTORS [*]													
Volume converting zone, m³								18.3	183.1	110.3	179.9	80.3	70.4
Length converting zone, m								4.5	12.3	10.0	12.7	9.7	8.5
PROJECTED BLOWING RATE [**], Nm³/h													
CONVERTERS (using $R_s = 550\ h^{-1}$)													
Based on volume correlation	30,300	37,000	43,700	57,200	47,600	48,700	62,000						
Based on McKerrow (16) correlation (Eqn.2)	31,500	37,000	43,700	57,200	45,900	47,300	57,800						
BATH SMELTING REACTORS													
Based on converting bath area (Eqn. 3)								(16,200)	85,200	59,700	85,800	50,100	44,000
Reported blowing rate, Nm³/h	30,000	37,000	50,400	51,300	45,000	45,000	47,600	7,100	76,500	28,920	62,000	34,000	30,000
Actual bath surface area, m³ [***]	24.3	26.7	31.5	41.0	31.0	32.4	36.5	22.4	83.9	60.0	87.0	52.5	46.1
Bath area, smelting/converting zone, m³ [***]								10.1	50.3	36.0	52.2	31.5	27.7
Superficial gas velocity, m/s	1.9	2.2	2.5	1.9	2.3	2.2	2.0	1.1	2.4	1.2	1.8	1.7	1.7

Refractory lining assumed to be 0.38 m (15 in.). Calculated data have been rounded.

[*] Smelting and converting zone in bath smelting reactors assumed to be 60% of internal volume (except Home pilot where factor is 45% due to vessel configuration).

[**] Basis: Blowing rate for "standard" 13 ft. by 40 ft. Peirce-Smith converter was assumed to be 37,000 Nm³/h; for bath smelting reactors, 50,400 Nm³/h for 13 ft. by 35 ft. vessel.

[***] At average melt level for reported blowing rate.

In both of these cases, reactor scale-up should be done on the basis of converting bath surface area, i.e.,

$$v_{gas} \propto Ld \qquad (3)$$

where L here refers to the length of the converting bath zone.

It is believed that Equation (3) provides a realistic, and the most conservative, means for projecting the injection capacity of larger bath smelting reactors. The use of this equation is illustrated in the following example.

Actual performance data for converters and various bath smelting reactors are shown in Table I and are also plotted in Figure 3. In bath-smelting reactors, which, unlike converters, are used for smelting plus converting duties, only part of the vessel length is used for converting. For instance, in the case of the Noranda reactors, it is estimated that only 60% of the reactor length is active for smelting and converting, while the rest provides a quiescent zone under the reactor mouth, and for settling of high-grade matte from the slag (slag zone). Assuming that other bath-smelting vessels, such as the El Teniente converters, also use 60% of their length for converting, we have also plotted in Figure 3 the reported gas injection data for bath-smelting reactors on this basis. This plot indicates that the gas injection rate in bath smelting reactors, such as the Noranda Process and the El Teniente converters, could potentially be increased appreciably before reaching the levels projected by blowing rate correlations for the Peirce-Smith converter. It is also of interest to note that the projected operating range of a proposed continuous converter (19) falls within the range for bath smelting vessels.

Let us now estimate the projected injection rate in a vessel the size of the Noranda reactor at the Horne, which would correspond to the performance data for the Mount Isa converter. On the basis of the data reported in the literature (13, 21), this converter (3.2 m I.D. x 9.91 m inside length) has one of the best blowing rate performances in the world, with an R_v (on a copper blow) equal to $(50,400 \ Nm^3/h)/79.8 \ m^3 = 630 \ h^{-1}$. On the basis of Equation (3) and using the above R_v value, the corresponding projected injection rate in the above-mentioned Noranda reactor (4.35 m ID x 12.4 m converting zone) would then be:

$$V_g = 50,400 \times \left(\frac{12.3}{9.91} \right) \times \left(\frac{4.35}{3.2} \right)$$

$$= 85,200 \ Nm^3/h$$

If one now assumes a number of reactors of the same diameter (4.35 m I.D.) but different lengths of converting zone, the corresponding linear relationship between gas injection rate and reactor volume is shown by the dotted line in Figure 3. It can be seen that on the basis of either volume or converting bath surface projection, the large bath smelting reactors have potentially room to increase gas injection rate to the levels attained by the best operating converter. It is also apparent from the large scatter of reported gas injection rates in Figure 3, that other design factors, such as tuyere positioning and maintenance, vessel holding capacity, blower specifications, mouth size and location, off-gas handling system, matte grade (and slag quality) and, above all, good operating practice all have a considerable effect on converter and bath smelting reactor performance.

Table II - Typical Operating Data - Converters Bath Smelting and Fluid Bed Reactors

Parameter	Peirce-Smith Converter (11,21)	Noranda Process Reactor (4)	El Teniente Converter (6)	Anode Refining Furnace (22)	Slag Fuming Furnace (39)	Zinc Fluid Bed Roaster (24)	Nickel-Copper Fluid Bed Roaster (24)
Unit or Plant Reference	Typical 30ft x 30ft (11,21)	Horne (4)	El Teniente (6)	Horne (22)	Trail (39)	CEZ (24)	Falconbridge (24)
Vessel shell dimensions, (dia.xlength), m	3.96 x 9.14	5.11 x 21.34	5 x 22	3.96 x 10.67	3.05(w) x 7.32(l) x 3.05(h)*	9.6 m dia. Bed height: 1.6 m Reactor height: 16 m	5.5 m dia. Bed height: 1.5 m Reactor height: 9.1 m
Vessel inside dimensions, (dia. x length), m	3.20 x 8.38	4.35 x 20.58	4.24 x 21.24	3.20 x 9.91			
Vessel volume, m³	68	305	300	80	68	---	---
Nominal melt volume, m³	31	84	101	40	18	---	---
Capacity, tonnes/day							
Concentrate	120 t matte/cycle	2,660	1,660	305 t/charge	70 t/cycle	550	1176
Total charge	---	3,500	2,000	---	---	(550)	(1176)
Blowing rate, Nm³/h	37,000	76,500	59,500	850	24,000	37,000	35,000
Blowing rate, Nm³/h/tuyere	770	1,420	1,080	850	~400	400	---
Oxygen, %	21.28	37	32	Natural gas	21	21	21
Calculated data							
Specific blowing rate: Nm³/h t of melt	24.0	193.2	120	3	340	---	---
Nm³/min t of melt	4.0	3.2	2.0	0.05	5.7	---	---
Superficial gas velocity (converting area)**, m/s	2.2	2.4	1.8	0.3	1.7	0.6	1.3
Specific smelting rate (total charge):	---	---		---			
Tonnes/day m³ (furnace volume)	---	11.4	6.7	---	---	---	---
Tonnes/day m³ (melt volume)	---	41.7	19.8	---	---	---	---
Total heat input rate, kw/m³ (furnace vol.)	222	266	196	109	244	---	---
Mixing power, KWh/mt of total melt	30	23	16	0.8	24	---	---

Note: Some calculated data have been rounded.
* As designed and lengthened in 1931 (39).
** Based on smelting and converting zone as 60% of total melt surface area.

360

GAS-LIQUID INTERACTION

Superficial Gas Velocity

In the operation of the "standard" size of converter, as described above, the superficial gas velocity, defined as,

volumetric rate of injected gas at melt temperature/surface area of bath

is calculated to be approximately 2.2 m/s. In order to put this value into perspective, it is instructive to compare it with velocity parameters for other gas-stirred reactors. Table II shows that the converter velocity is nearly three times higher than the superficial velocity in a zinc fluid bed roaster where roasted zinc concentrates are maintained in suspension by the injected gas. Also, it is interesting to note that the superficial velocity in the converter is more comparable to the superficial gas velocity in a typical zinc fuming furnace (39,42).

In the Noranda Process reactor at the Horne smelter (4), the average tuyere blowing rate is 76,500 Nm^3/h of oxidizing gas at 37% oxygen, injected through 54 operating tuyeres (Table II). While this injection rate is the highest of all bath smelting reactors, the superficial velocity of the rising gas in the smelting and converting zone, as defined earlier, is 2.4 m/s, which is close to converter operations. The data for other bath smelting vessels are in a similar range. As in the case of fluid bed reactors (Table II), where the carryover of particles increases with superficial velocity, a lower velocity in bath smelting reactors signifies less intensive splashing in the "freeboard" space above the bath. It can be seen that the superficial velocity is directly related to the blowing rate per unit bath surface area of the reactor, as discussed in the previous section. Of course, the number of tuyeres used for injecting the gas (and the blowing rate per tuyere) is also important, as will be discussed later under plume velocity.

Gas-Liquid Interfacial Area

When a gas jet is injected into the melt, the bubbles formed at the orifice are hydrodynamically unstable and shatter to smaller bubbles shortly above the point of injection. This phenomenon depends on the injected gas flowrate and the degree of mixing, or turbulence intensity, within the melt; therefore, it can be observed only at high injection flowrates. Several investigators (25-28) have studied bubble characteristics and have reported evidence of bubble break-up close to the orifice even at low flowrates. Thus, Fruehan (25) injected gas bubbles into molten silver or copper, and observed that the bubbles emerging from the surface of the melt were much smaller than the sizes expected on the basis of bubble frequency observations at the orifice. For example, at an argon flowrate of 0.015 Nl/s into liquid copper, the expected bubble size was calculated to be 1.8 cm for a frequency of 22 bubbles/s at the 0.19 cm dia. orifice; however, the bubbles breaking the melt surface were much smaller than this.

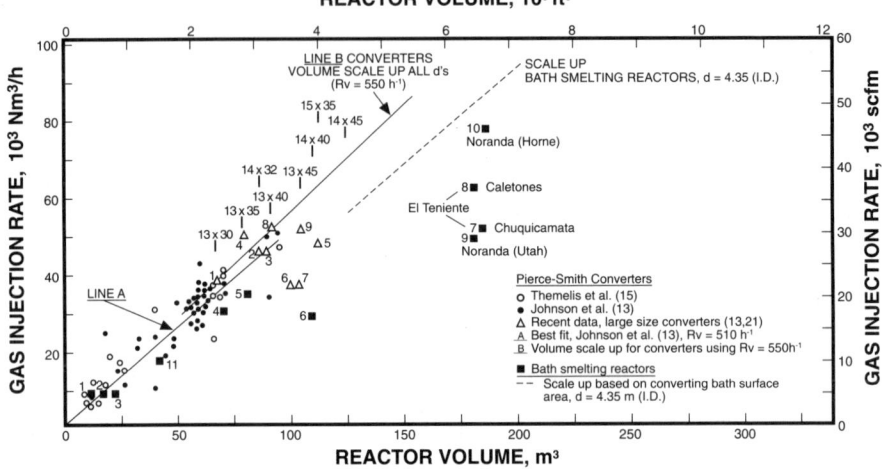

FIGURE 3

Relationship between gas injection rate and reactor volume for Pierce-Smith converters (Lines A and B), and bath smelting reactors (dotted line), (13,15,21). Assumed refractory lining 0.38 m (15 in.). The internal volume of bath smelting reactors here is assumed to be 60% of total volume. The numbered points refer to the following vessels (large bath smelting vessels also labeled):

Large-size Peirce-Smith converters (17, 21):

1. Standard, 13 ft. x 30 ft., 37,000 Nm3/h
2. Horne, 14 ft. x 32 ft.
3. Norddeutsche Affinerie, 4.2m x 10.3 m
4. Mt. Isa, 13 ft. x 35 ft.
5. San Manuel, 15 ft. x 35 ft.
6. Saganoseki, 4.2 m x 11.51 m
7. Toyo, 4.2 m x 11.9 m
8. Projected 13 x 40 ft. unit
9. Inco stretch, 13 ft. x 45 ft.

Bath smelting reactors (4, 6, 8, 19, 20):

(ET = El Teniente, N = Noranda)

1. Saganoseki converter smelting
2. Hitachi converter smelting
3. Horne pilot plant (N - copper mode)
4. Las Ventanas (ET)
5. Potrerillos (ET)
6. Southern Copper (N)
7. Chuquicamata (ET)
8. Caletones (ET)
9. Horne (N)
10. Kennecott (N)
11. Continuous converter (19)

362

On the basis of mass transfer measurements, Fruehan estimated that the average diameter of these secondary bubbles was about 1 cm in diameter. Sahai and Guthrie (26) also concluded that when argon was injected into molten iron at gas flowrates of about 0.1 Nl/s, the large bubbles formed were unstable, and disintegrated above the nozzle into an array of smaller bubbles. Sano and Mori (27) have shown that for bubble swarms in liquid metals (i.e., in the case of gas injection into a liquid column through a multitude of fine orifices), the average bubble diameter would be about 2 cm. The bubble size was found to be independent of the gas flow rate.

On the basis of the ideal gas law and assuming spherical shape bubbles, the overall bubble - melt interfacial area in the bath at any given instant can be calculated from the following equation:

$$A_i = v_{gas} (T_{melt}/273) (1/(P_{gas,ave})) t_b (6/d_p) \qquad (4)$$

where,

A_i = overall bubble - melt (or gas-liquid) interfacial area, m^2

v_{gas} = volumetric rate of air through tuyeres, Nm_3/s

$T_{melt}/273$ = temperature expansion factor for melt temperature, T_{melt}, K

$1/(P_{gas,ave})$ = pressure compensation factor for average pressure in gas bubbles; in the case of the Noranda reactor, $P_{gas,ave}$ = 1.25 atm.

t_b = average residence time of bubbles of diameter, d_b, in melt, s

$6/d_b$ = surface to volume ratio of a spherical bubble of diameter d_b

For the conditions in the Noranda Process reactor (Table II), and on the basis of the studies noted above, the average bubble diameter may be assumed, conservatively, to be 5 cm. Using a depth of immersion in the melt (at 1500 K) of 1 m, an average gas flowrate per tuyere of 0.40 Nm^3/s and the calculated plume velocity of 5.9 m/s (see below), the average residence time of the gas bubbles in the bath, t_b is estimated to be 0.17 s. For these values, A_i in Equation (4) is calculated to be 1,910 m^2. Even if bubbles of a different diameter are assumed, a large gas-liquid interfacial area still results (for example, if d_b = 2.5 cm, A_i is 3,820 m^2, while A_i is 955 m^2 for 10 cm bubbles).

Gas Holdup in the Bath

From the above calculations, it is evident that the injected gas stays in the melt for a very short period of time. The volume of "gas holdup" in the melt is expressed by the following equation:

$$V_{holdup} = v_{tuyere} t_b (T_{melt}/273) (1/(P_{gas,ave})) \qquad (5)$$

where,

v_{tuyere} = gas flow rate per tuyere, Nm^3/s

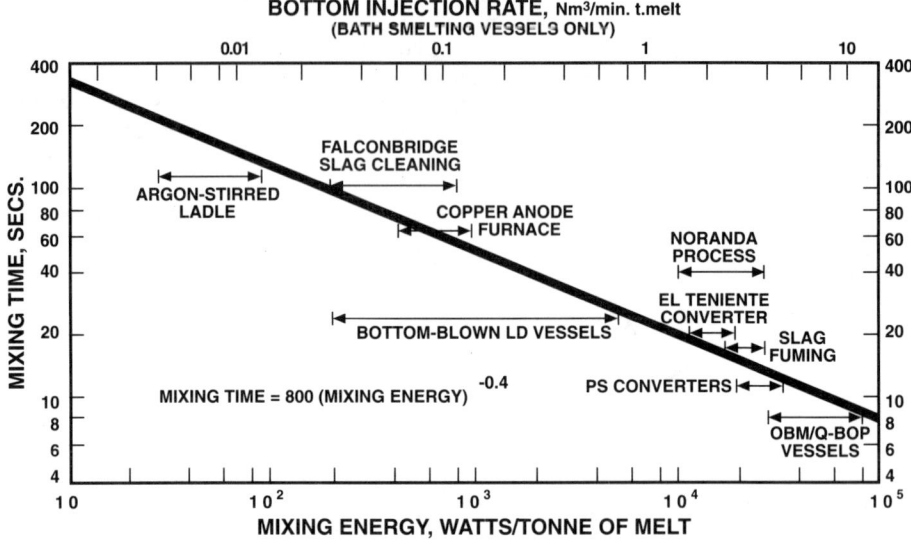

FIGURE 4

Range of mixing energy input level and mixing time for various ferrous and non-ferrous processes, after Mackey (8) with modifications by authors, Nakanishi *et al.* (30) and Themelis *et al.* (31). Note that the bottom injection rate (Nm³/min. tonne of melt) on the upper axis applies only to bath smelting vessels such as Noranda Process or Peirce-Smith converter.

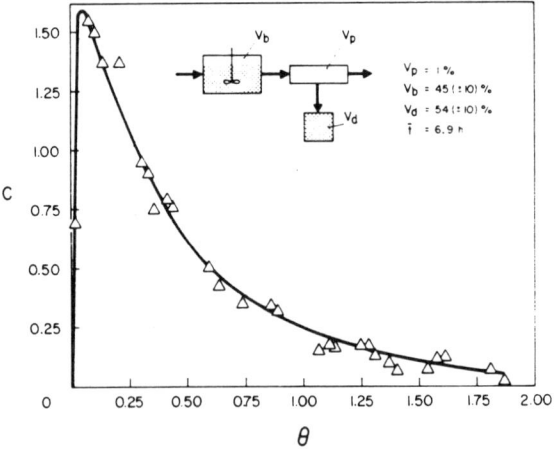

FIGURE 5 -   C-curve for flow in matte layer of Noranda Process reactor after Tarassoff (32). In the figure, the horizontal axis represents the dimensionless time ratio, while C is the dimensionless concentration ratio.

For the conditions described above,

$$V_{holdup} = 76,500/3,600 \times 0.17 \times 5.5 \times 1/1.25$$

or $\quad V_{holdup} = 15.9 \text{ m}^3$

Assuming that this gas holdup is uniformly distributed within the entire melt, it would, on the average, represent a 0.2 m expansion of the melt level in a Noranda reactor such as at the Horne. Actually, the gas bubbles move only through part of the melt, and such expansion across the full melt surface does not take place. The gas-liquid plume rises nearly vertically (38) forming a bubble (gas holdup) "dome" or "spout" (36) at the surface, where the bubbles disengage from the liquid phase. Further work is necessary to examine whether the projected holdup is as high as projected or, alternatively, the plume velocity is higher than estimated in this analysis.

Mixing Energy in Bath and Flow Pattern Through Reactor

The gas injection imparts an enormous amount of mixing energy to the melt due to the introduction, rise and expansion of the injected gas bubbles. For the case of gas injection in a melt of specific gravity ρ_m, this energy, P_m can be expressed as follows (1, 30, 31):

$$P_m = 0.74 v_{gas} T \ln (1+ (\rho_m L/P_a) \tag{6}$$

where,

P_m	=	mixing power, kW
v_{gas}	=	gas flowrate, Nm^3/s
T	=	bath temperature, K
ρ_m	=	specific gravity of the liquid bath, g/cm^3
L	=	depth of tuyere immersion, cm
P_a	=	atmospheric pressure, g/cm^2

The term P_m can also be expressed in terms of mixing power input per tonne of molten bath, or,

$$\dot{\epsilon} = P_m/W_m \tag{7}$$

where,

$\dot{\epsilon}$	=	mixing energy per tonne of melt, kW/tonne
W_m	=	weight of melt, tonne

According to the above equation, for an injection depth of about 1 m, a total gas flow rate of 21 Nm^3/s and an assumed average specific gravity of the bath of 4,700 kg/m^3, the mixing energy in the large size Noranda reactor is calculated to be about 21 kW/t of melt bath. The value for the Peirce-Smith converter is typically in the range 20-30 kW/t. For the smaller Noranda reactor in operation at Southern Copper (8), the mixing energy is about 12 kW/t of melt bath. These levels of mixing energy are comparable (8) to that achieved in the bottom-blown OBM/Q-BOP oxygen steelmaking process and other bath smelting systems, Figure 4. This indicates that in the tuyere zone, the Noranda reactor should behave as a nearly "perfect-mixer" reactor.

The mixing regimes in the Noranda Process reactor were discussed by Tarassoff in the 1984 Extractive Metallurgy Lecture (32). The tracer concentration test curve based on experimental measurements when a "pulse" addition of high silver concentrates was made is shown in Figure 5. Analysis of the tracer test data showed that the overall reactor melt volume (from feed end to slag taphole) could be considered to consist of three regimes of flow as presented in Table III (32).

Table III - Flow Regimes in Noranda Process Reactor(32)

| | | Volume Fraction, % | |
Zone	Plug flow	Perfect Mixing	"Deadwater"
Matte tracer test	1	45	54
Slag tracer test	21	53	26

The above data indicate that part of the matte and slag layers, probably in the lower layers of the quiescent zone of the reactor, is relatively inactive. The rest of the matte layer acts nearly as a perfect mixer, while nearly one-third of the slag "active" zone is in plug flow, probably in the slag settling zone.

MASS TRANSFER RATE BETWEEN GAS AND LIQUID

As discussed above, the injection of gas in a bath smelting reactor provides a very large interface area for reaction between the oxidizing gas and the matte-slag emulsion within the bath. This large surface area results in very high reaction rates. Because of the relatively high concentration of sulphur in the matte (approximately 20% S, even in the case of direct production of copper in the Noranda Process reactor (2)), and the high chemical reaction rates at elevated temperatures, it is expected that the rate of oxidation is controlled solely by the supply of oxidizing gas. A simple calculation of the maximum rate of reaction of oxygen in a Noranda Process reactor can be made by assuming that it is equal to the formation of sulphur dioxide by "combustion" of the sulphur in the melt (i.e. excluding the oxidation of iron in the matte, which, in high-grade matte production, consumes only 20% of the input oxygen):

$$\dot{m}_o = N_s \; A_i$$
$$= k_d \; A_i \; \rho_m \; (X^1_{s,b} - X^1_{s,e}) \tag{8}$$

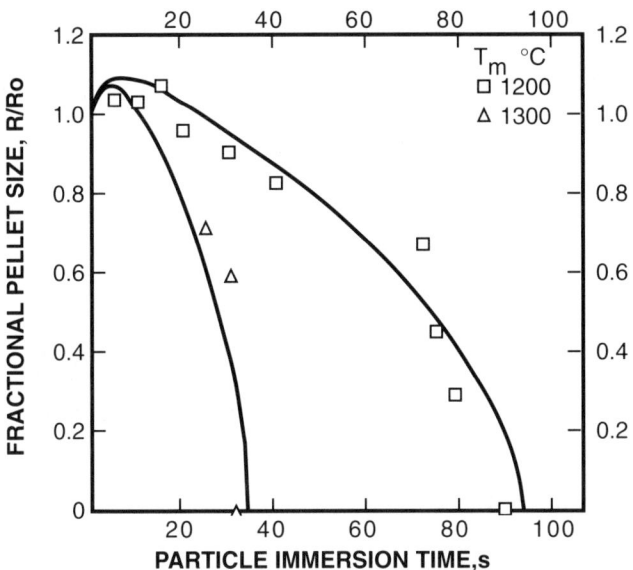

FIGURE 6 -   Comparison of observed and calculated melting time of Cu-Ni ore pellets immersed in Cu$_2$S melt, as determined by the fractional pellet size vs. pellet immersion time, after Jiao and Themelis (33), experimental data determined by Mwansa and Warner (34). In the figure, R$_o$ = initial pellet radius, R = pellet radius after time, t.

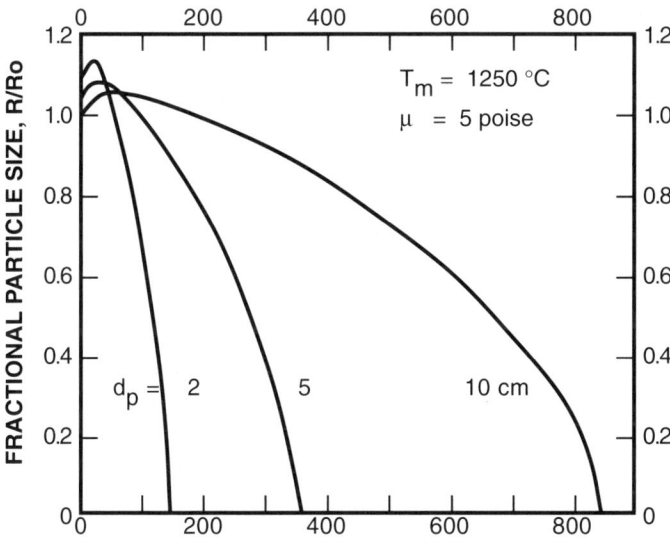

FIGURE 7 -   Calculated rate of melting determined by fractional particle size change as a functional particle diameter (d$_p$) and particle immersion time for slag particles immersed in slag melt under natural convection, after Jiao and Themelis (33). In the figure, R$_o$ = initial particle radius, R = particle radius after time, t.

where,

$\dot{m}_o$ = moles of oxygen (as O_2) reacted/second
N_s = molar flux of sulphur reacted
A_i = overall gas-liquid interface area, m^2
k_d = mass transfer coefficient of sulphur at the gas-liquid interface, m/s
ρ_m = density of matte, kg/m^3
$X^1_{s,b}$ = fractional mass concentration of sulphur in the matte bulk
$X^1_{s,e}$ = fractional mass concentration of sulphur in the matte at equilibrium with oxygen (in this case ~ 0)

Substituting in the above equation the value of k_d = 0.0002 m/s (0.02 cm/s), established by experimental work on the oxygen-carbon reaction in iron melts and other gas-liquid metal systems (25, 30, 31), the above calculated value of A_i (1,910 m^2), a matte density of 6,000 kg/m^3 and a fractional mass concentration of sulphur in the melt of 0.18, we obtain:

$$\dot{m}_o = 0.0002 \times 6,000 \times (0.18/32) \times 1,910 \qquad (9)$$
$$= 12.9 \text{ kmol/s}$$
$$= 290 \text{ Nm}^3/\text{s}$$

By comparison, the actual rate of oxygen injection in the Noranda Process reactor, for example, is about 8 Nm^3/s (76,500 Nm^3/h at 37% oxygen). This very approximate calculation shows that the bath has the mass transfer potential to react a much larger quantity of oxygen than that which is actually injected in a Noranda Process reactor; the same potential exists for other bath smelting reactors. The very large difference between the projected mass transfer potential and current operating practice indicates that it may be possible to design much more intensive bath smelting reactors.

HEAT TRANSFER BETWEEN SOLID PARTICLES AND SLAG/MATTE BATH

Injection-forced Heat Transfer from Melt to Charge

When concentrates or other charge materials are introduced onto the melt surface in a bath smelting vessel such as the Noranda Process reactor, a solid shell or crust is first formed around the colder charge, and grows around the particle, soon reaching a maximum thickness. This crust then starts to melt when the heat transfer by convection from the melt to the crust becomes greater than the rate of heat conduction out through the particle.

Jiao and Themelis (33) developed a mathematical correlation for the time of complete melting of crust and particle immersed in a relatively quiescent bath. Figure 6 shows that there was good agreement between the calculated values (33) and the experimental results by Mwansa and Warner (34) on the melting time of Cu-Ni ore pellets immersed in a copper sulphide melt. Figure 7 (33) shows the calculated time for melting of slag particles immersed in a non-ferrous slag; if the slag bath is at least 100° C above its melting point, a 5 cm solid slag particle would melt fully in 360 seconds and a 2 cm particle in 140.

However, for both Figures 6 and 7, the heat transfer coefficient between slag and particle depended only on natural convection, as described by the following equation (33, 35):

$$Nu = 2.0 + 0.6 \ (Gr^{0.25} \ Pr^{0.333}) \tag{10}$$

where Nu, Gr and Pr are the Nusselt, Grashof and Prandtl numbers respectively, and are defined as,

$$Nu = h \ d_p \ /k \tag{11}$$

$$Gr = d^3_p \ \rho^2 g \ \beta \ (T_b - T_m)/\mu^2 \tag{12}$$

$$Pr = C_p \ \mu/k \tag{13}$$

where for the slag phase, the following parameters and values apply (33):

Symbol	Term	Units	Value
C_p	slag specific heat	J/kg°C	920
d_p	solid particle diameter	m	0.02
g	gravitational constant	m/s^2	9.81
h	slag/particle heat transfer coefficient	w/m^2°C	-
k	slag thermal conductivity	W/m°C	10.6
T_b	melting point of slag	°C	1170
T_m	temperature of melt	°C	1250
β	thermal expansion coefficient	°C^{-1}	0.0001
μ	slag viscosity	kg/m s	0.2
ρ	slag density	kg/m^3	2890

According to the above equations, with Gr=131 and Pr=17.3, the value of the Nusselt number using Equation (10) was found to be equal to 7.3.

However, as discussed above, injection of enriched air through the tuyeres results in intense mixing of the bath. Therefore, one would expect a much faster rate of heat transfer to the solid particles than predicted by the above Equation (10) which is based on natural connection only. In order to obtain a rough estimate of the enhancement of heat transfer due to mixing of the bath, which we have termed **injection-forced heat** (and mass) **transfer**, the authors proceeded as follows:

Using the Sahai and Guthrie correlation (36) for plume velocity (u_p) and the recirculation velocity (u_{ave}) in a bottom injected bath, the plume velocity above each tuyere can be expressed by:

$$u_p = (4.4 \ L^{0.25} \ v_{tuyere}^{\ 0.33})/R^{0.33} \tag{14}$$

while the average recirculation velocity can be expressed as follows:

$$u_{ave} = 0.18u_p/R^{0.33} \tag{15}$$

where,

$$L \quad = \text{ tuyere depth immersion, m}$$
$$v_{tuyere} \quad - \text{ gas flow rate per tuyere, Nm}^3/\text{s}$$
$$R \quad = \text{ radius of ladle; in the present case, this parameter was assumed to be}$$
equal to the tuyere "influence" radius (taken to be equal to the center-to-center distance between tuyeres, 0.165 m)

The nominal air flow through each tuyere in the Noranda Process reactor is 0.4 Nm^3/s. For a tuyere depth of immersion of 1 m, and assuming that the zone affected by each tuyere extends to a radius of 0.165 m from the tuyere, the plume velocity was calculated from the above correlations (36) to be 5.9 m/s, while the corresponding recirculation velocity was 1.93 m/s. If it is assumed that the latter is an effective measure of the relative velocity between the bath fluid and the charge feed particles, then the particle Reynolds number, Re = $d_p u_{ave} \rho / \mu$ is calculated to be approximately equal to 560. Finally, incorporating this value into the well known Ranz-Marshall correlation (35) for forced flow heat convection to the particles due to the recirculation velocity,

$$Nu = 2 + 0.6 \ (Re^{0.5} \ Pr^{0.333}) \tag{16}$$

the value of the injection-forced Nusselt number is calculated to be 29.4 which is about four times larger than that calculated for natural convection (Equation 10). Therefore, these admittedly rough calculations show that in the presence of injection-forced heat transfer, i.e. due to gas injection in the bath smelting reactor, a 5 cm particle (Figure 7) would melt completely in about 90 seconds. For comparison, a 2 cm particle would melt in only 35 seconds, and a 10 cm particle in 210 seconds.

Smelting Capacity of Bath

If we now consider the operation of a bath smelting reactor such as the Noranda Process reactor at the Horne (Table II), the above estimates can be used to compare the projected rate of smelting relative to the actual rate of feeding particles into the reactor. Considering an average instantaneous total feed rate of 3,000 tonnes per day (including concentrates, flux and other materials), and an average particle diameter of 2 cm, the number of particles fed is calculated to be approximately 2,800 particles per second. If the feed is evenly distributed over the surface of the smelting and converting zone, the number of 2 cm spherical particles that can be accommodated per square meter of melt surface area in a monolayer is calculated to be 2,500. Considering that the total surface available at the average melt level is about 50 m^2, this leads to the conclusion that, on a statistical basis, a particle would melt in 35 seconds, as calculated above, while it would take,

$$2,500/2,800 \times 50 = 45 \text{ seconds}$$

for another particle to be fed to the exact location.

The above calculation was made for a monolayer of particles. However, in the intensely mixed bath, a particle is enveloped instantly in the recirculating flow of the slag and matte plumes. Therefore, such an intensely mixed bath may be able to "absorb" much more than a monolayer of particles. This calculation indicates that the smelting capacity of bath smelting reactors can be increased substantially.

CONCLUSIONS

The factors influencing gas injection rates and rate phenomena in bath smelting reactors and in converters have been examined. On the basis of all the phenomena analyzed and discussed above, it is evident that bath smelting reactors have not as yet reached their full potential, in terms of rate of production per unit volume of reactor. In order to reach this potential, further research and development are required. For example, actual measurements of the holdup volume during operation can be used to verify the plume velocity correlation used in this study. Another need for research is to understand better the limits to the flow of process gases through the space above the melt and through the mouth in such vessels (and potential for dust carry-over). Of course, the real limits to productivity will also depend on overcoming technological barriers, such as wear and tear on tuyeres and reactor lining, and operating factors relating to close monitoring and control of feed input and distribution, matte grade and temperature (37) within the reactor.

ACKNOWLEDGEMENTS

The authors wish to express their thanks to Noranda Inc. for the opportunity to publish this paper. The continuing support of Noranda Technology Centre to academic research is gratefully acknowledged.

REFERENCES

1. H.H. Kellogg and N.J. Themelis, "Principles of Sulfide Smelting" in, <u>Advances in Sulfide Smelting</u>, (H.Y. Sohn, D. George and D. Zunkle, eds.) *TMS-AIME*, Vol. 1, 1983, pp. 1-29.

2. N.J. Themelis, G.C. McKerrow, P. Tarassoff and G.D. Hallett, "The Noranda Process for the Continuous Smelting and Converting of Copper Concentrates", *J. of Metals*, Vol. 24, No. 4, April 1972, pp 25-32.

3. R.J. Anderson, R.R. Beck and A.J. Weddick, "The Utah Smelter as Modified for Environmental Compliance, in, <u>Copper Smelting - An update</u>, (D.B. George and J.C. Taylor, eds.), *TMS-AIME*, 1982, pp.173-199.

4. D.G. Pannell and P.J. Mackey, "Noranda Process Operations 1988 and Future Trends", Paper presented to the Copper Committee meeting of the GDMB, Antwerp, Belgium, 27-29 April, 1988.

5. Hudson Bay Mining and Smelting Co., Limited, Annual Report, 1990, Winnipeg, Manitoba, Canada, 1990; "A Renaissance at Hudbay", *Canadian Mining Journal*, Vol. III, No. 11, November 1991, pp. 14-17.

6. R.J. Campos, V.O. Achurra and O.C. Rojas, "El Teniente Converter: A Leading Pyrometallurgical Technology", <u>Pyrometallurgy of Copper</u>, (C. Diaz, C. Landolt, A. Luraschi and C.J. Newman, eds.) Proceedings of the Copper 91 - Cobre 91 International Symposium, Volume IV, August 18-21, 1991, Ottawa, Ontario, Canada, The CIM, Montreal, 1991, pp. 229-246.

7. "Chilean Technology Updates Nkana Smelter, Zambia", *Mining Journal*, London, February 1992.

8. P.J. Mackey, "A Look at the New Copper and Nickel Metallurgy", Proceedings of the Elliott Symposium on Chemical Process Metallurgy (P.J. Koros and G.R. St-Pierre, eds.), The Iron and Steel Society, 1991, pp. 51-99.

9. P.J. Mackey, "Trends in Copper Processing", paper presented at Metals Week Copper Conference, Orlando, FL, January 7-8, 1991.

10. R.J. Fruehan, "Iron Bath Smelting - Current Status and Understanding", Ibid. Ref. 8, pp. 1-10.

11. J.A. Innes, J.P. Moodie, I.R. Webb, and K. Brotzmann, "Direct smelting of Iron Ore in a Liquid Iron Bath", Process Technology Conference Proceedings, Vol. 7, 1988, Toronto, Ontario, Canada, pp. 225-231.

12. R.J. Fruehan, K. Ito and B. Ozturk, *Trans. Iron and Steelmaker*, November, 1988, p. 83.

13. R.E. Johnson, N.J. Themelis and G.A. Eltringham, "World-wide Survey of Copper Converting Practice", *J. of Metals*, Vol. 31, No. 6, June 1979, pp. 28-36; also Ibid. Ref.17, pp. 1-32.

14. F.E. Lathe and L. Hodnett, "Data on Copper Converter Practice in Various countries, *Trans. TMS-AIME*, Vol. 212, 1958, pp. 603-617.

15. N.J. Themelis, P. Tarassoff and J. Szekely, "Liquid Momentum Transfer in a Copper Converter", *Trans. TMS-AIME*, Vol. 245, 1969, pp. 2425-33.

16. G.C. McKerrow, "A 14 ft. x 32 ft. Converter at the Noranda Smelter", Pyrometallurgy Processes in Non-Ferrous Metallurgy , (E. Anderson and P.E. Queneau, eds.), Metallurgical Society Conferences, Vol. 39, The Aime, New York, 1965, pp. 247-258.

17. D.W. Rodolff and I.A. Rana, "Converter Practice at Magma Copper Company in 1978", Copper and Nickel Converters, (R.E. Johnson, ed.), the Metallurgical Society of AIME, 1979, pp. 81-109.

18. J.D. McCain, K.D. Driggs, and T.M. Gonzales, "The Magma Smelter Modernization and Commissioning", Paper presented at the American Mining Congress, San Francisco, September, 1989.

19. P.J. Mackey and J.B.W. Bailey, "Process and Apparatus for Continuous Converting of Copper and Non-Ferrous Mattes", U.S. Patent 4,544,141, (1 October, 1985).

20. T. Tsurumoto, "Copper Smelting in the Converter", *J. Metals*, Vol. 20, No. 11, Nov. 1961, pp. 820-824.

21. World Survey of Non-Ferrous Smelters, J.C. Taylor and H.R. Traulsen, eds., The Metallurgical Society, Inc., 1987, 399 pp.

22. J.A. Vogt, P.J. Mackey, and G.C. Balfour, "Current Converter Practice at the Horne Smelter", Ibid. Ref. 17, pp. 357-390.

23. "Noranda Smelter Contracts", *Mining Journal, London*, Vol. 318, No. 8173, May 15, 1992, p. 353.

24. N.J. Themelis and G.M. Freeman, "Fluid Bed Behaviour in Zinc Roasters", *J. Metals*, Vol. 36, No. 8, August 1984, pp. 52-57.

25. R.J. Fruehan, "Mass Transfer Between Liquid Metals and Injected Gases", *Metals Technology*, Vol. 7, Part 3, 1980, pp. 95-101.

26. Y. Sahai and R.I.L. Guthrie, Int. Symposium on Steelmaking, Jamshedpur, India (1981).

27. K. Mori and M. Sano. SCANINJECT I: First International Conference on Injection Metallurgy, Jernkontoret, Sweden, 1977; M. Sani and K. Mori, *J. Iron Steel Inst. Japan*, Vol. 60, 1974, pp. 348-354.

28. G.A. Irons and R.I.L. Guthrie, "Bubble Formation at Nozzles in Pig Iron", *Met. Trans. B.*, Vol. 9B, 1978, pp. 101-110.

29. A.V. Vanyukov *et al.*, "Pyrometallurgical Method for Processing Heavy Nonferrous Metal Raw Materials", U.S. Patent 4,252,450, (24 February 1981).

30. K. Nakanishi, T. Fujii and J. Szekely, "Possible Relationship between Energy Dissipation and Agitation in Steel Processing Operations", *Ironmaking and Steelmaking*, Vol. 3, 1975, pp. 193-197.

31. N.J. Themelis and P. Goyal, "Gas Injection in Steelmaking: Mechanism and Effects", *Can. Metall. Quart.*, Vol. 22, No. 3, 1983, pp. 313-320.

32. P. Tarassoff, "Process R and D - The Noranda Process", *Metall. Trans. B.*, Vol. 15B, 1984, pp. 411-431.

33. Q. Jiao and N.J. Themelis, "Mathematical Modeling of Heat Transfer During the Melting of Solid Particles in a Slag or Metal Bath", *Can. Metall. Quarterly*, in press.

34. J.C. Mwansa and N.A. Warner, "Natural Convective Heat Transfer between Single Ore Pellets and Molten Copper", in <u>African Mining</u>, The IMM London, 1987, pp. 253-266.

35. J. Szekely and N.J. Themelis, <u>Rate Phenomena in Process Metallurgy</u>, J. Wiley and Sons, New York, 1971, 750 pp.

36. Y. Sahai and R.I.L. Guthrie, "Hydrodynamics of Gas Stirred Melts: Part I. Gas/Liquid Coupling", *Met. Trans. B*, Vol. 13B, June 1982, pp. 193-202; R.I.L.Guthrie, <u>Engineering in Process Metallurgy</u>, Clarenden Press, Oxford, England, 1989.

37. Y. Prevost, "Approaching Twenty Years of Operation of the Noranda Process Reactor at the Horne Division of Noranda Minerals Inc.", Savard-Lee International Symposium on Bath Smelting, TMS, Warrendale, PA, 1992.

38. E.O. Hoefele and J.K. Brimacombe, "Flow Regimes in Submerged Gas Injection", *Metall. Trans. B*, Vol. 10B, 1979, pp. 631-648.

39. R.R. McNaughton, "Slag Treatment for the Recovery of Lead and Zinc at Trail, British Columbia", *Trans. AIME*, Vol. 121, 1936, pp. 721-737.

40. P.E. Queneau and R. Schulmann, Jr., "The QS Continuous Oxygen Converter", *J. of Metals*, Vol. 26, No. 8, 1974, pp. 14-16.

41. P. Arthur, A. Siegmund and M. Schmidt, "Operating Experience with QSL Submerged Bath Smelting for Production of Lead Bullion", Savard-Lee International Symposium on Bath Smelting, TMS, Warrendale, PA, 1992.

42. N.J. Themelis, J.G. Peacey and Q. Jiao, "Recovery of Zinc in a Slag Resistance Electric Furnace", H.H. Kellogg International Symposium, <u>Quantitative Description of Metal Extraction Processes</u> (N.J. Themelis and P.F. Duby, eds.), TMS-AIME, Warrendale, PA, 1991, pp. 331-347.

Fluid Flow and Mixing in Two-Phase

Channel Reactors with Bottom Gas Injection

H. Y. Sohn,[1] D. G. C. Robertson,[2]
A. K. Agrawal,[3] and K. M. Iyer[1]

[1]Utah State Center of Excellence for Advanced Pyrometallurgical Technology
and Department of Metallurgical Engineering
University of Utah
Salt Lake City, Utah

[2]Center for Pyrometallurgy
Department of Metallurgical Engineering
University of Missouri-Rolla
Rolla, Missouri

[3]USS Technical Center
Monroeville, Pennsylvania

Abstract

The paper describes two studies on flow and mixing in reactor models, carried out in different flow regimes. In the first study, relatively low gas strength was used. The high-temperature reactor was simulated using a 200-cm long and 20-cm wide channel, in which tetrachloroethylene and water were used to simulate the metal and the slag, respectively. Bottom gas blowing was carried out to enhance interphase mass transfer.

A novel tracer technique, using a thermal tracer, was developed to measure the eddy diffusivities in the two phases and to simulate mass transfer using the heat and mass transfer. On the system using one liquid phase, the effects of liquid velocity, gas flowrate, bubbler separation, and scale were studied. It was found that, for constant values of other parameters, there existed a characteristic bubbler separation, which resulted in the highest longitudinal mixing. Successful scale-up criteria were developed, using dimensional analysis.

In the countercurrent system, the overall heat-transfer coefficients were found to be a linear function of the mixing-energy input per unit volume and were independent of the bubbler separation. Characteristic bubbler separations existed for both the upper phase and the lower phase, when the respective eddy diffusivities were maximum. The eddy diffusivities for the upper phase were of the same order as those for the lower phase. The mass-transfer coefficients were deduced from the heat transfer data, which allowed the prediction of reactor performance using a computer model. A design of a large-scale steel refiner was proposed.

In the second study, a much higher gas strength was used (jetting regime), and the model geometry was based on the QSL reactor, with cylindrical cross-section. The model was 0.3 m in diameter and 2.4 m in length. The liquid depth was varied between about one-sixth and one-half the diameter. The gas injection rates were high enough to cause the two phases

Proceedings of the
Savard/Lee International Symposium on Bath Smelting
Edited by J. K. Brimacombe, P. J. Mackey,
G. J. W. Kor, C. Bickert and M. G. Ranade
The Minerals, Metals & Materials Society, 1992

to form an emulsion.

A mathematical model which describes the reactor as a combination of ideal reactors has been developed for studies in the high gas-strength regime. The model incorporates plug-flow reactors with recycle to represent the flow between the gas injection points and a perfectly mixed reactor to represent the flow in the gas plume. The data on exit-age distribution functions were obtained using a conductivity tracer. There was an excellent agreement between model predictions and the observed experimental exit-age distribution. A correlation has been developed between the recycle ratio and the experimental conditions.

I. Introduction

Many smelting and molten metal refining processes are characterized by slag/metal or matte reactions. The metal/slag interfacial area is an important factor in enhancing these processes. This is true of both ferrous and nonferrous processes. To ensure good metal/slag mixing, processes using gases both as oxidant and mixing media have been developed. The ferrous processes were the first to adapt this technology, leading to the development of such processes as BOF and Q-BOP steelmaking. The nonferrous industry is just beginning to capitalize on the potential of this technology (1-4). The first nonferrous process to do so is the Q-S oxygen process (QSOP for short), which uses submerged injection of tonnage oxygen.

There is at present a great emphasis on the development of continuous metal extraction and refining processes. Out of the various proposed designs of a continuous extraction or refining process, the countercurrent system is potentially the most efficient in terms of purity of the product, refining reagents requirement, and control of the process. Previous pilot-plant work has shown that this is a very promising concept (5-9).

The main advantages of a continuous process include simpler operation under steady-state conditions, consistent quality of product, and economic benefits. The capital costs can be greatly reduced because of the reduction in size of the furnace and holding units and a reduced need for elaborate material-handling equipment. Continuous processes are more suitable for automation as compared to their batch counterparts, resulting in reduced labor costs.

In the ideal operation of a countercurrent system for metal refining, plug flow should be attained in the horizontal direction and "complete mixing" in the vertical direction. This would allow transfer of impurities from metal to slag at a high rate and still maintain the concentration gradients along the reactor that are required for efficient operation. The extent to which this ideal situation can be approximated by the judicious selection of the system geometry and operational parameters such as liquid velocity, gas flow rate, and injector placement, is of major interest.

The QS reactor is a long circular cylindrical reactor lying on its side and inclined at an angle of 1° with the horizontal (2). Flash burners in the roof of the vessel serve to flash smelt the ore and fluxes. In the reactor, the matte or metal and slag are in countercurrent flow, and tonnage oxygen is bottom-blown into the reactor using the Savard-Lee injectors. The bottom injection serves to stir the bath, thus mixing the matte or metal and slag, and also to oxidize the impurities in the matte or metal. There is a staged control of the oxygen partial pressure along the length of the reactor, by the use of shallow suspended baffles, making it possible to make metal at one end while cleaning slag at the other. The slag cleaning is carried out by bottom injection of oxygen, sulfur dioxide, and pulverized coal.

The cold model studies presented in this paper can be classified into two categories: (1) Low injection-gas-strength studies, in which the extent of longitudinal mixing and interphase mass transfer in a countercurrent system with two liquid phases was studied under low injection gas strength, and (2) high injection-gas-strength studies, in which the overall fluid-flow behavior and mixing characteristics were studied in a QS-type reactor.

II. Experimental Technique

The studies were carried out in two distinct regimes of gas injection into one or two-phase liquid baths. In the low gas-strength studies the gas flow is predominantly in the bubbling regime, and the two liquid phases remained essentially separate with the formation of a small amount of emulsion in the interfacial area between the two liquids. The high gas-strength studies were carried out in the jetting regime, and there was extensive emulsification between the two liquid phases in the plume region with some air entrainment leading to the formation of a foamy emulsion.

II.1. Low Gas-Strength Studies

The high-temperature continuous metal/slag system was simulated in the laboratory using a low-temperature physical model (10). Tetrachloroethylene (TCE) and water were used to simulate the dense phase (metal or matte) and the slag phase, respectively. A thermal tracer technique was developed in which a disturbance was introduced in the system in the form of a temperature signal, and the resulting temperature changes were recorded at various locations in the system. The analogy between heat transfer and mass transfer was used in this simulation.

The experiments were first conducted on a single-phase system, which is simpler than the countercurrent two-phase system. Experiments were conducted at different widths, while maintaining a constant width/height ratio, with different liquids (TCE and water), by changing the gas flow rate, liquid velocity, and injector separation. The results were interpreted using the longitudinal dispersion model (11).

The countercurrent experiments required separate measurements of overall heat-transfer coefficient for the TCE/water system under various experimental conditions. The coefficient for heat loss from the system to the surroundings was also measured. This was followed by tracer studies on the countercurrent system to obtain the residence time distribution (RTD) for the two liquid phases.

A computer model of the countercurrent system was developed. The laboratory data were fed to the computer model, and the RTD predicted by this model was compared with the actual RTD (from the laboratory data) to obtain the eddy-diffusivity in a particular liquid phase. These measurements were done for different gas flow rates, bubbler separations, liquid velocities, and liquid depths.

The tank used for the experiments was rectangular, made of 1.27-cm thick plexiglas plates. The tank was 200 cm long, 20 cm wide, and 40 cm in height. Two baffles were installed on each end of the tank so that the liquid entering the experimental length of the reactor approached streamline flow.

It was necessary when running the experiments at different scales to be able to alter the tank width. For this a temporary wall made of 0.62-cm thick plexiglas plate was installed in the tank when needed. Experiments were run at 10-cm and 12.8-cm tank widths, in addition to the full 20-cm width of the tank.

Bottom blowing of gas using injectors was simulated using bubblers made of 1-mm bore pyrex capillary tube bent in an L-shape. The bubblers were connected to a nitrogen gas cylinder using flexible tubing. The gas was injected from a point source.

Two immiscible liquids were required with physical properties (density, viscosity, surface tension) very similar to high-temperature metal/slag systems. Unfortunately, it is not easy to find liquids that satisfy all the criteria for an ideal simulation. Since most organics are only slightly less dense than water, the density difference could be maximized by choosing a halogenated hydrocarbon that was more dense than water. Of the various liquids considered, tetrachloroethylene (C_2Cl_4) was found to be compatible with plexiglas. TCE is a nonflammable, colorless liquid at room temperature. It has a density of 1.614 g/cc and viscosity of 0.84 cP at 25°C, and it approximated the desired physical constants. The schematic diagram of the arrangement made to obtain countercurrent flow of TCE and water is shown in Figure 1.

Accurate measurement of very small temperature changes was essential for the experiments. Considerable time and effort was required to develop a temperature measurement system with high sensitivity, which could be used for direct data logging using a data acquisition system.

In the single-phase system, only distilled water or only TCE flowed through the reactor. The stimulus-response technique was used to obtain the RTDs under different conditions of injector separation, gas flow, liquid velocity, and reactor width.

At the start of the experiment, a disturbance was introduced at the inlet end of the system in the form of liquid at a different temperature than the bulk liquid. The warmer (or colder) liquid addition was made just before the inlet baffle, in between two baffles. This was done to minimize turbulence at the time of warmer liquid addition. The addition was done with the help of a funnel with its lower end submerged in the flowing liquid.

Four to six probes were located at different points in the channel for obtaining the temperature-vs.-time curves. The hot liquid addition between the baffles resulted in a maximum temperature rise of 3°C to 4°C at the reactor inlet, since the hot liquid quickly mixed with the surrounding liquid. Figure 2 shows the input and output curves for a single-phase experiment with water.

The experiments with TCE and water flowing countercurrent to each other were conducted by adding heat to one of the phases by using an immersed tubular heater. Temperature changes with time at different points in both the phases were recorded by using six temperature probes. The procedure for temperature measurement and data storage was similar to that used for the single-phase experiments. Figure 3 shows an example of temperature/time curves at the TCE and water outlets when the signal was added to TCE.

For each countercurrent experiment, the heat-transfer coefficient from TCE to water was separately measured. A stationary system was used for the measurement of the heat-transfer

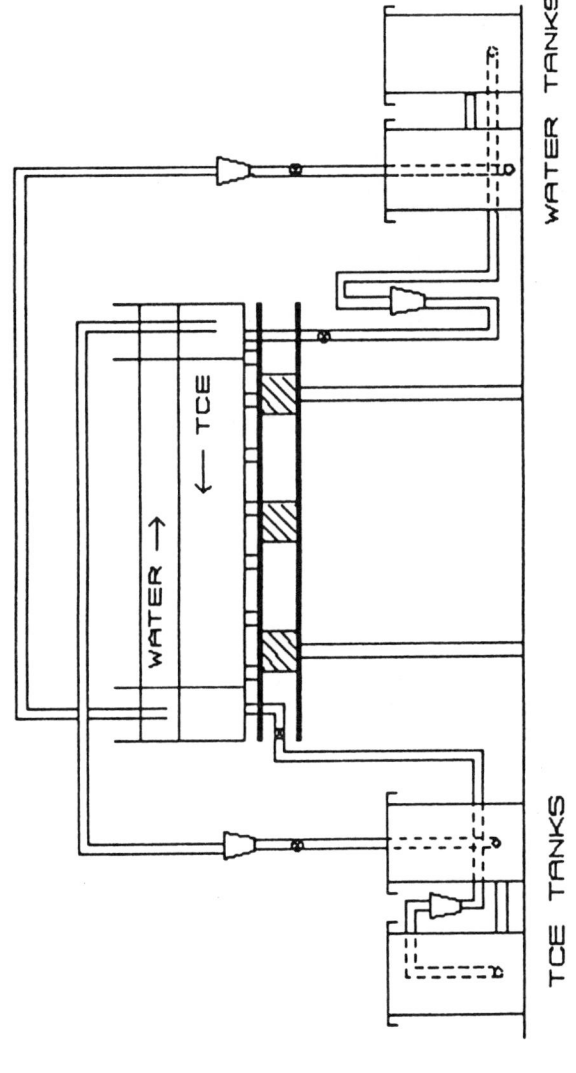

Figure 1.   Schematic diagram showing the arrangement made for countercurrent flow of TCE and water.

381

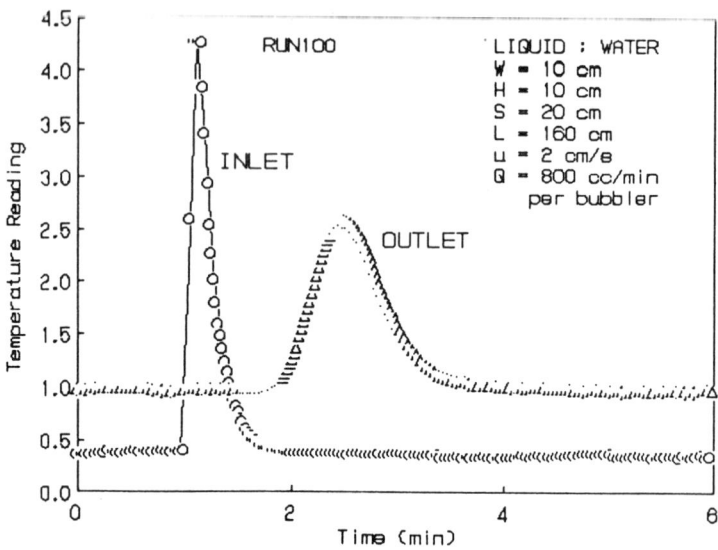

Figure 2. Temperature/time curves at the inlet and the outlet of the single-phase system.

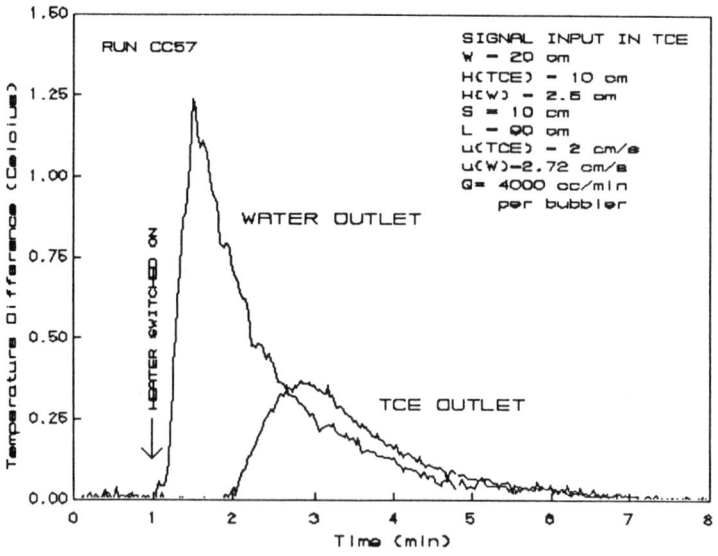

Figure 3. The temperature/time curves obtained at the TCE and water outlets from a countercurrent experiment.

382

coefficient. At the start of the experiment, the temperature of the lower phase (TCE) was raised instantly along the length of the reactor. TCE cooled down and water temperature went up until both phases reached the same temperature. The temperature of each phase was measured by the probes. A program was written to treat the data acquired to obtain the values of the overall heat-transfer coefficient.

II.2. High Gas-Strength Studies

These studies involved cold model experiments in a QS-type reactor, in which the overall fluid-flow behavior and mixing characteristics in a shallow bath with high-strength bottom gas blowing were determined.

Residence time distribution (RTD) studies were performed for various conditions of overall flow rate of the liquid and gas phases. The apparatus was a cylindrical plexiglas vessel, 0.3 m ϕ × 2.4 m L, with a slope of 1° to the horizontal, shown schematically in Figure 4. The vessel is provided with holes along the bottom to accommodate a variable distance between the submerged gas injectors and a different total number of injectors. Compressed air was used as the injection gas, and water and kerosene were chosen to represent the matte or metal and slag phases, respectively. The gas injection rates were varied over a range around to the sonic velocity. The range of gas flow rates includes the regime where the jet breaks up into a bubble plume on exiting the injector to the regime where the jet shoots right through the bath and impacts the roof of the reactor. The gas Mach number cannot exceed unity under experimental conditions because the injector only has a converging section. Thus, the gas is compressed, increasing its density to maintain the mass flow rate at high rates of gas injection. The flow rates for the two liquids were also varied independently between 0 and 5 liters/min.

The experiments that were carried out involved two types: Single-liquid experiments, using water, and two-liquid experiments, using water and kerosene.

(a) Single-Liquid Experiments. The single-liquid experiments were carried out with water. The submerged gas injectors were turned on at the desired gas injection rates, prior to flowing the liquid into the vessel. Water was run continuously through the reactor. After steady state was reached at the desired bath height and flow rate of water, a pulse tracer was injected into the inlet water stream. A concentrated solution of potassium chloride was used as the tracer. The RTD for the tracer was monitored by detecting the variation in the conductivity of the water at the exit, using an on-line conductivity probe.

(b) Two-Liquid Experiments. The two-liquid experiments were carried out using water and kerosene. The start-up procedure for the two-liquid experiments was similar to that for the single-liquid system. Once steady state was reached for the water flow through the vessel, the desired quantity of kerosene was added. The kerosene was either left stagnant (0 bulk flow rate) or recirculated at the desired flow rate to set up the countercurrent flow. The tracer RTD studies were conducted similarly to the single-liquid system, using an on-line conductivity probe and a concentrated potassium chloride solution as the tracer. The bath heights of the individual phases were varied to determine their effect on the RTD behavior.

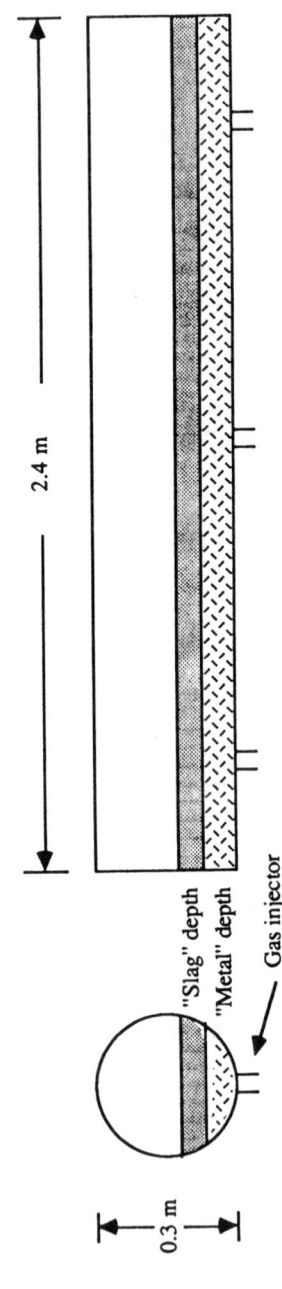

"Slag" depth
"Metal" depth
Gas injector

2.4 m

0.3 m

Figure 4. Schematic diagram of experimental setup for high gas-strength studies.

III. Mathematical Modeling for High Gas-strength Systems

In order to simulate and describe the RTD response of the channel reactor, the reactor was modeled as a series combination of ideal reactors. Ideal reactors were chosen because their RTD responses are well known and they are easy to model. The various ideal reactors that were used are the continuously stirred tank reactor (CSTR), the plug flow reactor (PFR) and the plug flow reactor with a second PFR in the recycle stream (this combination being referred to henceforth as the recycle reactor) (11).

III.1. Model Basis

The plume region above the injector was modeled as a CSTR. Since the gas is injected with high energy and the bath is rather shallow, the fluids in the plume region are very well mixed and would be expected to behave as a CSTR.

Analyzing the fluid flow around each plume, it was evident that a recirculating flow is set up on either side of the plume. Thus, these regions were modeled as recycle reactors. A schematic diagram for the combination of ideal reactors used to model the region around each plume, including the plume itself, is shown in Figure 5(a). Since each gas plume has a similar arrangement around it, it was decided that the recycle reactors between two plumes would not overlap but would equally share the distance between the two plumes. The transport of fluid between two such units would be accomplished by the bulk flow. A schematic diagram of the model setup for a system with two injectors is shown in Figure 5(b).

III.2. Model Analysis

The model is based on a single parameter, α (shown in Figure 5), the recycle ratio in the recycle reactor section. All other model variables are determined as functions of α.

The values of the time constants for the PFR and recycle PFR units are assumed the same. Because of the turbulent nature of the flow set up by the gas plume, the velocity profiles in both the PFR and recycle PFR sections are relatively flat and have the same magnitude. Since the average length of both units is the same, the time constants should be same. The value of the time constants τ_P ($= V_1/Q_1$) and τ_R ($= V_2/Q_2$) is determined as follows. Let Q_1 be the flow through the PFR section and Q_2 the flow through the recycle PFR section, and V_1 and V_2 their respective volumes. Then, we have

$$Q_1 = Q_2 + Q_L \tag{1}$$

$$V_1 = V - V_2 \tag{2}$$

$$V_1/Q_1 = V_2/Q_2 \tag{3}$$

$$\alpha = Q_L/Q_1 \tag{4}$$

Solving these equations simultaneously, we get

$$\tau_P = \tau_R = V \alpha /((2 - \alpha) Q_L) \tag{5}$$

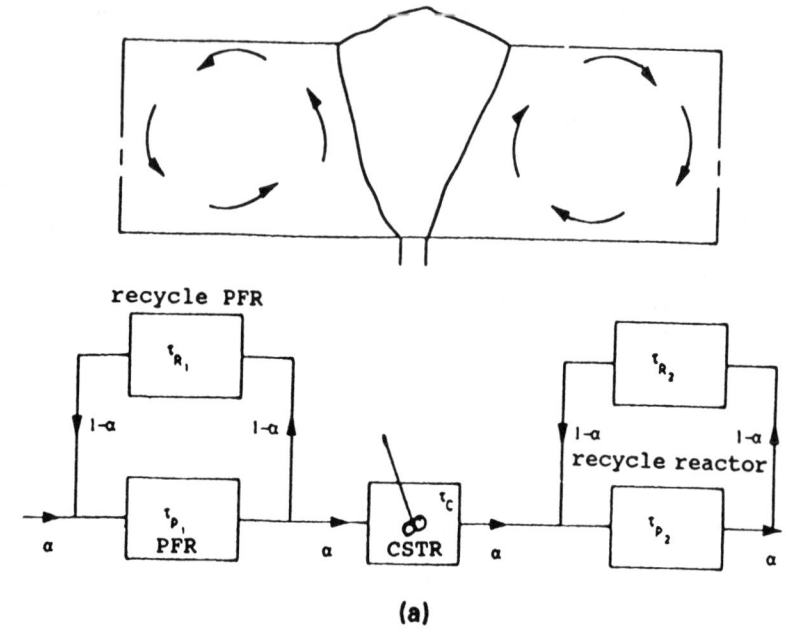

(a)

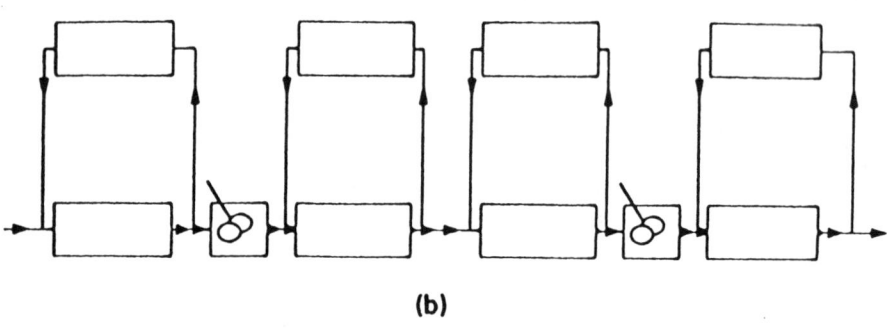

(b)

Figure 5. (a) Schematic diagram of mathematical model. (b) Schematic diagram of reactor with two injectors.

where Q_L = throughput rate of the heavy liquid

            V = total volume occupied by the recycle reactor section

            τ_P = nominal residence time in the PFR

            τ_R = nominal residence time in the recycle reactor

The volume of the CSTR section is a function of the plume cone angle. The angle is constant at 21° for the gas flow rates used in our experiments. Since the gas plume forms a cone at the injector, the volume occupied by the gas plume can be written as

$$V_c = \pi r^2 h/3 \qquad (6)$$

where V_c = volume occupied by gas plume

            h = bath height at injector location

However, r can be related to h as follows

$$r = h \tan \theta \qquad (7)$$

where, θ is half the plume cone angle. Therefore,

$$V_c = \pi h^3 \tan^2 \theta/3 \qquad (8)$$

and

$$\tau_c = V_c/Q_L \qquad (9)$$

where τ_c = nominal residence time in the CSTR

This time constant is independent of α.

III.3. Model Equations

The model is based on the RTD response characteristics of the various ideal reactors. The PFR response to a pulse input can be written as

$$E(t) = \delta(t - \tau_P) \qquad (10)$$

where $E(t)$ = response function

            δ = Dirac-delta function

            t = time

The CSTR response to a similar pulse input is

$$E(t) = \tau_c^{-1} \exp(-t/\tau_c) \qquad (11)$$

and that for the recycle reactor is

$$E(t) = \alpha (1 - \alpha)^n [\delta\{t - \tau_P - n(\tau_P + \tau_R)\}] \qquad (12)$$

387

where $\quad n = $ Integer $[(t - \tau_P)/(\tau_P + \tau_R)]$

Let us now consider a CSTR with a normalized input function $F(\theta)$ which satisfies

$$\int_{\theta=0}^{\infty} F(\theta)\, d\theta = 1 \tag{13}$$

The exit response to the input function introduced between time 0 and $d\theta$ is given by

$$\delta E = \tau_c^{-1} [F(0)\, d\theta]\, \exp\, (-t/\tau_C) \tag{14}$$

The exit response to the input function introduced between any time θ and $\theta + d\theta$ is given by,

$$\delta E = \tau_c^{-1} [F(\theta)\, d\theta]\, \exp\, (-\{t - \theta\}/\tau_C) \tag{15}$$

Therefore, the response at any time t is given by

$$E(t) = \int_{\theta=0}^{t} \tau_C^{-1}\, F(\theta)\, \exp\, -\big((t - \theta)/\tau_C\big)\, d\theta \tag{16}$$

The model equations for the various combinations of ideal reactors can be obtained by combining the responses of the individual reactors. The response of the first combination of the recycle reactor and CSTR can be expressed as follows:

$$E_1(t) = \sum_{i=0}^{n} \left[\alpha\, (1 - \alpha)^{n-i} \left\{ \int_{\theta=0}^{t-\tau} \tau_{C_1}^{-1}\, F(\theta)\, e^{-(t-\theta)/\tau_{C_1}}\, d\theta \right\} \right] \tag{17}$$

where $\quad \tau = t - (2i + 1)\, \tau_{P1}$

$\quad\quad\quad n = $ Integer $[(t - \tau_{P1})/(2\tau_{P1})]$

$F(\theta)$ equals the normalized input function at reactor inlet, as in Equation (13), while that for the recycle section following the CSTR can be written as

$$E_2(t) = \sum_{i=0}^{n} \alpha\, (1 - \alpha)^{n-i} \left[\int_{\theta=0}^{t-\tau} E_1(\theta)\, d\theta \right] \tag{18}$$

where $\quad \tau = t - \tau_{P1} - (2i + 1)\, \tau_{P2}$

$\quad\quad\quad n = $ Integer $[(t - \tau_{P1} - \tau_{P2})/(2\tau_{P2})]$

These equations can be extended similarly to further identical sections down the length of the reactor, by using the response of the previous section as the input function to the next section.

IV. Results and Discussion

IV.1. Low Gas-Strength Studies

(a) Single-Liquid Work. For the first set of experiments, the reactor width was reduced to 10 cm by installing a temporary wall, to reproduce the conditions used by Sahai (12). The experiments were carried out at three bubbler spacings, three liquid velocities, and two gas flow rates.

Figures 6 and 7 show plots of inverse Peclet number vs. dimensionless bubbler separation (S/W, where S is the distance between two adjacent bubblers and W is the reactor width) for different velocities at Q = 200 and 800 cc/min per injector, respectively. The longitudinal mixing first increased as the bubblers were brought close to each other. However, below S/W = 2, the longitudinal mixing began to drop as the bubblers were brought closer. Thus the D/uW-vs.-S/W curves show a peak at S/W = 2 under most conditions of velocities and gas flow rates. The occurrence of a maximum D/uW at a characteristic bubbler separation was also noticed by Sahai (12).

Figure 8 shows the plot of D/uW vs. Reynolds number (uW/ν) for two gas flow rates. D/uW initially dropped quickly as the value of Reynolds number increased, but as Re was increased further, the D/uW value became less sensitive to this change.

An important part of the present work was to study the effect of changes in scale. The next set of runs was aimed at doing the single-phase experiments on a different scale, but keeping the dimensionless groups the same. Single-phase experiments were thus conducted in the 20-cm wide reactor.

The important dimensionless groups for this system were shown to be (10):

D/uW	Inverse Peclet number.
Q/uW^2	Dimensionless gas flow rate.
u^2/gW	Froude number (Fr).
uW/ν	Reynolds number (Re).

where D is the eddy diffusivity, u is the liquid velocity, W is the reactor width, Q is the gas flow rate per tuyere, and ν is the kinematic viscosity of the liquid.

If the Froude number, dimensionless gas flow rate, Reynolds number, and the reactor geometry are kept constant, the inverse Peclet number obtained for two systems at different scales would be expected to be the same. However, it is impossible to achieve similarity with respect to all three groups without changing the kinematic viscosity of the liquid. Since water was used for this set of experiments, the similarity with respect to at least one group had to be ignored. Since this system was already reasonably turbulent, it was concluded that a change in Re would have the least effect on the system. The experimental parameters were then chosen to impose similarity with respect to the other two groups. The reactor geometry was not changed.

Figure 9 shows the plot of inverse Peclet number vs. dimensionless bubbler spacing for the two scales at the lower gas flow rate. For the two curves marked u = 0.25 cm/s and

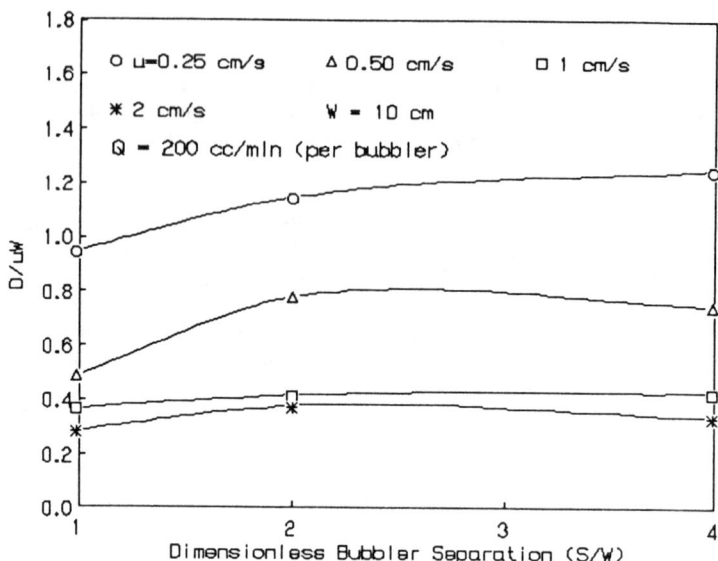

Figure 6. Inverse Peclet number vs. dimensionless bubbler separation for 10-cm wide tank at Q = 200 cc/min.

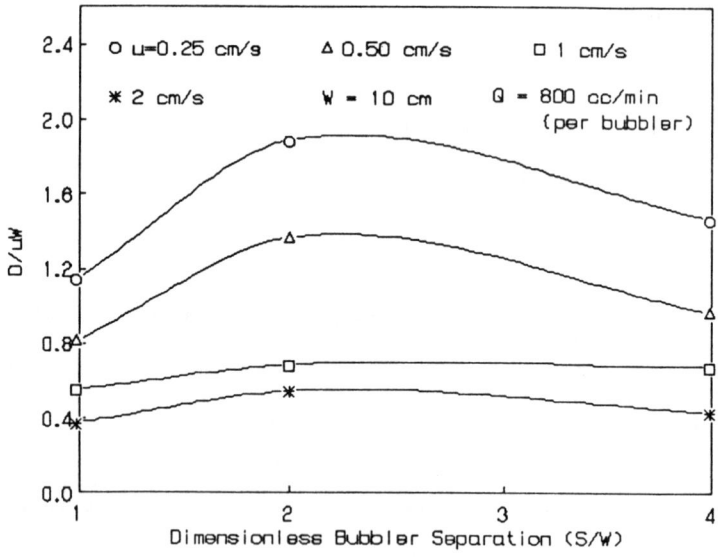

Figure 7. Inverse Peclet number vs. dimensionless bubbler separation for 10-cm wide tank at Q = 800 cc/min.

390

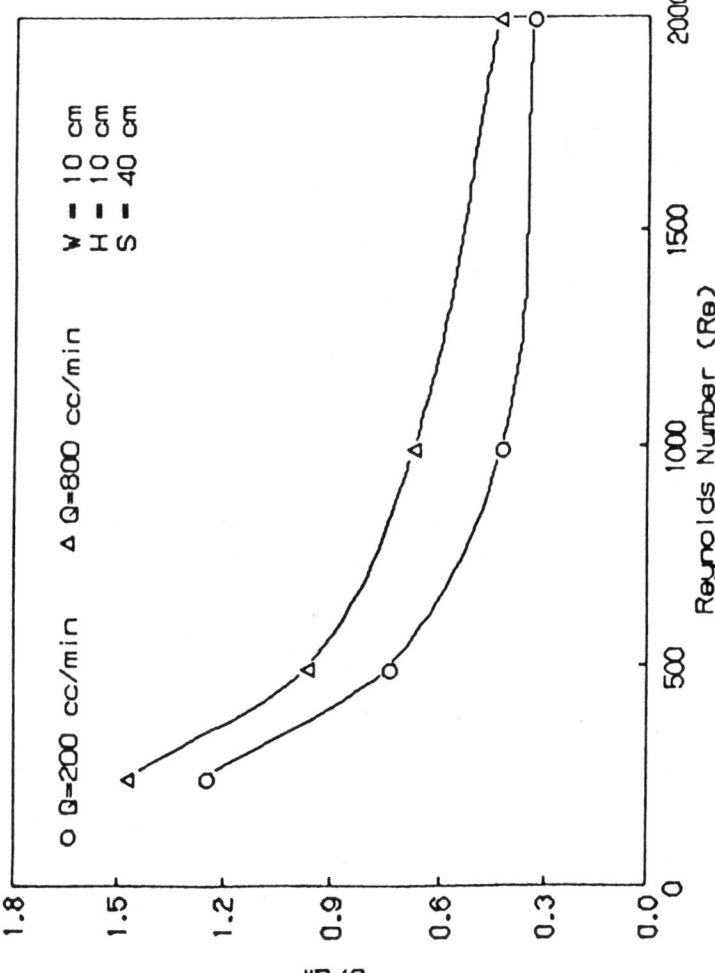

Figure 8. Inverse Peclet number vs. Reynolds number for single-phase experiments with water.

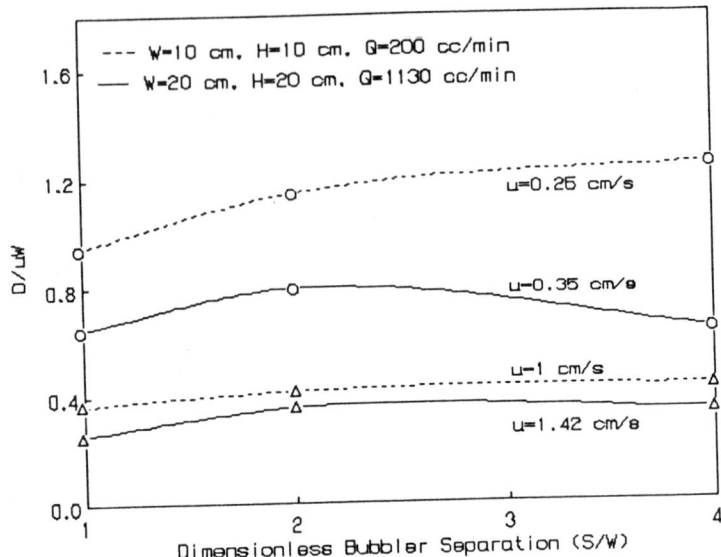

Figure 9. Comparison between D/uW obtained from 10-cm and 20-cm wide tanks at the lower gas flow rates.

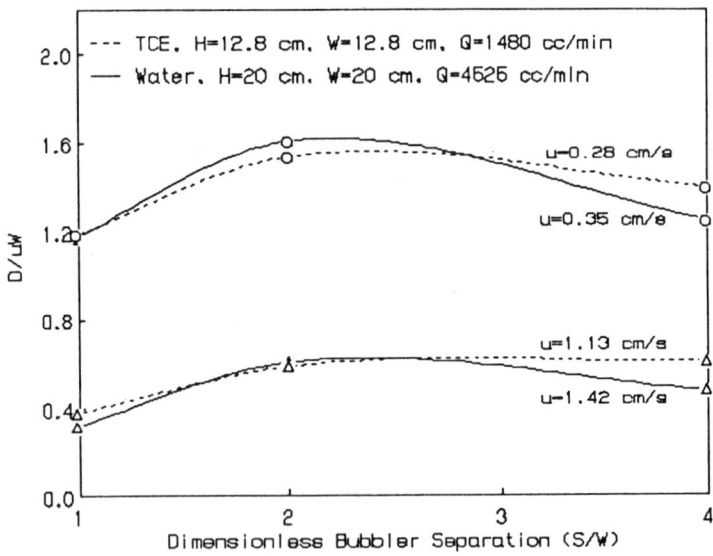

Figure 10. Comparison between dynamically similar systems at the higher gas flow rates.

u = 0.35 cm/s, the values of the Froude number and the dimensionless gas flow rate were the same. However, the inverse Peclet numbers in the case of the 10-cm width were higher. This result can be explained by noting that complete dynamic similarity between the two systems could not be imposed. The Reynolds numbers in the 20-cm wide tank were 2.8 times higher than the corresponding values for the 10-cm wide tank. From Figure 8 it was seen that a higher Reynolds number results in a drop in inverse Peclet number. The Re value for the 10-cm wide tank with 0.25 cm/s velocity was 250, whereas that for the 20-cm wide 0.35 cm/s velocity was 700. These two values fall on the lower end of Re values on Figure 8 where D/uW is sensitive to a change in Re.

Similarly, the dimensionless gas flow rate and Froude number for the curves marked u = 1 cm/s and u = 1.42 cm/s are the same. In this case the difference in inverse Peclet number between the two curves is much smaller than the lower velocity cases. This can again be explained by noting that the Re values for the 10-cm and 20-cm wide tanks were 1000 and 2800, respectively. At these higher values of Re, the inverse Peclet number was less sensitive to a change in Re (as shown in Figure 8), which would result in the behavior shown in Figure 9.

Single-phase experiments were also conducted with TCE to change the kinematic viscosity of the liquid. The kinematic viscosity of TCE is 0.0052 cm²/s compared to 0.01 cm²/s for water. The reactor width, gas flow rates, and liquid velocities were selected so as to obtain complete dynamic similarity with single-phase water experiments in the 20-cm wide tank.

The channel width was reduced to 12.8 cm by installing a temporary wall. The injector separations selected for the experiments were such that the geometry of the system was the same as the 20-cm wide tank experiments.

In Figure 10 the inverse Peclet number vs. dimensionless bubbler spacing for both TCE experiments and the corresponding water experiments (in 20-cm wide tank) are plotted for the higher gas flow rates. Since the values of the three important dimensionless groups were the same, the D/uW values would be expected to be in agreement. This was found to be approximately true at the higher gas flow rate. Figure 10 shows that, although the liquids, scales, liquid velocities, and gas flow rates are quite different in the two cases, the results of one system can predict those for the other system (of the same geometry) as long as the dimensionless groups have the same value. This is an important consideration in the scale-up of the system.

(b) Two-Liquid Work. The overall heat-transfer coefficient was measured with no bulk flow of liquids for two gas flow rates and three injector separations. The h_{ov} data were interpreted by plotting it against the mixing energy input to the reactor. The mixing energy input per unit volume of the liquid (γ) is given by:

$$\gamma = \frac{Q[(\rho g H)_{TCE} + (\rho g H)_{water}]}{WS(H_{TCE} + H_{water})} \tag{19}$$

The term in the numerator gives the energy input to the reactor per second (watts) when gas is blown at the rate of Q (m³/s) per bubbler through the volume of liquid given by the denominator.

The overall heat-transfer coefficient was plotted against these values. This is shown in Figure 11 for the 12.8-cm wide tank in which the preliminary countercurrent experiments were carried out. This figure shows that the overall heat-transfer coefficient follows a straight line relationship with the mixing energy input per unit volume. The mixing energy input to the system can be increased in two ways:

— by reducing the injector separation and keeping the gas flow rate per injector constant.

— by increasing the gas flow rate per injector and keeping the injector separation constant.

Figure 11 shows that the overall heat-transfer coefficient fell on the same straight line, irrespective of the way the mixing energy input was changed.

This was followed by experiments in the 20-cm wide tank. For this set of experiments, the depth of the TCE layer was 20 cm, and the depth of the water layer was 5 cm. Experiments were carried out at two gas flow rates, two liquid velocities, and four injector spacings. Measurements of h_{ov} and $h_{ov,l}$ were first carried out. The mixing energy input to the reactor per unit volume of the liquid was calculated for each of the experiments, and the h_{ov} data were plotted against it. This plot is shown in Figure 12. This figure again shows a straight line relation between h_{ov} and mixing energy input per unit volume.

Figure 13 shows inverse Peclet number vs. dimensionless bubbler separation for $Q = 1000$ cc/min per bubbler. Figure 14 shows a similar plot for $Q = 4000$ cc/min per bubbler. The variation of D/uW with S/W was found to be similar to that in the single-phase system. As the bubblers were brought closer together, D/uW first reached a maximum at S/W = 2 and then dropped as the bubbler spacing was reduced further.

A regression analysis of the data showed that the eddy diffusivity in the lower phase was proportional to $Q^{0.11}$, and it was also proportional to $u^{0.09}$. The values of exponents of Q and u in the one-phase work were found to be 0.25 and 0.45, respectively. Thus in the counter-current reactor, the dependence of eddy diffusivity on Q and u was found to be comparatively lower.

For the next set of experiments, the layer heights of TCE and water were reduced to half their values (i.e., 10 cm and 2.5 cm, respectively) compared to those used for the previous set of experiments. The gas flow rates and the injector separations were not changed. As in the previous cases, the mixing measurements were preceded by h_{ov} measurements. This is shown in Figure 15. The relationship was again a straight line.

Thus in all the cases studied, the value of overall heat-transfer coefficient was found to depend only on the mixing energy input per unit volume of the liquid. Even if the same mixing energy as a given case was obtained by quadrupling the gas flow rate and injector separation, the resultant h_{ov} values were very close to each other.

The overall heat-transfer rate for a given set of layer heights of the two liquids was thus independent of the injector spacing and gas flow rate per injector, as long as the mixing energy input per unit volume was kept constant.

To measure the eddy diffusivity, the gas flow rates per bubbler were kept the same as

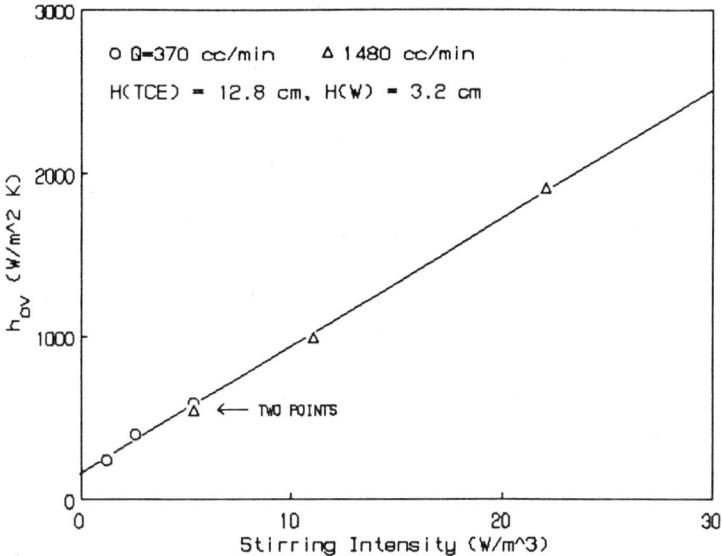

Figure 11. Overall heat-transfer coefficient vs. stirring energy per unit volume for the TCE/water system in a 12.8-cm wide channel.

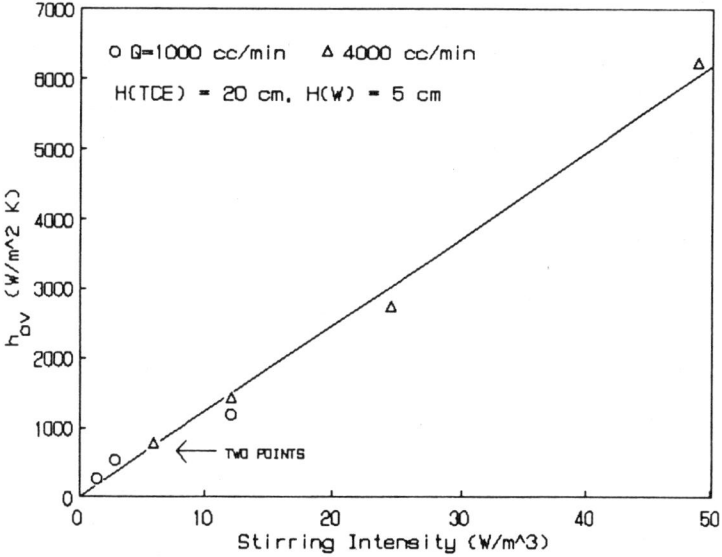

Figure 12. Overall heat-transfer coefficient vs. stirring energy per unit volume for the TCE/water system in a 20-cm wide channel.

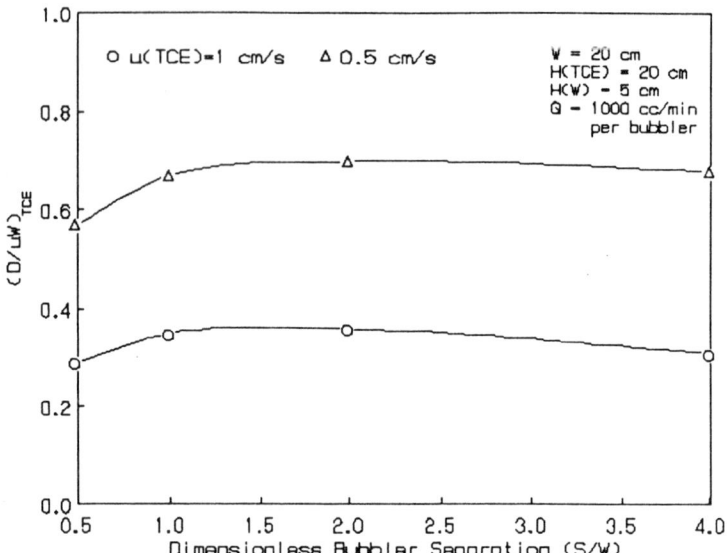

Figure 13. Inverse Peclet number in the TCE vs. dimensionless bubbler
separation for Q = 1000 cc/min per bubbler.

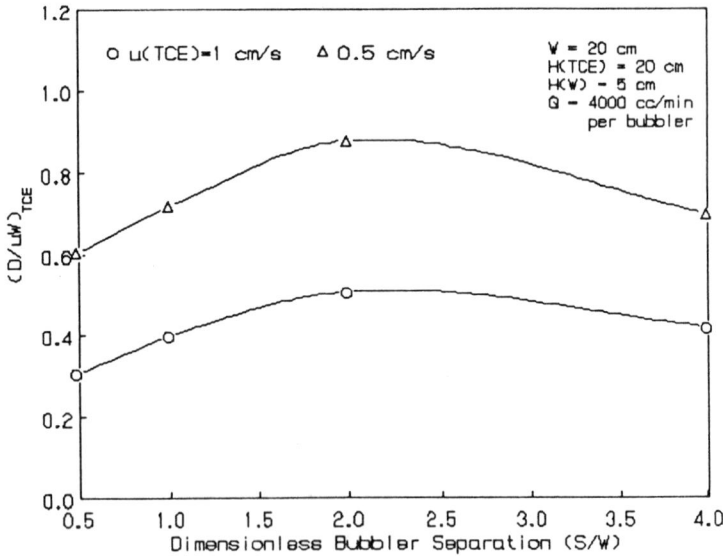

Figure 14. Inverse Peclet number in the TCE vs. dimensionless bubbler
separation for Q = 4000 cc/min per bubbler.

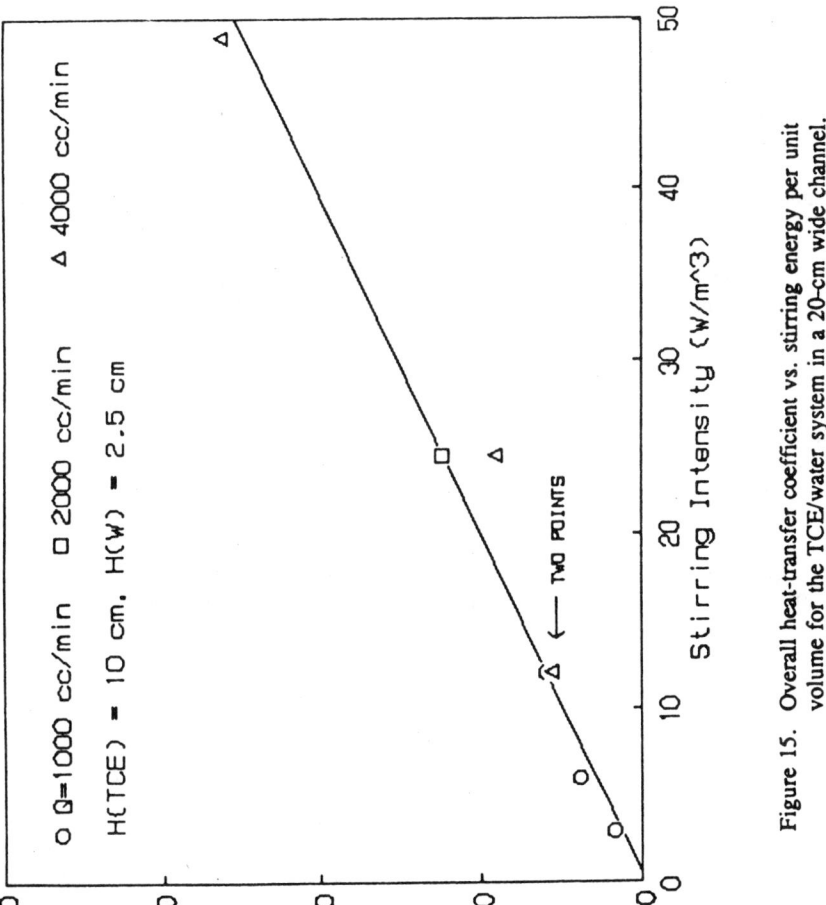

Figure 15. Overall heat-transfer coefficient vs. stirring energy per unit
volume for the TCE/water system in a 20-cm wide channel.

those in the previous set of experiments. Experiments were conducted up to a maximum S/W value of 2. Mixing measurements were also conducted at a higher liquid velocity compared to the previous set of experiments. The data for the gas flow rate of 1000 cc/min per bubbler were plotted for the three liquid velocities and are shown in Figure 16. The data for the gas flow rate of 4000 cc/min are plotted in Figure 17. These two Figures show that as S/W is decreased, the mixing intensity goes to a maximum before dropping down. However, in this case the mixing intensity reached a maximum at S/W = 1. This is in contrast with the previous cases (both single-phase and counter-current arrangements) where the maximum occurred at S/W = 2. In the previous cases, the lower phase height was equal to the reactor width, whereas in the present case the lower phase height was half the reactor width. Thus the maximum value of D/uW appears to occur when H/(S/2) is equal to 1, i.e., with a square cell.

A regression analysis of the data showed that in D $\propto$ Q^m u^n, the average values of the exponents of Q and u were 0.1 and 0.23, respectively.

The drop in TCE height from 20 cm to 10 cm resulted in a drop in the inverse Peclet numbers between 25% and 54%. This significant reduction in longitudinal mixing at lower H/W ratios was also noticed by Gassel (13) in the single-phase stationary system.

The longitudinal mixing in the slag phase was found to be much higher than that in the metal phase in the Lulea steel refiner (7). The D/uL value for the slag phase was 1.0 and that for the metal phase was 0.064. Thus it was considered important to investigate the magnitude of mixing in the upper phase of the TCE-water system.

The mixing measurements for the upper phase were carried out using a 10-cm TCE layer and 2.5-cm water layer in the 20-cm wide tank. Experiments were run at two liquid velocities, two gas flow rates, and three bubbler separations. The overall heat transfer coefficients corresponding to this geometry were measured earlier. The longitudinal mixing data were interpreted in the same way as that for the lower phase.

The variation of inverse Peclet number with dimensionless bubbler separation for Q = 1000 cc/min is shown in Figure 18. Figure 19 shows a similar plot for Q = 4000 cc/min. These Figures show that the mixing intensity varies with the bubbler separation in a manner similar to the lower phase. A maximum in D/uW occurred in all cases at a dimensionless bubbler spacing of 1.

The eddy diffusivity in the upper phase showed a greater dependence on a change in the gas flow rate or liquid velocity than the lower phase. In the expression D $\propto$ Q^m u^n, the average values of m and n were found to be 0.50 and 0.48, respectively.

It can be seen that the mixing in the upper phase was not very high. A comparison of water mixing data with the TCE mixing data revealed that the eddy diffusivity in water was either lower than or very close to that in TCE in most of the cases.

The height to width ratio for the upper phase is one fourth that of the lower phase. From the earlier lower phase measurements in this work, and also from Gassel's work (13), it was found that the lower the height to width ratio, the lower is the longitudinal mixing. To some extent, this finding also holds good for the upper phase even though the nature of the bubble plume in the upper phase is somewhat different from that in the lower phase (or in a single-phase system). The bubbles originate in the upper phase from the plume at the top

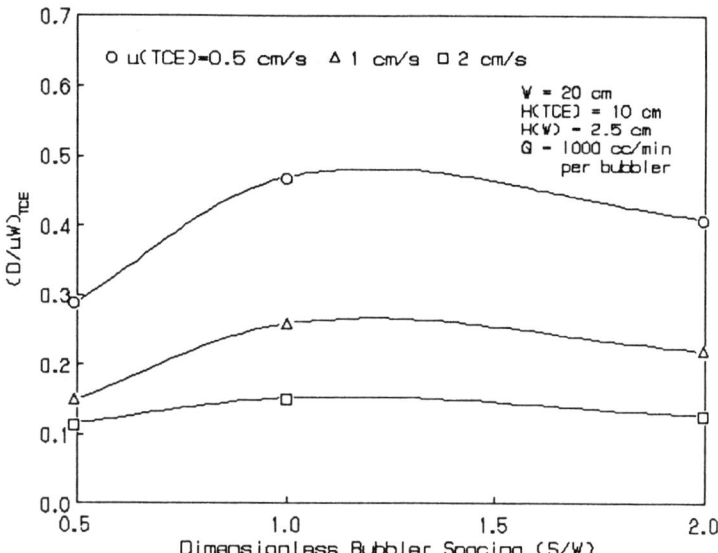

Figure 16. Inverse Peclet number in the lower phase vs. dimensionless
bubbler separation for Q = 1000 cc/min per bubbler.

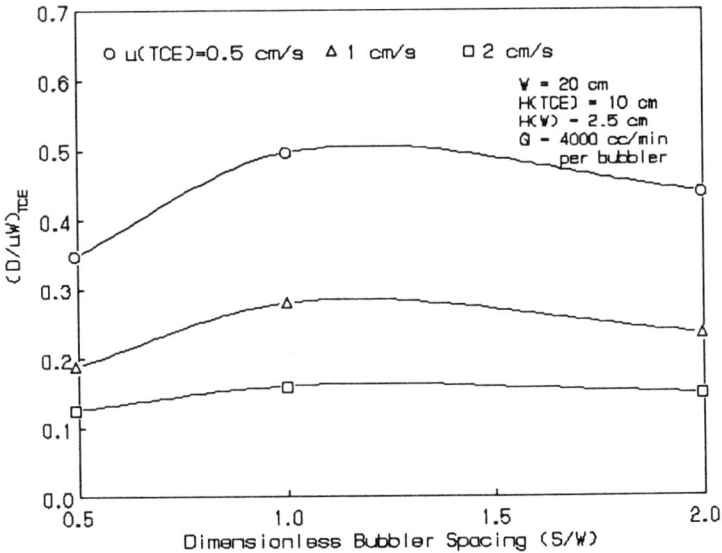

Figure 17. Inverse Peclet number in the lower phase vs. dimensionless
bubbler separation for Q = 4000 cc/min per bubbler.

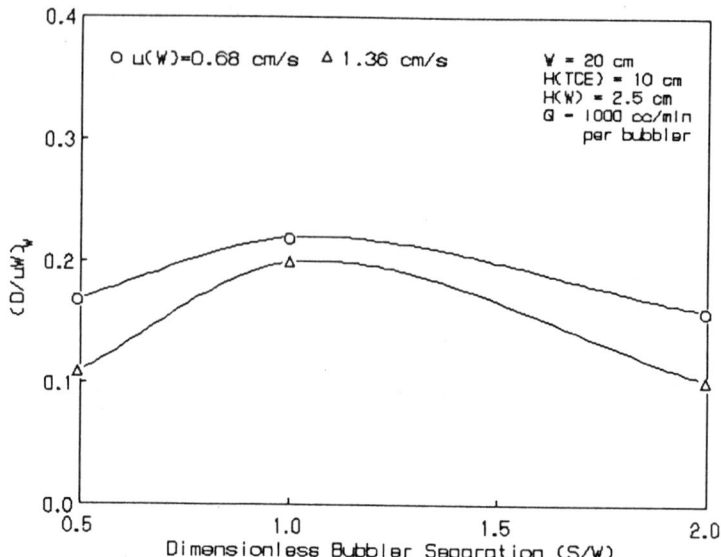

Figure 18. D/uW vs. S/W for the upper phase for Q = 1000 cc/min per
bubbler.

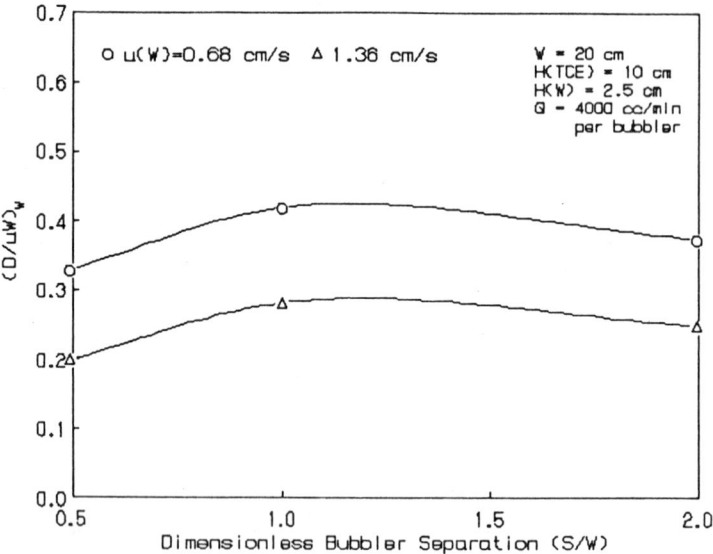

Figure 19. D/uW vs. S/W for the upper phase for Q = 4000 cc/min per
bubbler.

surface of the lower phase. Engh et al. (9) found that D/uL for the slag phase in the Lulea refiner was close to 1.0 compared to 0.064 for the metal phase. The reason for such high longitudinal mixing in the slag phase was presumably that the gas blowing was carried out by top jetting using inclined lances. This transferred most of the mixing energy to the upper phase, and inclination of the lances would result in extremely high longitudinal mixing. For the same reason the turbulence in the lower phase was not very high, resulting in a large difference between the mixing intensities between the two phases. The situation was quite different in the present case, and there are no particular conditions that would impart a larger longitudinal mixing to the upper phase.

The upper phase eddy diffusivity cannot be predicted and must be measured. The lower height-to-width ratio of the upper phase suggests that it should have a lower eddy diffusivity than the lower phase, but the more distributed nature of the gas flow from the plume should increase D above the value it would have from a single bubbler. The mobility of the lower interface of the upper phase will also tend to increase the eddy diffusivity in the upper phase.

While mixing in vertical gas/liquid contactors has been studied extensively by chemical engineers, the information on horizontal bottom blown reactors is quite limited. The mixing and mass-transfer data in reactors with two liquid phases with flow are almost absent, particularly because of the complications involved in modeling such systems.

No systematic modeling studies on a horizontal counter-current reactor with bottom gas blowing have been reported. Some tracer studies were conducted on the WORCRA steelmaking reactor. Comparison with that system is particularly difficult since very little data were published. In addition, the system configuration was quite different. Top jetting using inclined lances was employed, which results in a very different fluid flow behavior compared to bottom blowing.

In the present study, countercurrent system studies were made at two different scales (W = 12.8 cm and 20 cm) and two different H/W ratios of the phases. In all the cases, a linear relationship between the overall mass-transfer coefficient (k_{ov}, derived from the overall heat transfer coefficient data) and the mixing-energy input was found.

A number of studies have been carried out on the measurement of mass-transfer coefficient for liquid-liquid mass transfer in different systems. The experiments have mostly been carried out in cylindrical reactors.

Robertson and Staples (14) worked with (i) a lead/molten salt system and (ii) an amalgam/aqueous system. They found for both these systems,

$$k_{ov} \propto Q^{0.5}$$

Patel and coworkers (15) studied mass transfer in a bubble-agitated aqueous/organic system, and they found the effect of gas flow rate on mass-transfer coefficient to be,

$$k \propto Q^{0.72}$$

Asai, Kawachi, and Muchi (16) have shown, from a comprehensive study of the previous works, that the value of the exponent of Q varied in a wide range, between 0.25 and 3.0. The value of this exponent increased substantially in the higher range of q (gas flow rate per unit mass of the liquid). There was an abrupt change in dependence of the mass-transfer coefficient on the mixing energy input, which was explained by noting that the liquid/liquid interfacial area was constant at lower mixing energy. The agitation was rather quiet in this region. However, the surface area increased substantially at high mixing energy, increasing the mass-transfer rate rapidly. In a separate study on the transfer of benzoic acid between tetraline and water, they found a similar abrupt change in dependence of k on mixing energy. Entrapment of drops of tetraline in the water phase was observed at nearly the same q, at which the abrupt change of mass-transfer rate appeared.

When the height of the lower phase was halved (i.e., H/W was reduced from 1 to 1/2), the eddy diffusivity dropped by between 25% and 54%. This can be compared with Gassel's results from stationary single-phase reactor with single central bubblers. His results showed that the eddy diffusivity dropped by between 40% and 52% when H/W was reduced from 1 to 1/2.

IV.2. High Gas-Strength Studies

An example of the comparison of model predictions and experimental results for a single-liquid experiment is shown in Figure 20. Those for a two-liquid experiment are shown in Figure 21. As can be seen from these figures, there is good agreement between model predictions and experimental results. The model predictions were calculated using the model equations, by selecting the value of α which gives the best fit for the RTD response of the experiments.

(a) Single-Liquid Experiments. The variation of α with various experimental conditions is shown in Table 1. It is seen from the table that α varies with Q_l, Q_g, h and L, and it is

Table 1. Variation of α with Experimental Conditions

Q_L (lpm)	Q_2 (lpm)	Q_g (scfm)	h_1 (cm)	h_2 (cm)	L (cm)	d_{inj} (cm)	α
2.84	0	4.5	9.0	0	120	0.2117	0.150
0.60	0	2.4	10.5	0	120	0.2117	0.025
2.75	0	2.4	9.6	0	80	0.2117	0.160
2.19	0	2.4	9.6	0	40	0.2117	0.210
2.40	0	3.75	5.2	0	80	0.2117	0.200
3.50	0	4.0	5.1	0	80	0.2117	0.245
3.03	0.152	4.0	4.6	5.6	80	0.2117	0.295
0.75	0.780	4.0	5.5	6.2	80	0.2117	0.167
0.75	1.560	4.0	5.5	6.2	80	0.2117	0.173
0.75	2.340	4.0	5.7	6.2	80	0.2117	0.165
0.75	3.510	4.0	5.7	6.2	80	0.2117	0.162
0.75	4.680	4.0	5.7	6.2	80	0.2117	0.159

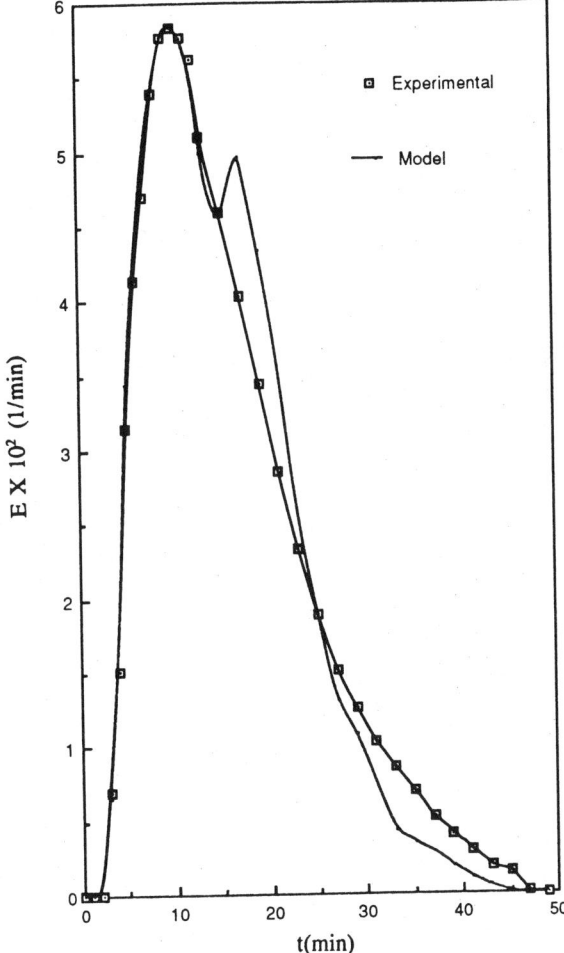

Figure 20. Comparison of experimental results with model predictions for single-liquid experiment. ($t_{expt} = 16.1$ min, $t_{model} = 16.1$ min.)

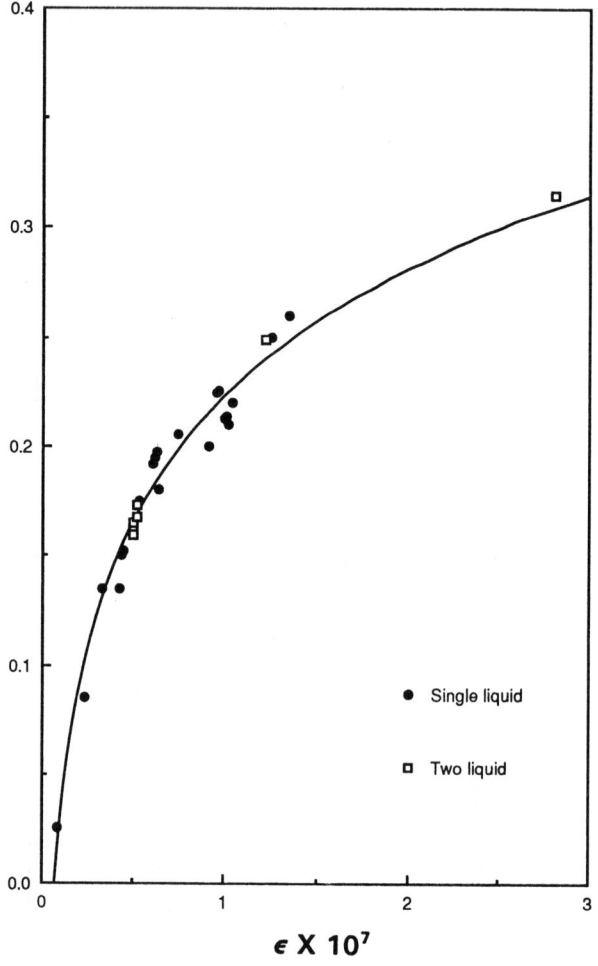

Fig. 21. Variation of α with ϵ

404

expected that α is also a function of d_{inj} and D by affecting among others the velocities of liquid and gas. Thus,

$$\alpha = f(Q_L, Q_g, h, L, d_{inj}, D) \tag{20}$$

where Q_g = volumetric flow rate of gas at 25°C and 1 atm

h = bath height

L = distance between injectors

d_{inj} = injector diameter

D = reactor diameter

Based on these parameters, we can define the following additional variables,

$$U_L = Q_L/A \tag{21}$$

$$U_g = 4 Q_g/\pi d_{inj}^2 \qquad (Ma < 1) \tag{22}$$

$$ = a \text{ (constant)} \qquad (Ma > 1) \tag{23}$$

$$\rho_g = \rho_g^{\circ} \qquad (Ma < 1) \tag{24}$$

$$ = \rho_g^{\circ} Q_g /\pi d_{inj}^2 a \qquad (Ma > 1) \tag{25}$$

where U_L = nominal heavy-liquid flow velocity

A = average cross-sectional area of bath

U_g = average gas velocity

a = speed of sound in air at room temperature

ρ_g = average gas density

ρ_g° = gas density at STP

Ma = apparent gas Mach number

$(\equiv U_g/a$, where U_g is defined as in Equation (22))

While the gas Mach number cannot exceed unity under experimental conditions, the Mach number in these calculations is based on calculations of gas velocity at the injector tip at standard conditions of temperature and pressure.

Performing a dimensional analysis, it was determined that the most appropriate dimensionless group is the modified energy ratio (ϵ) which is defined as

$$\epsilon = \rho_L U_L^2 h (Q_g/Q_L)/(\rho_g U_g^2 L) \tag{26}$$

where ρ_L = heavy-liquid density

A plot of α vs. ϵ is shown in Figure 22. As can be seen, there is an excellent correlation between α and ϵ.

(b) Two-Liquid Experiments. Plotting the values of α vs. ϵ (as defined in Equation 26) for two-liquid experiments, it was evident that there is a deviation in the value of α from the modified energy number. However, when the value of h in Equation (26) was modified to include the effect of the added weight due to the second liquid that the injected gases have to overcome, as follows:

$$h = h_1 (1 + (h_2 \rho_2)/(h_1 \rho_L)) \tag{27}$$

where h_1 = height of heavy liquid

h_2 = height of light liquid

ρ_2 = density of light liquid

it was found that there was good agreement between α and ϵ for two-liquid flows. Thus the incorporation of this correction in this correlation makes it applicable to countercurrent two-phase flows. It has been observed that the flow rate of the light liquid is not an important parameter.

V. Design of a Steel Refiner

In the following, the design of a full-scale steel refiner is proposed, based on the low gas-strength study. The input steel is assumed to be low in carbon content, and hence the primary reactions will be slag-metal reactions, enhanced by bottom gas blowing using tuyeres. It should be emphasized that this design does not pertain to a smelting reactor (for steelmaking from hot metal), which would involve oxygen blowing to oxidize the impurities and would result in considerable gas evolution from the carbon/oxygen reactions.

(a) Design Calculations. A desired capacity to refine 400 tons of steel per hour was initially assumed.

The width and length of this reactor were decided based on practical considerations. A conventional continuous casting tundish is approximately 10 meter long and 1 meter wide with a metal layer height of up to 1 meter. For the countercurrent reactor, a wider and more shallow metal bath was selected. A 10-meter long and 1.5-meter wide channel was assumed, with the H/W ratio of metal equal to 1/2. For this geometry, the maximum D/uW occurred at S/W = 1; however, the lowest value of S/W at which experiments were carried out was 1/2, and so it was decided to keep S/W at 1/2, which means that the tuyere separation was 75 cm.

For the above set of parameters, the metal velocity in the channel was calculated to be 1.37 cm/s. The laboratory model experiments, which corresponded to these parameters were the ones with $(H/W)_{TCE}$ = 1/2 in the 20-cm wide tank with TCE flow velocity of 0.5 cm/s.

If the values of dimensionless gas flow rate (Q/uW^2), Fr (u^2/gW), and Re (uW/ν) for the full-scale reactor were equal to the corresponding values for the model reactor, then (for

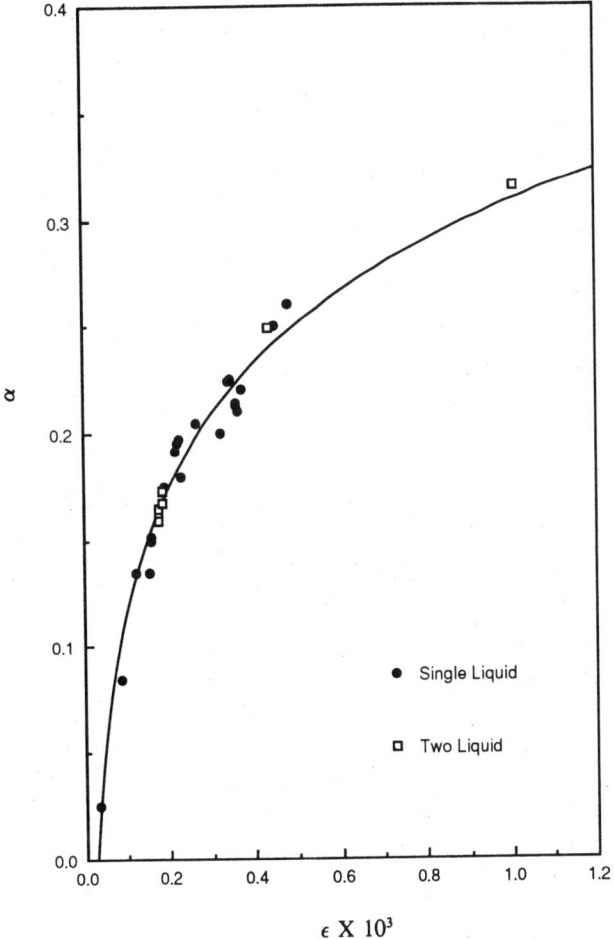

Figure 22. Variation of α with ϵ.

the given geometry) the D/uW values in the two cases would be the same. In the following discussion, the model values are represented by subscript "m", and the full-scale values are represented by subscript "f".

i. $(u^2/gW)_f = 1.27E\text{-}5$, $(u^2/gW)_m = 1.27E\text{-}5$.

Thus, the Fr number had the same value in the two cases.

ii. For the model experiments with Q = 4000 cc/min per bubbler, $(Q/uW^2)_m = 0.33$. For the dimensionless gas flow rate for the full-scale reactor to have the same value, Q = $0.616 \text{ m}^3/\text{min}$ (at the operating temperature) per tuyere. For an operating temperature of 1600 C, this amounts to 90 N liters/min per tuyere).

iii. Similarity with respect to Re could not be achieved, since $u_f = 2.7 \ u_m$, $W_f = 7.5W_m$, and $f = 2f_m$. Thus, Re in the full-scale reactor was much higher than the model value. However, it was shown in Figure 8 that D/uW is much less sensitive to Re at high Reynolds numbers. In the present case, the Re value for the model was 1938, which is close to the range in which Reynolds number did not have any significant effect. The dissimilarity with respect to Re was neglected in this scale-up study.

Under these conditions, the D/uW values for the two systems are the same. Thus from the laboratory measurements:

$(D/uW)_f = 0.35$, or $D_{metal} = 72 \text{ cm}^2/\text{s}$.

(b) Mass-Transfer Coefficient. Following Robertson and Staples (13), Gaye, Gatellier and Riboud (17) found that the mass-transfer coefficient for sulfur could be expressed by,

$$k = B \ (D_s \ Q/A)^{0.5}$$

where D_s is the diffusion coefficient of sulfur in the metal (cm^2/s), B is an empirical parameter $(\text{cm}^{-1/2})$, and Q/A is the specific volumetric gas flow rate across the interface $(\text{cm} \cdot \text{s}^{-1})$.

On the basis of their desulfurization data, they showed that $B = 50 \ \text{cm}^{-1/2}$. Taking the value of $D_s = 2 \ E\text{-}5 \ \text{cm}^2/\text{s}$, the value of k was calculated to be 0.21 cm/s.

The value of the distribution coefficient (β) has been reported in the literature to be 200 (or even higher). A conservative value of 50 was initially selected.

$\beta m_s/m_m$ is an important parameter for the reactor, where m_m and m_s are the mass flow rates of metal and slag, respectively. It is the ratio of the amount of an impurity that can be retained in the slag to the amount of that impurity coming in with the metal. The value of this parameter is normally kept well above one to obtain sufficient slag capacity for impurity removal.

For $\beta m_s/m_m = 5$, and $H_m/H_s = 4$, the slag velocity was calculated to be 0.98 cm/s (using a value of 4000 kg/m^3 for the slag density). Since the D/uW value for the upper phase was found to be close to that for the lower phase in the model system, a value of 0.35 was also used for the slag phase.

For the above set of parameters, the computer model was used to calculate the reactor performance. The percentage impurity removal was found to be 80%.

Normally, an impurity removal in the vicinity of 95% would be desirable. With a value of $\beta m_s/m_m$ of 5, the reactor performance can only be improved by (i) increasing the mass-transfer rate and/or (ii) increasing the residence time in the reactor.

Usui (15) has reported that mass-transfer coefficient values in the range of 0.24 cm/s to 0.46 cm/s were measured for a 250 t ladle desulfurization process, using top slag with intense stirring. If a value of k_{ov} = 0.46 cm/s could be achieved by stronger gas blowing, and the resultant increase in longitudinal mixing were neglected, then the model predicted a 96% impurity removal.

The reactor performance was also calculated for a lower through-put rate of 200 tons of steel per hour through a reactor of the same size. The values of H_m/W (= 1/2), S/W (= 1/2) and H_s/H_m (= 1/4) were kept at their earlier values.

The metal velocity in this case was calculated to be 0.68 cm/s, giving a residence time in the reactor of 24.3 min (which is close to the residence time of approximately 20 minutes in the WORCRA steelmaking reactor (6)).

The previous set of D_{metal} and D_{slag} values were used for these calculations also. (Again, exact dynamic similarity with the model system was lost when the velocities were lowered. In this case, a better performance than the one predicted would be expected).

For k_{ov} = 0.21 cm/s and $\beta m_s/m_m$ = 5, the model predicted a 92% impurity removal. When the k_{ov} was increased to 0.46 cm/s, the impurity removal predicted was 98%.

The calculations presented above are summarized in Table II. These calculations indicate that good refining can be achieved at reasonably high production rates, using a reactor that was not much larger than present day continuous casting tundishes. The weight of the metal in the proposed reactor at any time would be 81 tons. This can be compared with the 70 ton tundish in Bethlehem Steel's Burns Harbor, Indiana plant, which is the largest tundish in North America.

VI. Conclusions

The modeling work on the continuous countercurrent reactor, at low gas strengths, showed that longitudinal mixing in such reactors could be controlled by a judicious selection of injector separation. Injector separation below a characteristic value provides the possibility of keeping the longitudinal mixing low, while obtaining a high mass-transfer rate.

In an effort to understand the overall fluid-flow patterns and mixing taking place in a shallow bath channel reactor with countercurrent fluid flow and high-energy bottom gas injection, such as a QS oxygen reactor, a basic approach has been adapted by modeling the reactor as a combination of ideal reactors. The analysis is based on a single parameter α. α has been correlated to various operating parameters as well as physical properties of the different fluids involved, thus making it possible for a priori prediction of RTD behavior of the channel reactor.

With the adjusted bath height as defined in Equation (27), a correlation has been developed for α based on a modified energy ratio (ϵ, as defined in Equation (26)). It has been

Table II. Summary of the Design Calculations.

Reactor Dimensions and Geometry

Length of the reactor	(L)	10 m
Width of the reactor	(W)	1.5 m
Metal layer height	(H_m)	75 cm
Slag layer height	(H_s)	18.5 cm
Bubbler Separation	(S)	75 cm

S/W 1/2
H_m/W 1/2
H_m/H_s 4

Case 1: Throughput = 400 TPH of steel

Metal velocity	(u_m)	1.37 cm/s
Slag velocity	(u_s)	0.98 cm/s
D/uW (metal, slag)		0.35
D/uL (metal, slag)		0.053
D_{metal}		72 cm²/s
D_{slag}		51 cm²/s
Diffusivity of the impurity		2 E-5 cm²/s
Distribution coefficient		50
$m_s\beta/m_m$		5

Case 1(a)	Gas flow rate per bubbler	(Q)	90 N l/min
	Mass-transfer coefficient	(k_{ov})	0.21 cm/s
	Percent impurity removal		80
Case 1(b)	Higher gas flow rate		
	Mass-transfer coefficient		0.46 cm/s
	Percent impurity removal*		96

Case 2: Throughput = 200 TPH of steel

Metal velocity	(u_m)	0.68 cm/s
Slag velocity	(u_s)	0.49 cm/s
D_{metal}	72 cm²/s	
D_{slag}	51 cm²/s	
D/uW (metal, slag)		0.70
D/uL (metal, slag)		0.105
Diffusivity of the impurity		2 E-5 cm²/s
Distribution coefficient		50
$m_s\beta/m_m$		5

Case 2(a)	Gas flow rate per bubbler	(Q)	90 N l/min
	Mass-transfer coefficient	(k_{ov})	0.21 cm/s
	Percent impurity removal**		92
Case 2(b)	Higher gas flow rate		
	Mass-transfer coefficient		0.46 cm/s
	Percent impurity removal		98

* Eddy diffusivity increase due to higher gas blowing neglected.
** Eddy diffusivity decrease due to lower velocities neglected.

found that the same correlation applies to both the single- and two-liquid systems.

In the low gas-strength work, D/uW values did not vary greatly with S/W (injector spacing) at a constant gas flow rate per injector. However, the values of D/uW were quite different for the same γ, defined by Equation (19), if they were obtained in one case by doubling the flow rate and in the other by halving the injector spacing. The heat-transfer coefficient (and presumably mass-transfer coefficient) increased with increasing γ, independent of injector spacing.

Acknowledgement

The low gas-strength work was supported by the Department of the Interior's Mineral Institute program administered by the Bureau of Mines through the Generic Mineral Technology Center for Pyrometallurgy under grant number USDI G1105129 2923. The high gas-strength work was supported in part by the State of Utah Centers of Excellence Program and by the Utah Mineral Leasing Fund. During the course of this work K. Iyer received the University of Utah Research Committee Fellowship and the Geneva Steel Fellowship.

References

1. Queneau, P. E., "Oxygen Technology and Conservation," *Met. Trans. B*, Vol. 8B, No. 3, Sept. 1977, pp. 357.

2. Queneau, P. E. and Schuhmann, R. Jr., "The QS Oxygen Process," *Journal of Metals*, Vol. 26, No. 8, August 1974.

3. Queneau, P. E., "Coppermaking in the Eighties — Productivity in Metal Extraction from Sulfide Concentrates," *Journal of Metals*, Vol. 33, No. 2, February 1981, pp. 38.

4. Queneau, P. E., "Innovation and the Future of the American Primary Metals Industry," *Journal of Metals*, Vol. 37, No. 2, February 1985, pp. 59.

5. H. K. Worner, "Extra Clean Steel Production Via the WORCRA Steelmaking Process," *Iron and Steelmaker*, January 1988, pp. 23-24.

6. H. K. Worner, F. H. Baker, I. H. Lassam and R. Siddons, "WORCRA (Continuous) Steelmaking," *Journal of Metals*, Vol. 21, June 1969, pp. 50-56.

7. T. A. Engh, C. E. Grip, L. Hansson and H. K. Worner, "A One Dimensional Model of the WORCRA Steel Refining Launder," *Jernkontorets Annaler*, Vol. 155, 1971, pp. 553-564.

8. T. W. Jenkins, N. B. Gray and H. K. Worner, "Application of the Levenspiel Dispersion Model to Metal Flow in Pilot Plant WORCRA Continuous Steelmaking Furnace," *Metallurgical Transactions*, Vol. 2, 1971, pp. 1258-1259.

9. T. A. Engh, L. Hansson, H. K. Worner and K. Wulff, "Measurement of the Residence-Time Distribution for the WORCRA Steel Refiner," *Jernkontorets Annaler*, Vol. 155, 1971, pp. 93-99.

10. A. K. Agrawal, "A Model Study of Mixing and Mass Transfer in a Counter-current Reactor for Metal Refining," Ph.D. thesis, University of Missouri-Rolla, 1991.

11. O. Levenspiel, "Chemical Reaction Engineering," Second Edition, John Wiley and Sons, 1972.

12. Y. Sahai, "Mixing in Channel Reactors Stirred by Bubbles," Ph.D. thesis, Imperial College, London, 1979.

13. J. B. Gassel, "Characterization of the Fluid Dynamic Mixing of a Bottom Gas Injected Channel Reactor Using Low Temperature Modeling," M.S. Thesis, University of Missouri-Rolla, 1987.

14. D. G. C. Robertson and B. B. Staples, "Model Studies on Mass Transfer Across a Metal-Slag Interface Stirred by Bubbles," Process Engineering of Pyrometallurgy, The Institute of Mining and Metallurgy, ed. M. J. Jones, 1974, pp. 51-59.

15. P. Patel, M. G. Frohberg and K. Biswas, "The Effect of Rising Bubbles on the Mass Transfer between Liquid Iron and FeO-Bearing Slags," Kinetics of Metallurgical Processes in Steelmaking, ed. W. Dahl, K. W. Lange and D. Papamantellos, 1975, pp. 180-191.

16. S. Asai, M. Kawachi and I. Muchi, "Fluid Flow and Mass Transfer in Gas Stirred Ladles," Foundry Processes—Their Chemistry and Physics, ed. S. Katz and C. F. Landefeld, 1988, pp. 261-292.

17. H. Gaye, C. Gatellier and P. V. Riboud, "Physico-Chemical Aspects of the Ladle Desulphurization of Iron and Steel," Foundry Processes—Their Chemistry and Physics, ed. S. Katz and C. F. Landefeld, 1988, pp. 333-356.

18. T. Usui et al., "On the Study of Best Desulphurization Conditions by Gas Injection through Lance," Tetsu-to-Hagane, Vol. 66, 1980, pp. S259.

PROCESS DEVELOPMENT IN PILOT PLANTS

Eric Burström[1], Theo Lehner[2]

[1]MEFOS (The Foundation for Metallurgical Research)
P.O Box 812
S-951 28   Luleå, Sweden

[2]Boliden Mineral
S-932 00   Skelleftehamn, Sweden

Abstract

The feasibility of utilizing pilot plants is seldom publicly discussed but is of major concern for MEFOS (MF) and its users.

The authors elaborate on why, how, when and by whom pilot plant facilities for pyro-metallurgical research should be used. Aspects on the feasibility of laboratory, pilot and/or plant investigation are presented and also what the customer should think of to make efficient use of pilot plants.

Studies related to bath smelting are reported to illustrate investigations on pilot plant scale. A major issue has been the introduction of gases and powders through the bottom, the sidewall and the top of vessels. The increase of knowledge from injection technology trials in pilot plant at MEFOS is exemplified.

Proceedings of the
Savard/Lee International Symposium on Bath Smelting
Edited by J. K. Brimacombe, P. J. Mackey,
G. J. W. Kor, C. Bickert and M. G. Ranade
The Minerals, Metals & Materials Society, 1992

Description of MEFOS

MEFOS - The Foundation for Metallurgical Research - owns and runs pilot
plant facilities in Luleå, Sweden. MEFOS is a research institute for the
Nordic Steel and Metal Industries and consists of two research departments,
The Metallurgical Research Plant (MF) and The Metal Working Research Plant
(BTF).

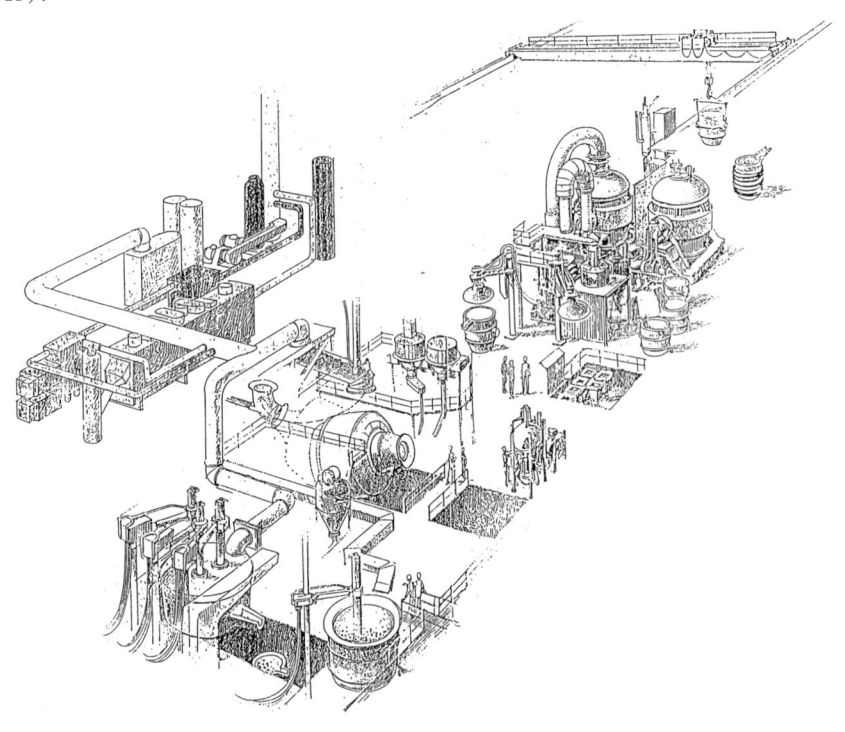

Figure 1 - The Metallurgical Research Plant (MF)

MEFOS conducts metallurgical and metal working research and development,
both jointly and on a contract basis for individual clients. Out of the actual research budget about 50 % is joint research where government funding
is about 50 % (approx. 25 % of the total turnover).

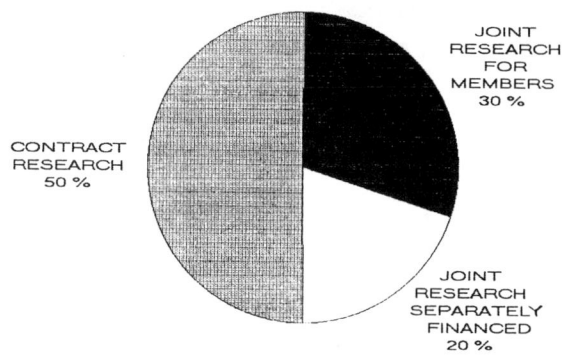

Figure 2 - MEFOS's type of research.

The other half of the budget is contract research projects for members and clients from all over the world.

In the following only the pyro-metallurgical research will be discussed. The pyro-metallurgical research at the Metallurgical Research Plant (MF) concerns:

Type of research
Pilot Plant research
Plant implementation
Consulting/Engineering

Field of research
Reduction processes
Oxygen processes
Electro steel processes
Refining
Ladle metallurgy
Casting
Non-ferrous metallurgy
Ferro-alloys metallurgy

In recent years we carried out studies on bath smelting and associated technologies in the field of smelting reduction, converter technology, coal gasification and injection in ladles.

The normal procedure is to utilize our existing equipment which is modified for the actual task. But also separate major demonstration units have been erected and run at MEFOS.

The staff consists of experienced steelworkers and engineers with well balanced theoretical and practical knowledge.

The organization and equipment is "slimmed" so that normal research and development work can be performed and so that we always have enough key persons. However, for more extensive work external personnel is engaged temporarily.

Research for non-members

MEFOS is an independent research foundation. This enables us to carry out secret research projects for anyone who is interested.

Recently the board of MEFOS has expressed the willingness to extend the group of members outside the Nordic countries. The conditions for such a membership is not yet completely settled, but in principle everyone interested can join certain joint research projects. An example is our strip casting project with participants from both Japan and Europe. Another example is the so called "multi-client" projects for which MEFOS has achieved participation worldwide.

415

Development in Pilot Scale

According to Johnstone and Thring /1/ the aim of the pilot plant is to provide design data for the ultimate large one and that of the model to exhibit
the effect of change on operation conditions more quickly and economically
than would be possible on the full scale. Baekeland, the inventor of bake-
lite, already in 1916 stated /2/ "Commit your blunders on a small scale and
make your profits on a large scale".

But in order for the research to be valuable, the pilot scale should be as
close to production size as possible. Similar to get real "piloting" from
tests it is important that all big scale physical and chemical phenomena occur in the same way in the pilot equipment. This also means that only
long-term factors like definite figures on refractory consumption, maintenance, manning etc. should remain after a test campaign.

Proper scale is not the same for all kinds of parameters to be studied /3/.
When studying mass transfer one has to make sure that fluid flow patterns
are alike. In the study of rate phenomena one can define a scale-up number
N_r where

$$N_r = \frac{\text{area of "local phenomena"}}{\text{area of "overall phenomena"}}$$

For gas stirred melts the local phenomena is equal to gas plume and overall
phenomena include all other reactions occurring at top surface between
atmosphere and melt, interface between liquid and refractory etc. Here one
observes the following.

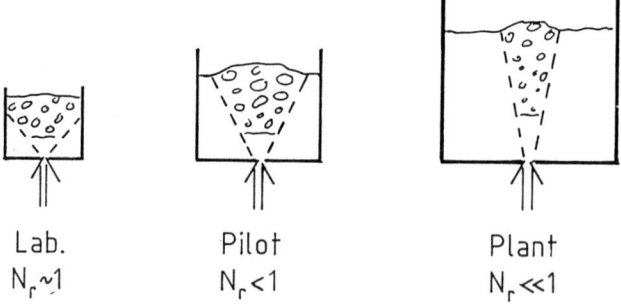

Figure 3 - Scale effect at lab, pilot and plant for injection trials

Some of the bath smelting phenomena that have successfully been studied in
pilot scale at MEFOS are illustrated in the table below.

416

Table I Examples of phenomena studied in pilot scale at MEFOS.

Process	Phenomena	Knowledge
Materials handling	Pneumatic transport	Powder characteristics
Tuyere technology	Blockage	~ Mach Number
	Wear	Rotating jet flow
	Design	Position, high pressure
Transfer Phenomena	Mixing Foaming	$\tau \sim \dot{\epsilon}$ [1]), $u_v \sim \dot{V}_g$ [2]) $h = k_1 + k_2 \cdot dC/dt$ [3])
Refining	ΔS	Transitoric, permanent reactions
	ΔO	Deoxidation, reoxidation rates,
	Inclusion control	Ca-additions
	ΔP	Slag treatment to prevent
	Δ	PH_3 formation
	ΔCu	Contact surface, temp.
Non-ferrous	Smelting	New smeltinging technology (BOLD)
	Slag refining	$\Delta As/As_0$, matte fuming
Energy/ Smelting	Coal gasification	Kinetics and process technology (MIP)

1) τ = mixing time, $\dot{\epsilon}$ = stirring effect
2) u_v = surface velocity, $\dot{V}_g$ = gas volume
3) h = foam weight dC/dt = decarburization rate, k_1 and k_2 = constants
K_2 = f(carbon reactivity, slag comp.)

A large number of different powders have been tested and the information collected into an "Injection hand book". The well-known "Geldart"-diagram has proved to be valid in several occations.

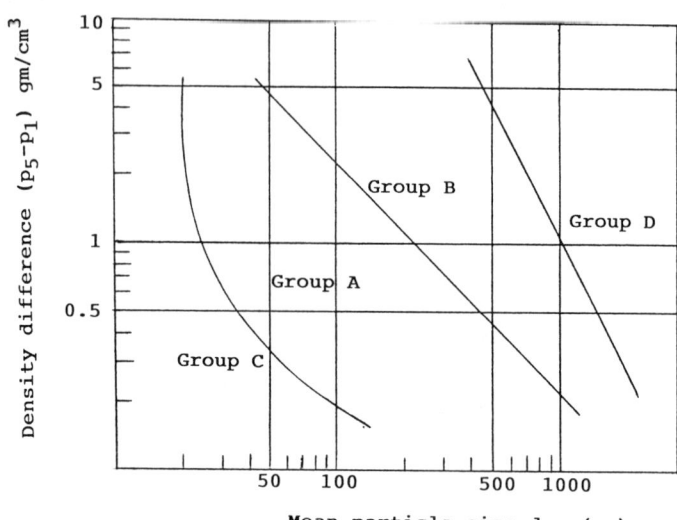

Figure 4 - Particle size versus density difference and classification
of powders in different groups with certain injection
characteristics. From Geldart /4/.

Groups A + B powders are normally easy to handle but lately group C powders
have been more interesting for injection purpose (steelworks dust etc.). Instead of using expensive flow-aids MEFOS has developed a mechanical stirring
device that enables us to inject most steelworks dust without flow-aids or
micro-agglomeration /5/.

Tuyere technology has been an issue almost since the start-up of MEFOS. Prevention of blockage is easiest done by high Mach Numbers. Wear can be reduced by rotating jet flow in tuyeres /6/.

High pressure tuyeres give other flow patterns and indications of better
slag-bath reactions /7/. A further important feature is the better regulation possibility compared to porous plugs.

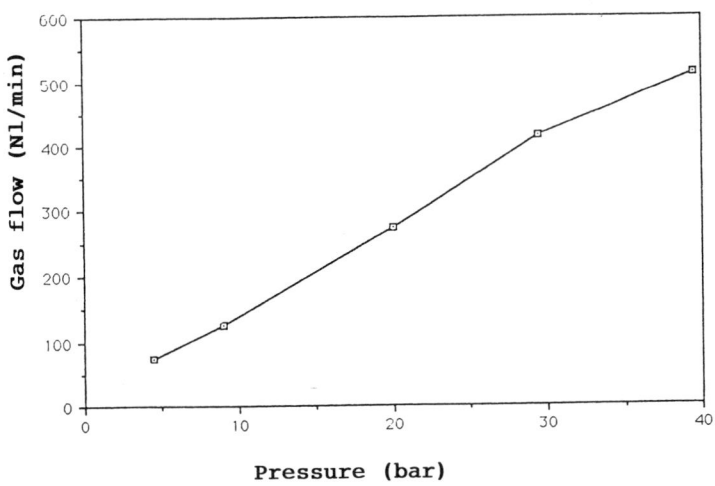

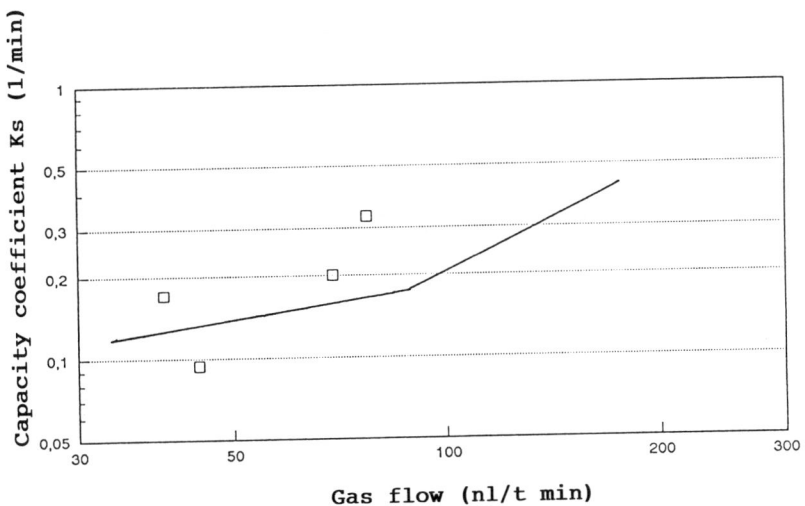

Figure 5 - Characteristics of high pressure tuyeres.
a) Flow versus pressure. b) Desulphurization rate
with gas purging and top slag.

419

Mixing has been studied by tracer additions and related to mass transfer by measurement of dissolution rates and to measured melt velocities /8/.

Conclusion: By proper understanding of the phenomena involved we are able to

- predict full scale operation
- advice on process modification
- gain a basis to new processes.

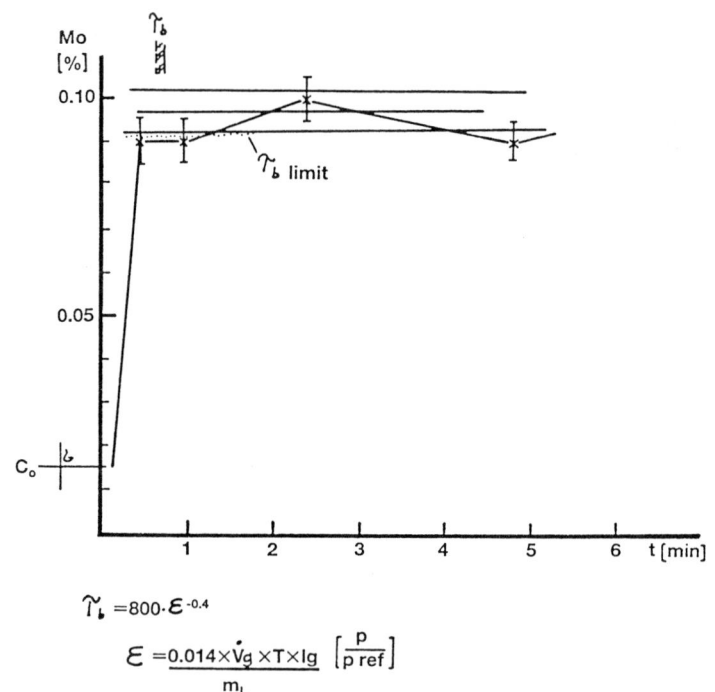

$$\tau_b = 800 \cdot \varepsilon^{-0.4}$$

$$\varepsilon = \frac{0.014 \times \dot{V}_g \times T \times lg}{m_L} \left[\frac{p}{p\,ref}\right]$$

Figure 6 - Mixing of matte in pilot converter.

Removal of sulphur, phosphorous etc. can easily be studied in pilot scale. Ritakallio et.al. showed the influence of transitoric versus permanent reactions during powder injection /9/.

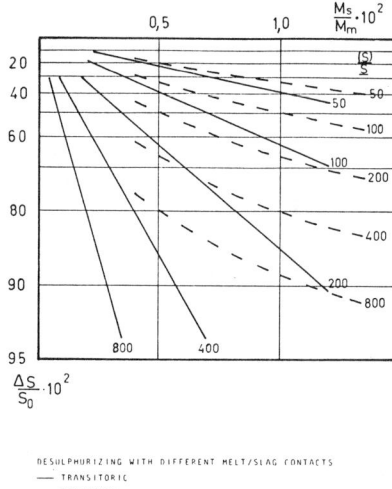

Figure 7 - Degree of desulphurization vs. specific reagent
supply (M_s/M_m) and equilibrium sulphur distribution
for transitoric phase contact.

Deoxidation and reoxidation reaches an equilibrium during injection ($k_{deox} = k_{reox}$). The oxygen transport in a ladle reaches very high values /10/.

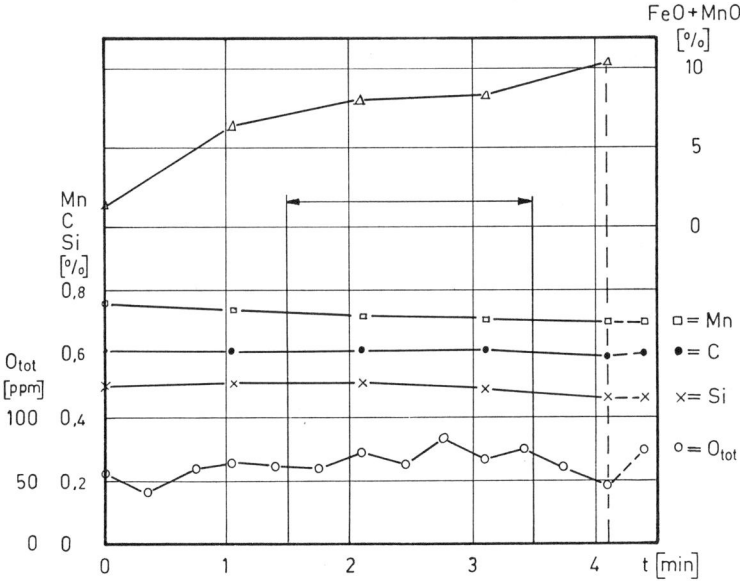

Figure 8 - Melt composition during reoxidation experiment. Iron-ore and
lime-fluorspar- deoxidant are continuously added to the system.

Modification of inclusions by different Ca-additives has been studied. Generally it can be concluded that these type of investigations should preferably be made on full scale especially if trial inclusion morphology is the subject. This is primarily due to the sampling problems and/or the need for
statistics for safe evaluation of the results and that the plant specific
solidification metallurgy is difficult to simulate.

Dephosphorization has been studied under reducing conditions with the injection of CaC_2-mixtures /11/. The problem of slag destruction (prevention
of formation of phosphines) has in principle been solved by a technique with
an oxidating burner on the ladle without rephosphorization of the steel.

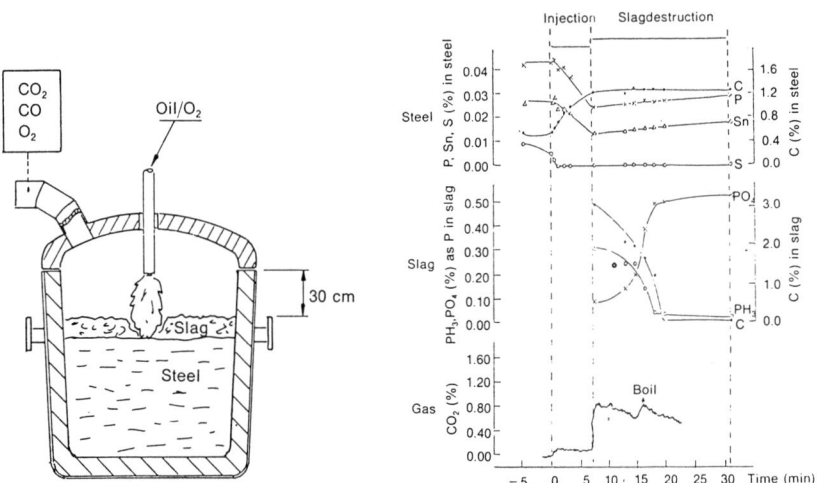

Figure 9 - Dephosphorization during reducing conditions.
a) Procedure for destruction. b) Results from
injection and slag destruction.

Removal of copper from liquid steel is for several reasons something very
suitable for pilot plant scale, especially because of the safety-risk with
sodium reagents.

The general conclusion is that these methods are too costly but also that
there is a high potential for improvement since the used sodium mattes are
far from saturated.

422

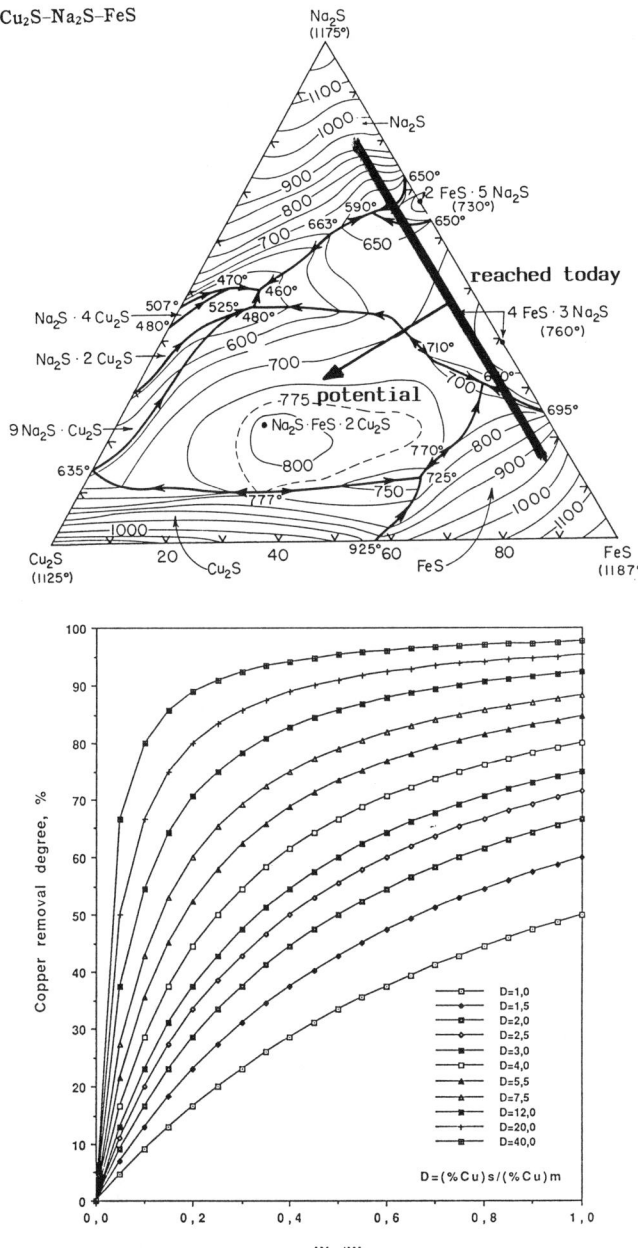

Figure 10 - Cu-refining of steel with Na₂S-FeS mattes. a) Potential for improvements of Cu₂S-content in Na₂S mattes. b) Degree of removal at different distribution ratio, $D = (\% \, Cu)_s / (\% \, Cu)_m$ and slag-metal ratio $C = W_s / W_m$.

In non-ferrous metallurgy bath smelting and converting of copper concentrates has been developed in the BOLD-process.

The general feature is a combination of ferrous and non-ferrous processes:
BOF technology and bath smelting. The process has a great potential for
handling a complexity of feed materials and high impurity removal capacity.
The capital cost barrier (including the learning curve) is dampening its introduction.

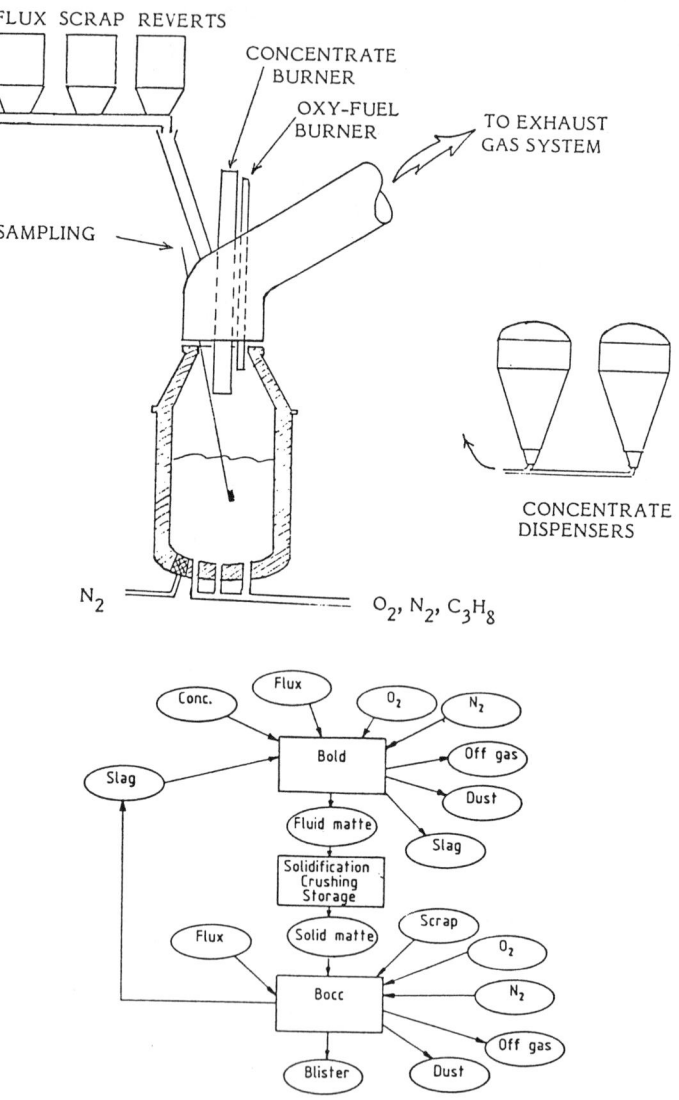

Figure 11 - a) BOLD layout.   b) Process flow sheet.

Another example in the non-ferrous field is the refining of As, Sb and Bi from matte by fuming and from copper by slagging /8/.

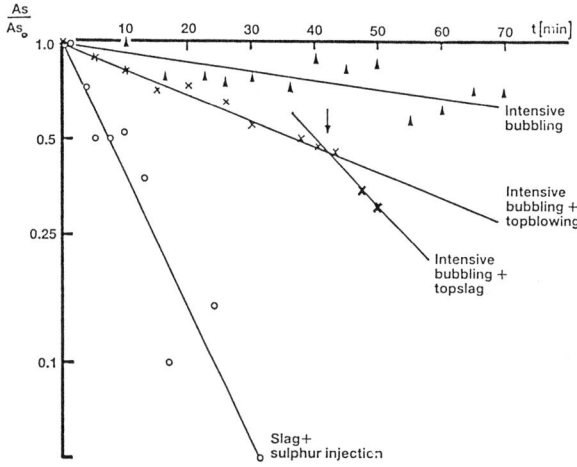

Figure 12 - Comparison of methods tested for refining, As removal.

Smelting reduction has been studied in MEFOS´s equipment (ELRED) in a special demonstration plant (INRED) and also the final step in converter type smelting reduction in the two big coal gasification projects run at MEFOS (MIP and P-CIG).

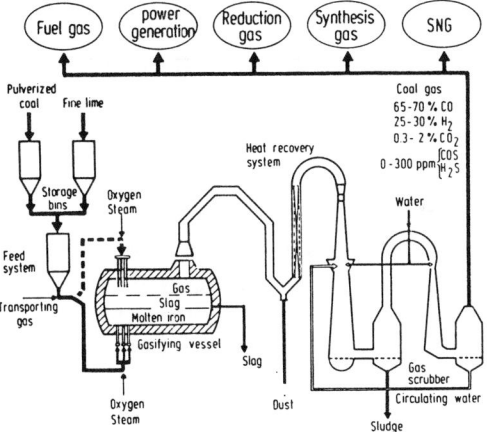

Figure 13 - MIP (Molten Iron Pure gas) process. One of two demonstration plants for coal gasification at MEFOS.

These two projects with the concept of bottom injection of coal/oxygen into an iron bath has also increased our knowledge for application in another emerging area namely environmental technologies for the metal industry.

The following chapter summarizes the experience from several years of research and development at MEFOS and might give the audience some ideas of
the advantages and drawbacks for the utilization of pilot plants.

In Table II some factors when choosing the proper development path are listed.

Table II Using proper scale (++ very ample, + ample, ± viable,
+/- difficult, - not recommended, () depends on circumstances)

Phenomena to be studied	Lab	Pilot	Demo	Plant
Thermodynamics	++	+-	-	-
Kinetics	(+)	++	-	+
Process techniques	-	++	+(+)	-+
Optimization	-	+-	+	++
Knowledge	++	+(+)	-	+-

When looking at the total cost this has to be related to the number of parameters that can be checked and also the time from idea to a definite answer.
The time factor has become much more important since it is very essential to
be the first in the world to earn money. Slow development work means
"leakage" of ideas to competitors which means that the tendency is to take a
chance to test in full scale without any "disturbing" long term research.
When looking at new process concepts the following is roughly valid.

Table III Scope, time and costs in different scale.

	Lab	Pilot	Demo	Plant
No test variables	10^2	10^1	10^0	10^0
Dev. time, year	2-5	0.2-1	2	3
Costs, MUSD	0.1-1	0.2-2	20	100

The demand for fast development thus often should lead to the use of pilot
plants! Other factors that have changed the development route during the
last decade are

-         better theoretical models (thermodynamic, kinetic programmes)
-         improved measuring technique, mobile measuring units and process
          models
-         experimental design and multivariate analysis

The above shows that very efficient research can be done by plant investigations. The tendency for MEFOS is also that some 30 % of our activities are
at the member plant sites but almost 100 % of contract research is in pilot
scale and concerns new process technologies.

The earlier mentioned new circumstances are changing the development route
for new technologies according to:

Table IV Influence from changing factors and development route.

	Lab	Pilot	Plant
Time-cost	-	++	+
Theoretical models	-	+-	+
Measuring technique	-	+-	++
Multivariate analysis	-	-	++

Efficient Use of Pilot Plants

Below we have listed some important aspects from experiences to run pilot plants during almost thirty years.

1. ### Definition of the problem
    Some clients are so enthusiastic about having the possibility to test their ideas on bigger scale that they try to solve all problems instantly. So, concentrate on solving only the key aspect and not for example some material feeding system that can be purchased on the market.

2. ### Potential time factor and cost
    What is the potential of the new technology, especially the market situation. How important is it to get fast answers and thereby being the first to earn money? What is the most cost effective way for development?

3. ### Choice of pilot plant
    If you need to test in a pilot plant, should we then build up something ourselves or utilize an external pilot plant? The triple Q (Quality, Quick service, Quietness = secrecy) is very important.

4. ### Planning of pilot tests
    Plan with purpose and not with purse! If your idea cannot take the cost of proper pilot set-up then you should probably not realize your idea.

5. ### Evaluation of results
    Involve your potential clients to take part in the test terms! The demonstration effect is very strong and often makes your client to come up with bright new ideas.

New Research Projects at MEFOS

Among the projects MEFOS will carry out in the coming years there are some that could be of special interest to bath-smelters.

-   Annular Laval Nozzle for BOF
-   Injection of liquids into liquid metals
-   Development of a high pressure gas-tuyere for ladles.
-   Measuring technique
-   Environmental technologies for treatment of steelworks dust.

The first project "Annular nozzle for BOF´s" is a multi-client project where the normal multi-hole tuyere is replaced by an annular laval shaped nozzle with other behaviour than normal BOF-tuyeres. Pilot plant tests have shown

for example less slopping and lower necessary back-pressure besides the possibility for simple manufacturing.

Injection of liquids into liquid metals is highly interesting from an environmental point of view but means also the possibility for new process technologies.

The high pressure tuyere is very interesting since it gives the possibility
to vary the gas flow in a very wide range but gives also better refining
characteristics. We have even managed to make the tuyere so that it can be
closed and re-opened.

Measuring technologies include work on ultrasonic detection for inclusion
control in molten metal, application of emission, mass spectrometry and XRF
for direct analysis among other things.

Among the environmental projects, inexpensive upgrading of low grade Zn-con-
taining dusts will be made with injection into liquid steel slags and is an
example of a project that needs both pilot plant and plant development phases.

In Figure 14 we have indicated the correlation between knowledge and plant
applicability and indicated how the above projects fit to this model and the
different development phases.

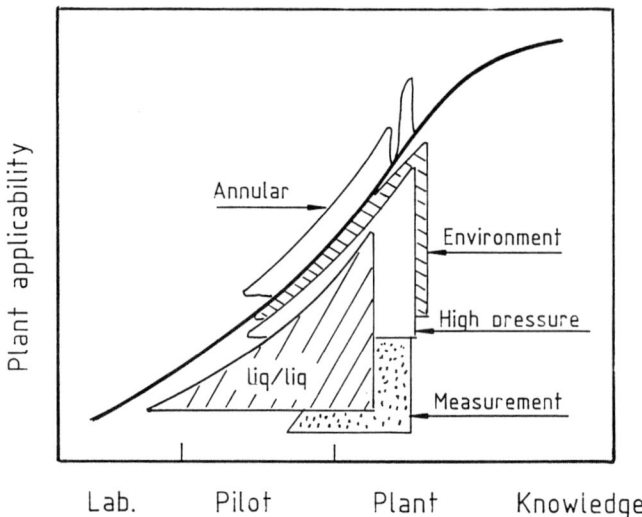

Figure 14 - Knowledge versus plant applicability and development
phases for some new projects at MEFOS.

A couple of comments to Figure 14: Due to safety reasons liquid-liquid has
to be run at first in small scale and gradually increased up to final plant
implementation. Most of the measurement technology will on the other hand be
developed on a lab scale and tested on plant scale due to the small impact
on plant conditions and the enormous amount of results that can be gained at
a low price.

428

Conclusions

Pilot plants should be used for:

- Development and demonstration of new technologies.
- Testing of extremes and projects with safety hazards.
- Clients worldwide at pilot plants with triple Q.
- Fast answers.

References

1. R.E. Johnstone, M.W. Thring: "Pilot Plants, Models and Scale-up Methods in Chemical Engineering", 1957, New York, McGraw Hill.

2. Baekeland: Ind. Eng. Chem. 1916, I, 177-190.

3. B. Berg, T. Lehner: "A Strategy for the Utilization of Pilot Plants", Scand. Journal of Metallurgy 10(1981), pp 99-107.

4. D. Geldart: Powder Technology 7 (1983), pp 285-292.

5. D. Bergman: "Dispenser Reduces Flue Dust Treatment", Steel Times vol 220 (No. 3), March 1992, p 6, 23.

6. E. Burström, G. Carlsson: "Alternative Tuyere Design - Prolonging the Life of the Tuyere", Scaninject IV, June 11-13 1986, Luleå, Sweden.

7. H. Gripenberg: "High Pressure Injection of Inert Gases Into Metal Melts", Int. Symp. on Injection in Process Metallurgy, New Orleans, 17-21 February 1991.

8. T. Lehner, G Lindkvist: "Injection Metallurgy for the Refining of Matte & Blister Copper", CIM Meeting 1984.

9. P. Ritakallio: "Desulphurization and Nitrogenization as Examples for Powder Injection", Scaninject I, June 9-10 1977, Luleå, Sweden.

10. T. Lehner, G. Carlsson "On Fluid Flow and Metallurgical Reactions in Gas Stirred Melts", Scaninject II, June 12-13, 1980, Luleå, Sweden.

11. E. Burström, G. Ye: "Refining of Impurities From Steel Using Ladle Metallurgical Methods", Scand. Journal of Metallurgy 20(1991), pp 126-134.

12. S. Ångström, L. Hedlund: Continuous Converting of Copper Matte to Blister Copper", Technological Advances in Metallurgy, September 20-21, 1988, MEFOS, Luleå, Sweden.

REACTION AND HEAT TRANSFER IN BATH SMELTING OF IRON

AND FERROALLOYS

Hiroyuki Katayama*, Tetsuharu Ibaraki**, and Masaki Fujita***

   * Technical Development Bureau, Nippon Steel Corporation
     20-1 Shintomi, Futtu, Chiba-ken, Japan, 299-12
  ** Sakai Works, Nippon Steel Corporation
     1 Yawata, Sakai, Osaka-fu, Japan, 590
 *** Takaoka Ferroalloy Works, Japan Metals & Chemicals
     1 Yoshihisa, Takaoka, Toyama-ken, 933

Synopsis

In bath smelting, in which a thick layer of slag separates an oxygen jet from the metal bath, consistency of high post combustion and the progress of reducing reaction is possible. As bath stirring was thought to be effective for accelerating reducing reaction and heat transfer, a top and bottom blowing converter was used. Bath smelting of ferrochrome, ferromanganese and iron was conducted with furnaces of various scale by changing bottom blowing condition.

(1) The appropriate intensity of stirring by bottom blowing is rather low (1-3 kW/t) for the following two reasons:
  (i) Under excessive stirring intensity of the metal bath by bottom blowing, the increased amount of metal droplets in the slag layer disturbs the separation of an oxygen jet and metal, and the decrease of post combustion and increase in dust generation occurs.
 (ii) The acceleration of reducing reaction and heat transfer is accomplished by the circulation of carbonaceous materials in the slag layer.

(2) In the tested composition range ([Cr]≦58%, Mn≦64%), the influence of metal composition on characteristics is rather small except for the rate of reduction. Therefore, this bath smelting with a thick layer of slag can be used commonly, both for the production of iron and for alloys of chromium and manganese.

Proceedings of the
Savard/Lee International Symposium on Bath Smelting
Edited by J. K. Brimacombe, P. J. Mackey,
G. J. W. Kor, C. Bickert and M. G. Ranade
The Minerals, Metals & Materials Society, 1992

Introduction

The smelting reduction process, in which an oxygen jet is separated from the metal bath with a thick layer of slag, is now under research.[1-13] Experiments were performed with a top-and-bottom blowing converter of various scale. The reason why the top-and-bottom blowing converter type furnace was chosen is that stirring by bottom bubbling was thought to be useful in accelerating reducing reduction and heat transfer.

In this paper, we will discuss the influence of stirring intensity on reaction and heat transfer, based on the experimental results from bath smelting of ferrochrome, ferromanganese, and iron.

Concept of smelting reduction process with a thick layer of slag

Fig. 1 is a schematic drawing of the smelting reduction furnace with a thick layer of slag. An oxygen jet is separated from the metal bath by a thick layer of slag. This state can be attained by the combination of (1) the suppression of vigorous slag foaming with carbonaceous materials, which makes it possible for a large amount of slag to exist stably in a furnace, and (2) soft blowing of oxgen gas through a lance with multi-flow nozzles. In this process, consistency of high post combustion, progress of reducing reaction, and suppression of dust generation are possible at the same time in a stirred bath.

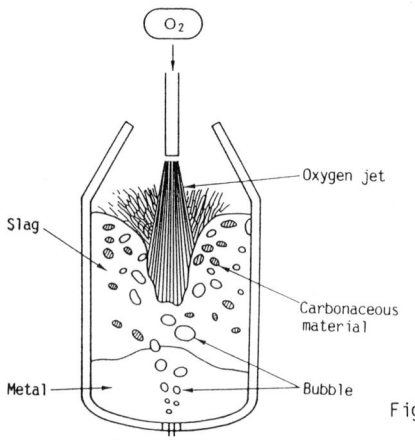

Fig. 1 Image of smelting reduction with thick layer of slag.

This process concept was obtained in research on the production of ferrochrome by smelting reduction of chromium ore[1], in which about 1000 kg of slag is formed for 1 t of ferrochrome, as chromium ore contains a large amount of gangue matter. It is possible to apply it in a case where the amount of gangue matter is not great, by leaving the necessary amount of slag from the preceding heat, without increasing the slag ratio.

Experimental method

Experimental furnaces of various scales are shown in Fig. 2 and the experimental condition in Table 1. With these furnaces, experiments in bath smelting of ferrochrome, ferromanganese, and iron were conducted step by step as shown in Fig. 3. In those experiments, 1 t and 5 t scale experiments of bath smelting of ferrochrome were performed by the Research Association for New Smelting Technologies and the 100 t scale experiment of bath smelting of iron was performed by the R & D task force of the Committee for R & D on the New Iron-ore Smelting Reduction Process, the Japan Iron and Steel Federation.

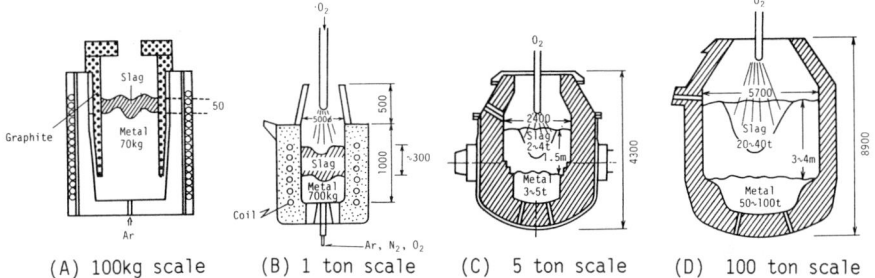

(A) 100kg scale   (B) 1 ton scale   (C) 5 ton scale   (D) 100 ton scale

Fig. 2 Profile of experimental converter.

Table 1 Experimental conditions.

Scale	1 ton	5 tons	100 tons
Oxygen blowing rate (top) (Nm³/h)	48	700~900	20000~30000
Bottom bubbling (Nm³/h)	Ar : 10.4 O_2 : 0.42	N_2 :100~150 O_2 : 0~300	· CO_2 :1800 · O_2-N_2, · N_2
Amount of slag (t)	0.2	2 ~ 4	20~40
Average thickness of slag (m)	0.3	1 ~ 2	3 ~ 4

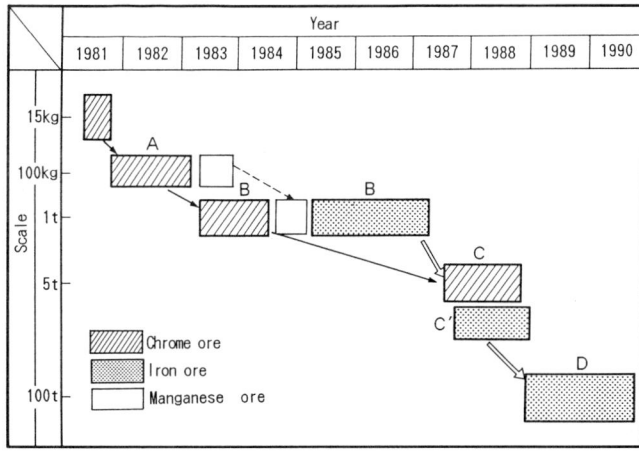

Fig. 3 Procedure of stepped up experiment for smelting reduction with thick layer of slag.

The materials used in the experiments are summarized in Table 2. There were two types of bottom blowing, oxygen gas blowing through an annular nozzle, and nitrogen gas blowing through a single nozzle. The intensity of stirring by bottom blowing gas was calculated with the equation presented by Mori et al.[14], on the assumption that 2CO was generated in oxygen gas blowing. The

433

main characteristics are the rate constant of reducing reaction, the rate of dust generation, post combustion, and heat efficiency. The definition of each characteristic is shown in Table 3.

Table 2 Composition of Raw materials.

· Chromium ore.

	T.Cr	T.Fe	CaO	SiO$_2$	Al$_2$O$_3$	MgO	P	S	C	$\frac{M.Cr}{T.Cr} \times 100$
Partially reduced pellet (standard) (ϕ8 ~12mm)	30.3	16.9	1.2	10.3	13.5	11.8	0.015	0.371	6.0	55.8
Fine ore (< 1mm)	23.4	13.1	1.1	8.1	9.8	9.6	0.017	0.292	15.5	—

· Manganese sintered ore.

T.Mn	Mn$^{4\cdot}$	T.Fe	CaO	SiO$_2$	Al$_2$O$_3$	MgO	P	S
52.4	14.2	5.3	9.3	5.9	5.7	1.2	0.084	0.090

· Iron ore.

T.Fe	SiO$_2$	Al$_2$O$_3$	CaO	MgO
66.8	3.00	0.51	0.05	0.01

· Carbonaceous materials.

	VM	Ash	FC	S
HVM Coal	36.5	8.3	55.2	0.50
MVM Coal	21.7	6.9	70.1	0.52
Coke	0.6	12.1	84.9	0.36

Table 3 Definitions of rate constant, post combustion and heat efficiency of post combustion.

Apparent rate constant of reduction	$R = A \cdot K \{(\%M) - (\%M)f\}$ (M:Fe, Mn, Cr) ········ (1) K: (kmol-O$_2$/(T.Fe)/m^2/min)
Post combustion	$PC = \dfrac{\{(\%CO_2) + (\%H_2O)\} \times 100(\%)}{(\%CO_2) + (\%CO) + (\%H_2O) + (\%H_2)}$
Heat efficiency of post combustion	$\eta_{PC} = [\ 1 - \dfrac{(\text{Super heat of exhaust gas})}{\text{Heat generated by post combustion}}\ ] \times 100(\%)$

Experimental results

Bath smelting of chromium ore

The experiments were performed by using furnaces A, B, and C in Fig. 2. In the experiment with furnace A, the influence of [%Cr] and the amount of carbonaceous material on the rate of reducing reaction was investigated, without oxygen blowing.[1] The apparent equilibrium: $(\%Cr)_f$ in eq.(1) is negligible, if carbonaceous material coexists. The apparent rate constant: k increases with the increasing amount of carbonaceous material (coke in this case) (Fig. 4)[16], and decreases with increasing [%Cr] up to 20%, while above 20% nearly constant (Fig. 5).

Also in the experiment with furnace B with oxygen blowing, $(\%Cr)_f$ is as low as less than 0.4%, if there is a sufficient amount of slag and carbonaceous material, independent of top-and-bottom blowing of oxygen and high post combustion (Fig. 6).

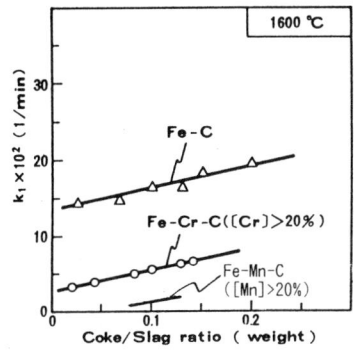

Fig. 4 Influence of coke/slag ratio on the rate constant : k_1

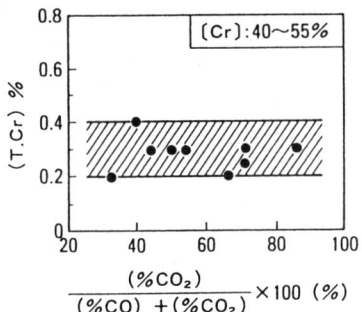

Fig. 6 Relation between (%T.Cr) and post-combustion ratio in 1 ton smelting reduction with thick layer of slag without chrome ore addition.

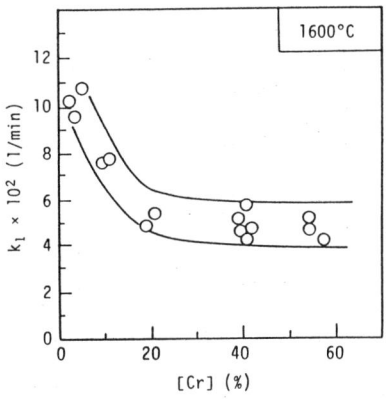

(a) Chromium reduction

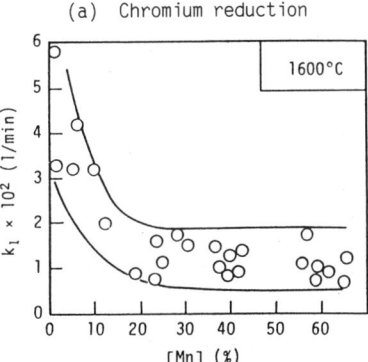

(b) Manganese reduction

Fig. 5 Influence of metal composition on the reduction rate constant of each oxide (Furnaces B in Fig. 3).

Influence of bottom blowing condition on various characteristics was investigated with furnace C (Table 4). In the standard bottom blowing condition (bottom blowing gas: 200 Nm^3/h, the stirring intensity by bottom blowing: 2-3 kW/t), stable operation with a thick layer of slag was possible. Next, bottom blowing was intensified by increasing bottom bubbling gas to 290 Nm^3/h. Then, the increase of slopping and dust formation, and the decrease of post combustion and metal yield were observed. Therefore, in order to obtain the best condition for bottom blowing, the stirring intensity was decreased step by step. When the amount of bottom blowing was decreased to 67 Nm^3/h (N$_2$), no trouble was observed and some characteristics were somewhat improved. When the bottom blowing gas was decreased to 42 Nm^3/h, the composition of exhaust gas became unstable. It means that materials added intermittently from the top of the furnace formed an island, which sometimes intercepted an oxygen jet.

Finally, an experiment without bottom bubbling was performed. Then, only

435

the part directly under an oxygen jet formed a molten bath, where vigorous dust generation continued, and no molten alloy was produced.

Production process of high carbon ferrochrome by bath smelting is composed of the cycle: smelting reduction period, finishing reduction period, and tapping. In the smelting reduction period. (%Cr) is in the range of 4% and 12%, which varies with operational conditions. In the finishing reduction period after the stop of chromium ore addition, (%Cr) is decreased to the appointed value. In tapping, some amount of slag and metal is left behind for the succeeding heat.

High carbon ferrochrome of 58%Cr and slag, in which (T.Cr) is less than 1%, were produced with ordinal chromium ore.[2] It is easy to produce it in the same manner as an alloy, in which [Cr] is less than 58%, by adding an iron source (e.g., pig iron).[15]

Table 4   Influence of stirring intensity.

	Standard	Strong	Weak 1	Weak 2	Weak 3	No bubbling
Amout of gas (blown from bottom)	200Nm³/h	290	110	67	42	0
Operation	stable	unstable	stable	stable	critical	impossible
Post combustion	50 − 70%	20 − 50	50 − 90	50 − 80	45 − 85	—
Dust generation (ratio)	1.0 (standard)	1.1	0.65	0.46	0.64	—

Bath smelting of manganese ore sinter

Behaviour of the content of manganese oxide in the slag was studied with furnace A.[3] The rate of reduction of manganese oxide was also expressed by the equation of first order reaction (1), but the apparent equilibrium (final) manganese content in the slag: $(T.Mn)_f$ is high (Fig. 7). Therefore, in the bath smelting of manganese ore, it is necessary to decrease $(T.Mn)_f$ by controlling slag composition and temperature. On the other hand, the influence on metal composition and the amount of carbonaceous materials on the rate constant was almost the same as in the case of the reduction of chromium ore (Figs. 4, 5).[4]

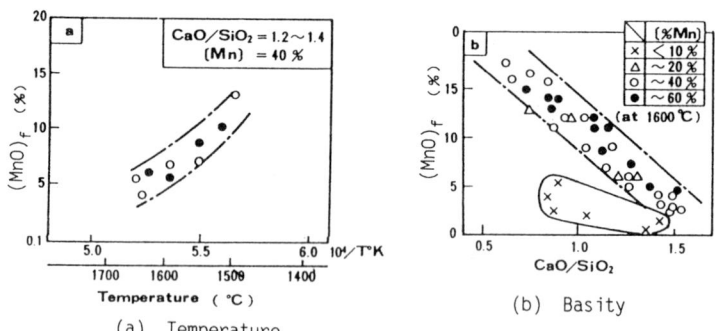

(a) Temperature

(b) Basity

Fig. 7   Influence of temperature and slag composition on apparent efuiblium : $(MnO)_f$.

In the experiment with furnace B with oxygen blowing, oxygen gas in the bottom blowing increases $(T.Mn)_f$ (Table 5), as well as dust generation (Fig. 8). This means that oxygen gas bottom blowing is harmful in the bath smelting of manganese ore. [%Mn] was increased to 64% in the experiment. $(T.Mn)_f$ and dust generation increased with the increasing [%Mn], but it is expected that this process can be used for the production of ordinary high carbon ferromanganese ([Mn]=75%) by adjusting operational conditions.

Table 5　Influence of oxygen bottom blowing on apparent equilibrium manganese content of slag ; $(Mn)_f$ in 1t scale experiment.

Bottom bubbling	$(Mn)_f$
Oxygen	7.5 %
Ar	3.0 %

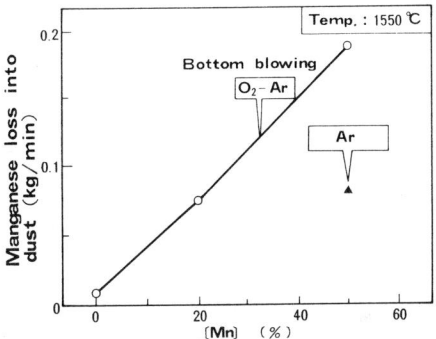

Fig. 8　Influence of [Mn] and bottom blowing condition on the manganese loss into dust (600kg scale).

Bath smelting of iron ore

Experiments were conducted with furnaces A, B, C' (similar but different from C) and D. The rate constant obtained in experiments with furnace A was higher than those of chromium and manganese alloys (Figs. 4, 5).

Based on the data obtained in experiments with furnaces B, C', D, the influence of the amount of slag per cross area on the rate constant is shown (Fig. 9).[5] The rate constant seems to increase with the amount of slag. The influence of the stirring intensity by bottom bubbling on post combustion and heat efficiency is shown in Fig. 10. Heat efficiency increases by increasing the stirring intensity, but it saturates above 3 kW/t. On the other hand, post combustion decreases smoothly with increasing stirring intensity.

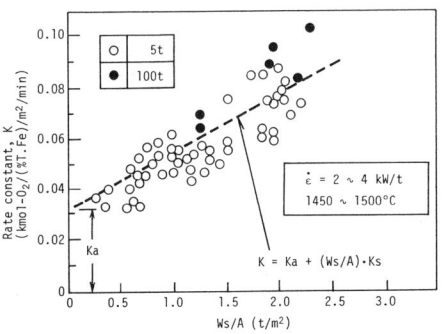

Fig. 9　Relationship between rate constant per cross area and the amount of slag.

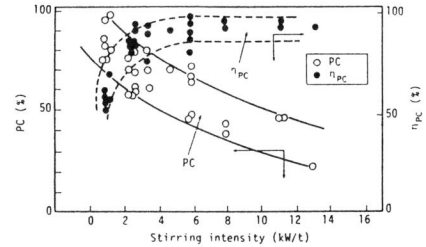

Fig.10　Effect of stirring intensity on post-combustion ratio and heat efficiency (5t experiment). (Carbonaceous material : coke)

437

Fig. 11 shows the data obtained in the experiment, where stirring intensity was changed in a wider range.[12] The relation between various characteristics and the intensity by bottom bubbling is clear, independent of injection of fine materials, and coincides with the trend shown in Fig. 10. When the stirring intensity is high, (T.Fe) level during the smelting reduction period decreases, but productivity decreases, because the heat generation decreases with the decrease of post combustion if the oxygen flow rate is the same.

The relation between coal consumption and bottom bubbling condition is shown in Fig. 12. When coal with high volatile matter is used, bottom bubbling of oxygen gas decreases post combustion and increases coal consumption.[6] Therefore, it is better to use an inert gas such as nitrogen for bottom bubbling rather than to use oxygen, particularly when the carbonaceous material used is coal with high volatile matter.

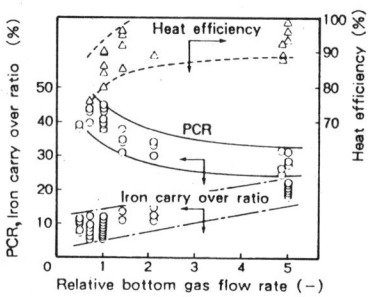

Fig.11 Operational properties of bottom injection.

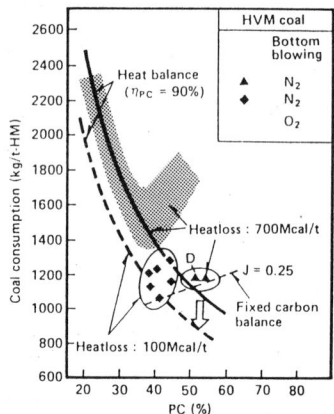

Fig.12 Influence of heat loss on coal consumption.

Consideration

Phenomena in excessive stirring. Fig. 13 represents the distribution of metal droplets in the slag layer. When the stirring intensity is moderate, the amount of metal droplets in the slag layer is about 10wt% or less except in the lower 30% of slag layer.[10] According to the tracer test, 90% or more of metal droplets come from the metal bath. When the stirring intensity is increased, the amount of metal droplets in the slag layer increases and finally a lump of molten metal is jetted into the atmosphere through the slag layer. When stirring by bottom bubbling is excessive, the contact of an oxygen jet or oxidizing atmosphere with metal droplet occurs. Then, reactions

$$O_2 + 2C \rightarrow 2CO \qquad (2)$$
$$CO_2 + \overline{C} \rightarrow 2CO \qquad (3)$$

lead to the decrease of post combustion and the increase of dust generation by the bursting of bubbles.

The influence of oxygen gas bottom bubbling. When oxygen is injected into the high carbon molten bath, oxidation of carbon chromium, manganese, or iron and the reduction of generated oxide by carbon in the bath occur. In the case of chromium alloy or iron, oxide can be reduced rapidly by carbon and the influence of oxygen gas bottom blowing on the oxide content in slag is small. But in the case of manganese alloy, apparent equilibrium: $(T.Mn)_f$

is high, so the oxygen gas bottom blowing increases (%T.Mn) in slag.

The reason why oxygen gas bottom blowing increases coal consumption when coal contains a great amount of volatile matter (Fig. 13) is as follows: Existence of solid carbon in slag is necessary for stable operation without vigorous slag foaming in bath smelting with a thick layer of slag. Fixed carbon in coal added in the furnace is consumed by the following three routes:
  (i) For reduction of oxide in slag
 (ii) For carburization of metal
(iii) Burned by oxidizing atmosphere.

Item (iii) contains indirect burning of solid carbon, that is, consumption of solid carbon for reduction and carburization of melt oxidized by atmosphere. After being consumed for (i) - (iii), solid carbon must be left to suppress vigorous slag foaming.[17]

Therefore, coal consumption is determined not only by heat balance, but also by solid carbon balance. When oxygen gas is injected into the melt, the fixed carbon balance line is lifted, as shown in Fig. 14.

In order to decrease coal consumption, high post combustion and high heat efficiency are desirable. In Fig. 14, it is possible to operate at point P temporarily, but fixed carbon remaining in the slag layer decreases gradually and finally vigorous slag foaming occurs. If the amount of carbonaceous materials supplied is increased, the operational point moves from P to Q. It means that the stable operation of minimum coal consumption is determined as the cross point of the heat balance line and fixed carbon balance line.

Therefore, oxygen gas bottom blowing decreases post combustion and increases coal consumption (from Q to R), when coal containing high volatile matter is used.

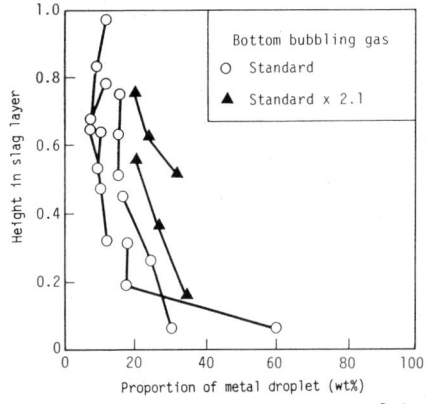

Fig.13 Distribution of metal droplets in slag layer.

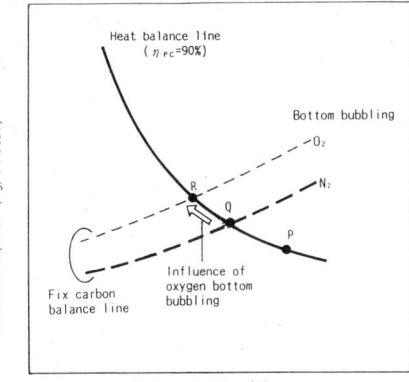

Fig.14 Schematic drawing of influence of oxygen gas liottom bubbling on coal consumption in smelting reduction with high volatile matter coal.

Mechanism of reducing reaction and heat transfer. The reduction site of oxide can be divided into the following:
(a) Interface between slag and metal bath
(b) Interface between slag and metal droplets

(c) Interface between slag and carbonaceous materials.

In the case of iron oxide, the amount of reduction in each site was calculated for the various conditions by using values on rate constant obtained
in experiments (Fig. 15).[5] In a small furnace experiment, the proportion of
reduction at the slag-metal bath interface is predominant. But the proportion of reduction at the slag-metal droplets interface and the slag-carbonaceous materials interface increases with the increasing experimental scale
and the amount of slag.

In order to increase the rate constant of reducing reaction, stress should
be given to an increase of the slag-carbonaceous materials interface, because an increase of metal droplets in the slag accompanies the bad effect
described.

Fig. 16 shows a sketch of an X-ray transmission image of the fundamental
experiment on the reaction site of manganese oxide in slag. Bubble generation by the reduction of manganese oxide by carbon can be observed at the
slag-metal bath interface when [Mn] of the bath is 7%. The amount of bubble
generation at the interface decreases with the increasing [%Mn]. Only a
small number of bubbles were observed at [%Mn]: 22%, and no bubbles at [Mn]:
50%.

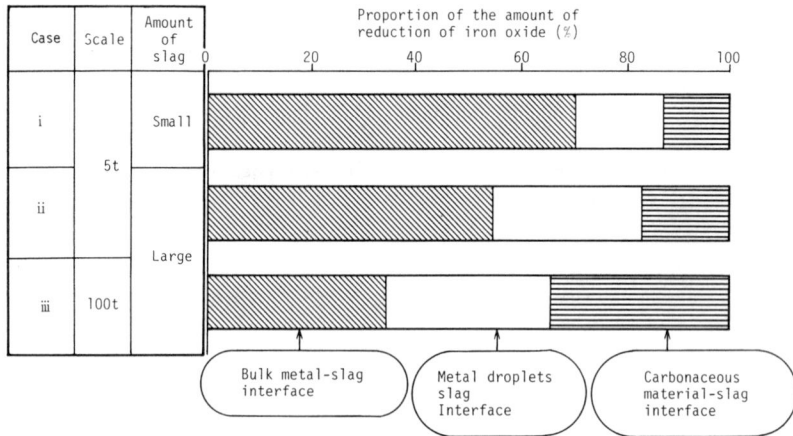

Fig.15 Calculated values of the amount of reduction
occurring at each site.

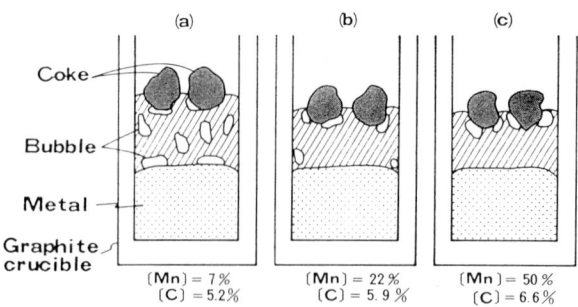

Fig.16 Sketch of X ray transmission image of reduction of manganese oxide.

440

This means that the amount of reduction at the slag-metal bath interface decreases with [%Mn] and in the bath smelting of manganese alloy, almost all the reduction occurs at the slag-carbonaceous material interface. It is possible to explain Fig. 5 by combining the idea of the three reaction sites described above and an observation of the X-ray transmission image.

Ordinary radiation and convection is thought to be predominant as the mechanism generated by post combustion in the furnace. A mathematical model was formulated to estimate the heat transfer by radiation and convection[18], and the results calculated were compared with the experimental results (Fig. 17). When exhaust gas temperature is higher than 2000°C (Case C), heat transfer can be explained by radiation. But this operational condition is not practical due to the low heat efficiency. On the other hand, when heat efficiency is as high as 90% (gas temperature: 1705°-1765°C: Case A, B), the calculated results of heat transfer by radiation and convection is only 20-30% of real heat transfer. Therefore, another mechanism, as well as radiation and convection, must be taken into account.

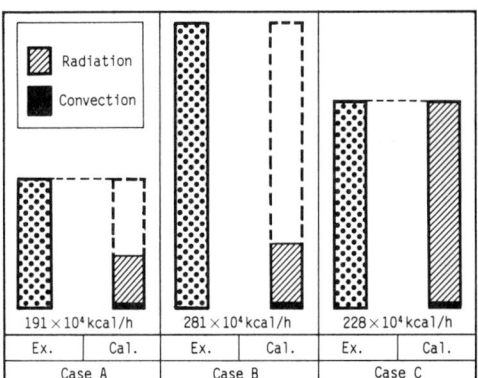

	Case A	Case B	Case C
Coal	Coal	Coke	Coal
Post combustion (%)	30%	55%	60%
Temerature of exhaust gas (°C)	1707	1765	2171

Ex. : Necessary heat transfer in experiment
Cal. : Calculated results

Fig.17 Comparison of calculated heat transfer amounts with experimental amounts.

The heat transfer by circulation of carbonaceous materials within the slag layer was considered by assuming the following cycle: (1) carbonaceous materials are rolled up in the stirred slag, (2) bubbles attach to the surface of the carbonaceous material in the slag layer, and then (3) carbonaceous materials float on the slag layer and loose bubbles.

In this cycle, if super heated carbonaceous materials react with oxide in slag, heat is effectively transferred. As this mechanism concides with various phenomena[18], it is possible to consider that 70-80% of total heat transfer is caused by the circulation of carbonaceous materials when gas temperature is lower than 1765°C.

Therefore, in the practical condition of bath smelting, the role of carbonaceous materials rolled up in slag in reducing reaction and heat transfer should be stressed.

Necessary intensity of stirring and appropriate operational condition. The ideal condition in bath smelting with a thick layer of slag is (1) the slag layer is nearly uniform by stirring, (2) carbonaceous materials circulate

441

actively by being rolled up in the slag layer, and (3) the amount of metal droplets in the slag where an oxygen jet invades is small. An appropriate intensity of bath stirring by bottom bubbling was 1-3 kW/t-metal. If the amount of slag increases, the density of metal droplets can be decreased, even if the amount of formed metal droplets is the same. This condition will improve various characteristics.

As separation of an oxygen jet and metal is essential in this process, excessive bottom bubbling, particularly, oxygen gas bottom bubbling, is harmful. In the HISMELT process, in which high post combustion is also the target, the stirring condition is different from that of this process. It is expected that there will be discussion on an appropriate bottom bubbling condition by considering the differences in operational conditions (e.g., oxygen gas and hot air).

Summary

Various scale experiments (100 kg-100 t) of bath smelting with a thick layer of slag were conducted with a top-and-bottom blowing furnace for producing ferrochrome, ferromanganese, and pig iron. The influence of stirring by bottom bubbling and of metal composition on reducing reaction and heat transfer is summerized as follows:
(1) When bottom bubbling is too strong, a decrease in post combustion and metal yield occurs. On the other hand, when bottom bubbling is too weak, ununiformity becomes a problem. The appropriate stirring intensity by bottom bubbling was 1-3 kW/t, independent of the experimental scale.
(2) Too strong bottom bubbling increases the contact between the oxygen jet and metal by increasing the amount of metal droplets in the slag layer, and disturbs the essential condition of this process.
(3) Though strong stirring increases the rate constant of reducing reduction, it decreases productivity due to a decrease in post combustion, because the amount of reduction depends on heat supply.
(4) Reducing reduction at the slag-metal (bath and droplets) interface depends on metal composition. The proportion of reduction at the slag-carbonaceous materials interface increases with the increasing experimental scale. When %Cr or %Mn is higher than 20%, the reducing reaction at the slag-carbonaceous materials interface becomes predominant and the rate constant becomes almost constant.
(5) 70-80% of heat transferred at high heat efficiency operation is explained by the circulation of carbonaceous materials.
(6) The ideal condition is the coexistence of a uniform slag layer, active circulation of carbonaceous materials, and the small amount of metal droplets in the slag layer. An increase in the slag amount will improve various characteristics by decreasing the density of metal droplets.

References

1) M. Fujita et al., "Smelting reduction of chrome ore pellet in stirred bath, " Tetsu-to-Hagane, 74 (1988), 680-687.
2) H. Katayama et al., Production of ferrochrome by smelting reduction with top-and-bottom blowing converters," Proc. of International Conference on New Smelting Reduction and Near Net Shape Casting Technologies for Steel, Pohang (1990), 297-306.
3) M. Fujita et al., "Smelting reduction of manganese sintered ore in stirred bath," Tetsu-to-Hagane, 74 (1988), 816-624.
4) H. Katayama et al., "Production of ferroalloys by smelting reduction," Proc. of Shenyang International Symposium on Smelting Reduction, Shenyang (1986), 290-302.

5) M. Matsuo et al., "Smelting reduction of iron ore with top-and-bottom blowing converter," Tetsu-to-Hagane, 76 (1990), 1871-1878.
6) M. Matsuo et al., "Relationship between post-combustion, heat efficiency and coal consumption in smelting reduction of iron ore with top-and-bottom blowing converter," Tetsu-to-Hagane, 76 (1990), 1879-1886.
7) M. Yamauchi et al., "Operation results of in-bath smelting reduction in the 100t furnace," CAMP-ISIJ, 3 (1990), 1074.
8) M. Yamauchi et al., "Phenomena of slag in the in-bath smelting reduction furnace," CAMP-ISIJ, 3 (1990), 1075.
9) T. Ibaraki et al., "Scale-up rules of in-bath smelting process," CAMP-ISIJ, 3 (1990), 1076.
10) T. Ohno et al., "Mechanism of iron dust formation during smelting reduction," CAMP-ISIJ, 4 (1991), 1170.
11) M. Yamauchi et al., Method of ore addition into the smelting reduction furnace," CAMP-ISIJ, 4 (1991), 1171.
12) M. Yamauchi et al., "Bottom injection of iron ore and coal in the in-bath smelting reduction furnace," CAMP-ISIJ, 4 (1991), 1172.
13) T. Ibaraki et al., "Development of smelting reduction of iron ore," I & SM (12) (1990), 30-37.
14) K. Mori et al., "Process kinetics in injection metallurgy," Tetsu-to-Hagane, 67 (1981), 672-695.
15) S. Kitamura et al., "Quantitative Estimation about the influence of various factors on the smelting reduction rate of Cr-ore by top-and-bottom blowing converter," Tetsu-to-Hagane, 74 (1988), 672-679.
16) M. Matsuo et al., "Comparison of smelting reduction of iron ore and chrome ore with 1t top-and-bottom blowing converter," Tetsu-to-Hagane, 72 (1986), 970.
17) Y. Ogawa et al., "Slag foaming in smelting reduction and its control with carbonaceous materials," ISIJ International, 32 (1991), 87-94.
18) T. Kawamura et al., "Mechanism of heat transfer in smelting reduction with a thick layer of slag," Tetsu-to-Hagane, (1992), 367.

443

On the Formation of Thermal Accretions ('Mushrooms') in Steelmaking Vessels

R.I.L. Guthrie[*], H.C. Lee[*], Y. Sahai[**]

[*] McGill Metals Processing Centre
Department of Mining and Metallurgical Engineering
McGill University, 3450 University Street, Montreal, Canada H3A 2A7

[**] Department of Materials Science and Engineering
Ohio State University, Columbus Ohio, USA 43210-1179

Abstract:

The Savard-Lee shrouded tuyere technology for oxygen injection relies on the protection of the underlying refractory from rapid chemical erosion at the oxygen jet-refractory interface. The inventors accomplished this task by causing an accretion (mushroom) of solid steel to form and thereby prevent FeO fluxing of adjacent nozzle refractory materials. A simple heat balance model is given to describe the formation and growth of these solid accretions. It is shown that the endothermic cracking of hydrocarbon gases together with heating of gaseous hydrogen to steel bath temperatures provides cooling rates compatible with estimated convective heat input rates from the steel bath. Furthermore, the thermal efficiency of their original process is estimated to be 60 percent, based on data deriving from initial tests at U.S. Steel on Q-BOP heats.

Low temperature experiments in aqueous systems demonstrated that the passage of cold gases (vaporized liquid nitrogen) could result in the formation of ice accretions that are similar in morphology, and sometimes aspect ratio, to those observed in actual steel melts.

Subsequent to that work, laboratory experiments at McGill University, involving the injection of carbon dioxide into steel melts were carried out. Again, the results confirmed that carbon dioxide can also be a useful gas for top and bottom blowing operations, in that the endothermic cracking of carbon dioxide in the presence of dissolved carbon in a steel melt can again lead to the formation of a protective accretion. Furthermore, the reaction

$$CO_2 + \underline{C} \rightarrow 2CO$$

doubles the gas flowrate through the melt, at the same time as enhancing carbon removal, in such processes as the STB (Sumitomo Top and Bottom Blowing) Operation.

Proceedings of the
Savard/Lee International Symposium on Bath Smelting
Edited by J. K. Brimacombe, P. J. Mackey,
G. J. W. Kor, C. Bickert and M. G. Ranade
The Minerals, Metals & Materials Society, 1992

Introduction

Robert Lee and Guy Savard made a unique and fundamental contribution to the world's steel industry with their demonstration of the 'shrouded tuyere.' This tuyere makes it possible to blow a pure oxygen jet into a bath of molten steel without eroding refractory adjacent to the penetrating submerged jets of oxygen.

Submerged gas injection into molten iron and steel has been of great interest ever since the early days of Bessemer's bottom blown steelmaking operations in the 1850's when air was blown through an array of nozzles placed in the bottom surface of a refractory lined vessel. However, one of the problems of Bessemer steel (apart from sulphur and phosphorus) was the excessive nitrogen pickup from the submerged air blast. Consequently, despite its attractive kinetic features, and the introduction of the Thomas or Basic Bessemer converter for high phosphorus European operations, the process was eventually superseded by the versatile, but slower, Open Hearth Process, first demonstrated by Siemens in about 1865[1].

By the late 1940's - early 1950's, the Open Hearth process had become dominant and Bessemer 'teapot' reactors were virtually extinct in North America. Nonetheless, experiments continuing in Germany on Basic Bessemer, or Thomas, type vessels showed that while oxygen injected through the tuyeres led to excessive refractory erosion around the nozzle, oxygen could be used through a top blown lance to produce satisfactory grades of steel at extremely competitive rates of production *vis à vis* the Open Hearth process. Dofasco was the first company to licence and introduce this new technology, known as the LD (Linz-Donovitz) Process in Europe, into North America, with the pioneering efforts of John Francis McMulkin in 1955-56, using two ladles welded together for the pilot scale furnace. What he, and others, had not anticipated were the copious amounts of iron oxide fume that formed during the refining as compared to the regular Open Hearth operations. Dismay on the part of the citizenry of Hamilton at the fine coating of red dust appearing every morning following their nightly trials, led to Dofasco's rapid introduction of bag-houses for the new process![2]

In Sydney, Nova Scotia, our two celebrity inventors, who were in the gas business with *Air Liquide*, were trying to solve a similar fuming problem in a steel plant that needed to lower the level of silicon in a hot metal by introducing oxygen into the transfer ladle through a submerged tuyere rather than overhead lances.

To protect the refractory in the vicinity of a submerged oxygen jet, two approaches proved fruitful. The first method, studied by Lee and Savard[3], was to inject the oxygen under very high pressure into the melt and to rely upon the Joule-Thomson cooling effect resulting from adiabatic gas expansion to 'neutralize' the exothermic reaction between oxygen and iron, and subsequent fluxing of adjacent refractory. The second method[4], leading to refractory protection by an accretion of solid steel, was to introduce a coolant gas into a concentric nozzle surrounding the oxygen nozzle.

Unfortunately for Savard and Lee (and steelmakers in general), Dofasco and many other North American steel plants were in engaged in the further development of Top Blown Processes, and their concepts fell on infertile ground. Their French patent was nearing expiration before further work in Germany on their shrouded tuyere technology led U.S. Steel to license the technology for the United States, under the acronym Q-BOP steelmaking. Figure *1*, shows a cross section of the injector, and a metallic mushroom, for an early 200 ton Q-BOP operation. As pictured, the annular spacing between the inner and outer tubes is very small,

Figure 1 (top)
Section of a thermal accretion formed in a 200 ton Q-BOP vessel

Figure 2 (right)
Apparatus used for vertical gas injection

Figure 3 (bottom)
Two types of Nozzle used; for dimensions, see Table I

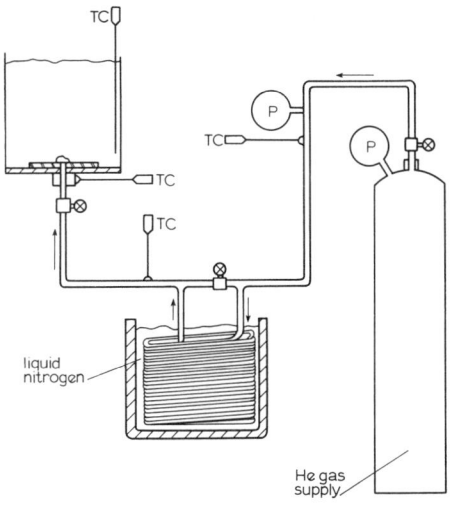

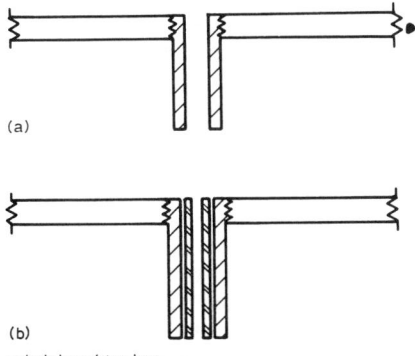

(a)

(b)

a single bore; *b* two bore

measuring about 1.5 mm. The great advantage of the second solution was the availability of oxygen at low (~ 1,000 kPag), versus high, pressures (~ 8,000 kPag).

When this early sample of a Q-BOP 'mushroom' first appeared at McGill University in about 1980 for Professor Williams to analyze metallurgically, one of the authors (R. Guthrie) was invited by the inventors to provide a decent 'academic' explanation for its appearance.

Believing that this invention could be explained in terms of thermal phenomena, a Master's student, by the name of Leo Matikainen, was set to work on a low temperature analogue system. Thus, in order to simulate the injection of a cold gas into a liquid metal and to observe possible phenomena at work, it was decided that a low-temperature aqueous equivalent might provide a simple, but effective, analogue. Consequently, cold (helium) gas was injected through bottom- and side-blown orifices into water, and freezing phenomena taking place around the penetrating jet were observed.

Figure 2 shows the equipment constructed for this purpose. Helium gas passed through a helical wound copper pipe immersed in liquid nitrogen. This pipe then delivered cooled gas to one of the two nozzle configurations shown in Figures 3a and b. Figure 3a shows a single-bore nozzle, 3-6 mm dia., into which cold gas was passed at flowrates between 0.3 and 1.5 litres/sec. Figure 3b shows the second configuration, in which cold gas could be injected through an outer annular bore while warmer gas passed up through the central bore. The glass tank, filled with water to a height of 0.3 m, was 0.3 m square. A mass flow-meter (Turbine flow-meter, Cox Instruments), provided information on gas flowrates from the nozzle or orifice, while thermocouples, suitably placed, monitored bulk-gas and water temperatures.

Ice Accretions

Single Bore and two-bore nozzles

Figure 4 shows a series of typical plaster replicas of ice accretions formed around a single nozzle during the course of 20 min gas injection. These plaster replicas were produced by pouring molten wax around the frozen ice accretions, emptying the water from the cavity formed, filling the cavity with plaster of Paris, and finally melting away the surrounding wax, the work being performed by Prof. C. Xu, on sabbatical leave from Chongquing University.

Figure 5 shows a typical series of accretions formed around a single nozzle after 5, 10, 15, and 20 min of gas injection. The shapes of these accretions differ from the tall conical shapes obtained by Boxall et al.,[5] who used a very similar cold-gas technique; their taller shapes resulting from the use of a higher gas nozzle velocity.

In the second series of experiments, the concentric two-bore nozzle was used in an effort to simulate, physically, the type of accretion formation occurring when a combination-type tuyere, such as the Savard-Lee oxygen injector, is used. Figure 6 illustrates typical accretion shapes associated with the concentric two-bore nozzle. As seen, there was a marked difference in the physical characteristics and effects of accretions formed with the single and concentric nozzles; for the former, the ice slowly grew up, around, and then over the nozzle entrance, producing the icy equivalent of a porous sponge. This 'sponge' grew during the course of an experiment, disrupting the jet's coherence and forward momentum and caused it to act more in the manner of a gas sparger or distributor. This is shown in Figure 7, which shows gas entering the water through many holes.

a   b   c   d

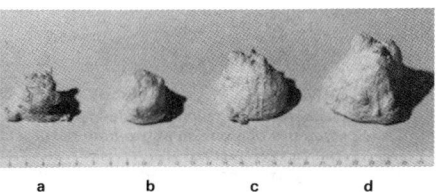

a   b   c   d

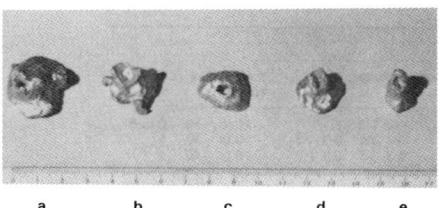

a   b   c   d   e

a   b

Figure 4   Plaster replicas of thermal accretions formed at single nozzle; various water temperatures, gas temperature -40 ± 5°C, gas flowrate 0.9 litres/sec., time of gas flow 20 min; scale in cm; *a 6°C; b 4°C; c 2°C; d 0°C*

Figure 5   Plaster replicas of thermal accretions formed at single nozzle; various times of gas flow, water temperature 0°C, gas temperature -40 ±5°C, gas flowrate 0.9 litres/sec.; scale in cm; *a 5 min; b 10 min; c 15 min; d 20 min*

Figure 6   Plaster replicas of thermal accretions formed at two-bore nozzle; various air/gas flowrate ratios, water temperature 1°C, gas flowrate 0.3 litres/sec., time of gas flow 20 min; scale in cm; *a 1.2; b 3.0; c 3.4; d 3.5; e 4.0*

Figure 7   Gas flow through typical accretions at single nozzle; flow is significantly reduced through large accretion to illustrate that gas is entering water through many holes *a small accretion; b large accretion*

Figure 8 (bottom, left)   S e c t i o n s through plaster replicas of thermal accretions formed at two-bore nozzle; different combinations of gas flowrate and air/gas flowrate ratio, water temperature 1°C, time of gas flow 20 min; scale in cm; *a,b 0.6 litres/sec, 1.5; c 0.3 litres/sec, 3.0*

Figure 9 (bottom, right)   Plaster replicas of thermal accretions formed at two-bore nozzle; various air/gas flowrate ratios, water temperature 1°C, gas blown *horizontally* at flowrate of 0.3 litres/sec., time of gas flow 20 min; scale in cm; *a 0; b 1.0; c 2.0; d 3.0*

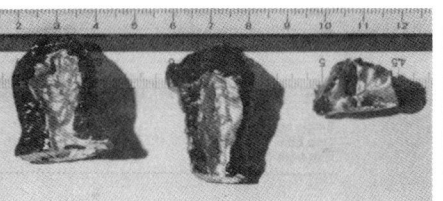

a   b   c

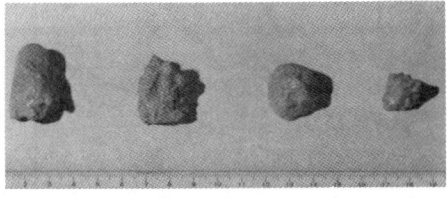

a   b   c   d

449

In the two-bore concentric nozzle, however, the jet's integrity and forward (vertical) penetration remained more in evidence, as ice slowly grew over the annular gas exit but not around the warm central jet of air. This is shown in Figure 8, where the core region of the accretions remains hollow. The figure shows the inside surfaces of accretion replicas sectioned vertically. The walls of the accretion, which are seen to be relatively thick, were in fact slightly porous.

Side-blown vessels

Figure 9 illustrates a series of typical accretions which grew around a jet directed horizontally into water. It is interesting to observe that the shape of the accretion reflected the trajectory followed by the penetrating jet, which bent sharply upward toward the free surface. The mode of formation was interesting and resembled that of bottom-blown accretions from single nozzles. A conical shape of high aspect ratio first formed around the tuyeres, reminiscent of the shapes described by Boxall et al.[5] . This trumpet-shaped tube bent upwards within the first two minutes of blowing, and then divided into many tree-like branches. Gas flowing through these icy branches caused them to thicken, until a large, hemispherical porous mass of ice was formed. Suitable conditions of gas flow and nozzle design allowed the accretion shape to be controlled, but a discussion of this is beyond the scope of the present paper.

For ease of reference, Table I gives a summary of the range of experimental conditions studied (nozzle sizes and type, and flowrates and temperatures of the gases and water), including data for Figures 4-9.

Table I   Summary of Experimental Values

Nozzle Type	O.D., mm	I.D. of inner bore, mm	Annular width of outer bore, mm	Flow rate, He, l/sec.	Flow rate, Air, l/sec.	Temp., °C, He	Temp., °C, Water
Single Bore	3-6	...	...	0.3-1.2	...	-10 to -70	0 to 6
Two Bore	...	2.5-3.0	0.4-1.0	0.3-0.6	0.3-1.2	-10 to -40	1
Two Bore (side)	...	3.0	0.4	0.3-0.6	0.3-1.2	-10 to -40	1

As well as recording the physical characteristics of these accretions, their masses were measured as a function of relevant parameters.

Figure 10 shows how the mass of the accretion increased with time during the course of gas injection. The increases were found to be linear, the slopes depending of the bath

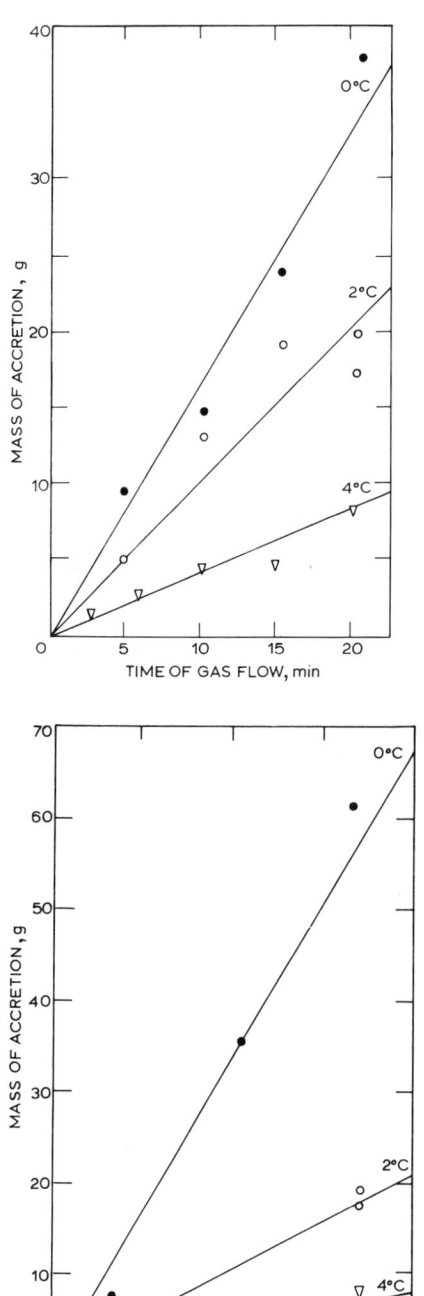

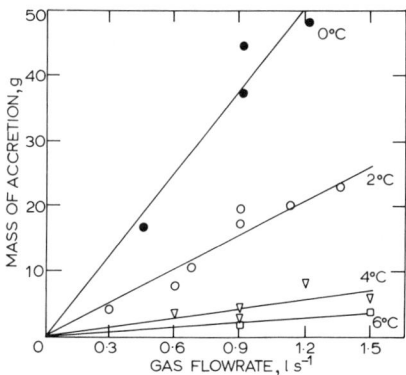

Figure 10 (top, left)

Variation of mass of ice accretions formed at single nozzle with time of gas flow, for different water temperatures; gas temperature -50±10°C, gas flowrate 0.9 litres/sec

Figure 11 (below, right)

Variation of mass of ice accretions formed at single nozzle with gas flowrate, for different water temperatures; gas temperature -40±5°C, time of gas flow 20 min

Figure 12 (bottom left)

Variation of mass of ice accretions formed at single nozzle with gas temperature for different water temperatures; gas flowrate 0.9 litres/sec., time of gas flow 20 min

451

superheat temperature. Figure *11* shows that the mass of the accretion increased linearly with increasing gas flow, the slope again depending on bath superheat conditions.

Figure *12* shows that the mass of the accretion also increased linearly with gas 'supercool' temperature, i.e. the difference between the temperature of the incoming cold gas and the freezing point of the accretion.

One may therefore conclude from these studies that the mass m of accretion formed is directly proportional to time of injection, t, gas flowrate, Q, and gas 'supercool' temperature, $\Delta\theta_G$, and inversely proportional to liquid superheat temperature, $\Delta\theta_L$ as one might have reasonably expected. Similarly, the linear plots of m versus t showed that conditions were far from thermal equilibrium in the system studied.

Structure and Chemical Composition of Steel Accretions

Figure *13* shows polished and mounted sections of the accretion shown in Figure *1*. This was chemically analyzed for carbon and manganese at the marks indicated on the sample. The accretions were found to consist essentially of partially refined hot metal with a gradation in carbon content. The carbon content at the base of the accretion was about 2.5 percent C, dropping to 0.1 percent C near the outer rim. The manganese content varied in a random manner from approximately 0.2 to 0.4 percent Mn.

Figure *14* shows some early results from the Nippon Steel Corporation where, in the work of Okohira et al.,[6] refractory faced concentric tuyeres were dipped below the surface of a bath of molten steel during steelmaking operations in a 1 ton electric arc furnace to evaluate the cooling ability of several kinds of coolants. This figure shows the result of blowing carbon dioxide into a low-carbon bath of steel at $1700 \pm 20°C$ for three minutes via an annular tuyere. A flowrate of 300 Nlitres/min through the inner pipe ($d_i = 4$ mm, $d_o = 7$ mm) and 429 Nlitres/min through the outer pipe ($d_i = 9$ mm, $d_o = 12$ mm), resulted in the hemispherical accretion shown. Figure *14a* provides a side view of the frozen mass of steel, which is somewhat similar to a sponge-like structure, and through which gas must bubble or jet through a large number of small holes. Figure *14b* shows a vertical cross-section through the accretion in which aqueous picric acid solution has been used as an etchant to reveal the microstructure of the accretion. It is interesting to note the layered, onion-like structure of this example, suggesting discrete, but incremental, size changes with time.

Similar work was carried out at McGill in order to explore the potential use of carbon dioxide as an alternative to (expensive) argon for post blowing practices in combination blown vessels, or for bubbling in ladles containing aluminum killed steels. Figure *15* illustrates some typical results when a 'Masrock' lance, $d_i = 0.8$ cm and $d_o = 4$ cm, was immersed into a covered inductively stirred melt of low carbon steel of about 30 Kg mass, and carbon dioxide blown at a rate of 50 Nlitres/min., into the bath which was maintained at a superheat temperature of 10°C. One will observe that the dissolved aluminum burned off rapidly following CO_2 introduction, presumably according to the highly exothermic reaction:

$$2\underline{Al} + 3CO_2 \rightarrow Al_2O_3 + 3CO \qquad \Delta H^\circ_{298} = -200 \text{ Kcal/mole} \qquad (1)$$

However, this did not prevent the formation of an accretion through the strongly endothermic reaction of CO_2 and dissolved carbon, until a lower limiting carbon value of 0.018 wt percent was reached. At carbon levels below this, one must conclude that CO_2 reacted with molten iron according to

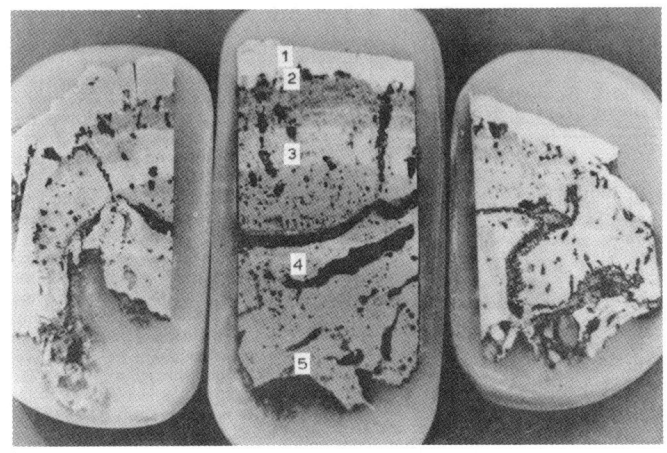

Figure 13

Polished section of thermal accretion formed in a 200 ton Q-BOP vessel

a

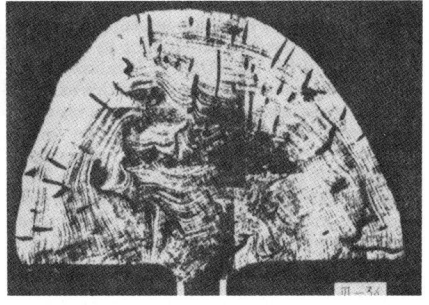

b

Figure 14

Thermal accretions formed in electric furnace steelmaking using carbon dioxide in both central and annular tuyeres (Nippon Steel Corporation)

$$CO_2 + Fe \rightarrow FeO + CO \qquad \Delta H^o_{298} = 1.5 \text{ Kcal/mole} \qquad (2)$$

even though thermodynamic calculations suggest carbon levels as low as 0.001 wt percent C should be possible before iron is oxidized preferentially to dissolved carbon. Above this threshold, much of the CO_2 reacted according to

$$CO_2 + \underline{C} \rightarrow 2CO \qquad \Delta H^o_{298} = +41.2 \text{ Kcal/mole} \qquad (3)$$

The fact that accretions could be formed in the presence of high levels of dissolved aluminum, suggests that the reaction between carbon dioxide and aluminum takes place away from the accretion.

From such experimental investigations, it is evident that 'mushrooms,' or accretions, can be formed when either reducing gases (e.g. CH_4), neutral gases (e.g. Ar), or mildly oxidizing gases (e.g. CO_2) are blown into steel at relatively high flow rates and/or low superheats. Furthermore, the accretions are essentially metallic in nature and can grow, or shrink, during the course of a heat.

Thermal Model of Steel Accretions

On the basis of this various information, gathered from water model, pilot-scale, and plant experiments, a simple thermal analysis of an accretion's growth can be carried out, in which one argues that the heat absorbed by the protective gas in heating, and cracking is set equal to the heat supplied by the hot metal and to the heat required for solidification of the metal. Thus, referring to Figure 16,

$$\varrho \, Q \, \Delta H = h \, A_M \, (\theta^B_L - \theta^*_L) + \dot{m}L \qquad (4)$$

where ΔH represents the net enthalpy change per unit mass of hydrocarbon or other protective gas, Q is the gas flow rate, h a convective heat transfer coefficient from the melt, L the latent heat and $\dot{m}$ the mass growth rate of accretion.

Initially, when the accretion/liquid metal surface area, A_M, is very small, convective heat transfer from liquid metal to accretion is small, and the cooling caused by the protective gas will result in freezing more metal. Later, towards the end of the blow, steady-state conditions are achieved, that is, zero growth and $\dot{m} = 0$. Equation (4) can then be rewritten as

$$\varrho \, Q \, \Delta H = h \, A_M \, (\theta^B_L - \theta^*_L) \qquad (5)$$

For liquid metals, h can be calculated from the relationship

$$Nu = (RePr/\pi)^{\frac{1}{2}} \qquad (6)$$

in which R_M, the characteristic length dimension, is taken over the radial surface width of an accretion, and V, the mean velocity of steel flow over the mushroom, is estimated on the basis of previous work[7]. The value of h was thereby calculated to be $0.98 \pm 0.15 \text{ cal cm}^{-2}\text{s}^{-1}\,^o\text{K}^{-1}$.

This analysis ignores the cooling of adjacent refractory by gases ascending the tuyere pipes, and treats the mushroom/refractory interface as being adiabatic. However, the latter represents a reasonable first approximation in view of the insulating properties of refractories in comparison to liquid metals.

To test this simple analysis, consider the specific conditions relating to the mushroom formed in Figure 1, in which the oxygen flow rate per tuyere was $\sim 32 \text{ Nm}^3/\text{min}$. Since the

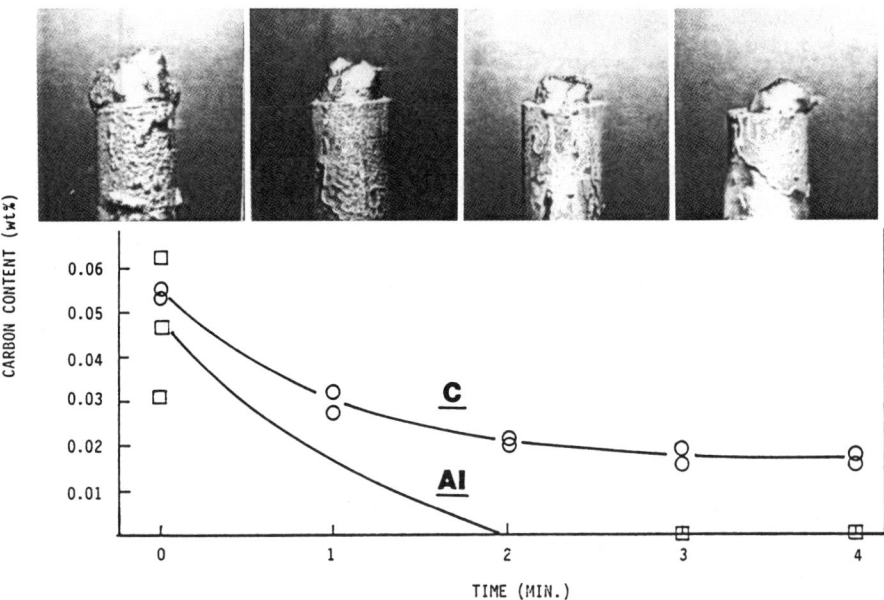

Figure 15 (above)

Variation in dissolved carbon and total aluminum levels in a low carbon melt (0.06 wt%) versus CO_2 bubbling time (min.s). Low superheat.

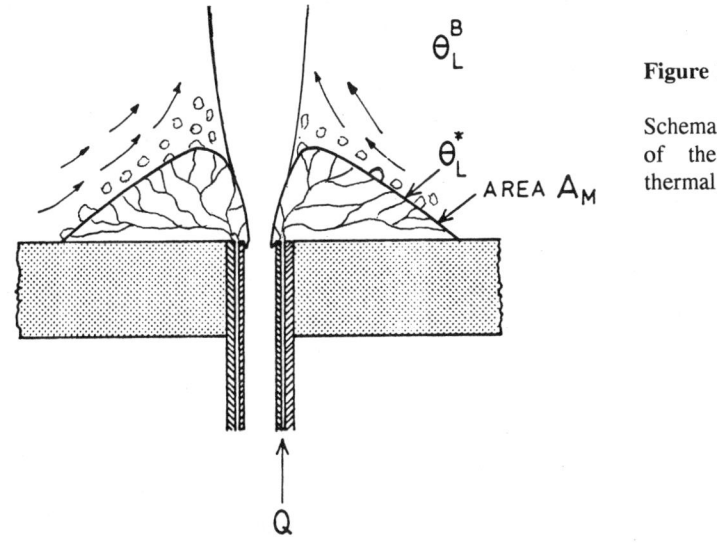

Figure 16 (left)

Schematic representation of the section of a thermal accretion

natural gas flow was 8 percent that of the oxygen, the net enthalpy change in cracking CH_4 to $C + H_2$ and heating the products to accretion temperature would have been 88 Kcal/second.

From equation 2, we may write

$$Q_{cooling} = h \, A_M \, (\theta_L^B - \theta_L^*) = 0.98 \times 460 \, (1620 - 1500) \rightarrow 54 \; Kcal \; s^{-1} \qquad (7)$$

From this simple analysis it can be concluded that a balance between heating and cooling factors was established, wherein approximately 60 percent of the hydrocarbon's maximum cooling capabilities were available for freezing a protective accretion of steel around the tuyere. This would seem reasonable since another function of the hydrocarbon is to burn with the outer rim of the oxygen jet, so as to shield the steel cavity close to the injector from direct contact with oxygen. Furthermore, some of the hydrocarbon may not fully pyrolyse within the mushrooms.

Predictions for Industrial Applications of Protective Accretions

Considering the order of magnitude agreement, it is worthwhile to study those industrial conditions needed for obtaining thermal accretions around submerged gas jets in steelmaking operations. Table II shows the cooling that can be obtained as a result of the complete endothermic dissociation of the hydrocarbon, together with the sensible heat requirements for raising the products to an average accretion temperature of 1623 K (1350°C). Also shown are the heat requirements for argon, water, nitrogen and carbon dioxide. In typical Q-BOP operations, however, only natural gas, propane and butane are used.

Assuming that 60 percent of the maximum cooling capabilities of these gases are available for freezing a protective accretion around a tuyere, the cooling that can be produced by natural gas, propane, carbon dioxide and butane as a function of oxygen flow rate (Nm^3/min) is as shown in Figure 17 In practice, it is normal to quote the protective gas flow rate as a percentage of the oxygen flow. Taking a typical case of 32 Nm^3/min/tuyere of oxygen, a second abscissa scale can be drawn showing the percentage cooling gas. We see from Figure 17, therefore, that an 8-10 percent flow of natural gas could be replaced by a 3-5 percent flow of butane, propane or CO_2 to provide the same cooling effect. These percentages are approximately observed in industrial practices.

The left side of Figure 17 shows the radial width, R_M, of the protective mushroom that is predicted to form for the various conditions of liquid steel superheat. For example, 8 percent natural gas, 4.25 percent propane gas or 3.3 percent butane gas is expected to form a protective mushroom of 10 cm radial width at 100°C superheat or 15 cm radial width at 50°C superheat. In these estimates, the accretion was assumed to have the form of a mushroom, or spherical cap, with an aspect ratio (height/radial width) of 0.5, or similar to that observed in Figure 1.

Returning to Table II, note the far greater cooling potential of a hydrocarbon gas versus, for example, argon. As seen, methane has a cooling capacity some 15 times greater than argon, about half of which is due to the sensible heat requirements for expanding hydrogen up to 1623 K. Similarly, hydrocarbons higher in the homologous series are expected to have even greater endothermic cooling effects. As mentioned, propane, on a molar basis, should be twice as effective as methane as a coolant.

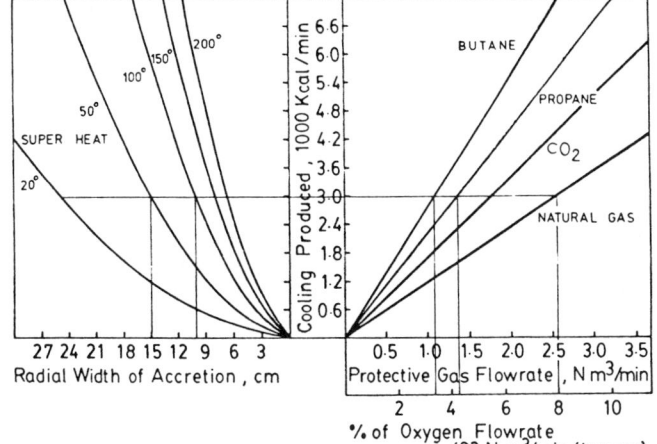

Figure 17 (left)

Predicted size of accretions forming at various superheats, as a function of varying gas flow rate and different protective gases, assuming 60 percent of hydrocarbon's cooling efficiency

Table II   Maximum Cooling Capabilities of Various Protective Gases

Protective Fluid	Heat of decomposition at 298 K, Kcal/mole	Sensible heat requirement, 298 - 1623 K, Kcal/mole (figure in brackets is H_2 heating component)		Total heat required Kcal/mole	Total heat required per Nm^3, Kcal
Natural Gas	17.8	25.7	(19.6)	43.6	1948
Ethane	20.2	41.7	(29.4)	61.9	2765
Propane	24.8	57.6	(39.2)	82.4	3682
Butane	29.8	73.6	(49.0)	103.4	4617
Pentane	35.0	89.5	(58.8)	124.5	5561
Hexane	39.9	105.5	(68.6)	145.4	6495
Heptane	44.8	121.4	(78.4)	166.3	7427
Octane	49.8	137.4	(88.2)	187.2	8359
Argon	...	2.9		2.9	132
Nitrogen	...	9.9		9.9	442
CO_2 (+ $\underline{C}$ → 2CO)	41.2	20.5		61.7	2757

Turning to the mildly oxidizing gas, carbon dioxide, the situation would appear to be more complex. Carbon dioxide may either react with dissolved carbon to form carbon monoxide or may react with iron to form iron oxide, as already noted.

$$CO_{2(298K)} + \underline{C}_{(1623K)} \rightarrow 2CO_{(1623K)} \qquad \Delta H \simeq 61.75 \; Kcal/mole \qquad (8)$$

$$CO_{2(298K)} + Fe_{(1623K)} \rightarrow CO_{(1623K)} + FeO_{(1623K)} \qquad \Delta H \simeq 32.07 \; Kcal/mole \qquad (9)$$

Comparison of N.S.C.'s results in Figure *14*, with the predictions of Figure *17*, show that their combined flow rate of 0.73 Nm^3 min^{-1} should lead to a steady-state accretion of about 4.5 cm radius. This is very close to the results observed and shown in Figure *14a*, based on reaction with dissolved carbon. It is also worth noting that Sumitomo engineers are able to control the size of their accretions by using a suitable mix of CO_2 and O_2 in the outer and inner nozzle bores respectively, and that by providing for about 5 percent of the net oxygen flow in top blown BOF's, are able to match the enhanced mixing provided by the submerged jets in Q-BOP, or OBM, operations.

Returning to the question of gradations in phosphorus and carbon contents that are observed between the base and rim of the accretions, these may be explained in terms of early growth of the accretion in only partially refined hot metal, followed by later growth in raw steel towards the end of the blow. The solidus carbon content of 2.5 percent carbon observed in Figure *1* would correspond to a bath composition of about 4 percent C, while the 0.1 wt percent carbon near the accretion's surface would correspond to raw steel.

Discussion of Analysis and Results

While the heat transfer analysis just presented would seem plausible, and fits observations nicely, there are certain difficulties associated with this simple analysis, the most problematic being the question of pyrolysis of the shrouding gas, into its components, within the annular slot and passageways of the accretion. One may cite the work of Chen et al.[8] on the mechanism of thermal decomposition of methane which was shown to proceed extremely slowly, as depicted in Figure *18*. At a reactor temperature of 1038 K, the yield of hydrogen was 9 x 10^{-7} moles/litre of methane fed after 500 seconds of reaction time. Similarly, a study of propane pyrolysis by Dunkleman and Albright [9] is shown in Figure *19*, where the abscissa, space-time, is defined as the reaction vessel volume divided by the volumetric feed rate (at vessel temperature and pressure). Assuming a tuyere annulus, 1.5 mm thick, 1 m long and 55 mm in diameter, with a methane flow of 0.5 Nm^3/s, the space-time condition would be about 0.6 milliseconds. Again, one would have little or no pyrolysis under these conditions, according to such data. Evidently, if the thermal analysis presented previously is correct, then the minute passageways of the growing accretion must have an unexpected catalytic effect. Similarly, the much higher temperatures of 1873 K would, on the basis of an Arrhenius relationship, lead to significantly higher cracking rates than those indicated by Figures *18* and *19*. The alternative is to suppose, as claimed by L. Matikainen,[10] and various converts, that thermal accretions are, in fact, artifacts caused by pools of liquid steel freezing in refractory sumps around each of the Q-BOP nozzles, the 'bumps' over the nozzles being caused by deformation of the pool from gas pressure within the tuyeres!

While much of the present paper summarizes earlier work by the authors[11,12] on accretions in steelmaking, similar phenomena occur in the formation of accretions in copper making.[13,14,15] For these, Brimacombe and co-workers have developed a finite difference model for copper accretions, similar to that presented by Ohguchi and Robertson for steel accretions,[15] and based on equivalent thermal principles to those presented here. Further, the

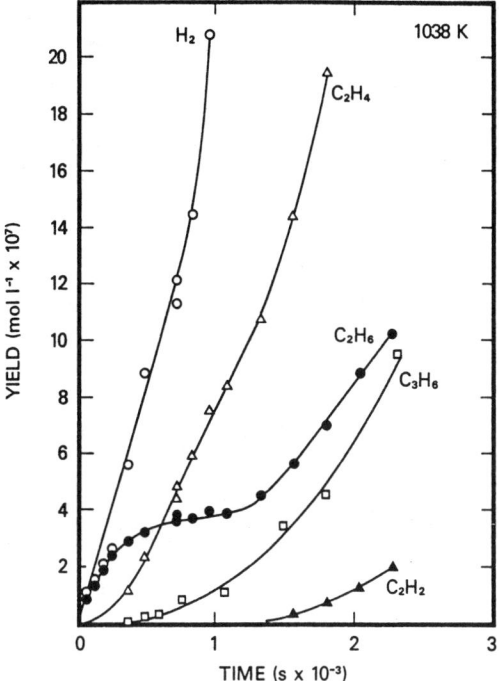

Figure 18 (top, left)

Methane Yield versus Reaction Time (from Chen *et al.*)

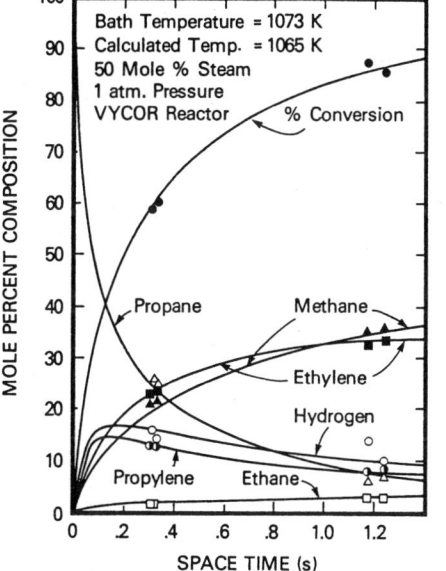

Figure 19 (bottom, left)

Propane Yield versus Reaction Time (from Dunkleman and Albright)

459

work of Bustos *et al.*[13] has shown that high pressure operations (~276 kPag) in copper converters causes air to 'jet' rather than 'bubble' into the melt from the tuyere, in a manner equivalent to Q-BOP operations. As such, accretions remained open while at 207 kPag, tuyeres became blocked. Historically, conventional Pierce-Smith Converters operate at 100 kPag to 250-300 kPag, and this accounts for the need for tuyere punching in conventional operations. In steelmaking, relatively high operating pressures, high velocities, and radiant heat from the reacting gas envelope of the oxygen jet above the tuyeres, keep it open (Figure *1*). For combination blowing operations, at lower flows, accretions can either cover the whole of the nozzle (*e.g.* Figure), or not, as desired. A recent summary of general work in the area of submerged gas injection into liquid metals is provided by J.K. Brimacombe in "Basic Aspects of Gas Injection in Metallurgical Processes."[17]

Conclusions

In conclusion, accretions can be formed during submerged gas injection operations in steelmaking and their formation and growth depend upon a heat balance between the entering shroud gas and heat transfer from the bath. They are metallic in nature, consisting of partially refined hot metal, with carbon compositions ranging from pig iron to steel in the case of typical steelmaking operations. These accretions protect adjacent refractory from excessive erosion when jetting oxygen into molten steel. As such, they provided one of the key factors to the successful implementation of Q-BOP, OBM, K-BOP, *etc.*, steelmaking technologies, for which Savard and Lee's original development work proved pivotal.

Acknowledgements

The authors are grateful to Dr G. Savard and Mr R. Lee for providing early samples of Q-BOP thermal accretions, and posing the question, to Dr K. Okohira of Nippon Steel Corporation for providing the photographs for Figures *14 a* and *b*, and to Mr J. Eisenwasser of Liquid Carbonic, USA for initiating the CO_2 injection work.

Nomenclature

A_M	mushroom/liquid steel interface area
h	average convective heat transfer coefficient
ΔH	net enthalpy change per unit mass of hydrocarbon or other protective gas
l	characteristic length (radial width of accretion)
L	latent heat of freezing
m	mass of accretion
$\dot{m}$	freezing rate of accretion
q_c	cooling parameter
Q	gas flow rate
R_M	radial width of accretion, or mushroom
t	time
V	mean velocity of steel flow over surface of accretion
α	thermal diffusivity of liquid steel
θ_G^f	temperature of gas on leaving accretion
θ_G^o	temperature of cold gas on entering accretion
$\Delta\theta_G$	gas 'supercool' temperature
θ_L^B	bulk liquid temperature
θ_L^*	temperature of liquid at accretion/liquid interface
$\Delta\theta_L$	liquid superheat temperature
θ_{mp}	melting temperature of accretion
μ	dynamic viscosity of liquid steel
ν	kinematic viscosity of liquid steel
ρ	density of gas
R_e	Reynolds number, $\rho V R_{M/\mu}$
Nu	Nusselt number, hl/k
Pr	Prandtl number, ν/α

461

References

1. J.R. Stubbles, "The Original Steelmaker," Monograph, ISS of AIME (1984), 23-24

2. J.F. McMulkin, private communication, Dofasco, 1975.

3. G. Savard and R. Lee, US Patent #2,855,293, October, 7, 1958.

4. G. Savard and R. Lee, French Patent #1,450,718, July 18, 1966.

5. G. Boxall, A.K. Sabharwal, T. Robertson and R.J. Hawkins, Symposium on "Injection phenomena in extraction and refining," University of Newcastle upon Tyne, England, 1982, pp.B1-B18, Editor A.E. Wraith.

6. K. Okohira, private communication, Nippon Steel Corporation, Yawata Works, August, 1982.

7. Y. Sahai and R.I.L. Guthrie, "Hydrodynamics of Gas Stirred Melts: Part II. Axisymmetric Flows," Metallurgical Transactions, B, 13B (1982), 203.

8. C.J. Chen, M.H. Back, and R.A. Back, Industrial Laboratory Pyrolyses, L.F. Albright and B.L. Crynes, ed.s, p.1, Amer. Chem. Soc., Washington, 1976.

9. J.J. Dunkleman and L.F. Albright, ibid., p.261

10. L. Matikainen, "Thermal Aspects of Subsurface Gas Injection" (M.Eng. thesis, McGill University, 1982) 76-80.

11. C. Xu, Y. Sahai and R.I.L. Guthrie, "Thermal Accretions in Submerged Gas Injection Processes," Injection Phenomena in Extraction and Refining, University of Newcastle upon Tyne, England, 1982, pp., ed. A.E. Wraith.

12. Y. Sahai, R.I.L. Guthrie, "The Formation and Growth of Thermal Accretions in Bottom-Blown/Combination Blown Steelmaking Operations," IJSM, April, 1984, pp.34-38.

13. A.A. Bustos, J.K. Brimacombe and G.C. Richards, "Accretion Growth at the Tuyeres of a Pierce-Smith Copper Converter," Canadian Metallurgical Quarterly, 27 (1988), 7-21.

14. J.K. Brimacombe, S.E. Meredi and R.G.H. Lee, "High Pressure Injection of Air into a Pierce-Smith Copper Converter," Metallurgical Transactions, B, 15B (1984), 243-250.

15. A.A. Bustos, J.K. Brimacombe, G.C. Richards, A. Vahed and A. Pelletier, "Development of Punchless Operation of Pierce-Smith Converters," Copper '87, Vol. 4, ed.s C. Diaz, C. Landolt and A. Luraschi (Santiago, Chile: Universidad de Chile, 1987), 347-373.

16. S. Ohguchi and D.C.G. Robertson, "Formation of Porous Accretions Around Tuyeres During Gas Injection," Ironmaking and Steelmaking, 10(1983), 15-23.

17. J.K. Brimacombe, "Basic Aspects of Gas Injection in Metallurgical Processes," International Symposium on Injections in Process Metallurgy, ed.s T. Lehner, P.J. Koros and V. Ramachandran, TMS of AIME, 1991, 13-29.

Session 5

Injection into Baths

Fundamentals of Gas Injection in Refining Processes

Masamichi Sano and Kazumi Mori

Department of Materials Processing Engineering
Nagoya University
Furo-cho, Chikusa-ku, Nagoya 464-01
Japan

Abstract

Gas injection into liquid metal is utilized in many high temperature metallurgical processes. To clarify the mechanism of rate phenomena, fundamental knowledge on the behavior of injected gas in the bath is essential. The present paper is concerned with fundamentals of various phenomena occurring in refining processes with gas injection. Correlating equations for the size of bubbles are classed according to gas-flow rate and gas chamber volume. Transition of behavior of gas jet from "bubbling" to "jetting" is discussed and advantages of gas injection in the "jetting" regime are emphasized. Dispersion behavior of the injected gas in the bath is investigated to elucidate the structure of jet and plume. Fluid flow and mixing characteristics in the gas-stirred bath are briefly mentioned. Gas-metal and slag-metal reactions in the refining processes are also discussed.

Proceedings of the
Savard/Lee International Symposium on Bath Smelting
Edited by J. K. Brimacombe, P. J. Mackey,
G. J. W. Kor, C. Bickert and M. G. Ranade
The Minerals, Metals & Materials Society, 1992

I. Introduction

Techniques of gas injection into molten iron have made great progress in the pretreatment of hot metal, bottom-blowing and combined-blowing oxygen steelmaking, ladle metallurgy, etc. Consequently, advantages of gas injection into molten iron have been well recognized.

Many fundamental studies have been made on various aspects of gas injection into melts, such as bubble formation, behavior of gas jet near a tuyere (bubbling and jetting), accretion formation, tuyere erosion, dispersion behavior of injected gas in a bath, trajectory of side-blown jet, circulating flow and mixing in metallurgical reactors, bubble-metal reactions, slag-metal reactions with bubble stirring, emulsification, splashing, etc.

The present paper is concerned with selected fundamental research works performed mainly in the authors' laboratory during the last over two decades.

II. Bubble formation from nozzle

Knowledge on the bubble size is indispensable in analyzing kinetic data of bubble-liquid reaction systems. Many studies have been made on the bubble formation from nozzles in aqueous solutions [1]. It has been shown that factors controlling the bubble size are volumetric gas-flow rate V_g, nozzle diameter d_n, surface tension σ, liquid density ρ_l (or $\Delta\rho = \rho_l - \rho_g$, ρ_g: gas density), gas chamber volume V_C, etc.

There are three regions of bubble formation according to the gas-flow rate [2].

(I) At low gas-flow rates, the bubble size is independent of the gas-flow rate and determined by the nozzle diameter, liquid physical properties and gas chamber volume.

(II) At intermediate gas-flow rates, the bubble size depends on the gas-flow rate and nozzle diameter.

(III) At high gas-flow rates, small bubbles of various sizes are formed at about 10 cm above the nozzle by disintegration of large irregular bubbles formed at the nozzle.

As the gas chamber volume increases, three regions of bubble formation are obtained [2, 3].

(A) Provided that the chamber volume is smaller than a critical volume, it does not affect the formation of bubbles.

(B) For the chamber volume larger than the critical volume, the size of bubbles increases with increasing chamber volume.

(C) When the chamber volume increases sufficiently, the formation of bubbles becomes to be unaffected by the volume.

Here, it is to be noted that the effect of the gas chamber volume on the bubble formation results from the surplus gas accumulated in the chamber between bubble detachments.

Wettability of the nozzle to a liquid has a definite effect on the bubble formation [2]. Figure 1 schematically illustrates difference in bubble formation between wetted and non-wetted nozzles, orifices and porous plugs.

466

In the case of bubble formation from a wetted nozzle or orifice, the circular line of contact of the liquid with the surface of the nozzle or orifice remains at its inner edge. On the other hand, in the case of bubble formation from a non-wetted nozzle or orifice, the base of a forming bubble spreads much larger than the cross sectional area of the gas outlet . Therefore, the nozzle diameter controlling the bubble size is the inner diameter d_{ni} for the wetted nozzle and the outer one d_{no} for the non-wetted nozzle. Here, it should be emphasized that although a wetted porous plug can produce very tiny bubbles, a non-wetted porous plug can not produce small bubbles, but does rather large bubbles, as shown in Fig. 1.

Correlations obtained for the bubble formation from wetted and non-wetted nozzles are classed according to the gas-flow rate and the gas chamber volume, and shown in Table 1. In applying the equations in the table, the outer diameter should be used for the non-wetted nozzle, and the inner diameter for the wetted one. N_G in the table is a dimensionless gas chamber volume defined by Eq.(1) [3], and should be used for the wetted nozzle. If the nozzle is non-wetted and temperature in the gas chamber T is not constant, a modified dimensionless gas chamber volume N_G' defined by Eq. (2) should be used instead of N_G [3].

Equation (10) is applied to calculate the bubble size formed from nozzles in various liquids. The calculation is made for the nozzle of 0.6 cm in diameter. As seen from Fig. 2, at low gas-flow rates (region (I)), the bubble size decreases in the order of decreasing σ/ρ_l, that is, molten iron, water, molten silver, and mercury. As the gas-flow rate increases, the difference in the value of bubble size becomes smaller, and the region, where the bubble size no longer depends on the liquid physical properties, is attained.

Here is mentioned the bubble formation in highly viscous liquids such as slags. At very low gas-flow rates, the liquid viscosity has no influence on the bubble size. This is because the bubble size is determined by a balance of buoyancy and surface tension forces, so that the bubble size is given by Eq. (3) for region (I)-(A).

With increasing gas-flow rate, the viscous resistance becomes to have an influence on the bubble formation. Assuming that the bubble is moving with the Stokes velocity according to its size, Davidson and Schuler [11] obtained the following equation

$$V_B=(4\pi/3)^{1/4}(15\mu_l V_g/2\rho_l g)^{3/4} \qquad (16)$$

where V_B is the bubble volume.

At higher gas-flow rates, the momentum of the surrounding liquid mass (virtual mass) should be considered. The bubble formation in this region has been analysed by Davidson, Schuler [11], and Kumar, Kuloor [1].

III. Bubbling and jetting

A full understanding on behavior of gas jets injected into liquids is helpful in solving serious process problems such as refractory erosion due to jet action and clogging of tuyeres by frozen metal. Hence, many studies have been made on bubbling and jetting phenomena in energetic gas injection into liquids.

Considering the physical properties of water are much different from those of molten metals, the present authors [12] have used mercury. As shown in Fig. 3, nitrogen was injected into a mercury bath through an orifice located at the transparent vessel bottom. Behavior of the gas jet was photographed through the bottom plate by a high speed cinecamera.

Two flow regimes were observed. At low gas-flow rates, the injected gas was shown to expand immediately upon discharging from the orifice and formed discrete bubbles. This proves the presumption stated earlier that the base of the bubble formed from a non-wetted orifice spreads much larger than the cross sectional area of the orifice. This phenomenon is called "bubbling". At gas-flow rates higher than a critical rate, an apparent coincidence between the base and the orifice diameters begins to occur over various time ranges. In these time ranges, the injected gas is considered to leave the orifice as a continuous jet. This phenomenon is called "jetting".

In Fig. 4, time fractions for both bubbling and jetting are plotted against the gas-flow rate. As clearly seen in the figure, jetting is initiated at a critical gas-flow rate for each orifice diameter. A gradual change from bubbling to jetting takes place in a transitional range of gas-flow rate. Figure 5 shows bubbling-jetting time fraction plotted against the nominal Mach number M'. Here, M' is defined by the ratio of nominal gas-flow velocity to sonic velocity at room temperature, and the nominal gas-flow velocity is the volumetric gas-flow rate divided by the cross sectional area of orifice. It is found that the critical value of M' at which the transition from bubbling to jetting begins to occur is the same for the three diameters of orifice. The critical gas-flow velocity is a little bit larger than but very close to the nominal sonic velocity (M'=1) irrespectively of the orifice diameter.

The bubbling-jetting phenomena have been studied by many researchers. Hoefele and Brimacombe [13] visually studied the behavior of gas horizontally injected into water, molten zinc-chloride solution, and mercury. They distinguished two regimes of flow, bubbling and steady jetting, and represented the critical condition for bubbling-jetting transition on the diagram shown in Fig. 6. Their results are shown by the hatched area, which indicates that the steady jet behavior is observed at high modified Froude number and high ratio of gas to liquid densities. Here, the modified Froude number Fr' is given by

$$Fr' = \rho_g u_o^2 / \rho_l d_o g \qquad (17)$$

The critical values of Fr' for the initiation of jetting observed by Ozawa et al. [12, 14, 15] and other researchers [16~18] are also shown in Fig. 6. As seen from the figure, bubbling-jetting transitions are quite different among the researchers. This suggests that various definitions for the transition are adopted by the researchers.

Under the condition of high gas-flow rates, detachment of a bubble is followed by the growth of a new bubble which frequently touches and links with the former bubble. After the linkage, the lower bubble becomes a lengthened stem through which the gas once again flows to the upper bubble. This behavior of gas jets was reported by Wraith and Chalkley in detail [16]. Ozawa and Mori [15] measured "no linking" time, which is the period of time between the severance of the first bubble from the gas supply and the

468

subsequent linkage of the severed bubble to the forming second one. As shown in Fig. 7, they found that the time fraction for "no linking" is expressed as a simple function of $M'(\rho_g/\rho_l)^{1/2}$. Here the time fraction for "no linking" is defined as the ratio of the sum of "no linking" times to the total time in a series of bubble formation.

From the examination of the results by Hoefele and Brimacombe [13] shown in Fig. 6, Ozawa and Mori [15] clarified that the critical condition for the transition from bubbling to steady jetting observed by Hoefele et al. can be expressed by

$$M'(\rho_g/\rho_l)^{1/2}=0.009 \tag{18}$$

This critical condition is indicated by an arrow in Fig. 7. It is found that the critical condition adopted by Hoefele et al. is the time fraction of 30% "no linking". Ozawa and Mori stated that when the time fraction for "no linking" is smaller than about 30%, the severance of the forming bubble from the gas supply occurs less frequently and the jet becomes to look just like a continuous jet of gas.

Farias and Robertson [17] studied the bubbling-jetting phenomena in terms of the injection number $N_I(=3.3Fr'^{2/5}(\rho_g/\rho_l)^{3/5})$, and Wraith, Chalkley [16] and Leibson et al. [9] in terms of $Re_0(=\rho_g u_0 d_0/\mu_g)$. Again Ozawa et al. [15] could have shown that those critical conditions correspond to the time fraction of 30% "no linking". Recently Zhao and Irons [19] applied a combined Kelvin-Helmholtz and Rayleigh-Taylor instability theory to the surface of a forming bubble at an orifice and determined the critical condition for the bubbling-jetting transition. They showed that the critical condition is given by the following equation.

$$W_e=10.5(\rho_g/\rho_l)^{-1/2} \tag{19}$$

This condition is shown to be in reasonable agreement with the time fraction of 30% "no linking".

Thus through the above described discussion, apparently different views on the gas jet characteristics proposed by various investigators could be correlated with each other.

Physical behavior of injected gas in the initial jet formation zone has important connection with serious process problems such as refractory erosion due to jet action and clogging of tuyeres by frozen metal.

Sakaguchi et al. [20] made an important investigation on the erosion of a double-tuyere for injection of oxygen into molten steel. From their data giving the injection condition under which the tuyere refractory is little eroded, Mori et al. [12] have estimated the critical Mach number for erosion of the tuyere refractory used for injecting oxygen into a steel bath. In Fig. 5 the value of this critical Mach number is shown by a dot-dash line. It is seen that the estimated value almost coincides with the critical velocity for the bubbling-jetting transition in the N_2-Hg model. Thus, it has been cleared that the bubbling-jetting transition is of considerable significance in the practical problem of erosion of tuyere refractory.

Studies on the mechanism of the penetration of liquid back into the tuyere during gas injection are very few. Engh et al. [21] , from model studies of gas injection in water and glycol, obtained a correlation equation for the formation rate of droplets entering into the nozzle.

However, a criterion for predicting liquid penetration leading to nozzle clogging was not given. Marukawa et al. [22] found, from studies of gas injection in a 250 t ladle, that gas injection at sonic velocity is necessary to prevent nozzles from clogging. Recently, Sharma [23] has done experiments of injecting inert gas into molten steel. He concluded that the controlling factor for metal penetration in the tuyere is the jet Froude number, and not the gas velocity (Mach number). According to him, injecting gases at sonic velocity may not be sufficient to prevent metal penetration.

Bustos et al. [24] discussed the blockage of tuyeres in a copper converter by accretion formation. They showed that bubbling behavior is related to the tuyere blockage and recommended that the pressure in the tuyere should be raised to operate gas injection in the "choked" flow regime, which was the necessary condition for jetting.

IV. Dispersion behavior of gas injected into liquid

Many studies have been made on the dispersion behavior of injected gas in liquid, such as bubble frequency, gas holdup, bubble rising velocity, and size of jet and plume zones. In these studies, the electroresistivity probe technique was used to detect the gas in the bath.

The present authors [25] measured the distribution of gas holdup in mercury to study the behavior of gas jet in the vicinity of gas exit (bubbling and jetting) and the size of jet and plume zones. The vessel containing mercury was a stainless steel column of 15 cm inner diameter and 40 cm height. Nitrogen was injected through an orifice positioned at the center of the vessel bottom. A hole in the bottom plate was used as the orifice. The diameters of the orifice , d_o, were 1 and 2 mm. The bath depth was $20 \sim 140$ mm. The gas-flow rate V_g was $140 \sim 2200$ cm^3/s under the conditions at the gas exit. The nominal Mach number M' was $0,5 \sim 3.5$. The local gas holdup, ϕ_{100} was measured at vertical distances, h, of $5 \sim 120$ mm.

Contour maps of gas holdup are plotted in Fig. 8 to examine the effect of the gas velocity (or the inertial force). In the range of $h < \sim 30 \sim 40$ mm, the contour map for d_o=1 mm is different from that for d_o=2 mm. The local gas holdup ϕ_{100}=0.8 is found in the case of d_o=2 mm. In the range of $h > \sim 30 \sim 40$ mm, the two contour maps are almost the same. In order to investigate the phenomena more closely, the gas holdup on the center axis $\phi_{r=0}$ is plotted against h in Fig. 9. At a gas-flow rate, V_g, of 550 cm^3/s and in the range of h $> \sim 30$ mm $\phi_{r=0}$ for d_o=2 mm (③) tends to increase with decreasing h. But $\phi_{r=0}$ for d_o = 1 mm (④) in the same range shows a reverse tendency. This tendency is observed in the gas injection under the sonic conditions and reflects the "jetting" phenomenon. In the range of $h > \sim 30$ mm, $\phi_{r=0}$ increases with increasing gas-flow rate (①$>$②$>$③), and decreases with increasing h.

The motion induced by injection of a fluid into another fluid is roughly classified into two regimes: jet and plume. The jet is a fluid motion induced by the inertial force of the injected fluid. The plume is used to represent a fluid motion induced by the buoyancy force acting on the injected fluid. As shown in Figs. 8 and 9, while the distribution of gas holdup for $h < \sim 30 \sim 40$ mm is affected by the gas inertial force, that for $h > \sim 30 \sim 40$ mm is dependent only on the gas-flow rate. From the viewpoint of the gas dispersion, transition of flow behavior from jet to plume occurs at $h \approx 30 \sim$

470

40 mm under the present injecting conditions.

Figure 10 shows the effect of bath depth H on the dispersion of injected gas. As H becomes smaller, the gas tends to be dispersed in a smaller region of the bath and the gas holdup decreases as a whole. It is clear that the gas holdup for H=20 mm is very small. The sound produced by the gas injection for $H < \sim 40$ mm was quite different from that for $H > \sim 40$ mm.

Several investigators directed their attention to the size of the plume zone [26 ~ 29]. It is shown that the radial distribution of gas holdup obeys the Gaussian distribution.

$$\phi_{100} = \phi_{r=0} \exp\{-r^2/(r^2_{1/2}/\ln2)\} \tag{20}$$

where, $r_{1/2}$: the plume half radius, i.e., the radial distance at which ϕ_{100} becomes $\phi_{r=0}/2$.

Dispersion characteristics such as $\phi_{r=0}$, $r_{1/2}$, etc. are experessed by correlating equations [27~ 29].

V. Mixing characteristics in a gas-stirred molten metal bath

Fluid flow phenomena and mixing characteristics in refining processes have decisive influences on the refining reaction rate. Many experimental and theoretical studies have been made on circulating flow and mixing time in the molten metal bath.

Nakanishi et al. [30] have first introduced the concept of mixing time to evaluate the mixing characteristics in the steelmaking processes and obtained a quantitative relation between the mixing time t_m and the input stirring power ε_M.

Mori and Sano [31] noticed that the relations between t_m and ε_M obtained from experiments for different bath sizes scattered largely. They made a theoretical analysis on circulating flow in a bubble stirred bath [32]. At a steady state, the rate of energy dissipation due to the liquid circulation in the bath and the bubble slip in the plume zone is equal to the input power of injected gas. The analytical result obtained by Sano et al. is given by

$$t_m = 100 \ \{(D^2/H_0)^2/\varepsilon_M\}^{0.337} \tag{21}$$

where t_m is the mixing time (s), D the bath diameter (m), H_0 liquid depth (m), and ε_M the stirring power per liquid mass (W/ton). This stirring power ε_M is expressed as

$$\varepsilon_M = (V_g RT_1/V_N W_1) \ln(P_1/P_2) \doteqdot (V_{gM} \rho_1 g H_0)/W_1 \tag{22}$$

where V_g is the gas-flow rate (Nm^3/s), R the gas constant ($W \cdot s/mol \cdot K$), T_1 the liquid temperature (K), V_N the normal molar volume of gas (Nm^3/mol), W_1 the liquid mass (ton), P_1 and P_2 the static pressures at the nozzle exit and the ambient pressure (atm), V_{gM} the gas-flow rate at T_1 and logarithmic mean pressure of P_1 and P_2, and g the acceleration of gravity.

Here it is to be noted that the actual work done by the bubble during its rise given by Eq. (22) is exactly half the stirring power overestimated by several investigators who adopted the sum of work done by the buoyancy force and the volumetric work to obtain the stirring power.

Equation (21) is plotted by using a solid line in Fig. 11, in which the experimental results obtained using water models and 6, 60 ton ladles are also shown. Close agreement between the calculation and the experiment is seen in the figure.

Asai et al. [36] made a dimensional analysis of mixing time on the basis of Navier-Stokes equation and obtained correlating equations corresponding to the flow regimes. If the mixing length is equal or proportional to the characteristic length of the bath, their equations for laminar and turbulent flow regimes are similar to those obtained by Sano and Mori.

Although important studies have been made on fluid flow and mixing phenomena by many researchers [37~40], they are not cited here.

VI. Gas-metal reactions through injected bubble interface

The present authors have made a series of studies on rates of nitrogen absorption and desorption [41~44] in molten iron through bubble interface. They used a Tammann furnace to melt iron in a magnesia crucible. Nitrogen or argon was injected into the melt through an immersed alumina nozzle of 0.1 cm I.D. and 0.3 cm O.D., the tip of which was faced upward. The immersion depth of the nozzle was 3.3~4.7 cm. The gas-flow rate was 70~117 Ncm^3/min. The oxygen and sulfur concentrations in the metal were varied widely. During each experiment the frequency of bubble formation was measured to calculate the volume of the formed bubble at the nozzle.

Figure 12 shows the experimental result of nitrogen absorption. A mixed-conrol model of mass transfer and chemical reaction is developed to explain the rate. In the model calculations, the nitrogen absorption is considered to take place during both bubble formation at the nozzle tip and bubble ascent. The mass transfer coefficient is estimated by using the Higbie model. The chemical reaction rate constant k_C is determined by the trial and error method. As seen from Fig. 12, the kinetic model can explain the absorption rate. The value of k_C is shown in Fig. 13. It is seen that k_C obtained for the bubble-metal interface is in good agreement with that obtained from an isotope exchange reaction [45]. Figure 13 indicates that k_C decreases with increasing [%O]+[%S]/2.

The experiment on nitrogen desorption from molten iron was made by injecting argon. During each experiment, an argon-nitrogen mixture whose nitrogen pressure was controlled to be in equilibrium with the nitrogen concentration in the melt, was blown onto the melt, so that the reaction through the bath surface could be neglected.

The experimental result is shown in Fig. 14 and compared with that calculated from a mixed control model, which is developed with considering mass transfer in both liquid and gas phases and chemical reaction. The chemical reaction rate constant k_C determined from the desorption experiment is plotted in Fig. 13 by open circles. As can be seen from Fig. 14, k_C determined from the desorption experiment agrees well with that from the absorption experiment and also with that of the previous study [45]. Since the value of k_C obtained from the experiments in wide ranges of nitrogen concentration in molten iron and nitrogen pressure in the gas phase can be correlated by a unified function of [%O]+[%S]/2, one can conclude that both nitrogen absorption and desorption in molten iron are controlled by mass transfer and chemical reaction.

In the earlier stage of the study of nitrogen desorption, Takahashi et al. [43] got a preliminary result shown in Fig. 15. The open circle represents the data for nitrogen desorption by Ar injection together with blowing onto the melt. The closed circle represents the experimental result with Ar blowing onto the melt. The dotted line shows the change of nitrogen concentration calculated for the desorption through the bubble interface. It is clear that the rate of nitrogen desorption with Ar injection is enhanced markedly by Ar blowing onto the melt. Possibly, a high degree of turbulence is produced from the appearance of bubbles at the surface of the melt, leading to the increase of the effective free surface area (A_S).

On the basis of this preliminary result, Takahashi et al. [43] made a systematic study on the enhancement of the rate of nitrogen desorption from molten iron by Ar injection together with blowing onto the melt. It has been learned from the comparison between calculation and experiment that the effective free surface area of the melt is about 2.1 times as large as the cross sectional area of the crucible (A_C) (Fig. 16).

In most refining processes, the size of dispersed bubbles is not known, so that the rate data are usually analyzed in terms of volumetric coefficients defined by kA, kA/V, kA/H. Here, k is the apparent first order rate constant, V the metal volume, and H the bath depth.

Figure 17 shows the relation between kA/H and gas flow rate. It is seen that the volumetric coefficient for oxygen absorption in molten silver obtained by Mori et al. [47] is in good agreement with those obtained in molten silver and copper by Fruehan [48]. The volumetric coefficient obtained for nitrogen absorption in molten iron and stainless steel are considerably smaller than those for oxygen absorption in molten silver and copper.

In case that the absorption rate is controlled by liquid phase mass transfer, k is equal to the mass transfer coefficient k_l given by

$$k_l = 2(D_l v_B / \pi \, d_B)^{1/2} \tag{23}$$

where D_l and v_B are the liquid-phase diffusivity and the bubble rising velocity.

Using the following equations for the bubble-metal interfacial area, A, and bubble rising velocity, v_B,

$$A = (6V_g/d_B)H/v_B \tag{24}$$
$$v_B = (0.5d_B g)^{1/2} \tag{25}$$

and assuming $D_l = 10^{-4}$ cm^2/s, one can obtain

$$kA/H = 0.0144 V_g/d_B^{7/4} \tag{26}$$

where g is the acceleration of gravity, and the units are cm and s.

In Fig. 17, the relation between kA/H and V_g is plotted for various bubble diameters. Provided that d_B does not change with V_g, kA/H is proportional to V_g. Actual dependence of kA/H on V_g observed experimentally is smaller than that predicted by Eq. (26). This smaller dependence of kA/H on V_g is due to the change in d_B with V_g. The bubble size in the experiment of Mori et al. [47] is estimated to be 0.69 and 0.98 for

473

V_g=3.95 and 14.2 cm^3/s. Considering the increase in d_B with V_g, one can conclude that fairly good agreement is obtained between experiment and calculation.

Sano et al. [41] reported that the diameter of bubbles formed from a nozzle in molten iron was in the range of 0.92 ~ 1.08 cm under their experimental conditions. They also showed that the nitrogen absorption is mixed-controlled and the apparent rate constant, k, for the nitrogen absorption shown in Fig. 17 is equal to 0.78k$_1$. Hence, it is found that their value of kA/ H agrees with Eq. (26). On the basis of the above discussion, the bubble size in the experiments of Fruehan [49] and Kawakami et al. [50] is presumed to be as large as 3~4 cm.

Fruehan analyzed the data of nitrogen absorption in Q-BOP. Nitrogen was injected into molten iron at a high flow rate (V_g=0.7 m^3/s). In this case, nitrogen bubbles were dispersed spatially in the bath. Then, the volumetric coefficient is given by

$$kA/H= kA_m(6\phi/d_B) \tag{27}$$

where ϕ is the gas holdup and A_m the cross sectional area of the bath.

Using the rate data (kA/V=0.051 min^{-1}, k=0.2 cm/min [52] and ϕ =0.33 [53]), Fruehan estimated that the average bubble diameter in the bath was 8.4 cm. This bubble diameter agrees well with that calculated from Eq. (15).

VI. Mass transfer between molten slag and metal under gas injection stirring

In recent years, kinetics of the reactions between slag and molten iron with gas injection in steel refining processes have become of increasing interest to the metallurgists. Previously studies were made by Richardson and co-workers on elucidating the fundamental relation between the slag-metal reaction rate and gas injection stirring conditions with using model reaction systems at low temperatures, i.e., aqueous solution-amalgam system [54, 56, 57] and molten salt-molten lead system [55, 57]. From the practical viewpoint, model studies have been made frequently with using organic solvent-water system. This kind of model studies may produce questionable results for the application to actual slag-metal systems because of the low interfacial tension of organic solvent-water systems.

These years, model experiments have been done in the authors' laboratory to investigate the role of gas injection stirring in slag-metal reactions at high temperatures with using a molten slag-Cu system where physical properties are similar to those of molten slag-iron system [58]. The explored reaction is oxidation of Si by FeO

$$Si\ (in\ Cu)+2FeO\ (in\ slag)=2Fe\ (in\ Cu)+SiO_2\ (in\ slag) \tag{28}$$

at 1250 °C, taking place under the condition of rate controlling by Si transport in the metal phase. The slag-metal bath was stirred by Ar gas injected through a nozzle located at the crucible bottom. The gas-flow rate (V_g), crucible diameter (d_C) and metal depth (h_M) were varied.

The rate of reaction (28) was found to be described by

$$-d[\%Si]/dt=k'_{Si}(A/V)[\%Si] \tag{29}$$

where A is the interfacial area, V the metal volume, k'_m the apparent mass transfer coefficient of silicon and t the time.

Figure 18 shows typical experimental results for the relation between the logarithm of Si concentration and time. Here, h_{SI} is the slag depth and V_g is expressed under the conditions of 1250°C and 1 atm pressure. All the data are represented by linear lines predicted from Eq. (29). From the slope of the linear line the apparent mass transfer coefficient of Si, k'_{Si} is calculated.

Figures 19 and 20 present the relation between k'_{Si} and V_g for $d_C=4$ and 7.5 cm, respectively. The dependence of k'_{Si} on V_g at constant d_C changes at transitional gas-flow rates denoted by V_g^* and V_g^{**}. A similar result is obtained for $d_C=3$ cm. The whole range of gas-flow rate explored is divided into three regions:

(i) Region I $(V_g < V_g^*)$ where $k'_{Si} \propto V_g^{1/2}$,
(ii) Region II $(V_g^* < V_g < V_g^{**})$ where the effect of the increase in V_g on k'_{Si} is considerably smaller than in Region I, and
(iii) Region III $(V_g > V_g^{**})$ where the extent of the increase in k'_{Si} with V_g increases again.

From the experiments of varying metal depth, interesting results presented in Fig. 21 are obtained, where h_I is the distance between the slag-metal interface and the nozzle tip. Here, $h_I = h_M - 0.5$(cm). As seen from Fig. 21, k'_{Si} is proportional to $h_I^{1/2}$ below a transitional depth, h_I^*, while it becomes independent of h_I larger than h_I^*.

Another phenomenon almost unexpected is that in Region II k'_{Si} increases with increasing h_{SI} (Fig. 22), while in region I k'_{Si} is independent of h_{SI}.

One could realize, from the above presentation of the experimental data for mass transfer coefficient (k'_{Si}), that the nature of mass transfer between metal and slag phases affected by gas injection stirring is very complicated. Hirasawa et al. [59] have made analytical approach to the metal-side mass transfer in Region I by use of the theory of turbulence phenomena developed by Davies [60]. This theory has previously been applied to the problem of slag-metal mass transfer by Robertson and Staples [56]. However, the assumption made by Hirasawa et al. for "the scale of energy containing eddies" is different from that by Robertson et al.

Hirasawa et al. [59] have developed equations for the metal-side mass transfer coefficient as a function of gas injection stirring conditions. The theoretical equations are rearranged to dimensionless equations.

Hirasawa et al. have obtained dimensionless equations for the four cases of gas injection stirring conditions: $V_g < V_g^*$ and $h_I < h_I^*$, $V_g < V_g^*$ and $h_I > h_I^*$, $V_g > V_g^*$ and $h_I < h_I^*$, and $V_g > V_g^*$ and $h_I > h_I^*$. These equations are summarized in Table 2, together with transitional conditions for V_g and h_I. As an example, Eq. (31) is plotted in Fig. 23 to compare with mass transfer data for $V_g < V_g^*$ and $h_I > h_I^*$. The mass transfer data obtained from model studies for molten slag-Cu reaction system at 1250°C and the available results of model studies at room temperature are correlated successfully by Eq. (31). It is also found that the mass transfer data of practical ladle desulfurization refining could be explained reasonably by Eq. (31).

Ⅵ. Concluding remarks

The performance of steel refining processes has greatly improved by the development of injection metallurgy, which was enhanced by the emergence of bottom-blowing oxygen steelmaking. The subjects discussed in the present paper have been concerned with only some aspects of injection metallurgy. Although important, accretion formation, fluid flow in gas-stirred systems, splashing, slag-metal emulsification, powder injection, etc., have not been dealt with in this paper. The most complicated phenomena are those occurring in the smelting reduction process, which involves gas-liquid-solid dispersion together with transport phenomena and chemical reactions. Efforts should be made further on elucidating the complex dispersion processes and on establishing the fundamentals of 'dispersion engineering'.

Nomenclature

A : Total bubble-metal interface
A_C : Cross sectional area of crucible
A_m : Cross sectional area of bath
A_S : Effective surface area of metal bath
D : Bath diameter
D_l : Metal phase diffusivity
d_B : Bubble diameter
$d_{B,Av}$: Average bubble diameter
d_C : Crucible diameter
d_{ni} : Inner diameter of nozzle
d_{no} :Outer diameter of nozzle
d_o : Orifice diameter
d_P : Diameter of plume zone
d_{vs} : Volume-surface mean diameter
Fr : Froude number ($=u_o^2/d_og$)
Fr' : Modified Froude number ($\rho_gu_o^2/\rho_ld_og$)
g : Acceleration of gravity
H, H_o:Bath depth
h : Vertical distance from gas exit
h_l : Vertical distance from nozzle tip to metal surface
h_M : Metal depth
h_{Sl}: Slag depth
K : Equilibrium constant or constant in Eq. (8) (K=10)
k : First order rate constant
k' : Apparent mass transfer coefficient
k_C : Chemical reaction rate constant
k_l : liquid phase mass transfer coefficient
M' : Nominal Mach number
N_C : Dimensionless gas chamber volume defined by Eq. (1)
N'_C : Dimensionless gas chamber volume defined by Eq. (2)

N_1 : Injection number $(=3.3Fr'^{2/5}(\rho_g/\rho_l)^{3/5})$
N_w : Dimensionless parameter $(=W_e \cdot Fr^{0.5})$
W_e : Weber number $(=\rho_g u_o^2 d_o/\sigma)$
Pe : Peclet number $(=(4V_g/\pi\, d_C^2)d_C/D_l)$
Re : Reynolds number $(=(4V_g/\pi\, d_C^2)d_C/\nu_l)$
Re_o: Orifice Reynolds number $(=\rho_g u_o d_o/\mu_g)$
R : Gas constant
r : Radial distance from center axis
$r_{1/2}$:Radius for $\phi_{r=0}/2$
Sh : Sherwood number $(=k_l d_C/D_l)$
t : Time
t_f : Bubble formation time
t_m : Mixing time
u_o : Gas velocity at gas exit
V : Bath volume
V_B : Bubble volume
V_b : Flow rate of gas blown onto the bath surface
V_C : Chamber volume
V_g : Gas-flow rate
V_{N2} : Nitrogen flow rate
V_s : Superficial gas velocity
v_B : Bubble rising velocity
W_{Fe} : Mass of iron
β : A_S/A_C
ε_M : Input stirring power per metal mass
μ_g : Gas viscosity
μ_l : Liquid viscosity
ν_l : Liquid kinematic viscosity
ρ_g : Gas density
ρ_l : Liquid density
$\Delta\rho$: $\rho_l - \rho_g$
σ : Surface tension
ϕ : Gas holdup
$\phi_{r=0}$: Local gas holdup at r=0
ϕ_{loc}: Local gas holdup

References

1) R.Kumar and N.R.Kuloor, Advances in Chemical Engineering Vol. 8, edited by T.B.Drew et al., Academic Press, (1970), p256.
2) M.Sano, and K.Mori, Trans. Jpn. Inst. Met. 17 (1977), 344.
3) T.Tadaki and S.Maeda, Chem. Eng. (Japan), 27 (1963), 147.
4) C.G.Maier, U.S.Bur, Mines Bull., (1927), 260.
5) J.F.Davidson and B.O.G.Schuler, Trans. Inst. Chem. Engrs., 38 (1960), 335.
6) L.Daividson and E.H.Amick, Jr., A.I.Ch.E. Journal, 2 (1956) 337.
7) A.Mersman, V.D.I. Forschungsheft, 491 (1962), 1.
8) K.Mori, et al., Trans. Iron Steel Inst. Jpn., 19 (1979), 553.
9) I.Leibson et al., A.I.Ch.E. Journal, 2 (1956) 296.
10) M. Sano and K.Mori, Trans. Iron Steel Inst. Jpn., 20 (1980), 675.

11) J.F.Davidson and B.O.G.Schuler, Trans. Inst. Chem. Engrs., 38 (1960), 144.

12) K.Mori et al., Trans. Iron Steel Inst. Jpn., 22 (1982), 377.

13) F.O.Hoefele and J.K.Brimacombe, Metall. Trans. B, 10B (1979), 631.

14) Y.Ozawa and K.Mori, Trans. Iron Steel Inst. Jpn., 23 (1983), 764.

15) Y.Ozawa and K.Mori, Trans. Iron Steel Inst. Jpn., 26 (1986), 291.

16) A.E.Wraith and M.E.Chalkely, Advances in Extractive Metallurgy, ed. by M.J.Jones, IMM, London, (1977), 27.

17) L.Farias and D.G.C.Robertson, Injection Phenomena in Extraction and Refining, compiled by A.E.Wraith, Univ. of Newcastle upn Tyne, 1982, E1-25.

18) M.J.McNallan and T.B.King, Metall. Trans. B, 13B (1982), 165.

19) Y.F.Zhao and G.A.Irons, Proc. Sixth Int. Iron Steel Congress (Nagoya Japan) Vol. 1 (1990), 452.

20) S.Sakaguchi et al. Tetsu-to-Hagane, 63 (1977), S534.

21) T.A.Engh et al., Scand. J. Metall., 5 (1976), 21.

22) K.Marukawa et al., Tetsu-to-Hagane, 65 (1979), S154.

23) S.K.Sharma, Iron Steelmaker, 14 (1987), No. 3, 73.

24) A.A.Bustos et al., Copper 87 (Santiago, Chile), Vol. 4, (1987), 347.

25) M. Sano et al., Trans. Iron Steel Inst. Jpn., 26 (1986), 298.

26) M.Kawakami et al., Tetsu-to-Hagane, 68 (1982), 774.

27) T.H.Tacke et al., Metall. Trans. B, 16B (1985), 263.

28) A.H.Castillejos and J.K.Brimacombe, Metall. Trans. B, 18B (1987), 659.

29) M.Iguchi et al., Tetsu-to-Hagane, 78 (1992), 407.

30) K.Nakanishi, et al., Ironmaking Steelmaking, 2 (1975), 193.

31) K.Mori and M.Sano, Tetsu-to-Hagane, 67 (1981), 672.

32) M. Sano and K.Mori, Trans. Iron Steel Inst. Jpn., 23 (1983), 169.

33) O.Haida et al., Tetsu-to-Hagane, 66 (1980), S253.

34) T.C.Hsiao et al., Scand. J. Metall., 9 (1980), 105.

35) L.H.Lehrer, I & EC Process Design & Development, 7 (1968), 226.

36) S.Asai et al., Trans. Iron Steel Inst. Jpn., 23 (1983), 43.

37) J.H.Grevet et al., Int. J. Heat Mass Transfer, 25 (1982), 487.

38) Y.Sahahi and R.I.L.Guthrie, Metall. Trans. B, 13B (1982), 203.

39) F.Oeters et al., Steel Research, 59 (1988), 192.

40) A.H.Castillejos et al., Metall. Trans. B, 20B (1989), 603.

41) M. Sano et al., Trans. Iron Steel Inst. Jpn., 24 (1984), 825.

42) M. Takahashi et al., Trans. Iron Steel Inst. Jpn., 27 (1987), 626.

43) M. Takahashi et al., Trans. Iron Steel Inst. Jpn., 27 (1987), 633.

44) M. Takahashi et al., Trans. Iron Steel Inst. Jpn., 27 (1987), 866.

45) M.Byrne and G.R.Belton, Metall. Trans. B, 14B (1983), 441.

46) K.Mori and K.Suzuki, Trans. Iron Steel Inst. Jpn., 10 (1970), 232.

47) K.Mori et al., Tetsu-to-Hagane, 55 (1969), 1142.

48) R.J.Fruehan, Metals Technology, 7 (1980), 95.

49) R.J.Fruehan, Metall. Trans. B, 6B (1975), 573.

50) M.Kawakami et al., Tetsu-to-Hagane, 73 (1987), 661.

51) R.J.Fruehan, Ironmaking Steelmaking, 3 (1976), 33.

52) R.D.Pehlke and J.F.Elliot, Trans. Metall. Soc. AIME, 227 (1963), 844.

53) F.Yoshida and K.Akita, A.I.Ch.E.J., 11 (1965), 9.

54) W.F.Porter et al., Heat and Mass Transfer in Process Metallurgy, ed. by A.W.Hills, Inst. Min. Met., London, (1967), 79.

55) J.K.Brimacombe and F.D.Richardson, Trans. Inst. Min. Metall. Sect. C, 82 (1973), C63.

56) D.G.C.Robertson and B.B.Staples, Process Engineering of Pyrometallurgy, ed. by M.J.Jones, Inst. Min. Met., London, (1974), 51.
57) F.D.Richardson et al., Proceedings Darken Conference on Physical Chemistry in Metallurgy, US Steel Corp. Research Lab., Monroeville, (1976), 25.
58) M. Hirasawa et al., Trans. Iron Steel Inst. Jpn., 27 (1987), 277.
59) M. Hirasawa et al., Trans. Iron Steel Inst. Jpn., 27 (1987), 283.
60) J.T.Davies, Turbulence Phenomena, Academic Press, New York, (1972).
61) H.Gaye and J.C.Grosjean, Steelmaking Proceedings, AIME, 65 (1982), 202.
62) T.Usui et al., SCANINJECT Ⅱ, Lulea (Sweden), (1980), 12:1.
63) K.Umezawa and H.Kajioka, Tetsu-to-Hagane, 63 (1977), 2043.
64) K.Mori et al., Int. Conf. "Secondary Metallurgy", Verein Deutscher Eisenhuttenleute, (1987).

Table 1. Size of bubbles formed from nozzle

V_g \ N_c, N'_c	(I) Low V_g	(II) Intermediate V_g	(III) High V_g
(A) $N_c, N'_c < 1$	Maier[4] , $f < 100/min$ $d_B = (6\sigma d_n/\Delta\rho g)^{1/3}$ (3)	Davidson and Schüler[5] $d_B = 1.381 V_g^{2/5}/g^{1/5}$ (6) Davidson and Amick[6] $d_B = 0.54(V_g d_n^{0.5})^{0.289}$ (7)*	Leibson[9] , $Re(=\rho_g d_n u_{go}/\mu_g) > 10^4$ $d_{VS} = 0.71 Re^{-0.05}$ (13)* Tadaki[3] , $N_w > 16$ $d_{B,AV}^3(\Delta\rho g/\sigma d_n) = 1730 N_w^{-1.3}$ (14) Sano and Mori[10] $d_{B,AV} = 0.091(\sigma/\rho_l)^{0.5} V_S^{0.44}$ (15)*
	Mersman[7] $d_B = \{(3\sigma d_n/\Delta\rho g) + (9\sigma^2 d_n^2/\Delta\rho^2 g^2 + KV_g^2 d_n/g)^{1/2}\}^{1/3}$ (8) Tadaki[3] $d_B = (\sigma d_n/\Delta\rho g)^{1/3}(6 + 2.5 N_w)^{1/3}, N_w < 16$ (9) Sano and Mori[8] $d_B = \{(6\sigma d_n/\Delta\rho g)^2 + 9.5(V_g^2 d_n/g)^{0.867}\}^{1/6}$ (10)*		
(B) $1 < N_c, N'_c < 9$	Tadaki[3] , $N_w < 2.4(N_c - 1)$ $d_B = (6\sigma d_n/\Delta\rho g)^{1/3} N_c^{1/3}$ (4) Sano and Mori $d_B = \{(6\sigma d_n/\Delta\rho g)^2 N_c^2 + 9.5(V_g^2 d_n/g)^{0.867}\}^{1/6}$ (11)*	The same as (II)-(A).	
(C) $N_c > 9$	Tadaki[3] , $N_w < 16$ $d_{B,AV} = 3.8(\sigma d_n/\Delta\rho g)^{1/3}$ (5) Sano and Mori $d_B = \{3030(\sigma d_n/\Delta\rho g)^2 + 9.5(V_g^2 d_n/g)^{0.867}\}^{1/6}$ (12)*	The same as (II)-(A).	

* C.G.S. units.

Wetted nozzle[3] $N_c \equiv 4V_c \Delta\rho g/\pi d_{ni}^2 P$ (1)

Non-wetted nozzle[2] $N'_c \equiv 4T_B[\int_0^{V_c}(1/T)dV_c(T)]\Delta\rho g \sin\theta/\pi d_{ni} d_{no} P$ (2)

480

Table.2 Dimensionless correlation equations and transitional conditions for metal-phase mass transfer coefficient.

	$V_g < V_g^*$ (Region I)
$h_I < h_I^*$	$Sh = 1.0 \times [Pe(\rho_\ell g d_C^2/\sigma)(h_I/d_C)]^{1/2} \cdots\cdots\cdots\cdots\cdots\cdots\cdots\cdots$ (30)
$h_I > h_I^*$	$Sh = 6.0 \times [Pe(\rho_\ell g d_C^2/\sigma)(d_p Re^{-1/3}/d_C)]^{1/2} \cdots\cdots\cdots\cdots\cdots$ (31)
	$V_g^* < V_g < V_g^{**}$ (Region II)
$h_I < h_I^*$	$Sh = 1.0 \times 10^2 \times [Pe^\alpha(\rho_\ell g d_C^2/\sigma)^\beta(h_I/d_C)(h_{Sl}/d_C)]^{1/2} \cdots\cdots\cdots$ (32)
$h_I > h_I^*$	$Sh = 6.0 \times 10^2 \times [Pe^\alpha(\rho_\ell g d_C^2/\sigma)^\beta(d_p Re^{-1/3}/d_C)(h_{Sl}/d_C)]^{1/2} \cdots$ (33)

$\alpha = 0.425 \quad \beta = 0.654$

Transitional conditions :
h_I^* : $(h_I^*/d_C) = 36 \times (d_p Re^{-1/3}/d_C)$
V_g^* : $(h_{Sl}/d_C) = 10^{-4} \times Pe^{*(1-\alpha)}(\rho_\ell g d_C^2/\sigma)^{(1-\beta)}$
V_g^{**} : $V_g^{**} = 50 \ cm^3/s$

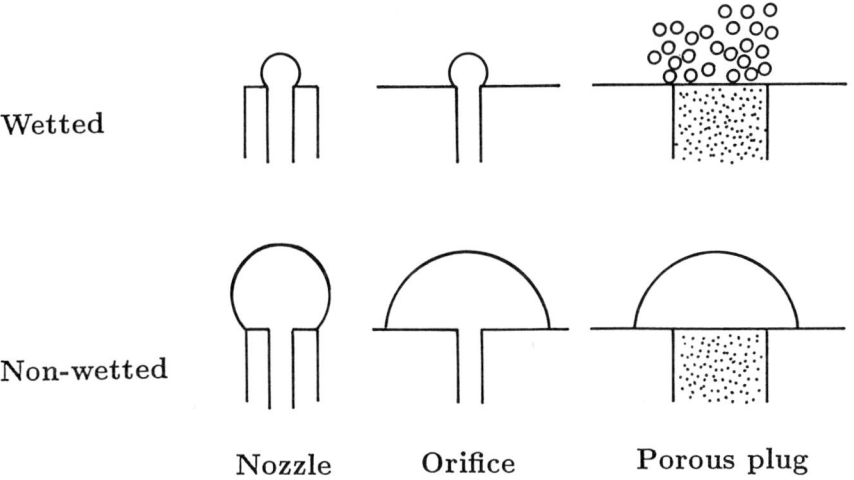

Wetted

Non-wetted

Nozzle   Orifice   Porous plug

Fig.1 Bubble formation at wetted and non-wetted nozzle, orifice and porous plug.

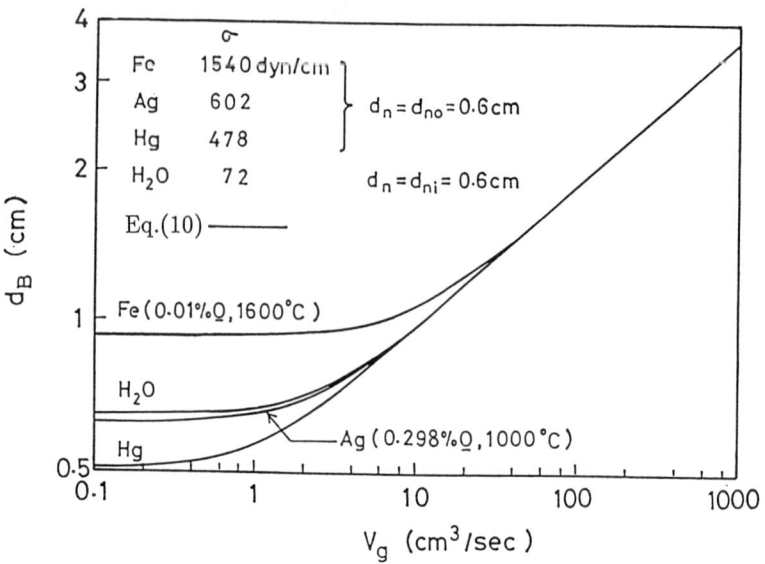

Fig.2 Comparison of sizes of bubble formed in various liquids [8].

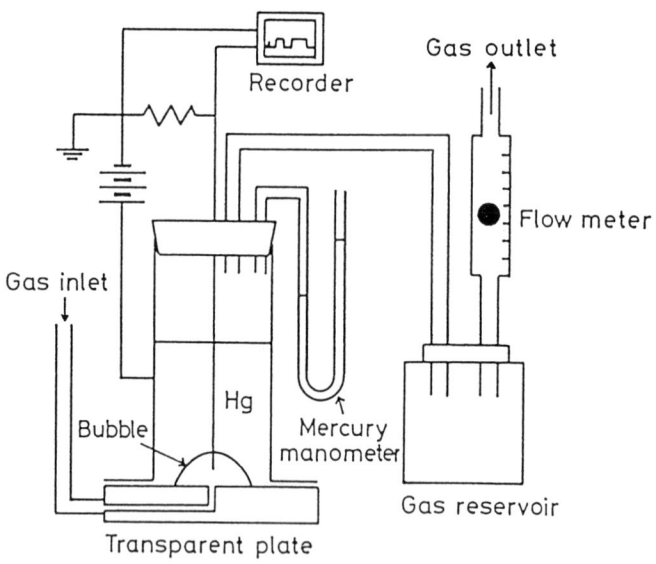

Fig.3 Schematic drawing of experimental apparatus [12].

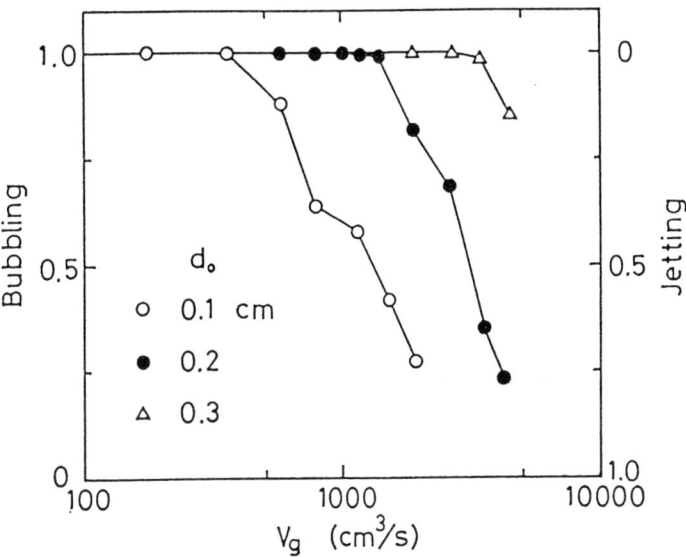

Fig.4 Time fractions for bubbling and jetting plotted
against gas-flow rate [12].

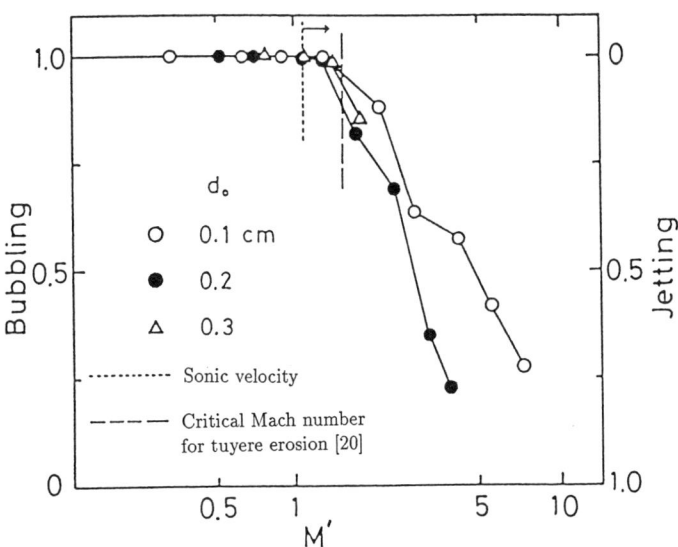

Fig.5 Time fractions for bubbling and jetting plotted
against nominal Mach number [12].

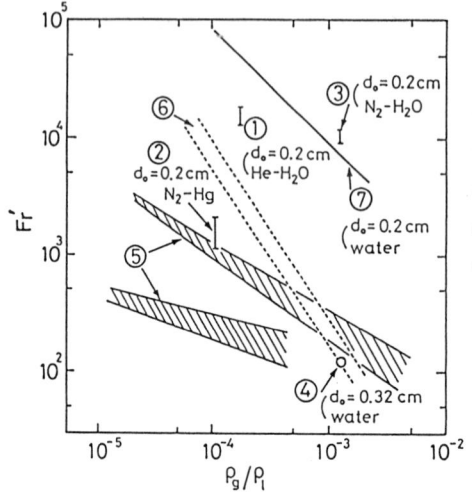

①	Ozawa and Mori [15]
②	Mori, Ozawa and Sano [12]
③	Ozawa and Mori [14]
④	Wraith and Chalkley [16]
⑤	Hoefele and Brimacombe [13]
⑥	Farias and Robertson [17]
⑦	McNallan and King [18]

Fig.6 Critical condition for bubbling-jetting transition of submerged gas jets[15].

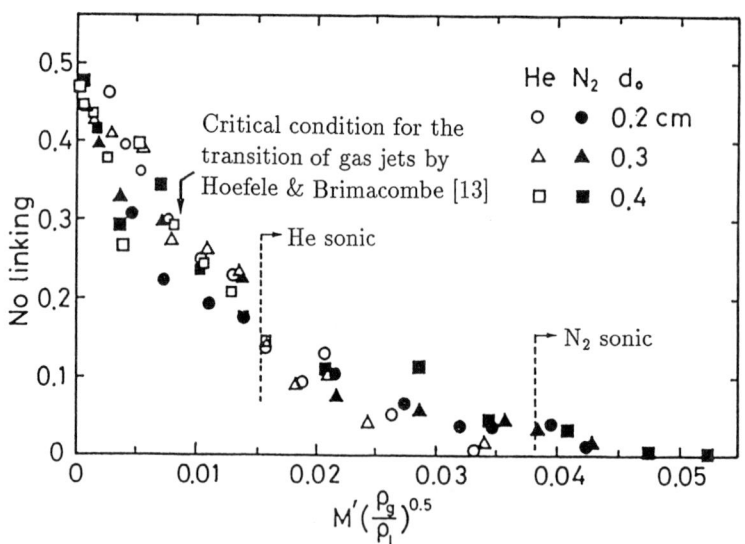

Fig.7 Time fraction for "no linking" plotted against $M'(\rho_g/\rho_\ell)^{0.5}$ [15].

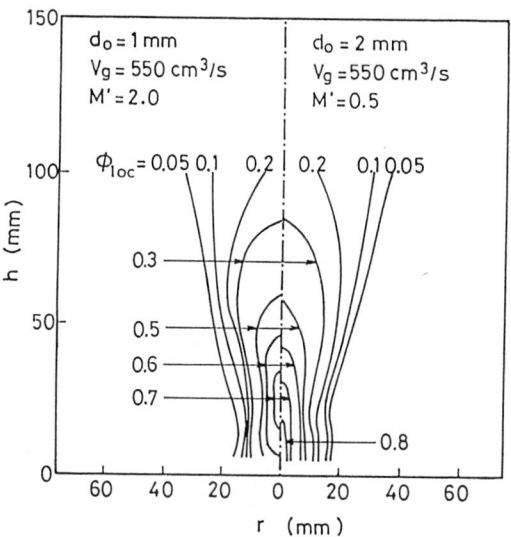

Fig.8 Contour map of gas holdup [25].

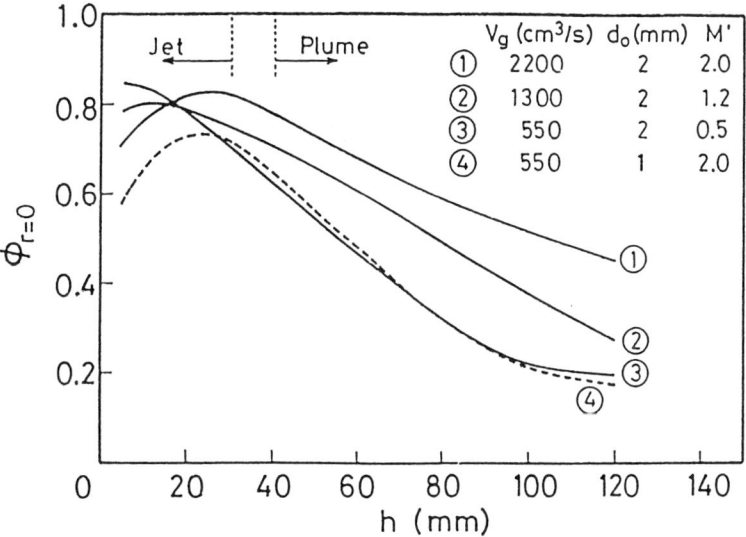

Fig.9 Relation between gas holdup on center axis and
vertical distance from orifice [25].

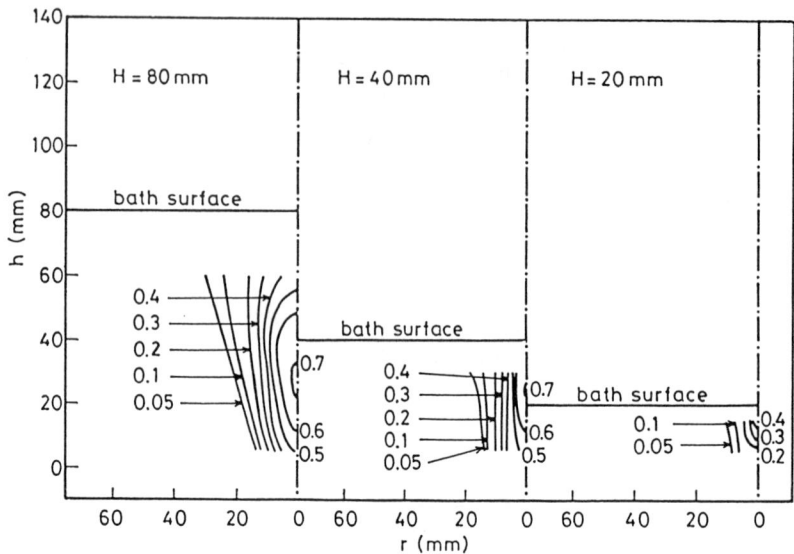

Fig.10 Effect of bath depth on dispersion of nitrogen injected into mercury (Nozzle, $d_o = 1mm, V_g = 960cm^3/s,$ $M' = 3.5$) [25].

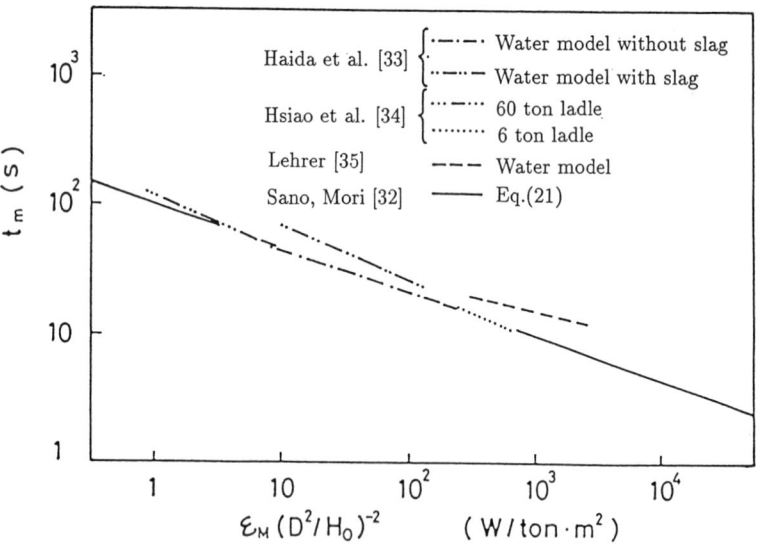

Fig.11 Relation between mixing time and $\varepsilon_M(D^2/H_o)^{-2}$ [32].

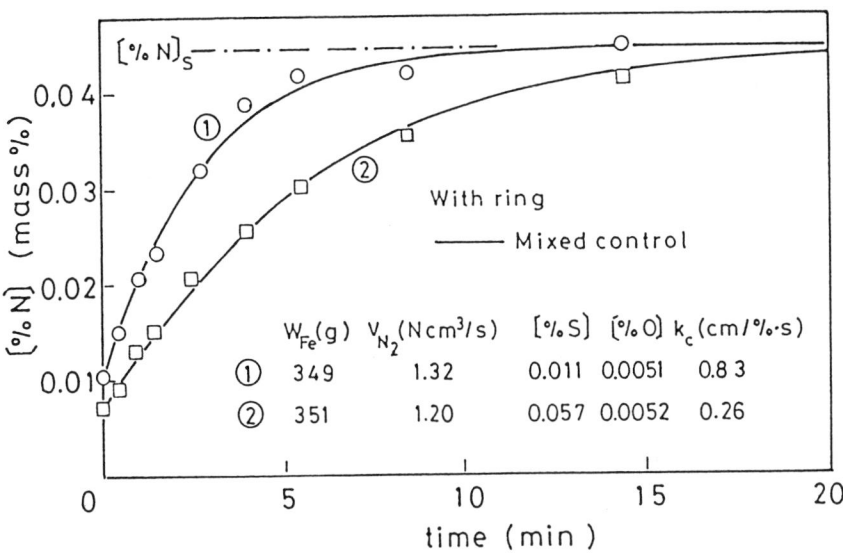

Fig.12 Change in nitrogen concentration with time [41].

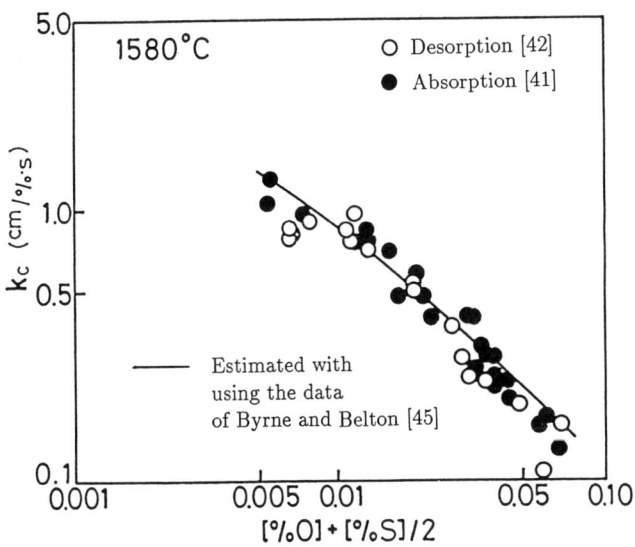

Fig.13 Chemical reaction rate constant plotted against
$[\%O] + [\%S]/2$ [42].

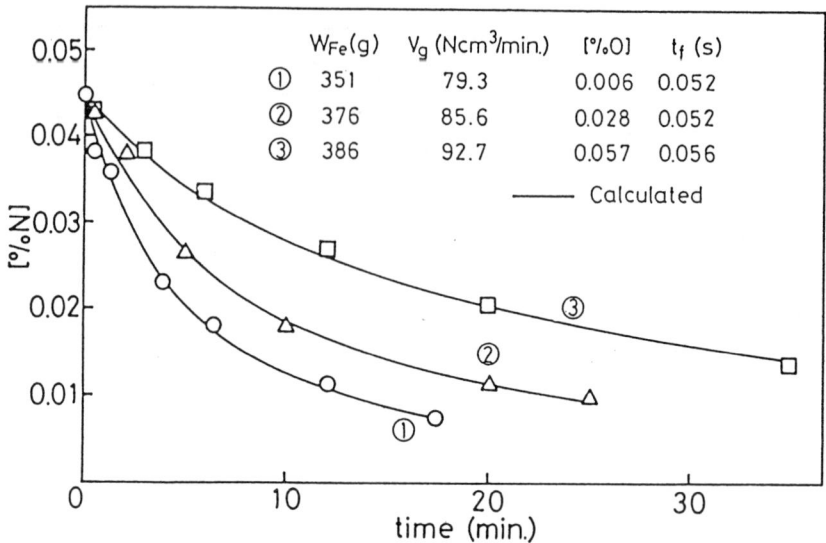

Fig.14 Change in nitrogen concentration with time [42].

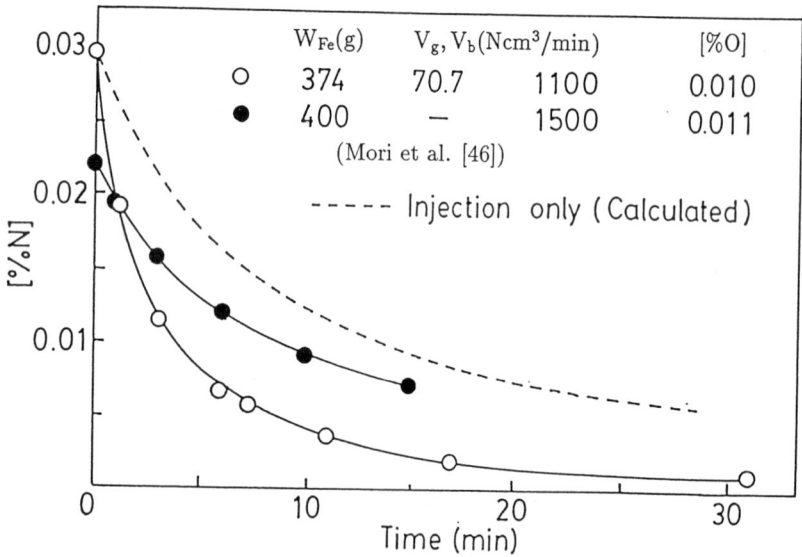

Fig.15 Comparison of changes in nitrogen concentration
with time by Ar injection together with blowing,
Ar blowing and Ar injection [43].

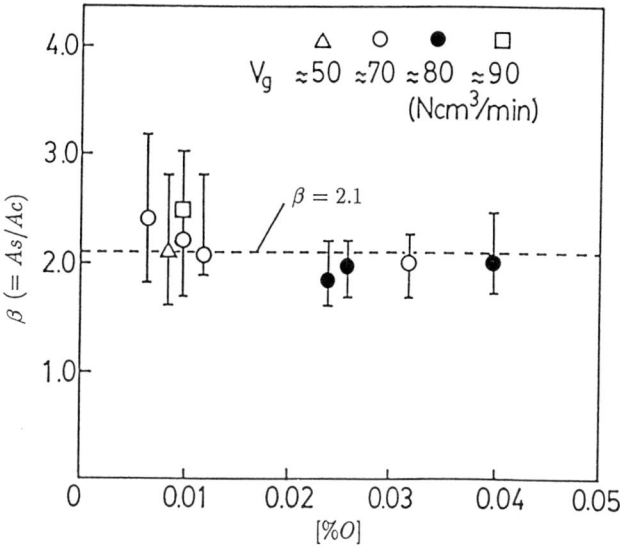

Fig.16 Ratio of the effective free surface area of the melt to the cross sectional area of the crucible [43].

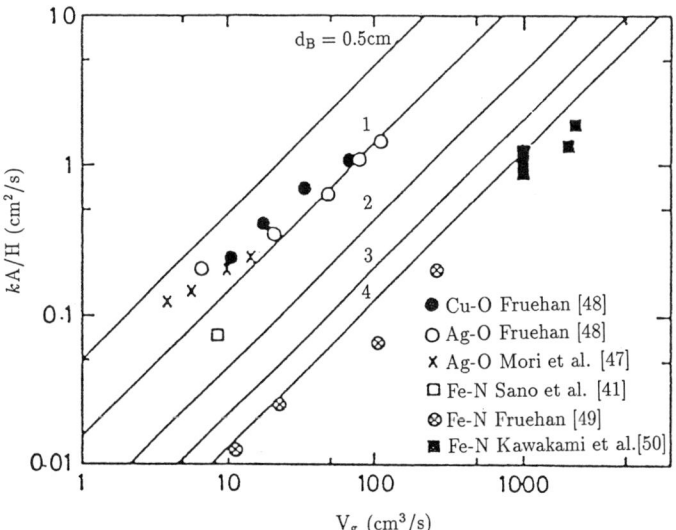

Fig.17 Relation between volumetric mass transfer coefficient and gas flow rate.

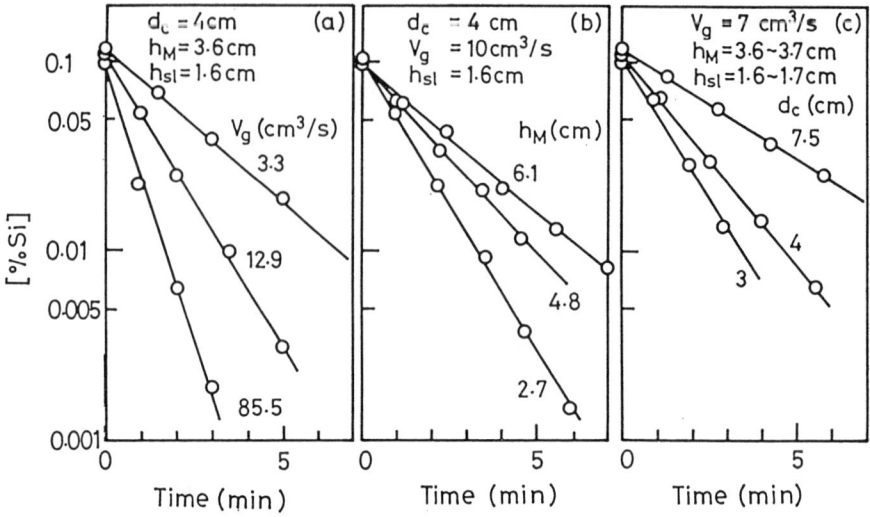

Fig.18 Typical relation between log[% Si] and time [58].

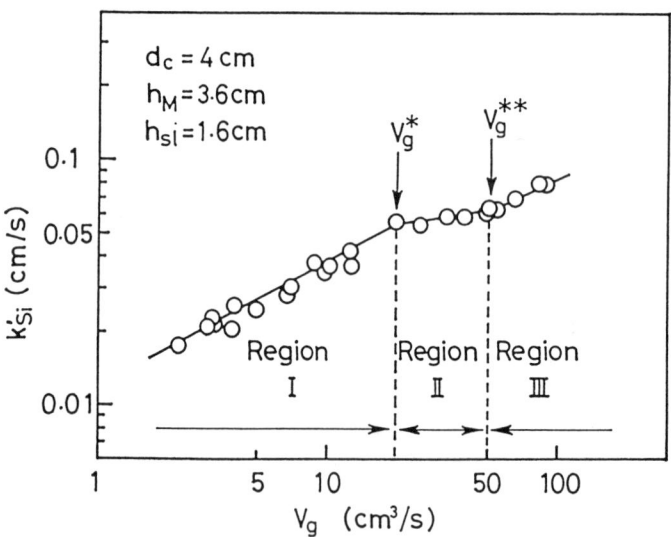

Fig.19 Relation between k'_{Si} and gas-flow rate
(d_c=4cm) [58].

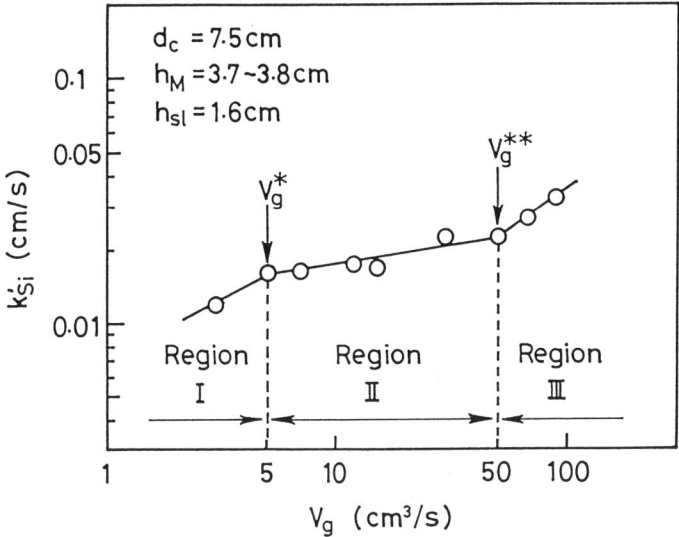

Fig.20 Relation between k'_{Si} and gas-flow rate
(d_c=7.5cm) [58].

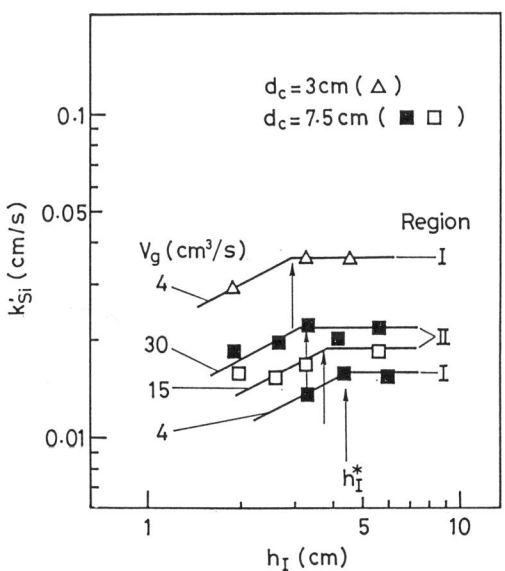

Fig.21 Relation between apparent mass transfer coefficient of
Si, k'_{Si}, and injection depth of metal phase, h_I [59].

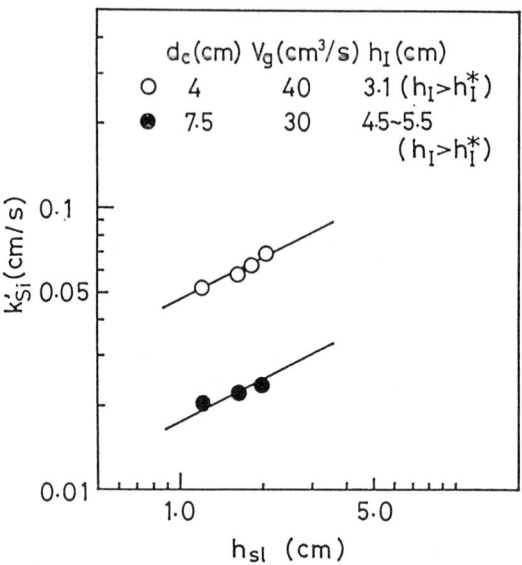

Fig.22 Relation between k'_{Si} and slag depth, h_{Sl} [59]

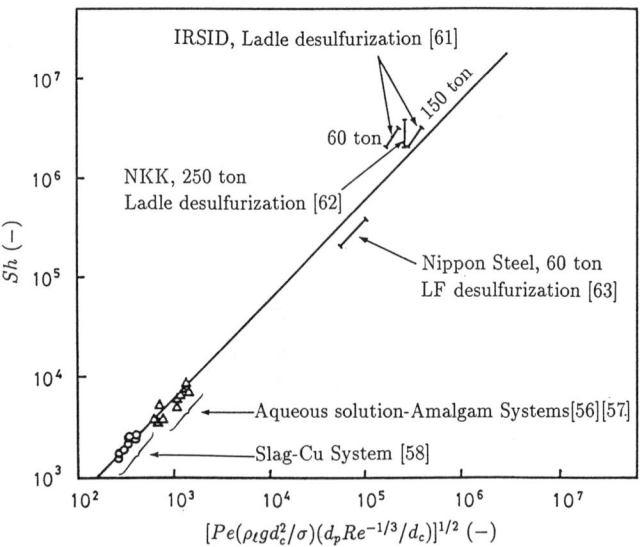

Fig.23 Dimensionless correlation of mass transfer data on metal-slag reactions [64].

FUNDAMENTAL ASPECTS OF SOLIDS INJECTION

FOR BATH SMELTING

G.A. Irons

Department of Materials Science & Engineering
McMaster University
Hamilton, Canada

Abstract

In bath smelting, the method of injection of ore, concentrates, reductants and oxidants is one of the few variables available to the process designer for control of the process. In this paper, aspects relevant to the use of powder injection in smelting processes will be reviewed, including: gas and particle trajectories, lance clogging, particle reaction rates, gas evolution, and mixing.

Proceedings of the
Savard/Lee International Symposium on Bath Smelting
Edited by J. K. Brimacombe, P. J. Mackey,
G. J. W. Kor, C. Bickert and M. G. Ranade
The Minerals, Metals & Materials Society, 1992

Introduction

In bath smelting, for both ferrous and non-ferrous processes, the method of addition or injection of the ore, concentrates, reductants and/or oxidants is one of the few variables which the process designer can manipulate to advantage. Often the solids are very fine which precludes simple top addition of the material; injection must be employed. The trajectories of the particles and carrier gas are very important, as are the contacting patterns of the particles with the gas and the melt into which they are injected. In bath smelting processes there is usually a large amount of gas evolution as a result of the smelting reactions. This phenomenon further complicates the flow patterns and reaction mechanisms in the vessel. In this paper, an attempt is made to develop a fundamental understanding of the influence of the injection practice on contacting patterns and reaction rates. Much of this work was originally performed in support of ladle metallurgy process modelling.

Powder injection technology is well-established in ladle metallurgy, and a fundamental engineering basis has been developed to support this technology. In this paper, aspects relevant to in-bath smelting will be reviewed, including: the flow regimes for powder injection (gas and particle trajectories), lance clogging, particle reaction rates, mixing and gas evolution.

Flow Regimes for Powder Injection

There has been considerable work performed with water models of injection processes, and a lesser amount with liquid metals. From this work, a unified picture of the various flow patterns has emerged [1]. Typical flow regimes that are expected in full scale injections are shown in Figure 1. For fine particles (approximately 100 μm or less) at very low loading, the carrier gas forms bubbles which are penetrated by the particles (Figure 1a). At higher loading for the same particles, continuous jets of gas and particles form (Figure 1b). For coarser particles, it is difficult to form continuous jets at any level of loading (Figures 1c and d). Figure 2 is a schematic representation of the various possible regimes which are described in detail in the original paper [1].

For the injection of gas alone (the left side of Figure 2), small bubbles are produced at low flow rate, in the "constant-volume" regime where the balance between surface tension and buoyancy governs bubble size. In the intermediate flow rate range, the "constant-frequency" regime, the bubble size is controlled by a balance between buoyancy and the inertia of the liquid around the growing bubbles. At high flow rates, the bubbles break down into steady jets, due to surface instabilities [2].

The major difference between the coarse and fine particles in Figure 2 is that the flow of fine particles can be coupled with the carrier gas flow, producing jets more readily. Most pneumatic conveying systems operate with medium flow rates according to the classification in Figure 2 (0.03 < Ma < 0.5). As will be shown in a following section, bubbling usually results in lance clogging. Therefore, the use of fine particles which form jets with the gas is preferred. In such cases, the gas and particles exit the lance with essentially the same velocity, and travel together along the length of the jet. The orientation of the lance has little effect on the flow regime immediately adjacent to the lance, but, of course, has a dramatic effect on the particle and gas trajectory, as discussed in the next section.

Models for Jet Penetration

As shown in the previous section, fine particles flow with almost the same velocity as the gas in an homogenous jet. For downward injection, the depth of penetration can be calculated with models which essentially balance the momentum of the jet against the buoyant force [3, 4, 1]. These models have recently been extended [5]; the model formulation and nomenclature are illustrated in Figure 3. The model is quite general in that the jet can composed of gas alone or gas and fine particles, and the jet can be injected at any downward angle into any fluid: gas, metal or slag.

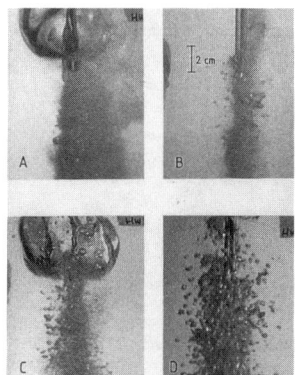

Figure 1: Photographs from high-speed cinematography of sand injection into water with helium carrier gas. (A) uncoupled flow, d_p = 115 μm, d_o = 7 mm, Q_R = 0.124 Nm^3/min, W = 0.636 kg/min. (B) coupled flow, d_p = 115 μm, d_o = 7 mm, Q_R = 0.042 Nm^3/min, W = 2.4 kg/min. (C) uncoupled flow, d_p = 450 μm, d_o = 7 mm, Q_R = 0.150 Nm^3/min, W = 0.702 kg/min. (D) partially coupled, d_p = 450 μm, d_o = 7 mm, Q_R = 0.217 Nm^3/min, W = 4.91 kg/min.

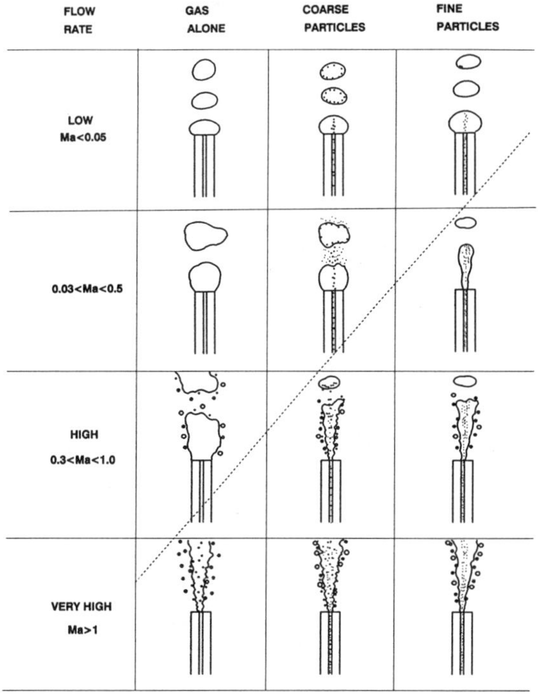

FLOW RATE	GAS ALONE	COARSE PARTICLES	FINE PARTICLES
LOW Ma<0.05			
0.03<Ma<0.5			
HIGH 0.3<Ma<1.0			
VERY HIGH Ma>1			

Figure 2: Summary of the different possible regimes in injection of gas and gas-powder mixtures into liquids [1]. The Mach Numbers give general guidelines concerning the flow rates for each regime, however, the transitions do not depend on Mach Number in a fundamental manner. The dotted line indicates in a general way the transition from bubbling to jetting; above the line bubbling occurs and jetting below. Similar flow patterns in the vicinity of the lance would be observed for downward injection.

495

Momentum in the vertical direction is dissipated by buoyant forces:

$$\frac{d(\rho A U^2 \sin\alpha)}{d\partial} = -gA(\rho_a - \rho) \tag{1}$$

whereas the momentum in the horizontal direction is unopposed:

$$\frac{d(\rho A U^2 \cos\alpha)}{ds} = 0 \tag{2}$$

The jet continuously entrains surrounding fluid into the jet. The local rate of entrainment is proportional to the local axial velocity [6]:

$$\frac{d(\rho A U)}{ds} = K(\rho_j \rho)^{1/2} U d \tag{3}$$

The entrainment coefficient, K, for single-phase jets is 0.25 [7], whereas a value of 0.06 is appropriate for gas-particle jets [1]. The volume fraction of the entrained fluid can be calculated from the area of the jets, corrected for local pressure:

$$\theta_a = 1 - \frac{U_o d_o^2}{U d^2} \tag{4}$$

and the local jet density becomes:

$$\rho = (1 - \theta_a)\rho_o + \theta_a \rho_a \tag{5}$$

Solving these equations simultaneously, one obtains the jet trajectories; typical results are shown in Figure 4 as a function of the modified Froude Number, N_{Fr}'. The model results are compared with experimental results from several groups of investigators in Figure 5. This data was acquired in a wide variety of systems including water and liquid lead; in some instances the lances were submerged, whereas in others it was above the bath surface, and some of the injections were at inclined angles. Therefore, such codes can be used to design the injection conditions to achieve penetration to precise locations. It is important to note that good penetration of jets can be obtained with non-submerged lances which is convenient for very exothermic reactions or aggressive slag environments.

Lance Clogging

One of the major operational headaches of powder injection equipment is lance clogging which occurs near the lance tip due to interactions of the flow with the melt. This is different from clogging of the dispenser or conveying lines; these are pneumatic handling problems. The lance clogging problem was studied with a liquid lead model of an injection system [8]. It was found that two conditions must be met simultaneously for lance clogging:

(1) the system must be operating in the bubbling regime, and

(2) the temperature on the inside of the lance must be below the melting point of the liquid alloy.

In the bubbling regime, liquid can enter the end of the lance between bubbles. If the inner circumference of the lance is cold enough, some liquid will freeze, and after a few bubbles, the lance will become clogged. Therefore, operating continuously in the jetting regime will avoid lance clogging because no liquid has the opportunity to enter the lance tip. A detailed heat transfer analysis of the lance revealed that powder flow substantially increases lance cooling, thus condition (2) is easily met. Consequently, the most effective way to prevent clogging is to use fine particles at high loadings to maintain jetting.

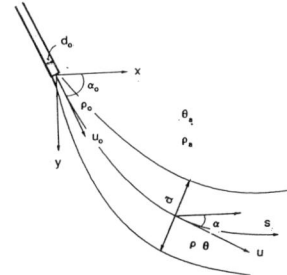

Figure 3: Schematic representation of a jet showing the nomenclature for the model. The jet expands from the lance diameter, d_o, to d because of entrainment and deceleration of the jet.

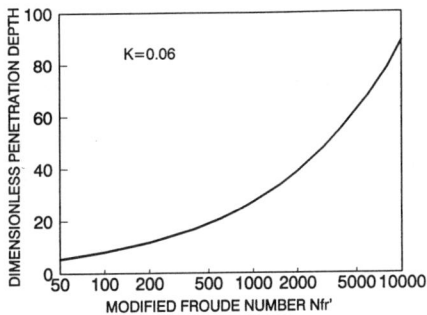

Figure 4: The dimensionless jet penetration (depth/lance diameter) plotted against the modified Froude Number, N_{Fr}'.

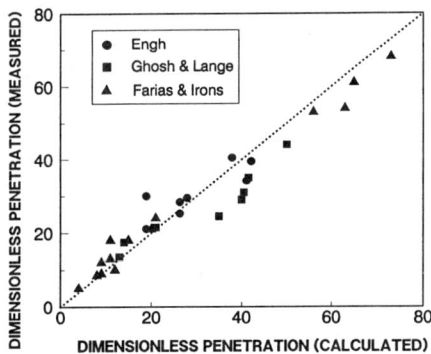

Figure 5: Comparison of the penetration of gas-particle jets into various liquids compared with the model. Engh [3,4] injected polystyrene particles into water; Ghosh and Lange [17] injected silica, graphite, and soda into water, ethanol, and tetrabromethane; Farias and Irons [1] injected Q-cel and silica into water and liquid lead, both submerged and non-submerged.

497

Reaction Rates

The flow regimes for the powder, the carrier gas, slag, metal and reaction products are very complex for powder injection in bath smelting processes, as the first section demonstrated. For heterogenous systems with several phases and many possible reactions, the interfacial area between the various phases has a dominant influence on which reactions will proceed, thus the flow regimes or contacting patterns are very important. At the present time, our understanding of contacting patterns in smelting systems is limited, thus, it is instructive to examine work on metal refining by powder injection. This work will be discussed in this section, where it will be shown that the contacting patterns between the injected particles, carrier gas and metal have a overwhelming influence on the reaction rate. Refining reactions are generally carried out at rates several orders of magnitude slower than smelting reactions. Furthermore, smelting reactions generally evolve huge quantities of gas which has an enormous effect on contact, and reaction rates; these factors will be discussed in the last section of this paper.

Figure 6 shows a schematic representation of the reaction sequence for a powder refining process, in this particular case calcium carbide hot metal desulphurization [9]. Nevertheless, it illustrates the general possible rate controlling steps for many refining operations when only one element is involved. There are several consecutive steps; any one or more could be rate-controlling:

(1) The rate at which sulphur-rich liquid is pumped into the plume by entrainment; this is termed pumping or mixing control. (Mixing is discussed more fully in the next section).

(2) The rate at which sulphur diffuses through the boundary layers around the particles. This rate depends on whether the particles are located on the bubble interfaces or in the liquid, thus it is called contact control.

(3) The rate at which sulphur penetrates the product layer which depends on the nature of the diffusive process through the layer. In the case of calcium carbide desulphurization, it is due to calcium vapour diffusion through the solid product layer of calcium carbide and graphite. In other systems, a liquid reaction product slag may form, or the layer may be stripped off.

As will be shown in the next section, the rate of mixing is often fast enough not to be rate-controlling. Furthermore, the time for reaction as the particles rise through the melt is short, thus the extent of reaction is limited, and the product layers are generally quite thin. Consequently, the third step is usually not rate controlling. The most important rate-controlling step is contact control.

Contact control was investigated with a mathematical model to represent refining at particles in the melt and those positioned on the bubble interfaces, as shown schematically in Figure 6 [9]. The reaction rate constant determined from the model is shown in Figure 7 as a function of the solids injection rate. If all the particles are in the liquid, the rate increases as the solids rate increases because more particles are in the melt at any instant. This also represents the maximum possible reaction rate. On the other hand, if all the particles are positioned on the bubble interfaces, then increasing the solids flow rate will have little effect on the reaction rate because the bubbles will just become more loaded with particles; the rate is limited by diffusion to the bubble interfacial area. (For a particular carrier gas flow rate, the bubble interfacial area will not change substantially). Thus, one can see that the contact pattern has a dramatic effect on reaction rates and therefore reagent utilization. Actual experimental data for the injection of calcium carbide into 3 Tonne heats of hot metal is also shown in the diagram. From this observation, it was deduced that only 30% of the particles enter the liquid. This is reasonable since liquid iron does not wet calcium carbide.

Further experiments were performed in 60 kg heats of iron to investigate the contacting problem [10]. It was again found from the reaction rates that only 30% of the particles contact the melt. In these experiments the bath cooling rate was measured during the injection. If the melt were acting as a

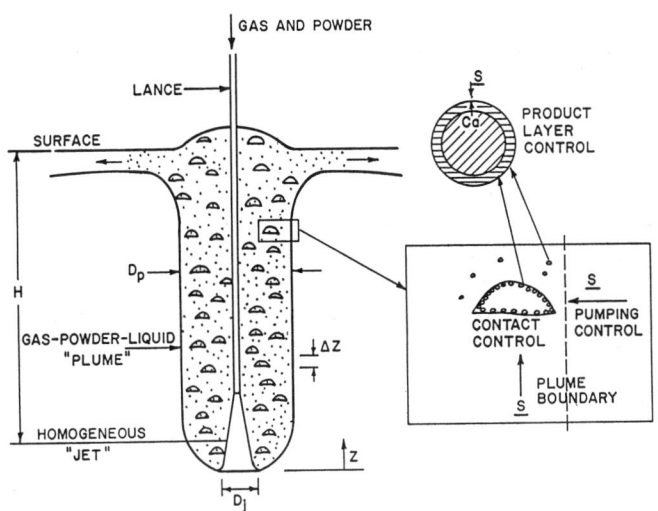

Figure 6: Schematic representation of the physical and chemical phenomena in the rising plume. The carrier gas and injected particles rise and react through the plume. The inset indicates that the particles may be on the bubble interfaces or in the liquid. At individual particles, calcium vapour diffuses through the product layers.

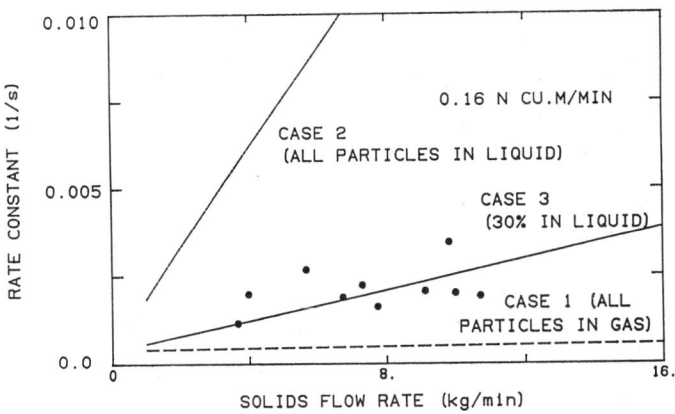

Figure 7: The experimental first-order rate constants for desulphurization in a 3 Tonne ladle as a function of the calcium carbide injection rate at 0.16 Nm^3/min of argon compared with the model cases.

calorimeter, then the theoretical bath cooling rate can simply be calculated from the heat capacities of the melt, the powder and the powder injection rate. The measured cooling rate was only 30 to 50% of this theoretical limit which confirms the conclusion that particle-liquid contact is quite limited.

Refining reactions in practice can seldom be limited to one element, in fact calcium carbide hot metal desulphurization also involves deoxidation. Techniques and models to understand multi-component refining have been developed, but are beyond the scope of this review [11, 12].

Mixing Phenomena

In oxygen steelmaking, it is well-known that mixing is very important; it has been found that conventional top blowing practices can be improved by the injection of relatively small quantities of bottom-stirring gas. The aim of this section is to examine the effect that powder injection may have on mixing phenomena.

For gas-only injection, the time to reach a particular degree of mixing after the addition of a tracer is usually determined by an equation of the form:

$$t_1 = B\epsilon^{-c}$$

(6)

where ϵ is the specific energy input rate, and A and B are parameters which depend on the particular geometry of the vessel, the number of injectors, the presence of a slag-simulating phase, and the particular degree of mixing chosen [13, 14].

The specific energy input rate for gas injection is usually taken as the buoyant potential energy of the gas in rising from its point of injection to the bath surface:

$$\epsilon_G = \frac{2 P_R Q_R}{M_1} \ln\left(1 + \frac{\rho_1 g Z}{P_R}\right)$$

(7)

The specific kinetic energy input rate of an injected phase is:

$$\epsilon_k = \frac{U^2 W}{2 M_1}$$

(8)

Not only is the kinetic energy input rate important, but so is the degree of coupling of this input energy with the liquid to move it around in the vessel. A simple calculation was performed in which it was assumed all of the particle momentum is transferred to the liquid; this is the best possible coupling. The results in Figure 8 indicate that the kinetic energy transfer is poor [15]. The energy transferred from the particles to the liquid ϵ_1, is two orders of magnitude smaller than the buoyant energy from the carrier gas, ϵ_G. Therefore, contrary to intuition, particles cannot directly contribute to improved mixing.

However, particles can exert an indirect contribution to mixing which is illustrated with some experiments performed in a water model of a torpedo car [15]. The results are shown in Figure 9. For gas-only injection, the mixing time decreases as the lance is moved more deeply, as Equations 6 and 7 would suggest. For the gas and powder injection in this model, the flow was in the jetting regime, and due to the particles, the jet was powerful enough to penetrate to the bottom of the torpedo car. Consequently, the gas rose from the bottom, and the mixing times were comparable to the deepest injection with gas-only. This illustrates the fact that the flow regime and particle trajectories need to be carefully considered when designing new processes.

As mentioned in the previous section, mixing is one of the consecutive steps in the mass transfer process for refining. This concept is illustrated in Figure 10 with an electrical analog for the resistances to mass transfer [15]. This diagram was originally prepared for hot metal desulphurization in

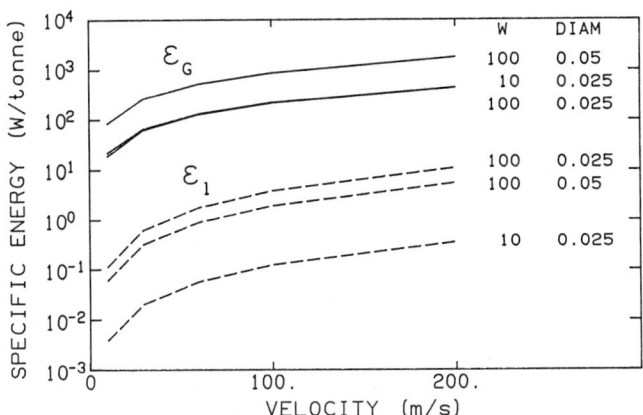

Figure 8: Specific potential energy input of gas, ϵ_G, and specific kinetic energy of particles which is transferred to the liquid, ϵ_l, as a function of gas and particle injection velocities, showing that the latter quantity is relatively small under a variety of conditions; calcium carbide flowrate W (kg/min) and lance diameter are indicated for each condition.

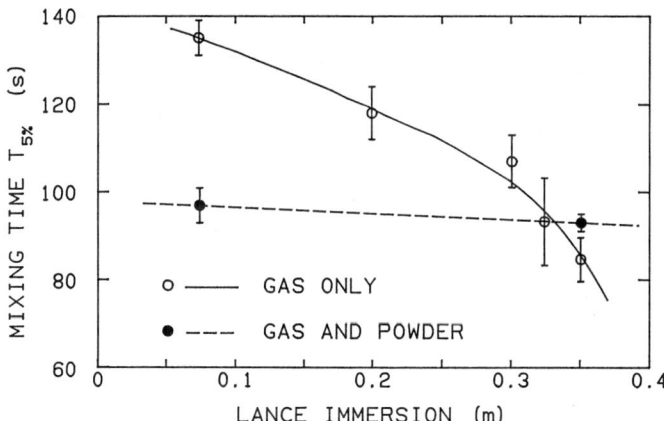

Figure 9: Mixing times at 5% from perfect mixing in 240 tonne torpedo car model as a function of lance immersion depth for air alone and Q-cel injection; gas and solids flow rates were 0.1 Nm^3/min and 0.3 kg/min, respectively.

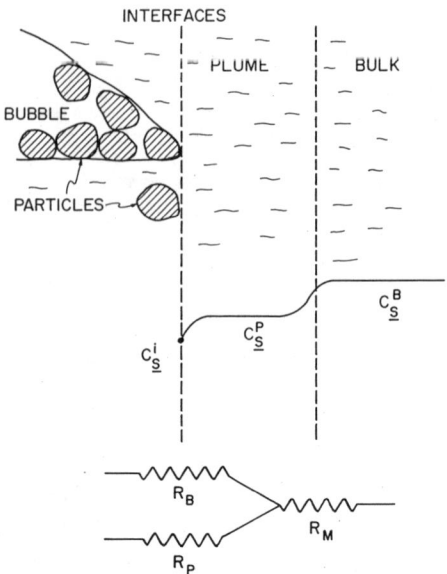

Figure 10: Schematic representation of model in which sulphur is pumped into the plume region from the bulk of the torpedo car, and then diffuses to the particles which may be on the carrier gas bubble surfaces or in the liquid. The expected concentration profile along with the electrical circuit analog for resistance to mass transfer are also shown. Under most circumstances resistance due to mixing R_M is small and resistance due to diffusion to particles in the liquid, R_P, is large, so that resistance due to diffusion to the bubbles, R_B, primarily controls the rate.

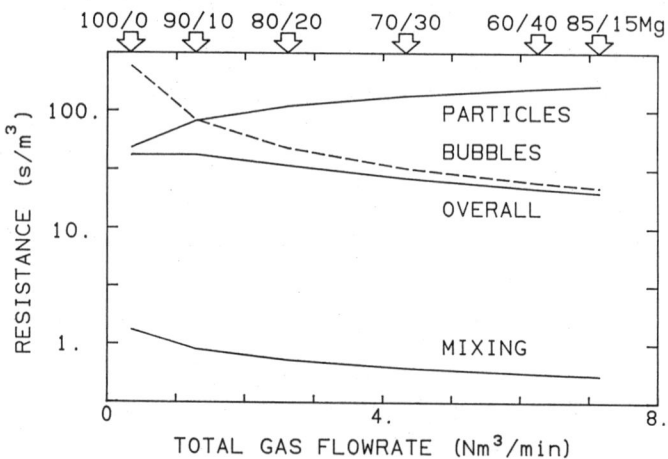

Figure 11: Resistances to mass transfer plotted as a function of the total gas flow rate, including any gas releasing effects. The basis for the diagram is the injection of 38 kg/min of calcium carbide with various limestone mixtures at 180 kg solids/Nm³ of nitrogen into 150 tonne car having metal depth of 2.1 m. Across the top the corresponding calcium carbide/limestone ratios are given, as well as the gas of 85% calcium carbide-15% magnesium. Note that R_M due to mixing is almost two orders of magnitude smaller than other resistances.

a torpedo car. If the mixing rate is fast, then R_M will be small and there will be a small concentration difference between the bulk and the plume. The mixing resistance can be estimated from the mixing time measurements. The contact resistances for particles in the liquid, R_p, and particles on the bubble interfaces, R_B, were estimated from the model in the previous section. The results of the calculation are summarized in Figure 11. It shows that the resistance due to mixing is approximately two orders of magnitude smaller than the other resistances, indicating that mixing is fast enough to keep up with the rate at the particle and bubble interfaces. Refining rates are generally several orders of magnitude smaller than smelting rates, therefore good mixing is even more crucial in bath smelting processes. The gas evolution rates in smelting are also several orders of magnitude higher, thus mixing should also be faster. Most of the gas in smelting processes is generated by the smelting reactions themselves which has a dramatic effect on the contacting patterns and transfer rates. This subject is discussed in the next section.

The Effects of Gas Evolution

In bath smelting processes, the particles often evolve considerable gas volumes as they enter the melt. In smelting-reduction processes for ironmaking, coal is combusted to produce carbon monoxide and carbon dioxide, and carbon monoxide is also generated by the iron ore reduction. In non-ferrous smelting large volumes of sulphur dioxide are produced. Obviously, this creates a great deal of turbulence, but it is not obvious what effect this gas evolution has on transport phenomena. In order to simulate the effects of gas evolution from powder injection, aluminum hydroxide was injected into liquid lead [16]. Aluminum hydroxide decomposes according to:

$$2\,Al(OH)_3 \rightarrow Al_2O_3 + 3\,H_2O \tag{9}$$

This reaction occurs between 200 and 350°C, whereas the temperature of the lead in the experiments was between 400 and 500°C. The injection lance was placed adjacent to a quartz window so that the flow patterns could be seen. High speed photographs are reproduced in Figure 12 where it can be seen that the water vapour is evolved as very small bubbles, in contrast to the large carrier gas bubbles. It would be expected that reactive gases evolved in this manner would react very quickly because of their high surface area.

During the injection of the aluminum hydroxide, the bath temperature was monitored. It decreased very rapidly; the bath cooling rate is shown in Figure 13. There is very good agreement between the measured cooling rates and those calculated assuming that the melt acts as a calorimeter, and that all the energy necessary to heat and decompose the aluminum hydroxide comes from the lead. This is in marked contrast to the cooling rate if non-reactive particles are injected, such as the glass spheres also shown in Figure 13. For non-reactive particles, the cooling rate is only 30 to 50% of the calorimetric limit. As mentioned in the section on Reaction Rates, the bath cooling rate for calcium carbide injection is also only 30 to 50% of the calorimetric limit, and the rate of mass transfer is only 30% of that expected if all the particles enter the melt. The conclusion from this finding is that the gas evolution substantially improves particle-liquid contact. It may be postulated that the gas release creates stirring and turbulence which pushes other particles in contact with the melt, which in turn generates more gas and more contact. In this manner virtually complete particle-liquid contact is achieved. Similar results could be anticipated for mass transfer rates.

Summary

In this paper, it has been shown that the method in which powders are injected into melts has a profound influence on flow patterns, heat transfer rates and mass transfer rates. There is also a great opportunity to optimize the design of bath smelting processes by judiciously using this information for selecting the injection conditions and configurations. It is also apparent that there are still significant gaps in our understanding of the detailed phenomena associated with bath smelting processes. It is expected that as these processes develop further, fundamental understanding will advance along side these processes.

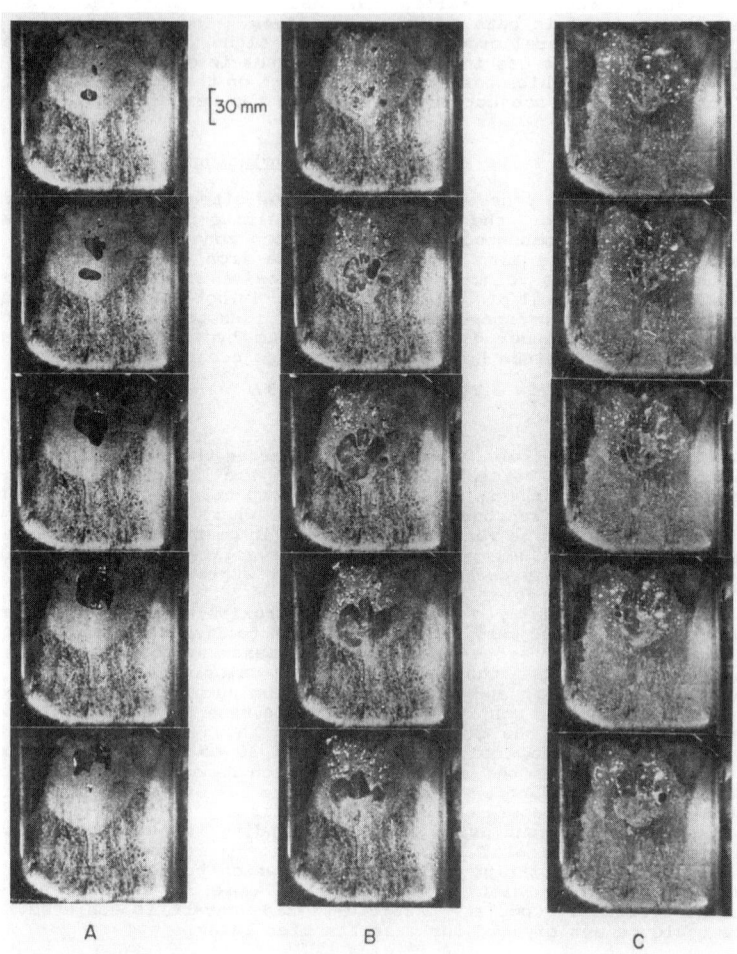

Figure 12: Frames from high speed photography for the aluminum hydroxide injections. The time between frames is 12.5 ms. (A) Nitrogen alone at 0.040 Nm^3/min. (B) 80% Glass shot, 20% Al(OH)$_3$ at 1.84 kg/min and nitrogen at 0.039 Nm^3/min. (C) 100% Al(OH)$_3$ at 0.64 kg/min and nitrogen at 0.029 Nm^3/min.

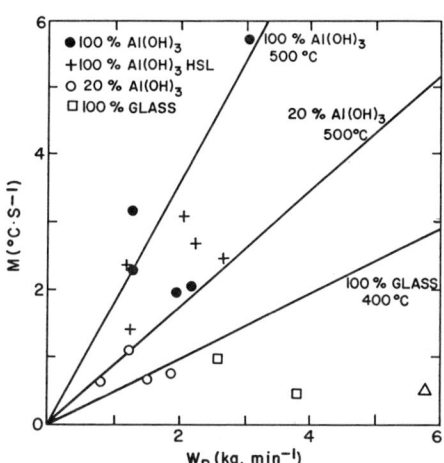

Figure 13: Lead bath cooling rate as a function of the solids injection rate for Al(OH)₃ and glass shot mixtures. The solid lines show the cooling rates expected if all the particles were heated to the indicated bath temperature. HSL stands for hockey-stick lance.

Nomenclature

A	cross-sectional area of jet (m^2)
B	constant in Equation 6
C	constant in Equation 6
d	jet diameter (m)
d_o	lance diameter (m)
d_p	particle diameter (m)
g	gravitational constant (m/s^2)
K	entrainment coefficient (-)
Ma	nominal Mach Number, U/velocity of sound (-)
M_l	mass of liquid (kg)
N_{Fr}'	modified Froude Number, $\rho U_o^2/(\rho_a - \rho)/g/d_o$ (-)
P_R	pressure at reference conditions (Pa)
R_B	mass transfer resistance at bubble interfaces (s/m^3)
R_M	mass transfer resistance due to mixing (s/m^3)
R_p	mass transfer resistance at particle interfaces (s/m^3)
s	distance along jet trajectory (m)
Q_R	gas flow rate at reference conditions (m^3/s)
t_i	time to reach i% away from complete homogenization (s)
U	velocity (m/s)
W	mass flow rate (kg/s)
Z	vertical distance (m)
α	jet angle from the horizontal
ρ	jet density (kg/m^3)
ρ_a	density of ambient fluid around jet (kg/m^3)
ρ_l	liquid density (kg/m^3)
ϵ	specific energy input (W/kg)
θ_a	volume fraction of ambient phase entrained in jet (-)

References

1. L.R. Farias and G.A. Irons: <u>Metall. Trans. B</u>, 1985, Vol. 16B, pp. 211-225.

2. Y.F. Zhao and G.A. Irons: <u>Metall. Trans. B</u>, 1990, Vol. 21B, pp. 997-1003.

3. T.A. Engh and H. Bertheussen: <u>Scan. J. Met.</u>, 1975, Vol. 4, pp. 241-49.

4. T.A. Engh, K. Larsen and K. Venas: <u>Ironmaking and Steelmaking</u>, 1979, no. 6, pp. 268-73.

5. H. Gou, G.A. Irons and W.K. Lu: unpublished research, McMaster University, 1992.

6. G.K. Batchelor: <u>Quart. J. Roy. Meteorol. Soc</u>, 1954, Vol. 80, pp. 339-58.

7. F.P. Ricou and D.B. Spalding: <u>J. Fluid Mech.</u>, 1961, Vol. 11, pp. 21-32.

8. G.A. Irons: <u>Metall. Trans. B</u>, 1987, Vol. 18B, pp. 105-117.

9. L.K. Chiang, G.A. Irons, W.K. Lu and I.A. Cameron: <u>Trans. ISS</u>, Jan. 1990, I&SM, pp. 35-52.

10. Y.F. Zhao and G.A. Irons: <u>Proc. Sixth International Iron and Steel Congress</u>, 1990, Nagoya, ISIJ, Vol. 3, pp. 65-69.

11. S. Ohguchi and D.G.C. Robertson: <u>Ironmaking and Steelmaking</u>, 1984, Vol. 11, No. 5, pp. 261-273.

12. Y.F. Zhao: <u>Ph.D. Thesis</u>, McMaster University, 1992, in preparation.

13. K. Nakanishi, J. Szekely and C.W. Chang: <u>Ironmaking and Steelmaking</u>, 1975, Vol. 2, pp. 115-124.

14. D. Mazumdar and R.I.L. Guthrie: <u>Metall. Trans. B</u>, 1986, Vol. 17B, pp. 725-733.

15. G.A. Irons: <u>Ironmaking and Steelmaking</u>, 1989, Vol. 16, No. 1, pp. 28-36.

16. G.A. Irons and L.R. Farias: <u>Can. Metall. Quart.</u>, 1986, Vol. 25, no. 4, pp. 297-306.

17. D.N. Ghosh and K.W. Lange: <u>Ironmaking and Steelmaking</u>, 1982, Vol. 19, no. 3, pp. 136-41.

IMPROVED TECHNIQUES FOR GAS FLUXING IN ALUMINUM FURNACES

Guy Béland, François Tremblay

Alcan International Limited,
Arvida Research and Development Centre
1955, Mellon Blvd., P.O. Box 1250
Jonquière, Québec, Canada G7S 4K8

and

Jean-Claude Pomerleau

Société d'électrolyse et de chimie Alcan Ltée,
Complexe Jonquière,
1955, Mellon Blvd. P.O. 1500
Jonquière, Québec, Canada G7S 4L2

Abstract

Prior to being cast into ingots, molten aluminum alloys must be treated to meet inclusion and alkali element content specifications. A widespread approach to molten aluminum alloy treatment in furnaces consists of injection of gas mixtures (usually containing chlorine) through stationary lances. This method has limited efficiency caused by restricted gas/liquid contact area and contact time inherent to large gas bubbles, poor metal circulation and inadequate geometry of furnaces dictated by heat transfer considerations.

Recent developments in ceramic materials have led to the introduction of gas injection through porous plugs installed in the floor of furnaces. Laboratory evaluations and plant production information available to date are discussed; design and operational problems identified indicate that this approach does not constitute a viable alternative to static lance injection at this time.

This paper discusses improved gas fluxing approaches for aluminum treatment in furnaces based on coupled metal circulation/gas injection systems developed by Alcan and presents results obtained in production-scale facilities.

Proceedings of the
Savard/Lee International Symposium on Bath Smelting
Edited by J. K. Brimacombe, P. J. Mackey,
G. J. W. Kor, C. Bickert and M. G. Ranade
The Minerals, Metals & Materials Society, 1992

INTRODUCTION

Prior to being cast into ingots, molten aluminum must be treated to meet certain specifications related to its composition and cleanliness. One of the methods used to carry out this partial or total treatment is to inject a reactive gas such as a mixture of chlorine and an inert gas (N_2, Ar) into the molten aluminum held in casting furnaces. This technique is used more specifically for the removal of dissolved or dispersed impurities. In the case of aluminum and its alloys, the principal dissolved impurities are hydrogen and alkaline metals such as sodium, calcium and sometimes lithium. Dispersed impurities are mainly composed of solid inclusions such as oxides, carbides and borides.

The efficiency of such a treatment process is related to obtaining a maximum contact surface between the liquid and gaseous phases. A large contact surface maximizes mass transfer which then improves reaction kinetics. When a gas is injected into a liquid, the contact surface can be increased by the generation of gas bubbles with the smallest possible diameter so as to maximize the ratio of contact surface to gas volume.

In industrial furnaces, the conventional technique consists of injecting the reactive gas from stationary lances. Although widespread, this method has very limited efficiency [1]. Gas injected with a lance generates columns of large bubbles within a confined region near the injection lance (Figure 1). Moreover, due to the large diameter of the bubbles and the enormous difference of density between gas and aluminum, the gas rises very rapidly to the bath surface, thus limiting the contact time between gas and liquid.

Figure 1
Presence of Surface Turbulence Generated by Gas Bubbles
Produced with Stationary Lance Injection

Another disadvantage of this practice is the high level of oxidation generated during the treatment. When they rise out of the aluminum bath, the gas bubbles produce turbulence of the metal at the surface thus producing an excessive amount of dross by oxidation of the aluminum in contact with air. The generation of dross is undesirable for economic, environmental and metal cleanliness reasons. Moreover large amounts of unreacted chlorine are also released into the furnace stack [1].

Due to the geometry of the aluminum bath in industrial furnaces, characterized by a small depth/length or surface ratio, the use of multiple injection lances (2 to 10) is often essential to ensure treatment of the whole metal volume.

Given the low efficiency of the practice described above and because of environmental considerations (dross generation, atmospheric emissions), some new fluxing in furnace techniques have been developed over the past several years.

One of these new techniques proposes the injection of reactive gas into the metal bath using porous plugs installed in the furnace refractory. Alcan's evaluation of this method shows that this practice does not improve, in a satisfactory manner, the overall treatment efficiency. More details concerning the evaluation of fluxing with porous plugs are given in the next section.

An efficient treatment method for aluminum fluxing is the injection of a reactive gas using a rotary stirrer. The shearing produced by the stirrer results in the formation of small gas bubbles in conjunction with strong metal circulation. This method is very well adapted to the in-line treatment of molten aluminum during casting and is the basis of commercial treatment units (Union Carbide SNIF, Pechiney Alpur, Alcoa 622). Although it corrects most of the lance injection problems, its application to metal treatment in furnaces, having capacities greater than 40 tonnes, is very limited and not very practical. These limitations are mainly due to the diameter of the stirrer, the number of stirrers required for large-capacity furnaces and the compatibility of moving parts with the furnace heating system [1, 2].

Given that neither of these proposed solutions have been adapted to replace or improve the present method used for chlorine fluxing, Alcan has developed a new gas injection method. This method is based on the use of the Alcan Jet Stirrer developed by Alcan to ensure molten metal stirring on most types of furnaces [4].

METAL TREATMENT WITH POROUS PLUGS IN HOLDING FURNACES

Recently an approach to metal treatment in furnaces based on the use of porous plugs installed in the floor refractory lining of furnaces has been proposed [3]. Claimed improvements included a reduction of maintenance costs associated with broken graphite fluxing tubes and a potential increase in fluxing efficiency due to reduction of reactive gas bubble size in furnaces, leading to a reduction of Cl_2 usage.

During the fall of 1989, an evaluation of porous plug assemblies built in a furnace floor was instigated. Five porous plugs were cast in the floor of a 4-tonne furnace (Figure 2). Objectives were to observe the physical integrity of the plug assemblies in the furnace floor, their restart ability and measure their metallurgical efficiency in removing alkalies and inclusions. However, experiments were cancelled due to a chlorine leak from the plugs on the first attempt to run Cl_2 through them (on the 21st day of service life). The plugs had been cycled continuously (20 minutes ON, 2 hours OFF) to simulate holding furnace applications using 35 sLpm (standard liter per minute) of N_2. Their restart ability was excellent. The floor of the furnace was then dismantled and the plugs were removed for failure analyses. The postmortem showed that chlorine leaked under the plug between the steel shell of the furnace and the refractory.

Later, three newly designed porous plug assemblies were tested. In addition, two modified older designs were tried. A general arrangement of the new porous plugs in the experimental 4-tonne furnace is illustrated in Figure 3. A metallurgical experiment showed that at a flow rate of 17 sLpm (maximum flow at the time of the test), the Ca and H_2 removal efficiencies of a porous plug and a stationary lance were equal. This proves that gas bubbles generated in aluminum by porous plugs are of the same size as the ones produced by a lance at equal flow rate, i.e. 17 sLpm. The high surface tension of aluminum prevents the easy formation of small gas bubbles. The formation of small gas bubbles in water is a lot easier than in aluminum. To date, the trials have been halted since only one plug functions without a leak; further metallurgical testing were cancelled due to an insufficient flow rate, i.e. 7 sLpm. The other four plugs stopped functioning within the first 13 days of service. Complete blockage of the gas flow and/or gas leakage were identified as the two causes of failure.

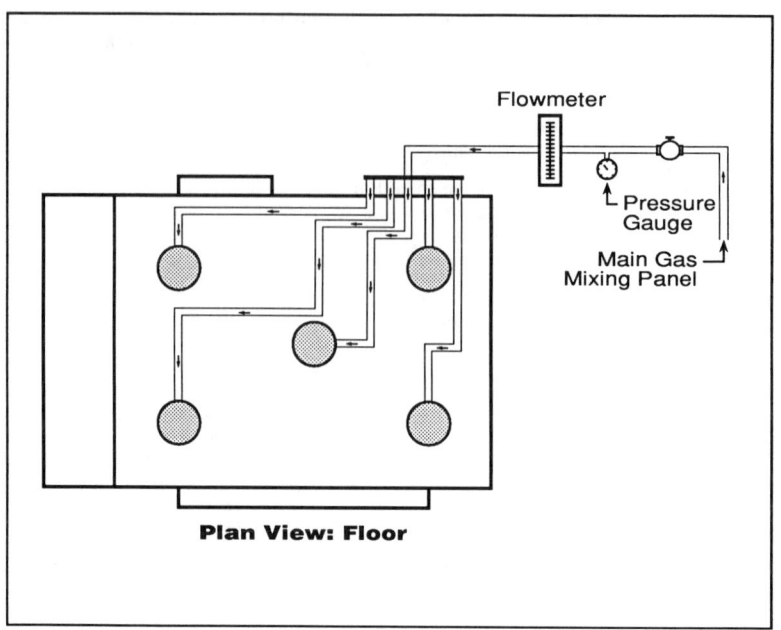

Figure 2
Initial Installation of Porous Plugs

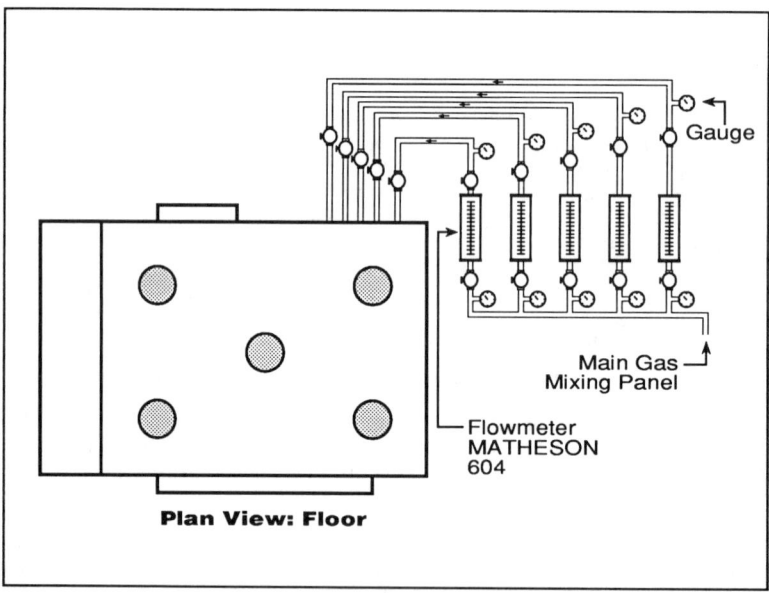

Figure 3
Second Installation of Porous Plugs

Figure 4 illustrates how the flow through the only remaining porous plug varied with time. It also shows that the flow resistance increased with service time (i.e. to maintain a metallurgically efficient gas flow, applied gas pressure had to be continuously increased). The flow rate gradually diminished from a starting value of 35 sLpm to a low value of 12 sLpm at day 34 and 7 sLpm one year after installation.

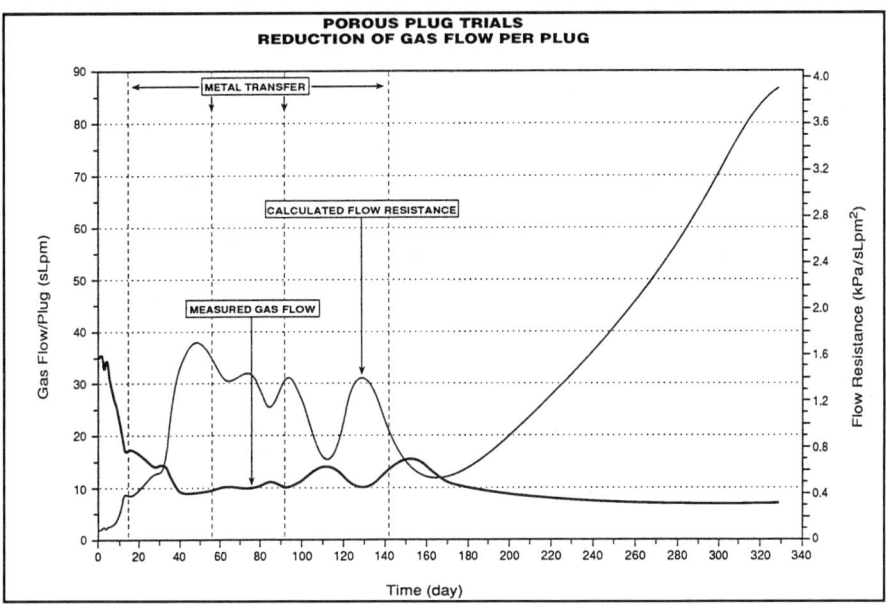

Figure 4
Variation Over Time of Gas Flow and Resistance in Porous Plugs

In conclusion, Alcan has elected not to proceed with the installation of porous plugs in the floor of holding furnaces for the following reasons:

1. From the information and experience gathered to date, porous plug potential service life (12 months) is not in the same range as the refractory of a furnace floor, i.e. 10 years.

   Heterogeneities in a furnace floor will always increase the chances of molten metal penetration. This is corroborated by the 4-5 cm metal penetration found at the plug-refractory interface.

2. The restriction by the porous media to the gas flow increased continuously throughout the period during which the plugs were operating. Each plug required its own valve and piping system.

   If more than 10 plugs are to be installed, it will be very difficult to detect which one of them is leaking.

   Undetected leaks, between the refractory and the steel shell of a furnace, can lead to disastrous consequences for its structure integrity and safety. Chlorine is very corrosive at high temperature.

Given that the porous plug option has not been proven superior to the stationary lance fluxing method, Alcan continued its efforts to find an improved treatment system. The next section describes the physical modelling of Alcan's Jet Fluxing system.

511

DESCRIPTION OF THE JET FLUXING TECHNIQUE

The treatment technique developed by Alcan is called Jet Fluxing (patent pending). It consists of injecting a reactive gas into a 2 to 5 m/s liquid aluminum jet. The aluminum jet allows the dispersion of gas into a large number of small bubbles and their transport throughout the liquid metal bath.

The gas is injected perpendicularly to the axis of the jet so as to be sheared into small bubbles. The direction of the metal jet created in the furnace is parallel to the bath surface. Gas bubbles are entrained by the jet through the bath.

The metal jet is generated by the Alcan Jet Stirrer (AJS) [4]. The Alcan Jet Stirrer is an electro-pneumatic device for subsurface metal stirring that can be used on any type of furnace. Figure 5 shows a schematic view of the stirrer. The operating principle is based on the periodic operation of two distinct phases. The first is the filling or depressurization phase during which a certain amount of molten aluminum is drawn into the reservoir. This is followed by the expulsion or pressurization phase during which the accumulated amount of metal is expelled back into the molten metal furnace through the same orifice. The operating mode of the AJS consists of repeating these two phases alternately. The furnace stirring is ensured by the intermittent metal jet which lasts 5 seconds with a 5 second pause time (filling phase). The conduit, linking the furnace and AJS reservoir, is about 14 cm in diameter.

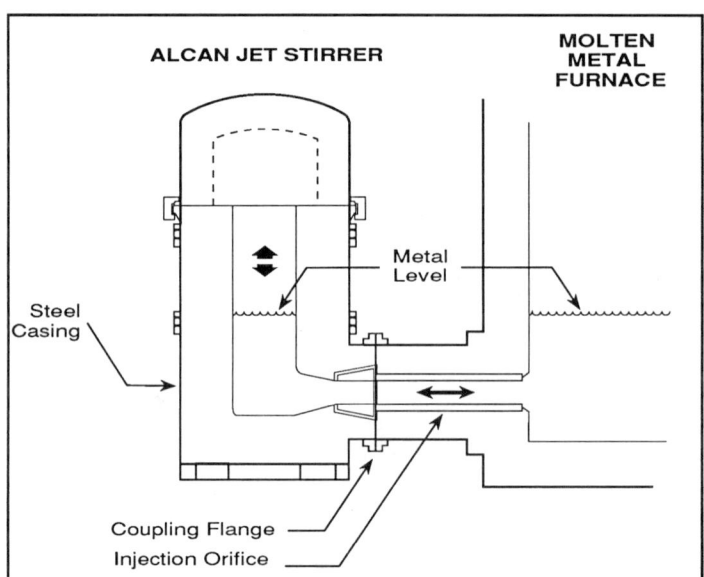

Figure 5
Schematic of the Main Elements of the Alcan Jet Stirrer

Gas injection into a liquid metal jet improves the treatment efficiency in many ways when compared with the conventional stationary injection method:

- The gas/liquid contact surface is larger for a similar volume of gas. This is caused by the generation of smaller bubbles through shearing.

- The retention time of the gas in the metal is increased due to the smaller diameter of the gas bubbles and also because the metal jet induces a horizontal speed component which increases the length of their trajectory in the liquid.

512

- The metal jet of the AJS, in addition to carrying along gas bubbles, stirs the metal bath ensuring that the whole metal volume is in contact with the injected reactive gas.

- Because gas dispersion is improved, the movement of the surface metal caused by the escaping of large bubbles is reduced, thus decreasing the dross generation due to oxidation of aluminum at the metal/air interface.

Figure 6 shows where and how the gas is physically injected into the stirrer's metal jet. The gas is injected near the AJS orifice. Other options are possible in order to obtain the shearing and distribution of gas in the aluminum bath by the stirrer's jet. One of these methods would be to inject the gas directly into the AJS conduit and another is to inject the gas into the stirrer's reservoir. Nevertheless, for maintenance and operation reasons the gas injection by a lance located at the exit of the AJS aluminum jet (in the furnace) is preferred.

To maximize efficiency, the optimal position of the gas injection in the metal jet has been determined in a 1/3 scale water model. Removal of dissolved oxygen in water by nitrogen gas was used to measure efficiency and to establish the best position to inject the reactive gas. The position of the gas injection point was defined along the longitudinal axis of the jet, with the gas injection perpendicular to the liquid jet stream.

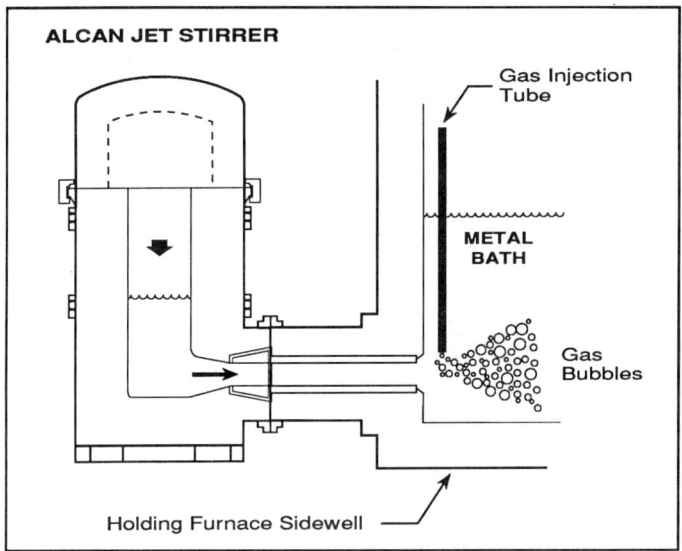

Figure 6
Preferred Position for the Injection of Gas in the Metal Jet

The best location to inject gas (Figure 7) is at a distance from the orifice outlet along the longitudinal axis of the jet equivalent to 3 to 5 times the diameter of the stirrer's orifice (Do). This position takes advantage of the maximum shearing and entrainment power given by the jet stream.

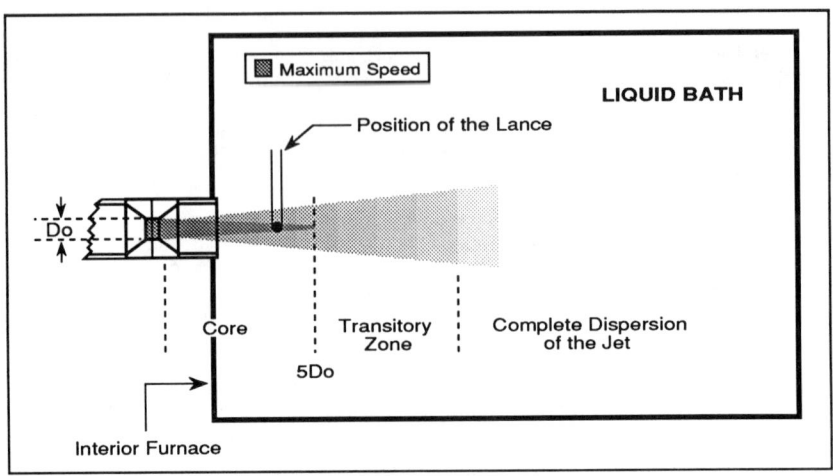

Figure 7
Dispersion of the Liquid Jet

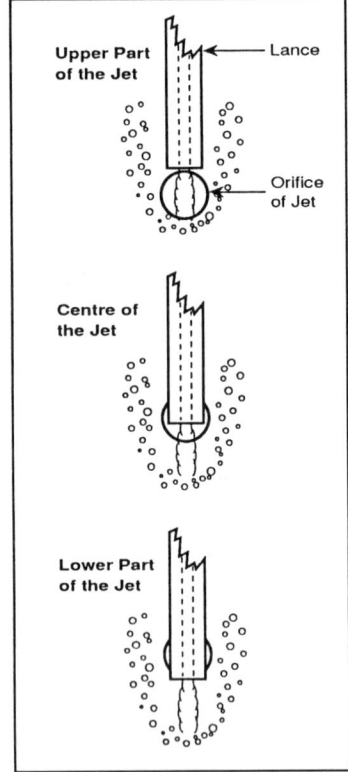

Figure 8
Vertical Position Possibilities for Gas Injection in the AJS

Figure 7 shows a schematic of the jet profile. The free jet is characterized by three distinct zones. In the first zone, located immediately after the outlet orifice, the maximum speed reached in the orifice is maintained at the center of the jet. The length of this zone has been measured to be equivalent to five times the diameter of the stirrer's orifice. Therefore, the position of gas injection has to be within the distance where the speed is high in order to assure maximum gas shearing.

For maintenance reasons, reactive gas must not enter the stirrer's reservoir. During the vacuum phase of the stirrer, it was observed that, when the gas tube was placed at a distance smaller than three times the orifice's diameter, the gas from the tube penetrated the stirrer.

On the axis perpendicular to the jet, three possible gas outlet positions have been studied: in the upper part of the jet, in the center or in the lower part (Figure 8). The optimal position of the gas outlet is at the center where the maximum liquid velocity is found.

Finally, the physical model has been used to compare the stirring efficiency between jet fluxing and standard fluxing technique. The total mixing time of the liquid bath was measured to quantify the stirring efficiency. The mixing time of the bath was reduced from 4 min to 1 min when Jet Fluxing (Figure 10) was used instead of standard fluxing (2 lances: Figure 9). The mixing efficiency of Jet Fluxing is superior because of the use of the AJS. The mixing effect of stationary lances is very intense near the tubes, due to ascending movement of the gas bubbles. However, it is non-existent (diffusion only) outside the cone formed by the rising bubbles.

INDUSTRIAL EVALUATION

A series of tests were performed at an Alcan production facility to:

- compare the conventional fluxing technique to the Jet Fluxing technique;
- quantify the benefits of Jet Fluxing;
- validate the physical model.

The conventional fluxing technique consists of using two graphite gas injection lances placed as shown in Figure 9. Their positioning was executed manually.

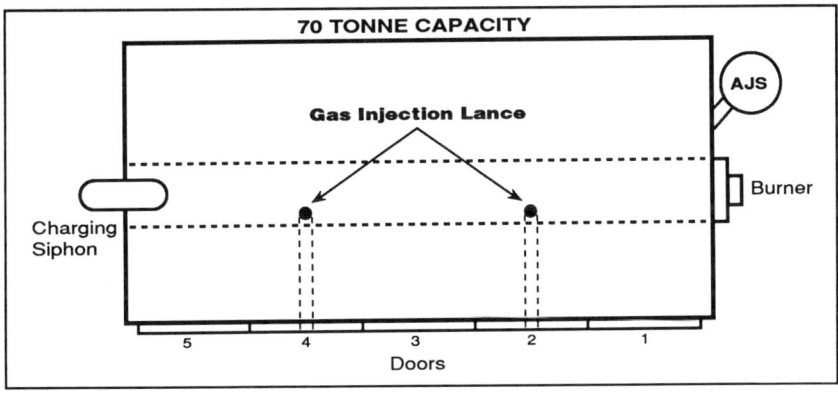

Figure 9
Positioning of the Gas Injection Lances for the Conventional Technique

515

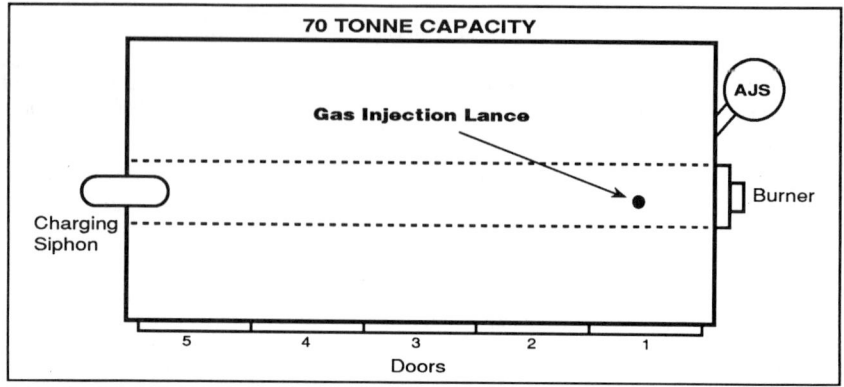

Figure 10
Positioning of the Gas Injection Lances for the Jet Fluxing Technique

On the other hand, the Jet Fluxing technique consists of automatically positioning one graphite gas injection lance (Figure 10). This new technique provides:

- Increased chlorine treatment efficiency.
- Decreased gas consumption (N_2-Cl_2).
- A reduction of emissions to the atmosphere (Cl_2, HCl).
- Decreased skim generation.

Chlorine Treatment Efficiency

The chlorine treatment efficiency was evaluated as a function of the alkali removal rate (e.g. $[Ca]/[Ca]_0$ as a function of time).

The first order of reaction constant is used to compare removal efficiencies.

$$[Ca] = [Ca]_0 \, e^{-Kt}.$$

Referring to Table I, the reaction constant is increased by approximately 30 % as compared to the conventional method when chlorination is carried out in the metal jet.

Table I Comparison of First Order Reaction Constants

	$K_{Standard}$ (min^{-1})	$K_{Jet\ Fluxing}$ (min^{-1})	Increase (%)
Lithium	.016	.021	31
Sodium	.026	.034	31
Calcium	.032	.043	34

Figure 11 shows removal curves of each alkaline element for the reaction constants shown in Table I.

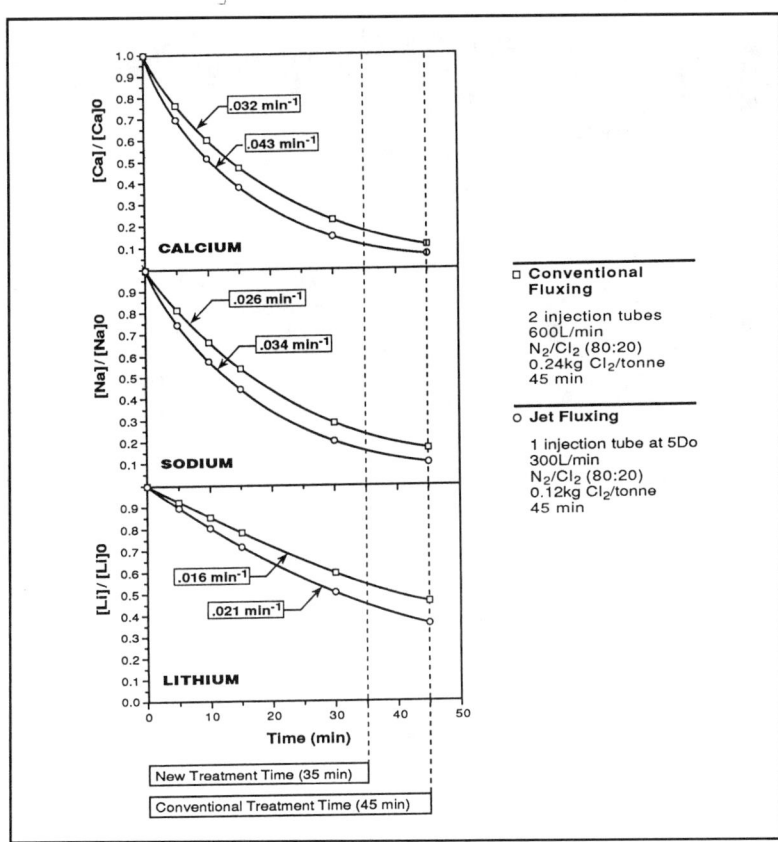

Figure 11
Comparison Between Conventional and Fluxing Treatments
Using to Ca, Na and Li Removal Curves

Chlorine Consumption

The gas (80 % N_2, 20 % Cl_2) consumption was 600 L/min with the standard method compared to 300 L/min for Jet Fluxing (50 % reduction). In addition, with the increased efficiency, the treatment time can be reduced from 45 to 35 minutes, i.e. a reduction of about 30 % while achieving the same final alkali concentration (Figure 11). Therefore, with the use of the new fluxing treatment, chlorine consumption can be reduced from 0.24 kg/t to 0.09 kg/t.

As a consequence, Jet Fluxing increases the chlorine treatment efficiency and reduces the chlorine consumption by 62.5 %.

Dross Reduction

As shown with the physical model (water), Jet Fluxing allows the use of only one lance for the treatment of molten aluminum bath without massive disturbance of its surface.

This decrease, in the number of lances used and in the surface turbulence, greatly reduces the quantity of skim generated (Figure 1 vs Figure 12).

517

Figure 12
Minor Surface Turbulence During the AJS Pressure Cycle (Bubble Shearing)

The average dross generated per batch with the conventional chlorine treatment is 750 kg. When the new chlorine treatment is used, the average dross generated is 380 kg (same furnace, same alloy). This represents a dross reduction of 370 kg, or approximately 50 % of the dross quantity generated by the conventional furnace treatment.

OPERATING PRINCIPLE

General Assembly

As the efficiency of Jet Fluxing depends on the fractionation and the distribution of the gas bubbles in the liquid metal bath, it is essential that the lance be positioned very precisely in the metal jet generated at the outlet of the AJS orifice.

The Jet Fluxing mechanical system possesses maximum rigidity and flexibility which guarantee the stability of the lance in the molten aluminum jet stream.

This system consists of a mobile carriage supporting a graphite lance. This carriage, which is located on top of the furnace, travels on a set of tracks. Traction from a cable controls the movement of the carriage. This traction is produced by the winding and unwinding of the cable which is attached to a cylinder (Figure 13).

The cylinder is connected to an electric motor by a chain and cogwheels. When the motor is activated, the carriage moves up or down.

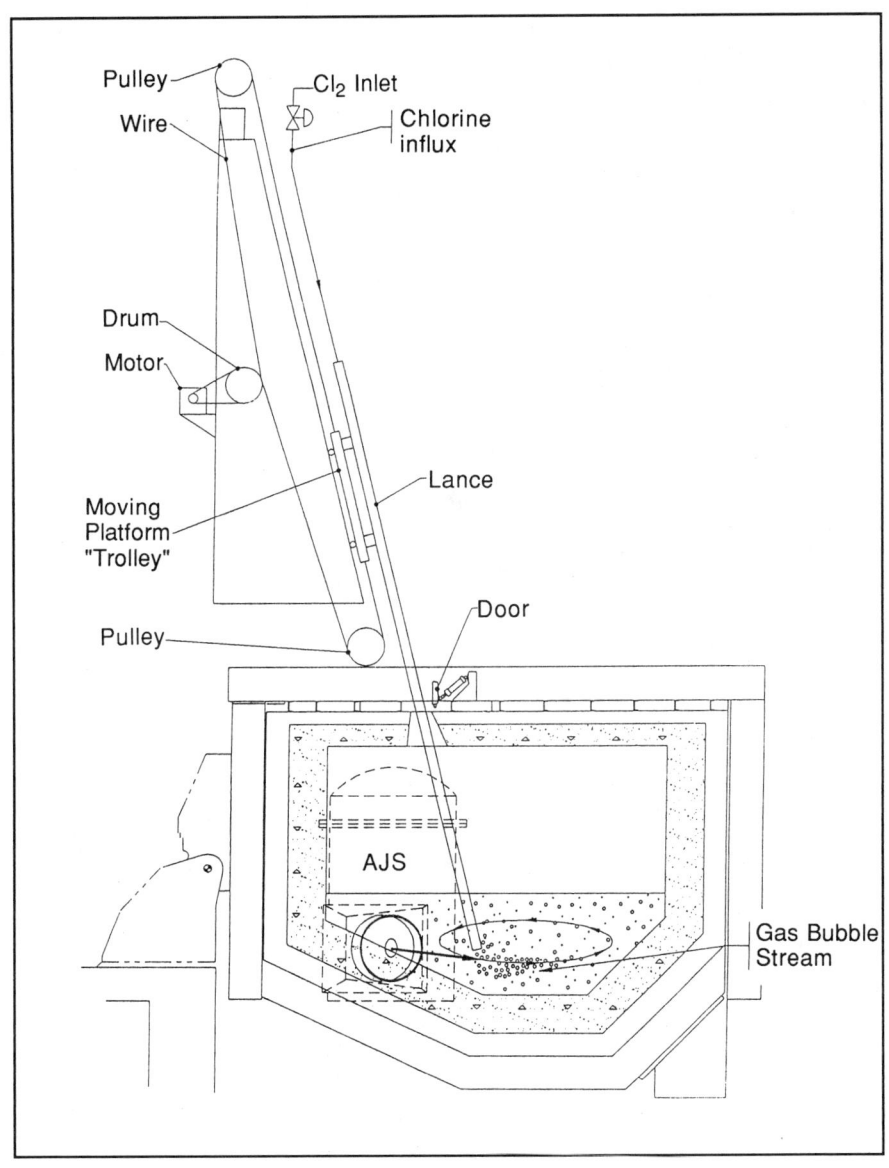

Figure 13
Lance Insertion Phase

Insertion of the Lance

When the proper temperature of the metal bath is reached, the operator initiates the automatic control system. This system sequentially opens the furnace's access port (located on the roof) and activates the descent of the carriage. As the carriage moves, the lance, attached to the carriage by guides, is directed towards the orifice of the AJS.

519

When the lance is inserted into the furnace, the control system triggers the arrival of the gaseous mixture (nitrogen-chlorine) just before entry of the tube into the molten metal. The injection of the gas into the liquid metal jet, released by the AJS, results in the production of tiny bubbles which disperse in the furnace bath.

The duration of the gas mixture injection is controlled by a Programmable Logic Controller (PLC). The duration is determined by the standard practice established for furnace treatment.

Lance Removal

At the end of the programmed fluxing period, the PLC triggers the lance to be raised. Gas injection continues until the lance exits the metal bath therefore avoiding any obstruction by solidified metal.

When the carriage has reached the top of the tracks, a detector transmits the position of the carriage to the PLC. The PLC initiates the closing of the furnace access port and shuts off the electric motor. The Jet Fluxing system is then ready to be reactivated whenever necessary.

Control System

All the components of the injection system are located on top of the furnace. This makes access to the components difficult. Therefore, the intervention by the operator is limited to the start-up of the system (by pressing a button located in the local control panel) and the replacement of the graphite lance.

Position of the Lance

The extremity of the lance must be in the center of the metal jet. Metallurgical results show that any other position considerably reduces the Jet Fluxing efficiency.

Life Span of the Lance

For both treatment systems (stationary or Jet Fluxing), graphite lances deteriorate at the metal/air interface. Their cross-section being reduced at this location, breakage occurs. The less the number of lances, the less their consumption. The Jet Fluxing system uses only one lance.

Specifically to the tests performed at an Alcan production facility, additional differences between the two techniques exist:

- With stationary fluxing, the operator must manipulate each lance manually. When removed from the furnace, the extremity of the lance (which is very hot) falls to the floor. The life span of the lance will depend on its resistance to mechanical shock rather than the deterioration at the metal/air interface.

- With Jet Fluxing, the life span of the lance depends solely on the deterioration at the metal/air interface.

At present, a casting center uses, on average, 6 fluxing lances for 36 batches produced. With the automatic system, the estimated consumption would be 1 lance for every 30 batches produced.

When comparing the two systems on an annual production basis, the savings would be approximately 136 lances, or a reduction of 80 % (Table II).

Table II Comparison Usage of Fluxing Lances

System	Number of Lances Used	Reduction
Stationary Fluxing	170	80 %
Jet Fluxing	34	

Reliability of the Jet Fluxing System

As the mechanical system is to be located on top of the furnace, its reliability is very important.

The results of several tests have demonstrated the reliability of the industrial prototype system as pertains to the positioning of the extremity of the lance in the center of the metal jet and to the different mechanical components.

CONCLUSION & DISCUSSION

The efficiency of furnace fluxing, using reactive gases, can be increased to a maximum by increasing the contact surface between liquid aluminum and reactive gases. To do so, an optimal gas fluxing system should provide small gas bubbles in a continuously renewed aluminum environment i.e. the system should continuously destroy the concentration gradient of dissolved and dispersed impurities that occur around the gas injector(s). The fluxing system should also leave the bath surface as quiescent as possible to prevent the formation of additional oxide impurities. As illustrated throughout the present document, experiments performed with the Alcan Jet Fluxing system show that these objectives are met.

In addition to the Jet Fluxing efforts, Alcan smelter plants are using TAC [5] units to reduce alkali and carbide inclusions before they reach the furnaces. Instead of using a reactive gas (Cl_2) to generate salts and remove unwanted alkalies, the TAC units use AlF_3 which reacts readily with the impurities to remove them. Usage of TAC units on pot room metal before it reaches the furnace coupled with the use of in-line treatment units, is part of an integrated approach to decrease chlorine usage in furnace treatments.

In fact this integrated approach has allowed Alcan to significantly reduce its overall chlorine consumption (Figure 14) and to tend towards stoichiometric use, thereby minimizing atmospheric emissions.

Answers to the aluminum industry goal of reaching zero Cl_2 furnace emissions will be yielded mainly by the research and development of mechanically sound machines; such machines will enable us to achieve a 100 % efficient usage of reactive gaseous, liquid or solid fluxes.

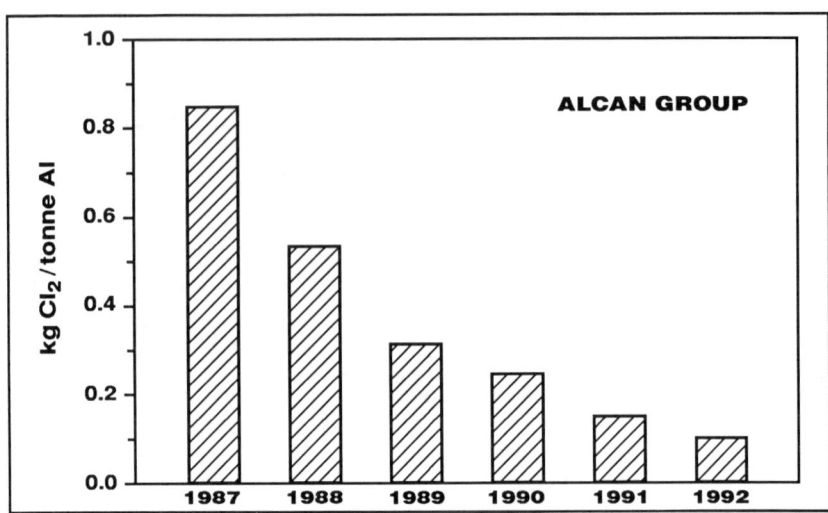

Figure 14
Chlorine Consumption

REFERENCES

[1] Celik, C.; Doutre, D.; "Theoretical and Experimental Investigation of Furnace Chlorine Fluxing"; Light Metals 1989.

[2] Boeuf, F.; Baud, O.; "Rotating Injectors in Casthouse Furnaces"; Proceedings Light Metals 1989; pp. 815-818; March 1989.

[3] Nelson, B.R.; "AL CLean-Porous Refractories for Clean Aluminum"; Ingot and Continuous Casting Process Technology Seminar for Flat Rolled Products, Vol. 1, New Orleans, 10-12 May 1989, pp. 149-180.

[4] Thibault, M.-A.; Tremblay, F.; Pomerleau, J.-C.; "Molten Metal Stirring: The Alcan Jet Stirrer", Light Metals 1991, pp. 1005-1012, February 1991.

[5] Gariépy, B.; Dubé, G.; Simoneau, C.; Leblanc, G.; "The TAC Process: a Proven Technology"; Proceedings Light Metals 1984, pp. 1267-1279, March 1984.

INJECTION IN MATTE CONVERTING AND METAL REFINING

N.B. Gray and M. Nilmani

G.K. Williams Centre for Extractive Metallurgy
Department of Chemical Engineering
The University of Melbourne
Parkville, Victoria, 3052, Australia

Abstract

Matte converting operations have traditionally been carried out in horizontal cylindrical reactors. Few changes have been made to the original design and method of operation of these reactors to overcome operating and environmental problems. In the last two decades intensively stirred vertical cylindrical reactors have been developed involving submerged gas injection through lances with and without powder injection to obtain more efficient gas/liquid and solid/liquid contacting. In this paper previous matte converting studies are briefly reviewed followed by discussion of recent developments in converting and metal refining.

Application of powder injection technology is illustrated through the development of a novel nickel matte fluxing technique where the flux has been injected through the tuyeres of a P-S converter. The potential of this approach to improve productivity, process chemistry, the environment and minimise tuyere punching are discussed. Mathematical and physical modelling studies relating to heat transfer and pressure drop through a Sirosmelt lance are discussed. Experimental data for turbulent swirling flow in a heated annulus are presented. Investigations carried out on the causes and effect of slopping at the free surface in horizontal cylindrical reactors are discussed. Recent developments have focussed on damping the slop or operating under conditions where slopping does not occur.

In the area of metal refining, plant investigations on the copper anode furnace have shown that the rate of copper deoxidation is governed by mass transport of oxygen from the bulk of the bath to the gas-liquid interface at the bath surface. Recent development of an optimal control strategy (involving both the oxidation and reduction cycles) has resulted in significant improvements to current practice. Finally, reference is made to the "Helical Plug Lance", a novel gas injection device being developed for melt treatment. The lance has been tested in aluminium holding furnaces both in primary cast houses and foundries. Significant improvements in refining efficiency with respect to dissolved gases and alkalis have been obtained.

Proceedings of the
Savard/Lee International Symposium on Bath Smelting
Edited by J. K. Brimacombe, P. J. Mackey,
G. J. W. Kor, C. Bickert and M. G. Ranade
The Minerals, Metals & Materials Society, 1992

Introduction

Injection of gases and solids into molten baths has been practiced in both the non-ferrous and ferrous industries for many years. Traditionally matte converting has been carried out in horizontal cylindrical reactors with gas injection through a row of tuyeres into the molten matte contained in the converter. Recently, intensively stirred vertical cylindrical reactors have been developed involving submerged gas injection with powder additions to obtain more efficient gas/liquid and solid/liquid contacting. Injection has been carried out through vertical self cooled lances protected by a slag coating on the lance. Such developments show that the processing vessels and the methods of injection used have not been optimally designed to carry out their functions of converting and metal refining. Investigations continue to be made to overcome key operating problems such as tuyere blockage, refractory wear at the tuyere line and environmental problems arising from air pollution.

In this paper, previous matte converting studies are briefly reviewed followed by discussion on recent developments in both matte converting and metal refining at the G.K. Williams Cooperative Research Centre for Extractive Metallurgy at the University of Melbourne. These developments are showing that it is vital to understand more fully the process chemistry as well as the process dynamics of each step in the processes involved through a combination of mathematical modelling and measurements made in the laboratory and on the plant.

Matte Converting

Matte converting of an iron-copper or iron-nickel sulphide concentrate is an oxidation process where iron and sulphur are gradually oxidised and removed in a slag and gas phase. Under certain conditions solid magnetite may also be formed. The process is carried out in a Peirce-Smith converter by injecting air or oxygen enriched air through a row of tuyeres into molten matte contained in the converter.

Although the converter has been in operation for almost a century, little change has been made to the design and operation of this reactor. The problem of tuyere blockage and the practice of tuyere punching for removal of accretions has remained essentially unchanged except for some improvements to the design of mechanical punching devices. Concern over air pollution has lead to the development of tight fitting water cooled hoods but gases containing sulphur dioxide can still escape during the charging and pouring sequences. Flux additions are still made by batch additions through the mouth. The operation is dusty and uncontrolled and has not changed over the years. Both punching and fluxing operations affect the process efficiency and are bad for the working environment. As the industry today faces ever increasing economic and environmental challenges it is essential to review practices which may not be acceptable in the near future and would need alternate solutions and modifications to meet these challenges.

The matte converter consists of a circular cylinder laid on its side. When viewed from the circular cross section, the geometry is known as a "circular canal" geometry or "horizontal cylinder". Few studies have been carried out to investigate the complex fluid dynamics of gas injection in liquids in such a geometry. During injection of gases into liquids various zones are formed as follows:

(a)   large bubble zone
(b)   dispersed bubble zone
(c)   disengagement zone (slopping and spout region) and
(d)   splashing zone.

As the gas discharges from the tuyeres of the matte converter, large bubbles are formed. The size and shape of these bubbles are a function of gas/liquid density ratio, gas momentum, tuyere diameter and liquid viscosity. Matte converting is usually operated at shallow tuyere seal depths and consequently the large bubble zone is dominant in providing the gas/liquid interfacial area (see Figure 1). In addition the free surface and spout region may play important roles in providing a high gas/liquid interfacial area. Each of these zones interact with the surrounding liquid. Most of the experimental work has focussed on the large bubble zone (Hoefele and Brimacombe[1], Liow and Gray[2] and Engh and Nilmani[3]), the small bubble dispersion zone (Castillejos and Brimacombe[4], Sheng and Irons[5]) interaction between these zones and mixing in the liquid phase. Recently the disengagement zone (Sahajwalla et al[6]) and the slop and splash zone (Liow and Gray[7,8]) has begun to receive attention and these will be referred to later.

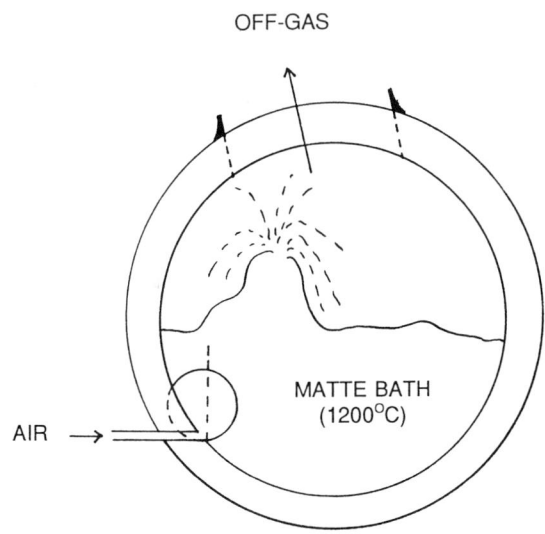

OFF-GAS

AIR →

MATTE BATH
(1200°C)

Figure 1:   Vertical section through a copper converter.

525

Investigation of the large bubble zone in matte converting has led to an increased understanding of the critical parameters involved in tuyere blockage and refractory wear with proposals to overcome these problems. Tuyere blockage in copper and nickel converting requires frequent clearing by forcing a steel bar through each tuyere to dislodge built up accretions. It is considered that formation of accretions and tuyere blockage involves several possible mechanisms as follows (Nilmani et al[9]):

(a)   liquid metal surging back into the tuyere
(b)   melt entering the tuyere as a splash of droplets, and
(c)   bath material solidifying in front of the tuyeres.

To investigate the large bubble zone Hoefele and Brimacombe[1] set up a physical model in which different gases (air, argon, helium) were injected into baths of water or mercury. The modified Froude number was used as the major variable together with the gas to liquid density ratio. The dynamic pressure inside the tuyere and at the tuyere exit was measured simultaneously with observations of the gas discharge using high speed cinematography. At low Froude numbers, characteristic of matte converting, regular pulses were observed with a frequency of about $10s^{-1}$. Subsequently industrial trials were carried out on operating converters to measure the dynamic pressure at the tip of the tuyeres and again pressure pulses were observed. This indicated that in matte converting gas also discharges continuously from the tuyeres into the bath and the liquid periodically flows in against the tuyere. Bustos et al[10] attached a tuyerescope to the back of a given tuyere on operating copper converters to observe the build up of accretions. During the slag blow it was found that the accretions grew upwards by a freezing mechanism from the bottom of the horizontal tuyere. The major constituent of the accretions was found to be $Cu_{1.96}S$ with various other copper-iron sulphides. No magnetite was detected in the accretions.

To overcome tuyere blockage in matte converting various alternatives have been proposed such as:

(a)   higher momentum gas injection
(b)   extended tuyeres
(c)   oxygen enrichment of the air, and
(d)   injection of fluxes to increase the gas momentum.

Applications of the first two alternatives in matte converting and metal refining will be highlighted here while the potential of flux injection will be discussed later.

Gray et al[11] investigated high pressure gas injection in the copper anode furnace to show that by installing critical orifices upstream of each tuyere to dampen pressure fluctuations in the tuyere, the blockage problem could be eliminated. No blockages occurred when the orifices were operated at critical flow conditions with an upstream pressure of 530 kPa. Replacement of tuyere inserts after each deoxidation cycle was also found to eliminate the blockage problem.

Brimacombe, Meredith and Lee[12] used high pressure air/oxygen supply operated at 414 KPa on four tuyeres on a copper converter at the Tacoma Smelter of ASARCO. The high pressure tuyeres did not require punching over a period of 89 days in which 88 charges were processed showing that operating the converter without punching is achievable.

Refractory wear in matte converting occurs mainly around the tuyeres and at the melt contact line at the bath surface. Various causes have been proposed for wear around the tuyeres. Brimacombe et al[13] proposed that the punching action in combination with local bath motion causes severe refractory wear at the tuyere line. Brimacombe et al and more recently Fountain[14] monitored bubble frequencies throughout the compaign life of a matte converter. Figure 2 shows that there is a significant change in bubble frequency with increasing number of charges. Initially the bubble frequency is in the range of 12-14 per second but this rapidly decreases to 4-6 per second after 20-25 charges. As erosion of the refractory lining develops at the tuyere mouth the effective tuyere diameter will increase leading to interaction between tuyeres and the bubble frequency would be expected to decrease. It is also noticeable from Figure 2 that the bubble frequency in the Port Kembla converter decreased more rapidly than that observed in the Utah Smelter. Nilmani et al[9] suggested that in the case of the Utah converter the use of a magnetic wash to provide an intial lining on the furnace gave lower refractory consumption. Liow and Gray[15] estimated the wear profile due to the formation of standing waves at the bath surface in a Sirosmelt reactor. The wear profile was found to have a maximum just below the bath surface showing that wave erosion may be a significant cause of refractory wear in matte converting.

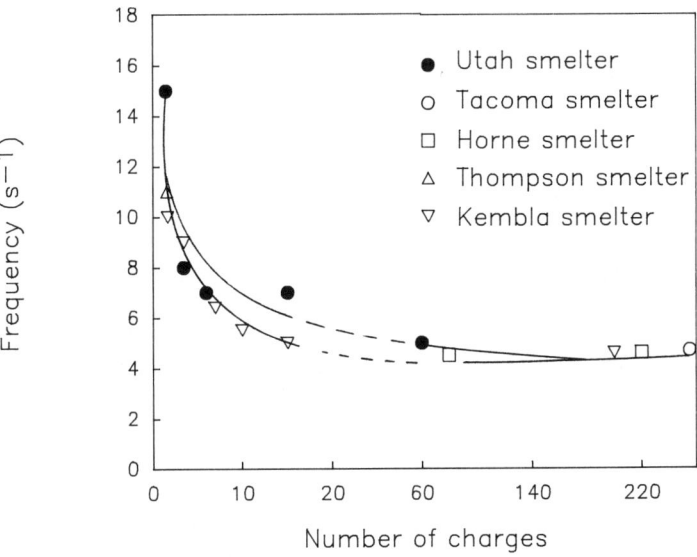

Figure 2: Tuyere bubble frequency versus the number of converter blows after the relining of the P-S converter. The diagram shows a decrease in th bubble frequency as the tuyere lining aged.

Injection Fluxing In Matte Converting

Although fluxing of melts by submerged powder injection technique is an established practice and has been successfully incorporated in a number of steel refining operations, its use in the non-ferrous industry has not been realised to any appreciable extent, except for zinc fuming and more recently in QSL and Sirosmelt processes. Many opportunities still lie in the non-ferrous industry to improve the processes utilising solid-melt contacting by controlled injection of fine solids in place of lump additions to the melts. This section outlines one such example where the fluxing step in P-S converter practice has been closely examined by Nilmani and Collins[16] and Park[14].

Flux Injection Tests   Currently at the Kalgoorlie Nickel Smelter of WMC the method of fluxing of the nickel converter is by overhead cranes, charging the sand in batches of two or four tonnes through the mouth of the converter. To save on time during the flux addition operation the blowing air is turned down to 10% to 20% of the normal blowing rate rather than turning off altogether. This batch method of fluxing the converter causes two main problems; an environmental problem due to sand being blown out of the converter during charging, throwing dust into the working environment and loss of sand to the flue and poor control of slag chemistry. This method of charging also adds to the problem of fugitive sulphur dioxide in the working environment of the smelter. Injecting the sand through the tuyeres instead of adding through the mouth was seen to be a workable solution to these problems. Therefore an experimental program of investigations was undertaken at Kalgoorlie Nickel Smelter. The trials were conducted on one of the available nickel converters and were designed to assess the merits of injection of sand flux through a selected tuyere. Three campaigns of trials were conducted.

The general layout of the test rig is shown in Figure 3. The tests were performed on a 3.6 m x 7.01 m nickel converter, blowing air through a bank of 28 tuyeres of 65 mm bore. For sand injection the 7th tuyere from one end was selected. A 35 mm diameter mild steel flux injection lance was inserted into the tuyere and secured in position using a specially designed bayonet mount arrangement to lock it onto the tuyere faceplate. Sufficient clearance was made available between the lance and the main tuyere to allow the primary air flow to continue during the lance insertion. A 4 mm diameter steel pipe running inside the lance relayed the pressure fluctuation from the tuyere - lance tip to a pressure transducer mounted at the lance entry point. Pressure signals after amplification were displayed on an oscilloscope and either taped on a data cassette recorder or collected on a U-V chart recorder (See Figure 4).

Sand of two size fractions was used to investigate the effect of sand penetration efficiency in the molten nickel matte. The dried sand was stored in a fifty tonne capacity hopper. A conveyor belt system fed the sand into a 20 tonne capacity surge bin, installed directly above the pressurised dispenser. In the early tests sand conveying air was provided by a commercially hired compressor, however plant compressed air at 600 kPa was made available in the latter tests. Air and sand was conveyed through a 50 mm "shot and blast" hose, 20 m long.

Between each subsequent skim, flux was injected through the tuyere at a rate of

200-300 kg/min with conveying air at 14.4 Nm^3/min. After some initial difficulties in sand conveying, later tests were conducted successfully and the converter was totalling fluxed by sand injection through one tuyere.

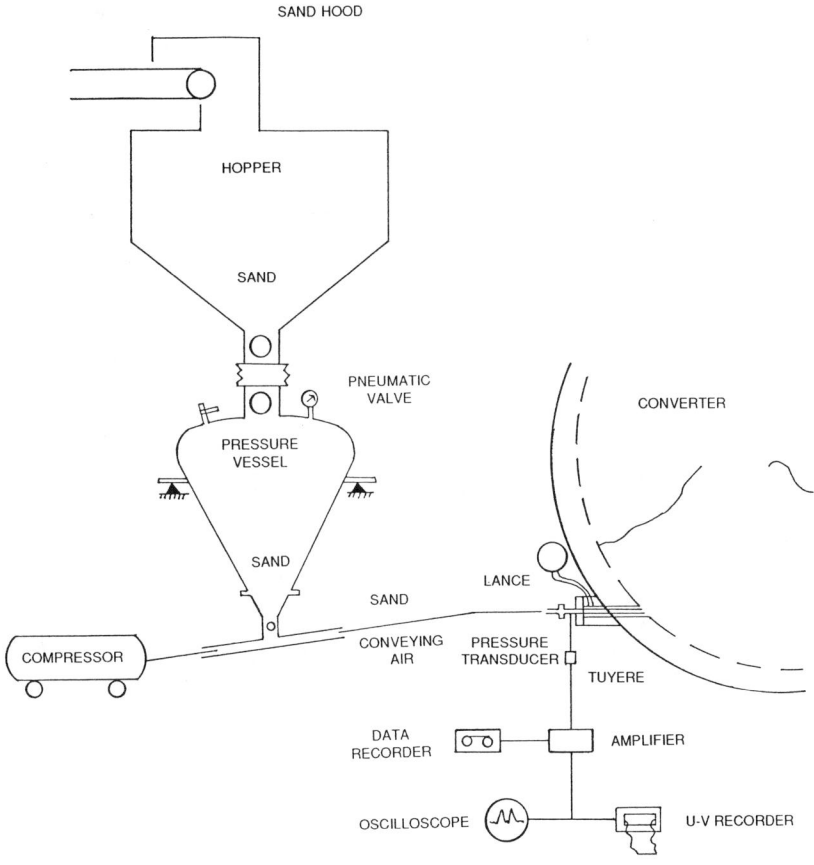

Figure 3:   Arrangement for flux injection tests through tuyeres at the Kalgoorlie Nickel Smelter.

<u>Efficiency of Flux Injection</u> One of the important aspects of the plant investigation was to assess the degree of sand flux incorporation in the matte during injection trials. The following approaches were used to analyse the efficiency of flux injection, i.e. total quantity of sand penetrating through the gas/melt interface.

In the first approach tuyere pressure fluctuations as a result of unsteady gas discharge[3,17] into the matte were analysed. Figure 4 compares the two pressure-time traces with and without sand injection when the sand flow is smooth without any slugging. The frequency of pressure peaks which relates to bubbling indicated that in both cases the frequency is about 10-11 Hz, and is not altered by the injection of sand.

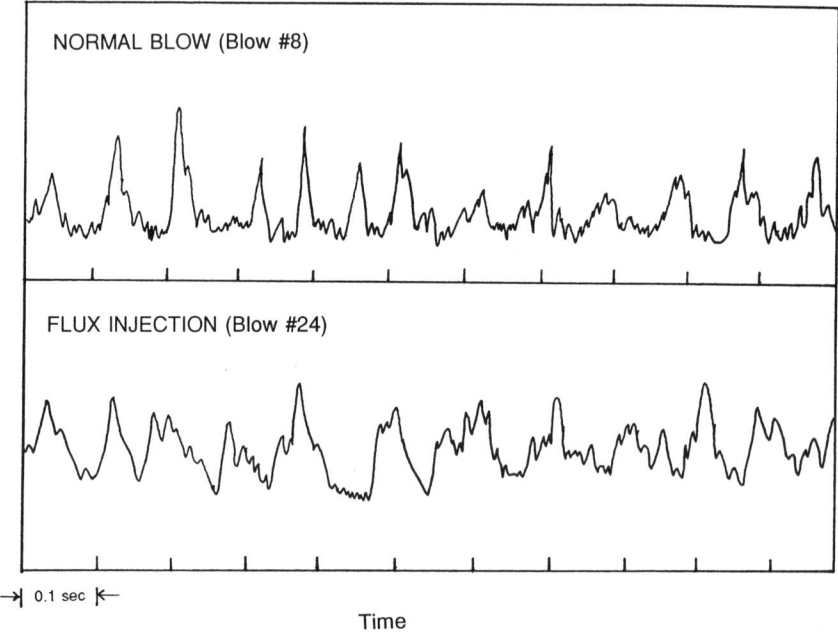

Figure 4:   Tuyere-lance pressure fluctuation traces with and without flux blowing

According to earlier work by Nilmani et al[18] and Farias and Robertson[17], in the bubbling regime, the bubbling frequency has been reported to be influenced by the efficiency of particles penetrating through the bubble/melt interface. In the limiting condition, if particles do not penetrate and transfer their total momentum to the growing bubble frontal interface, premature bubble detachment will occur and this will increase the bubble frequency. On the other hand, complete particle penetration will not transfer the kinetic energy to the bubble/melt interface and bubbling will continue unhindered. Therefore, in the latter case the bubbling frequency with and without particle injection should be identical. Of course there will be an intermediate range where only some of the particles will penetrate. Based on this argument it appears that under the given operating conditions all sand particles did penetrate across the gas/matte interface since the frequency of 10-11 Hz was maintained under conditions with and without sand injection.

In the plant test the average size of particles employed was ~ 250 microns, the gas velocity was 56.6 m/s. Let us assume the particle velocity to be half of the gas velocity. For a matte density of 4500 kg/m^3 and surface tension of 0.5N/m the particle Weber number for plant tests can be calculated as 900. According to Ozawa and Mori[19] under these conditions the critical Weber number for particle penetration is 35 and any particles above this size should penetrate. However this is applicable to a single particle situation which often does not represent a practical situation where a multitude of particles are used.

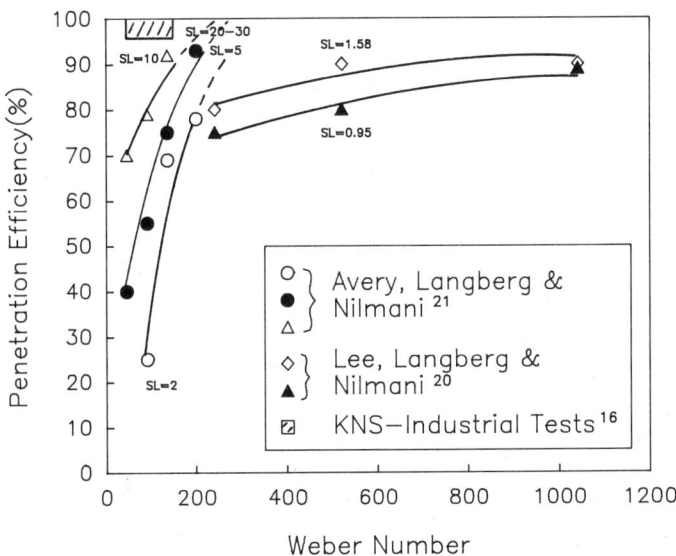

Figure 5:   Effect of particle Weber number and solids loading (SL) on sand flux penetration efficiency across the gas/liquid interface.

Most of the published work so far is on single particle penetration with very little work addressing penetration efficiency of multiple particles through a gas/melt interface. Figure 5 shows some experimental data by Lee et al[20] and Avery et al[21]. They have shown that increased solids loading and Weber number both increase penetration efficiency. The KNS plant data for a solids loading in the range of 20-30 with the lowest Weber number of 90 falls in the 100% efficiency range.

All the above analyses of flux injection tests shows that 100% sand penetration should be achieved in the matte bath. Although not conclusive, some initial tests showed a saving of up to 20% in sand consumption when injected through the tuyere compared to normal fluxing.

Benefits From the test campaigns Nilmani and Collins[16] conclusively found that the converter "out of wind" time required for flux charging was saved. The converter blow time was decreased which could be directly translated to increased productivity per converter.

Environmentally there were definitely some improvements as there was less dust in the air. The limited dust emission samples collected from the converter balloon flue indicated that the total dust concentration was substantially reduced from 340 mg/m^3 to 158 mg/m^3.

It was observed that the tuyeres stayed open throughout the injection period indicating that sand injection can eliminate the need for tuyere punching.

Flux injection through tuyeres does indicate a number of benefits to the P-S converting practice in terms of increased productivity, improved working environment, flux saving, and possibilities of elimination of tuyere punching. There are further benefits to be gained in terms of matte/slag chemistry.

Lance Injection With Swirled Gas Flow

With the development of more intensively stirred vertical cylindrical reactors other types of gas injection devices have been developed including lances and porous plugs. The SIROSMELT process involves a simple lance made from mild steel consisting of two concentric pipes. Helical vane swirlers are placed in the annular section at various intervals to promote turbulence in the air and improve heat transfer across the lance wall. Heat transfer from the bath to the lance and pressure drop within the lance depend critically on lance design. Previous work (Dave and Gray[22,23]) under isothermal conditions has shown that pressure losses within the lance are largely due to entrance effects and the internal design and placement of swirlers.

Further work has been carried out on the heat transfer and pressure drop characteristics of the flow within a SIROSMELT lance, involving an experimental and mathematical modelling study of swirling flow in a heated annulus (Solnordal[24]).

Mean velocities, Reynolds stresses, mean and fluctuating air temperatures, wall shear stresses, heat transfer coefficients, and the heat transfer per unit pumping power were all determined. For turbulence measurements, non-isothermal hot-wire techniques were developed and employed.

Experimental Ambient air was passed through a bellmouth inlet into the 3 m long test section of an open circuit wind tunnel. Swirl was imparted to the flow using either guide vanes that were equispaced around the bellmouth, or by helical vane swirlers inserted at the entrance to the test section. On leaving the test section, the working fluid passed through a conical diffuser and into an 11 kW Richardson Type PP blower. It was then exhausted to atmosphere.

The annular test section was made from two concentric pipes. The outer pipe had an inner diameter of 82.8 mm and the inner pipe had an outer diameter of 33.4 mm. This provided a nominal annular gap of 24.7 mm.

Thermocouples and static pressure tappings were positioned along both the inner and outer walls of the test section. Ports in the outer wall allowed access to the air stream for velocity and temperature measurement, using hot-wire sensors, and "cold-wire" resistance thermometers.

A maximum heat input of 3.6 $kW.m^{-1}$ of tube was generated using three 240 V, 15 A Pyroheat heating coils wrapped helically around the outer tube. The entire test section was wrapped in a 40 mm layer of Fiberfrax insulation to minimise heat losses.

Experiments were performed covering the conditions summarised in Table I. Each experiment consisted of a series of traverses at 5 or 6 axial stations along the test section. For each axial position, a hot-wire or hot-film sensor was traversed radially across the annular gap to measure both velocity and temperature. This data allowed the direct calculation of axial and tangential velocity, Reynolds normal stresses, and air temperature. In conjunction with the wall temperature and static pressure measurements, heat transfer coefficients and shear stresses could also be determined.

Table I: Range of Parameters

Variable	Value	
Reynolds Number	85000 - 200000	
Swirl generator	Helical vane swirler	$p/d_o = 0.91 - 1.21$
	Guide vane entry	$\psi = 0°, 30°, 45°, 60°$
Heat input (kW)	0, 9.45, 10.5	
Swirl Number (S) (=tangential/axial momentum)	0 - 1.39	

Two different types of swirl generator were used. The guide vanes produced a free vortex swirl, where the tangential velocity was inversely proportional to radial distance. The helical vanes generated forced vortex swirl, where tangential velocity was directly proportional to radial distance. The nature of these flows could then be compared.

Five heated swirling flow runs were performed. The details of each are summarised in Table II.

Table II: Details of Experimental Runs

Experiment	Re	Heat Input (kW)	Swirl Generator	Decay Length (L/d_h)
1	125000	10.5	20 x Guide Vanes, $\psi = 45°$, $S = 0.42$	64.7
2	175000	10.5	20 x Guide Vanes, $\psi = 45°$, $S = 0.38$	64.7
3	150000	10.5	20 x Guide Vanes, $\psi = 60°$, $S = 0.82$	64.7
4	100000	9.45	Helical vane swirler, $p/d_o = 1.21$, $L = 2p$, $S = 1.27$	60.7
5	85000	9.45	Helical vane swirler, $p/d_o = 0.91$, $L = 2p$, $S = 1.39$	61.7

By comparing the average tube friction factors and average heat transfer coefficients, a graph was made of heat transfer per unit pumping power (Figure 6). This graph shows the ratio of heat transfer on friction factor for a decaying swirl flow over the same coefficient for an axial flow. The horizontal axis is a measure of the decay length of tube. Thus, from the graph in can be seen that a one metre length of lance with a short helical vane swirler at the inlet (Experiment 4) has a heat transfer per unit pumping power only 0.35 times that of the equivalent Reynolds Number axial flow.

Since the pressure drop for this configuration is mainly associated with the entrance losses and losses through the short helical vane at the inlet, a longer length of decay will increase the average performance compared with axial flow.

For guide vane generated swirl, it is possible for the heat transfer per unit pumping power to increase over the axial flow case (Experiments 1 and 2). This is because the pressure drop associated with the guide vanes is very small. However, the actual increase in heat transfer is not adequate for protecting industrial scale lances.

The overall conclusion from the graph is that, although the helical vanes do not perform as well as the guide vanes using the heat transfer per unit pressure drop criterion, the helical vanes must be used to enhance the heat transfer enough to protect the lance.

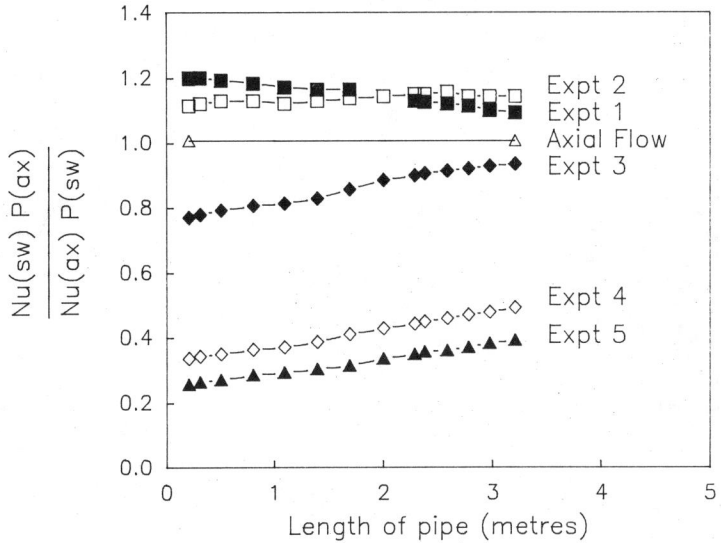

Figure 6: Heat transfer per unit pumping power versus length of lance. Nu(sw) and Nu(ax) are the swirling flow and axial Nusselt numbers respectively. P(sw) and P(ax) are the swirling flow and axial pressure drop respectively.

<u>Mathematical Modelling</u> The turbulent flow experimental data were modelled mathematically using the computer package FLUENT. A turbulence model was used to provide conditions to close the turbulent Navier-Stokes equations. Two models were available for use, these being the k-ε model, and an algebraic stress model (ASM).

The measured profiles of velocity, air temperature, and turbulence were used to verify the applicability of each turbulence model to swirling flow predictions.

Figures 7 and 8 show the predictions of each model for the tangential velocity profiles of experiments 1 and 4. The velocity is presented as W/U_{av}, i.e. the tangential velocity normalised with respect to the average axial velocity. This is plotted against radial distance from the inner wall. Each profile represents the tangential velocity at successive distances downstream of the inlet. The profiles are staggered by a factor of $W/U_{av} = 0.2$ so that they are easily distinguished.

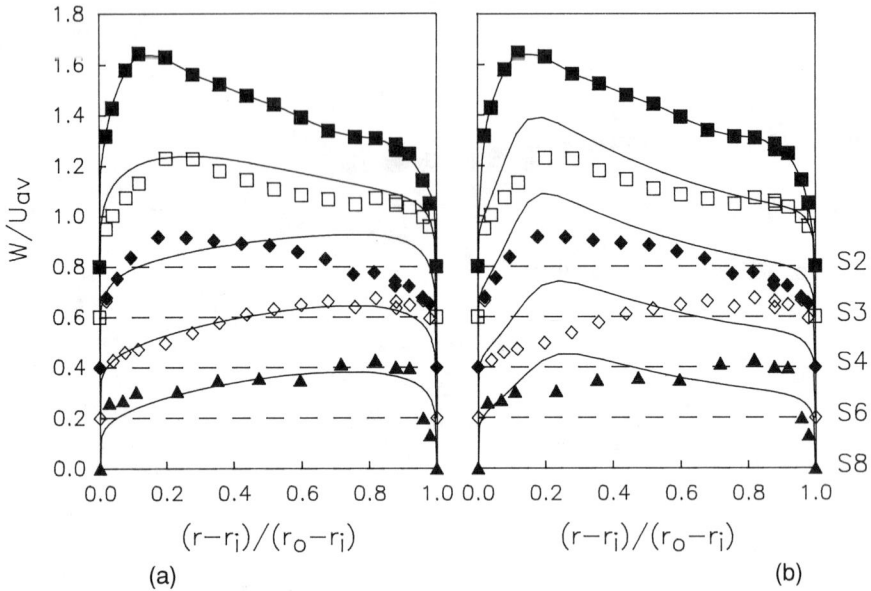

Figure 7:   Plot of W/U$_{av}$ versus radial distance from inner wall showing results from Experiment 1 and predictions based on (a) k-ε model and (b) ASM model.

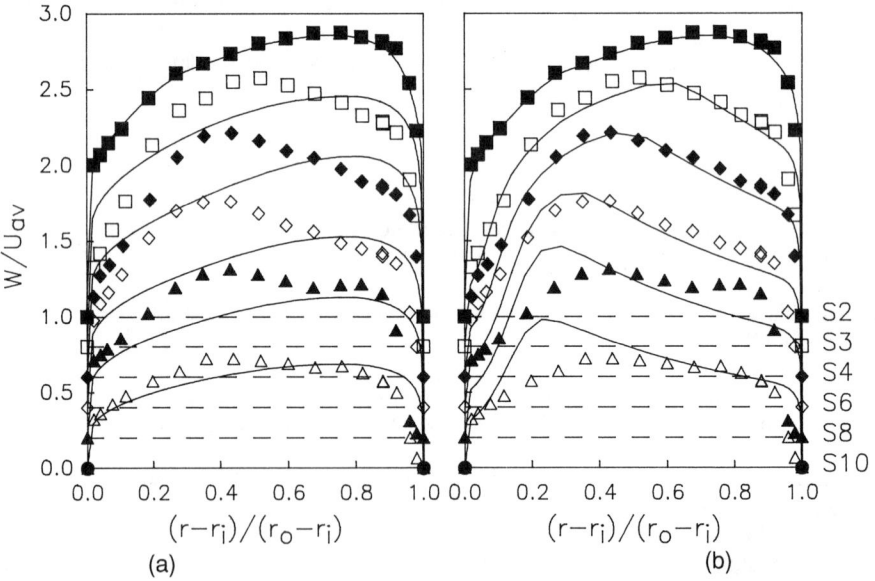

Figure 8:   Plot of W/U$_{av}$ versus radial distance from inner wall showing results from Experiment 4 and predictions based on (a) k-ε model and (b) ASM model.

In Figure 7, it can be seen that both turbulence models simulate the decay of the guide vane generated free vortex to a weak forced vortex. While the k-ε model fits the experimental data well, it is the ASM that seems to model the structure of the free vortex. The profile is almost linear near the inner wall, as would be expected for a forced vortex. The remaining profile is predicted to be concave up, which is in agreement with experimental data, and is expected for a free vortex flow. This combined vortex structure has been observed extensively in the literature (Yeh[25]; Scott and Rask[26]; Scott and Bartelt[27]; Morsi[28]; Yowakim and Kind[29]).

In Figure 8 (a) it can be seen that the much stronger forced vortex swirl of experiment 4 is not predicted at all well by the k-ε model. This is because the k-ε model assumes the turbulence structure to be locally isotropic. Turbulence measurements from this study and elsewhere (Morsi[28]; Kitoh[30]) reveal the three components of normal stress to be unequal, this effect being greater for higher swirl strengths. The weak forced vortex structure represented by the downstream profiles for both figures 7 and 8 is a swirl that has an approximately isotropic turbulence structure. Therefore, any swirling flow simulated using the k-ε model will probably quickly decay to this type of swirl. The simulations in this study revealed this behaviour under all circumstances.

The ASM does not assume an isotropic turbulence. By using simplified algebraic forms of the transport equations for turbulence, the ASM modelled the decay of forced vortex swirl to a combined vortex (Figure 8(b)). Further decay to a weakly forced vortex is not predicted using this model. This suggests that the algebraic equations in the ASM do not approach the isotropic turbulence condition for weakly swirling flows.

However, for the first 20 hydraulic diameters downstream of the helical vane swirler, the ASM models the decay of swirl accurately.

Splashing and Slopping in Matte Converting

Splashing and slopping often occur during matte converting. Slopping which is the creation of standing waves, can be created by the action of gas injection into a melt, the motion of the vessel during transportation or during the pouring of melt into a vessel. The standing wave formed occur in multiples of half a wavelength based on the bath length in the direction of the slop. Asymmetric standing waves have wavelengths as 1/2, 1 1/2, 2 1/2, ..., times that of the bath length, while symmetric standing waves have wavelengths as integer multiples of the bath length. As slopping is a natural phenomena, the presence of a force exerted on the melt can cause slopping to occur.

Work has been carried out by Liow and Gray[7,8] to understand slopping and to find means of damping the slop or operate under conditions where the slop frequency is not in resonance with the exciting forces. The formation of slop occurs when the kinetic and potential energy of the gas imparted to the melt and acted on by gravity is sufficient to overcome the viscous forces and turbulent dissipation occurring within the melt. Work on Peirce-Smith converters have shown that the first asymmetric wave is the dominant slop frequency. Higher order wave modes can be obtained but their amplitudes are normally so small that they hardly disrupt the process at all, except in very large vessels or when separation between interfaces is small.

Liow and Gray[7,8] studied a water model of the Peirce-Smith converter in detail and found that the power input required to create slopping is higher than the power input required to sustain the wave. The gas plume formed by the gas injected resonates with the slop. The tuyere angle and bath depth plots presented in Figure 9 show several regions. Region S is where spitting occurs. Region A1 is where the first asymmetric standing wave occurs with no spitting. Region B is where the first symmetric wave mode occurs. Region C is where the first symmetric wave was not strong enough to propagate across the bath resulting in a half formed symmetric standing wave near the gas injection end. No first symmetric standing waves were formed in region D. There is a region between S and A1 where both spitting is not continuous and slopping does not form over a large range of gas flow rates. This region is suitable for process operation, as high gas flow rates can be used.

Liow and Gray proposed that the wave steepness is responsible for the changes in the type of standing waves found in the water model. The wave steepness can contribute to the splashing of liquid from the first asymmetric standing wave if the amplitude of the standing wave is large. However, the onset of the slop creates little splashing, if any, when the tuyere is adequately submerged in the bath. The results showed that the lower gas flow limit of the first asymmetric standing wave can be calculated from the power per unit mass required to sustain the standing wave. An experimental value of 0.7 to 1.0 watt/kg was obtained for the power per unit mass from the potential power supplied. If the kinetic energy of the gas flow is large, its contribution to the power per unit mass to form a first asymmetric standing wave can be obtained as a function of the tuyere angle.

Work at the G.K. Williams CRC (Parker and Rodgers[31]) with the top submerged injection model in a cylindrical vessel has shown that there are regions in which slopping does not appear irrespectively of the gas flow rate used. The effect of placing a thin layer of lighter viscous oil over the water bath also results in a significant reduction in the region where slopping can be formed. Figure 10 compares the effect of no oil layer and a 1 cm oil layer on the region where slop are formed. Thus the large amount of slag in a metallurgical vessel can have a stabilising influence on the slopping phenomena which allows processes to use higher gas flow rates than if there was no slag layer. In these cases, the gas flow rate is limited by splashing that occurs rather than the formation of slop. However, in cases where there is little slag left in the vessel, for example, the copper blow during copper converting, slopping will be important in limiting the gas flow rate.

Splashing formed during gas injection has been found to be caused by two phenomena (Nielsen et al[32]). The first is the collapse of cavities resulting in liquid being ejected from the jet formed due to Rayleigh-Taylor instabilities. The second is the shearing of liquid sheets by gas moving upwards which results in both Rayleigh-Taylor and Kelvin-Helmholtz instabilities in the liquid sheets. Studies in furnaces at the G.K. Williams CRC have shown that the second phenomenon dominates the formation of splashing in large furnaces. However, in the top submerged vessels, the large amount of gases discharged relative to the surface area of the melt at the top of the vessel also gives rise to large cavities opening

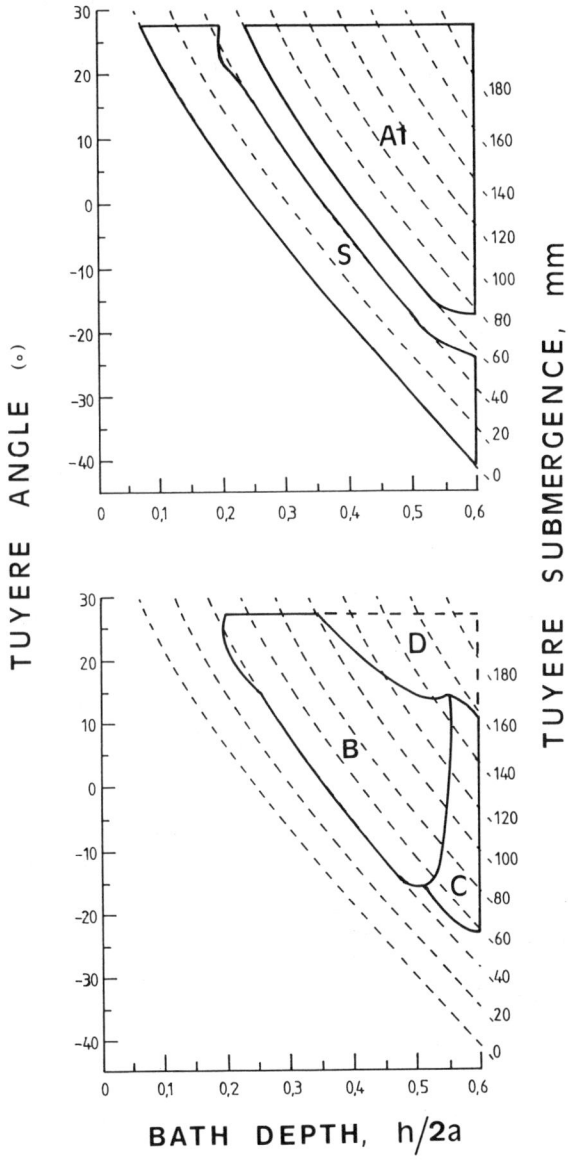

Figure 9: Effect of the tuyere angle/tuyere submergence and bath depth on the presence or absence of standing waves in the water model of a Peirce-Smith converter: (A1) first asymmetric standing wave, (B) first symmetric standing wave, (C) half first symmetric standing wave, (D) no first symmetric standing wave; (S) spitting.submerged vessels, the

and closing intermittently. Current work at the GKW CRC is directed towards understanding the mechanism of splash formation and the physical criteria which determine the instabilities formed during the break up of the liquid sheets and jets.

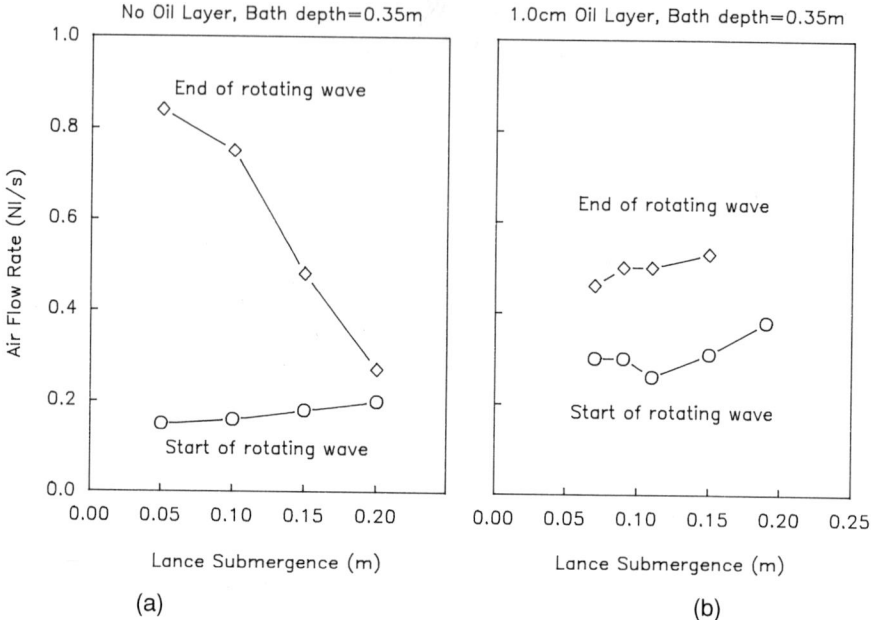

Figure 10: Effect of an oil layer in reducing the region where slopping occurs for a cylindrical vessel with top submerged gas injection.

Metal Refining

Melt refining prior to casting is an important step practiced in most base metal industries. Many examples can be cited. The steel industry treats massive tonnages of molten steel through the degassing process and controls other impurities through well known ladle metallurgy processes. Similarly, in copper processing excessive dissolved gases and unwanted elements are removed by oxidation/reduction stages depending on the chemistry of the impurity. In the last two decades improved understanding and applications of the principles of thermodynamics, kinetics and process dynamics have led to significant improvements to the existing processes and development of radical new techniques. The contribution of Savard Lee in introducing the annular tuyeres to bottom blown steelmaking is the most significant example.

In this section two recent examples of metal refining developments in the base metal industry are highlighted.

Anode Furnace

Crude blister copper from the converters contains about 0.05 percent dissolved sulfur and 0.7 percent dissolved oxygen. At these levels, if the metal was cast into anodes for electrorefining without prior treatment, on solidification the oxygen and sulfur would combine to form blisters of sulfur dioxide in and on the newly cast metal. The levels of oxygen and sulfur must therefore be reduced before the casting operation. Molten blister copper is commonly fire refined in a rotary cylindrical furnace in two stages. In the first stage or oxidation cycle, air is injected through one or two tuyeres to oxidise the sulfur to sulphur dioxide, leaving 0.001-0.002 percent sulfur in the copper. In the second stage or reduction cycle, hydrocarbons are injected to combine with the dissolved oxygen to lower its level to 0.1-0.5 percent. To optimise the operation with respect to minimising cycle times and consumption of raw materials requires a basic understanding of the total refining process.

Oxidation Cycle Shibasaki et al[33] investigated the rate of desulphurisation during the oxidation cycle in the anode furnace at the Naoshima Smelter and Refinery. The rate of desulphurisation was found to be proportional to the sulfur concentration for sulfur levels less than 0.2 percent. Improvements to the oxidation cycle time were obtained by addition of steam to the blowing air.
They proposed that the sulfur removal step was a liquid phase mass transfer controlled process but have not determined whether this occurs at the upper bath surface or at the bubble stirred interface. The sensitivity of the model to the sulphur dioxide partial pressure above the bath was not adequately evaluated.

Reduction Cycle Gray et al[11] carried out plant trials on the rate of removal of oxygen during the reduction cycle in the rotary anode furnaces at Mount Isa Mines Ltd. During several reduction cycles involving vaporised naphtha injection, the oxygen partial pressure in the gas phase above the copper was measured with solid-electrolyte probes. The results are plotted in Figure 11 which shows log weight percent oxygen in copper versus time. Most cycles conformed to a linear relationship showing that the deoxidation process can be described by a first order kinetic model with respect to oxygen in the molten copper. The results suggested that the rate of deoxidation was controlled by liquid phase mass transfer of dissolved oxygen to the gas-liquid interface at the bath surface. The change in bath oxygen content with time can be written as follows:

$$\ln\left(\frac{Co}{Co_i}\right) = \frac{-k_L At}{V} \qquad (1)$$

provided that the concentration of oxygen in the bath at the gas-liquid interface approaches zero. Values of $\dfrac{k_L A}{V}$ for each cycle can be obtained from the slope of the lines on Figure 11. A mass and heat transfer process model was developed for the deoxidation of copper with naphtha based upon the above observations. The model consisted of maintaining a constant oxygen potential difference for the removal of oxygen between the gas and the liquid. Ideally the oxygen potential in the gas phase should be buffered so that ingress of a small amount of air will not have a significant effect on the direction or rate of mass transfer. Values of the oxygen potential in the gas phase were predicted at a known temperature. It was found that apart from the extremes at which soot formation and stoichiometry are approached, the oxygen potential for these gas mixtures does not vary greatly with composition i.e. they are buffered with respect to oxygen addition. Over the range of process conditions considered $\log_{10} P_{O_{2(g)}}$ was usually within the range of -11 to -14. This was consistent with experimental

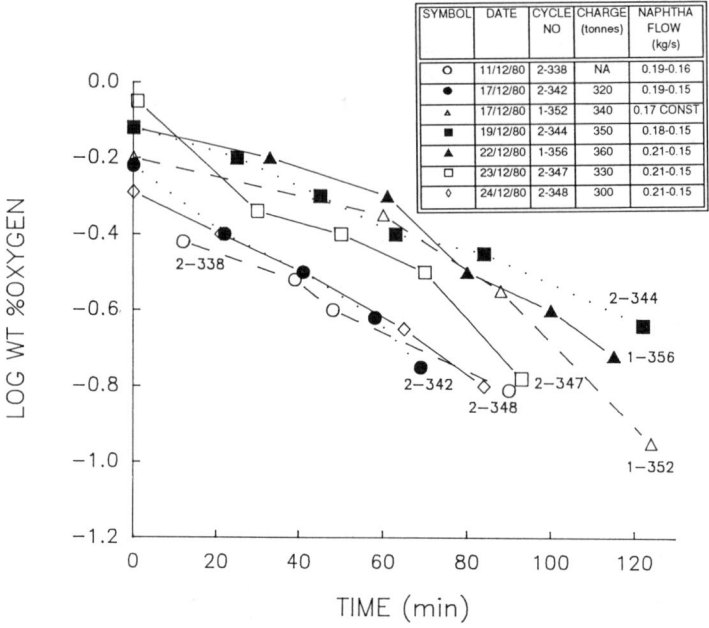

SYMBOL	DATE	CYCLE NO	CHARGE (tonnes)	NAPHTHA FLOW (kg/s)
O	11/12/80	2-338	NA	0.19-0.16
●	17/12/80	2-342	320	0.19-0.15
△	17/12/80	1-352	340	0.17 CONST
■	19/12/80	2-344	350	0.18-0.15
▲	22/12/80	1-356	360	0.21-0.15
□	23/12/80	2-347	330	0.21-0.15
◇	24/12/80	2-348	300	0.21-0.15

Figure 11: Semilogarithmic plot of bath oxygen content against time for several deoxidation cycles in an anode furnace[11].

results when the naphtha flowrate and damper setting were carefully adjusted to maintain a positive pressure within the furnace to minimise ingress of air. If the oxygen partial pressure above the bath exceeded $10^{-10} - 10^{-11}$ atm. the rate of deoxidation was reduced. This suggests that under these conditions the deoxidation conditions may be controlled by mass transfer in the gas phase.

The depth of immersion of tuyeres between 0 and 500 mm was observed to have a significant effect on the rate of reduction. This presumably results from an increase in buoyancy energy imported to the bath which enhances mixing. This observation is consistent with liquid phase transport control. Increasing the tuyere submergence beyond 500 mm did not enhance the deoxidation rate.

The mathematical model was used to determine the optimum naphtha injection rate through a deoxidation cycle assuming the oxygen partial pressure in the gas phase is kept below 10^{-11} atm. Predictions of the model are compared with initial and adjusted naphtha practices in Figure 12 for a charge of 320 t, initial oxygen content of 6500 ppm and final oxygen content of 1500 ppm. The model predicts approximately 30% reduction in naphtha consumption if the rate can be continuously reduced through the deoxidation stage. The lower limit to naphtha addition was imposed by tuyere blockage problems. The blockage problem was solved by installing critical flow orifices upstream of each tuyere to dampen pressure fluctuations in the tuyere.

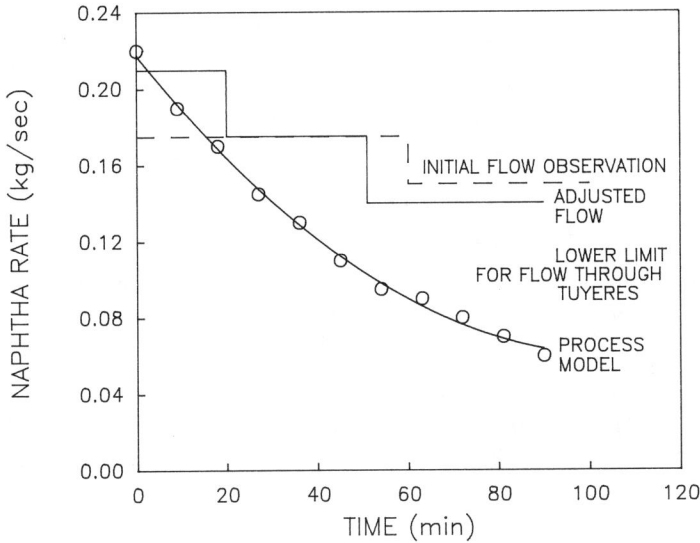

Figure 12:   Comparison of initial and adjusted naphtha injection rates with model predictions for a deoxidation cycle[11].

Optimal Control Strategy   Bateman[34] has extended the previous work to develop an optimal control strategy for the anode furnace.   This strategy aims to minimise batch times and consumption of raw materials for the oxidation and reduction cycles while operating within the physical constraints of the system to produce a consistent product.   Dynamic models were developed for the sulphur and oxygen removal processes as well as the temperature of the copper bath.   An objective function was defined containing the parameters mentioned above.

After developing a general method for optimising the oxidation and reduction cycles, the two cycles were then linked using a "smart" searching method.   The optimal control technique was found to generate significant improvements to current practice.

Helical Plug Lance

In the melt refining the most fundamental step is the effective contacting between the melt and the gas which is provided by a suitable gas distributor.   A number of devices are commonly used such as lances/tuyeres, diffusers and rotors. Although rotors are very effective means to breakup gas into fine bubbles, their use is restricted at high temperatures and their adaptability to the different designs

543

of furnaces is also limited. Diffusers are prone to clogging, wearing out and are mechanically weak. Lances are very simple and easy to use devices, popularly employed across the metallurgical industry in almost any shape and size of furnace.

The plain lance, often called a flux wand or flux tube in the aluminium industry is commonly employed both in the primary and in the foundry industry for gas fluxing and degassing operations. The simplicity of the lance operation, ruggedness and low cost have favoured its popular use over the years. Since the lance does not have a rotating part its maintenance cost is also low. However the lance has its limitations in terms of its ability to disperse gas as fine bubbles. In an effort to improve the efficiency of a plain lance used in the aluminium industry a helical plug lance was developed[35].

The plug is designed in such a way that it could be incorporated into the existing lance currently in use in the industry. The pressure losses and entry effects due to the plug are optimised so that it can be used on the existing gas line pressure. The second design aspect is to provide multiple angled gas outlets to provide a uniform gas dispersion of finer bubbles. The lance has been successfully tested in the aluminium industry in the refining and degassing of aluminium in the holding and crucible furnace[40].

Plain lances produce large bubbles and their volume(V_b) can be estimated by the relationship[36];

$$V_b = K \, Q^{6/5} \, g^{-3/5} \qquad (2)$$

Where K is a constant, g is the gravitational constant and Q is the gas flowrate. This relationship is valid when the gas momentum is not significant[37]. With increasing gas momentum at a constant gas flow rate (Q) the actual bubble size (V) decreases. The effect of increasing gas momentum has been shown in Figure 13 as injection number, N_I, versus the ratio of the actual bubble size V to the bubble size V_b suggested by equation 2, where N_I is the ratio of the gas momentum and bubble buoyancy. With increasing N_I, the bubble size is drastically reduced. Plain lances generally operate at N_I <0.1 producing large bubbles according to equation (2), whereas the helical plug lance operates at very high momentum (N_I>1) producing very fine bubbles. Figure 13 shows that there is a marked difference in the actual bubble size produced from a helical plug lance compared to a plain lance.

This difference is given more explicitly in Figure 14 as contours of bubble frequency when measured[39] by an electrical contact probe in water for a plain and a helical plug lance. Comparison of the contours clearly points out the two main benefits of the helical plug lance when compared with the plain lance. First the frequency of interfaces is higher by a few orders of magnitude and second the lateral spread of the bubble zone increases.

In an attempt to assess the performance of a helical plug lance Nilmani et al[40] carried out laboratory experiments using a water oxygen desorption analog test. They compared the performance of a plain lance (25 mm id), a helical plug lance (25 mm lance with helical plug insert) and a diffuser (3 disc type) in a large water tank (1.22 x 1.22 x 1.3 m). Water was first saturated with oxygen by blowing air,

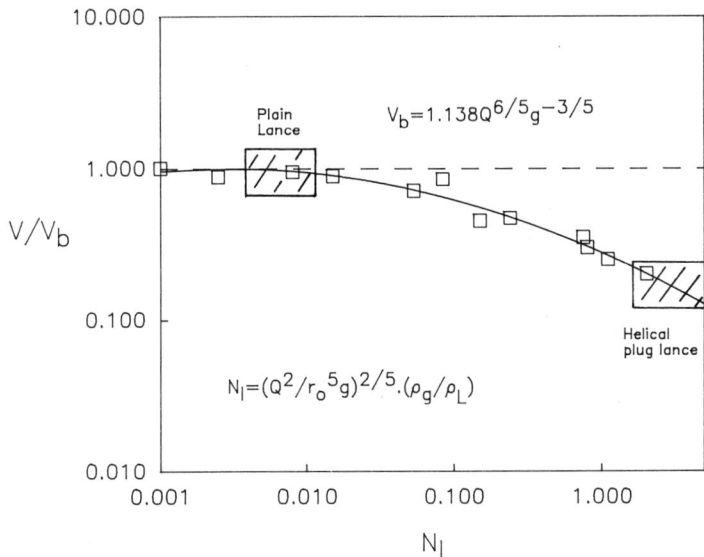

Figure 13: Dimensionless bubble volume plotted against injection number. Experimental points are for room temperature modelling work[37] Q-gas flowrate, V_b-lance radius, ρ_g and ρ_L gas and liquid densities, V-measured bubble volume, V_b-calculated volume from equation[2].

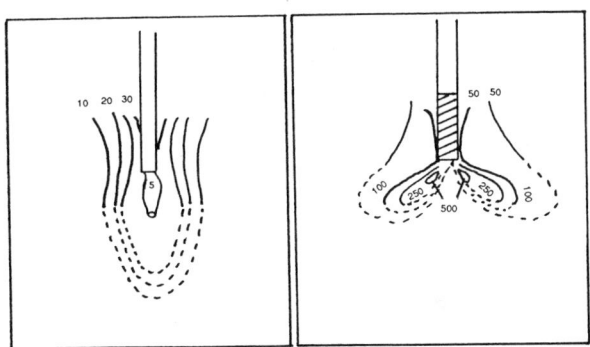

Figure 14: Contours showing frequency of interfaces passing (a) plain lance and (b) helical plug lance[39].

followed by nitrogen purging with a selected device to desorb the dissolved oxygen. The drop in oxygen concentration was monitored by a dissolved oxygen probe. The relative performance of the three devices were tested at 5×10^{-3} Nm^3 /sec, a typical flowrate for a holding furnace. From the gradient of the log-linear relationship of oxygen drop with time as shown in Figure 15, it is evident that the helical plug lance achieves much higher desorption rates compared to the other two devices.

Further tests carried out in the aluminium smelter on the sodium removal by Nilmani et al[40], showed a similar trend. They tested a helical plug lance and a plain lance in a 52 tonne holding furnace and a 3.5 tonne transfer crucible. Figure 16 demonstrates that the helical plug lance reduces the sodium levels faster than the plain lance. The reduction of the sodium level in a 52 tonne holding furnace was 65% in 20 min using a helical plug lance, compared with 56% reduction in 30 min when a plain lance was used. This was despite the fact that the helical plug lance was operating at only 95% of the gas flowrate used in the plain lance.

Tests conducted in a 3.5 tonne open transfer crucible also revealed similar results (Figure 17). The plain lance took 10 minutes to register 52% drop in sodium level, whereas the helical plug lance recorded a 65% sodium reduction in just eight minutes and with a gas flowrate of 83% of that used in the plain lance.

This example indicates that the current refining practices of gas lancing can be significantly improved and in the melt pretreatment operations, the newly developed helical plug lance holds considerable promise.

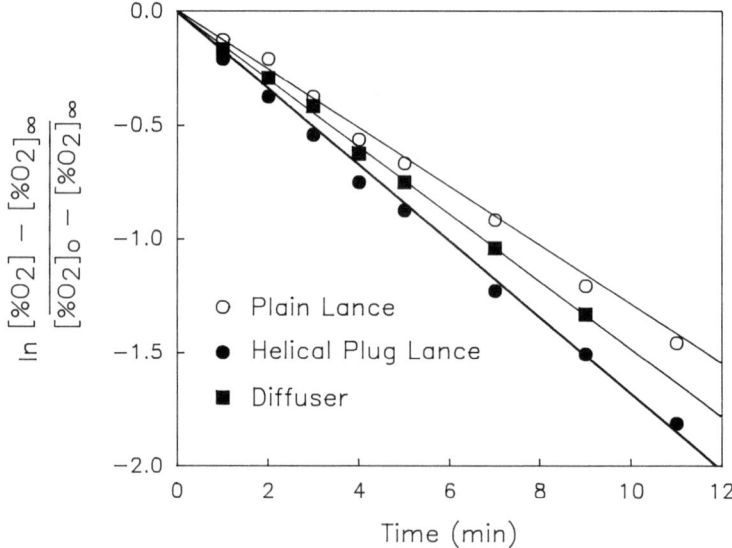

Figure 15:   Oxygen desorption from water comparing three injection devices.
        $[\% \, O_2]$ - O_2 concentration wt %
        $[\% \, O_2]_o$ -   initial O_2 concentration
        $[\% \, O_2]_\infty$ -   steady state O_2 concentration

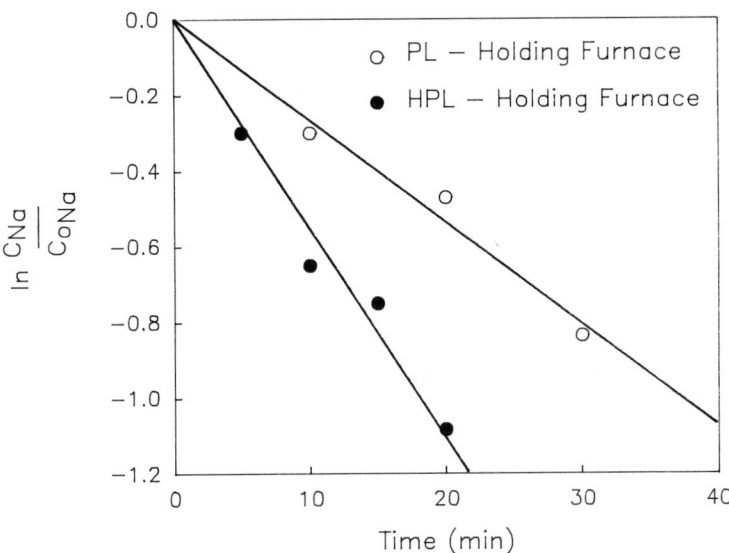

Figure 16: Sodium removal from an aluminium holding furnace C_{Na} and C_{oNa} are the sodium concentration at a given time and initial sodium concentration respectively.

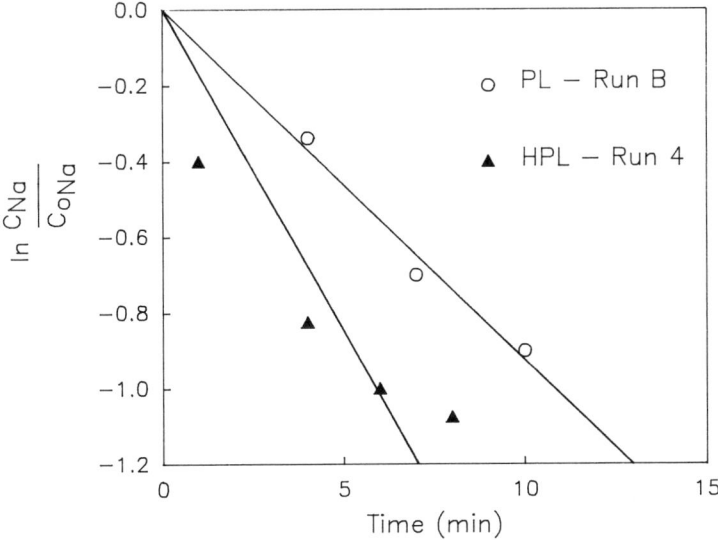

Figure 17: Sodium removal by argon and chlorine mixture injection from a transfer crucible. C_{Na} and C_{oNa} are the sodium concentration at a given time and the initial sodium concentration respectively.

Summary and Conclusions

In matte converting and metal refining, changes have been slow to emerge to minimise tuyere punching, improve refractory wear and overcome environmental problems. Few attempts have been made to optimally design such processes and develop appropriate control strategies. To make such changes and develop alternate solutions it has been necessary to critically examine each step in the processes involved using the techniques of mathematical modelling coupled with laboratory and plant measurements. Alternate solutions have begun to emerge through improvements to the operation and control of existing processes and by the development of specifically designed intensive reactors with specially designed lances.

Acknowledgments

Financial support for this work has been provided by The Australian Research Council and the Australian Mineral Industries Research Association. Madhu Nilmani is thankful to KNS for funding and support provided for the flux injection work.

548

References

1. E.O. Hoefele and J.K. Brimacombe, "Flow regimes in Submerged Gas Injection", Metall. Trans. B, 10B (1979), 631-48.

2. J.L. Liow and N.B. Gray, "A Model of Bubble Growth in Wetting and Non-wetting Liquids", Chem. Eng. Sci 43(12) (1988), 3129-3139.

3. T.A. Engh and M. Nilmani, "Bubbling at High Flow Rates in Inviscid and Viscous Liquids (Slags)" Metall. Trans. B, 19B (1988), 83-94.

4. A.W. Castillejos and J.K. Brimacombe, "Measurement of Physical Characteristics of Bubbles in Gas-Liquid Plumes. Part I: An Improved Electroresistivity Probe Technique", Metall. Trans. B, 18B (1978a), 649-658.

5. Y.Y. Sheng and G.A. Irons, "A Combined Laser Doppler Anemometry And Electrical Probe Diagnostic For Bubbly Two-Phase Flow", J. of Multiphase Flow, 17 (1991), 585-598.

6. V. Sahajwalla, A.H. Castillejos and J.K. Brimacombe, "The Spout of Air Jets Upwardly Injected Into A Water Bath", Metall. Trans. B, 21B (1990), 71-80.

7. J.L. Liow and N.B. Gray, "Slopping Resulting From Gas Injection In A Peirce-Smith Converter: Period of the Standing Wave", Metall. Trans. B, 21B (1990a), 657-664.

8. J.L. Liow and N.B. Gray, "Slopping Resulting From Gas Injection In A Peirce-Smith Converter: Water Modelling", Metall. Trans. B, 21B (1990b), 987-996.

9. M. Nilmani, T.A. Engh, N.B. Gray and J.M. Floyd, "Plant And Laboratory Investigations On Injection And Related Phenomena In Some Smelting And Refining Processes", Scandinavian Journal of Metallurgy, 19, (1990), 278-287.

10. A.A. Bustos, G.G. Richards, N.B. Gray and J.K. Brimacombe, "Injection Phenomena in Non-ferrous Processes", Metall. Trans. B,15B (1984), 77-89.

11. N.B. Gray, M.J. Hollitt, R.G. Henley and J. Pritchard, "Investigation And Modelling Of Anode Furnace And Casting Operations At Mount Isa Mines, Ltd., Queensland, Australia", Trans. Inst. Min. Metall., (Sect C: Mineral Process. Extr. Metall.) 91 (1983), C54-C63.

12. J.K. Brimacombe, S.E. Meredith and R.G.H. Lee, "High Pressure Injection Of Air Into A Peirce-Smith Copper Converter", Metall. Trans. B, 15B (1984), 243-250.

13. J.K. Brimacombe, A.A. Bustos, D. Jorgensen and G.G. Richards, "Towards A Basic Understanding Of Injection Phenomena In The Copper Converter" Kellogg Symposium, AIME Annual Meeting, New York, February 25-28, (1985), 327-351.

14. C.R. Fountain, "The measurement and analysis of gas discharge dynamics in metallurgical converters" (Ph.D. thesis, The University of Melbourne 1988)

15. J.L. Liow and N.B. Gray, "Modelling of Slopping and Its Effects on Refractories In Metallurgical Vessels" TMS Annual Meeting. International Symposium on Refractories for Process Vessels, New Orleans, USA (1991), 295-307.

16. M. Nilmani and D. Collins, "Plant Trials on Flux Injection in a P-S Nickel Converter", CIM, Conference, Montreal, (1988), 24.

17. L. Farias & D.G.C. Robertson, "Physical Modelling of Gas-Powder Injection into Liquid Metals", Proceedings of 3rd Process Technology Conf., Pittsburgh, PA, ISS and AIME, (1982), 206-220.

18. M. Nilmani, A.K. Das, A. Johnson and T.A. Engh, "Modelling Gas Powder Discharge into Melts", The 14th Australian Chemical Engineers Conference, Adelaide, S.A. (1986), 107-112.

19. Y. Ozawa and K. Mori, "Critical Conditions For Penetration Of Solid Particles Into Liquid Metal" Trans ISIJ, 23 (1983), 769-774.

20. B.W. Lee, D. Langberg and M. Nilmani, "Physical Modelling of Particle Penetration Relevant to Injection Metallurgy", Int. Symp. Injection in Process Metallurgy, ed. Lehner et al, TMS New Orleans (1991), 129-142.

21. S. Avery, D. Langberg and M. Nilmani, "Particle penetration at high loading", Unpublished work. University of Melbourne, 1992.

22. N. Dave and N.B. Gray, "Modelling of Annular Swirled Flow Lances with Helical Inserts", Trans. Inst. Min. Metall. (Sect C: Mineral Process Extr. Metall) 98 (1989), C178-184.

23. N. Dave and N.B. Gray, "Fluid Flow Through Lances with Constant and Variable Pitch Swirled Inserts", Metall. Trans. B, 22B (1991), 13-20.

24. C.B. Solnordal, "Modelling of Fluid Flow and Heat Transfer in Decaying Swirl Through a Heated Annulus" (Ph.D. thesis, The University of Melbourne, 1992).

25. H. Yeh, "Boundary layer along annular walls in a swirling flow", Trans ASME, 80(1958), 767-776.

26. C.J. Scott and D.R. Rask, "Turbulent viscosities for swirling flow in a stationary annulus", J. Fluids Eng, Trans ASME, 95 (1973), 557-566.

27. C.J. Scott and K.W. Bartelt, "Decaying annular swirl flow with inlet solid body rotation", J. Fluids Eng Trans ASME, 98 (1976), 33-40.

28. Y.S.M. Morsi, "Analysis of turbulent swirling flows in axisymmetric annuli" (Ph.D thesis, London University, 1983).

29.   F.M. Yowakim and R.J. Kind, "Mean flow and turbulence measurements of annular swirling flows", J. Fluids Eng Trans ASME, 100 (1988), 257-263.

30.   O. Kitoh, "Experimental study of turbulent swirling flow in a straight pipe" J. Fluid Mech, 225 (1991), 445-479.

31.   L. Parker and J. Rodgers, "Wave Formation And Damping In An Upright Cylindrical Vessel During Submerged Top Gas Injection" Final year project report, G.K. Williams CRC, Department of Chemical Engineering, University of Melbourne, 1990.

32.   R.D. Nielsen, M.R. Tek and J.L. York, "Mechanism Of Entrainment Formation In Distillation Columns" Symposium on 2-phase flow. University of Exeter, Ed. Lacey P.M.C., 1965, F201-225.

33.   T. Shibasake, T. Shimizu and N. Oguma, "Analysis of Process Dynamics And Improvement Of Actual Operation At Anode Furnace" International Symposium on Injection in Process Metallurgy, 1991, TMS, 266-276.

34.   I.R. Bateman, "Mathematical Modelling and Optimal Control of Copper Anode Furnace" (PhD thesis. The University of Queensland, 1992).

35.   M. Nilmani and J.M. Floyd, "A novel gas injection device for liquid metal treatment", Aust Provisional Patent PI 7760, 1988.

36.   M. Nilmani, "Gas injection into melts: Relevant to holding furnace pretreatment" Course proceedings on Aluminium Melt Refining & Alloying, (ed. M. Nilmani), University of Melbourne, 1991, pp. F1-13.

37.   M. Nilmani and D.G.C. Robertson, "Model studies of gas injection at high flowrates using water and mercury" Trans. Inst. Min. Metall. (Sect C: Mineral Process. Extr. Metall.), 89 (1982), C42-C53.

38.   A.E. Wraith and M.E. Chalkley, "Tuyere injection for metal refining". IMM Symposium, 1977, 27-33.

39.   M. Nilmani and D.S. Conochie, "Gas dispersion with swirled lances" Scaninject IV, Part 1, Lulea, Sweden, 1986, pp.7:1-7:19.

40.   M. Nilmani, P.K. Thay, C.J. Simensen and D.W. Irwin, "Gas fluxing operation in aluminium melt refining laboratory and plant investigations", Light Metals, (1990), 747-754.

41.   S.J. Park, "Injection of flux Into Nickel Converter", Western Mining Internal Technical Conference/Metallurgy, 20-21 March, 1991, pp12.

SUBMERGED INJECTION IN NON-FERROUS PROCESSES

G.G. Richards

Centre for Metallurgical Process Engineering
University of British Columbia
Vancouver, B.C., Canada V6T 1Z4

Abstract

The non-ferrous smelting industry is critically dependent on the submerged injection of a wide range of different gases and solids in a variety of processes. Virtually all routes for the extraction of non-ferrous metals use this bath smelting technology in some way, including a number of the hydrometallurgical leaching operations. The purpose of this paper is to review our current understanding of submerged injection and the potential benefits it offers for future process development.

Matte converting is the most wide spread non-ferrous bath smelting process. Examples include the converting of copper and nickel mattes and the oxidation stage of the QSL process and in these cases the gas is introduced through submerged injectors or tuyeres. Not surprisingly, the ability to use oxygen or air in this way is dependent both on process constraints as well as available technology. In a number of processes such as Mitsubishi process and Sirosmelt/Isasmelt a lance is used to introduce the reactants into the bath. The lance has certain advantages over the use of submerged tuyeres. A number of unit operations rely on the ability to inject solid reagents or feed material into the bath. For example, in zinc fuming use is made submerged tuyeres to introduce a solid reductant into the smelting system.

The design and optimum operation of bath smelting processes requires a thorough understanding of submerged injection phenomena. Submerged injection of gases and solids allows for an intimate interaction of the reactants with the bath giving a large contact area which promotes rapid reaction. There are several key problems which must be overcome however, if the process is to be used effectively. These include tuyere blockage which results from the formation of an uncontrolled accretion at the end of the injector. Under certain circumstances the erosion of the tuyere and surrounding refractory can also be a serious problem. The injection of fine solids through tuyeres is very problematic and systems have to be carefully designed to the characteristics of solid material.

Although submerged injection has served the industry well there is substantial room for future improvements, and in this regard there are lessons that can be learned from ferrous metallurgy. There are a number of examples that can be cited such as the use of high oxygen enrichment in Peirce-Smith converting and the use of gases for stirring in electric furnaces and metal refining.

Proceedings of the
Savard/Lee International Symposium on Bath Smelting
Edited by J. K. Brimacombe, P. J. Mackey,
G. J. W. Kor, C. Bickert and M. G. Ranade
The Minerals, Metals & Materials Society, 1992

Introduction

For a number of reasons bath smelting processes will remain a vital part of the nonferrous smelting industry. They are very intense and are able to treat a wide variety of feed materials. Furthermore, the presence of a bath of molten material contributes to process inertia which aids in the damping out unavoidable fluctuations in operating variables. For a bath smelting process to operate successfully close attention must be paid to a number of factors including coupling of the vessel to off-gas handling systems and bath-container interactions. Critical, however, is the submerged injection of reactants and fluxes through submerged injectors.

The use of submerged injection in the non-ferrous industry followed closely its introduction into ferrous metallurgy. The most wide spread non-ferrous bath smelting process is converting which is directed to the removal of sulphur and iron from sulphide feed streams. Examples include the converting of copper and nickel mattes and the oxidation stage of the QSL process. In these cases the gas is introduced through submerged injectors or tuyeres. The traditional process uses air or enriched air while oxygen has found application in the QSL. Oxygen is also used in lead softening to remove impurities such as arsenic and antimony. Not surprisingly, the ability to use oxygen or air in this way is dependent both on process constraints as well as available technology. A lance is used in the Mitsubishi and Sirosmelt/Isasmelt processes to introduce reactants into the bath. The lance has certain advantages over the use of submerged tuyeres. Finally, the submerged injection of solid reagents or feed materials is an important aspect of a number of processes including zinc slag fuming.

Injection Dynamics

The fundamental aspects of gas injection in metallurgical systems have been reviewed in some detail recently(1) and it is beyond the scope of this paper to reiterate this information in detail. It is appropriate however to briefly deal with the major aspects of injection dynamics as it affects the basic injection phenomena discussed in the subsequent section.

Gas Injection Dynamics

The submerged injection of gas into a liquid bath is a subject that has received considerable study in recent years. A large number of physical phenomena are involved in determining the behaviour of the gas once it enters the liquid. However, the primary factor is the energy of the gas flow. Under high pressure injection conditions the gas tends to enter the liquid in a continuous stream. Velocities of the order of Mach one or greater are necessary to ensure that the bath does not enter the tuyere or injector at any time. As the injection pressure is decreased, to what is standard in nonferrous converting, a considerable change in the phenomena occurs. At low pressures the gas energy is low and incapable of sustaining a jet at the injector tip. In this regime bubbles are formed which grow and detach periodically. During the detachment process liquid washes back against the tuyere.

This process is influenced by the general injection conditions as shown in Figure 1 by back-pressure traces.(2) What is termed 'classic' bubbling is what would be seen for single (or isolated tuyeres) at a relatively deep submergence. At bubble detachment liquid bath washes back against the tuyere and results in a steep increase in pressure. This is followed by a gradual decline in pressure indicative of the greater ease with which gas flows into the bubble as it grows. At some point however the buoyancy force causes the bubble to detach from the tuyere and liquid moving into the vacated space results in the pressure spike and the process is repeated. A distinctly different behaviour is seen when multiple tuyeres are involved and are close enough to interact as is the case in the Peirce-Smith converter. Here the regime might be termed: 'unstable envelope'. Instead of separate bubbles the pressure trace shows a constant injection pressure interrupted by spikes of short duration. During the periods of constant pressure injection the gas is flowing into a envelope or tube of gas which runs down the length of the tuyereline. Periodically this envelope breaks down in places as local region detaches and liquid washes in toward the tuyere. This appears as the steep vertical forward side of the spike. The gas however quickly punches through to the gas envelope which is as close as the adjacent tuyere. There is therefore no period of prolonged bubble growth. Instead the pressure rapidly drops as the gas flows easily into the envelope. Refractory erosion along the tuyereline likely stabilizes the envelope and contributes to this behaviour. A third form of bubbling seen in nonferrous vessels is 'channelling'. Here the important additional factor is the tuyere submergence. During injection at shallow submergences the gas bubbles can grow right through to the surface while they are still connected to the tuyere. A low pressure pathway for the gas is thereby opened up and it passes with little resistance through the bath. Periodically the gas flow is interrupted by an actual bubble detachment and liquid flows back against the tuyere causing a pressure spike.

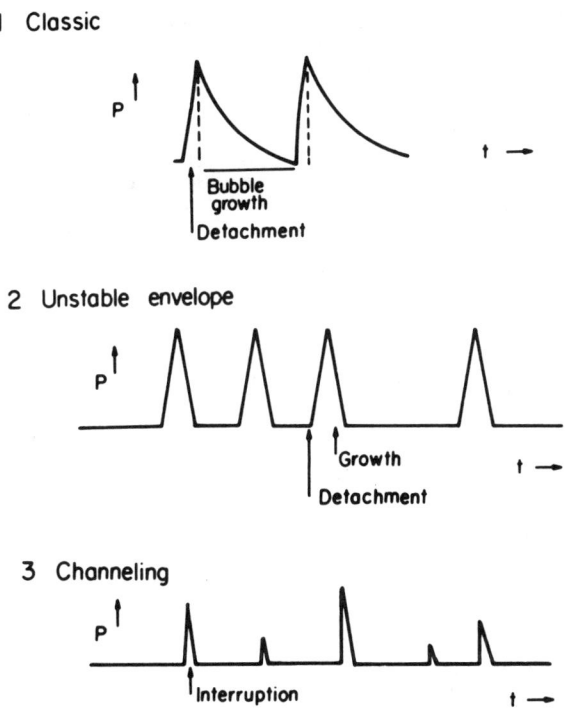

Figure 1. Idealized tuyere pressure traces for three regimes of gas discharge from horizontal, closely spaced tuyeres.(2)

Lance Injection

The injection of gas and feed materials through lances is an important part of nonferrous smelting processes such as the Mitsubishi and Isasmelt or Sirosmelt technologies. Depending on the submergence of the lance in the bath various phenomena are important. If the lance is above the surface then the interaction of the jet with the liquid is a function of the kinetic energy of the stream. At low gas velocities the impinging jet simply causes 'dimpling' of the surface. As the velocity is increased penetration of the gas and splashing become dominant. If the lance is submerged below the surface then the questions of bubbling and jetting discussed above must be considered. The major difference is the direction of the buoyancy force which in the case of the lance acts against the injection direction. At low pressures the gas will enter the bath as bubbles and lance blockage will be likely.

Gas-Particle Injection

The presence of solid particles in the injection stream can have a significant impact on injection dynamics. If the gas and particles are flowing in a closely coupled manner they will tend to behave as a 'single phase' jet on entry into the liquid. The presence of the particles will also cause jetting rather than bubbling to occur at lower injection velocities because the particles contribute a significant amount of energy to the injection stream. In general, the gas will be largely entrained with the particles in these circumstances. However, if the flow is uncoupled, then the particles will behave independently and their entrainment in the liquid will depend more on the dynamics of single particle-liquid interaction. The slip velocity will therefore be a more important parameter. When a single particle impacts a liquid surface a range of outcomes are possible. If the velocity of the particle is low then the particle may simply remain at the interface. At higher velocities the particle can penetrate into the liquid and may even form a cavity and draw a gas bubble into the liquid with it.(3)

It is worth noting that there are differences in vertical and horizontal gas-particle flow. In horizontal flow non-homogeneous conditions can develop at high loadings if the gas velocity is not kept above the saltation velocity. The presence of a dense, slower moving suspension on the bottom of the pipe could dramatically reduce particle entrainment in the liquid. Finally, heat and mass transfer phenomena cannot be ignored. The injection of a relatively cold gas-particle mixture into the liquid results in local cooling and the potential formation of accretions which may block the injector.

Injection Phenomena

The submerged injection of gas into a bath has two direct consequences: local cooling which may or may not result in an accretion and/or blockage of the injector and local refractory wear. These two phenomena represent the major aspects of submerged injection which must be dealt with in the operation of a practical process. It should be noted in passing that the submerged injection of gas results in several indirect consequences such as bath slopping and splashing which may have important effects on the process.(4) These questions however are considered in detail elsewhere in the Proceedings.

Gas-Bath Reactions

Bath smelting processes are efficient reactors for high temperature gas-liquid reactions because a very large contact area can be generated. This is important because the residence time of the gas is short (of the order of 0.1 to 1 second) and for effective use of the gas a large area must be available. The reaction between a reactive gas such as air and a sulphide matte has been shown to be gas-phase mass transfer controlled(5,6) and hence submerged injection phenomena are critical to efficient operation. As noted above, under low pressure injection conditions the gas first enters the bath in the form of bubbles, Figure 2, which detach from the tuyere and probably break up as they rise through the bath.

Given that bubble frequencies are of the order of 5 - 10 per second it is obvious that the time the gas spends at the tuyere during bubble formation can be a significant fraction of its total residence time. Figure 3 shows the results from a physical modelling study of the converter in which SO_2 was absorbed from the injected gas into a peroxide solution.(6) Under dynamically similar injection conditions somewhat more than 50% of the SO_2 reacted during bubble formation (t = 0 in the figure). It is important therefore to consider the implications for process kinetics and local refractory wear. It is worthwhile to distinguish this case

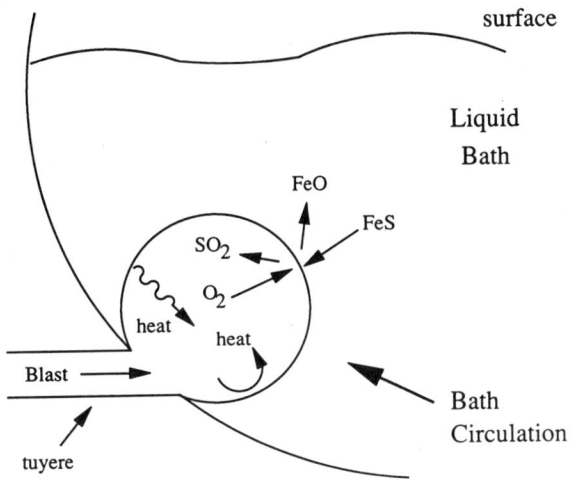

Figure 2. Schematic diagram of bubble formation at the tuyere of a Peirce-Smith converter.

from that of high pressure injection as seen in ferrous converters where the gas effectively issues form the injector as a jet and spends virtually no time confined to the region immediately in front of the tuyere.

Accretions and Injector Blockage

Accretion formation is a common phenomena in submerged injection processes. There are generally three types: hemispheres, hollow-cored hemispheres and pipes. The former two are found in the ferrous industry and result respectively from the injection of inert gases through a single injector and the shrouded tuyere injection of oxygen. In the non-ferrous industry the pipe type accretion is most common and forms during the injection of a reactive gas through a single pipe injector at relatively low pressures.

Whatever there form, accretions are basically a thermal phenomena and result from the freezing of bath material, either slag or matte.(7,8) Accretions themselves are found to be beneficial *so long as they do not impede the flow of the gas.* They form as a result of local cooling and therefore are indicative of less severe thermal conditions in the vicinity of the tuyereline. They also serve to move the gas-liquid reaction zone away from the vessel wall and therefore also reduce refractory wear. In the case of injection in the ferrous industry high injection pressures are used to ensure that accretions are non-blocking. Figure 4 shows a predicted accretion growth sequence at a tuyere in a Peirce-Smith copper converter treating a 30% matte grade for a 5 K superheat.(9)

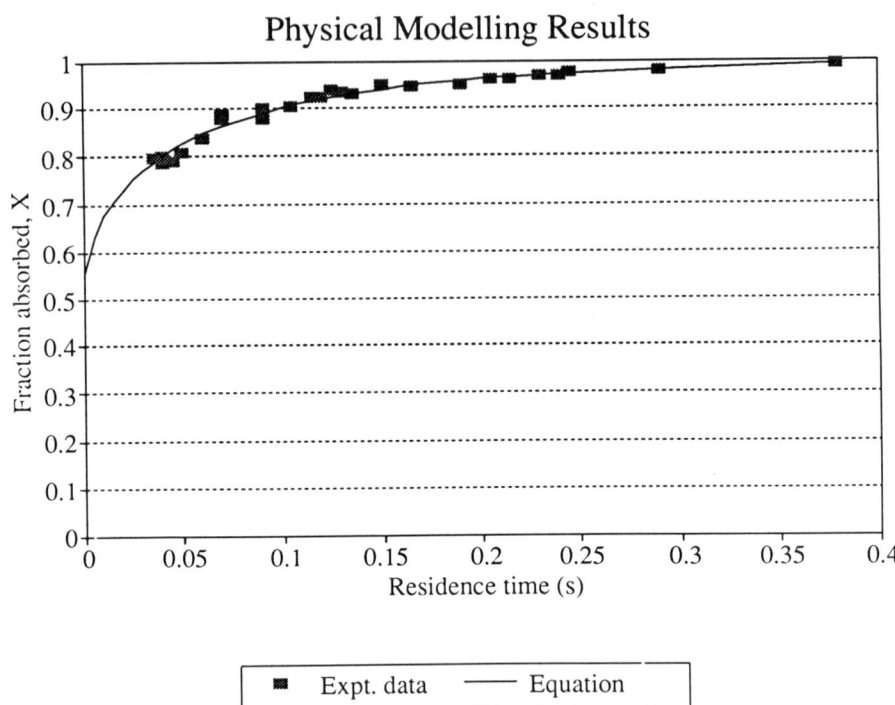

Figure 3. SO$_2$ adsorption into a hydrogen peroxide solution during a physical modelling simulation of copper converting.(6)

The accretion reaches a steady state size in about six minutes. The critical variables which influence accretion size and tendency toward blocking are bath superheat and circulation velocity. The latter is important because it is very much a factor in controlling the flow of heat from the bath to the accretion. As the velocity increases the flow of heat is increased substantially and the accretion melts back as shown in Figure 5. In this regard wear of the tuyereline and the development of a 'notch' which would shield the accretions and encourage their formation. The presence of accretions in turn would increase the tendency to blockage.

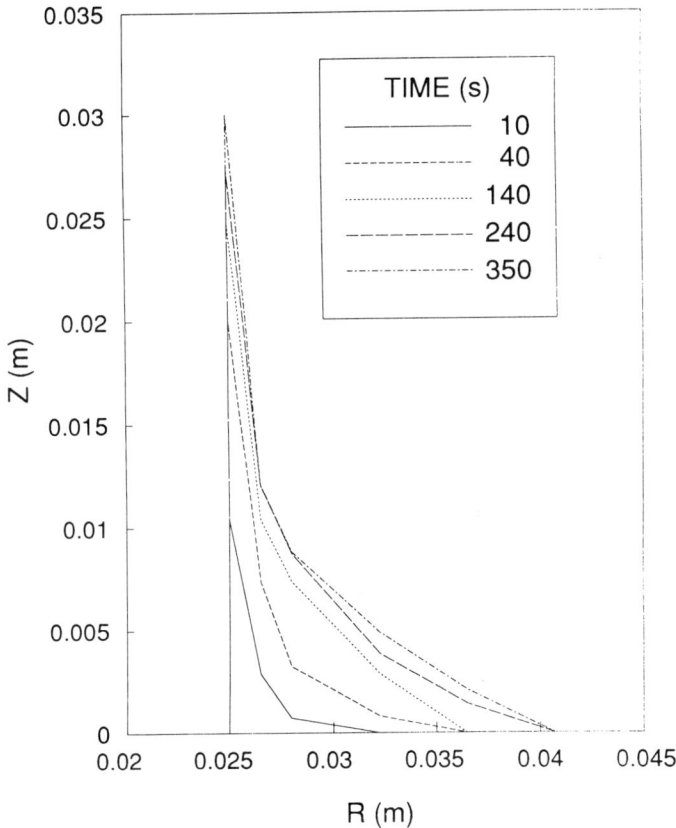

Figure 4. Predicted growth profile for a slag accretion during converting of a 30% copper matte.(9)

In the nonferrous industry however accretions formation often results in a gradual blocking of the injectors. It has therefore been found necessary to install punching devices to clear the blockages by mechanical means. This is a far from satisfactory solution as it leads to high noise levels, additional maintenance work and accelerated tuyereline refractory deterioration. It would be quite unacceptable in the modern steel plant.

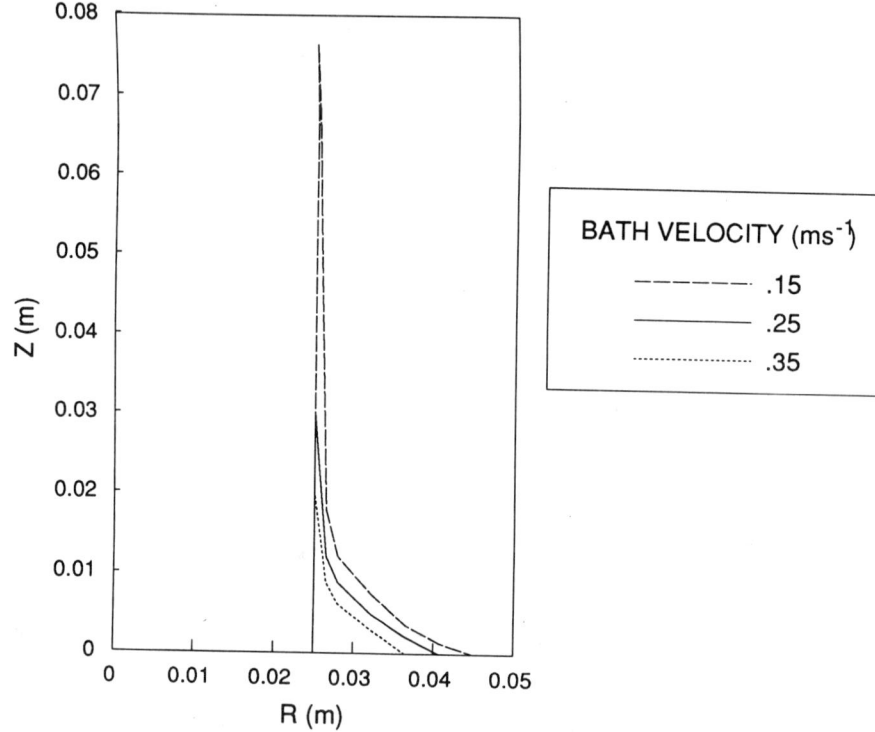

Figure 5. The effect of local bath velocity on the predicted accretion size for a slag
accretion during converting of a 30% copper matte.(9)

Refractory Wear

Refractory wear is a problem intimately connected to injection phenomena and it
would be remiss to ignore it. Tuyereline refractory deterioration and the downtime
it causes testify to the need to take this issue seriously. Wear at the tuyereline or
in the vicinity of injectors can result from a combination of several phenomena.
In the first place there are thermal stresses leading to refractory spalling which
can be generated by differential heating and cooling of the refractory by the bath
and gas flowing through the injector pipe. It is interesting to note that the presence
of an accretion can have a profound effect on the thermal profiles at the tuyere
which may exacerbate the thermal stress problem.(9) Thermal stresses at the
tuyereline are also affected by radiative cooling during out-of-stack time. Those
tuyeres which are directly opposite the mouth suffer the greatest heat loss and
temperature decline, and hence the largest thermal stresses. Furthermore, if the
tuyereline cools below the freezing point of the matte there may be severe accretion
growth problems immediately after blowing is started in the stack.   There are
therefore compelling reasons to cover the converter mouth during any out-of-stack
time which will last longer than a few minutes.(10) Blockage of the tuyeres resulting
from uncontrolled accretion growth is an indirect cause of tuyereline refractory
wear. The use of 'punching' to clear the blockages can damage the refractory when

the blocking accretion is frozen to the refractory adjacent to the tuyere. The violent mechanical removal of the accretion then tends to knock parts of the brick work away.

The other causes of refractory wear in the vicinity of the injectors stem from the effects of gas-bath reactions and fluid flow. If the reactive gas spends a significant amount of time at the tuyere then the heat generated by reaction can act directly on the local brick work. High local temperatures are generated which enhance thermal and chemical attack of the refractory. The motion of the bath at the injector is also important. If the gas enters the bath in a bubbling regime then there is a periodic washing of the bath back at the injector and surrounding refractory. This is likely to accelerate local erosion.

Future Developments

It is clear from this discussion that an appreciation of submerged injection phenomena is intimately tied to proper operation and development of non-ferrous processes. It is therefore appropriate at this conference to discuss future possibilities for submerged injection in non-ferrous processes because of the technological ideas which can be adapted from the ferrous industry. The obvious technologies are high pressure gas injection and the use of shrouded tuyeres. Both of these offer the opportunity to increase the capacity and intensity of nonferrous metallurgical reactors by increasing the flow of oxygen. This can be accomplished one of two ways: through the use of tonnage oxygen and by increasing the flow of air. The latter route has the disadvantage of increased off gas volumes and splashing as alluded to above. It is therefore only logical to discuss the injection of oxygen enriched air or tonnage oxygen itself.

High Pressure Injection

High pressure injection is routinely applied in the ferrous industry. Under high pressure injection conditions the gas tends to enter the bath as a jet with little or no bubble formation at the tuyere and little penetration of the bath into the tuyere pipe. This situation is complicated of course when an accretion forms at the end of the injector. However high pressure injection generally ensures that even when the accretion forms it is porous to the gas flow.

High pressure injection can then benefit non-ferrous processes in two ways. In the first place it acts to change the injection regime towards jetting rather than bubbling. This ensures that more of the gas-liquid reaction takes place in the bulk of the bath and reduces the degree of local heating at the tuyereline. This is demonstrated by a comparison of the accretions formed in during copper converting for low and high pressure injection at the same matte, bath flow and superheat conditions as shown in Figure 6. For the same gas flow the accretion for high pressure injection is roughly a factor of four or five longer. The second result of high pressure injection is a substantial reduction of bath penetration into the tuyere and hence the tendency to blockage.

These advantages offer a potential solution to questions of increases in reactor capacity and intensity through injection of enriched air or tonnage oxygen. Using this type of technology oxygen levels can be increased without severe injection or refractory problems. There do however appear to be limits to what can be done solely with high pressure injection.(8) There are likely limits on oxygen enrichment levels unless extremely high pressures are used. It does, nonetheless, represent a worthwhile consideration in the retrofitting of existing vessels especially when the potential savings in tuyereline refractory and punching equipment are considered.(11)

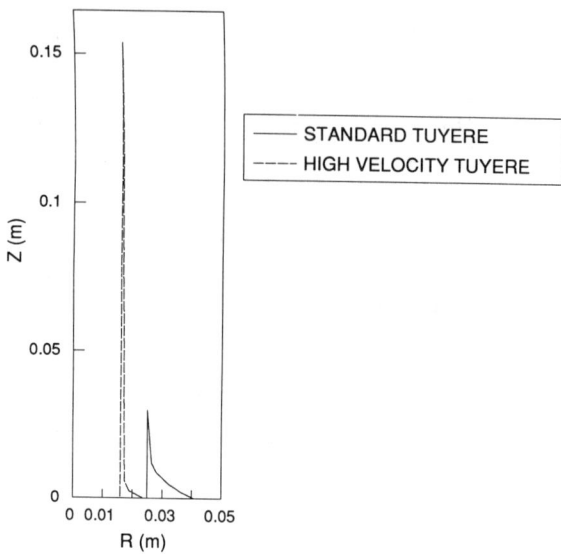

Figure 6. A comparison of predicted accretion size for a slag accretion during converting of a 30% copper matte under standard (solid line) and high pressure (dashed line) injection conditions.(9)

Shrouded Tuyere Injection

A modern non-ferrous converting process would seek to take advantage of injection of tonnage oxygen and the Savard-Lee shrouded tuyere offers this possibility. There is no reason to believe that a shrouded tuyere would not be feasible in these applications and it simply a question of developing the proper design.(13) Clearly it would be necessary to address the consequences for the process heat balance. This could be done a number of ways including the development of higher temperature refractories, the use of the excess heat for melting scrap, injection of cold SO_2 or reactor cooling.

References

1. J.K. Brimacombe et al., "Process Dynamics: Gas-Liquid" Proceedings of the Elliot Symposium, (Warrendale, PA, The Metallurgical Society, 1990), 343-412.

2. A.A. Bustos et al., "Injection Phenomena in Nonferrous Processes," Metall. Trans. B, 15B (1984), 77-89.

3. G.G. Richards, "Fundamental Aspects of Gas-Particle Injection" Materials Handling in Pyrometallurgy, ed. C. Twigge-Molecy and T. Price (Toronto, Canada, Pergamon, 1990), 3-13.

4. G.G. Richards et al., "Bath Slopping and Splashing" The Reinhardt Schuhmann International Symposium, ed. D.R. Gaskell et al. (Warrendale, PA, The Metallurgical Society, 1987), 385-402.

5. D.W. Ashman, J.W. McKelliget and J.K. Brimacombe, "Mathematical Model of Bubble Formation at the Tuyere of a Copper Converter", Can. Metall. Quarterly, 20 (1981), 387-97.

6. E. Adjei and G.G. Richards, "Physical Model of Mass Transfer in a Peirce-Smith Converter", Copper 91, ed. C. Diaz, C. Landolt, A. Luraschi and C.J. Newman (Pergamon, Elmsford, New York, 1991), 377-88.

7. A.A. Bustos, J.K. Brimacombe and G.G. Richards, "Accretion Growth at the Tuyeres of a Peirce-Smith Copper Converter," Can. Metall. Quarterly, 27 (1988), 7-21.

8. T. Kimura et al., "Protection of Refractory by High-Speed Blowing in P.S. Converter," J. Metals, 38 (9) (1986), 38-42.

9. A.K. Kyllo and G.G. Richards, "Accretion Growth at Single Pipe Tuyeres" submitted to Metall. Trans. B, July, 1992.

10. A.A. Bustos, J.K. Brimacombe and G.G. Richards, "Heat Flow in Copper Converters" Metall. Trans. B, 15B (1984), 77-89.

11. J.K. Brimacombe, S.E. Meredith and R.G.H. Lee, "High-Pressure Injection of Air into a Peirce-Smith Converter" Advances in Sulphide Smelting, ed. H.Y. Sohn, D.B. George and A.D. Zunkel (Warrendale, PA, The Metallurgical Society, 1990), 839-54.

12. G.R. Garrido and R.G.H. Lee, "Optimizing Oxygen Enrichment of High-Pressure Air Injection into Copper Converters" Copper '87, ed. C. Diaz, C. Landolt and A. Luraschi (Universidad de Chile, Santiago, 1987), 375-93.

13. A.K. Kyllo and G.G. Richards, unpublished research, 1992.

INJECTION INTO THE ELECTRIC ARC FURNACE

- THE K-ES PROCESS -

W. Ballandino *, F.G. Hauck **, K. Klintworth **

* Acciaierie Venete SpA Padova, Italy
** Klöckner Contracting und Technologie GmbH, Hamburg, Germany

Steel production on scrap basis in the electric arc furnace has a steadily increasing share. Today nearly 30% of the total world raw steel production are manufactured in the electric arc furnace. Of course, also in this process the technologies initiated by Mr. Savard and Mr. Lee have a field of application.

The electric arc furnace showed, specially during the last twenty years, to be a very efficient tool in scrap melting. Additionally, melting of direct reduced iron is actually realized by 100% in the EAF.

During the two last decades a very important progress was obtained in productivity of the EAF. It was the result of a constant improved combination of electric design and electrical furnace operation with metallurgical operation practice.

The power of the transformers was constantly increased as soon as an improved operation with foamed slag allowed to increase secondary voltages and power factor.

Tap to tap times of one hour are not seldom today. Considering that during this short melting time scrap pieces of different size have to be melted down and superheated it is a necessity to complement this process by methods which basically contribute to a melt down homogenization, bath

Proceedings of the
Savard/Lee International Symposium on Bath Smelting
Edited by J. K. Brimacombe, P. J. Mackey,
G. J. W. Kor, C. Bickert and M. G. Ranade
The Minerals, Metals & Materials Society, 1992

temperature homogenization and optimized melt/slag equilibrium. Melt down homogenization and acceleration can be obtained by oxygen use in order to take advantage of the exothermic heat of oxidation reactions. Used initially for slag foaming purposes, the oxygen application was constantly increased by making use of injection pipes and burners.

This increase showed to be limited to max. 35 m^3/t by these methods as with increasing oxygen use its efficiency decreases.

Figure 1 shows a statistical evaluation of Inagaki for Japanese furnaces using high oxygen rates.

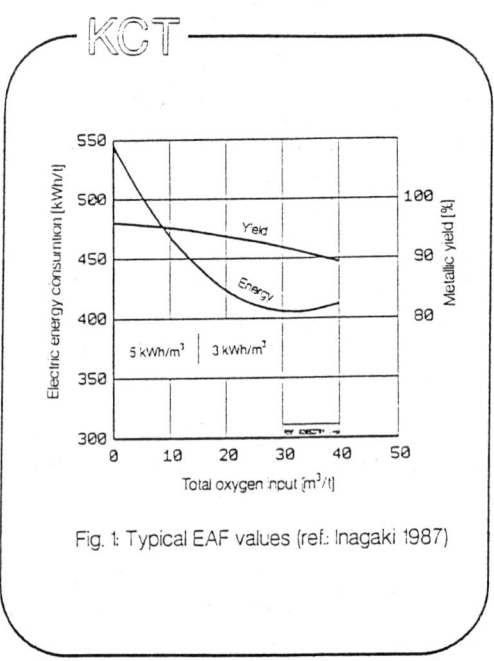

Fig. 1: Typical EAF values (ref.: Inagaki 1987)

It is shown, that at 35 m^3/t no gain, expressed in electrical energy savings, is finally obtained. Higher oxygen rates lead even to an increased electrical energy consumption due to the fact, that the metallic yield is drastically reduced.

The experience gained in converter process technologies as OBM, K-OBM (Q-BOP) and KMS indicate that there is still a field in the electric arc furnace to overcome this limit.

Figure 2 shows a K-OBM converter, the characteristical top and bottom oxygen blowing are also beneficial for the EAF as they allow to improve it on the following points:

1) Whereas the CO emerging from the melt normally is postcombusted in the exhaust system of an EAF, top blowing of oxygen should allow an important rate of postcombustion in the furnace. The released heat must be transferred into the melt.

2) The EAF has characteristically not the bath agitation rate of a converter as it normally is operated with scrap containing by far less dissolved carbon than hot metal.

   Additionally the reduced carbon content of the melt during superheating and the typical absence of electromagnetic stirring effects in the EAF make necessary an assistance for bath temperature homogenization and improved melt/slag equilibrium. As faster the furnace this necessity increases. Therefore, implementation of bottom blowing would permit this improvement with the additional advantage of a better interaction of the melt to be heated with the heat generated by the postcombustion.

3) As already mentioned, the charge of an EAF normally consists of scrap with typically low carbon content.

   Additional carbon input for example lumpy coal or injected coal, would therefore be advantageous as in case of high postcombustion rates the profit obtained by carbon combustion is a considerable amount of heat.

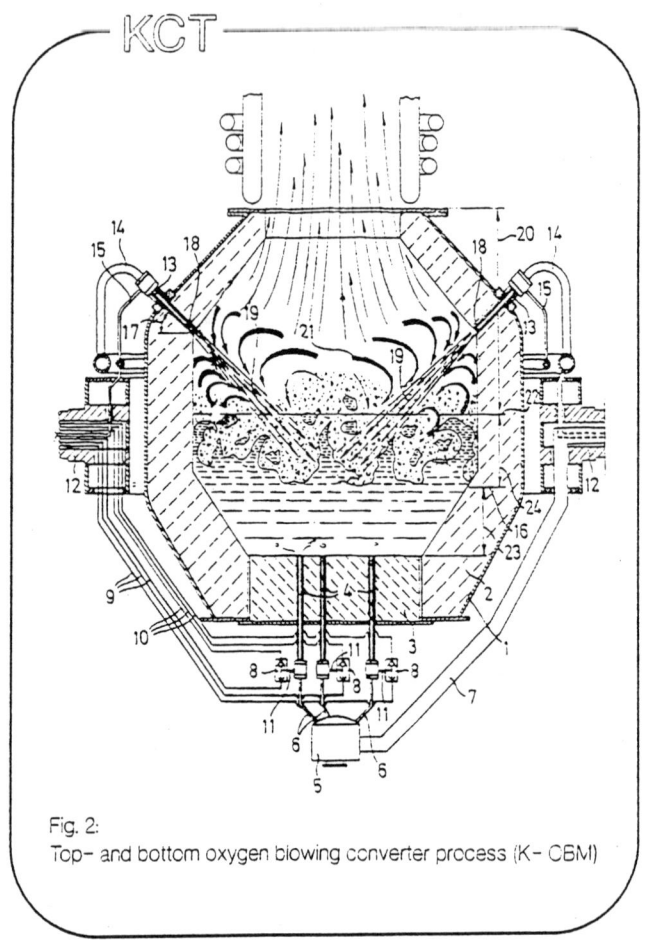

Fig. 2:
Top- and bottom oxygen blowing converter process (K- CBM)

Figure 3 shows a "ranking' of heats gained by oxidation reactions taking place in the furnace.

The high amounts of heat obtained by oxidation of aluminium, silicon and manganese are already exploited at a maximum in the EAF. Their combustion reduces considerably the electrical energy consumption contributing to the characteristical savings of even 5 kWh per m^3 of oxygen at low oxygen injection quantities as expressed in figure 1.

KCT

$2\langle Al\rangle + 1.5\,(O_2) = \{Al_2O_3\}$ 298K 298K 1800K	-7.1 kWh/kg Al -11.4 kWh/m^3 O_2
$\langle Si\rangle + (O_2) = \{SiO_2\}$ 298K 298K 1800K	-9.0 kWh/kg Si -11.3 kWh/m^3 O_2
$\langle Fe\rangle + 0.5\,(O_2) = \{FeO\}$ 298K 298K 1800K	-0.8 kWh/kg Fe -3.9 kWh/m^3 O_2
$\langle C\rangle = [C]$ 298K 298K	-0.5 kWh/kg C "-0.6 kWh/ m^3 O_2"
$[C] = [C]$ 298K 1800K	$+1.4$ kWh/kg C "$+1.5$ kWh/m^3 O_2"
$[C] + 0.5\,(O_2) = (CO)$ 1800K 298K 1800K	-2.3 kWh/kg C -2.4 kWh/m^3 O_2
$(CO) + 0.5\,(O_2) = (CO_2)$ 1800K 298K 1800K	-5.9 kWh/kg C -6.3 kWh/m^3 O_2
$(CH_4) + 2\,(O_2) = (CO_2) + 2\,(H_2O)$ 298K 298K 1800K 1800K	-8.9 kWh/m^3 CH_4 -4.4 kWh/m^3 O_2

Fig. 3: Heats of oxidation

Additionally, aluminium and silicon are often used for fast superheating of the melt before tapping with the additional advantage of reduced FeO-content in the slag in combination with quick tapping procedures.

Also the high amount of heat gained by oxidation of iron is largely exploited in the conventional EAF practice. The oxygen blowed by pipes and burners (as superstoichiometric oxygen) generates a considerable heat due to the scrap oxidation. The created iron oxide is partially reduced from the slag. Even blowing carbon simultaneously, the amount of oxygen used with this target is limited as in other cases it would originate a reduced metallic yield, agressive and no foaming slags, etc.

Finally, the figure shows the great potential of the carbon combustion in case a high rate of postcombustion to CO_2 can be achieved and the heat transferred to the charge. All attempts to bring additional carbon into the melt in order to increase the injectable oxygen quantity cannot be successful if not enough profit of postcombustion to CO_2 is obtained.

As it can be deducted from the amount of evolved heat in the table, dissolution of carbon in the melt combined with 100% of CO generation leads to only 1,5 kWh/m^3 oxygen. In comparison, additional 6,3 kWh/m^3 oxygen are generated in case the carbon monoxide is totally postcombusted to carbon dioxide.

In comparison, the table shows that combustion of CH_4 by burners leads to "only" 4,4 kWh/m^3 oxygen in the hypotetical case of total postcombustion of this fuel to CO_2 and H_2O and 100%-heat transfer to the charge.

Conclusions are therefore:

a) The EAF must be equipped with postcombustors in order to take advantage of the enormous amount of heat evolved by this reaction.

b) During foamed slag operation in order to obtain an optimum profit from the electrical energy input, the generated postcombustion heat must be transferred to the melt through areas which are free from the thermal insulating slag. This can be realized by bottom blowing tuyeres, which simultaneously keep free of slag areas of the melt surface and increase their interaction surface with the postcombusted gas, generate exactly there a maximum amount of CO and finally contribute to the necessary bath homogenization and slag/melt equilibrium.

c) If the furnace is equipped with postcombustors, the carbon quantities should be maximized as the postcombustion heat considerably reduces the electrical energy consumption. Replacement by oxygen and carbon which normally are of lower cost than electrical energy saves considerable costs as soon as approx. 3 kWh can be replaced by one m^3 of oxygen.

These demands take the K-ES technology into account.

Implementation of the K-ES Technology at the 72t-EAF of Acciarierie Venete SpA

Figure 4 shows a model of the furnace at Acciaierie Venete. It is an oval shape bottom tapping furnace with a shell diameter of 5500 mm extended by 700 mm on the long axis.

In accordance with this design, 6 postcombustors were placed above the bath level fixed in the wall with a well distribution around the shell.

As the furnace was already equipped with 6 burners, the postcombustors were arranged in combination with them for achievement of an optimized homogeneous melt down.

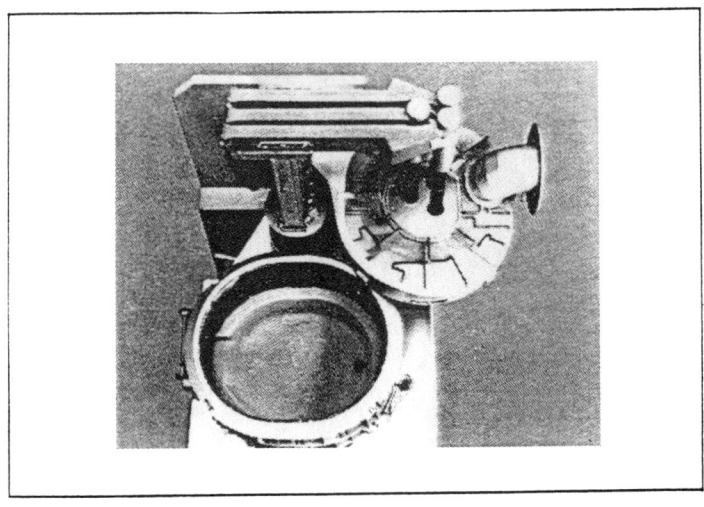

Figure 4 - OBT - Arc Furnace Model

For interaction with the postcombustors and the mentioned stirring purposes, 5 bottom tuyeres were placed in the furnace bottom. In order to not disturb the electric power input of the arcs, they were situated on an oval pitch circle in between electrode pitch circle and furnace wall (figure 5).

Additional carbon input in the EAF is possible by lumpy coal with the scrap charge. Nevertheless, as these quantities are not enough for an additional input till the end of the heat, they were complemented by a pneumatic injection device. One of the best places to realize carbon injection is the area of the electric arc in the EAF as this place is always the hottest in the melt and no unmelted pieces of scrap can obstruct the injection way.

Thus, a device for injection through a hollow electrode was developed. It is a system which allows to automatically connect the carbon supply line with the electrode column. As carbon injection is also necessary for slag foaming purposes in the conventional EAF practice, this new system allows to simultaneously do this work in a complete automized manner.

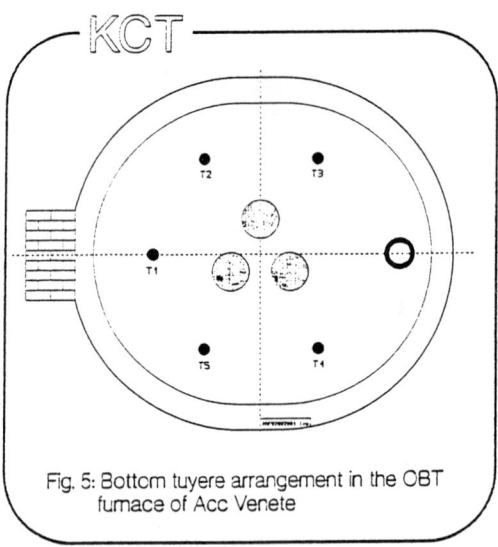

Fig. 5: Bottom tuyere arrangement in the OBT furnace of Acc Venete

Operation results

A. Electrical energy consumption and furnace productivity

Figure 1 showed the obtainable gains with oxygen by conventional injection methods. Figure 6 can be superposed to figure 1 in order to make visible what has been achieved.

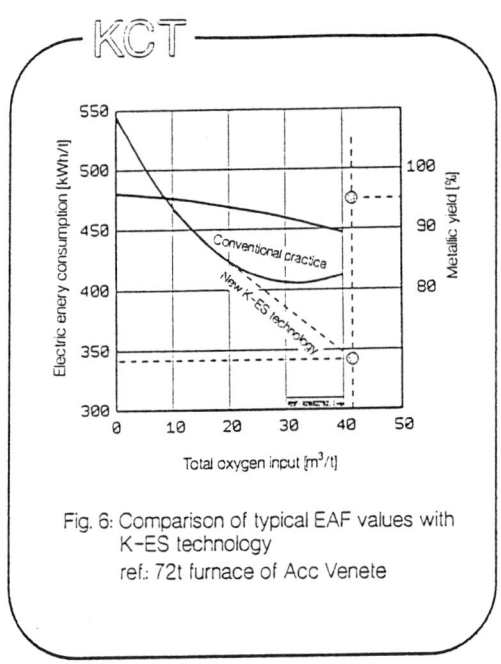

Fig. 6: Comparison of typical EAF values with
K-ES technology
ref.: 72t furnace of Acc Venete

It clearly reveals the improvement, namely that the increase in oxygen consumption from approximately 25m^3/t to now 42 m^3/t permitted to reduce the electrical energy consumption from approx. 400 kWh/t to below 350 kWh/t instead of leading to an electrical energy consumption increase as it would be the case applying conventional methods.

As the melting time of the EAF depends on the power input the transformer is able to bring per unit of time into the melt, the obtained reduction in electrical energy consumption signifies to have achieved simultaneously a considerable increase of productivity. Savings of 50 kWh/t are a significant reduction of approx. 6 minutes electrical time in the case of Acc. Venete.

The metallic yield is not altered by this practice. The furnace is now operated with typical yields of 95% which is even a slightly better figure than obtained by conventional practice.

B. Bottom blowing effects

In comparison to the classical converter which is mainly charged with hot metal, the EAF has to melt scrap. Therefore, in principle, there is no necessity in the EAF to reduce the carbon content of the melt in order to make steel.

The target of bottom blowing is consequently different and doesn't imply the use of oxygen. Nevertheless, oxygen is known as the best way to improve the slag/melt equilibrium in comparison to inert gases as their bubbles are reactive. Simultaneously, this reactivity leads to an improved CO-generation as the necessary supersaturation of carbon and oxygen in the melt is by far lower.

In case large quantities of oxygen would be blown through the bottom of an EAF in the same way as it is done in a converter, several problems arising from the consequent spitting would appear as the vessel of an EAF has a completely different design.

As conclusion, by far lower quantities of gas are necessary to fulfil at least the priority necessities of the applied technology. Interaction with postcombustors, activation of CO-generation, bath homogenisation and considerable improvement in the slag/melt equilibrium can already be obtained by blowing small amounts of inert gas through the tuyeres in the furnace hearth.

The well known OBM tuyere design is appropriated to do also the required work as it is no problem to simply operate it with closed central pipe. The remaining annulus is the tool to inject the small quantities of inert gas for the mentioned purpose.

Small additions of natural gas contribute to the buildup of the mushroom for optimum refractory protection.

The obtained figures of electrical energy consumption mentioned and represented in figure 6 were obtained by this type of operation.

The amount of injected bottom gas is approx. 1 m^3/t, i.e. only about 1 m^3/min operating the furnace with 5 small OBM bottom tuyeres.

The detailed quantities are depending on the necessities of each stage of the melt. During melt down of the scrap larger amounts of bottom gas are required for the interaction with the postcombustors than during fining and superheating of the melt when only a soft stirring for temperature homogenization is required.

Also the quality aspects have to be considered in a different manner for the EAF. Charging in steel scrap and not hot metal the CO-purging effect during melt down and fining is by far lower than in a converter. Injection of nitrogen and hydrocarbons can imply nitrogen or hydrogen pickup in case of large bottom gas quantities or during critical stages of the melt. For these periods the injection device switches automatically to Argon for nitrogen replacement, respectively reduces the quantity of the injected hydrocarbon. CO_2 has shown to be also very useful as it is cheaper than Argon and leads to an excellent slag foaming behaviour due to its reactivity with the melt.

Conclusions

Injection of oxygen for CO-postcombustion purposes in the EAF can be successfully applied. The obtained results indicate a postcombustion rate of approx. 40 to 50%. This is achieved in case only inert gases and no oxygen is injected through the furnace hearth.

Even better results can be obtained in case oxygen is injected through the bottom. In this case the bottom tuyere on which Mr. Savard and Mr. Lee are working since 1963 is one of the keys to increase the injectable quantities. The quantities realized till now are surely not the end of the story.

Larger quantities will allow to further improve the effectiveness of the oxygen in the EAF, probably eliminate hard hand work operated door pipes and finally further reduce the use of electric energy which can partially be replaced by cheaper fossil energy.

The EAF at Acciaierie Venete demonstrates that also a furnace with relatively low transformer power is able to achieve highest productivity by this way.

As the tap to tap time is now in the order of 54 minutes, this EAF has been the first all over the world who achieved a productivity higher than 2 tonnes per MWh.

Session 6

Emerging Bath Smelting Processes

STUDY ON DIRECT IRON ORE SMELTING REDUCTION
PROCESS (DIOS) AND FUTURE PROGRAM

KENJI SAITO

Manager, R & D Task Force, JISF DIOS Process
Japan Iron and Steel Federation
Kudan Plaza Building 2-3, 2-chome,
Kudan-minami, Chiyoda-ku, Tokyo, 102 Japan

Abstract

A study on the DIOS process has been undertaken by the Japan Iron and Steel
Federation as a 7-year project which began in April 1988, with a subsidy
provided by the Ministry of International Trade and Industry for the
promotion of coal-utilized technology. The focus of the study has been
placed on the smelting reduction process which is considered as an
effective process capable of making use of the strong points, as well as
compensating for the weak points of the blast furnace process. The first
three years were spent on the element study, which was envisaged and then
conducted at five leading Japanese steel plants. This iron bath smelting
reduction study primarily aims at the concurrent achievement of a high post
combustion ratio and a high level of heat transfer efficiency. It also
aims at the establishment of a continuous operation technology as well as a
technology to protect furnace shell refractories from wearing and to cool
the furnace shell. Studies on these technologies were performed at the
Sakai steel plant of the Nippon Steel Corporation, the Fukuyama steel plant
of the NKK Corporation and the Kashima steel plant of the Sumitomo Metal
Industries, Ltd. Through these studies it was clarified that the important
factors are that iron droplets and char need to be distributed in slag, the
slag bath needs to be stirred and the furnace needs to be pressurized, in
order to achieve a high post combustion ratio and a high level of heat
transfer efficiency simultaneously. At the Fukuyama steel plant, the
smelting reduction furnace, in its closed, pressurized condition, was
tested to ensure a prolonged, continuous operation, while at the Sakai
steel plant, tests involving the cooling of bricks with gas and of panels
with water were performed, thereby clarifying the applicability of them to
a smelting reduction furnace. Following the element studies, tests by
using a pilot plant were contemplated, and currently a 500-ton/day pilot
plant is under construction on the compound of the Keihin steel plant of
the NKK Corporation. The pilot plant is designed so that it allows for the
performance of tests in a range in which the prereduction rate of iron ore
in the prereduction fluidized bed, and the post combustion ratio of gas in
the smelting reduction furnace, are combined in several ways. Its test
operation will begin from October 1993 onward, and a feasibility study on
industrialized-scale DIOS process will be carried out in 1994.

Proceedings of the
Savard/Lee International Symposium on Bath Smelting
Edited by J. K. Brimacombe, P. J. Mackey,
G. J. W. Kor, C. Bickert and M. G. Ranade
The Minerals, Metals & Materials Society, 1992

Introduction

The blast furnace process, a contemporary, major ironmaking process, allows for a mass production of iron with an extremely high level of heat efficiency. On the contrary, it indispensably needs the pretreatment of iron ore and coal, hence requires enormous capital investment in sintering machines and coke ovens, as well as lacks productional flexibility. In order to make use of the strong points as well as to compensate for the weak points of the blast furnace process, we found out the advantages of the smelting reduction process that allow for the use of fine and granular non-coking coal and iron ore directly without their pretreatment such as coking and sintering which are required by the blast furnace process, iron ore is subsequently prereduced in a fluidized bed, and non-coking coal is treated directly or devolatized, and then charged in a smelting reduction furnace to produce hot metal. We therefore determined to carry out the necessary tests and research [1,2], which have since been continuously undertaken, as a 7-year project starting from fiscal 1988, with a subsidy provided by the Ministry of International Trade and Industry for the promotion of coal-utilized technology. The first three years were spent to program the element study, which aims at the concurrent achievement of a high post combustion ratio and a high level of heat transfer efficiency in smelting reduction, and the establishment of a technology for continuous operation as well as technologies to protect refractories from wearing and to cool the furnace shell. Up until now, the element studies have been completed, and at present the construction of a pilot plant is underway. This paper will report on the results of studies on iron bath smelting reduction, and the future program.

Iron Bath Smelting Reduction Study

Element studies on smelting reduction were conducted at five steel plants shown in Table 1 [3]. In this table the features of these steel plants and the main study tasks are given. The Sakai steel plant of the Nippon Steel Corporation, the Fukuyama steel plant of the NKK Corporation, and the Kashima steel plant of the Sumitomo Metal Industries, Ltd. concentrated their efforts on the study of iron bath smelting reduction. The main results which have been achieved through the element studies follow.

Table 1. Share of the element studies.

Testing place	Test facility and condition	Main task
Sakai steel plant, Nippon Steel Corp.	A 100-ton iron bath smelting reduction furnace (inner volume: 138m^3). Batch operation under normal pressure. Oxygen blown at top and bottom. Amount of oxygen: max 35,000 Nm^3/h	○ Achieving a high P.C. and a high η_{pc}. ○ Extracting problems involved in a up-scaled furnace, and examining the necessary countermeausres. ○ Examining for the possibility of using coal with high volatile matter. ○ Clarifying the in-furnace behavior. ○ Examining for a gas-reforming technology.
Fukuyama steel plant, NKK Corp.	A 5-ton iron bath smelting reduction furance directly connected to a prereduction fluidized bed. Operating pressure: max. 1.9 kgf/cm^{2}G. Oxygen blown at top and bottom. Amount of oxygen: max. 2,500 Nm^3/h. Tapping device available. Bubbling fluidized bed: 1mø × 8m high	○ Achieving a high P.C. and a high η_{pc}. ○ Examining for the possibility of using coal with high volatile matter. ○ Achieving prolonged, consecutive operation with the PRF directly connected to the SRF. ○ Examining for the steady operational conditions of the fluidized bed. ○ Assessing the reduction ability of the fluidized bed.

Kashima steel plant, Sumitomo Metal Industries, Ltd.	A 5-ton iron bath smelting reduction furnace. Batch operation under normal pressure. Oxygen blown at top, bottom and side. Amount of oxygen: max. 1,400Nm³/h	° Achieving a high P.C. and a high η_{PC}. ° Examining for the possibility of using coal with high volatile matter. ° Ensuring the effect of injecting oxygen into the slag bed from the side.
Chiba steel plant, Kawasaki Steel Corp.	A circulating fluidized bed (0.7mø × 7.3m high)	° Examining for the steady operational conditions of the fluidized bed. ° Assessing the reduction ability of the fluidized bed.
Kobe steel plant, Kobe Steel, Ltd.	A gas reforming test facility composed of a carbonizing furnace and a plasma heater.	° Examining for a gas reforming technology.

Concurrent Achievement of Technologies Involving a High Post Combustion Ratio and High Heat Transfer Efficiency. Improving the heat transfer efficiency and decreasing the unit coal consumption share parts of the large task involved in the production of molten iron. With the DIOS process, the main heat source is the post combustion heat resulting from oxygen, and carbon oxide gas which is generated in the process of smelting reduction. To improve the heat transfer efficiency and to reduce the unit coal consumption, technologies enabling the simultaneous achievement of a high post combustion ratio and a high level of heat transfer efficiency are indispensable. To achieve this technology, the individual steel plants carried out a survey on the role of slag, the effect of slag and iron bath stirring and the effect of pressurization.

The Role of Slag in a Smelting Reduction Furnace. With the DIOS process, the existence of a mass of slag in the smelting reduction furnace shares one of the DIOS process' large features. As well, slag is responsible for the roles to be played in its receiving post combustion heat, smelting and reducing iron ore, sustaining char, and preventing iron droplets and dust from being scattered. Therefore, in order to clarify these roles of slag, a survey on the density of slag, the distribution of the iron droplets in slag, the distribution of char, as well as the distribution of slag temperature, in the smelting reduction furnace, was conducted[4]. Figure 1 shows the relationship between the weight and layer thickness of the slag in the smelting reduction furnace, which resulted from a survey conducted at the Sakai steel plant. The layer thickness of slag remained in a uniform range of 2 to 3 meters, and unaffected by the weight of slag which fluctuated in a range of 20 to 40 tons. Figure 2 plots the relationship between the weight of slag per furnace barrel and the mean density of the slag layer, and that between the weight of slag per furnace barrel and the existent ratios

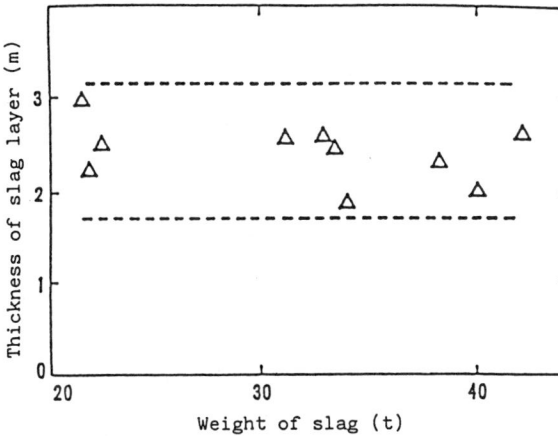

Figure 1. Relationship between weight and layer thickness of slag.

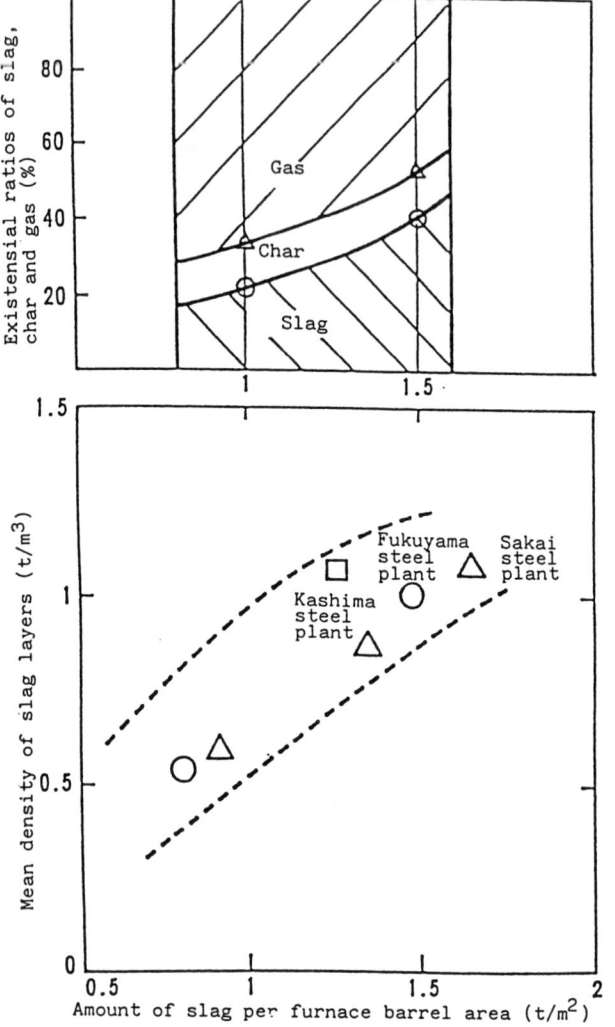

of slag, char and gas. In the figure, although information on the smelting reduction furnaces with individually different sizes and capacities at Sakai, Fukuyama and Kashima steel plants is displayed concurrently, one can see the weight of slag increasing along with an increasing mean density of slag layer, irrelevant to the sizes of the smelting reduction furnaces. This is due apprehensively to the fact that as plotted in the figure, the existential ratio of gas decreases as the weight of slag increases. Based on the findings mentioned above, with the smelting reduction furnace, a change in the weight of slag entails a varying gas content and mean slag density. Thus, the layer thickness of slag would remain approximately uniform. Figure 3 plot the result of temperatures measured at the various locations in the smelting reduction furnace at the Sakai steel plant.

Figure 2. Existential ratios of slag, char and gas.

All measurements indicated the similar temperature distributions, and that the temperature in the lower zone was identical to the temperature of molten iron, but that in the upper zone was higher than the lower zone by tens to hundreds of degrees in centigrade. This affirmatively suggests that the temperature in the lower zone was the same as that of molten iron due to the stirring effect of bottom-blown gas, while the temperature in the upper zone was greater due to the effect of the post combustion of top-blown oxygen. As well, this obviously shows that the heat transfer efficiency fully relates to a way in which post combustion takes place, and the pattern in which the upper slag layer zone is mixed with the lower slag layer zone. The results of survey

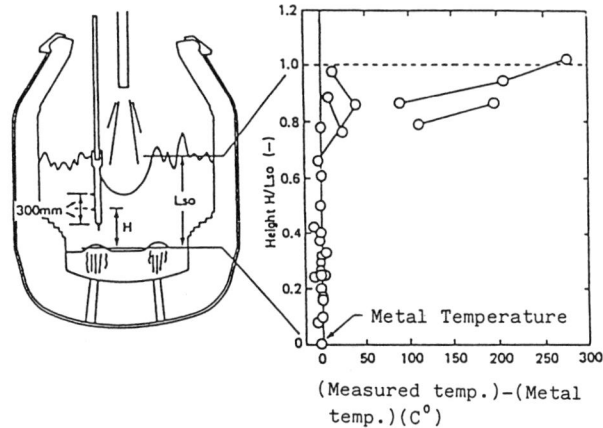

Figure 3. Temperature distribution in slag layer
(Sakai steel plant).

conducted on slag at the Sakai steel plant are shown in Figures 4 and 5. These surveys used various samplers required to collect accurate information, and enabled the clarification of iron droplet and char distributions in slag. Iron droplets are most in the lower slag zone, but also in existence in the upper slag zone. Char is also distributed throughout the slag layer, and this has led to the inference that the slag layer plays an important role as a zone in which reaction between iron droplets and slag, and iron droplets and char, takes place. It is known that the reductive reaction rate of molten iron oxide upon the carbon in molten iron is far greater than upon solid carbon, and that the existence of droplets in slag is very important for a smelting reduction process which aims to achieve a high level of productivity. Also, it is considered important that char is distributed throughout the slag layer in order to ensure the reaction of carbon on iron droplets or molten bulk iron. With the DIOS process, the method of operating it with iron droplets and char being properly distributed in slag is important in achieving a high level of productivity for molten iron.

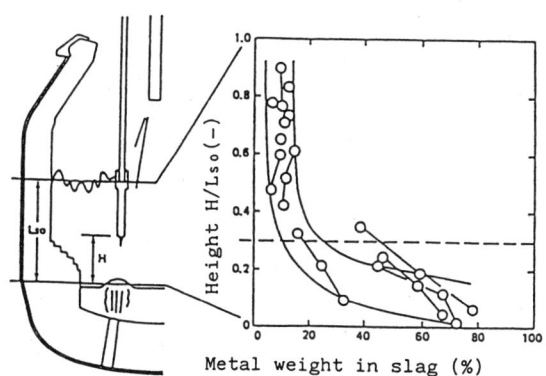

Figure 4. Metal droplet distribution in
slag layer (Sakai steel plant).

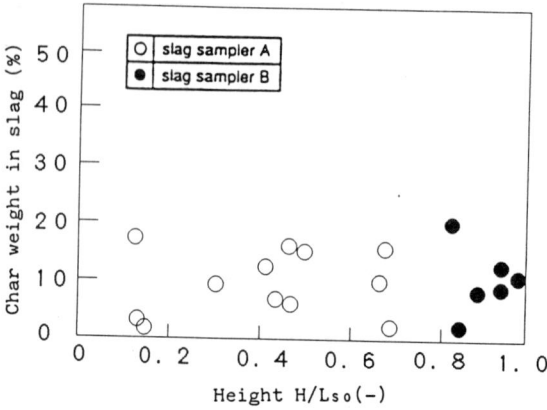

Figure 5. Char distribution in slag layer (Sakai steel plant).

Stirring Effects: Surveys on the effects of stirring the iron bath were conducted at the Sakai and the Kashima steel plants [5),6)]. Figures 6 and 7 show the results; at the Sakai steel plant, bottom-blown gas was increased

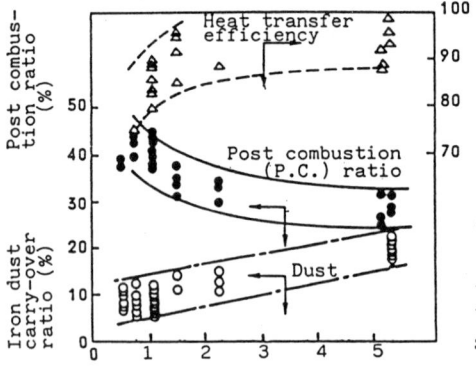

Figure 6. Effects of the amount of bottom-blown, stirring gas on post combustion ratio and heat transfer efficiency (Sakai steel plant)

in order to intensify the stirring of slag and molten iron. This entailed an increased loss of dust as well as a decreased post combustion ratio. The increased iron droplets in slag allowed for an increased ratio of the reaction of oxygen on the iron droplets, and this resulted in a decreased post combustion ratio. As shown in Figure 7, at the Kashima steel plant, side-blown gas stirred slag without the excessive inclusion of molten iron into slag, thereby enabling the increase of heat transfer efficiency by 5 ~ 10 percent. With the DIOS process, stirring the slag and metal in the smelting reduction furnace is a prerequisite, however excessive stirring will result in an increased iron loss as well as in a decreased post combustion ratio, hence constitutes one of the important operational conditions.

Effects of Pressurization. At the Fukuyama steel plant, provisions were made so that the smelting reduction furnace be directly connected to the prereduction fluidized bed. The coupled facility was pressurized to a maximum of 1.9kg/cm²G, in order to survey the effects of the pressurization [7)]. By pressurizing the smelting reduction furnace, the gas flow velocity was reduced, and as plotted in Figures 8 and 9, was found effective to reduce the iron ore and coal carry-over ratios. Furthermore, as shown in Figure 10, it became evident that along with an increasing pressure, the post combustion ratio could be increased while sustaining the heat transfer efficiency. As shown in Figure 11, by increasing the post combustion ratio, a decreased unit coal consumption could be achieved. From these findings, the pressurization would be necessary with the DIOS process.

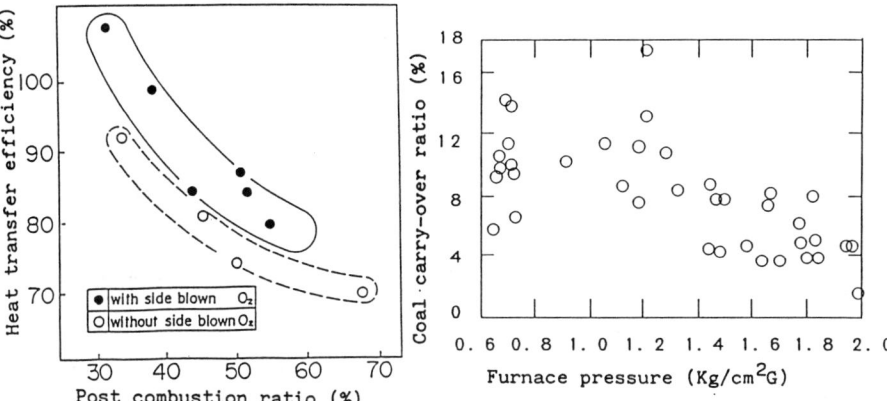

Figure 7. Effects of stirring with side-blown gas on post combustion ratio and heat transfer efficiency.

Figure 8. Relationship between furnace pressure and coal carry-over ratio.

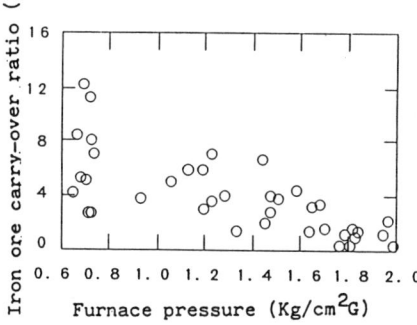

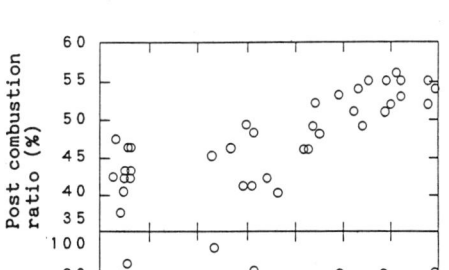

Figure 9. Relationship between furnace pressure and iron ore carry-over ratio.

Figure 10. Relationships between furnace pressure and post combustion ratio and heat transfer efficiency.

Continuous Operation of the Smelting Reduction Furnace

With the DIOS process, the smelting reduction furnace is provided with a taphole, a similar one provided with a blast furnace, which will be opened to discharge the molten iron and slag from inside the furnace following the elapse of a given time. At the Fukuyama steel plant, a continuous operation test of the smelting reduction furnace remaining pressurized and completely closed, was performed. The molten iron and slag were tapped by using the mudgun opener method, similar to the one traditionally used with the blast furnace process, while oxygen was being blown continuously into the furnace[8]. Thereby, a technology of controlling the tonnage of slag and molten iron in the furnace was established to ensure a prolonged, continuous operation[9].

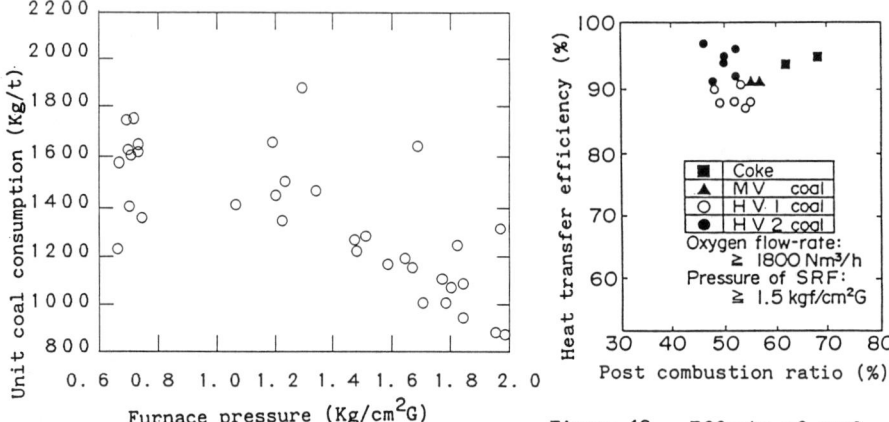

Figure 11. Operational results registered on furnace pressures and unit coal consumptions.

Figure 12. Effects of coal-contained volatile matter on post combustion ratio and heat transfer efficiency.

Technology to Use Non-coking Coal

The relationships between coal-contained volatile matter, the post combustion ratio and the heat transfer efficiency were studied[10]~[12]. Figure 12 plots the effects of coal-contained volatile matter on the post combustion ratio and the heat transfer efficiency, in which it is shown that the use of coal which contains high volatile matter ensures some 50% post combustion ratio and 86% or more heat transfer efficiency. Thus, we had the prospect of being able to use coal with high volatile matter contained to the extent that it can be used with the DIOS process.

Technologies to Protect Refractories from Wearing, and to Cool the Furnace Shell

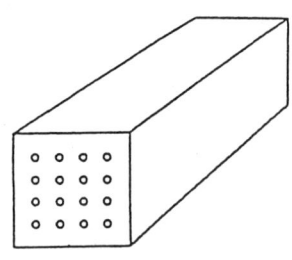

Figure 13. Schematic construction of gas-cooled bricks.

With the DIOS process, we aim at the reduction of the unit coal consumption as well as the concurrent achievement of a high post combustion ratio and a high level of heat transfer efficiency. Under the conditions being aimed at, the refractories of a smelting reduction furnace will be exposed to a very high temperature of molten iron oxide, and thus will be extremely worn. To reduce the wear of the refractories, improvement of refractory material, and tests of gas-cooled bricks as well as water-cooled panels, were performed. Bricks, the shape of which is as illustrated in Figure 13, were manufactured, and were given off-line tests, and then were tested by using them in a large

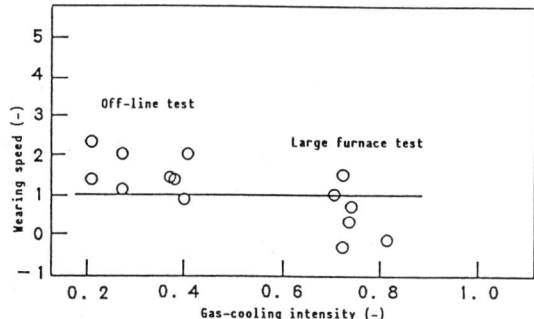

Figure 14. Relationship between gas-cooling intensity and wearing speed.

furnace. The results are plotted in Figure 14. It became evident that the wear of the refractories was reduced proportionately to the magnitude of intensity of gas being supplied, and through which the application of this gas cooling method in a smelting reduction furnace was found to be feasible. With water-cooled panels installed in the smelting reduction furnace located at the Sakai steel plant, the applicability of the water-cooled panels in the smelting reduction furnace was studied to evaluate the heat-resisting property of the panels and the thermal load of the furnace. The locations at which the water-cooled panels were installed are as illustrated in Figure 15. A total of 22 heats were tested, and all of the panels remained unaffected as shown in Photo 1. The test results indicated that the water-cooled panels are thoroughly resistant against the thermal load of the furnace.

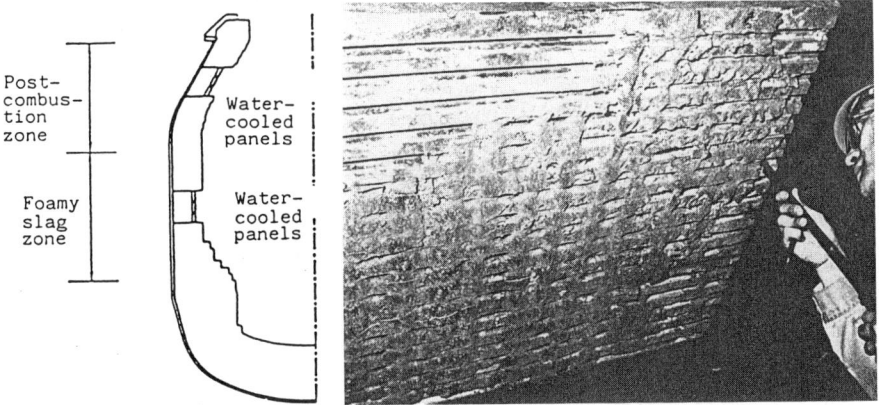

Figure 15. Water-cooled panels in place.

Photo. 1 View of water-cooled panels after 22 heats in the 4th campaign.

PILOT PLANT STUDY

Following element studies, tests by using a pilot plant have been planned, and the construction of a 500 tons/day pilot plant is currently underway at the Keihin steel plant, NKK Corporation. Its operation will commence in October, 1993. Figure 16 shows the ranges of tests being planned to be carried out at this pilot plant. These tests will be on the various combinations of prereduction degrees in the prereduction fluidized bed and of the post combustion ratios in the smelting reduction furnace [1,2], in order to establish practical system specifications. Table 2 shows the main system specifications involving the smelting reduction furnace of the pilot

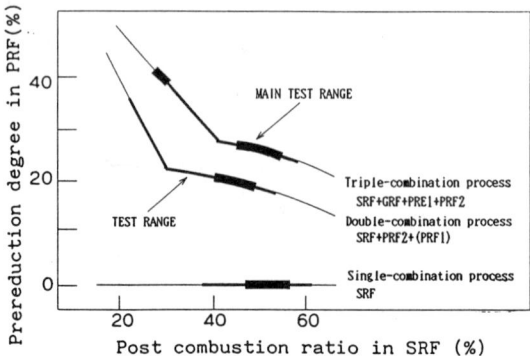

Figure 16. Ranges of testing to be performed by the pilot plant.

plant. The dimensions of the smelting reduction furnace have been determined based on the results of the element studies, with consideration given to the productivity of molten iron, the possibility of reforming gas with coal, and the flow velocity of gas. Figure 17 compares the shape of the smelting reduction furnace of the pilot plant with the shapes of the smelting reduction furnaces located at the Sakai and the Fukuyama steel plants, both of which had been used for the element studies. Based on the results of the element studies, the material balances involving all of the systems of the pilot plant under varying conditions were calculated. Figure 18 exemplifies the results of material balance calculations in terms of the triple-combination process (smelting reduction furnace (SRF) + gas reformer (GRF) + prereduction furnace (PRF) 1 + prereduction furnace (PRF) 2). Test operation of the pilot plant is intended to verify the abovementioned assumptions, and this will be followed with a feasibility study which will be performed in fiscal 1994 on an industrialized-scale DIOS process.

Table 2. Main service capacities.

Production facility	500 tons/day, 42~63 tons/tap (120 ~ 180 min-cycle)		
Gas reformer	Fine coal injection rate : max. 4 tons/h		
Smelting reduction furnace	Inside dimensions Production rate Inside pressure Bottom-blown rate Top-blown rate	: 3.7mø × 9.3mH : 21 tons/h (max. 25 tons/h) : less than 2.0kg/cm²G : N2; 500 ~ 6000Nm³/h : O₂; max. 20,000 Nm³/h	
Tapping device	Mudgun-opener type Taphole	: : max. 70mmø × 1	

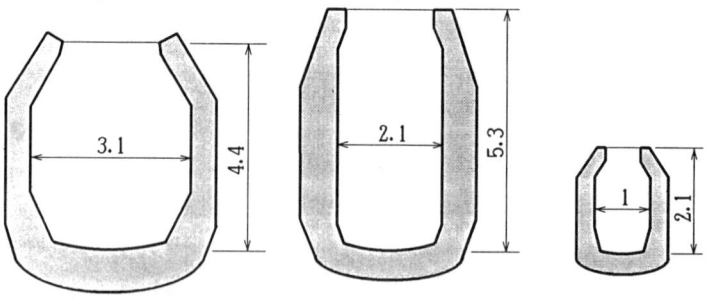

Sakai steel plant   Pilot plant   Fukuyama steel plant

Figure 17. Comparison between principal inside dimensions shown by ratio
of the smelting reduction furnace of the pilot plant and of the
smelting reduction furnaces of two steel plants at which
element studies were conducted.

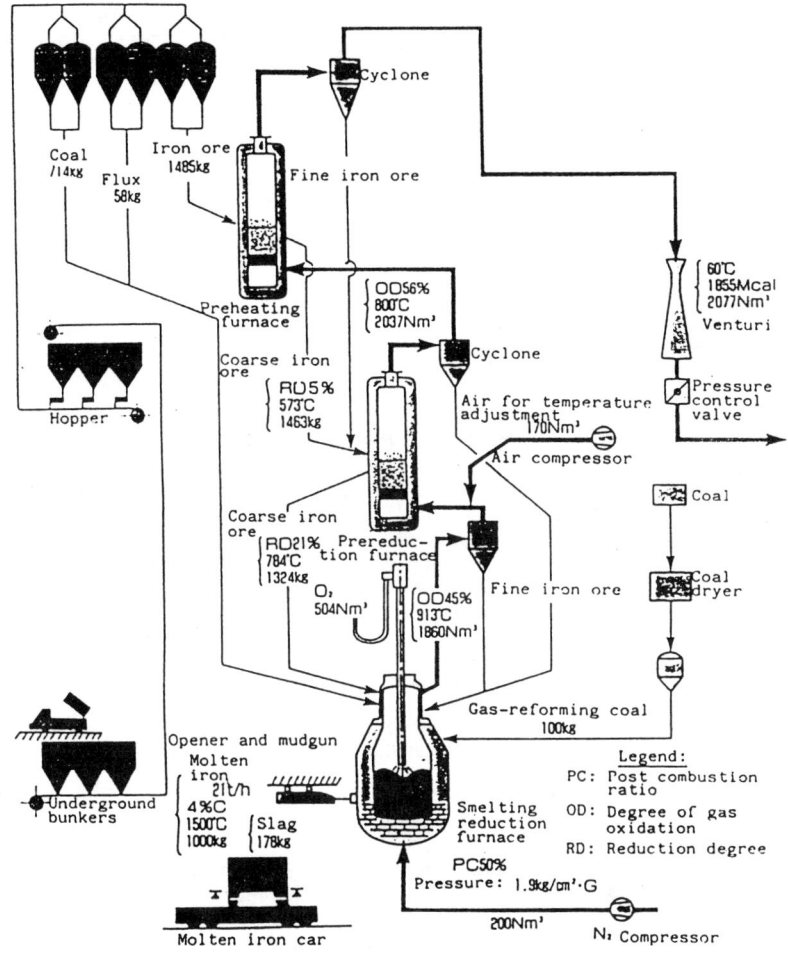

Figure 18. Process flow in the DIOS pilot plant.

CONCLUSIONS

At the Japan Iron and Steel Federation, study on the DIOS has been carried out with a subsidy provided by the Ministry of International Trade and Industry for the promotion of coal-utilized technology, as a 7-year project which began in April, 1988. In the first three years, element studies were envisaged, and the Sakai steel plant of the Nippon Steel Corporation, the Fukuyama steel plant of the NKK Corporation and the Kashima steel plant of the Sumitomo Metal Industries, Ltd. have shared the study on smelting reduction, thereby clarifying the importance of iron droplet and char distributions in slag, as well as bath stirring and furnace pressurization, in order to achieve a high post combustion ratio and a high level of heat transfer efficiency concurrently. At the Fukuyama steel plant, the continuous operation test of the smelting reduction furnace under a closed, pressurized condition was performed thereby achieving the prospect of prolonged, continuous operations. Through tests on gas-cooled bricks and water-cooled panels, their applicability to a smelting reduction furnace was confirmed. With the results of the element studies being reflected, the construction of a 500-ton/day pilot plant is underway on the compound of the Keihin steel plant of the NKK Corporation. The test operation of the pilot plant will begin in October, 1993, and a feasibility study on an industrialized-scale DIOS process will be carried out in fiscal 1994.

References

1) K. Kanamori et al., "Study on th DIOS Process in Japan," Quarterly Journal SEAISI, April 1992 issue.

2) T. Fukushima, "Smelting Reduction in Japan," The International Symposium on Future Ironmaking Process, June 1990.

3) T. Inatani, "The Current Status of JISF Research on the Direct Iron Ore Smelting Reduction Process," Ironmaking Conference Proceedings ISS-AIME, 50, April 1991.

4) M. Yamauchi et al., "Phenomena of Slag in an In-bath Smelting Reduction Furance," CAMP-ISIJ, vol. 3, 1990, 1075.

5) T. Hirata et al., "Improvement of In-bath Smelting Reduction Process through Side- and Bottom-blowing," ISS-AIME 50, April 1991.

6) M. Yamauchi et al., "Bottom-injection of Iron Ore and Coal into an In-bath Smelting Reduction Furnace," CAMP-ISIJ, vol. 4, 1991, 1172.

7) I. Kawata et al., "Integrated Operation of a Five-ton-scale Test Plant for Smelting Redeuction," CAMP-ISIJ, vol. 3, 1990, 1071.

8) S. Matsubara et al., "Development of Tapping Techniuqe for Smelting Reduction Process," CAMP-ISIJ, vol. 4, 1991, 45.

9) M. Kawakami et al., "Continuous Integrated Operation of Fukuyama Test Plant for Smelting Reduction," CAMP-ISIJ, vol. 4, 1991, 45.

10) I. Kikuchi et al., "Influence of Carbonaceous Materials on Post Combustion in In-bath Smelting Reduction Furance," CAMP-ISIJ, vol. 4, 1991, 46.

11) T. Ibaraki et al., "Scale-up Rules of Carbonaceous Materials on Post Combustion in In-bath Smelting Reduction Process (Experiment of In-bath Smelting Reduction in a 100t Experimental Furnace - Ⅲ)," CAMP-ISIJ vol. 3, 1990, 1076.

12) M. Yamauchi et al., "Phenomena of Slag in the In-bath Smelting Reduction Furance," CAMP-ISIJ, vol. 3, 1990, 1075.

RESULTS OF THE AISI/DOE DIRECT STEELMAKING PROGRAM

Egil Aukrust[1]

AISI Direct Steelmaking
640 Rodi Road
Pittsburgh, Pennsylvania   15235

Abstract

The AISI/DOE Direct Steelmaking Program has been in existence for more than three years. It is a multi-faceted program including a major pilot plant operation supported by process research at five universities and industrial laboratories, as well as full-scale steel plant tests to assess scale-up criteria. The program is projected to cost $54 million over five years, of which 77% ($41.6 million) will be funded by the federal government and the remainder will be paid by AISI.

The program has progressed to the stage where scale-up to a 350,000 tonnes per year demonstration ironmaking plant is being planned. An independent assessment by Mannesmann Demag has confirmed our findings and defined the key design parameters for the demonstration plant.

The information obtained at the pilot plant and backed by the research support programs has provided a basic understanding of the relationship among key process variables. Postcombustion, the sulfur path, utilization of the volatile matters from the coal, dust generation, and refractory performance have been carefully studied. The process capability in terms of production intensity and fuel efficiency has been analyzed.

A parallel path toward commercialization of this technology is now being pursued. This involves proceeding with the demonstration plant for ironmaking while the longer-term development of the overall direct steelmaking flowsheet continues.

---

[1]Senior Technical Director, LTV Steel Corporation, and Program Director, AISI Direct Steelmaking Program.

Proceedings of the
Savard/Lee International Symposium on Bath Smelting
Edited by J. K. Brimacombe, P. J. Mackey,
G. J. W. Kor, C. Bickert and M. G. Ranade
The Minerals, Metals & Materials Society, 1992

Introduction

The period of 1985 - 1990 saw a growing worldwide interest in direct ironmaking based on smelting of iron ore with coal and oxygen in a liquid bath. The basic assessments performed during that time period set the stage for industrial development programs in Europe, Japan, Australia, and the United States. These developments are now approaching commercialization.

In the 1987 Howe Memorial Lecture, Karl Brotzmann[1] traced his own extensive work on bath smelting and presented evidence of the recovery of high temperature heat from postcombustion within the smelting process. The following year Smith and Corbett[2] presented a comprehensive review of process developments to date. In the previous year the fundamentals of this technology had been examined by Oeters[3] and Tokuda.[4]

The AISI Direct Steelmaking Program was formulated during this time period. With aging coke ovens, stricter environmental regulation, and mounting capital requirements, the long-term economic viability of the existing iron and steelmaking technology became increasingly more questionable. In contrast to these negative economic trends, bath smelting had the potential of leading to a continuous, environmentally sound process with lower energy requirements and improved costs.

AISI realized the possibility for a major technological change and organized a collaborative development program which, in scope and rate of implementation, is unique for the modern American steel industry.

In July, 1987, AISI authorized a study of direct steelmaking, starting with a worldwide survey of ongoing projects and the formulation of a development program for the American steel industry. By July, 1988, a detailed cost proposal was submitted to DOE for support under the Steel Initiative legislation. DOE accepted the project in November, 1988, at which time AISI had the management team in place and was ready to release a major collaborative research program on the physical chemistry of smelting reduction with J. F. Elliott at MIT and R. J. Fruehan at CMU. Pilot plant work with HYLSA at Monterrey, Mexico, on preheating and prereduction of iron ore pellets proceeded in parallel with the design and construction of a 5 ton/hour pilot plant for in-bath smelting at Pittsburgh. During the same time, the supportive laboratory research program was expanded to include fluid flow and heat transfer studies at McGill and McMaster Universities and the Linde Division of Union Carbide. Scale-up studies were performed at Dofasco's #2 Q-BOF shop. Experimentation of continuous refining was undertaken, awaiting completion of construction of the pilot plant. The smelter was started up on June 14, 1990, less than two years after the original program was formulated.

This project is now in its fourth year of a projected five year duration. It is the objective of this paper to summarize the major findings and delineate the approach toward commercialization of this technology in North America.

Development Approach

At the beginning of the project, the basic decision was made to focus on the fundamental technological issues of each process unit in parallel and to study these processes simultaneously on laboratory, pilot plant, and full scale. The integration of the total flowsheet would be deferred to the end of the program. This approach would avoid the pitfalls of being bogged down in the excessive mechanical complexity of a fully-integrated pilot plant operation prior to having adequate knowledge of the unit operations involved. This approach of starting with the basics and adding sophistication and complexity as knowledge is developed has served us well.

A Technical Advisory Committee, consisting of senior professional personnel from the steel industry, was created to provide technical oversight for the program. The organization of the project and the Project Board is shown in Figures 1 and 2.

The overall direct steelmaking flowsheet was divided and studied in the following component parts: a) smelting, b) preheating and prereduction, c) gas cleaning and tempering of the smelter gases for prereduction, and d) continuous refining of the smelter metal.

It was realized that, to fully investigate the smelting process and to define the relationships among the key process variables, it might be necessary to test several furnace designs.

Smelter Design

The first installation, a BOF-type converter, was operated for 48 trials over a 14 month period for fundamental studies of the smelting operation. These investigations included the effects of:

a. Type of coals - Five types of coals were studied, including anthracite, low vol, mid vol, and two types of high vol coals,

b. Degree of prereduction - Blast furnace hematite pellets, wustite pellets, and DRI,

c. Blowing practices - Different types of oxygen blowing practices were investigated, including hard and soft blowing, top blowing vs. side injection, and various levels of bottom stirring, and

d. Amount of slag and char in the system.

This program, with the underpinning of the university and laboratory programs, provided a basic understanding of the thermochemistry of the system and allowed us to proceed to the second smelter, a stationary furnace, with a length-to-width ratio of 2.5, popularly referred to as the "horizontal smelter," which was installed in January, 1991. This larger aspect ratio allowed:

Figure 1

BOARD OF DIRECTORS

Milton Deaner, AISI

John M. Farley, AISI

Brian W. H. Marsden, Acme Steel

Alexander McLean, University of Toronto

Allan M. Rathbone, USS

Roger R. Regelbrugge, Georgetown Industries

Figure 2

PROGRAM MANAGEMENT

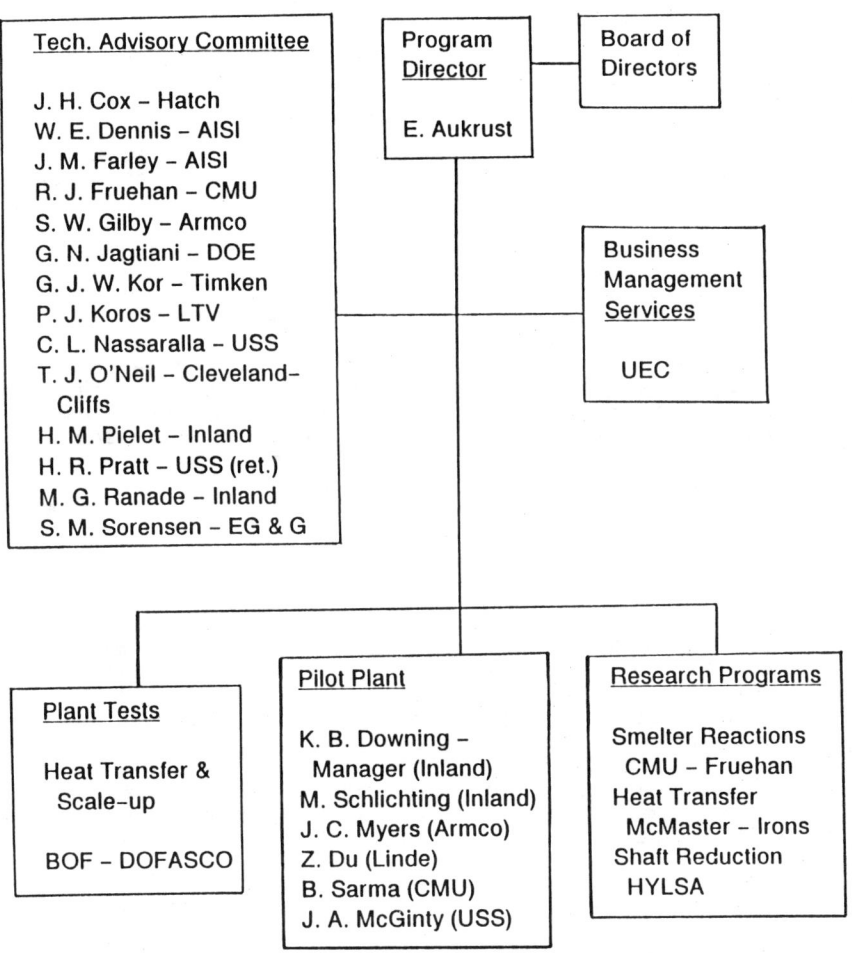

Tech. Advisory Committee

J. H. Cox – Hatch
W. E. Dennis – AISI
J. M. Farley – AISI
R. J. Fruehan – CMU
S. W. Gilby – Armco
G. N. Jagtiani – DOE
G. J. W. Kor – Timken
P. J. Koros – LTV
C. L. Nassaralla – USS
T. J. O'Neil – Cleveland–
  Cliffs
H. M. Pielet – Inland
H. R. Pratt – USS (ret.)
M. G. Ranade – Inland
S. M. Sorensen – EG & G

Program Director

E. Aukrust

Board of Directors

Business Management Services

UEC

Plant Tests

Heat Transfer &
Scale–up

BOF – DOFASCO

Pilot Plant

K. B. Downing –
  Manager (Inland)
M. Schlichting (Inland)
J. C. Myers (Armco)
Z. Du (Linde)
B. Sarma (CMU)
J. A. McGinty (USS)

Research Programs

Smelter Reactions
  CMU – Fruehan
Heat Transfer
  McMaster – Irons
Shaft Reduction
  HYLSA

a. Experimentation with multiple vs. single oxygen lances,

b. Larger freeboard volume with introduction of the top-charged feed material away from the high velocity gas stream in the duct and, thus, reduced dust losses,

c. Evaluation of water-cooled furnace structure at the slag line and for the furnace roof with the aim of reducing refractory consumption,

d. Development of an engineering solution to the slag metal separation and tapping practices from a stationery furnace, and

e. Examining whether a furnace with the larger aspect ratio of length-to-width is amenable for zoning with smelting at one end and refining at the other, thus providing the basis for continuous steelmaking.

After defining the principle issues of the two-zone smelter, <u>a third smelter will be installed</u>, incorporating the principle findings regarding smelter design from the two previous programs. This smelter, currently under design, will be installed during the first quarter of 1993 and will be the prototype for the direct ironmaking demonstration plant. This smelter will be operated under positive pressure and incorporate the gas cleaning and tempering system for delivery of the process gases at the prerequisite temperature and gas composition to the shaft prereducer.

To facilitate the initial process analysis, an overall heat and material balance model was provided by R. J. Fruehan et al[5]. It was concluded that prereduction of iron ore pellets to wustite would be possible utilizing offgases from the smelter with 40% postcombustion. The selection of these levels of prereduction and postcombustion would also enhance a stable controllable operation, as wustite is stable over a substantial range of gas compositions and temperatures and provides the basis for a reserve zone similar to that existing in the blast furnace. All work since then has confirmed the validity of this conclusion.

Design of a Moving-bed Shaft Reducer

Blast-furnace-type iron ore pellets is the raw material of choice for the majority of American steel plants. Although shaft reducers are used around the world to produce highly metallized pellets, the production of wustite represents a new technology. Accordingly, HYLSA of Monterrey, Mexico, who has a shaft pilot plant and good research and development laboratories and is well known in direct reduction technology, was contracted to develop a shaft furnace process to produce wustite pellets.

Tests were conducted at HYLSA's pilot plant at a production rate of 30 tonnes per day of wustite using simulated smelter gases at a 40% postcombustion level and at temperatures of 900 and 950 C. Tests were conducted over a wide range of gas flows ranging from 800 to 2000 NM^3 of gas per tonne of iron produced (NM^3/tFe). As shown in Figure 3, the conversion to wustite was reasonably

Figure 3

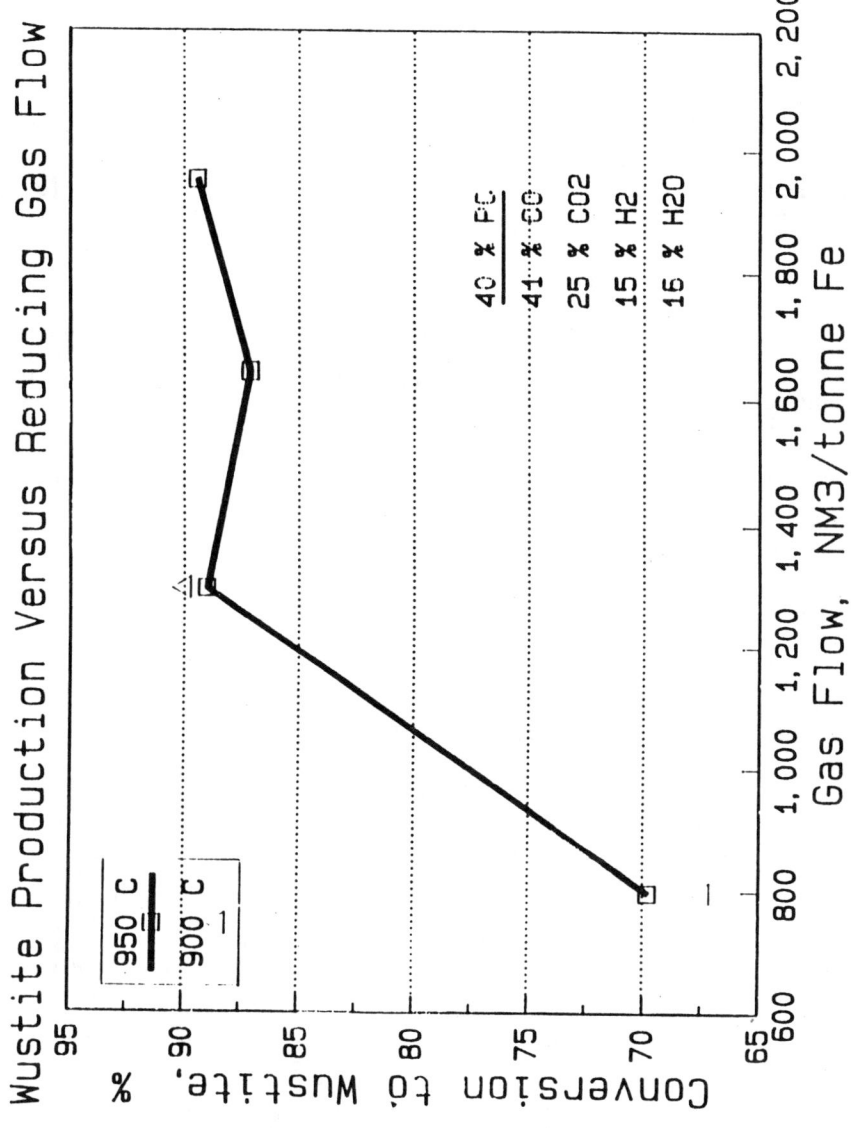

Wustite Production Versus Reducing Gas Flow

constant over a wide range of gas flows but decreased
dramatically at the 800 NM³/tFe gas flow rate. HYLSA estimates
that the minimum gas flow for good conversion to wustite is about
1050 NM³/tFe. This pilot plant information, in conjunction with
laboratory tests, has enabled HYLSA to develop a mathematical
model for a wustite shaft furnace that can be used to predict
the reduction level as a function of reducing gas composition,
temperature, and flow.

Based on information from the R & D program, HYLSA has developed
the functional engineering specifications for a reduction shaft
furnace to be used with the AISI smelter process. Furthermore,
HYLSA has provided the preliminary design specifications for a
shaft furnace to supply wustite to a proposed demonstration plant
to produce 50 tonnes per hour of hot metal. This information is
adequate to permit the detailed design of the shaft furnace to be
completed.

The Rotary Kiln Option

In mid 1990 the Technical Advisory Committee decided to explore
the use of a rotary kiln as an alternative to the shaft furnace
process for prereduction. It was perceived that a rotary kiln
might offer some advantages over the shaft process, including
operation at a lower pressure drop, thus decreasing the pressure
required in the smelter and offering more flexibility in startup
and shutdown. Furthermore, it was suggested that it might be
possible to use the smelter offgas directly in the kiln, thus
obviating the need for the gas cleaning and tempering loop
required for the shaft.

Following a literature review, it appeared that the kiln might be
a possible alternate to the shaft. To explore the viability of
the kiln in detail, consultations were held with a number of
noted kiln experts from academia and industry.

After extensive evaluation, the kiln approach was abandoned from
further consideration due to the following reasons:

1. The concept of directly using smelter offgas
   (temperature above 1500 C) in a kiln is not technically
   viable because the freeboard temperature in the kiln
   must be maintained below 1200 C to prevent accretions
   (rings) from forming. Even then, there is concern that
   the dust entrained in the smelter offgas could still
   produce accretions. In addition, the use of smelter
   offgas would produce a kiln gas rich in combustibles,
   which results in a process control problem should air be
   suddenly induced into the kiln. Finally, the volatiles
   produced from the charring of coal in the kiln are
   sufficient to provide the energy for the process, thus
   obviating the need for smelter offgas.

2. The cost of the kiln, including a second gas cleaning
   system for the proposed demonstration plant, is
   significantly higher than that for the shaft.
   Furthermore, the coal rate for the kiln is estimated to
   be at least 120 kilograms per tonne of hot metal higher

than that for the shaft, which is a direct indication
that the kiln operating cost would also be higher.

3. The chemical energy in the kiln offgas is too low to
    permit recovery. Thus, this loss in energy credit also
    increases the operating cost of the kiln.

Based on the aforementioned conclusions, it was recommended to
and accepted by the AISI Direct Steelmaking Board to discontinue
work on the rotary kiln.

Smelter Pilot Plant Results

Fuel Rates for Pilot Plant and Projections for the Commercial Plant

Table I provides fuel rate averages by coal type for a coke and
coal combination, an anthracite, Buchanan and Amonate (two
medium-vol coals), and Rowland and EKO (two high-vol coals).
Steady-state coal rates are calculated to account for non-steady-
state conditions with respect to temperature and char, which may
each be increasing or decreasing during a run segment or a series
of segments. The fuel rates are also corrected for the dust
losses as measured in the scrubber water. In an integrated
plant, the dust would be recycled. The data in the four rows at
the bottom of the table have been calculated on a dry basis,
correcting for the moisture in the coal. The raw data and the
net fuel rates are summarized in the bar chart, Figure 4.

The pilot plant model was used to calculate a projected fuel rate
for a 50 tonnes per hour commercial plant for producing hot
metal, Table II. Each of the steps corrects for a particular
difference between pilot plant operation and the expectations of
a commercial plant, and Table II presents the new value for each
of the process variables as calculated by the parametric model.

In step 1, correction is made for improved treatment of the
dust. With improved charging and offgas collection
techniques in a commercial plant, it should be possible to
reduce the net carbon losses after recycle to less than 1.5%
of the carbon charge.

In step 2, the heat loss per tonne of hot metal for the
larger vessel is assumed to be 80,000 kcal/t rather than the
138,250 kcal/t calculated for the pilot plant from energy
balances. This is approximately equivalent to a 0.75 power
scaling law.

In step 3, credit is taken for the fact that the wustite
from the shaft furnace supplying the smelter in the
commercial plant will be charged at 800°C, rather than at
25°C as in the pilot plant. Coal savings for this credit are
about 106 kg/t.

Thus, the fuel rate per tonne of hot metal for this Rowland coal,
about 29% volatile matter, would be approximately 760 kg/t.

Production rates of 5 tonnes per hour were achieved in the pilot
vessel but were limited by the oxygen supply and the offgas

Table I

Pilot Plant Fuel Rate Averages by Coal Type

Coal Type Volatile Matter (%)	Coke with Coal Injection 9.5	Anthracite 5.6	Buchanan 19.9	Amonate 22.9	Rowland 29.4	EKO 38.2
Raw Data Fuel Rate (kg/t)	1,014	1,226	1,297	1,270	1,482	1,828
Fuel Rate Adjustment to Steady State (kg/t)	-30	-42	-22	-5	-16	-9
Steady State Fuel Rate (kg/t)	984	1,184	1,275	1,265	1,466	1,819
Fuel Rate Adjustment for Dust Recovery (kg/t)	-35	-141	-161	-158	-337	-385
Net Coal Rate (kg/t)	949	1,043	1,115	1,106	1,129	1,434
Moisture (%)	2.74	3.33	2.15	2.53	3.85	4.61
Fixed Carbon (%)	80.29	84.54	69.86	68.76	64.13	53.59
Steady State Fixed C (kg/t)	768	968	872	848	904	930
Steady State VM (kg/t)	91	64	248	282	415	662
Net Fixed C (kg/t)	741	852	762	742	696	733
Net VM (kg/t)	88	56	217	247	320	522

Figure 4

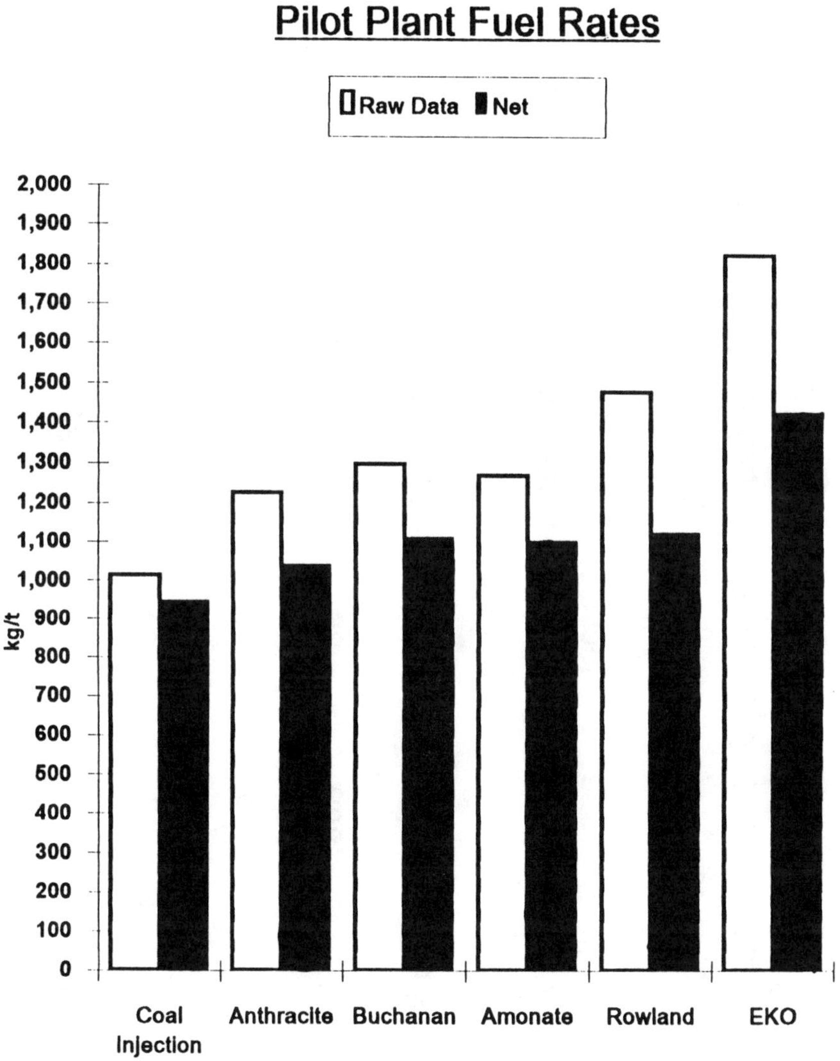
Pilot Plant Fuel Rates
Raw Data
Net
2,000
1,900
1,800
1,700
1,600
1,500
1,400
1,300
1,200
1,100
1,000
900
800
700
600
500
400
300
200
100
0
kg/t
Coal Injection
Anthracite
Buchanan
Amonate
Rowland
EKO

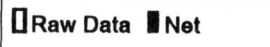

Table II

SCALE UP FROM PILOT PLANT TO COMMERCIAL

	Base Pilot Plant	Recycle of Dust	Reduced Heat Losses	Projected Demo Plant
Postcombustion %	40	40	40	40
Dust Loss kg/ton	309	26	26	26
Heat Loss kcal/ton	138.250	138.250	80.000	80.000
Ore Preheat °C	25	25	25	800
Coal Rates kg/ton	1176	893	867	761

handling system. Models based on reaction rates and foam volume acceptable in the vessel led to production intensities of 9.6 and 15 tonnes per cubic meter per day at operating pressures of 1 and 3 atmosphere, respectively.

Process control strategies and systems were developed, and data logging and information processing routines were established. Analytical and parametric models were developed based on energy and mass balances for use in analyzing the data. Considerable effort was devoted to the determination of dust and coal volatile matter bypass in the offgas system in order to establish the energy and material balances and to assess the gas cleaning and tempering requirements for coupling to the shaft furnace preheater and prereducer.

Refractory wear was controlled by keeping the process temperature to 1650°C maximum and by keeping the iron oxide content in the smelter in the 3 - 8% range. Despite the cyclic operation that stressed the refractories considerably more than would steady-state operation and despite occasional high temperature excursions in the cone or freeboard region, refractory performance was acceptable. High quality magnesia graphite refractories were used for the working lining and magnesia or alumina for the safety lining. The refractory wear for the demonstration plant has been projected to be 0.25 mm/operating hour.

The process intensity projections are given in Tables III and IV. A comparison of energy use with the COREX process is shown in Table V. These comparisons are for a single-zone smelter, making a hot metal at about 5% C. Upon success with the horizontal smelter in making a lower carbon liquid, we also foresee a major opportunity to serve the electric arc furnace user, in which the substitution of medium or low carbon hot metal for high quality scrap or for conventional offshore-produced DRI, could lead to very attractive savings. This is one of the incentives for the horizontal smelter.

The results from the research and experimentation thus far can best be summarized as follows:

- Bath smelting of wustite pellets is a manageable, high intensity process.

- Attainment of 40% postcombustion and use of this gas for prereduction of hematite pellets to wustite is an industrially attractive route.

Planning for the Demonstration Plant

To create a technical and economic basis for the further development of a commercial size plant, Mannesmann Demag was retained to prepare an independent and comprehensive feasibility study for the industrial-scale application of the AISI direct ironmaking process with the objective of constructing a 350,000 tonnes per year demonstration plant on the site of a host company still to be identified.

Table III

PROCESS INTENSITY COMPARISONS

MANNESMANN DEMAG STUDY
WORKING VOLUME FOR 50 t / HR

	COREX (m³)	AISI (m³)
SHAFT FURNACE	340	130
SMELTER VESSEL	1,010	130
	1,350	260

Table IV

PROCESS INTENSITY COMPARISONS

HOT METAL PER WORKING VOLUME

COKE AND BLAST FURNACE	COREX (SHAFT & MELTER)	AISI (SHAFT & SMELTER)
1.0	0.9	4.6

$$t / d / m^3$$

Table V

ENERGY CONSIDERATIONS

COREX AND AISI

	COREX	AISI
COAL (kg / t)	840	700
OXYGEN (Nm^3 / t)	500	430
EXCESS ENERGY (GJ / t)	11.3	7.4

COAL BLEND	%	
STEAM	80	VM = 31.6 %
LOW VOL	20	KJ / kg = 33,900
		ASH = 6.7 %

The basic study consisted of a critical survey of the results gained from the trial runs and from theoretical computations. It also included the determination of a preliminary plant concept with rough estimated costs and rough feasibility evaluation.

The scale-up criteria to prepare the basic study are based on the experimental results at Universal and Monterrey and on the assumptions of AISI and HYLSA where proven data have not yet been gathered from pilot plant operation.

The design basis for the proposed demonstration plant is a BOF-type smelter vessel and a soft blowing oxygen lance, which can secure a proper distribution of the oxygen over the entire cross section of the vessel.

The following design criteria were selected:

1. Capacity - 50 t hot metal/h
2. Postcombustion - 40%
3. Prereduction to wustite - 27.5%
4. Pressurized operation
5. Production rate - 4 t/h m^2
6. Production intensity - 10 t/day m^3
7. Continuous smelting
8. Batch tapping
9. Refractory wear 0.25 mm/h

AISI technoeconomic analyses indicated favorable variable operating costs and Mannesmann Demag's projections estimated significantly lower capital costs than for the coke oven-blast furnace route. The results of the basic study confirm a considerable potential for technoeconomic advantages of the AISI Direct Ironmaking Process and conclude that it would be a compact plant concept with high productivity.

The success of this AISI-DOE collaborative program on direct steelmaking has led us to proceed on two parallel paths:

1. Proceed with a 350,000 tonnes/year ironmaking demonstration plant for ironmaking at the same time as

2. Completion of the pilot plant program in support of the demonstration plant and further exploration of the direct steelmaking objective. The timing for these projects is shown in Tables VI and VII.

In conclusion, we believe that AISI has formulated with this project a creative approach for collaboration within the steel industry and with the government and that we now are at the forefront worldwide for implementation of an energy efficient and environmentally sound process for direct iron and steelmaking for the 21st century.

Table VI

SCHEDULE FOR IRONMAKING DEMONSTRATION PLANT

AMERICAN IRON AND STEEL INSTITUTE (AISI)

Determine host site for demonstration plant	June 1992
Complete proposal for demonstration plant	November 1992
Initiate plant construction with DOE FY 1994 funds	January 1994
Begin operation of demonstration plant	August 1995
Complete evaluation of process	August 1997

Table VII

AISI–DOE PROJECT SCHEDULE

A. Pilot Plant Program

    (1) Operation of the horizontal smelter   April –
October 1992

    (2) Engineering and procurement of the   April –
        gas cleaning and tempering loop   October 1992
        (Hatch) with recycle of the fines to
        the 10-ton/hr. pressurized smelter
        (Mannesmann Demag)

    (3) Installation of the third smelter and   November 1992 –
        gas loop   February 1993

    (4) Operations of the third smelter as   March –
        prototype for the ironmaking   September 1993
        demonstration plant

    Major Decision Point – June 1993

        Option 1 – Continue pilot plant operation
        through 1994 in support of the ironmaking
        demonstration plant and/or development
        of direct steelmaking.

        Option 2 – Close off the project by
        December 31, 1993.

B. Supporting Studies (Mannesmann Demag)

    (1) Feasibility study for the ironmaking   May – December
        plant   1992

    (2) Basic study continuous steelmaking   January – March
                                                1993

References

1.   K. Brotzmann, "New Concepts and Methods for Iron and Steel Production," _Steelmaking Proceedings_, ISS-AIME, _70_ (1987) 3.

2.   R. D. Smith and M. J. Corbett, "Coal Based Ironmaking," _Ironmaking and Steelmaking_, The Metals Society, _14_ (1987) 49.

3.   F. Oeters and A. Saatci, "Some Fundamental Aspects of Iron Ore Reduction with Coal," _PTD Proceedings_, ISS-AIME, _6_ (1986) 1021.

4.   M. Tokuda, "Conceptual and Fundamental Problems of Smelting Reduction Process for Ironmaking," _Transactions ISIJ_, _26_ (1986) B192.

5.   R. J. Fruehan, K. Ito, and B. Ozturk, "Analysis of Bath Smelting Processes for Producing Iron," _Transactions of the ISS_, (1988) 81.

Operational Results of 100 ton/day Test Plant

for Smelting Reduction of Iron Ore in NKK

T. Kitagawa [1], T. Hasegawa[2,3], K. Takahashi[2,3],
T. Ariyama [4] and N. Kimura[1]

R & D Task Force, JISF DIOS Process
Japan Iron & Steel Federation
2-3-2 Kudan-minami, Chiyoda-ku, Tokyo 102 Japan

NKK Corporation
1-1-2 Marunouchi Chiyoda-ku, Tokyo 100 Japan

Abstract

The core of the research program in NKK is to develop the process technology
that would achieve a high post combustion degree (50% to 55%) coupled with
high heat transfer efficiency of more than 85% in the smelting reduction
furnace (SRF). Such a highly heat efficient SRF makes the pre-reduction
unit compact (pre-reduction degree of less than 30%) and reduces the export
energy to the level equivalent to the blast furnace process.

Test operations of the 100 ton/day plant, which consists of SRF, bubbling
bed type pre-reduction furnace (PRF), tapping equipment and SRF dust recycling system have been carried out in NKK Fukuyama Works. The results
obtained with -8 millimeters sinter feed ore and high volatile matter coal
of maximum 37% volatility are summarized in this paper.

The reduction of iron ore in PRF coincides well with the theoretical analysis. The degradation of iron ore in PRF and size distributions of classified fine and coarse ore can be well predicted with the mathematical models.
To obtain a high heat transfer efficiency of more than 85%, submerged combustion and strong agitation of slag bath are ultimately important. The
oxygen lance nozzle design and the control of slag foaming are also
important for submerged combustion. The ultra soft blowing of oxygen with
the double flow lance nozzle and the increase in the operational pressure of
SRF considerably improve the post combustion degree. A post combustion
degree of more than 50% coupled with high heat transfer efficiency has been
attained with the high volatile matter coal.

Stable operation of the full system over the tapping operations has been
performed for several days. A good perspective of the operation of the
full smelting reduction system on a larger scale is obtained through the
tests. This paper contains the result of the DIOS collaboration research.

*1: Steel Division, NKK Corporation
*2: Fukuyama Laboratory, R & D Task Force, Japan Iron & Steel Federation
*3: Smelting Reduction R & D Team, Keihin Works, NKK
*4: Steel Research Center, NKK

Proceedings of the
Savard/Lee International Symposium on Bath Smelting
Edited by J. K. Brimacombe, P. J. Mackey,
G. J. W. Kor, C. Bickert and M. G. Ranade
The Minerals, Metals & Materials Society, 1992

Introduction

The R & D program of smelting reduction process was started in as early as 1978 in NKK, though several research projects are now on-going throughout the world. The basic concept of the process as an alternative to the current blast furnace process and the major research subjects were set by 1984.

The process should be furnished with the positive features of the blast furnace process, which are high energy efficiency, high productivity, and optimum energy supply to the downstream of the integrated steel works. It should also have the features that will eliminate the weak points of the blast furnace process, which are: direct use of various grades of raw materials, simple and compact production equipment, increased flexibility of production, and reduction in carbon dioxide emission.

The above process features will be realized by the highly heat efficient smelting reduction furnace (SRF), where the post combustion (PC) degree of higher than 50% coupled with a high heat transfer efficiency (HTE) of higher than 85% is attained. It allows the reduction of the pre-reduction degree of iron ore to less than 30%, which results in very compact iron ore pre-reduction and pre-treatment equipment. In this report the post combustion degree is denoted by the oxidation degree (OD) of SRF off-gas because it is directly introduced to PRF without change in its chemical composition.

A fundamental study was started in 1985[1],[2] to attain such a high PC degree coupled with a high HTE in the smelting reduction furnace. Promising results in the fundamental study lead to the installation of 5-ton SRF at NKK Fukuyama Works in 1986. Fundamental studies of the pre-reduction of iron ore were also carried out with the bench scale fluidization bed furnace in 1987 and 1988 [3], which were followed by the application of the bubbling bed type pre-reduction furnace (PRF) to the SRF in Fukuyama Works in 1988. The installation of the full units of the smelting reduction test plant in Fukuyama was completed by the attachment of the tapping equipment in 1989 and the SRF dust recycle system in 1990, which resulted in a 100 ton/day maximum production capacity.

NKK joined the collaboration research program of smelting reduction process (DIOS) when the Japan Iron & Steel Federation (JISF) organized it in 1988. The operation of the test facilities in Fukuyama has been partly funded by the DIOS collaboration research program since that time. The operational results of the full system in NKK Fukuyama Works, including the results obtained by the DIOS research program are described in this paper.

Brief Review of the Fundamental Study

Post Combustion[1],[2]

The PC degree in BOF type bath smelting furnace (SRF) can be increased relatively easily if the oxygen jet is effectively isolated to prevent the reaction with the carbon saturated metal including metal droplets generated in SRF. The oxygen blown in the SRF should be utilized for the combustion of the carbon monoxide evolved as the reduction reaction product as much as possible in this purpose. The combustion heat of carbon monoxide should be effectively transferred to the reaction site of iron oxide, which is a heat sink because of the endothermic reaction. The most effective way to recover the sensible heat of PC gas jets is to combust the carbon monoxide in the slag, which is well known as the submerged combustion [3]. An intensive study was then carried out to find the controlling factors in a submerged combustion to improve the heat transfer from the PC jets to the slag with a laboratory scale SRF of 100 and 400 Kg capacity. Data obtained with the various combinations of the experimental conditions are empirically summarized in Eq.(1).

$$\eta_{p,c} = \left\{ 1 - \frac{\text{Amount of Superheat of SRF Offgas}}{\text{Amount of Post Combustion Heat}} \right\} = 1 - a(OD)^{0.9} P^{-1} \quad (1)$$

where the constant "a" can be estimated 90 by a heat transfer model, and P represents the operational parameters of SRF as indicated in Eq.(2).

$$P = 350 \; nL/Vg \qquad (2)$$

where n is the number of the PC
nozzle, L is the length of the PC
characterized by the oxygen jet
trajectory in the foaming slag,
and Vg is the super- ficial velocity of the gas in the PC zone
of SRF. The results are represented in Fig. 1. It is suggested that the higher the PC degree,
the higher the P value should be
obtained for better heat transfer.
To obtain a high P value the number
of the lance nozzle for PC should be

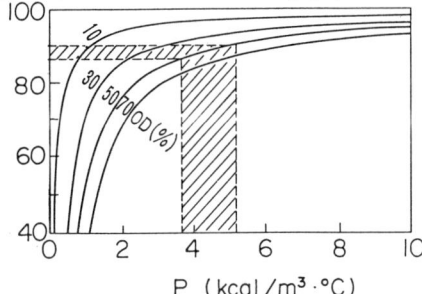

Fig. 1. Relationship between HTE and operational parameter, P.

increased and the oxygen jet trajectory should be extended in the foaming slag. To improve HTE intensive stirring of the slag should also be important.

Prereduction of Sinter Feed Ore[4]

Iron ore fines can be pre-heated and pre-reduced with <u>bubbling fluidizing bed</u>(BFB) furnace without difficulties if the pre-reduction degree is limited to less than 33%. However the <u>circulating fluidizing bed</u> (CFB) furnace with much higher gas velocity should be applied for higher pre-reduction degree to prevent the operational problems associated with the agglomeration of ore particles due to the metallic iron formation. Batch type and small scale fluidizing bed experiments were carried out to determine the specifications and the operational conditions of the test plant in Fukuyama.

<u>Fluidization Stability</u>. The minimum fluidi-
zation gas velocity, Umf and the criteria of
fluidization stability were studied with the
cold BFB model of 200 mm diameter. Obtained data of Umf were found to agree well with
those estimated by Wen-Yu's equation [5] . The
stable fluidization state was established by
the criteria shown in Fig.2, where the index,
Is, shows the size distribution of the ore.

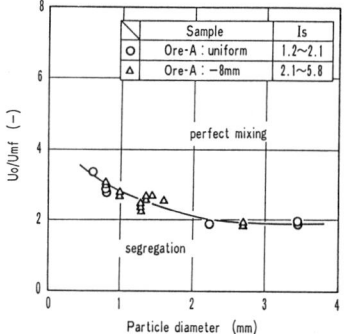

<u>Reduction Kinetics</u>. Hematite ore of 2.00 to
2.83 millimeters in diameter was reduced
with a batch type fluidizing bed furnace at
700 to 850℃ . The reactor was hung by a
balance so that the reduction rate was monitored during the experiment by the weight
change measurement. Microscopic analysis of
the reduced ore suggested that the reduction

Fig. 2. Stable fluidization criteria for iron ore fines. Is=(60% passing)/(10% passing)

proceeded as the topochemical state and the rate controlling step was not the diffusion step but the chemical reaction rate. The unreacted core model can be applied to simulate the reduction of the oreparticle. The reduction rate in the continuous operation in BFB was then analyzed with the reaction model under the the assumption of homogeneous single phase model. The results of the analysis are indicated in Fig. 3.

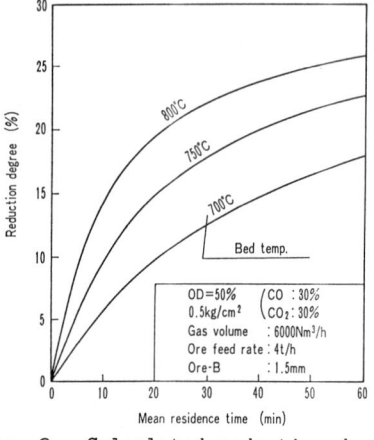

Fig. 3. Calculated reduction degree as a function of mean residence time.

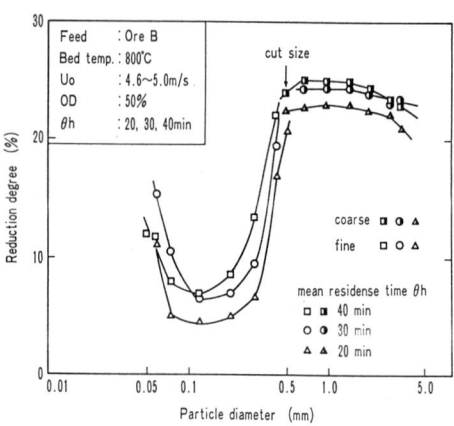

Fig. 4. Reduction degree observed in the small scale BFB.

The above analysis was verified with the small scale BFB reactor of 130 mm inner diameter and 4800 mm high, which enabled continuous operation and to circulate the classified fine ore. Observed reduction degree was well simulated by the analysis indicated in Fig. 3. The reduction degree observed in the experiments are shown in Fig. 4 as the function of particle diameter and the mean residence time of the coarse. It can be stated from the figure that the larger the mean residence time, the higher the reduction degree even for the classified fine ore.

Behavior of the Fine. Residence time of the fine ore to be classified from the bed was analyzed by the elutriation model. Analytical results showed that the residence time of the fine ore tended to increase with the particle diameter and reaches the same value as that for the coarse ore at the cut size. The reduction degree of sinter feed of wide size distribution can then be calculated for a given operational condition by combining the above analysis and the reaction kinetic model. The calculated results showed a similar tendency to those indicated in Fig.4. The effect of circulation of the classified fine ore on the reduction degree can be also evaluated by the analysis as indicated in Fig. 5. It is found that the effect of circulation on the reduction degree tends to saturate at a circulation rate greater than 3. This must be another feature of BFB.

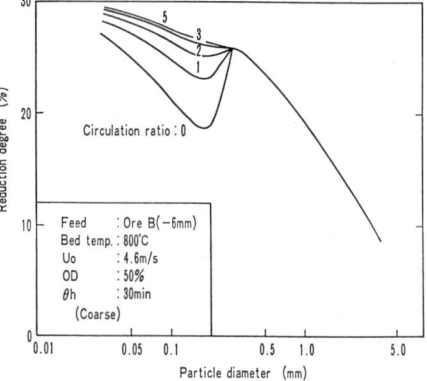

Fig. 5. Effect of circulation of the fine on the reduction degree.

<u>Operational Results of Fukuyama Plant</u>

<u>Test Facilities</u>

Specifications of the integrated test plant in Fukuyama Works
were determined based on the knowledge stated above. The final flow
sheet of the plant is shown in Fig.
6 and major specifications are
listed in Table I.
Features of the plant:
① The bubbling bed type PRF with
circulation system for fine ore
accommodates the sinter feed ore
with a wide size distribution.
② The reduction gas is directly
introduced to PRF via hot cyclone
by the gas evolution pressure from
SRF, which makes the gas introduction system simple and safe.
③ The SRF is operated at a pressure less than 2.0 Kg/cm² though
the necessary pressure for stable
bubbling bed formation in PRF is
0.7 Kg/cm².
④ Classified fine and coarse ore
in PRF are separately charged in
SRF to improve the yield.

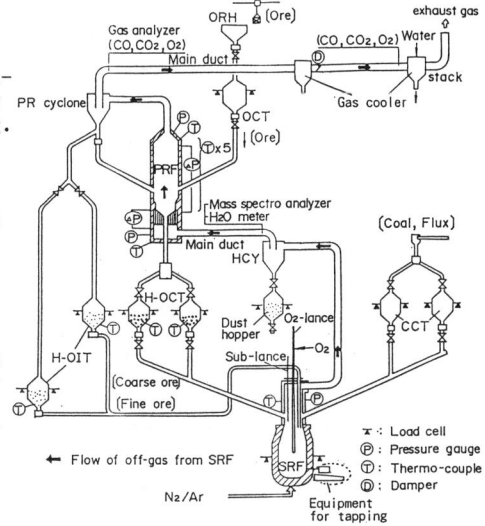

Fig. 6. Flow sheet of 100 ton/day
test plant in NKK Fukuyama Works.

⑤ Accurate operation control can be carried out based on the mass and heat
balance calculation by weighing every charging hoppers and dust collecting
hopper, sublance measurement of bath temperature, mass-spectrometer for gas
composition in the completely gas tight system.
⑥ Conventional type tapping equipment attached to SRF allowed to control
the amount of slag and metal without interrupting of the operation.

Table I Specifications of Test Plant in Fukuyama Works

PRF	Type	: Bubbling Bed with Circulation System
	Dimension	: 1.2 mφ × 9.8 m H
	Capacity	: Max. 6.5 ton/Hour
	Gas Flow Rate	: 4000-8000 Nm³/Hour
	Iron Ore Size	: Sinter Feed of −8 millimeters
SRF	Type	: In-bath Smelting Reduction Furnace
	Inner Volume	: 7 m³ as Lined
	Lance Nozzle	: Double Flow Type (Multi Holes)
	Oxygen Flow Rate	: Max. 2500 Nm³/Hour
	Max. Capacity	: 100 ton/Day

<u>Operational Procedure</u>

The atmosphere of the entire system was completely exchanged by nitrogen
after charging the initial hot metal of around 6 tonnes to SRF. After
charging the hot metal oxygen top blowing was started to heat up the metal
and the initial slag was produced. Iron ore was gradually charged into PRF
according to the gas evolving rate from SRF during the heat up period.
After enough amount of gas was obtained in PRF the entire system was
pressurized for steady operation of SRF. Sinter feed ores of −5 or −8

millimeters listed in Table II were used in the operation. The effect of properties of coal on the performance of SRF was evaluated by using various grades of carbonaceous materials listed in Table III.

Table II Chemical Composition of Ore (mass %)

Brand	T.Fe	FeO	SiO$_2$	Al$_2$O$_3$	MgO	P	S	Ig.loss
A	62.4	0.2	4.11	2.28	0.05	0.07	0.04	3.20
B	67.4	1.8	1.08	0.65	0.13	0.05	0.02	1.28
C	68.1	1.6	0.87	0.46	0.08	0.03	0.01	0.21

Table III Chemical Composition of Carbonaceous Materials

Coal or Coke	Proximate Analysis			Ultimate Analysis				
	Ash	VM*	FC **	C	H	N	O	S
HV1	10.4	31.7	57.9	73.3	4.6	1.6	9.5	0.5
HV2	9.2	36.6	54.2	74.2	5.2	1.6	9.3	0.5
MV	12.3	22.0	65.7	77.7	5.4	1.4	2.9	0.3
Coke	12.9	0.5	88.6	85.4	0.3	0.9	0.2	0.2

*: volatile matter   **: fixed carbon   (dry base, mass %)

Operational Results and Discussion

Heat Transfer Efficiency. As already indicated, the interfacial area of the PC jets, which is represented by nL in Eq.(2) should be enlarged as much as possible for higher HTE. The lance nozzle geometry was well tuned for this purpose. For a given nozzle design and oxygen flow rate, the control of lance height relative to the foaming slag level, (Ll-Ls), is ultimately important as well as intensive stirring of the bath to attain high HTE as shown in Fig. 7.

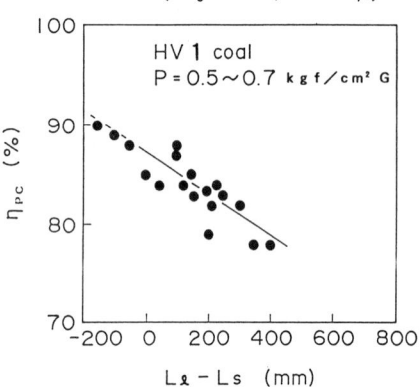

Fig. 7. Effect of lance height relative to the foaming slag level on HTE.

Post Combustion. The above mentioned operational conditions for improvement of the HTE were fixed to obtain high PC degree. PC degrees observed with the high volatile matter coal, HV1, are indicated in the relationship between slag foaming height and ore feeding rate as shown in Fig. 8. [6] Also indicated in the figure is the lance height at which those data were obtained. The HTE corresponding to the plot for PC degree is also indicated in the upper part of the figure. The PC degree can be raised relatively easily at higher slag level, which is obtained at higher ore feeding rate, even if high volatile matter coal is used. The reason is, at higher slag foaming level the oxygen lance can be kept high without any drop in HTE and the interaction between PC jets and carbon saturated metal is avoided. This is a closed relationship among the PC degree, the slag

616

foaming height, and the ore feeding rate. The relationship between PC degree and HTE observed with the various kinds of carbonaceous materials listed in Table III is shown in Fig. 9. For all of the carbonaceous materials, high PC degree and high HTE are simultaneously obtained though only a slight decrease in PC degree is observed with an increase in volatile matter content. [6]

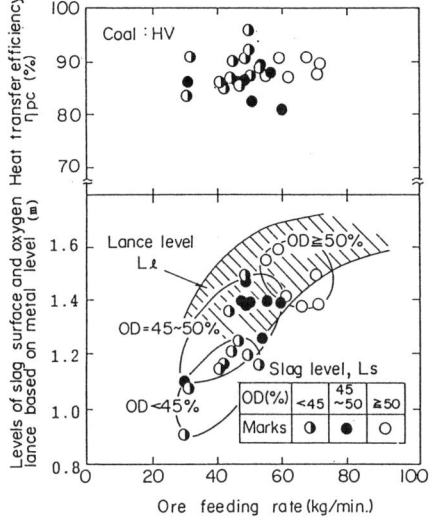

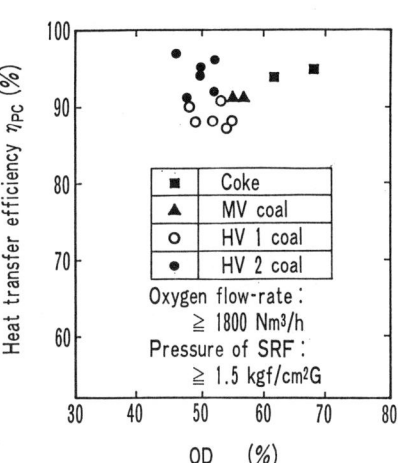

Fig. 8. Relationship between slag foaming level and ore feeding rate at the time specified post combustion degree was observed.

Fig. 9. Relationship between PC degree and HTE observed for various grades of carbonaceous materials.

The precise control of the charging rate of coal (carbon and hydrogen) and oxygen (from ore and gaseous oxygen) greatly affects the PC. PC degree should decrease drastically if the excess amount of coal is charged to SRF, and vice versa. Such charging rate control is also important for stable operation of SRF to be stated later. Pressurized operation of SRF carried out in NKK Fukuyama Works reduces the carry over ratio of coal and allows the precise operation control, which results in stable and highly heat efficient operation.

Gas Linkage to PRF. The change in SRF gas evolution rate may adversely affect the stability of fluidization since PRF is directly linked to SRF and a wide size distribution (ex.-8 mm sinter feed) of the iron ore. However, the gas evolution rate fluctuated by only 5 % at most in the steady state and any difficulties associated with this were not encountered even during start up and turn down operational periods. This can be regarded as one of the features of the bubbling bed type PRF, which has a gas distributor at the bottom of the furnace. Another subject for the smooth operation of the BFB type PRF was supposed to be nozzle clogging of the distributor. The nozzle tended to get choked as the reduction gas temperature rose. Controlling the working surface temperature of the distributor successfully solved the problem.

Classification of Ore in PRF. The classification point of the ore in the PRF was fairly consistent with the one estimated by the superficial velocity of reduction gas in the free board of the furnace and the terminal velocity of the particle. Typical size distributions of the classified fine and coarse ore are represented in Fig. 10. [7) The size distribution of the coarse ore diffuses beyond the cut size as shown in the figure. This can be explained by the elutriation model, in which the fine particle to be classified is well mixed with the coarse ore and dwells in bed before elutriation. The classified fine and coarse ores were charged into SRF individually to increase the charging yield. The fine ore was pneumatically transferred and was injected from the top of SRF, while the coarse ore was charged by gravity.

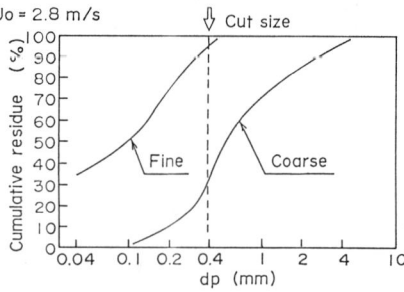

Fig. 10. Classification of sinter feed ore in PRF.

Degradation during Reduction. The degradation of ore could not be neglected with the increase in bed temperature. The degree of degradation should be correlated with the ore properties represented by the reduction degradation index, RDI, reduction temperature and reduction gas velocity. The former two factors may represent the structure change by reduction and the latter one may represent the mechanical effect applied to the particles. The degradation degree was expressed by Eq.(3) based on Rittinger's equation.

$$\Delta S = 1/dh - 1/dh° \qquad (3)$$

where dh° and dh are the harmonic mean diameter of the ore before and after degradation, respectively. The empirical equation was deduced statistically by using ΔS as a function of RDI, Tb and the superficial reduction gas velocity in PRF, U_0. [8) The degree of degradation was well simulated for various grades of ore with the model. The elutriation model stated above was then combined with the degradation analysis to estimate the size distributions of the fine and coarse ore. One of the results of the estimation is shown in Fig. 11.

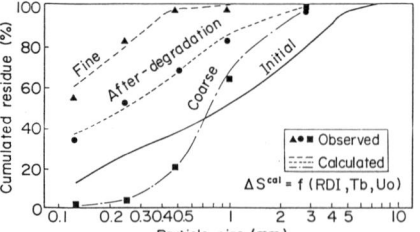

Fig. 11. Estimation of size distributions of fine and coarse ore obtained in PRF.

Reduction Characteristics. Pre-reduction degrees of both coarse and fine ore are shown Fig. 12, where the data for ore B (-8 mm) and the mean residence time in the range of 30 to 60 minutes are indicated. [9) The reduction degree almost solely depends on the bed temperature. The effect of the circulation of the classified fine ore on the reduction degree is not drastic, as is also indicated in the figure.

Continuous Operation. Conventional tap-hole opener and mud gun type tapping equipment was attached to SRF and the weighing system of tapped metal and slag was also equipped as shown in Fig. 13. Pressurized operation of SRF was found to be helpful for smooth tapping operation. The metal and slag can be tapped without change in the operational parameters though the slag in SRF is assumed to be foamed during the steady state operation. The amount of slag in SRF was determined by the mass-balance computation and the weighing of the tapped slag and was kept at the optimum level to maximize the heat efficiency of SRF. Examples of slag and metal weight control by the tapping of SRF are shown in Fig. 14.[10]

○   with circulation of fines
●   without circulation of fines

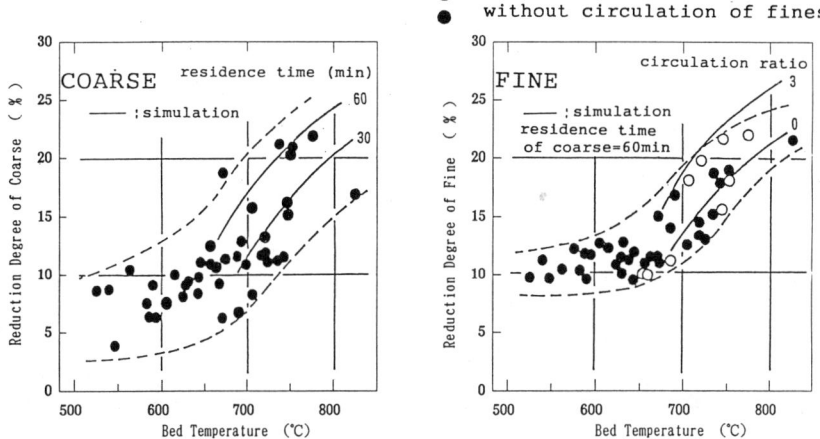

Fig. 12. Effect of bed temperature on the reduction degree of sinter feed ore B. (ore size:-8mm, gas OD: 40-50 %)

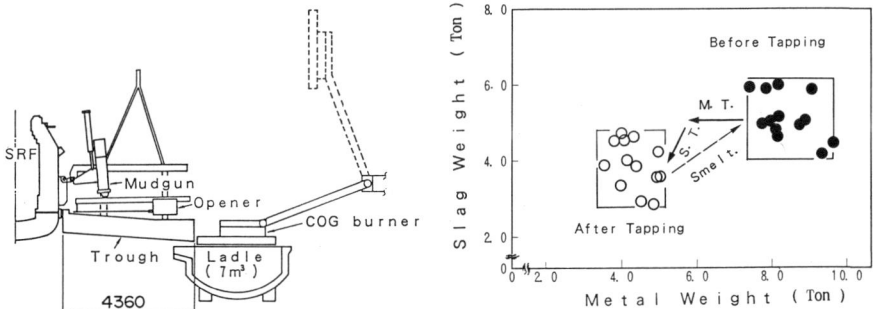

Fig. 13. Tapping equipments attached to SRF.

Fig. 14. Control of slag and metal weight in SRF by tapping operation.

The charging rate of coal should be balanced to that of ore for stable SRF operation. The amount of carbon to be supplied in SRF for a given operation time was determined by the evaluation of the carried-over coal and ore, the SRF off-gas analysis, and the amount of metal produced, which was calculated by the amount of pre-reduced ore charged and the estimated pre-reduction degree as shown in Fig. 15. Such calculation was carried out every one minute to control SRF operation. The full units of the integrated system were stably operated based on such operational control. A typical example is shown in Fig. 16.[11]

Coal Consumption.   Observed coal consumption for a unit weight of hot metal production is plotted against the production rate as indicated in Fig. 17. The coal consumption rate decreases with the production rate to attain a level as low as 850 Kg/ton of hot metal, which can be considered to be in the feasible range as an alternative to the blast furnace process. [11]   The effect of PC degree on the coal consumption rate is estimated as parameters

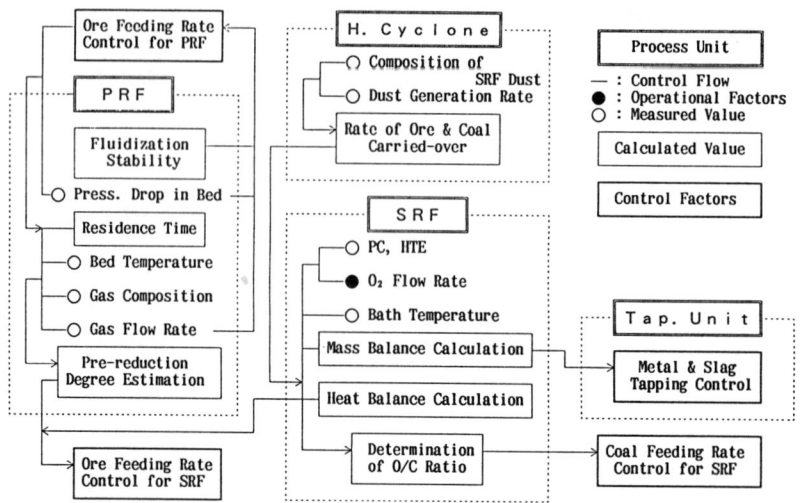

Fig. 15. Flow of operational control of the smelting reduction system.

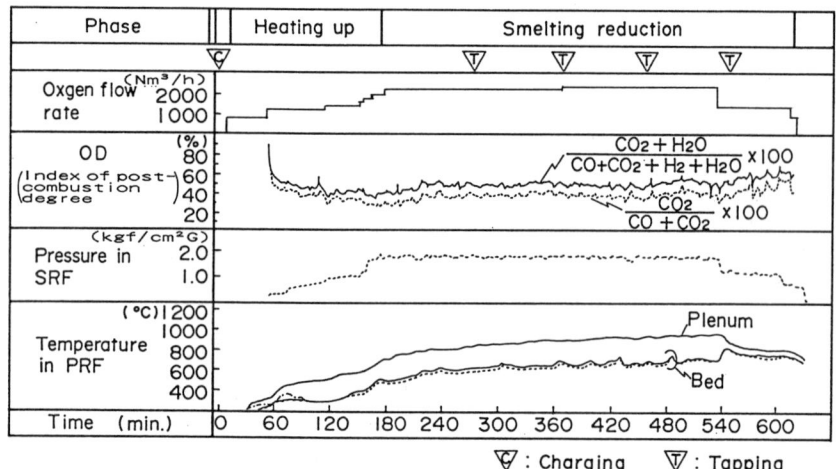

Fig. 16. Typical example of the continuous operation of the full system.

of the heat loss for a unit weight of hot metal produced and HTE as indicated in Fig. 18.[12] Major calculation conditions are, pre-reduction of ore; 11 %, moisture of coal; 4 %, carbon content of hot metal produced; 4.5 %. The results of heat and mass balance calculation are represented with solid lines for a coal grade HV1. For a given heat balance, the participation of the heat of fixed carbon combustion by the top blown oxygen was calculated hypothetically. It is expressed by the <u>fixed carbon combustion ratio</u>, FCR, which is defined by the amount of fixed carbon in the coal combusted by a unit amount of top blown oxygen, and is indicated by a set of broken lines in the figure. The observed data in the Fukuyama plant, which were taken under identical operational conditions as case "B", are also plotted in the figure. The calculated curve for case "B" coincides well with those observed. It can be stated that FCR is reduced to attain the low coal consumption rate because a large proportion of heat is generated by PC in a heat balance.

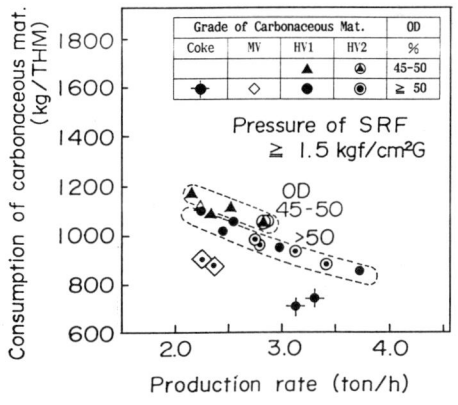

Fig. 17. Coal consumption rate attained with the integrated system in Fukuyama Works.

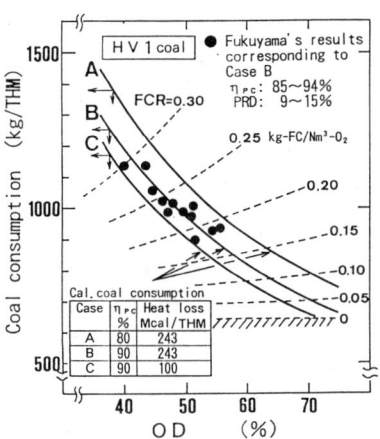

Fig. 18. Effect of PC degree on the coal consumption.

Conclusion

The following results have been obtained through operations of the 100 ton/day scale integrated smelting reduction plant at NKK Fukuyama Works.
1) The results of the fundamental study on the heat efficiency of SRF were verified. Submerged combustion in the foaming slag was realized to recover the PC heat effectively. Soft blowing of oxygen using a multi hole nozzle and intensive bath stirring are essential conditions for high HTE and high PC degree. Precise feeding rate control of ore and coal is also important for high PC degree.
2) Several operational factors affecting the reduction degree were evaluated. The behaviour of the sinter feed ore in the BFB furnace was studied to control the residence time in the furnace.
3) The BFB furnace was successfully linked to SRF and was proved to be suitable for pre-reduction of sinter feed ore of wide size distribution.
4) Metal and slag were tapped with a conventional type tapping device from the pressurized SRF without disturbing the steady operation.
5) The basic concept of stable and efficient operation of the full system was constructed and was verified through the continuous operation.

References

1) S. Sugiyama et al., "Heat Transfer Analysis of Post Combustion in LD Converter (Fundamental Study on Smelting Reduction Process IV)", Tetsu-to-Hagane, 72 (1986),1029.

2) S. Nishioka et al., "Fundamental Study on Post-combustion Technique in Strongly Stirred Iron Bath Reactor",Tetsu-to-Hagane,76 (1990), 2019-2024.

3) D. N. Rao et al., "Direct Contact Heat Transfer- A Better Way to High Efficiency and Compactness", Canadian J. Chem. Eng, 62(1984), 319-325.

4) T. Ariyama, et al., "Reduction Behavior of Iron Ore Fines in Fluidized Bed for Smelting Reduction Process", (to be presented at the 10th ISS PTD Conference, Toronto, Ontario, April 1992)

5) C. Y. Wen and Y. H. Yu, "A Generalized Method for Predicting the Minimum Fluidization Velocity", A. I. Ch. E. Journal, 12(1966), 610-612.

6) K. Takahashi, et al., "Post Combustion Behavior in In-bath Type Smelting Reduction Furnace", ISIJ Intern.,32 (1992), 102-110.

7) T. Ariyama, et al., "Operation and Analysis of Pre-reduction Fluidized Bed for Smelting Reduction (Reduction of Iron Ore in a Fluidized Bed III)", CAMP-ISIJ, 2(1989), 113.

8) S. Matubara et al., "Degradation and Classification Behavior of Iron Ore in Bubbling Fluidized Bed (Reduction of Iron Ore in a Fluidized Bed VI)", CAMP-ISIJ, 4(1991), 1162.

9) S. Matubara et al., "Reduction Characteristics of Iron Ore in Bubbling Fluidizing Bed (Reduction of Iron Ore in a Fluidized Bed VIII)", CAMP-ISIJ, 5(1992), 175.

10)S. Matubara et al., "Development of Tapping Technique for Smelting Reduction Process ", CAMP-ISIJ, 5(1992), 179.

11)M. Kawakami et al., "Continuous Intrgrated Operation of Fukuyama Test Plant for Smelting Reduction", CAMP-ISIJ, 4(1991), 45.

12)M. Muroya et al., "Relationship between Post Combustion and Carbon Combustion of Coal in Iron Bath Based Smelting Reduction Furnace", CAMP - ISIJ, 5(1992), 178.

622

Adaptation of Injection Technology for the

HIsmelt™ Process

G.J. Hardie[1], I.F. Taylor[2], J.M. Ganser[1], J.K. Wright[2], M.P. Davis[3], C.W. Boon[4]

1 : HIsmelt Corporation, Kwinana, Western Australia
2 : CSIRO, Clayton, Victoria, Australia
3 : CRA ATD, Perth, Western Australia (Formerly 4)
4 : CHAM Ltd., London, United Kingdom.

ABSTRACT

Top and bottom injection techniques play a critical role in the creation of process conditions that facilitate efficient, direct smelting of iron ore in a molten bath reactor. Conventional injection techniques, originally trialed on a K-OBM converter and a small scale direct smelting pilot plant, did not achieve the desired result in terms of process performance and suitability of the engineering.

In the OBM steelmaking application of the Savard-Lee tuyere, the formation of a mushroom or knurdle of frozen metal and slag at the tip of the tuyere plays a critical role in prolonging tuyere and refractory life. Adaptation of bottom injection technology for high solids and gas injection rates, utilised in the HIsmelt Process, also requires the formation of stable mushrooms. The morphology of these mushrooms are quite different to those observed in bottom blown steelmaking vessels and can have a significant influence on process operating performance and reactor design. The difference between steelmaking and direct smelting mushrooms is discussed in the context of the different injection and bath conditions.

Development of the top injection technology involved a combination of pilot plant trials, laboratory tests and extensive mathematical modelling. Identification of a suitably balanced reaction and transfer of heat between the hot post combustion jet and the bath was achieved through successive improvements to the top tuyere design and changes in bottom injection operating practices.

Proceedings of the
Savard/Lee International Symposium on Bath Smelting
Edited by J. K. Brimacombe, P. J. Mackey,
G. J. W. Kor, C. Bickert and M. G. Ranade
The Minerals, Metals & Materials Society, 1992

INTRODUCTION

A historical perspective of the HIsmelt Process development clearly demonstrates that the technology utilised to achieve efficient direct smelting of iron ore has evolved from components of the technology developed for conventional oxygen steelmaking and blast furnace ironmaking[1,2]. In almost all cases, the process and engineering challenges confronted in the development of new process technologies have required successive adaptation of the existing engineering and operating practices. Successful developments have consistently involved determining how new processes function and identifying the key features which provide a differential advantage. Adjusting the engineering and operating practices to take advantage of the desirable features is perhaps the most challenging and costly step.

Throughout the HIsmelt development, efforts have focused on utilising the beneficial features associated with top and bottom injection techniques utilised in steelmaking and coal gasification processes. This approach has resulted in a new direct smelting technology which is unique amongst the emerging processes. In this paper, the key attributes of the smelting process are explained in terms of the injection technologies devised for the HIsmelt $\underline{S}$melt $\underline{R}$eduction $\underline{V}$essel (SRV) and the approach utilised to modify the process technology to achieve the desired operating conditions.

THE HIsmelt PROCESS

The incentive to pursue a simple direct smelting process, capable of treating a wide range of raw materials with limited pretreatment steps and commercially viable on a substantially smaller scale than modern blast furnace installations, has been the driving force behind the HIsmelt development program[1-7]. The need to continually adapt the process technology and engineering in order to realise this vision was recognised very early in the development program. Early work with a direct smelting process, based on the use of an K-OBM converter as the reaction vessel, has been progressively taken through a series of improvement stages. These have resulted in modifications to the vessel geometry and injection technology and have permitted the the achievement of significantly higher levels of post combustion and heat transfer efficiency than was found to be achievable during the initial testing phase.

As a result of an extensive research program, conducted over a ten year period from late 1981, the HIsmelt Process has evolved into the high intensity direct smelting process described in Figure 1[1-7]. To reach this point, inputs have been required on many fronts and from many research scientists/metallurgists/engineers. Particular attention has been paid to the operation of a $\underline{S}$mall $\underline{S}$cale $\underline{P}$ilot $\underline{P}$lant (SSPP), which has been backed up by extensive laboratory investigations and an ongoing commitment to develop process mathematical models.

The next phase of the development involves the construction and operation of a 100,000 t/y plant in Kwinana, Western Australia. The $\underline{H}$Ismelt $\underline{R}$esearch and $\underline{D}$evelopment $\underline{F}$acility, (HRDF) portrayed in Figure 2, is seen as an interim step to investigate scale-up of the process to a commercial size plant with an annual capacity of approximately 500,000 t/y hot metal.

Key Features of the HIsmelt Direct Smelting Technology

Key features of bath smelting processes are the vessel design (shape) and the injection technologies utilised. These attributes determine the way smelting processes will function and what performance characteristics can be achieved. The pivotal impact injection technology has on almost all aspects of the process design and operation can be seen from the emphasis placed on developing this component of the smelting technology, as illustrated in Figure 3 and outlined in this paper.

624

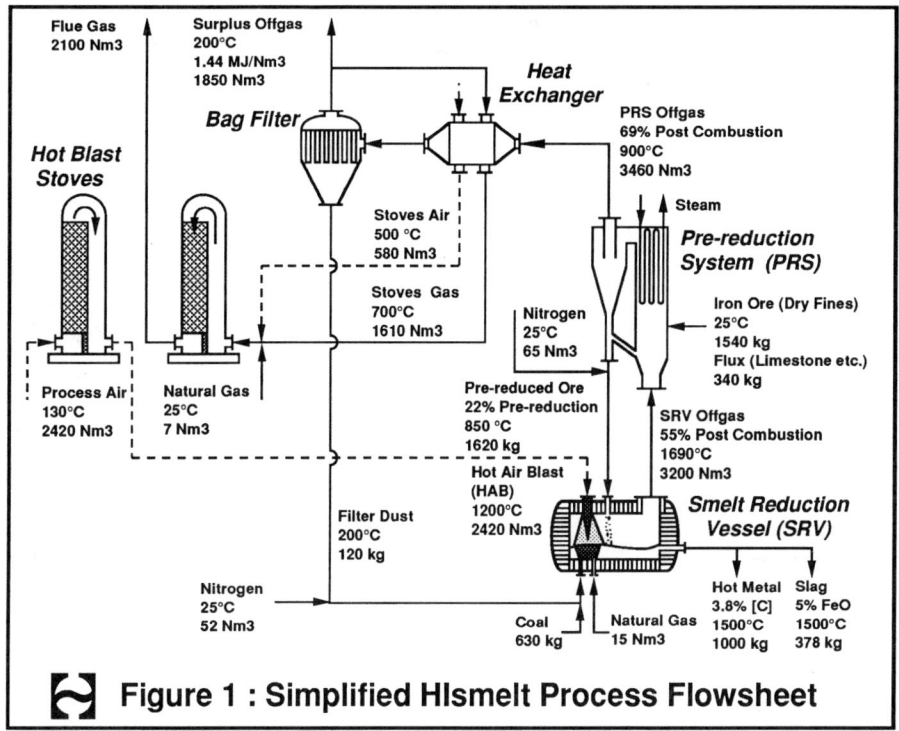

Ⓗ Figure 1 : Simplified HIsmelt Process Flowsheet

HIsmelt, in common with other iron bath based smelting, gasification and allothermic steelmaking technologies, shares the following desirable features associated with top and bottom injection technology :-

1) High Carbon Recoveries to the Iron Bath

The highest recovery of carbon to the iron bath can be achieved by bottom coal injection[4]. Submerged injection into a smelt reduction vessel not only permits recovery of the fixed carbon content of the coal, but a substantial proportion of the carbon released as volatile matter during the rapid pyrolysis experienced. Recent studies[8-10] have shown that the yield of carbon, as volatile matter, can increase between 10% and 30% when coals are rapidly pyrolysed at iron bath and slag bath temperatures when compared with normal proximate analysis procedures. Carbon recovery is an important characteristic, as carbon that does not dissolve in the bath can react with the offgas to reduce post combustion or cause excessive dust losses. Both problems result in significant reductions in fuel efficiency[4,5] or a reduction in process intensity.

Submerged injection is complementary to the other processes occurring in the SRV in that the high injection rates utilised also release large volumes of hydrogen and carbon monoxide which produce substantial bath turbulence. Turbulence generates increased metal and slag surface areas which affords high heat transfer efficiency and maintains low slag FeO levels through intimate mixing with the carbon containing hot metal. Containment of the intense bath turbulence has required the use of a horizontal cylindrical vessel similar to that depicted in Figure 3.

625

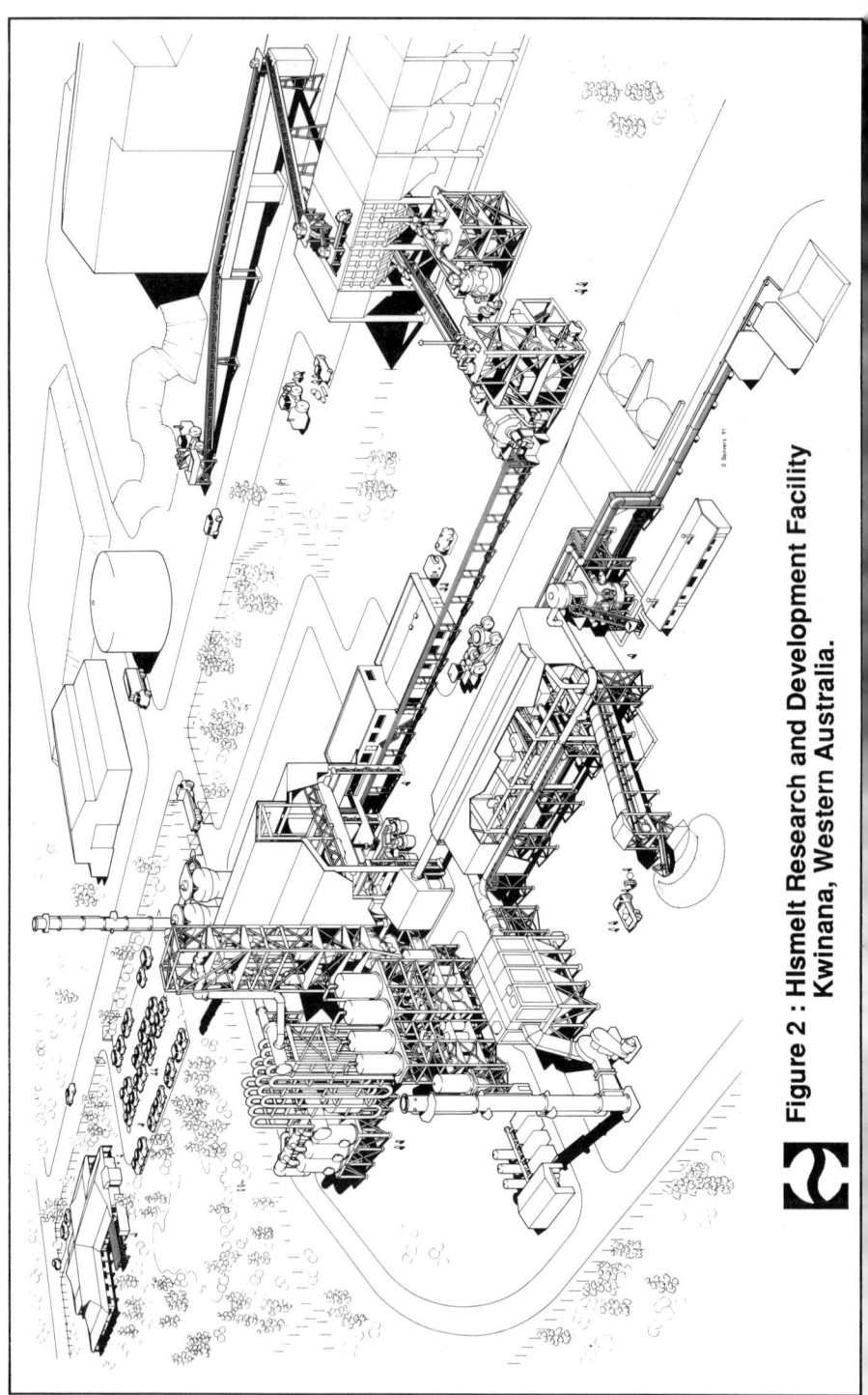

Figure 2 : HIsmelt Research and Development Facility Kwinana, Western Australia.

626

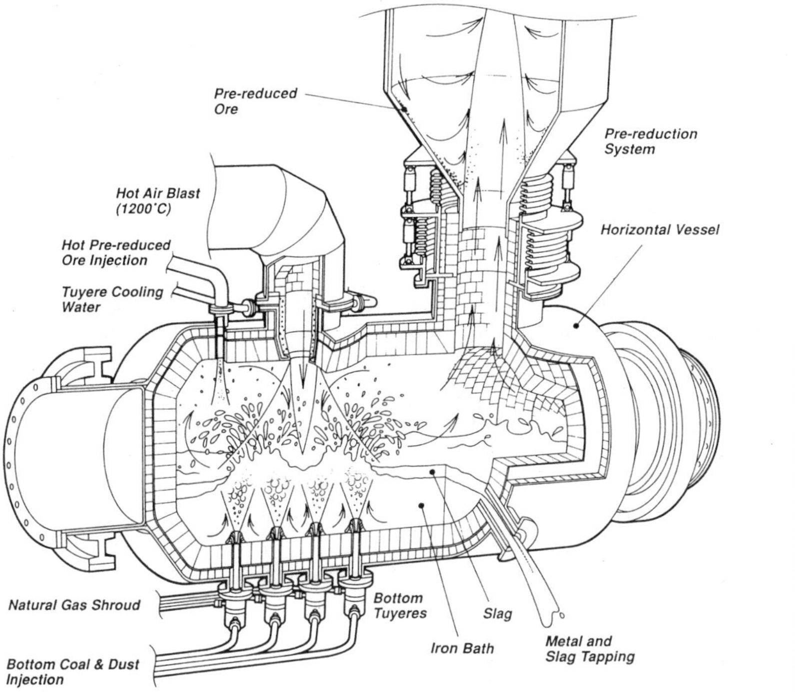
Pre-reduced Ore
Pre-reduction System
Hot Air Blast (1200°C)
Hot Pre-reduced Ore Injection
Horizontal Vessel
Tuyere Cooling Water
Natural Gas Shroud
Bottom Tuyeres
Slag
Iron Bath
Metal and Slag Tapping
Bottom Coal & Dust Injection

Figure 3 : Conceptual HIsmelt Smelt Reduction Vessel

2) Rapid Smelting in an Iron Bath

Smelting of the iron ore directly with a carbon containing iron bath affords the highest possible reaction rates and therefore high process intensity[11]. Submerged injection of iron ore, or top injection directly into a highly turbulent bath, leads to rapid reaction with the metal and generates additional bath turbulence due to the carbon monoxide released during reduction. Direct reaction with the iron bath minimises the slag FeO content as it avoids the steps of firstly dissolving the ore in the slag and subsequently reducing the FeO bearing slag with charred coal particles and hot metal droplets[12-14]. The intensity of iron bath based direct smelting processes is not governed by the available reduction sites and the concentration of FeO in the slag.

3) Intense Top Space Reaction Zone

The high bath turbulence generated by the intense bottom injection processes creates an ideal environment for high heat transfer above the bath surface. Droplets ejected into the top-space, in a fountain like fashion, rapidly remove heat from the hot post combustion jet and return this sensible heat to the bath.

The remainder of this paper examines some of the results and developments associated with top and bottom injection technology that have successfully utilised the features listed above and permitted the simultaneous achievement of post combustion levels in the SRV in excess of 60% and heat transfer efficiencies greater than 85%.

BOTTOM INJECTION TECHNOLOGY

One of the key technologies concerned with bottom blown, iron bath based processes was the successful application of the shrouded tuyere. The classical example of this type of technical development was the use of the Savard-Lee, shrouded, annular, tuyere in the development of the OBM steelmaking process[2]. This technique permitted pure oxygen to be bottom injected into a steelmaking converter, without burn-back or excessive refractory wear, by providing supplementary cooling at the tuyere exit. Endothermic cracking of hydrocarbons (natural gas, LPG etc.) injected through the outer annulus of the tuyere is the chemical cooling mechanism responsible for the formation of the protective "mushroom", or "knurdle".

Mushrooms are formed as a result of the cooling effect occurring in the vicinity of the bottom tuyere exit, freezing some of the liquid products in the bath to the refractory lining of the vessel. Control of the interface between the injection tuyere and the bath is crucial to maintain acceptable bottom refractory wear rates. As with bottom oxygen steelmaking, high bottom solids injection processes can experience reduced refractory life if the condition of the mushroom is not controlled. A number of ferrous and non-ferrous processes such as OBM (Q-BOP), KMS, KS, MIP, QSL etc. have successfully adapted the original concept to meet the challenges presented by the unique conditions confronting each process. A similar technical development was required to establish the viability of the HIsmelt Process. A typical example of a solids injection mushroom can be seen in Figure 4.

The Influence of the Local Environment on Mushroom Shape

The size and shape of the mushroom is a consequence of its thermal history. Heat is lost to endothermic processes such as the cracking of shroud gas or the pyrolysis of injected coal. A mushroom gains heat from direct contact with the bath and from exothermic reactions if they are occurring. If, at a given instant, the heat loss is greater than the gain, the mushroom will grow by the accumulation of frozen bath liquids. This process is reversible, if heat gain exceeds loss then frozen materials will be lost from the mushroom through melting. It is a plausible conjecture that in an otherwise static situation, the competing processes of accumulation and loss of material are self-regulating.

A simplified heat balance model suggests that the ultimate size of a mushroom is an outcome of dynamic equilibrium, but provides little insight into the shape which the accumulated material is likely to form. Because of the difficulty of studying mushroom growth in liquid metal, investigators have resorted to cold models. The typical experimental approach is to inject nitrogen, chilled well below 0°C, into water and observe the growth of the ice accretion which forms around the point of injection. In a series of experiments reported by Boxall et al[15], the ice accretion begins as an extension of the tubular geometry of the injection nozzle. Cracks may form in the ice and the escaping chilled gas will then form side branches. Eventually the initially simple tubular geometry is lost and the shape of the accretion may resemble a typical steelmaking mushroom. In similar experiments performed at the CSIRO Division of Mineral and Process Engineering, the same pattern of mushroom growth was observed. The actual pattern followed showed a large variation from experiment to experiment, but some general trends could be discerned :-

- Initial growth was tubular
- As the accretion grew, cracked and broke, the shape tended towards the typical flat gaussian or fried egg cross-section.

The rate at which a mushroom advances from its initial to mature shape is determined by the environment in which it grows. The transition from tubular to gaussian shape is a result of successive breakage and re-growth events. In a relatively peaceful environment conditions favour mushroom stability and tubular structures may be preserved for longer periods. A mushroom is subject to thermal and mechanical shock due to phenomena such as back

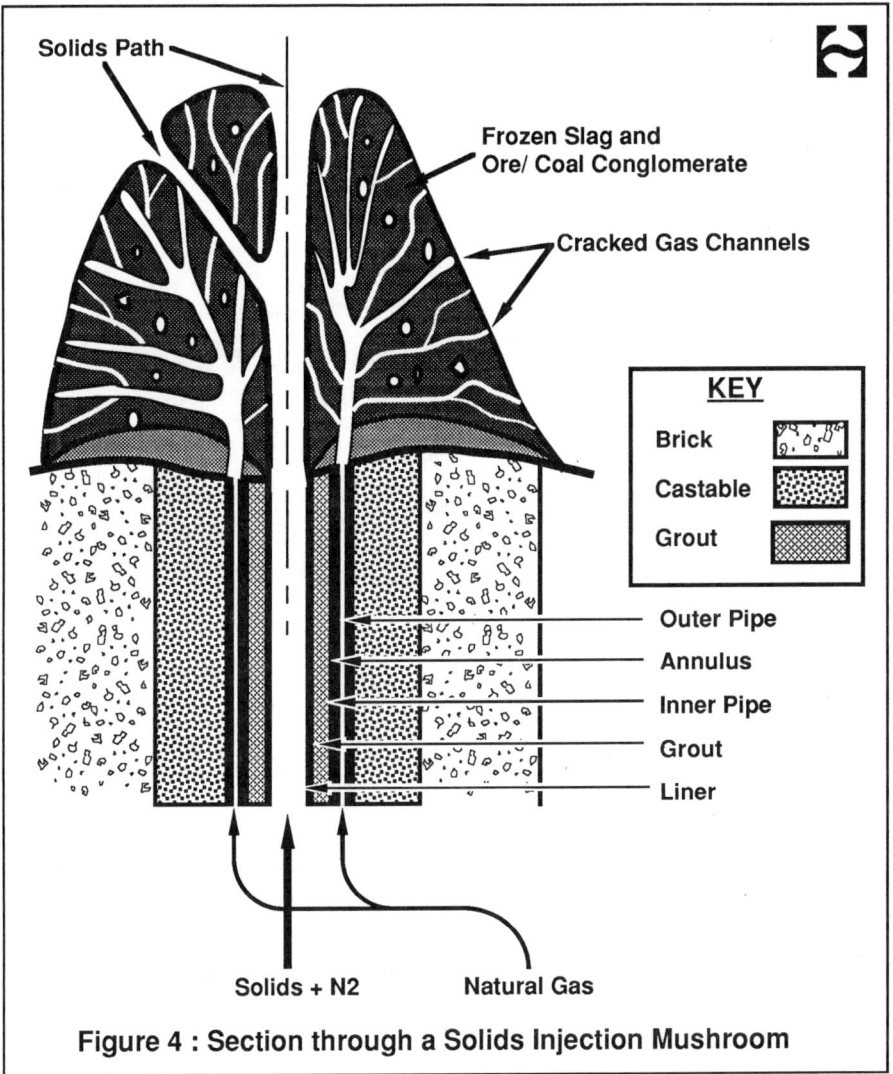

Figure 4 : Section through a Solids Injection Mushroom

attack[16], short term thermal fluctuations in the vicinity of the point of injection[17] and longer term thermal disturbances, due to process events such as steel converter turn downs[18].

<u>Mushroom Thermal Conditions.</u>   Prior to the commissioning of the SSPP, thermal conditions near steelmaking tuyeres were investigated to give an indication of what could be expected in a smelt reduction vessel. A series of experiments was undertaken by CSIRO, in conjunction with Klöckner Stahlforschung, to measure the temperature distribution within the refractory of a steelmaking converter. In addition to the data from the refractory temperature profiles, these experiments yielded much more valuable information when it was discovered that thermocouples continued to give meaningful readings after they had become frozen into mushroom metal. Figure 5 shows the configuration of a thermocouple embedded in a mushroom. A number of thermocouples were installed close

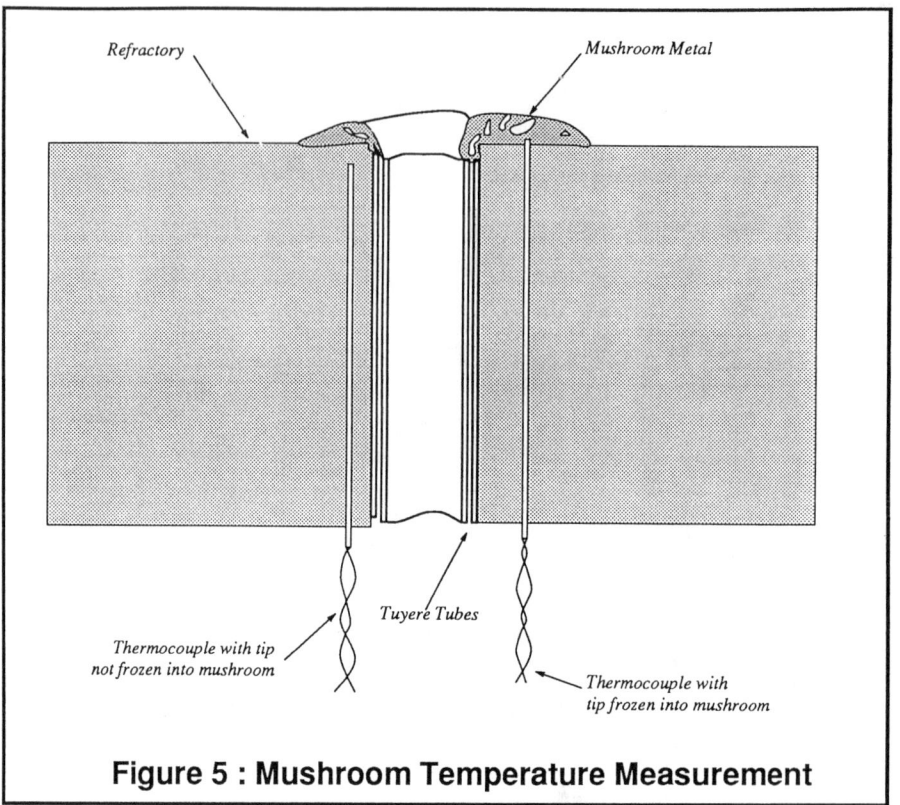

Figure 5 : Mushroom Temperature Measurement

to the tuyere tubes before the converter bottom was put into service. The tips of the thermocouples were placed at different depths relative to the top of the new refractory lining. As the refractory wore down in service, the thermocouples indicated the cyclic temperature pattern within the refractory due to the multi-phase, batch nature of the process. The top thermocouple is eventually exposed to the bath when the overlying mushroom breaks off. When the mushroom reforms, the thermocouple circuit is closed by a weld formed by mushroom metal. The thermocouple continues to function except for brief intervals when the mushroom is lost and reforms. Eventually several thermocouples become welded into the mushroom as the vessel bottom wears down. The incorporation of each successive thermocouple into the mushroom is signalled by the fact that its indications begin to track those of its predecessors.

Figure 6 shows the wide and rapid changes in mushroom temperature and the difference in mushroom temperature between a scrap melting phase, where lower bath temperatures and coal injection are employed, followed by a steelmaking phase involving bottom oxygen injection. The result of an interruption to shroud gas flow, under otherwise normal plant turned down conditions, can be seen after tapping, which commenced at approximately 17.9 hours. During the idle period the shroud gas supply was turned off and the mushroom temperature commenced to rise exponentially towards the temperature of the surrounding refractory. Shortly after, the measured temperatures rose through 1000°C and before equilibrium with the refractory was reached, the shroud gas was turned back on. The measured mushroom temperatures fell rapidly and the difference between the two temperatures was restored.

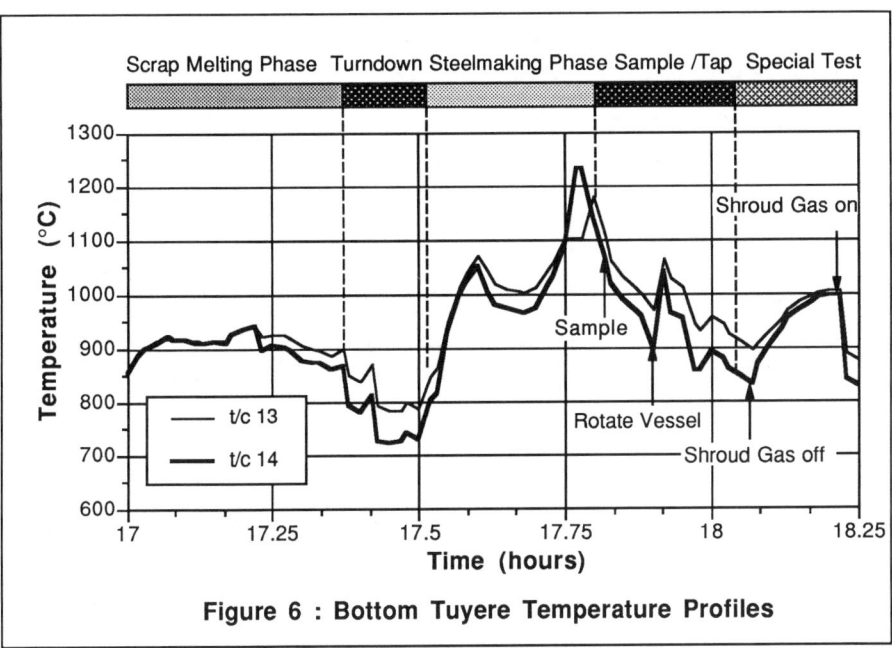

Figure 6 : Bottom Tuyere Temperature Profiles

HIsmelt Mushrooms

Although the initial 60t and SSPP trials were conducted with bottom oxygen injection, an analysis of process stoichiometry[4,5] and plant results quickly indicated that bottom oxygen injection was having a detrimental effect on the level of post combustion that could be achieved and was not necessary to generate sufficient bath turbulence to facilitate high heat transfer efficiency. Calculations, trials and mathematical modelling indicated that the gases used to inject coal, ore and lime, coupled with the gases liberated by coal pyrolysis and ore reduction, were sufficient to release enough buoyancy energy to generate the intense mixing of the bath and agitation of the surface required for high heat transfer efficiency. Operation of a completely endothermic bottom injection process had not been attempted previously. Special bottom injection practices were developed to cope with these new conditions. The essential differences between K-OBM steelmaking and HIsmelt bottom injection conditions are shown in Table I.

The substantial differences with the HIsmelt Process result from:-

- Lower bath temperatures
- Higher carbon content hot metal
- Different slag chemistry and melting point
- Higher slag quantities
- The absence of bottom oxygen injection
- Much higher solids injection rates
- The steady conditions encountered in direct smelting.

As the reactions occurring below the bath surface are all endothermic, burn-back of the tuyere is rarely observed. The mushroom is primarily required to provide protection against back attack and physical erosion problems associated with the high bottom solids injection rates utilised. The tuyere itself must be protected from the abrasive solids injected at very high velocities. Special durable lining techniques have been developed to cope with the highly

631

abrasive burnt lime, dust and iron ore injected through the bottom tuyeres. Figure 4 shows the use of a special ceramic lining technique.

Table I: Comparison of K-OBM and HIsmelt Bottom Injection Conditions

PROCESS	Bottom Blown Oxygen Steelmaking (K-OBM)			Direct Smelting Ironmaking (HIsmelt)		
PARAMETER	Max	Min	Steady	Max	Min	Steady
Metal Temp	1750	1200	NO	1550	1400	YES
Metal Composition [C]	4.8	0.0	NO	4.5	2.5*	YES
Slag Basicity (CaO/SiO$_2$)	3.5	2.0	NO	1.6	1.2	YES
Slag Quantity (kg/thm)	250	40	YES	***	200***	NO
Bottom O$_2$ (%)	100	30	YES	---	0	YES
Solids Inject. (kg/thm)	150	20	YES	***	700**	YES
Carrier Gas (Nm^3/t solids)	70	---	YES	70	40	YES
Shroud Gas (Type)	CH4, 10% of bottom O$_2$			Various gas mixtures eg. of CH$_4$ or N$_2$		

*   Lower carbon levels are possible
**   Not all materials are necessarily bottom injected
*** Depend on the gangue content of the raw materials.

Large mushrooms also present operational difficulties as the solids injection tuyeres exhibit higher pressure drop and block more readily. Long mushrooms can reduce the effective bath depth which results in reduced capture of injected materials (blowthrough), poor coal dissolution characteristics and reduced bath turbulence.

The morphology of the mushrooms are diverse as the temperature cycling and liquids present are also quite dissimilar for steelmaking and HIsmelt conditions. Steelmaking mushrooms are usually formed from frozen steel (low carbon, high melting point) and frozen high basicity slag (also high melting point), as these are the two liquids present at the end of the blow and most likely to remain solid during the high temperature phase of the process. High carbon content metal is only found in the mushroom where local carburisation occurs as a result of the shroud gas cracking and depositing carbon.

The mushrooms formed under HIsmelt conditions depend largely on the materials injected, as the mushroom is normally a conglomerate of the solids injected and the slag phase present. Frozen iron makes up an insignificant portion of the mushroom because the high carbon content results in a much lower freezing point than the slag phase. Sufficient cooling to freeze the iron in the vicinity of the tuyere would result in extremely large and problematic slag mushrooms. Slag chemistry is very similar to the blast furnace (except for the slightly higher FeO content) and is usually controlled to produce a non-aggressive, low temperature fluid slag suitable for lower bath temperature operation and easy tapping.

Close control of the cooling effect is required to maintain stable mushrooms under HIsmelt conditions. This is more critical if lower bath temperatures are utilised as the slag approaches freezing point and the cooling effect from coal cracking can cause long mushrooms. Injection of high volatile coals usually generates sufficient cooling to produce stable mushrooms without supplementary cooling, whilst lower volatile coals sometimes require a natural gas shroud. A diagram showing the morphology and shape of some typically observed mushrooms can be seen in Figure 7. Long tubular mushrooms, usually seen with high volatile coal injection, are often referred to as "Elephants Trunks" and can be controlled by changing the shroud gas composition by replacing natural gas with nitrogen or recycled process offgas.

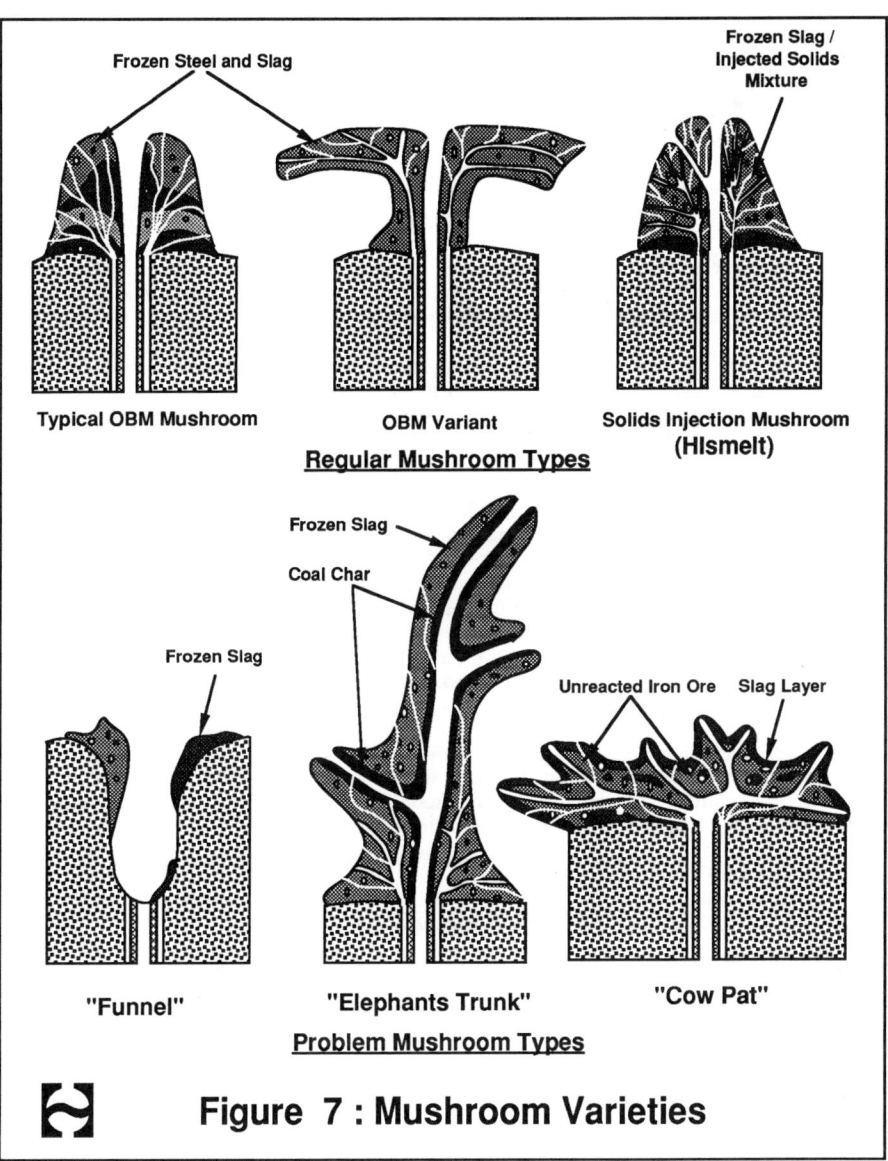

Frozen Steel and Slag

Frozen Slag / Injected Solids Mixture

Typical OBM Mushroom

OBM Variant

Regular Mushroom Types

Solids Injection Mushroom
(HIsmelt)

Frozen Slag

Coal Char

Frozen Slag

Unreacted Iron Ore   Slag Layer

"Funnel"

"Elephants Trunk"

"Cow Pat"

Problem Mushroom Types

Figure 7 : Mushroom Varieties

Ore and lime injection can also result in different mushroom shapes which may cause high pressure drop, blockages or blowthrough. Iron ore fines and burnt lime have relatively high melting points and can form large, flat, stable mushrooms when bound with a small quantity of slag. This type of mushroom is often known as a "Cow Pat" and can be controlled by injecting the ore with oxygen, which has a negative influence on the overall process fuel efficiency, or introducing some reductant into the shroud gas, which tends to form lower melting point oxides. Alternatively, coal and ore can be injected together because they react with each other at elevated temperatures and therefore avoid a localised concentration of one material which might have high temperature stability.

SSPP operation has shown that utilising high bottom injection of both gases and solids requires close control of the performance characteristics of the bottom tuyeres. For stable operation, the condition of the mushroom is critical to avoid :-

- Tuyere burn back
- Excessive refractory wear
- Metal run back
- Blockages
- Incomplete reaction of the injected materials (eg. blowthrough).

Mushrooms were observed to be a complex conglomerate of frozen liquid phases and unreacted solids. The morphology is a complicated matter, however, examination of numerous examples indicate the structure is determined by :-

- The type of materials being injected (solids, liquids and gases)
- Composition of the liquid phases in the bath
- Solidification temperature of the liquid phases in the bath
- Reactions between the injected materials and the mushroom
- Reactions between the injected materials and the bath.

The size and shape of the mushroom should be controlled to maximise the refractory lining life and the reaction of the injected materials with the bath. This can be accomplished by :-

- Implementing special tuyere designs
- Controlling the supplementary cooling (eg. shroud gas)
- Changing the injection conditions (eg pressure drop)
- Modifying the physical and chemical properties of the mushroom
- Modifying the bath condition (depth, temperature, composition, flow patterns, etc.)

Control of the bottom injection is critical to achieve process stability and low refractory wear. The process conditions and bottom injection techniques must be adapted to suit the raw materials and products. A sophisticated mushroom control technique was developed on the SSPP and will be further improved and automated in the HRDF.

TOP INJECTION TECHNOLOGY

Oxygen Steelmaking

Oxygen steelmaking processes employing combined blowing techniques have achieved post combustion levels up to about 35% with specially designed, height adjustable post combustion lances[19]. Usual steel plant operation involves more moderate post combustion levels in the 10% - 25% range due to the excessive converter hood refractory wear experienced at high post combustion levels.

Steelmaking processes are constrained from achieving significantly higher post combustion levels due to the nature of the steelmaking process and the design of the steelmaking converter. High post combustion levels are only practical when the heat liberated is returned to the iron bath. The consequence of low heat transfer efficiency, for oxygen based processes, is extreme offgas temperatures and accompanying refractory wear problems. High heat transfer efficiency can only be achieved when there is sufficient contact between the post combustion jet and the bath. Even for moderate post combustion levels, high bath turbulence is required to facilitate efficient heat transfer. This normally involves injecting at least 30%, and normally 50%, of the process oxygen through the bottom tuyeres. The carbon monoxide evolved from the bottom injection processes limits the level of post combustion that may otherwise be achieved[4,5].

There is little flexibility to change this situation, as additional bath turbulence is required to facilitate higher heat transfer efficiency. Supplementary inert gas injection does not offer a practical solution, as very large volumes would be required to make a significant change to an already turbulent system. The gains made by improved heat transfer are partially lost by the additional cooling of the bath by cold gas injection. Nitrogen injection is undesirable at the end of a steelmaking blow and would have to be substituted with oxygen or expensive argon. Additional bottom oxygen injection may be employed, but this has a strong adverse impact on the post combustion stoichiometry. Additional turbulence will also lead to unacceptable metal losses due to the ejection of liquid products into the offgas system. This was one of the principal results from the 60t K-OBM smelting reduction trials undertaken at Maxhütte in 1983-84 and led to the design of a horizontal cylindrical SRV used in the SSPP.

The second factor limiting the potential post combustion levels in conventional combined blowing converters is the back reaction of post combusted offgases with the metal and slag droplets in the top space. Plant results have indicated that by raising the centre lance[19,20], or adopting special multi-hole post combustion lances[21,22], improved thermal efficiency (increased scrap rates) can be achieved. The softer jets produced by long jet to bath distances, or special multi-hole post combustion lances, reduce the oxidising gas concentrations, gas velocities and temperatures close to the bath surface, thus reducing back reaction. Longer jets have the disadvantage of generating extreme temperatures at a great distance from the bath and thus provide limited opportunity for high heat transfer efficiency.

Investigation of this phenomena was undertaken during the early stages of the HIsmelt development[7,23]. Mathematical models of the processes occurring in the top space of a steel converter were used to identify and understand the various competing mechanisms. The results shown in Figure 8 for a multi-hole oxygen lance, with similar flow characteristics to those used in the SSPP indicate three important factors. Firstly, the complete reaction of the oxygen is being inhibited by the flame reaching disassociation temperature limitations as the jet is not releasing heat quickly enough to avoid temperatures in excess of 2700°C. Secondly, the individual divergent jets rapidly recombine to produce a single jet which exhibits similar properties to the well known free jet. Thirdly, the influence of combustion on free jet mechanics also tends to increase the effective length and reduce the entrainment and spreading of the flame. These factors lead to unreacted oxygen penetrating great distances into the converter top space and contacting the metal droplets present. As previously discussed[4,5,7], this will result in accelerated back reaction and thus lower post combustion levels. The concentration and residence time of other oxidising species (CO_2 and H_2O) is also high with oxygen based operation and will lead to increased back reaction and low post combustion levels.

Small Scale Pilot Plant Concept

Having completed the initial 60t converter smelting reduction trials at Maxhütte in 1984, it was apparent that changes would be required to achieve the aims of developing a simple, high intensity, direct smelting process. Results of these trials indicated that the SRV process development would need to concentrate on four principal areas to achieve simultaneous high post combustion levels and high heat transfer efficiencies :-

- Process containment
- Increasing effective bath surface area (high HTE)
- Improving the post combustion stoichiometry
- Reducing the back reaction without sacrificing HTE.

A 10 tonne bath size pilot plant was chosen to permit experimentation at a reasonable scale and allow rapid modification to accommodate the many new concepts developed during the initial testing phase. A horizontal cylindrical vessel was chosen as the reactor design to accommodate testing with substantially higher bottom injection rates and achieve acceptable containment of the reaction and bath products. Considerable flexibility

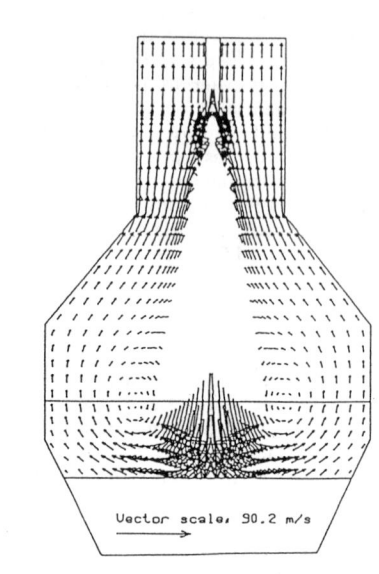

Oxygen Jet Velocity Vectors

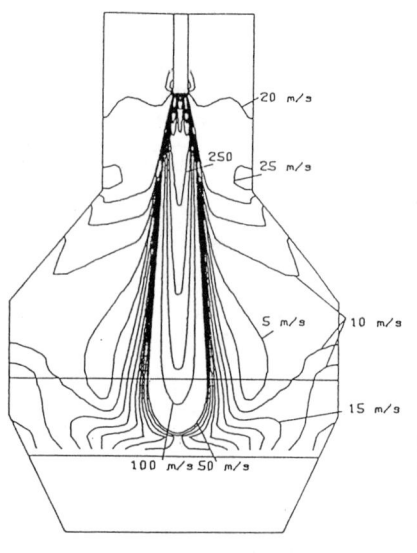

Contours of Jet Velocity

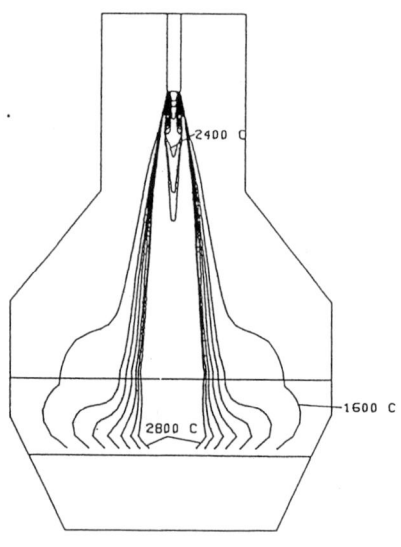

Contours of Gas Temperature

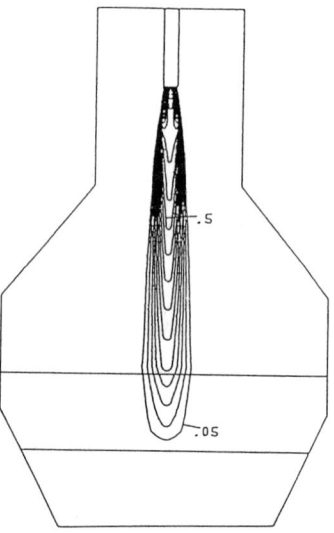

Contours of Oxygen Concentration

Figure 8 : Combined Blowing Oxygen
Steelmaking Simulation

CHAM

was included to accommodate a range of top and bottom injection tuyeres and locations. The vessel was designed to investigate both oxygen and hot air based operation.

Small Scale Pilot Plant Developments

Over a six year operating period an extremely large number of smelting reduction tests were completed, using a wide variety of raw materials, vessel configurations, tuyere types, operating pressures etc[3-7]. The following discussion summarises some the developments made to the injection technology and operating practices as a result of pilot plant trials, mathematical modelling and laboratory scale testing. The results discussed in the following section refer to steady state (stable bath temperature and carbon content) smelting reduction tests, where coal consumption and heat transfer efficiency can be reliably measured.

The pilot plant was commissioned using oxygen based operation and rapidly achieved a number of the initial design goals :-

- Acceptable yield achieved
- Steady state operation possible
- Practical operation demonstrated (tapping, sampling etc.)
- PC in the 15% - 25% level
- HTE in the 80% - 90% level
- Novel engineering successful (vessel, rotary joints etc.).

At the same time, mathematical models of the processes occurring in the top space of steelmaking and HIsmelt vessels were being developed at CHAM Ltd. to aid in process understanding, data analysis and to assist in future scale-up[7,23]. The modelling results assisted in tuyere placement and indicated that the opportunities to achieve high post combustion levels with oxygen based operation would be limited.

Much larger diameter tuyeres are required to inject the high volumes of hot air. It was suggested that the potential core of an air jet, containing unreacted oxygen, would contact the bath and result in drastically lower post combustion. The top space mathematical model predicted the opposite trend, involving an increase in post combustion and equivalent HTE levels.

SSPP results confirmed the model predictions as an immediate step change in post combustion was achieved. Post combustion levels up to 40% were obtained with a new tuyere and up to 45% as the tuyere eroded. These high post combustion levels could not be sustained as the tuyere rapidly enlarged and caused refractory wall damage as the soft jet attached to the lining and overheated the bricks. A new HAB tuyere was adapted from a blast furnace tuyere design, involving a refractory lined, water cooled jacket construction, to overcome the accelerated wear problems. In the first application of the water cooled tuyere design (see Figure 9), multiple tuyeres were adopted to reduce the tuyere diameter, as the jet to bath distance was correspondingly lower with a straight tuyere configuration. This concept also permitted the testing of multiple injection zones, where the ore and coal injection could be separated to assess the effect of the injected materials on transition zone generation and process performance.

Tuyere engineering proved to be successful and reliable results were obtained from the SSPP. But at levels of 30% to 35% post combustion, process performance was not improved when compared with previous designs. Problems were identified as poor combustion characteristics, jet stripping into the offtake, low gas residence time distribution and increased liquid losses. The HAB tuyere was changed to a "Bent Tuyere" design to reproduce the results obtained with the refractory tuyere. Results showed an immediate improvement, resulting in reliable post combustion levels between 35% and 40% and substantially higher heat transfer efficiencies. A long campaign was completed with this tuyere, involving extended operation and a series of parametric tests to study both top and bottom injection

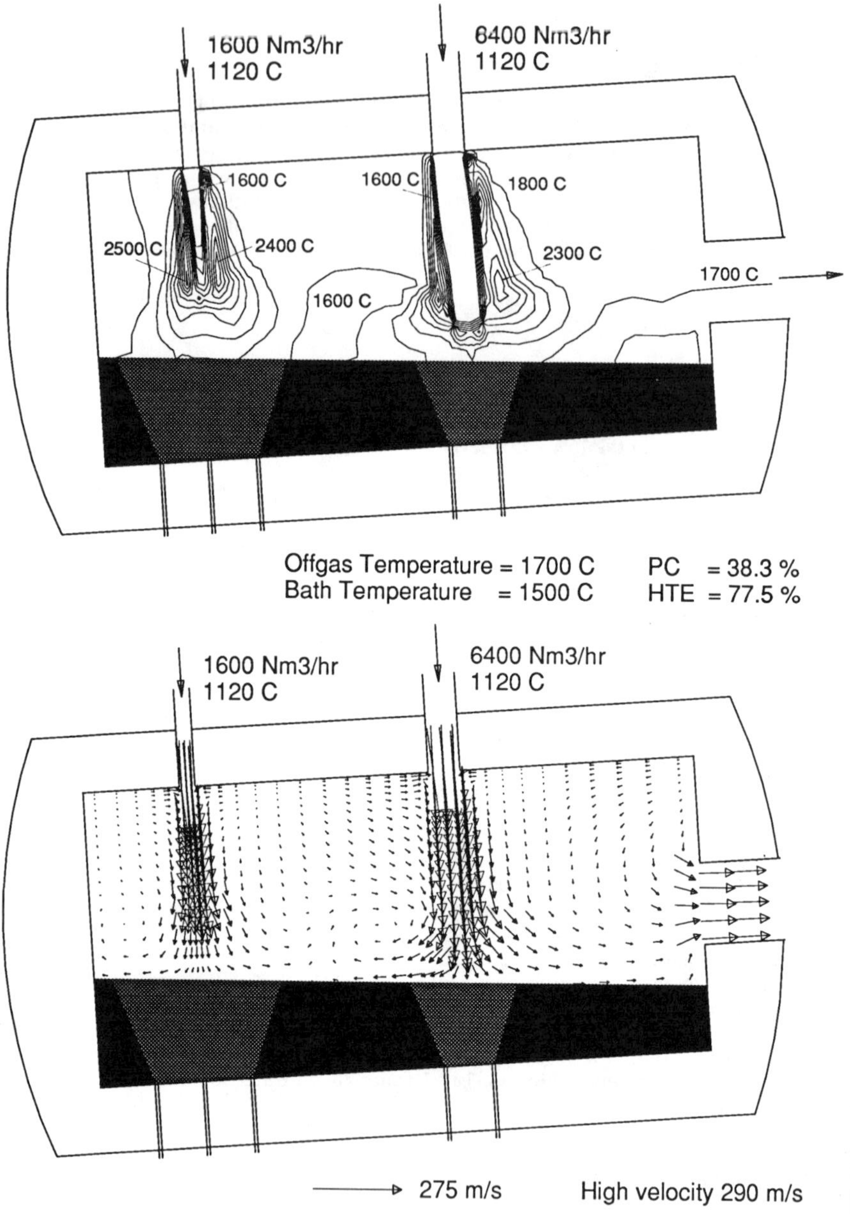

1600 Nm3/hr
1120 C

6400 Nm3/hr
1120 C

1600 C
1600 C
1800 C
2500 C
2400 C
2300 C
1700 C
1600 C

Offgas Temperature = 1700 C   PC   = 38.3 %
Bath Temperature   = 1500 C   HTE = 77.5 %

1600 Nm3/hr
1120 C

6400 Nm3/hr
1120 C

→ 275 m/s   High velocity 290 m/s

Figure 9 : SSPP Temperature Contours
and Velocity Vectors for Campaign 5

 CHAM

638

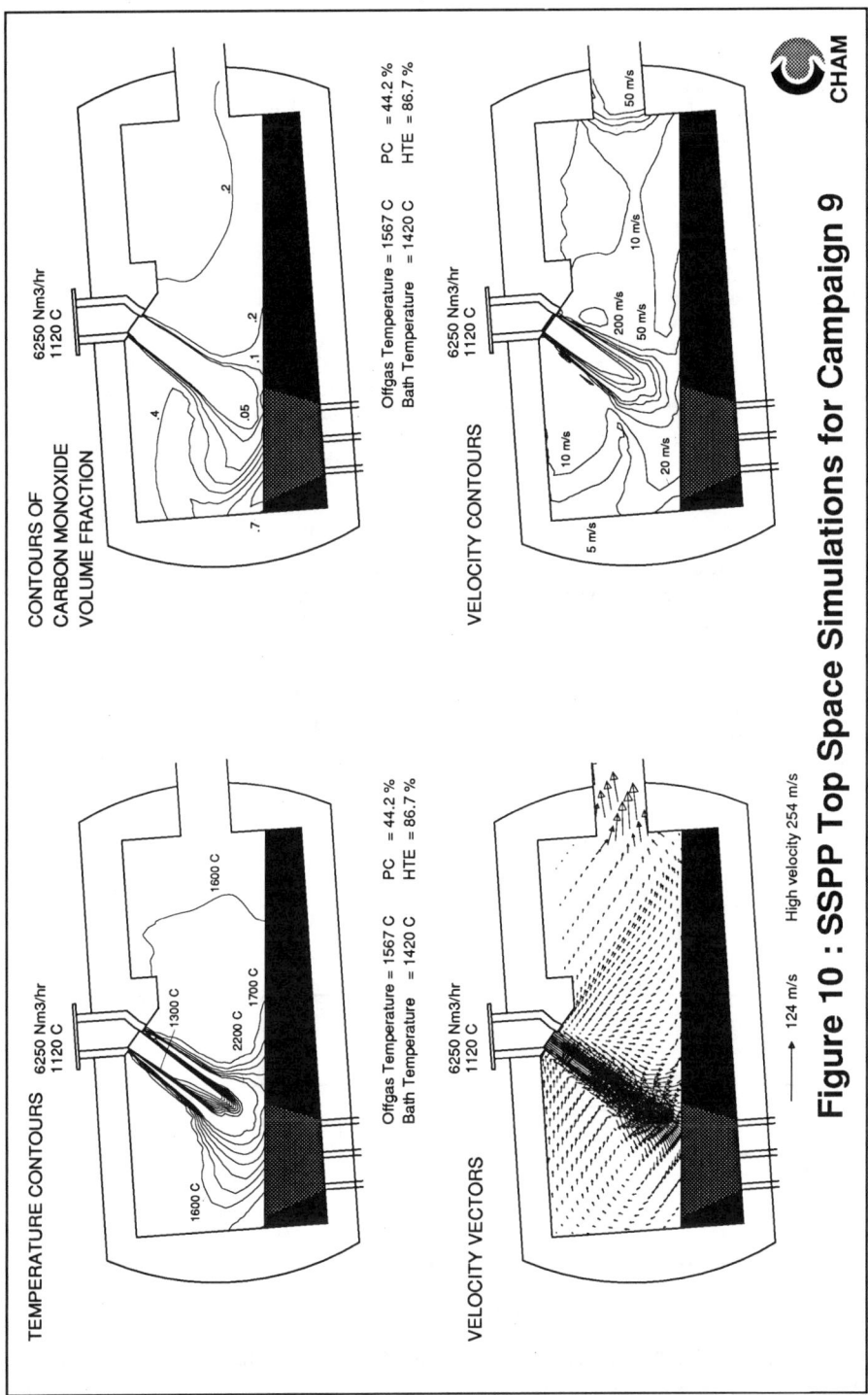

Figure 10 : SSPP Top Space Simulations for Campaign 9

phenomena[4,5]. Results of the simulation of these conditions can be seen in Figure 10. A larger diameter "Bent Tuyere" was tested to simulate the worn refractory tuyere that had previously produced higher post combustion levels. Lowering jet velocities by 30% was expected to reduce back reaction and permit higher post combustion levels. However, the consequence of a larger tuyere diameter is a longer unreacted potential core which results in additional oxygen penetrating further into the transition zone. Both plant results and modelling showed that the benefits of reduced jet velocity were cancelled by increased jet length. Only minor improvements in post combustion and a slight fall in heat transfer efficiency were obtained.

Scope for further improvement, utilising simple straight tuyeres and multiple tuyeres, involved major design changes and additional engineering and operational complexities for the SSPP. The following features were required to solve the problem of obtaining high post combustion and simultaneous high heat transfer efficiency :-

- Minimise oxygen penetration into the transition zone (high PC)
- Maximised gas mixing and jet entrainment (good combustion)
- Employ a short jet, close to the bath surface (high HTE)
- Reduced jet - droplet impingement velocities (high PC)
- Maximised contact between the combustion products and the transition zone (high residence time distribution = high HTE)
- Enhanced flame radiant heat transfer (high HTE)
- A degree of control over the above features.

An adjustable tuyere, in which the jet shape and "hardness" could be altered, was proposed to overcome the shortcomings of the original tuyere designs. The initial tuyere was designed to achieve control over the jet characteristics, so that a simple jet could be benchmarked against the previous satisfactory results obtained with the "Bent Tuyere".

Full top-space simulation also indicated a 10% to 20% increase in post combustion would result with only minor deterioration in the HTE. Full 3D models were used to assess the importance of different vessel geometries and tuyere configurations. The confidence gained in the model was sufficient for the new tuyere to be installed in the SSPP. The tuyere was commissioned and the results shown in Figure 11 were obtained. In these tests, the tuyere pressure drop and the jet axial velocity (hardness) was altered.

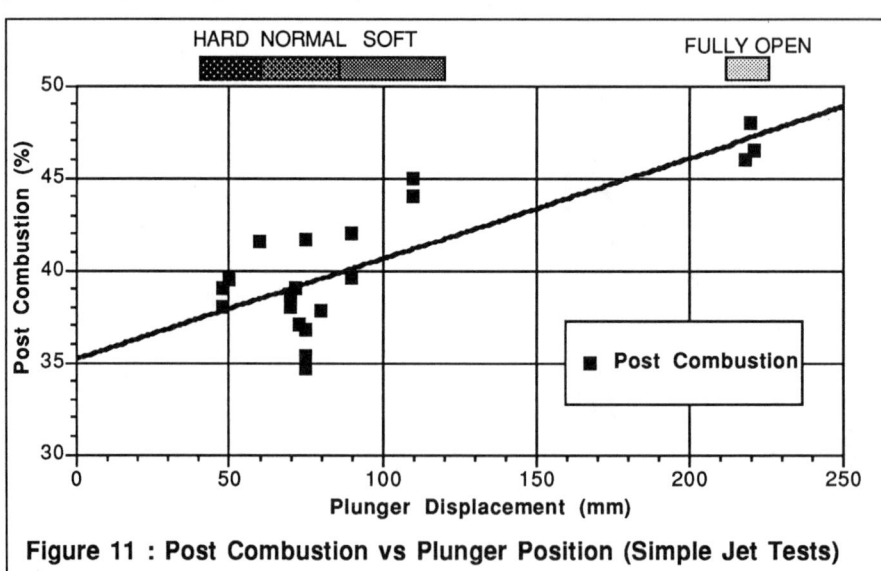

Figure 11 : Post Combustion vs Plunger Position (Simple Jet Tests)

The important result from this Figure is the small degree of control afforded by jet hardness. For these trials, jet velocity was altered by more than a factor of 3 and the post combustion levels only increased from 39% to 47%. The very soft jet could not be operated for extended periods due to instability of the post combustion (large fluctuations), which probably represents poor combustion characteristics. The normal operating range for jet hardness is indicated at the top of Figure 11.

Tests conducted to examine complex jet shapes involved normal reduction runs (where iron ore was smelted and the process was maintained at steady state) and some parametric tests to examine the influence of jet shape whilst all other conditions remained constant. The first experimental tuyeres permitted independent, on-line control over jet characteristics to facilitate rapid identification of optimum process conditions.

Parametric tests completed to study the influence of jet characteristics involved controlling top and bottom injection rates at the same level throughout the test (some difficulties were experienced maintaining all flowrates constant). The small fluctuations in HAB temperature and wind rate, as the pebble bed preheaters changed, during the test shown in Figure 12 is an example of the problems encountered. The influence of jet shape can be clearly seen as the test commenced with a simple jet and changed to more complex shapes when steady post combustion levels were obtained. To check for hysteresis, the jet was returned to the initial conditions the end of the test period. The small change in post combustion (~3%) can be attributed to small changes in the bath composition, temperature and HAB flowrate. This test

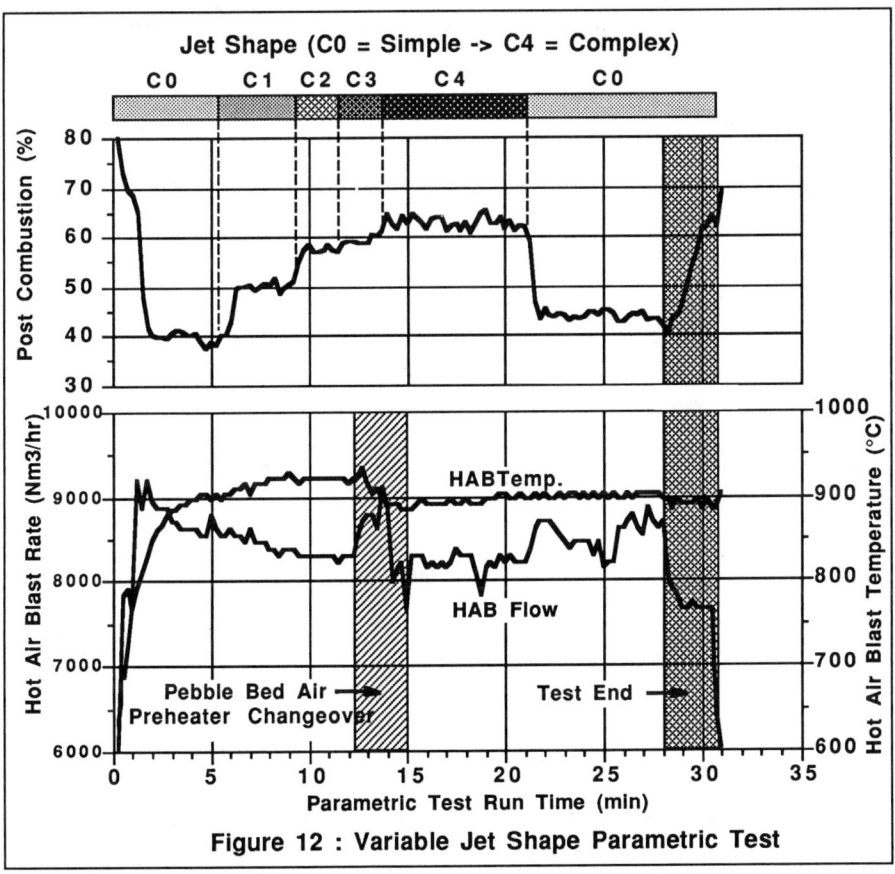

Figure 12 : Variable Jet Shape Parametric Test

demonstrates clear control over post combustion, but provided little information on heat transfer efficiency due to the changing process conditions and short duration of the test.

An extended series of smelting reduction trials were completed to investigate both post combustion and heat transfer efficiency. The result of one of these trials can be seen in Figure 13 where a campaign of long tests were completed with different jet characteristics. The results achieved under steady state conditions show an almost identical trend in post combustion to that achieved with the parametric test. More importantly, the results indicate that heat transfer efficiency does not deteriorate as the post combustion levels increase. Post combustion levels in excess of 60%, with heat transfer efficiencies greater than 85%, could be reliably obtained.

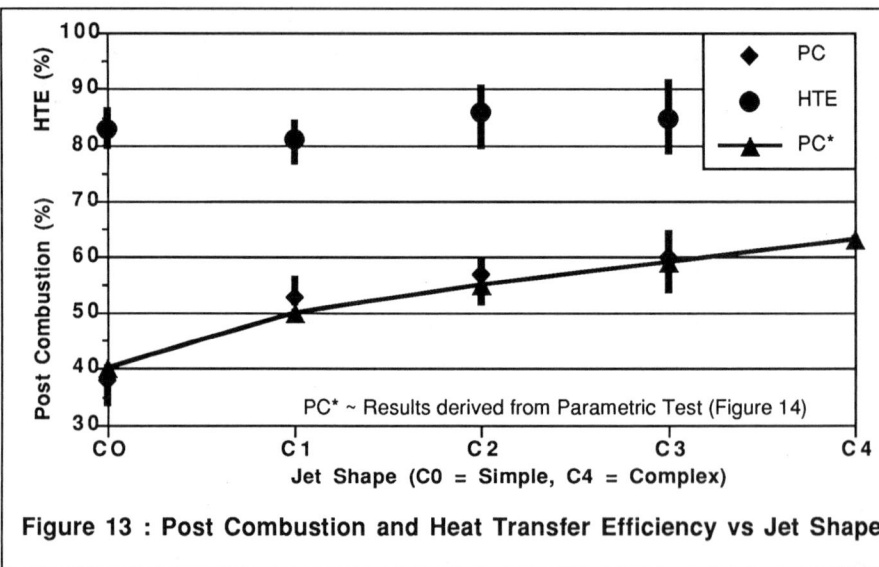

Figure 13 : Post Combustion and Heat Transfer Efficiency vs Jet Shape

An equally challenging development involved the engineering of the experimental variable jet characteristic tuyeres. Physical conditions experienced by this equipment are harsh and operation of the SSPP required reliable performance to achieve the results shown in Figure 13. Successive modification to the original tuyere design and engineering resulted in a robust and practical design which permitted on-line control over the post combustion level, in many ways superior to that achieved with a variable height centre lance.

CONCLUSION

To exploit the advantages offered by bottom injection into a molten iron bath, new process and engineering techniques were identified as playing key roles in the development of an alternative ironmaking technology. Having identified and studied the key process technologies, successive adaptations to existing steel converter and blast furnace technologies permitted the achievement of 60% post combustion and 85% high heat transfer efficiency in the smelting reduction vessel.

The end result of this development program is the novel and unique HIsmelt Process. The next phase of the development involves scale-up and continuing development of the engineering and process capabilities of the injection and smelting technology.

ACKNOWLEDGEMENTS

The authors acknowledge the pioneering (and continuing) efforts of a great many colleagues who have been closely involved with the development of the HIsmelt Process since the project was initiated in 1981. These include John Innes, the late Richard Turner, Don Philp, Robin Batterham, John Keogh, Richard Rusden, Jonathan Moodie, Ian Webb (all of CRA), Karl Brotzmann, Hans-Georg Faßbinder and Paul Mantey (members of the original Maxhütte Research team).

Others who have made important contribution to the work reported here include various CSIRO and CHAM staff members. Invaluable inputs have also come from Maxhütte Steelworks, Klöckner Stahlforschung, Mannesmann Anlagenbau (München), Thames Polytechnic and CRA Advanced Technical Development staff. In particular, the outstanding contribution of Nick Haller in the design and engineering of the most recent top and bottom injection technology is greatly appreciated.

REFERENCES

1. J.A. Innes, J.P. Moodie, I.D. Webb and K. Brotzmann, "Direct Bath Smelting of Iron Ores in a Liquid Bath - the HIsmelt Process," Process Technology Conference Proc, ISS/AIME, 17 (1988), 225-231.

2. K. Brotzmann, "Howe Memorial Lecture - New Concepts and Methods for Iron and Steel Production," Proc 70th Steelmaking Conference, ISS/AIME, (1987), 3-12.

3. J.A. Innes, R.E. Turner, R.C. Rusden, D.K. Philp and K. Brotzmann, "New Technologies for Efficient Utilisation of Coal in the Iron and Steel Industry," Proc. 13th CMMI Conference (1986), 87-92.

4. J.V. Keogh, G.J. Hardie, D.K. Philp and P.D. Burke, "HIsmelt Process Advances to 100,000 t/y Plant," 50th Ironmaking Conference Proceedings, ISS-AIME, (1991), 635-649.

5. B.L. Cusack, G.J. Hardie and P.D. Burke, "HIsmelt - 2nd Generation Direct Smelting," Second European Ironmaking Conference, Glasgow, Sept 1991.

6. J.K. Wright, I.F. Taylor and D.K. Philp, "A Review of the Progress of the Development of New Ironmaking Technologies," Minerals Engineering, 4 (1991), 983 - 1001.

7. G.J. Hardie, M. Cross, R.J. Batterham, M.P. Davis and M.P. Schwarz, "The Role of Mathematical Modelling in the Development of the HIsmelt Process," 10th Process Technology Conference Proceedings, ISS/AIME, 1992.

8. N. Suzuki, "Pyrolysis of Coal at High Temperature," Camp ISIJ, 2 (1989), 1047.

9. T. Kawamura, "Pyrolysis Products of Coal at Rapid Heating Conditions in a Wide Range of Temperatures," Camp ISIJ, 2 (1989), 1049.

10. T. Kawamura, "Coal Pyrolysis on a Molten Slag (Development of a Smelting Reduction Method with Direct Use of Coals -2," Camp ISIJ, 2 (1989), 162.

11. M. Tokuda and S. Kobayashi, "Process Fundamentals of New Ironmaking Processes," Process Technology Conference Proc, ISS/AIME, 17 (1988), 3-11.

12. E. Aukrust and K.B. Downing, "AISI Direct Steelmaking Program," 50th Ironmaking Conference Proceedings, ISS/AIME, (1991), 659-663.

13. T. Inatani, "The Current Status of the JISF Research on the Direct Iron Ore Smelting Reduction Project," 50th Ironmaking Conference Proceedings, ISS/AIME, (1991), 651-658.

14. K. Shiohara, "Research Program of the JISF's New Direct Iron Ore Smelting Reduction Process (DIOS Project)," Proc Second European Ironmaking Congress, (1991), 310-317.

15. G. Boxall, A.K. Sabharwahl, A.K. Robertson and R.J. Hawkins, "Modelling and Visualisation of Tuyere Accretion Growth," Injection Phenomena in Extraction and Refining, (A.E. Wraith ed, University of Newcastle upon Tyne, April 1982), B1-B18.

16. I.F. Taylor, J.K. Wright, and B.R. Baldock, "Instabilities in Gas Columns Resulting from Sonic Injections into Liquids," Presented at the CIM Metallurgists Conference, Winnipeg, Canada, (1987).

17. P. Moore, and A.E. Wraith, "Mathematical Modelling of the Refractory Temperature Fluctuations Generated Locally by Bubble Formation at an Annular Tuyere," Injection Phenomena in Extraction and Refining, (A.E. Wraith ed, University of Newcastle upon Tyne, April 1982), L1-L27.

18. G. Denier, J.C. Grosjean and H. Zanetta, "Heat Transfer for Tuyeres in Oxygen Bottom Blowing Converters," Iron and Steelmaking, (3) (1980), 123-126.

19. M. Nira, H. Take, N. Takashiba and F. Yoshikawa, "Development of Post Combustion Technique for the Top and Bottom Blowing Converter - Development of the Post Combustion in the LD Converter-II," Trans ISIJ, Vol 27, 1987, B74.

20. N. Takashiba, M. Nira, S. Kojima, H. Takeo and F. Yoshikawa "Development of the Post Combustion Technique in Combined Blowing Converter," Tetsu-to-Hagnaé, Vol 75(1), 1989, pp 89-96.

21. Y. Kato, J-C Grosjean, J-P Reboul, P. Riboud, "Influence of Lance Design and Operating Variables on Post Combustion in the Converter with Secondary Nozzles," Trans. ISIJ, Vol 28, 1988, pp 288-296.

22. H. Nakajima, S. Anezaki, Y. Tozaki, K. Ichihara, and K. Katohgi, "Improvement in the Heat Balance in the Combined Blowing Process, "Trans. ISIJ, Vol 26, 1986, pp 40-47.

23. J.P Moodie, M.P Davis and M. Cross, "Numerical Modelling for the Analysis of Direct Smelting Processes," 7th Process Technology Conference Proceedings, ISS/AIME, 1988, 55-64.

SUBMERGED OXYGEN INJECTION FOR PYROMETALLURGY

Guy Savard and Robert G.H. Lee

Abstract

This paper describes our numerous attempts to devise an injection system that would withstand the vigorous reaction created by the contact of oxygen with molten iron.

Most new ideas, products or processes are developed in response to a need and this technology is just such an example.

The process of development took place in several distinct phases.

Phase I. - A steelplant had a need to lower the level of silicon in the blast furnace iron to a constant level of less than 0.30%.

This was carried out in a hot metal transfer vessel, by injecting oxygen by the top through a one inch (25 mm) steel lance. The primary problem was solved, but it created two other problems: excess iron oxide fumes and a difficult working environment.

Phase II. - We reasoned that if the oxygen were introduced through the bottom of the vessel, desiliconizing would take place, the working conditions would be improved and the iron oxide fume generation would be reduced to an acceptable level. In practice, however the iron oxide fumes generated were still unacceptable. Another problem arose - that is the integrity of the submerged injector.

Phase III. - The previous test work (Phases I and II) gave some indication that burn-off of the submerged injector could be moderated by using high pressure oxygen. This was first observed during desiliconizing and in later tests when refining iron to steel with oxygen pressures as high as 8 270 kPa (1,200 psig).

Phase IV. - At this stage, attempts were made to improve the submerged injector design. Combinations of gases, including water additions, were tried to eliminate

Proceedings of the
Savard/Lee International Symposium on Bath Smelting
Edited by J. K. Brimacombe, P. J. Mackey,
G. J. W. Kor, C. Bickert and M. G. Ranade
The Minerals, Metals & Materials Society, 1992

or at least to reduce substantially the burn-off rate.

Phase V. - The reduction in the burn-off rate of the injector was found to be a direct function of the oxygen pressure which essentially relied on cooling brought about by the Joule-Thomson effect. These tests were a valuable learning exercise but it was realized that the high pressures that were needed would have to be reduced to levels that were commercially available. We also realized at this point that the solution to the injector burn-off would not be found in applying classic methods and materials.

Several reasons led us to consider shrouding the oxygen jet with a fluid that would be non reactive with molten iron but reactive with oxygen, i.e. an oxygen getter. The vigorous oxygen-molten iron reaction would then take place some distance removed from the tip of the injector. Hydrocarbons such as propane, methane and the like would meet this requirement.

Phase VI. - In 1967, Prof. Karl Brotzmann of Eisenwerk-Gesellschaft Maximilianshutte, mgH, Salzbach-Rosenberg, Germany, met with Canadian Liquid Air, Ltd. Within four months of this meeting, Prof. Brotzmann made the first industrial scale heat using the submerged oxygen injection technology.

INTRODUCTION

In the years following World War II there developed a great demand for steel to fill civilian needs.

At this time the most widely used steelmaking process in North America was the Open Hearth. In this process, hot metal from the blast furnace is charged into the open hearth together with scrap. The charge is then heated by burners to melt the scrap and to provide heat for refining the molten metal with iron ore which is the source of oxygen.

Because the charge ore must be heated to the temperature of dissociation, which is a lengthy process, it was not unusual to have tap-to-tap times of 12 hours.

The possibility of using gaseous oxygen as a replacement for the iron ore offered many advantages, notably an instant exothermic reaction between the oxygen and the metalloids which significantly shortened the heat times. The major limiting factor was the materials handling facilities in the melt shop.

Figure 1 - Desiliconization of blast furnace iron by means of a lance pipe. Oxygen at 700 k Pa. (100 psig) (1951)

647

Desiliconising blast furnace iron

In 1951, at a time when the use of oxygen was increasing rapidly, a major Canadian steel company requested our assistance to lower the silicon level in the blast furnace iron to a constant value.

It was decided to undertake a top lancing program consisting of a series of eighty heats of 40 tonnes each, using a one inch (25 mm) nominal steel pipe, as shown in Figure 1.

The lancing program produced a constant silicon analysis of 0.25%. This, in itself was quite successful, however the volume of iron oxide fumes generated was so dense that this desiliconising procedure was unacceptable.

Furthermore, the operating conditions for carrying out the "manual metallurgy" were too severe for the operators.

SUBMERGED OXYGEN INJECTION

Because of the difficult working conditions and because of the dense fumes generated, an alternative to top lancing had to be found. Subsequently, we reasoned that an oxygen injector placed in the bottom of a treating vessel would overcome these two problems. "Manual metallurgy" would be replaced by remote

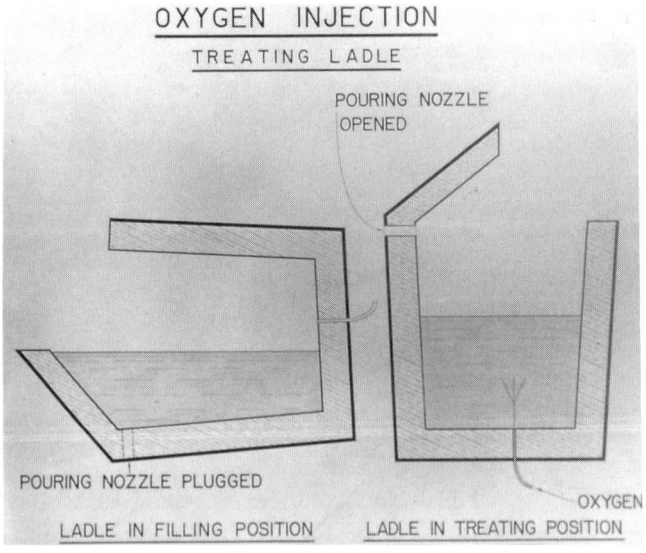

Figure 2 - Remote control for submerged oxygen injection. (1952)

oxygen control, and with the submerged oxygen injection system, less iron oxide fumes would be generated.

The solution is illustrated in Figure 2. This however created another problem: How to maintain the integrity of a submerged oxygen injector in contact with molten metal?

The destruction of the injector occurs at the tip where it is in contact with the vigorous high temperature oxidation reaction. The challenge then was to determine how to counteract this "hot spot".

<u>Joule-Thomson cooling of the single pipe injector</u>

We postulated that the Joule-Thomson cooling from an expanding jet could provide a solution. It was reasoned that the under-expanded jet would also serve to keep the high temperature reaction from contacting the tip of the injector.

From 1951 to 1956, seven series of high pressure oxygen injection tests were undertaken to determine the most practical design and operating conditions required to maintain the integrity of the injector.

During this period many companies graciously offerred the use of their facilities, manpower, metal, equipment and laboratories to carry out these tests. Heats ranging from 500 kg to 35 tonnes were carried out with oxygen pressures varying from 2750 kPa (400 psig) to 4825 kPa (700 psig).

Figure 3 - The first mushroom recovered intact with the
submerged injector. (1955)

Accretion formation at high pressure injection

The much sought after breakthrough in our research efforts occurred when refining semi-steel we discovered that an accretion (a mushroom) formed on the tip of the injector at an oxygen pressure of 4825 kPa. Figure 3 shows a mushroom that was recovered intact with the submerged injector.

We surmised that the integrity of the injector could be maintained as long as a mushroom formed at the tip.

Desiliconizing hot metal

The ability to deposit a mushroom rendered possible the continuation of the desiliconizing project by submerged oxygen injection. This program was carried out in a converted hot metal transfer vessel, shown in Figure 4.

Figure 4 - Hot metal transfer vessel modified for use as a
converter for submerged blowing. (1956)

Figure 5 shows the six submerged injectors located at the bottom of the converted vessel.

With the lower temperature needed for desiliconizing the hot metal, it was estimated that 4135 kPa (600 psig) oxygen pressure would be adequate to form a mushroom at the tip of the injector. Figure 6 shows an injector which is protected by a mushroom.

Figure 5 - Six injectors located at the bottom of the treatment vessel. (1956)

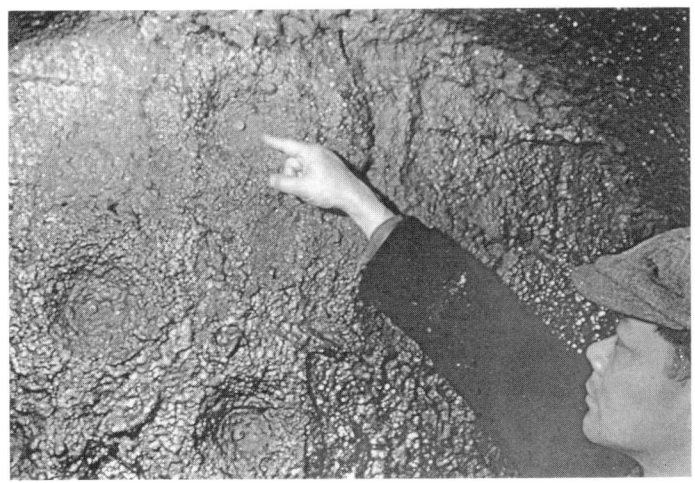

Figure 6 - A mushroom is shown at the tip of each injector. (1956)

Looking forward to the 21st century, Figure 7 shows the "state of the art" of the mobile gaseous oxygen storage system used for this desiliconizing program.

Figure 7 - Gaseous oxygen trailer. (1956)

Refining steel

The converted vessel along with the rotating mechanism was donated to Canadian Liquid Air Ltd. This made it possible to continue our research effort to develop a new process.

Since no handbook was available with the required data for molten steel temperature, it was estimated that 8270 kPa (1,200 psig) oxygen pressure would be required to produce the mushroom.

Following a series of successful tests on a 10 tonne scale, it was felt that a demonstration of the process was justified. A demonstration heat on a scale of 15 tonnes was undertaken in the presence of representatives of major North American steel companies.

The demonstration heat was successful. In spite of this, there was a feeling of disbelief that this technology was possible. - It was as though we had a secret that was not apparent. Consequently, no expression of interest followed.

In Europe at this time (1957) a substantial tonnage of steel was produced by the Thomas or Bessemer process. We considered, therefore, that there would be greater interest there for this new oxygen technology. However when a demonstration was made in Europe in 1961, the response was much the same as that in North America.

Refining low grade ferrochrome

Faced with this lack of interest, in 1962, the converted vessel which had been used for our tests was transferred to a ferrochrome smelter for the purpose of lowering the carbon and silicon contents. To achieve a minimal loss of chromium with pure oxygen, the metal temperature had to be extremely high.

As our original oxygen system was designed to operate at a maximum pressure of 8270 kPa (1,200 psig), it was felt that further cooling of the injectors would be needed. Therefore, water was introduced to produce an oxygen-water mist to supplement the Joule-Thomson cooling. This indeed improved the performance of the injector. With the refining of ferrochrome, however, another obstacle had to be overcome. It was noticed that the composition of the refractory surrounding the injector had a profound effect on injector life. To solve this refractory problem, an intensive co-operative development program was undertaken with a major Canadian refractory producer.

At the end of the program which extended over 370 heats, we succeeded in lowering the injector burnoff rate. With a higher rate of water injection, fume reduction was achieved.

Figure 8 shows the vessel in operation. As a thicker bottom refractory lining was required, the rotating mechanism was underpowered. Hence two counterweights had to be welded to the vessel to change its center of gravity making rotation possible.

THE CREATION OF THE RESEARCH DEPARTMENT

In 1963, Pierre Salbaing, president of Canadian Liquid Air Ltd., created a research group in Canada. This was a most important event because it enabled us to try many more ideas and to take many risks which could not have taken place in a metallurgical production plant.

Experimental Converter (the K-Vessel)

The first objective in continuing our metallurgical research was to build a vessel where bottom, top, and side oxygen injection could be carried out. The experimental K-Vessel shown in Figure 9 was designed and fabricated by the late Bob Kottmeier, a brilliant technician who was our partner in these projects since 1951.

Development continued on a scale of 150 kg of molten iron while seeking new solutions to the burn-off problem.

Figure 8 - Vessel is rotating to the blowing position.
Notice the counterweight. (1962)

Figure 9 - The K-Vessel designed with capabilities for top,
for bottom and for side oxygen injection was
built for $3,225.

Up to this time, only high oxygen pressures had been employed. It was decided that for practical reasons, the injector would have to <u>operate at commercially available oxygen pressures</u>.

Also, because no existing materials could withstand the high temperature reaction of oxygen with molten iron, it would be necessary to explore new techniques that would overcome the destructive effect of this high temperature, estimated at 3,900°C.

<u>Protective shield for the injector:</u>

We reasoned that if the oxygen could be shielded from immediate contact with the molten iron at the tip of the injector, the destructive oxidation reaction would then take place at a distance removed from the tip. It was our belief that an inert gas (e.g., nitrogen, argon, etc.) would provide this shield, a shield which would be more effective than that provided by the Joule-Thomson cooling.

<u>The "Getter"</u>

Pursuing this train of thought, if this inert gas shield were to be replaced by an oxygen "getter", the protection for the injector would be further improved. This getter would preferentially react with oxygen and be non-reactive with the molten iron. For this purpose, we reasoned that a hydrocarbon fluid would fulfill this requirement. The universal availability of methane and propane dictated that these be selected for the initial trials.

<u>Submerged combustion</u>

When injected oxygen is shielded with a hydrocarbon to prevent immediate contact with molten metal at the tip of the injector, it (oxygen) reacts first with the hydrocarbon getter.

In actual practice, for example, it was found that only 8% methane, based on oxygen, was required to protect the injector. According to the combustion equation (1) in the following table, the resulting reaction temperature is only 1400°C, contrary to popular belief that it is much higher than the normal temperature of molten iron.

Reaction	Temperature
(1) $8CH_4 + 100\ O_2 \rightarrow 8CO_2 + 16H_2) + 84O_2$	1400°C
(2) $2Fe_{(1425°C)} + O_2 \rightarrow 2FeO$	3900°C

Without the protective shroud, the injector would be subjected to the high temperature of 3900°C generated by the reaction of the molten iron and oxygen as shown by equation (2) in the table.

Therefore, even without the formation of a mushroom, the temperature exposed to the tip of the injector is lowered also about 2500°C by the oxygen getter. Because of the shroud, when a mushroom is formed at the tip of the concentric injector as shown in Figure 10, the injector is substantially protected against the molten metal and the vigorous refining reaction.

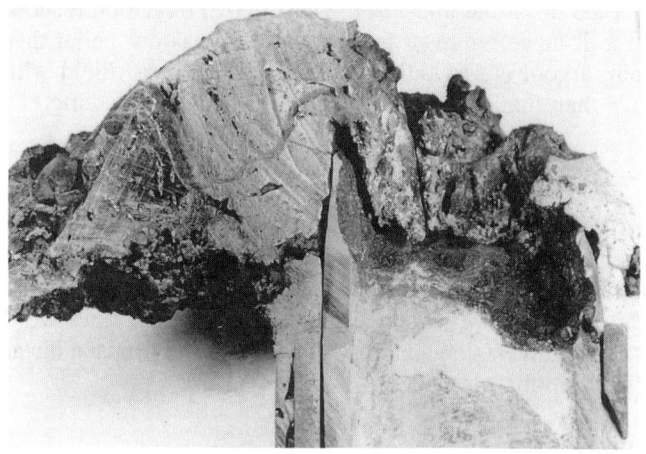

Figure 10 - Cross section of a typical mushroom formed on the tip of a concentric injector.

THE CONCENTRIC INJECTOR

Several injector designs were evaluated, bearing in mind the hazardous nature of the hydrocarbon in contact with oxygen. The final choice, as illustrated in Figure 11, was a simple, safe and effective injector design consisting of two coaxial pipes. The oxygen is injected through the central pipe. The hydrocarbon getter flows through the annulus creating the protective shroud which surrounds the oxygen pipe. Figure 12 records the first test employing the concentric pipe design for submerged injection of oxygen using propane as the shroud gas.

As the injected hydrocarbon is heated at the tip of the injector, cracking occurs. This reaction is endothermic thus contributing to the cooling of the injector and to the formation of the mushroom.

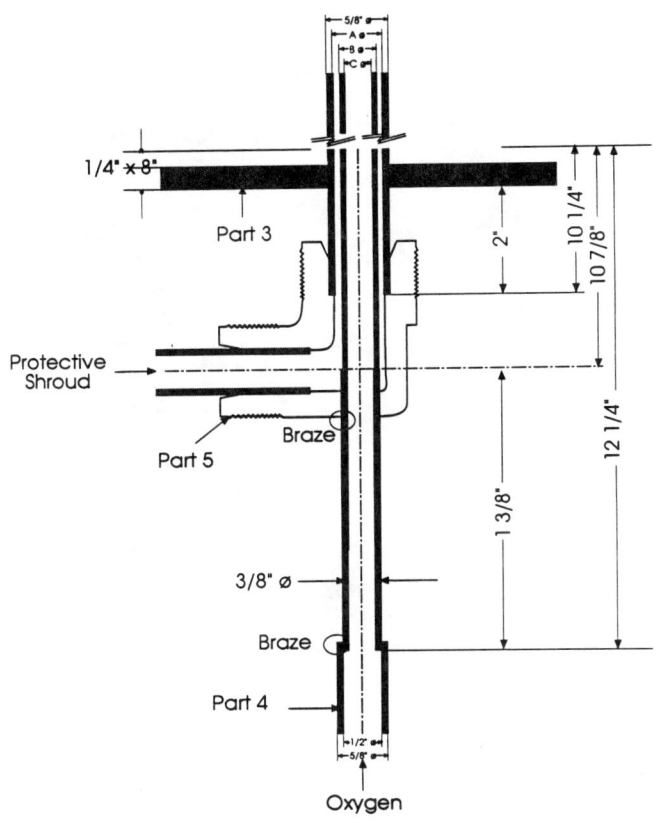

Figure 11 - Design of the concentric injector. (1964)

Getting the pig to market

There are many examples of new products or processes that were proven to be technically sound and commercially viable that never reached the market. This was our experience.

Contacts were made with many steel companies, and subsequently contacts with major producers of non ferrous metals. Neither was interested.

Finally in the Fall of 1967, Dr. Karl Brotzmann, Director of Research for

Eisenwerk-Gesellschaft Maximilianshutte mbH, having learned of our activities in bottom blowing, requested a meeting. This meeting took place in October, 1967.

Figure 12 - The first test with the concentric pipe design for submerged injection of oxygen using propane as the shroud gas. (1965)

Two demonstration heats were made. The injector showed no burn-off.

This meeting led to a license agreement with Maximilianshutte. This enabled them to develop the process to the full industrial scale - the OBM/Q-BOP Process.

The first industrial heat of 20 tonnes in a modified Thomas converter, shown in Figure 13, took place on December 17, 1967 - four months after the demonstration heats in Canada.

Figure 13 - The first industrial OBM/Q-BOP converter at
Eisenwerk-Gesellschaft Maximilianshutte,
mgH, Sulzbach-Rosenberg, Germany. (1967)

CONCLUSIONS

The main goal of this research project was to achieve a bottom blown oxygen refining process.

A project, such as the one described in this report, is best undertaken by a small group of innovative researchers.

Some important features which made the project possible were the material support, the keen co-operation from the steelmakers and related industries and the technical assistance of the Canadian Government.

It would be extremely difficult to develop a process of this nature today because of the emphasis on near term results, marketing and legal considerations.

Companies give high profile to research and technology, but in effect these receive low priority. It is urgent that this adverse situation be corrected.

ACKNOWLEDGEMENTS

We are grateful to the numerous companies and to the Canadian Government which have contributed to the development of the submerged oxygen injection technology. In particular, we wish to acknowledge the following:

B and T Foundry, Richmond, Quebec.

Bureau of Mines, Ottawa, Ontario.

Canadian Refractory Ltd., Kilmar, Ontario.

Dosco, Sydney, Nova Scotia.

Eastern Electro Castings, Lachine, Quebec.

IRSID, Maizières-les-Metz, France.

National Research Council, Ottawa, Ontario.

Quebec Iron and Titanium Corp., Tracey, Quebec.

Sorel Industries Ltd., Tracey, Quebec.

Strategic-Udy, Niagara Falls, Ontario.

We commend the co-operation of Canadian Liquid Air, Ltd. and its numerous personnel who accommodated our high pressure industrial gas needs to carry out the many projects.

Also, we wish to acknowledge the assistance of our many friends and associates, in particular, the late Robert Kottmeier, the late Horace Freeman, Don MacPhail, Al Bachmeier, John Convey, Omar Smith and many others.

Our special thanks to Keith Brimacombe, the catalyst who encouraged us to tell our story; and, also special thanks to Phillip Mackey who assisted in researching the insight of the story.

Finally, we are grateful to Jeanne Gould and Catherine Lee for their patient assistance in the preparation of the manuscript.

Subject Index

Author Index